Handbook of Experimental Pharmacology

Volume 120

Springer
Berlin
Heidelberg
New York
Barcelona
Budapest
Hong Kong
London
Milan
Paris
Santa Clara
Singapore
Tokyo

Antipsychotics

Contributors

V.P. Bakshi, M.E. Bardgett, D.L. Braff, E.S. Brodkin, O. Civelli
J.G. Csernansky, S.G. Dahl, M. Davidson, A.Y. Deutch, H.S. Fatemi
W.O. Faustman, J.Gerlach, M.A. Geyer, J.Golier, A.A. Grace, D.Hartman
A.L. Hoff, P.W. Kalivas, A.R. Koreen, J.F. Leckman, J. Lieberman
C.J. McDougle, H.Y. Meltzer, S.A. Minchin, F. Monsma, B.H. Mulsant
P. O'Donnell S.-O. Ögren, L. Peacock, B.G. Pollock, R. Ranjan,
B.L. Roth, R.E. See B. Sheitman, N.R. Swerdlow

Editor

J.G. Csernansky

Springer

Professor JOHN G. CSERNANSKY, M.D.
Washington University
School of Medicine
Department of Psychiatry
4940 Children's Place
St. Louis, MO 63110-1093
USA

With 29 Figures and 47 Tables

CIP data applied for

Die Deutsche Bibliothek – CIP-Einheitsaufnahme
Handbook of experimental pharmacology/ed. board G.V.R. Born ... Berlin; Heidelberg; New York; Barcelona; Budapest; Hong Kong; London; Milan; Paris; Tokyo; Springer.
Früher u.d.T.: Handbuch der experimentellen Pharmakologie
NE: Born, Gustav V.R. [Hrsg.]
Antipsychotics: [with 44 tables]/contributors S. Anuras ... Ed. R.G. Cameron ... – Berlin; Heidelberg; New York; Barcelona; Budapest; Hong Kong; London; Milan; Paris; Santa Clara; Singapore; Tokyo; Springer, 1996
(Handbook of experimental pharmacology; Vol. 120)
ISBN-13: 978-3-642-64653-9 e-ISBN-13: 978-3-642-61007-3
DOI: 10.1007/978-3-642-61007-3

NE: Anuras, S.; Cameron, R.G. [Hrsg.]
Vol. 120. Antipsychotics. – 1996

Cover design: Springer-Verlag, Design & Production

Typesetting: Best-set Typesetter Ltd., Hong Kong

SPIN: 10476601 27/3136/SPS – 5 4 3 2 1 0 –

Preface

Antipsychotic drugs were first discovered in 1953, and not since the late 1970s has the *Handbook of Experimental Pharmacology* taken up this topic. A new treatment of this topic would be due under any circumstances; however, this is now particularly true, since remarkable progress has been made on several fronts in furthering our understanding of the mechanisms of antipsychotic drug action. First, we have learned that schizophrenia is an illness with particular neuroanatomical abnormalities, many of which suggest that the illness is caused by errors in neurodevelopment. These findings have helped to form a context for understanding neurochemical aberrations in the illness and suggest new approaches for pharmacological treatment. Propelled forward by rapid advances in neurochemical anatomy, current pathophysiological hypotheses of schizophrenia and antipsychotic drug action have taken on the appearance of complex electrical circuit diagrams. Second, molecular biology studies have now revealed that there is a multiplicity of dopamine receptors (i.e., D_1, D_{2short}, D_{2long}, D_3, D_4, and D_5), some of which may become entirely new targets for antipsychotic drug action. Ironically, the development of drugs that are selective for these receptors and that can be used to investigate their function lags behind; yet the discovery of these new receptors offers unparalleled opportunities for developing drugs with improved efficacy and fewer side effects. Third, the rediscovery of clozapine in the mid-1980s has been a potent stimulus for the development of other new antipsychotic drugs that are mixed dopamine and serotonin antagonists. Some of these new drugs have already reached the commercial market and represent true innovations, perhaps the first for the treatment of psychotic disorders since the original discovery of conventional neuroleptics in the 1950s. Also, because clozapine has actions at a large number of neuroreceptors other than dopamine (i.e., serotonin, acetylcholine, noradrenaline, histamine, and the sigma site), the ability to treat psychosis by affecting the functioning of neurotransmitters other than dopamine is being increasingly attempted. If and when such efforts succeed, new insights into the pathophysiology of schizophrenia, and other severe mental disorders, will develop.

It has been a privilege to edit this volume. In every case, the chapter authors are outstanding investigators in their areas and have made major contributions to our understanding of the neuropharmacology of antipsychotic drugs. Their contributions to this book are characteristically thorough and stimulating. The chapters may be roughly divided into basic and clinical. Among the basic chapters are those which describe new advances in the molecular pharmacology of dopamine receptors, as well as those which provide up-to-date reviews of more traditional topics, such as the neurophysiology, neurochemistry, and behavioral pharmacology of antipsychotic drugs. Among the clinical chapters are those which describe the types of patients and symptoms most successfully treated with antipsychotic drugs and the predictors of antipsychotic drug efficacy. In addition, other clinical chapters discuss the potential for antipsychotic drugs to alter the underlying cognitive deficits and other biological abnormalities now known to be associated with schizophrenia. In many chapters, however, the reader will see violations of the imaginary boundary between basic and clinical science, which only adds to the richness of the discussion.

In addition to the chapter authors, I would like to thank many others who have made the preparation of this book possible. Gina Redding-Dennison, my administrative assistant, and Doris Walker, Desk Editor for Springer-Verlag, have made important contributions to the organization and implementation of this task. I also appreciate the many suggestions of my colleagues and students at Washington University regarding the contents and ordering of the chapters. And finally, I would like to thank Cynthia Csernansky, my wife, and, Leo Hollister, my mentor, for their inspiration and support.

J. CSERNANSKY

List of Contributors

BAKSHI, V.P., Department of Neuroscience, UCSD School of Medicine, 9500 Gilman Drive, La Jolla, CA 02093-0804, USA

BARDGETT, M.E., Department of Psychiatry, Washington University School of Medicine, Box 8134, 4940 Children's Place, St. Louis, MO 63110-1093, USA

BRAFF, D.L., Department of Psychiatry, UCSD School of Medicine, 9500 Gilman Drive, La Jolla, CA 02093-0804, USA

BRODKIN, E.S., Yale University School of Medicine, Department of Psychiatry, Abraham Ribicoff Research Facilities, Connecticut Mental Health Center, 34 Park Street, New Haven, CT 06508, USA

CIVELLI, O., Pharma Research, F. Hoffmann-La Roche, CH-4002 Basel, Switzerland

CSERNANSKY, J.G., Department of Psychiatry, Washington University School of Medicine, Box 8134, 4940 Children's Place, St. Louis, MO 63110-1093, USA

DAHL, S.G., Department of Pharmacology, Institute of Medical Biology, University of TromsØ, N-9037 TromsØ, Norway. Present address: Institut de Recherche Jouveinal, 3–9, rue de la Loge, F-94265 Fresnes Cedex, France

DAVIDSON, M., Bronx Veterans Medical Center, Psychiatry Service 116A, 130 W. Kingsbridge Rd., Bronx, NY 10468, USA

DEUTCH, A.Y., Departments of Psychiatry and Pharmacology, Yale University School of Medicine, CMHC, 34 Park Street, New Haven, CT 06508, USA

FATEMI, H.S., Department of Psychiatry, Case Western Reserve University School of Medicine, 2074 Abington Road, Cleveland, OH 44106, USA

FAUSTMAN, W.O., Department of Veterans Affairs Medical Center, Stanford/VA Mental Health Clinical Research Center, Unit 4B2, 3801 Miranda Ave., Palo Alto, CA 94304, USA

Gerlach, J., Linda Peacock, St. Hans Hospital, Dept. P, DK-4000 Roskilde, Denmark

Geyer, M.A., Departments of Psychiatry and Neuroscience, UCSD School of Medicine, 9500 Gilman Drive, La Jolla, CA 02093-0804, USA

Golier, J., Bronx Veterans Medical Center, Psychiatry Service 116A, 130 W. Kingsbridge Rd., Bronx, NY 10468, USA

Grace, A.A., Department of Neuroscience and Psychiatry, Center for Neuroscience, University of Pittsburgh, 458 Crawford Hall, Pittsburgh, PA 15260, USA

Hartman, D., Pharma Research, F. Hoffmann-La Roche, CH-4002 Basel, Switzerland

Hoff, A.L., Biological Psychiatry Treatment and Research Unit, Napa State Hospital, 2100 Napa-Vallejo Highway, Napa, CA 94558, USA

Kalivas, P.W., Department of Veterinary and Comparative Anatomy, Pharmacology and Physiology, Washington State University, Pullman, WA 99164, USA

Koreen, A.R., Hillside Hospital, Division of Long Island Jewisch Medical Center, Albert Einstein College of Medicine, P.O. Box 38, Glen Oaks, NY 11004, USA

Leckman, J.F., Yale University School of Medicine, Child Study Center, and Department of Psychiatry, 34 Park Street, Room 333, New Haven, CT 06508, USA

Lieberman, J., Hillside Hospital, Division of Long Island Jewisch Medical Center, Albert Einstein College of Medicine, P.O. Box 38, Glen Oaks, NY 11004, USA

McDougle, C.J., Yale University School of Medicine, Department of Psychiatry, and Child Study Center, 34 Park Street, New Haven, CT 06508, USA

Meltzer, H.Y., Laboratory of Biological Psychiatry, Case Western Reserve University School of Medicine, University Hospital of Cleveland, 11100 Euclid Avenue, Cleveland, OH 44106-5000, USA

Minchin, S.A., Department of Psychiatry, Washington University School of Medicine, Box 8134, 4940 Children's Place, St. Louis, MO 63110-1093, USA

Monsma, F., Pharma Research, F. Hoffmann-La Roche, CH-4002 Basel, Switzerland

Mulsant, B.H., Department of Psychiatry, University of Pittsburgh School of Medicine, Western Psychiatric Institute and Clinic, 3811 O'Hara Street, Pittsburgh, PA 15213, USA

O'Donnell, P., Department of Neuroscience and Psychiatry, Center for Neuroscience, University of Pittsburgh, Pittsburgh, PA 15260, USA

Ögren, S.O., Karolinska Institute, Department of Neuroscience, Division of Cellular and Molecular Neurochemistry, S-171 77 Stockholm, Sweden

Peacock, L., Linda Peacock, St. Hans Hospital, Dept. P, DK-4000 Roskilde, Denmark

Pollock, B.G., Departments of Psychiatry and Pharmacology, University of Pittsburgh School of Medicine, Western Psychiatric Institute and Clinic, 3811 O'Hara Street, Pittsburgh, PA 15213, USA

Ranjan, R., Laboratory of Biological Psychiatry, Case Western Reserve University School of Medicine, University Hospital of Cleveland, 11100 Euclid Avenue, Cleveland, OH 44106-5000, USA

Roth, B.L., Department of Psychiatry, Case Western Reserve University School of Medicine, 2074 Abington Road, Cleveland, OH 44106, USA

See, R.E., Department of Psychology, Washington State University, Pullman, WA 99164-4820, USA

Sheitman, B., Hillside Hospital, Division of Long Island Jewisch Medical Center, Albert Einstein College of Medicine, P.O. Box 38, Glen Oaks, NY 11004, USA

Swerdlow, N.R., Departments of Psychiatry and Neuroscience, UCSD School of Medicine, 9500 Gilman Drive, La Jolla, CA 92093-0804, USA

Contents

CHAPTER 2

Molecular Models and Structure–Activity Relationships

CHAPTER 3

Interaction of Antipsychotic Drugs with Dopamine Receptor Subtypes

CHAPTER 4

Atypical Antipsychotic Drugs: Clinical and Preclinical Studies
HOSSEIN S. FATEMI, HERBERT Y. MELTZER, and BRYAN L. ROTH 77

CHAPTER 6

Basic Neurophysiology of Antipsychotic Drug Action

CHAPTER 7

Tolerance and Sensitization to the Effects of Antipsychotic Drugs on Dopamine Transmission

CHAPTER 8

The Behavioural Pharmacology of Typical and Atypical Antipsychotic Drugs

CHAPTER 9

Antipsychotic Drug Action After Lesions to the Hippocampus or Frontal Cortex

CHAPTER 10

An Animal Model of Sensorimotor Gating Deficits in Schizophrenia Predicts Antipsychotic Drug Action

CHAPTER 13

Antipsychotic-Induced Side Effects Related to Receptor Affinity

CHAPTER 16

Antipsychotic Drugs in Children and Adolescents

CHAPTER 17

Use of Antipsychotic Drugs in the Elderly

CHAPTER 1

Classification Schemes for Antipsychotic Drugs

SUSAN A. MINCHIN and JOHN G. CSERNANSKY

A. Introduction

The purpose of this chapter is to review classification systems for antipsychotic drugs and their utility for new drug development. Antipsychotic drugs may be defined as medications which alleviate delusions, hallucinations, and some aspects of formal thought disorder when these symptoms occur in a variety of illnesses, most notably schizophrenia. The term "neuroleptic" (i.e., affecting the nervous system) has also been used to describe this class of compounds because of the regular association between the use of antipsychotic medications and the development of extrapyramidal side effects. However, as scientists continue to search for new drugs that are both more effective and more tolerable, this latter term may become obsolete. Diverse classification schemes for antipsychotic drugs have been used throughout the literature – schemes based on chemical structure, potency, efficacy and side-effect profiles, and sites of neurotransmitter receptor action. The history of antipsychotic drug development and the development of these classification schemes are the subjects of this chapter.

I. Historical Perspective

The pain and suffering of severe psychiatric illness has been appreciated by mankind since the beginning of civilization. Attempts to alleviate this suffering through medications, in the form of psychotropic drugs derived from plants, were made as early as 3000 B.C. by the Sumerians in the Tigris-Euphrates valley. The Sumerians appreciated the mood-elevating effects of the residue and juice of poppies and cultivated these plants for this purpose (SPIEGEL 1989). The analgesic effects of opium led to its use as a healing agent in the second millennium B.C. in Asia Minor, Cyprus, Mycenae, and Egypt (SPIEGEL 1989).

The root hellebore was used as a psychopharmaceutical agent for centuries in European culture to treat a wide variety of neuropsychiatric disorders – mania, melancholy, violent temper, "crazy ideas," mental retardation, epilepsy, inflammation of the brain, and hydrophobia (WITTERN 1983). Hellebore belongs to the Ranunculaceae, a family of plants that includes marigold, buttercup, peony, and larkspur, and contains several toxic glycosides which have emetic, cathartic, and convulsive properties. In some cases, such drugs might have allowed the removal of psychopathogenic substances. However, the use of hellebore declined in the nineteenth century when the medical community better realized the dangers of using a substance which lowers the seizure threshold. References to the treatment of psychotic symptoms date back to the use of *Rauwolfia serpentina*, a root traditionally used by Indians as a remedy for snake bites and for the treatment of insanity in ancient Hindu ayurvedic writings (HOLLISTER 1977). Decoctions of *R. serpentia* have long been noted to have calming effects on agitated psychiatric patients.

Before the introduction of what is known today as the first antipsychotic drug – chlorpromazine – treatments for severely ill psychiatric patients consisted of a mixture of segregation, physical restraint, various "somatic" remedies, and pharmacological sedation (SCHNEIDER 1824). The "insane" were institutionalized, usually in buildings located on the perimeter of a city or town. Physical restraints took the form of metal chains or straightjackets. Internal remedies involved the use of various laxative and cathartic preparations, calming agents (for example, decoctions of hemlock, belladonna, opium, tobacco) and/or nerve-invigorating agents, such as turpentine oil, camphor, sage, naphthalene, and peppermint oil. Other remedies promoted nausea and vomiting. "External remedies" included placing the patient in a revolving machine or chair, whipping the patient using whips with nettles, applying mustard plasters or blistering plasters, or immersing the patient in cold baths or snow baths. Other external remedies included bleeding, cupping, magnetism, galvanism, and sneezing powder (SCHNEIDER 1824).

By the late nineteenth century, the use of emetics, laxatives, and the other somatic treatments for psychosis was declining, and the psychopharmaceuticals recommended for use by the leading psychiatrists of the day became restricted to narcotics and hypnotics. In his textbook published in 1899, KRAEPELIN listed only nine substances available for treatment of psychiatric illness – opium, morphine, scopolamine, hashish, chloral hydrate, sulonal, alcohol, chloroform, and bromine salts (SPIEGEL 1989). BLEULER's textbook (1916) contained a similar list; however, Bleuler considered the risk of habituation to alcohol too great, and that it made this pharmaceutical agent both ineffective and dangerous. BLEULER included two other hypnotics, paraldehyde and veronal. Prior to the 1950s, the above-mentioned therapeutic modalities were used in various combinations with other treatments such as chemically induced comatose states and/or seizures, and electroshock treatment (KALINOWSKY 1970; CALDWELL 1978). Electroshock treatment for psychosis is considered by many to be the first truly modern treatment for psychosis, and it still remains under investigation. However, this topic is beyond the scope of this chapter. In addition, nonspecific therapies such as antibiotics were utilized for psychosis and have proven beneficial in rare instances (for example, in the neuropsychiatric complications of syphilis).

Reserpine became the first principal chemical compound clearly described as having antipsychotic activity. In 1931, SEN and BOSE published a paper describing the induction of tranquilization and the reduction in psychotic symptoms by extracts from the shrub *R. serpentina* (SEN and BOSE 1931). However, the psychoactive component of the extract was not isolated until 1952 when MULLER, SCHLITTLER and BEIN tested one alkaloid component, reserpine, for its antihypertensive effects in a group of institutionalized patients. They found this drug to have a calming effect on these patients, transforming them from agitated, talkative, impulsive patients into slow-moving subjects who did not interact much with their environment and displayed little interest. These changes, although perhaps preferable to psychotic agitation, are now also recognized as the neurovegetative symptoms of melancholic depression.

The first modern antipsychotic drugs, i.e., the phenothiazines, were synthesized by BERNTHSEN during his investigation of the dyes Lauth's, violet and methylene blue (HOLLISTER 1977). Phenothiazine was first tested in humans in the 1940s as a possible antihelmintic agent, antihistamine, or adjuvant to surgical anesthesia. Derivatives of phenothiazine were being developed at Rhône-Poulenc Laboratories, and chlorpromazine was synthesized in 1950 by CHARPENTIER. At the same time, the French surgeon LABORIT was searching for a compound to use as an anesthetic with more central effects than the then-available antihistamine promethazine, and he tested chlorpromazine. LABORIT is credited with the observation that chlorpromazine induced "ataraxia" (i.e., indifference) in his study subjects, and therefore might be of some use in treating psychiatric disorders (DENIKER 1970). LABORIT's colleagues then tested the compound in manic patients at the military hospital of Val de Grâce, and found moderate responses (DENIKER 1970). In 1952, Pierre DENIKER and colleagues published the results of a trial of chlorpromazine in 38 psychotic patients resistant to all existing therapies (DELAY et al. 1952). In all patients, agitation and aggressiveness as well as the core psychotic symptoms of their illnesses improved with administration of chlorpromazine.

Thus, a fortuitous set of circumstances led to the observation of antipsychotic activity in phenothiazines and set into motion an era of investigation to test one of the fundamental theories of modern psychiatry – that it is possible to treat psychiatric disorders, as other medical illnesses, with specific chemical compounds. As the number of medications found to be efficacious in psychiatric illness then increased, various classification schemes were developed to aid in the understanding of patterns of clinical efficacy and side effects, in the development of new medications, and in the quest for an understanding of the pathophysiology of psychiatric disease.

B. Classification by Patterns of Efficacy and Neurological Side Effects

The behavioral effects of antipsychotics in humans were described in 1955 to the French Academy of Medicine by Jean DELAY and Pierre DENIKER as the "neuroleptic syndrome," a characteristic syndrome consisting of inhibition of spontaneous movements and operant behavior with preservation of spinal reflexes (CADWELL 1978). Interestingly, this clinical syndrome was well defined before the behavioral effects of antipsychotics had been investigated in animals. The neuroleptic syndrome also involved a lack of initiative on the part of the patient, with the patient displaying few emotions and showing disinterest in his or her environment. The patients were also noted to respond slowly to external stimuli and tended to appear drowsy; however, they were easily aroused, capable of answering questions correctly, and their intellectual functioning was not grossly impaired.

Extensive investigation into the behavioral effects of antipsychotics in animals, primarily rats and monkeys, then followed during the 1960s and 1970s with the hope of discovering the mechanisms of action of these clinical effects. In animals, antipsychotics were found to reduce or inhibit spontaneous movement, exploratory behavior, feeding behavior, operant conditioning behavior, and conditioned avoidance behavior. Also, amphetamine-induced hyperactivity, apomorphine-induced aggression, and dopamine agonist-induced stereotypies were all blocked by antipsychotic medications. As in patients, antipsychotic medications were observed to cause indifference in animals and to induce cataleptic immobility (resembling waxy flexibility) while not affecting spinal reflexes or the ability to withdraw from noxious stimuli. Each one of these effects along with many others were then investigated in an attempt to identify which behavioral paradigms would be most sensitive and specific with respect to predicting clinical efficacy.

While all antipsychotic agents were found to inhibit operant responses regardless of the type of reinforcement, little correlation was found between the potencies of these agents in these behavioral paradigms and their potencies as antipsychotics in humans (ZIRKLE and KAISER 1970). This was also true for exploratory and feeding behavior. However, the results of tests studying the ability of antipsychotics to block conditioned avoidance in both rats and monkeys, such as the pole-climbing test of conditioned avoidance-escape, demonstrated a high correlation with the potencies of the medications as antipsychotics in man (NIEMEGEERS et al. 1969a,b; PFEIFFER and JENNY 1957). Additionally, the ability of antipsychotic drugs to reduce amphetamine-induced stereotypies, hyperactivity, and aggression in animal models was also found to be specific and sensitive for predicting antipsychotic efficacy in humans (FIELDING and LAL 1978; FIELDING et al. 1975; LAL et al. 1976; NIEMEGEERS and JANSSEN 1975). Thus, the expectation developed that new antipsychotic agents should block conditioned avoidance, and reduce or inhibit amphetamine-induced stereotypies, hyperactivity, or aggression in animals.

Once a drug has been classified as an antipsychotic, it may then be further classified as either typical or atypical. Typical, or classical, antipsychotics have been defined as those drugs that, in addition to the behavioral effects summarized above, also induce catalepsy. Given the similarity between catalepsy in animals and pseudoparkinsonism in man, these drugs may be more likely to induce various acute extrapyramidal side effects (and tardive dyskinesia) in humans, and stimulate prolactin secretion as well. Such drugs have a uniformly high affinity for the dopamine D_2 receptor, and cause the development of depolarization block in nigrostriatal as well as mesolimbic dopamine neurons during chronic administration. Several phenothiazines, thioxanthenes, indoles, butyrophenones, and diphenylbutylpiperidines meet these criteria, and thus are classified as typical antipsychotics. The substituted benzamides and thioridazine are considered by some as atypical antipsychotics due to their low propensity to cause extrapyramidal symptoms in humans; however, several members of the benzamide class can give rise to substantial extrapyramidal side effects (LEWANDER et al. 1990; HARNRYD et al. 1984; GERLACH et al. 1985), and thioridazine's significant anticholinergic properties may well be responsible for this compound's favorable side-effect profile. In addition, these compounds cause marked elevation in serum prolactin levels in humans (MELTZER 1991), and are thus best classified as typical in this classification scheme.

Atypical antipsychotics produce similar, if not superior, clinical efficacy. However, atypical drugs produce few, if any, extrapyramidal side effects in patients and no catalepsy in animals. The atypical agents appear to differ from typical antipsychotics in a number of other ways: (a) no upregulation of D_2 receptors during chronic treatment of animals, (b) no development of depolarization block in nigrostriatal dopamine neurons, (c) little to no tardive dyskinesia in patients, and (d) transient or no elevation of serum prolactin levels (LIEBERMAN 1993).

Atypical antipsychotics may be further classified into type A or type B. Type A atypical drugs produce no extrapyramidal side effects nor elevations in prolactin levels; many of these compounds have been found to be potent $5HT_{2a}$ receptor antagonists as well as D_2 antagonists. Type B atypical drugs produce mild extrapyramidal side effects (at doses above those required for efficacy) and transient increases in serum prolactin. Some drugs in this class have been found to be selective D_2/D_3 antagonists or partial D_2 receptor agonists (MELTZER 1993). Clozapine and fluperlapine are examples of type A atypicals, while remoxipride and amisulpride may be classified as type B atypical antipsychotics. In the case of type B drugs, not all investigators would classify these drugs as atypical.

The typical/atypical drug classification scheme has achieved wide popularity among both clinicians and researchers in recent years, despite the difficulty of classifying some drugs in this manner. The usefulness of this scheme lies in its ability to unite basic and clinical observations on a drug's activity. It is widely used in the other chapters of this volume.

C. Classification by Chemical Structure

Antipsychotic drugs have long been classified according to their chemical structure. Chemical classification systems for antipsychotic drugs were originally devised to determine if certain structural features were related to particular patterns of clinical activity or side effects. Studies of structure–activity relationships have identified several critical variables in the activity profiles of these psychopharmacological agents. Psychopharmacological agents must possess certain chemical characteristics that allow the drug to be absorbed from the gastrointestinal tract, to be transported into the central nervous system, to interact with the membrane neurotransmitter receptors, and then to be metabolized and excreted from the body (Petracek et al. 1978). In this regard, most drugs have been modeled after the aromatic chain-amine model, a three-sectioned model which imparts important general chemical and physical properties onto each compound. The basic amine nitrogen of the amine group acts as a base, accepting protons from the physiological solvent system and allowing absorption to occur. In addition, the amine group often plays a key role in the receptor site–drug interaction by acting directly with membrane surfaces and enzymatic proteins. The aromatic portion of the compound confers solubility to this compound, allowing the compound to be absorbed and transported into the central nervous system. The third section of this model is the aliphatic chain of carbon atoms which connect the aromatic group to the amine group. This connecting chain can determine the geometry of the compound, sometimes creating a rigid structure. The geometry of the compound, especially with respect to the spatial relationships between the aromatic and the amine groups, impact the drug's functional activity.

Two components of this three-sectioned model are known to be critical features affecting drug activity. First, the aromatic group is usually the largest space-occupying portion of the compound and therefore dominates the overall steric arrangement. The steric pattern of all chemical compounds is known to be important in determining pharmacological activity. In addition, the overall geometry of the molecule is also a critical feature affecting drug activity. The overall geometry of the molecule is determined in part by the spatial relationship between the aromatic ring and the amine group. Aliphatic chains often set the distance between the aromatic and amine groups.

In sum, antipsychotics are heterocyclic compounds, and three types of structural modifications markedly affect their pharmacological activity: (a) the aromatic ring substitutions, (b) the configuration of the tertiary amine group, and (c) the length of the aliphatic side chain. There are seven main chemical classes of antipsychotic compounds currently in use: (1) phenothiazines, (2) thioxanthenes, (3) butyrophenones, (4) diphenylbutylpiperidines, (5) indoles, (6) dibenzapines, and (7) benzamides.

I. Phenothiazines

These drugs were the first major class of antipsychotics to be developed, and members of this class are still widely used today. In addition, this class is the largest, due to the large number of derivatives of phenothiazine synthesized since 1950. Phenothiazine is a three-ring heterocyclic compound in which two aromatic rings are linked by a third ring containing sulfur and nitrogen atoms (see Fig. 1). There are three subclasses in this family of antipsychotic drugs, based on differences in the composition of the side chain linked to the nitrogen atom in the phenothiazine nucleus (position 10): (1) aliphatics, (2) piperidines, and (3) piperazines. The agents in this category differ in terms of potency and side-effect profile. Several members of this class of compounds are commonly prescribed today: chlorpromazine, thioridazine, fluphenazine, perphenazine, and trifluoperazine.

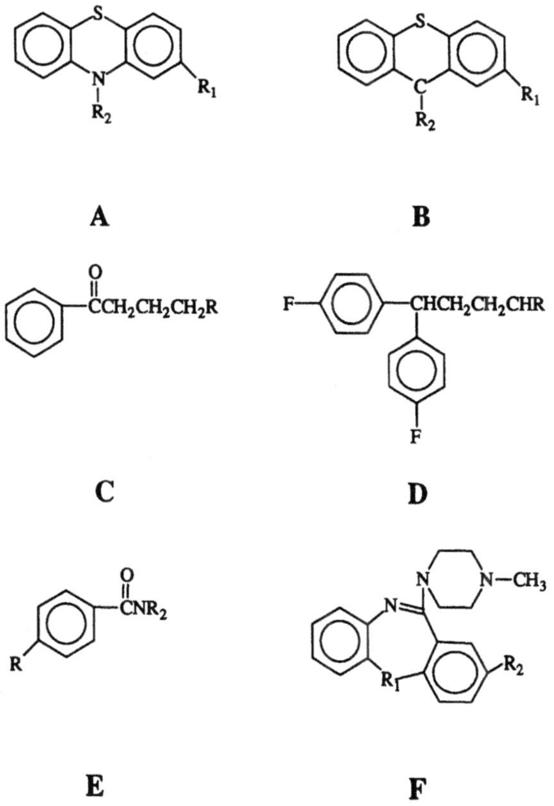

Fig. 1A–F. Structural formulas for the major classes of antipsychotics. **A** Phenothiazines. **B** Thioxanthenes. **C** Butyrophenones. **D** Diphenylbutylpiperidines. **E** Benzamides. **F** Dibenzapines

The efficacy of these compounds appears to be related to the presence of a substituent at the 2-position of the phenothiazine nucleus (BALDESSARINI 1980). Substitution of an electron-withdrawing group at the 2-position increases the potency of the phenothiazine – e.g., substitution of a trifluoromethyl group at the 2-position in place of the chloride atom in prochlorperazine yields trifluoperazine, an antipsychotic whose potency is much greater than the former drug. BALDESSARINI (1980) postulates that this increase in potency may be due to a conformational change in which the position of the substituent leans toward the substituted lateral ring. This structure is similar to the structure of endogenous catecholamines.

The spatial relationship of the nitrogen atom in the aliphatic, piperidine, or piperazine side chain to the position of the nitrogen atom within the phenothiazine nucleus also plays an important role in structure–activity relationships (ZIRKLE and KAISER 1970; BIEL et al. 1978). Aliphatic side chain substitutions at the 10-position yield two phenothiazines, chlorpromazine and trifluopromazine, that are relatively low in potency. A second class of phenothiazines, the piperidines, contains a substituted piperidine side chain at the 10-position. These compounds are similar in potency to the aliphatics but have a lower incidence of extrapyramidal side effects, presumably due to the acquisition of stronger anticholinergic effects. Thioridazine and mesoridazine are members of this subclass. The subclass of piperazines is comprised of eight compounds for which the 10-position nitrogen has a piperazine ring containing a side chain. Five of these compounds – prochlorperazine, trifluoperazine, perphenazine, fluphenazine, and acetophenazine – are among the most potent of the phenothiazines, and produce less sedation and autonomic side effects than other phenothiazines. However, these compounds carry a greater risk of acute extrapyramidal effects (BIEL et al. 1978).

II. Thioxanthenes

These drugs are closely related to the phenothiazine nucleus, differing only by the substitution of a carbon atom for a nitrogen atom in the central ring (see Fig. 1). The thioxanthenes are available with either aliphatic or piperazine substituents, but not piperidine substituents. The substitution of a carbon atom for the 10-position nitrogen atom confers a small decrease in general potency to this class of antipsychotics (HOLLISTER 1977). However, the clinical efficacy of the thioxanthenes is similar to that of the phenothiazines. Geometric isomers of the thioxanthenes exist secondary to the presence of a double bond between the 10-position carbon and the carbon in the side chain. The *cis* isomers are the more active of the isomers; BALDESSARINI (1980) postulates that this greater activity is due to a greater similarity between the *cis* structure of the thioxanthene and dopamine. FEINBERG and SNYDER (1975) demonstrated this concept with stereochemical models, computations, and crystallographic data. Chlorprothixene and thiothixene are the only members of this class available in the U.S., while others (e.g., *cis*-flupenthixol) are available in Europe.

The phenothiazines and the thioxanthenes share two characteristics: the amine is always tertiary and all the compounds have three carbon atoms interposed between the 10-position of the heterocyclic structure and the first amino nitogen atom of the side chain at the 2-position. When one or more of the substituents of the tertiary amine are removed, e.g., during metabolism, there is increasing loss of clinical efficacy of the compound (BALDESSARINI 1980). Similarly, clinical activity is also lost if the number of carbon atoms interposed between the 10-position and the first amino nitrogen atom is other than three atoms.

III. Butyrophenones

Haloperidol, the most commonly prescribed antipsychotic in the U.S. today, was developed using a novel approach in the early 1950s (JANSSEN 1970). After Paul JANSSEN joined his father's pharmaceutical company, he noted that while the company was manufacturing a number of classic prescription drugs, it was doing no research in chemical synthesis or drug development, and thus had no hope of obtaining patents or of expanding its market. With a team of young, highly motivated and industrious individuals, JANSSEN set out to synthesize new chemicals using simple methods and equipment, the cheapest possible intermediates, and efficient synthetic schemes. By the end of their first 5 years in this project, the company had synthesized over 2500 compounds. Butyrophenone antipsychotics were discovered in 1957, during this industrious period in the Janssen laboratories, following 4 years of tactical research into developing an analgesic more potent than morphine. After synthesizing a potent propiophenone, they began to alter the chemical structure to examine the effects of structural changes on potency systematically. Some of the compounds produced a morphine-like excitement in experimental mice, followed by sedation and mild catatonia (i.e., chlorpromazine-like activity). Derivatives of this compound which possessed both morphine-like and chlorpromazine-like activity were then synthesized. All the active derivatives contain a phenylpiperidine moiety (R; see Fig. 1) with the position of the nitrogen atom in the piperidine ring never disturbed. In 1958, haloperidol was synthesized. This compound was noted to be the most active neuroleptic drug known – it was more potent, longer-acting, and faster-acting than chlorpromazine. The compound was potent orally as well as parenterally and was well tolerated after long-term administration to experimental mice.

The synthesis of haloperidol has had a tremendous impact on the scientific and medical communities: over 1000 scientific papers on haloperidol were published between 1958 and 1969, and the amount of haloperidol consumed for medical purposes rose from 300 mg/year to over 2.3 million kg in the 10 years from 1959 to 1969 (JANSSEN 1970). Since the discovery of this psychoactive butyrophenone in 1958, over 5000 related compounds have been synthesized and pharmacologically investigated. Eight of these derivatives have become available for clinical and/or veterinary use: fluanisone, pipamperone, moperone, droperidol, benperidol, azaperone, spiperone, and trifluperidol. These butyrophenones were all discovered between 1957 and 1962 (JANSSEN 1974), and none of the subsequent derivatives of these compounds have been found to be more potent, longer-acting, more specific, or less toxic. Spiperone is one of the most potent antipsychotics synthesized and has become a major tool for labelling D_2 receptors in vitro and in vivo (BALDESSARINI 1985).

Structure–activity relationships for the butyrophenones have been extensively investigated. The general structure of a butyrophenone includes a tertiary amine containing at least one aromatic ring linked to the amine nitrogen by a keto group attached to an intermediate chain of three alkyl groups (see Fig. 1). The fluorine atom present in the *para* position of the aromatic ring is necessary for optimal potency of the butyrophenone (JANSSEN and VAN BEVER 1978). Placing a fluorine atom in the *meta* or *ortho* positions or placing other substituents in these positions or in the *para* position leads to compounds that are less potent than the compounds which contain fluorine in the *para* position. Compounds containing unsubstituted aromatic rings are two to eight times less potent that the fluoro-containing compounds.

The keto group of the butyrophenone compound is required for maximal potency; reduction of the ketone to an alcohol or other chemical modification of the carbonyl group results in less active compounds. There is one exception to this rule: the compound anisopriol, the alcohol resulting from the reduction of the keto group in fluanisone. Anisopriol is two times more potent than the carbonyl containing fluanisone. Isosteric replacement of the carbonyl moiety also leads to less active compounds. For example, replacement of the ketone group in spiperone by an oxygen atom results in a compound that possesses one-tenth the potency of spiperone and a shorter duration of action (JANSSEN 1966).

The length of the intermediate alkyl chain that links the keto group to the nitrogen atom of the tertiary amine is also critical for antipsychotic activity. Any alteration in the propylene chain, such as shortening, lengthening, branching, or incorporating it into a ring system, results in a decrease or loss of neuroleptic activity. BABOULENE and colleagues (BABOULENE et al. 1972; BABOULENE and STURTZ 1974) determined the antipsychotic potency of various compounds with cyclic intermediate alkyl chains or chains with less than or more than three alkyl groups and determined that their potency was significantly decreased.

Three other structural characteristics have also been found to be crucial to the antipsychotic activity of the butyrophenones. KOCH (1974) ascertained that the most potent of the butyrophenones must have the nitrogen atom of the tertiary amine incorporated in a piperidine ring with a substitution at position

4. Also, the plane of the piperidine ring must be perpendicular to the plane of the aryl group, that is, the substituent at position 4. The hydrogen atom attached to the piperidine ring needs to be located 3.5–6.5 Å from the nitrogen of the amine group and opposite to the lone pair of electrons on the nitrogen atom. The following compounds fulfill these requirements for optimal potency: haloperidol, spiperone, benperidol, and droperidol. In addition, fluspiperone, bromperidol, clofluperol, and moperone show optimal potency secondary to these structural characteristics.

IV. Diphenylbutylpiperidines

The diphenylbutylpiperidines are closely related in chemical structure to the butyrophenones (see Fig. 1). This class of antipsychotic compounds consists of butyrophenones in which the keto group (the alpha keto group) has been replaced by a 4-fluorophenylmethine moiety. The aromatic group is the same 4-fluorophenyl group as that present in the butyrophenones, as is the aliphatic chain attached to a tertiary amine incorporated in a 4-substituted piperidine ring. The diphenylbutylpiperidines are also similar to the butyrophenones with respect to the structural requirements for potency: the phenyl groups must contain (1) fluorine substituents at the *para* position, (2) an unbranched propylene intermediate side chain, and (3) a tertiary amine whose nitrogen atom is incorporated into a substituted piperidine ring.

Four compounds from this class have been developed: clopimozide, pimozide, penfluridol, and fluspirilene. All compounds have a long duration of action, have little sedative effects, and a low incidence of extrapyramidal side effects. Penfluridol and fluspiriline are particularly long-acting compounds, and pimozide has been suggested to be particularly effective in certain types of psychoses (Opler and Feinberg 1991; Stein and Hollander 1992; Tueth and Cheong 1993; Driscoll et al. 1993).

V. Indoles

The indoles are a small class of heterocyclic antipsychotic compounds (not shown in Fig. 1), and are structurally related to reserpine and lysergic acid. The main members of this class are oxypertine and molindone. These compounds exhibit typical clinical efficacy and relatively high potency. The antipsychotic activity of the indoles is dependent upon several different variables in this class. For compounds that are direct derivatives of the indole nucleus, the substituent of the nitrogen atom in the piperazine ring must be an aryl group in order for the compound to possess antipsychotic activity (Zirkle and Kaiser 1970). More distant derivatives of the indole nucleus, such as molindone, require alkyl or alkoxy groups attached to the central ring to retain antipsychotic activity.

VI. Benzamides

The benzamides comprise a large group of compounds that have structural similarities to the butyrophenones, and were developed by modification of the antiarrhythmic drug procainamide (see Fig. 1). Some of these compounds – sulpiride, remoxipride, raclopride, amisulpride, sultopride, tiapride, and DAN 2163 – are used in Europe as antipsychotics and others such as zacopride are undergoing clinical investigation. Metoclopramide is widely used as an antiemetic drug.

VII. Dibenzapines

The dibenzapines are derivatives of the piperazine tricyclic molecules (see Fig. 1). Compounds with two nitrogen atoms in the central ring are called dibenzodiazepines (clozapine and fluperlapine), while compounds with oxygen replacing the second nitrogen atom in the central ring are known as dibenzoxazepines (loxapine). When the same nitrogen atom is replaced by a sulfur atom, the compounds are known as dibenzothiazepines (metiapine and clothiapine).

VIII. Others

Further drug development will add new chemical classes of antipsychotic drugs to our armamentarium. For example, risperidone, a benzisoxazole, has recently been introduced throughout the world with success (CHOUINARD et al. 1993; CLAUS et al. 1992; BRESSA et al. 1991; MECO et al. 1989). Amperozide, a diphenylbutylpiperazine, has shown efficacy in the reduction of negative and positive symptoms (AXELSSON and NILSSON 1991). A dibenzothiazepine, ICI 204.636, is currently in phase II of clinical drug studies and has shown promise in small double-blind studies (FABRE et al. 1990). The tetracyclic cyano-dibenzoxepinoacepine derivative savoxepine has demonstrated antipsychotic efficacy in three open studies (BUTLER and BECK 1987; MOLLER et al. 1989; WETZEL et al. 1991). In each case, other drugs in these new classes may offer refinements of the advantages of the class.

D. Classification by Potency and Nonneurologic Side-Effect Profile

Antipsychotics have also been classified according to their potency, i.e., by dividing them into low- and high-potency compounds. Some authors also refer to drugs with intermediate potency. The potency of an antipsychotic has been traditionally expressed in equivalency units as compared to 100 mg of chlorpromazine, the prototypic antipsychotic drug. Low-potency drugs have chlorpromazine equivalencies in the range of 25–100 mg, with usual daily doses ranging from 40 to 1000 mg in adults. Triflupromazine, thioridazine, mesoridazine, chlorprothixene, and remoxipride are examples of low-potency antipsychotics with these characteristics (KANE 1993). Medium-potency drugs, such as prochlorperazine, perphenazine, acetophenazine, molindone, loxapine, and droperidol, exhibit chlorpromazine equivalencies in the range of 10–25 mg, with daily dosages ranging from 10 to 150 mg. Trifluorperazine, fluphenazine, thiothixene, haloperidol, and pimozide are all considered high-potency drugs. Their chlorpromazine equivalencies range from 1 to 5 mg and their daily doses range from 1 to 30 mg.

Nonneurologic adverse effects of antipsychotic preparations are inversely correlated with their potency; that is, with higher doses of lower-potency drugs, a variety of nonneurologic side effects are observed. This relationship between potency and nonneurologic side-effect profile is thus another way to classify antipsychotic compounds. For example, low-potency compounds tend to produce pronounced anticholinergic side effects, cardiovascular effects, including postural influences on blood pressure, sedative effects, and convulsant effects. Higher-potency drugs are relatively free of these nonneurologic problems.

Low-potency drugs such as chlorpromazine, mesoridazine, thioridazine, and chlorprothixene, at doses necessary to produce antipsychotic efficacy, block several different neurotransmitter receptors in addition to D_2 receptors: (1) alpha$_1$-adrenergic, (2) H_1 and H_2 histaminic, and (3) muscarinic receptors. Interactions with these receptor systems explain many of the adverse effects of these compounds: (1) orthostatic hypotension due to alpha$_1$-adrenergic blockade, (2) sedation due to antihistaminic and anticholinergic blockade, and (3) xerostomia, constipation, urinary retention, blurred vision, and increased resting heart rate due to blockade of muscarinic receptors. Other anticholinergic side effects include exacerbation of acute angle-closure glaucoma and adynamic ileus. At commonly used doses of these drugs, a variety of side effects not related to known receptor actions can also occur, e.g., photosensitivity reactions, simple allergic skin reactions, long-term pigmentary skin changes, weight gain, galactorrhea, amenorrhea, cholestatic jaundice, and seizures (PERRY et al. 1988).

High-potency agents – haloperidol, fluphenazine, trifluoperazine, pimozide, and thiothixene – also produce symptoms consistent with adrenergic, histaminic, and muscarinic blockade. However, the symptoms produced by this class of antipsychotics are less frequent and less severe than those produced by the low-potency compounds. The high-potency compounds are commonly associated with the development of acute extrapyramidal side effects including bradykinesia, a pseudoparkinsonian tremor of the upper extremities or perioral region, akathisia, and dystonia (CASEY 1993). Perphenazine, loxapine, and molindone are examples of intermediate potency compounds whose side effects are also intermediate between high- and low-potency drugs.

Potency does not predict the efficacy of a drug. For example, the atypical drug clozapine, although low in potency, has been shown to be effective in patients for whom other antipsychotics, even high-potency drugs, have not proven efficacious. Like other low-potency drugs, it is associated with strong anticholinergic side effects, anti-alpha$_1$-adrenergic effects, antihistaminic effects, and convulsant effects, making this antipsychotic difficult or impossible for some patients to tolerate. In addition, clozapine induces agranulocytosis in 0.8% of patients (LIEBERMAN and ALVIR 1992; LIEBERMAN et al. 1990), which further limits the use of this compound. It should be noted, however, that all psychotic compounds have been implicated in the development of rare cases of agranulocytosis (PERRY et al. 1988).

The association of disturbing autonomic side effects and sedation in the low-potency drugs and the association of increased incidence of extrapyramidal and potentially irreversible side effects of the high-potency drugs has guided the clinical use of these drugs for several decades, and has served as the driving force for an impressive wave of research efforts in the pharmaceutical industry. The main goal of this research has been to develop antipsychotics that are more effective and yet produce fewer side effects. Due to some recent successes of this strategy, the traditional system of clinical classification based on potency and side-effect profile may have lost its utility. For instance, the benzamide sulpiride has significant side effects on the autonomic system and the cardiovascular system, comparable to the low-dose compounds; however, this medication poorly penetrates the blood-brain barrier and thus must be given in high doses (GERLACH et al. 1985). Risperidone is another example of a compound which does not fit well into this classification system. Risperidone is a high-potency antipsychotic drug given at a chlorpromazine equivalency of 4–8mg. However, the side-effect profile of risperidone is more similar to the lower-potency drugs, i.e., it produces hypotension and has a low potential for producing extrapyramidal symptoms (CHOUINARD and ARNOTT 1993).

The late-developing extrapyramidal side effects of tardive dyskinesia and tardive dystonia are associated with the use of all antipsychotic drugs, as is the development of neuroleptic malignant syndrome (LIEBERMAN et al. 1991; CASSEY 1993; PEARLMAN 1986). These side effects are not well correlated with the high- or low-potency status of antipsychotics.

E. Classification by Pharmacological Mechanism

From the time that LABORIT observed that chlorpromazine had calming effects in psychiatric patients, the scientific community has sought to discover the mechanisms of action of antipsychotic compounds. CARLSSON and LINDQUIST (1963) first published experiments which demonstrated a link between antipsychotic drugs and dopamine. They reported that chlorpromazine and haloperidol increased the concentration of norepinephrine and dopamine metabolites. Later it was shown by ANDEN et al. (1964) and NYBACK and SEDVALL (1968) that antipsychotics preferentially increased dopamine metabolism. CARLSSON and LINDQUIST had hypothesized that the increase in catecholamine metabolites was the result of blockade of postsynaptic catecholamine receptors by antipsychotics, leading to secondary increases in catecholamine synthesis, release, and metabolism. Subsequent work involving in vitro labeling of dopamine receptors revealed that these receptors were bound by antipsychotic agents with high affinity (CLEMENT-CORMIER et al. 1974; SEEMAN et al. 1976). These studies, along with others showing that antipsychotic drugs blocked dopaminergic stimulation of adenylate cyclase and caused increases in the firing rates of dopamine neurons, led to formation of the dopamine hypothesis of psychotic disorders, particularly schizophrenia. This hypothesis asserted that psychosis was associated with an increase in dopamine transmission, and that neuroleptic-induced dopaminergic blockade corrected it.

Later, a strong direct correlation between the clinical potency of antipsychotic drugs and their affinity for dopamine D_2 dopamine receptors was noted (CREESE et al. 1976; SEEMAN et al. 1976; PEROUTKA and SNYDER 1980). This finding led to the proposal that blockade of dopamine D_2 receptors was essential for antipsychotic activity and emphasized the development of other D_2 antagonists as new antipsychotics. High affinity and specificity of dopamine D_2 receptor action was the goal. However, later findings in the neurobiology of schizophrenia suggested that other dopamine receptors and even other neurotransmitter systems probably play important roles in the pathophysiology of psychotic disorders. Molecular cloning along with receptor binding studies have demonstrated a family of dopamine receptors that are currently classified into two main groups containing five distinct receptor types and their variants. These advances have facilitated the current pharmacological classification system for antipsychotic drugs: (1) selective dopamine receptor D_2 antagonists; (2) combined D_2/D_3 antagonists; (3) combined D_2/D_1 antagonists, and (4) combined $5\text{-}HT_2\text{-}D_2$ antagonists.

I. Selective Dopamine Receptor D_2 Antagonists

Early hypotheses indicated that antipsychotic efficacy is wholly dependent upon dopamine D_2 receptor binding. In determining relative receptor affinities for typical and atypical antipsychotics, MELTZER et al. (1989) showed that phenothiazines, thioxanthenes, butyrophenones, indoles, pimozide, and loxapine all have high affinities for the D_2 receptor, possessing pK_i values of greater than or equal to 8.1. As expected, the affinity of these compounds for the D_2 receptor was positively correlated with their clinical efficacy with respect to resolution of psychotic symptoms. Also, SEEMAN et al. (1976) found a positive correlation between the dissocation constants for the D_2 receptor and the average clinical dose (in milligrams) of the antipsychotic drug. Haloperidol and pimozide have the highest affinities for dopamine D_2 receptor binding sites, with other members of the butyrophenones displaying similar activity. The phenothiazines fluphenazine and trifluoperazine also show high D_2 receptor affinity.

However, the atypical antipsychotics (i.e., clozapine, fluperlapine, melperone, and amperozide), although they bind to D_2 receptors, have lower affinities for this receptor compared to the typical compounds. In fact, the difference between affinities of the typical and atypical drugs, as measured by pK_i, is statistically significant and has been suggested to account for differences between typical and atypical drugs related to side-effect profile and patterns of symptom response (MELTZER et al. 1989).

II. Combined D_2/D_3 Antagonists

Several substituted benzamides with high affinity for the D_2 receptor have also been shown to bind to D_3 receptors. Sulpiride has long been used as an antipsychotic drug, and found to produce relatively mild extrapyramidal side effects (ALFREDSSON et al. 1984; GERLACH et al. 1985). Raclopride has more recently been shown to be a highly potent D_2 antagonist (KOHLER et al. 1985; ANDERSEN 1988), yet it produces little to no extrapyramidal symptoms. Basic pharmacological investigations have shown that both sulpiride and raclopride, along with remoxipride, are potent D_3 as well as D_2 receptor antagonists (GERLACH 1991). These substituted benzamides, along with amisulipride and emonapride, possess high affinity to D_2/D_3 receptors while maintaining antipsychotic efficacy and a low incidence of autonomic, cardiovascular, and extrapyramidal side effects. Interestingly, these combined D_2/D_3 antagonists exert behavioral effects in laboratory animals that differentiate them from selective D_2 antagonists. Sulpiride and remoxipride inhibit apomorphine-induced locomotion at doses below those required to cause catalepsy (ÖGREN et al. 1990). This differs from selective D_2 antagonists such as haloperidol which cause both behavioral effects at similar doses.

III. Combined D_2/D_1 Antagonists

The presence of more than one functional dopamine receptor in the central nervous system was demonstrated in a number of laboratories in the 1970s. Activation of the dopamine receptor located in the pituitary gland and the striatum was demonstrated to cause an inhibition of adenylate cyclase activity of neurons and to be involved behaviorally in the regulation of locomotion and low-intensity stereotypy (SPANO et al. 1979; DECAMILLI et al. 1979; STOOF and KEBABIAN 1981; WEISS et al. 1985). This class of dopamine receptor was named the D_2 receptor. The D_2 receptor was recognized to be a distinct entity from other dopamine receptors, first located in the parathyroid gland, which stimulated adenylate cyclase activity. Eventually, these receptors were named D_1 (KEBABIAN 1978; KEBABIAN and CALNE 1979). In animals, stimulation of D_1 receptors is linked to grooming behaviors and abnormal oral movements.

Most typical and atypical antipsychotics display some affinity for the D_1 receptor (SEEMAN and VAN TOL 1994; JACKSON et al. 1994). The thioxanthenes, particularly flupentixol and pitflutixol, have been shown to have affinities of equal relative intensity for both the D_1 and D_2 receptors (CLARK and WHITE 1987; MELTZER et al. 1989).

Several studies in rats and primates have suggested that selective D_1 receptor antagonists may have antipsychotic efficacy while producing fewer extrapyramidal side effects (WADDINGTON 1988; COFFIN et al. 1989; GERLACH 1991). This hypothesis has led to the development of relatively pure D_1 receptor antagonists, the first one developed being SCH-23390. This compound inhibited conditioned avoidance responses in rats and blocked hyperlocomotion, stereotypies, or grooming behavior produced by selective D_2 or D_1 agonists (CLARK and WHITE 1987). However, this compound was later also found to possess weak D_2 affinity and moderate affinity for $5HT_2$ receptors. SCH-39166, a compound more potent than SCH-23390 and more selective for the D_1 receptor, is currently undergoing preliminary testing prior to clinical trials (LIEBERMAN 1993).

IV. Combined 5-HT_2-D_2 Antagonists

As mentioned earlier, many of the antipsychotic agents currently available are both D_2 and $5HT_2$ antagonists. Analysis of the ratio of equilibrium constants for $5HT_2$ receptor binding to that of D_2 receptor binding has been performed in order to determine if this relationship is associated with important clinical characteristics of the drugs (MELTZER 1989; MELTZER et al. 1989). Typical antipsychotics have pk_i ratios ranging from 0.80 to 1.05, while atypicals such as clozapine, setoperone, and malperone have pk_i ratios greater than or equal to 1.15. Currently, studies are underway to investigate the possibility that refractory patients may respond only to compounds with a high $5HT_2$-D_2 ratio, since clozapine has a pk_i ratio of 1.19 (MELTZER et al. 1989) and is distinguished by its efficacy in this population. Risperidone also has high affinity for both D_2 and $5HT_2$ receptors. Preliminary evidence exists that suggests risperidone has antipsychotic properties with respect to both positive and negative symptoms (CASTELAS et al. 1989; MONFORT et al. 1989). Its efficacy in treatment-resistant patients is suspected (GELDERS et al. 1990; PEUSKENS et al. 1989), but is still under investigation. Zotepine, another combined $5HT_2$-D_2 antagonist, is currently undergoing clinical studies to determine its therapeutic efficacy.

F. Future Classification Schemes

There is little doubt that, in the future, new classes of antipsychotic drugs will be discovered and studied according to their pharmacological mechanisms at cellular and molecular levels. As our understanding of the neurobiology of schizophrenia and other psychotic disorders slowly increases, classification schemes based only on the empirical properties of drugs in animals or humans will become obsolete. As data continue to accumulate from investigations of the actions of antipsychotic drugs on a variety of neurotransmitter systems, many drugs, both old and new, are being found to have actions at other dopamine receptors (D_4 and D_5), serotonin receptors, sigma receptors, phencyclidine (PCP) receptors, glutamate receptors, and the benzodiazepine-gamma-aminobutyric acid (GABA) receptor complex. In addition, the ability of some drugs to act as partial dopamine receptor agonists may be of importance. These new findings have led to a multitude of hypotheses regarding the role of these receptor systems in the production of positive and negative symptoms in patients with psychotic disorders, and to a number of speculative new drug categories. A brief description of some of these categories is presented here.

I. Selective Serotonin Receptor Antagonists

The involvement of the serotonin neurotransmitter system in psychotic disorders has been inferred from observations that many antipsychotic drugs are serotonin antagonists, and that serotonin agonists such as lysergic acid diethylamide or dimethyl tryptamine have psychotogenic effects. Many typical antipsychotics, such as chlorpromazine, fluphenazine, haloperidol, pimozide, thiothixene, and loxapine, have been shown to possess affinities for the $5HT_2$ receptor (MELTZER et al. 1989). Molindone is an exception to this observation. Clozapine and other atypical antipsychotics possess relatively high affinities for the $5HT_2$ receptor (MELTZER et al. 1989). However, serotonergic antagonism alone does not appear to be adequate for antipsychotic activity (DEUTCH et al. 1991).

Risperidone, ritanserin, setoperone, melperone, olanzapine, ziprasidone, and R-79598 are all serotonin antagonists that display antipsychotic activity (LIEBERMAN 1993). Ritanserin and setoperone are the only pure $5HT_2$ antagonists in this group, and ritanserin has not been shown to definitely possess antipsychotic activity (LIEBERMAN 1993). Ritanserin has been shown to induce some improvement in the negative symptoms, mood symptoms (e.g., dysphoria) and to reduce extrapyramidal side effects in patients who were previously treated with typical antipsychotics (GERLACH 1991). In addition, it is unclear if the improvement seen in these patients was secondary to the serotonergic antagonism or simply due to the cessation of typical antipsychotic treatment. Preliminary studies have demonstrated a weak augmentation effect with regard to negative symptoms when ritanserin is added to typical antipsychotic therapy. Further investigations into the activity of $5HT_2$ antagonists will help elucidate the potential for these compounds to either potentiate or optimize the antipsychotic effect of other classes of compounds. Setoperone is a potent $5HT_2$ antagonist that displays efficacy for both positive and negative symptoms of psychotic disorders and is associated with a low incidence of extrapyramidal side effects (CEULEMAN et al. 1985). However, the poor bioavailability of this compound makes it pharmacologically impractical. Melperone possesses a moderate affinity for $5HT_2$ receptors and produces fewer extrapyramidal side effects, while providing equal antipsychotic coverage (BJERKENSTEDT et al. 1987; BJERKENSTEDT 1989).

Some antipsychotic compounds have been found to possess affinity for the $5HT_3$ receptor. For example, the dibenzoxazepine loxapine has a high affinity for the $5HT_3$ receptor as does the dibenzodiazepine clozapine (HOYER et al. 1989). However, $5HT_3$ antagonism alone may not be sufficient for antipsychotic activitiy, as the selective compound zacopride was found to have no significant antipsychotic effects in a recent pilot study (NEWCOMER et al. 1992).

II. Partial D_2 Agonists

MELTZER (1980) and CARLSSON (1988) have both postulated that partial agonists at dopamine D_2 receptors may be capable of modulating dopamine release in the central nervous system through stimulation of dopamine autoreceptors. Presumably, inhibition of dopamine activity would be achieved during periods of both excess and deficient dopamine release. Some of these compounds have been preliminarily shown to be efficacious with respect to both positive and negative symptoms while producing fewer extrapyramidal side effects (COWARD et al. 1989). Terguride, OPC-4392, SND-919, and EMD 49980 (roxindole) are all partial D_2 agonists and are currently undergoing clinical studies.

Dopamine agonist augmentation of standard antipsychotic treatment regimens has also been under investigation for several years. Amphetamine has been shown to improve mood and concentration in some schizophrenic patients concurrently under antipsychotic therapy (CESAREC and NYMAN 1985). L-DOPA has also been shown in several studies to improve negative symptoms is some patients maintained on antipsychotics (BUCHANAN et al. 1975; GERLACH and LUHDORF 1975; INANAGA et al. 1975; OGURA et al. 1976 and ALPERT et al. 1978).

III. Sigma Site Antagonists and Excitatory Amino Acid Agonists

Sigma site binding by opiate-like compounds and blockade of excitatory amino acid receptors by phencyclidine analogs have been associated with the development of psychotic symptoms. Antagonists of the sigma site have been investigated as potential antipsychotics. Remoxipride, a sigma site antagonist as well as a dopamine antagonist, have been shown to be efficacious in psychosis (HALL et al. 1986). However, three other sigma site antagonists – rimcazole, BWU-234, and BMY-14802 – have not been found to possess any antipsychotic activity (CHOUINARD and ANNABLE 1984; JAIN et al. 1987). Phencyclidine, which blocks the ligand-gated channel linked to several glutamate receptors, is clearly psychotogenic. Thus, compounds that facilitate glutamate receptor function may be helpful in the resolution of psychotic symptoms. Glycine, a synergist of glutamate, has been studied as a potential adjuvant to antipsychotic treatment, but with conflicting results (WAZIRI 1988; ROSSE et al. 1989).

IV. GABA-Mimetics and Partial Benzodiazepine Agonists

The role of the GABA-benzodiazepine neurotransmitter system in the pathogenesis of psychotic disorders is not well defined. ROBERTS (1976) proposed that the main function of this inhibitory neurotransmitter system is to preserve affective, cognitive, and behavioral activities through inhibition. Given this hypothesis, drugs that increase the inhibition of this neurotrans-mitter system may then improve the signs and symptoms assoicated with psychotic states. However, studies testing this hypothesis have been disappointing. GABA-mimetic drugs such as the agonist muscimol did not improve psychotic symptoms and in fact aggravated some psychotic symptoms (TAMMINGA and GERLACH 1987). Traditional benzodiazepines, however, have been used in double-blind, placebo-controlled trials and have been shown to be of some benefit with respect to the positive symptoms as well as symptoms of anxiety, tension, irritability, and agitation (LINGJÆRDE 1983; ARANA et al. 1986). The partial benzodiazepine agonist, bretazenil, has demonstrated efficacy in treating both negative and positive symptoms of schizophrenia (MERZ et al. 1988).

G. Conclusions

As new drugs are developed and their mechanism of action is elucidated, the number of categories in the pharmacological classification system of antipsychotics will increase. GERLACH and CASEY (1994) have proposed a pharmacological classification system for new antipsychotic drugs composed of three categories: (1) selective dopamine receptor antagonists, (2) multiple-receptor antagonists, and (3) nondopamine drugs. This system allows all old and all novel antipsychotics to be classified but does not give us any information on the clinical efficacy of the compound. Another system proposed by KANE and FREEMAN (1994) classifies the newer antipsychotics into four groups: (1) antipsychotics with selective dopamine receptor blockade, (2) partial D_2 receptor anonists, (3) nondopaminergic antipsychotics, and (4) antipsychotics blocking D_2 and other receptors.

An ideal classification system would aid in the development of new drugs by accurately predicting clinical efficacy and decreased side effects. Advances in molecular biology, pharmaceutical research, and clinical neuroscience may eventually provide enough information to develop such a classification system. It is hoped that new classification systems will not only allow the beneficial and adverse effects of antipsychotic drugs to be predicted, but will also provide scientists with new information about the underlying neurobiology of psychotic disorders such as schizophrenia.

References

Alfredsson G, Bjerkenstedt L, Edman G, Harnryd C, Oxenstierna G, Sedvall G, Wiesel FA (1984) Relationships between drug concentrations in serum and CSF, clinical effects and monoaminergic variables in schizophrenic patients treated with sulpiride or chlorpromazine. Acta Psychiatr Scand 311 [Suppl]:49–74

Alpert M, Friedhoff AJ, Marcos LR, Diamond F (1978) Paradoxical reaction to L-DOPA in schizophrenic patients. Am J Psychiatry 135:1327–1332

Andersen PH (1988) Comparison of the pharmacological characteristics of [3H]raclopride and [3H]SCH 23390 binding to dopamine receptors in vitro in mouse brain. Eur J Pharmacol 146:113–120

Anden NE, Roos BE, Werdinius B (1964) Effects of chlorpromazine, haloperidol and reserpine on the levels of phenolic acids in rabbit corpus striatum. Life Sci 3:149–158

Arana GW, Ornsteen ML, Kanter F, Friedman HL, Greenblatt DJ, Shader RI (1986) The use of benzodiazepines for psychotic disorders: a literature review and preliminary clinical findings. Psychopharmacol Bull 22:77–86

Axelsson R, Nilsson A (1991) Effects of amperozide in schizophrenia. An open study of a potent 5-HT_2 receptor antagonist. Psychopharmacology 104(3):287–292

Baboulene M, Sturtz G (1974) Aminomethyl-1-benzoyl-2-cyclopropanes. II. Etude stéréochimique; activité pharmacologique. Bull Soc Chim Fr 2929–2934

Baboulene M, Sturtz G, Hache J (1972) Synthèse d'aminomethyl-1-parafluorobenzoyl-2-cyclopropanes doués d'activité neuroleptique. Chim Ther 7:493–502

Baldessarini RJ (1980) Drugs and the treatment of psychiatric disorders. In: Gilman AG, Goodman LS, Gilman A (eds) Goodman and Gilman's the pharmaecological basis of therapeutics, 6th ed. MacMillan, New York, pp 391–418

Baldessarini RJ (1985) Chemotherapy in psychiatry. Principles and practice. Harvard University Press, Cambridge

Biel JH, Bopp B, Mitchell BD (1978) Chemistry and structure-activity relationships of psychotropic drugs. In: Clark WG, del Guidice J (eds) Principles of psychopharmacology, 2nd edn. Academic, New York, pp 140–168

Bjerkenstedt L (1989) Melperone in the treatment of schizophrenia. Acta Psychiatr Scand 352 [Suppl]:35–39

Bjerkenstedt L, Harnryd C, Grimm V, Gullberg B, Sedvall G (1987) A double-blind comparison of melperone and thiothixene in psychotic women using a new rating scale, the CPRS. Arch Psychiatr Nervenkrkh 226:157–172

Bleuler E (1916) Lehrbuch der Psychiatrie. Springer, Berlin

Bressa GM, Bersani G, Meco G, Boucique E (1991) One year follow-up study with risperidone in chronic schizophrenia. New Trends Exp Clin Psychiatry 7:169–177

Buchanan FH, Parton RV, Warren JW (1975) Double blind trial of L-DOPA in chronic schizophrenia. Aust NZ J Psychiatry 9:269–271

Butler B, Bech P (1987) Neuroleptic profile of cipazoxapine, a new tetracyclic dopamine antagonist: clinical validation of the hippocampus versus striatum ratio model of dopamine receptors in animals. Psychopharmacopsychiatry 20:122–126

Caldwell AE (1978) History of Psychopharmacology. In: Clark WG, del Giudice J (eds) Principles of psychopharmacology. Academic, New York, pp 9–40

Carlsson A (1988) The current status of the dopamine hypothesis. Neuropsychopharmacology 1:179–186

Carlsson A, Linquist M (1963) Effect of chlorpromazine or haloperidol on the formation of 3-methoxytyramine and normetanephrine in mouse brain. Eur J Pharmacol 99:103–105

Casey DE (1993) Neuroleptic-induced acute extrapyramidal syndromes and tardive dyskinesia. Psychiatr Clin North Am 16(3):589–610
Castelas JF, Ferreira L, Gelders UG, Heylen SLE (1989) The efficacy of the D2 and 5HT2 antagonist risperidone (R 64766) in the treatment of chronic psychosis. An open dose-finding study. Schizophr Res 2:411–415
Cesarec Z, Nyman AK (1985) Differential response to amphetamine in schizophrenia. Acta Psychiatr Scand 71:523–538
Ceulemans DLS, Gelders YG, Hoppenbrouwers MLJA, Reyntjens AJM, Janssen PAJ (1985) Effect of serotonin antagonist in schizophrenia: a pilot study with setoperone. Psychpharmacology 85:329–332
Chouinard G, Annable L (1984) An early phase II clinical trial of BW 234U in the treatment of acute schizophrenia in newly admitted patients. Psychopharmacology 84:282–284
Chouinard G, Arnott W (1993) Clinical review of risperidone. Can J Psychiatry 38 [Suppl 3]:S89–S95
Chouinard G, Jones B, Reminton G, Bloom D, Addington D, MacEwans GW, Labelle A, Beauclair L, Arnott W (1993) A Canadian multicenter placebo-controlled study of fixed doses of risperidone and haloperidol in the treatment of chronic schizophrenic patients. J Clin Psychopharmacol 13:25–35
Clark D, White FJ (1987) Review: D1 dopamine receptor – the search for a function: a critical evaluation of the D1/D2 dopamine receptor classification and its functional implications. Synapse 1:347–388
Claus A, Bollen J, De Cuyper H, Eneman M, Malfroid M, Peuskens J, Heylen S (1992) Risperidone versus halperidol in the treatment of chronic schizophrenia inpatients: a multicentre double-blind comparative study. Acta Psychiatr Scand 85:295–305
Clement-Cormier YC, Kebabian JW, Petzold GL, Greengard P (1974) Dopamine-sensitized adenylate cyclase in mammalian brain: a possible site of action of antipsychotic drugs. Proc Natl Acad Sci USA 71:1113–1117
Coffin Vl, Latranyi MB, Chipkin RE (1989) Acute extrapyramidal syndrome in Cebus monkeys: development mediated by dopamine D_2 but not D_1 receptors. J Pharmacol Exp Ther 249:769–777
Coward O, Dixon K, Enz A, Shearman G, Urwyler S, White T, Karobeth M (1989) Partial brain dopamine D2 receptor agonists in the treatment of schizophrenia. Psychopharmacol Bull 25:393–397
Creese I, Burt DR, Snyder SH (1976) Dopamine receptor binding predicts clinical and pharmacological potencies of antischizophrenic drugs. Science 192:481–483
DeCamilli P, Macconi D, Spada A (1979) Dopamine inhibits adenylate cyclase in human prolactin-secreting pituitary adenomas. Nature 278:252–255
Delay J, Deniker P, Harl JM (1952) Traitement des états d'excitation et d'agitation par une méthode médicamenteuse derivée de l'hibernothérapie. Ann Med Psychol 110:267–273
Deniker P (1970) Introduction of neuroleptic chemotherapy into psychiatry. In: Ayd FJ, Blackwell B (eds) Discoveries in biological psychiatry. Lippincott, Philadelphia
Deutch AY, Moghaddam B, Innis RB, Krystal JH, Aghajanian GK, Bunney BS, Charney DS (1991) Mechanisms of action of atypical antipsychotic drugs. Implications for novel therapeutic strategies for schizophrenia. Schizophr Res 4:121–156
Driscoll MS, Rother MJ, Grant-Kels JM, Hale MS (1993) Delusional parasitosis: a dermatological, psychiatric, and pharmacologic approach. J Am Acad Dermatol 29(6):1023–1033
Fabre L, Slotnick V, Jones V, Murray G, Malick J (1990) ICI 204.636, a novel atypical antipsychotic: early indications for safety and efficacy in man. Abstracts of the 17th congress of CINP, vol II. Kyoto, Japan, p 223
Feinberg AP, Snyder SH (1975) Phenothiazine drugs: structure activity relationships explained by a conformation that mimics dopamine. Proc Natl Acad Sci USA 72:1899–1903

Fielding S, Lal H (1978) Behaviorial actions of neuroleptics. In: Iversen LL, Iversen SD, Snyder SH (eds) Handbook of psychopharmacology. Plenum, New York, pp 91–128

Fielding S, Marky M, Lal H (1975) Elicitation of mouse jumping by combined treatment with amphetamine and L-dopa: blockade by known neuroleptics, Pharmacologist 17:210

Gelders YG, Hcylen SL, Vanden Bussche G, Reyntjens AJ, Janssen DA (1990) Pilot clinical investigation of risperidone in the treatment of psychotic patients. Pharmacopsychiatry 23:206–211

Gerlach J (1991) New antipsychotics: classification, efficacy, and adverse effects. Schizophr Bull 17(2):289–309

Gerlach J, Casey DE (1994) Drug treatment of schizophrenia: myths and realities. Gurr Opin Psychiatry 7:65–70

Gerlach J, Luhdorf K (1975) The effect of L-DOPA on young patients with simple schizophrenia, treated with neuroleptic drugs. Psychopharmacology 44:105–110

Gerlach J, Behnke K, Heltberg H, Munk-Andersen E, Nielsen H (1985) Sulpiride and haloperidol in schizophrenia: a double-blind cross-over study of therapeutic effect, side effects and plasma concentrations. Br J Psychiatry 147:283–288

Hall H, Sallemark M, Jerning E (1986) Effects of remoxipride and some related new substituted salicylamides on rat brain receptors. Acta Pharmacol Toxicol 58:61–70

Harnryd C, Bjerkenstedt L, Bjork K, Gullberg B, Oxenstierna G, Sedvall G, Weisel F-A, Wik G, Aberg-Wistedt A (1984) Clinical evaluation of sulpiride in schizophrenic patients – a double-blind comparison with chlorpromazine. Acta Psychiatr Scand 69 [Suppl 311]:7–20

Hollister LE (1977) Antipsychotic medications and the treatment of schizophrenia. In: Barchas JD et al. (eds) Psychopharmacology from theory to practice. Oxford University Press, New York

Hornykiewicz O (1982) Brain catecholamines in schizophrenia – a good case for noradrenaline. Nature 299:484–486

Hoyer D, Gozlan H, Bolanos F, Schechter LE, Hamon M (1989) Interaction of psychotropic drugs with central 5HT3 recognition sites: fact or artifact? Eur J Pharmacol 171:137–139

Inanaga K, Nakasawa T, Inoue K, Tachibana H, Oshima M, Kotorii T, Tanaka M, Ogawa N (1975) Double blind controlled study of L-DOPA therapy in schizophrenia. Folia Psychiatr Neurol Hpn 29:123–142

Jackson DM, Ryan C, Evenden J, Mohell N (1994) Preclinical findings with new antipsychotic agents: what makes them atypical? Acta Psychiatr Scand 89 [Suppl 380]:41–48

Jain AK, Kelwala S, Moore N, Gerson S (1987) A controlled clinical trial of tiosperone in schizophrenia. Int Clin Psychopharmacol 2:129–135

Janssen PAJ (1966) The chemical anatomy of neuroleptic drugs. Farm Revy 65:272–295

Janssen PAJ (1970) The butyrophenone story. In: Ayd FJ, Blackwell B (eds) Discoveries in biological psychiatry. Lippincott, Philadelphia, pp 165–179

Janssen PAJ (1974) Butyrophenones and diphenylbutylpiperidines. In: De Stevens G (ed) Psychopharmacological agents, vol 3. Academic, New York, pp 129–158

Janssen PAJ, Van Bever WFM (1978) Structure-activity relationships of the butyrophenones and diphenylbutylpiperidines. In: Iversen LL, Iversen SD, Snyder SH (eds) Handbook of psychopharmacology, vol 10. Plenum, New York, pp 1–36

Kalinowsky LB (1970) Biological psychiatric treatments preceding pharmacotherapy. In: Ayd FJ, Blackwell B (eds) Discoveries in biological psychiatry. Lippincott, Philadelphia

van Kammen PP, Peters J, Yso J et al. (1990) Norepinephrine in acute exacerbations of chronic schizophrenia: negative symptoms revisited. Arch Gen Psychiatry 47:161–170

Kane JM (1993) New antipsychotic drugs. A review of their pharmacology and therapeutic potential. Drugs 46(4):585–593
Kane JM, Freeman HL (1994) Towards more effective antipsychotic treatment. Br J Psychiatry 165 [Suppl 25]:22–31
Kane JM, Honigfeld G, Singer J, Meltzer HY (1988) Clozapine for the treatment-resistant schizophrenic: a double-blind comparison versus chlorpromazine/benzatropine. Arch Gen Psychiatry 45:789–796
Kebabian JW (1978) Dopamine-sensitive adenylate cyclase: a receptor mechanism for dopamine. Adv Biochem Psychopharmacol 19:131–154
Kebabian JW, Calne DB (1979) Multiple receptors for dopamine. Nature 277:93–96
Koch MHJ (1974) The conformation of neuroleptic drugs. Mol Pharmacol 10:425–427
Kohler C, Hall H, Ogren SO, Gawell L (1985) Specific in vitro and in vivo binding of 3H-raclopride. Biochem Pharmacol 34:2251–2259
Kraepelin E (1899) Psychiatrie, vol I: allgemeine Psychiatrie. Leipzig
Lal H, Mark M, Fielding S (1976) Effect of neuroleptic drugs on mouse jumping induced by L-dopa in amphetamine-treated mice. Neuropharmacology 15:669–671
Lewander T, Westerberg S-E, Morrison D (1990) Clinical profile of remoxipride – a combined analysis of a comparative double-blind multicenter trial programme. Acta Psychiatr Scand 82 [Suppl 358]:27–36
Lieberman JA (1993) Understanding the mechanism of action of atypical antipsychotic drugs. A review of compounds in use and development. Br J Psychiatry 163 [Suppl 22]:7–18
Lieberman J, Alvir J (1992) A report of clozapine-induced agranulocytosis in the United States. Drug Safety [Suppl 7]:1–2
Lieberman JA, Yunis J, Egea E, Conoso RT, Kane JM et al. (1990) HLA B38, DR4, DQW3 and clozapine-induced agranulocytosis in Jewish patients with schizophrenia. Arch Gen Psychiatry 47:945–948
Lieberman JA, Saltz BL, Johns CA, Pollack S, Borenstein M et al. (1990) The effects of clozapine on tardive dyskinesia. Br J Psychiatry 158:503–510
Lingjærde O (1983) Effect of the benzodiazepine derviative estazolam in patients with auditory hallucinations. Acta Psychiatr Scand 65:339–354
Marder SR, Ames D, Wirshing WC, Van Putten T (1993) Schizophrenia. Psychiatr Clin North Am 16(3):567–587
Meco G, Bedini L, Bonifati V, Sonsini U (1989) Risperidone in the treatment of chronic schizophrenia with tardive dyskinesia. Curr Ther Res 46:876–883
Meltzer HY (1980) Relevance of dopamine autoreceptors for psychiatry: clinical and preclinical studies. Schizophr Bull 6:456–475
Meltzer HY (1989) Clinical studies on the mechanism of action of clozapine: the dopamine-serotonin hypothesis of schizophrenia. Psychopharmacology 99:S18–S27
Meltzer HY (1991) The mechanism of action of novel antipsychotic drugs. Schizophr Bull 17(2):263–287
Meltzer HY (1993) Serotonin receptors and antipsychotic drug action. In: Gram LF, Balant LP, Meltzer HY, Dahl SG (eds) Clinical pharmacology in psychiatry. Strategies in psychotropic drug development. Psychopharmacology Series, volume 10. Springer, Berlin Heidelberg New York, pp 70–81
Meltzer HY, Matsubara S, Lee JC (1989) Classification of typical and atypical antipsychotic drugs on the basis of dopamine D-1, D-2 and serotonin pKi values. J Pharmacol Exp Ther 251: 238–246
Meltzer HY, Bastani B, Kwon K, Ramirez L, Burnett S, Sharpe J (1989b) Clozapine: new research on efficacy and mechanism of action. Eur Arch Psychiatry Neurol Sci 238:332–339
Merz WA, Alterwain P, Ballmer U, Bechelli L, Capponi R, Galeano-Munoz J, Marquez C, Nestoros J, Rivero-Almanzor L, Ucha-Udabe R, Versiani M (1988)

Treatment of paranoid schizophrenia with the partial benzodiazepine agonist, RO 16-6028 (abstract). Psychopharmacology 96 [Suppl]:237
Moller HJ, Kissline W, Dietzfelbinger T, Stoll K-D, Wendt G (1989) Efficacy and tolerability of a new antipsychotic compounds (Savoxepine): results of a pilot study. Pharmacopsychiatry 22:38–41
Monfort JC, Manus A, Bourguignon A, Bouhours D (1989) Risperidône: treatment of schizophrenic patients with negative symptoms. Abstracts of the 8th world congress on psychiatry, Athens, Greece. Elsevier, Amsterdam, p 440 (Excerpta Medica international congress series 899)
Muller JM, Schlittler E, Bein HJ (1952) Reserpin, der sedative Wirkstoff aus *Rauwolfia serpentina* Benth. Experientia 8:338
Newcomer JW, Faustman WO, Zipursky RB, Csernansky JG (1992) Zacopride in schizophrenia: a single-blind serotonin type 3 antagonist trial. Arch Gen Psychiatry 49:751–752
Niemegeers CJE, Janssen PAJ (1975) Differential antagonism to amphetamine-inducing oxygen consumption and agitation by psychoactive drugs. In: Fielding S, Lal H (eds) Antidepressants. Futurea, Mount Kisco, New York, pp 125–141
Niemegeers CJE, Verbruggen FJ, Janssen PAJ (1969a) The influence of various neuroleptic drugs on shock avoidance responding in rats. I. Nondiscriminated Sidman avoidance procedure. Psychopharmacologia (Berl) 16:1161–1174
Niemegeers CJE, Verbruggen FJ, Janssen PAG (1969b) The influence of various neuroleptic drugs on shock avoidance responding in rats. II. Nondiscriminated Sidman avoidance procedure with alternate reinforcement and extinction periods and analysis of the inter-response times (IRT's) (1). Psychopharmacologia (Berl) 16:175–182
Nyback H, Sedvall G (1968) Effect of chlorpromazine on accumulation and disappearance of catecholamine formed from tyrosine C-14 in brain. J Pharmacol Exp Ther 162:294–301
Ogren SO, Florvall L, Hall H, Magnusson O, Angeby-Moller K (1990) Neuropharmacological and behavioral properties of remoxipride in the rat. Acta Psychiatr Scand 82:21–26
Ogura C, Kishimato H, Nakao T (1976) Clinical effects of L-DOPA on schizophrenia. Curr Ther Res 20:308–318
Opler LA, Feinberg SS (1991) The role of pimozide in clinical psychiatry: a review. J Clin Psychiatry 52(5):221–222
Pearlman CA (1986) Neuroleptic malignant syndrome: a review of the literature. J Clin Psychopharmacol 6(5):257–273
Peroutka SJ, Snyder SH (1980) Relationship of neuroleptic drug effects at brain dopamine, serotonin, alpha-adrenergic, and histaminic receptors to clinical potency. Am J Psychiatry 137:1518–1522
Perry PJ, Alexander B, Liskow BI (1988) Psychotropic drug handbook. Harvey Whitney Books, Cincinnati
Petracek FJ, Biel JH, Bopp B, Mitchell BD (1978) Chemistry and structure-activity relationships of psychotropic drugs. In: Clark WG, del Giudice J (eds) Principles of Psychopharmacology. Academic, New York, pp 133–168
Peuskens J, Claus A, DeCuyer H, Bollen J, Eneman M, Wilms G (1989) Risperidone: a new approach in the treatment of schizophrenia. Abstracts of the 8th world congress on psychiatry athens, Greece, Elsevier, Amsterdam, p 347 (Excerpta Medica international congress series 899)
Pfeiffer CC, Jenny EH (1957) The inhibition of the conditioned response and the counteraction of shizophrenia by muscarinic stimulation of the brain. Ann NY Acad Sci 66:753–764
Roberts E (1976) Disinhibition as an organizing principle in the nervous system: the role of the GABA system. Application to neurologic and psychiatric disorders. In:

Chase TN, Tower DB (eds) GABA in nervous system function. Raven, New York, pp 515–540
Rosse RB, Theut SK, Banay-Schwartz M, Leighton M, Scarcella E, Cohen CG, Deutsch SI (1989) Glycine adjuvant therapy to conventional neuroleptic treatment in schizophrenia: an open-label, pilot study. Clin Neuropharmacol 12:416–424
Schneider PJ (1824) Entwurf zu einer Heilmittellehre gegen psychische Krankheiten – oder Heilmittel in Beziehung auf psychische Krankheitsformen. Tubingen
Seeman P (1980) Dopamine receptors. Pharmacol Rev 32:229–313
Seeman P (1992) Dopamine receptor sequences. Therapeutic levels of neuroleptics occupy D_2 receptors, clozapine occupies D_4. Neuropsychopharmacology 7(4):261–284
Seeman P, Van Tol HHM (1994) Dopamine receptor pharmacology. Trends Pharmacol Sci 15(4):264–270
Seeman P, Chau-Wong M, Tedesco J, Wong K (1975) Brain receptors for antipsychotic drugs and dopamine: direct binding assays. Proc Natl Acad Sci USA 72:4376–4380
Seeman P, Lee T, Chau-Wong M, Wong K (1976) Antipsychotic drug doses and neuroleptic/dopamine receptors. Nature 261:717–719
Seeman P, Bzowej NH, Guan HC, Bergeron C, Reynolds GP, Bird ED, Riederer P, Jellinger K, Tourtelotte WW (1987) Human brain D_1 and D_2 dopamine receptors in schizophrenia, Alzheimer's, Parkinson's and Huntington's diseases. Neuropsychopharmacology 1:5–15
Sen G, Bose KC (1931) *Rauwolfia serpentina*, a new Indian drug for insanity and high blood pressure. Indian Med World 2:194–201
Spano PF, Biggio G, Casu M, Gessa GL, Bareggi SR, Govoni S, Trabucchi M (1978) Interaction of metergoline with striatal dopamine system. Life Sci 23:2383–2391
Spano PF, Frattola L, Govini S, Tonon GC, Trabucci M (1979) Dopaminergic ergot derivatives: selective agonists of a new class of dopamine receptors. In: Fuxe K, Calne DB (eds) Dopaminergic ergot derivatives and motor function. Permagon, Oxford, pp 159–171
Spiegel R (1989) Psychopharmacology. An introduction. Wiley, London, pp 25–47
Stein DJ, Hollander E (1992) Low-dose pimozide augmentation of serotonin reuptake blockers in the treatment of trichotillomania. J Clin Psychiatry 53(4):123–126
Stoof JC, Kebabian JW (1981) Opposing roles for D-1 and D-2 dopamine receptors in efflux of cAMP from rat neostriatum. Nature 294:366–368
Sunahara RK, Seeman P, Van Tol HHM, Niznik HB (1993) Dopamine receptors and antipsychotic drug response. Br J Psychiatry 163 [Suppl 22]:31–38
Tamminga C, Gerlach J (1987) Neuroleptics and experimental antipsychotics in schizophrenia. In: Meltzer HY (ed) Psychopharmacology: the third generation in progress. Raven, New York, pp 1129–1140
Tueth MJ, Cheong JA (1993) Clinical uses of pimozide. South Med J 86(3):344–349
Waddington JL (1988) Therapeutic potential of selective D-1 dopamine receptor agonists and antagonists in psychiatry and neurology. Gen Pharmacol 19:55–60
Waziri R (1988) Glycine therapy of schizophrenia. Biol Psychiatry 23:210–211
Weiss S, Sebben M, Garcia-Senz JA, Bockaert J (1985) D2-dopamine receptor-mediated inhibition of cyclic AMP formation in striatal neurons in primary culture. Mol Pharmacol 27:595–599
Wetzel H, Wiedemann K, Holsboer F, Benkert O (1991) Savoxepine: Invalidation of an "atypical" neuroleptic response pattern predicted by animal models in an open clinical trial with schizophrenic patients. Psychopharmacology 103:280–283
Wiesel FA (1994) II. The treatment of schizophrenia. Neuroleptic treatment of patients with schizophrenia. Mechanisms of action and clinical significance. Br J Psychiatry 164 [Suppl 23]:65–70

Wittern R (1983) Die Geschichte psychotroper Drogen vor der Ära der modernen Psychopharmaka. In: Langer G, Heimann H (eds) Psychopharmaka – Grundlagen und Therapie. Springer, Berlin Heidelberg New York
Zirkle CL, Kaise C (1970) Antipsychotic drugs. In: Berger A (ed) Medicinal chemistry, 2nd edn. Wiley, New York, pp 1410–1469

CHAPTER 2
Molecular Models and Structure-Activity Relationships

SVEIN G. DAHL

A. Introduction

Classical structure-activity relationship (SAR) studies of compounds with a potential antipsychotic effect have examined the relationship between topological and physical-chemical properties of the drug molecules, and their activities in relevant biological systems. One aim of such SAR studies has been to define a specific pharmacophore model describing the spatial distribution of three or four key atoms or chemical groups in the drug molecule which are believed to be essential to antipsychotic potency.

The present era of cloning and molecular modelling of neurotransmitter receptors has taken SAR studies of antipsychotic drugs one step further. In addition to modelling of drug molecules and postulation of pharmacophore models, entire drug-receptor complexes may now be modelled and used to simulate drug-receptor interactions. Such simulations have added to our understanding of the relationship between the molecular structure of antipsychotic drugs and their biological effects.

Drug-receptor interactions have traditionally been explained by "lock and key", "zipper" or "induced fit" mechanisms. Simulations of ligand-receptor interactions, using three-dimensional receptor models, have demonstrated that ligand-receptor interactions should be regarded as dynamic processes, where the ligand may move between different molecular conformations as it approaches and binds to the receptor (DAHL et al. 1991a; EDVARDSEN and DAHL 1991, 1992; EDVARDSEN et al. 1992).

B. Structure-Activity Relationships of Antipsychotic Drugs

Since chlorpromazine was introduced in 1952, a number of potential antipsychotic compounds have been synthesized and their activities examined in various biological systems. After the postulation of the dopamine hypothesis of schizophrenia and antipsychotic drug action (CARLSSON and LINDQUIST 1963), biological assays and animal behavioural tests of antidopaminergic action were developed and used to assess the antipsychotic potential of new compounds, and to postulate pharmacophore models (HUMBER et al. 1975, 1979; PHILIPP et al. 1979).

The crystal structure of chlorpromazine was reported in 1969 (McDowell 1969), and the three-dimensional structures of several antipsychotic drugs have since been examined by X-ray crystallographic techniques (Tollenaere et al. 1977; McDowell 1974; Dahl et al. 1986). As in the structure of chlorpromazine, the crystal structures of other psychoactive phenothiazine and thioxanthene derivatives have the tricyclic ring systems folded with a 135°–145° angle between the phenyl ring planes, and the central ring in a "boat" conformation. The three-dimensional structures of butyrophenones do not closely resemble those of the tricyclic phenothiazine and thioxanthene drugs.

One prevailing hypothesis in the 1970s and early 1980s, based on the structures of chlorpromazine, haloperidol and various other antidopaminergic compounds, was that antipsychotic activity requires a three-carbon-atom bridge having an S-shaped conformation, between two nitrogen atoms or between a nitrogen and a carbon atom (Kaufman and Manian 1972; Harbert et al. 1980). Other pharmacophore models of antidopaminergic activity placed a nitrogen atom in a certain position above the plane of an aromatic ring system in the drug molecule (Humber et al. 1975; Philipp et al. 1979; Olson et al. 1981). Most neuroleptic drugs have a tertiary amino group that may act as a hydrogen bond acceptor, and the most potent ones also have a hydrogen bond donor. It has been proposed from conformational studies that the distance between these groups should be between 3.5 and 6.5 Å (Koch 1974).

The development of specific radioligand binding assays for dopamine receptors in the brain led to more precise assessment of dopamine receptor antagonist activity, and to more detailed pharmacophore models (Harbert et al. 1980; Asselin et al. 1986; Högberg et al. 1986; Manallack and Beart 1988; Rognan et al. 1990; Liljefors et al. 1990). It was proposed from one of these models that a phenolic hydroxyl group in dopamine or in an antagonist molecule may form a hydrogen bond with a serine residue in the dopamine receptor (Asselin et al. 1986). Site-directed mutagenesis experiments with dopamine D2 receptors later suggested that this may be the case, and that a pair of conserved serine residues in transmembrane helix (TMH) 5 may have such a function (Cox et al. 1992). Others have suggested that the catechol groups in dopamine form hydrogen bonds with the conserved Asp residue in TMH 2, further down in the central core of the receptor than the two conserved Ser residues in TMH 5.

However, the pharmacophore models of antidopaminergic activity that have been proposed have usually not taken into account that all currently used antipsychotic drugs have flexible molecular structures, and may attain various molecular conformations in solution and even in crystals. As shown in Fig. 1, the solid state molecular conformation of the free base of a psychoactive phenothiazine drug, methoxypromazine, was completely different from that of the maleate salt of the same drug (Viterbo et al. 1984; Dahl et al. 1986). The molecular conformation of any antipsychotic drug, determined by X-ray crys-

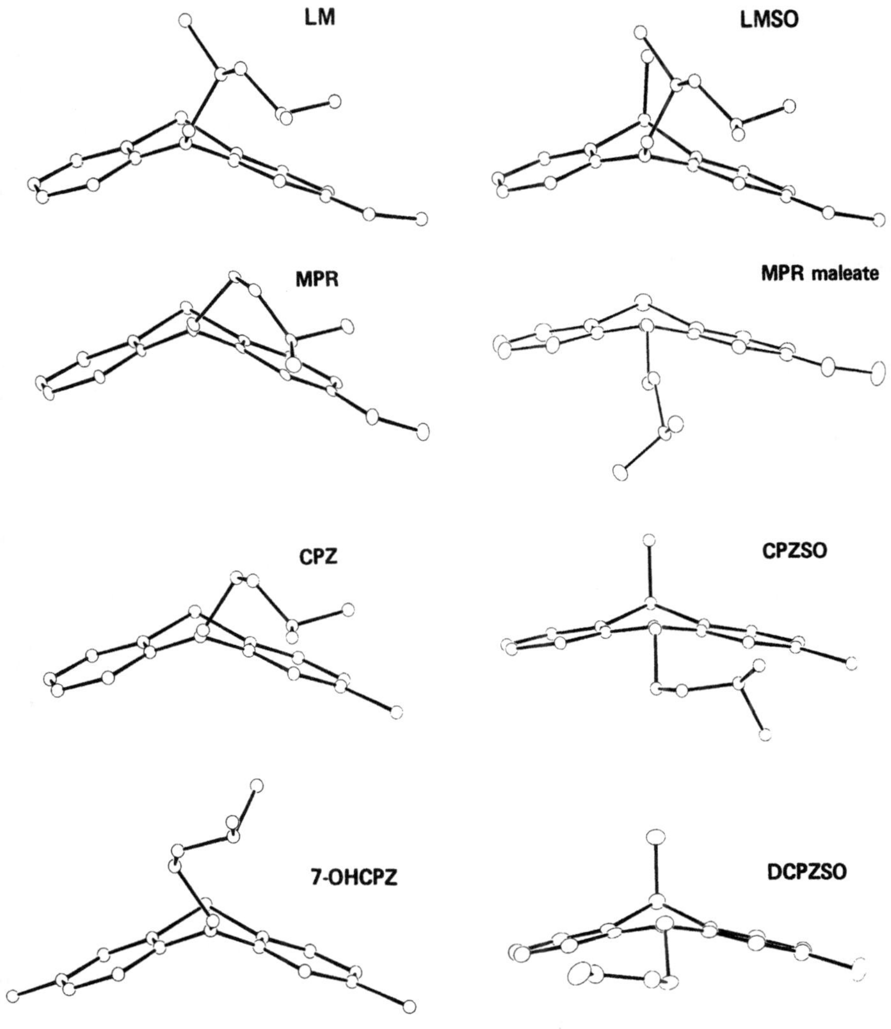

Fig. 1. Three-dimensional structures of levomepromazine (*LM*), levomepromazine sulphoxide (*LMSO*), methoxypromazine (*MPR*), methoxypromazine maleate (*MPR* maleate), chlorpromazine (*CPZ*), chlorpromazine sulphoxide (*CPZSO*), 7-hydroxy chlorpromazine (*7-OH CPZ*) and N-desmethyl chlorpromazine sulphoxide (*DCPZSO*), determined by X-ray crystallography. Reproduced from DAHL et al. (1986)

tallography, may therefore be different from its conformation when it is interacting with and binding to a receptor.

X-ray crystallographic and nuclear magnetic resonance (NMR) spectroscopic studies have provided detailed information about various side chain conformations of psychoactive phenothiazine derivatives relative to the ring system. This is illustrated in Fig. 1 for three different phenothiazine drugs and some of their metabolites. It has been postulated that each of the neuroleptic, antihistaminic and anticholinergic effects of phenothiazine drugs is produced by three distinct side chain conformations (BARBE et al. 1973; BARBE and BLANC 1976). This hypothesis, which was postulated long before any details regarding molecular structures of neurotransmitter receptors were known, has not been experimentally verified. Conclusive evidence concerning whether neuroleptics and other drugs have significantly different molecular conformations when they bind to different neurotransmitter receptors will probably only be available when detailed three-dimensional receptor structures become known from X-ray crystallographic experiments.

Fluctuations between various molecular conformations change not only the topology but also the electrostatic field around antipsychotic and other drug molecules (WEBER et al. 1986; DAHL et al. 1992). Different atoms have different electrostatic charges, and the net contribution from each atom to the electrostatic field around a molecule may depend on the molecular conformation. This is illustrated in Fig. 2 for chlorpromazine and one of its metabolites, chlorpromazine sulphoxide. As shown in Fig. 2, the degree of polarity of the metabolite depends on the side chain conformation. The sulphoxy (SO) group creates a strong electronegative field around this part of the molecule, which may be neutralized when the side chain has a conformation which places the protonated dimethylamino group near the SO group. As discussed below, the distribution of molecular electrostatic potentials around drug molecules may be as important as their topology for interactions with molecular targets.

C. Neurotransmitter Receptor Models

Up to 20 years ago, the term "receptor" was merely a theoretical concept used to explain cellular effects of drugs and endogenous signal transmitters. Since then receptors have been recognized as membrane proteins, and information about their molecular structures has accumulated rapidly, especially due to the development of molecular biology. Earlier studies of biological mechanisms related to antipsychotic drug effects focused mainly on dopamine receptors, but the interest in possible involvement of receptors for serotonin and other neurotransmitters has increased over the last years. Most receptor types and subtypes that have been linked to antipsychotic drug action belong to the superfamily of G protein-coupled receptors.

The three-dimensional structures of about 2000 different proteins have been determined by X-ray crystallography, but the detailed three-dimensional structures of the majority of proteins which have been cloned, including the G protein-coupled receptors, are still unknown. The classical experiments of ANFINSEN and his colleagues (ANFINSEN et al. 1961; ANFINSEN 1973) suggested that all information required to direct the folding of a protein into its tertiary structure lies in the amino acid sequence, and attempts to predict tertiary protein structures have since been made by a variety of experimental and computational methods. Accurate prediction of secondary and tertiary struc-

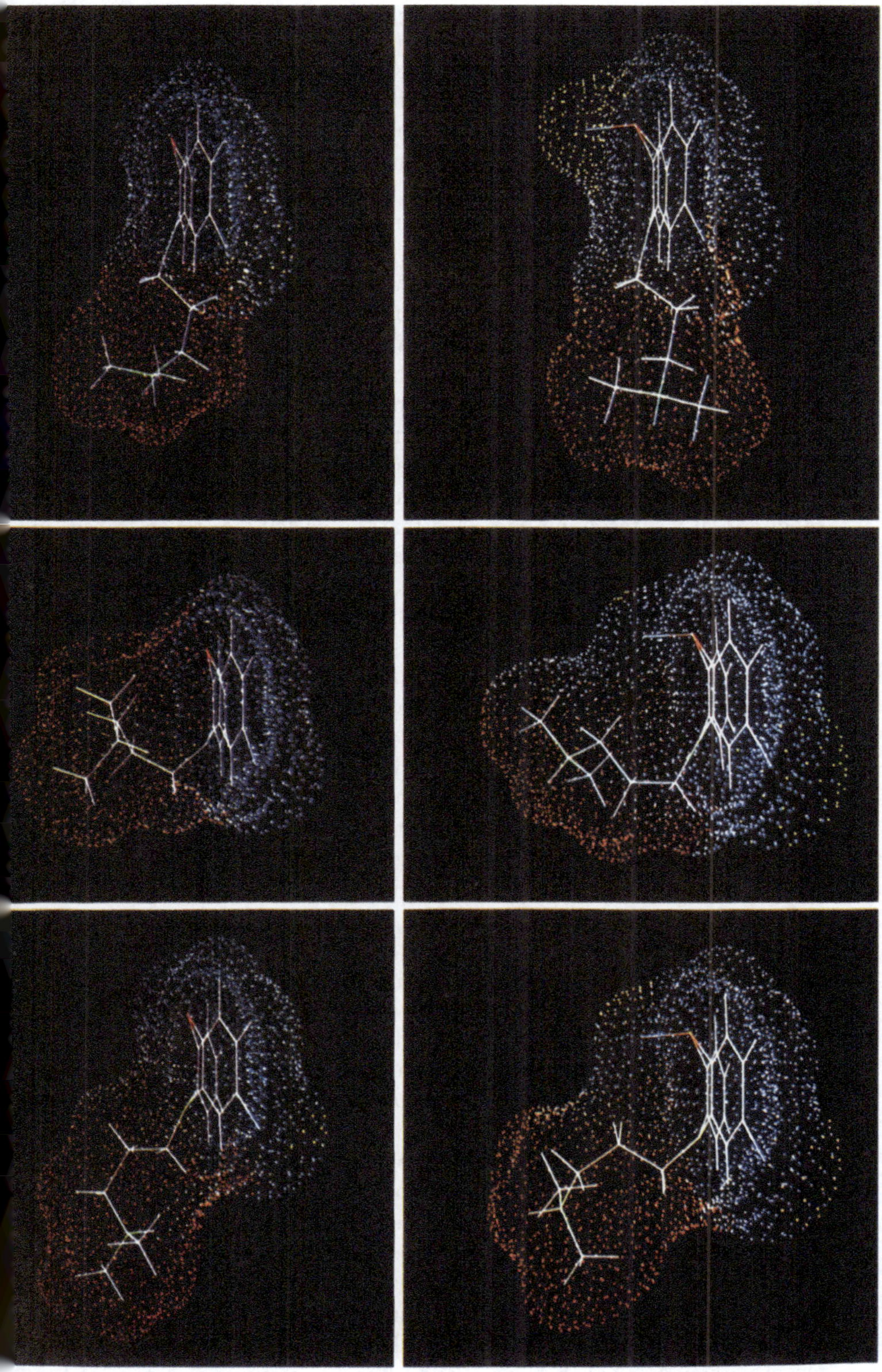

Fig. 2. Energy-minimized conformations of chlorpromazine (*left row*) and chlorpromazine sulphoxide (*right row*). The *dots* show the molecular surfaces, colour-coded according to the molecular electrostatic potentials (e, kcal/mol) 1.4 Å outside the surface: *yellow*, $e < o$; *blue*, $0 \leq e < 12$; *white*, $12 \leq e \leq 15$; *red*, $e > 15$. Reproduced from DAHL et al. (1992)

tures from the amino acid sequence has, however, proven to be extremely difficult, and no generally applicable method has yet been found.

Regardless of the limitations in protein models predicted mainly from amino acid sequences, we and others have constructed three-dimensional models of G protein-coupled dopamine and serotonin receptors, and used such models to simulate drug–receptor interactions (DAHL et al. 1989; DAHL and EDVARDSEN 1994; HUTCHINS 1994). These receptor models have been based on the amino acid sequences of the receptors, and also on postulated structural similarities in the membrane-spanning domains between G protein-coupled receptors and bacteriorhodopsin or visual rhodopsin.

A number of structural, biochemical and biological studies have suggested a seven α-helical membrane spanning topography as a general structure of G protein-coupled receptors. This has been confirmed by electron microscopic studies of the light receptor, rhodopsin, a chromoprotein which is linked to a G protein, transducin (G_t). A two-dimensional projection map of crystalline rhodopsin at 9 Å resolution indicated that the protein has seven transmembrane helices (TMHs), four of which are nearly perpendicular to the membrane plane (SCHERTLER et al. 1993).

Antagonist binding to G protein-coupled neurotransmitter receptors apparently takes place in the membrane-spanning domain, in a central core between the seven helices. This suggestion originated from site-directed mutagenesis studies of the β_2-adrenergic receptor (DIXON et al. 1986), and has since been confirmed by fluorescence emission spectroscopic experiments with the β_2 receptor (TOTA and STRADER 1990), and a number of site-directed mutagenesis studies with this and other neurotransmitter receptors. In dopamine receptors as in other G protein-coupled receptors, several different amino acid residues lining the central core of the receptor form a binding pocket and determine ligand binding specificity (KOZELL et al. 1994; HUTCHINS 1994).

Site-directed mutagenesis studies have shown that conserved aspartic acid residues in hydrophobic segments 2 and 3 have a functional role in G protein-coupled neurotransmitter receptors (SAVARESE and FRASER 1992). Replacement of the conserved Asp residue in hydrophobic segment 3 (Asp-114 in the rat dopamine D2 receptor) with Asn strongly reduces or completely abolishes the binding of both agonists and antagonists to various G protein-coupled receptors. Replacement of the conserved Asp residue in hydrophobic segment 2 (Asp-8 in the rat D2 receptor) with Asn abolishes agonist-induced signal transduction without significantly affecting antagonist binding.

The first dopamine D2 receptor model that was proposed had seven antiparallel membrane-spanning α-helices in a circular arrangement, and was refined by molecular dynamics simulations of the extra- and intracellular parts and by molecular mechanical energy minimization (DAHL et al. 1989, 1991b). The model placed Asp-80 in helix 2 further down in the central core of the receptor than Asp-114 in helix 3, and offered a steric explanation of how Asp-80 could be involved in signal transduction, and possibly also in agonist binding, without affecting antagonist binding. In this model both Asp-114 and

Asp-80 are required for high-affinity agonist binding, while only Asp-114 acts as a counter-ion in the binding of protonated antagonists.

A more recent dopamine D2 receptor model was built and energy-minimized with the seven transmembrane α-helices in an oval arrangement, and with a layer of water molecules around the synaptic and cytoplasmic parts and in the central core. When this receptor–water model was refined by molecular dynamics simulation, the seven helices moved into a rhodopsin-like arrangement with transmembrane helix 3 in a more central position in the seven-helix bundle (DAHL and EDVARDSEN 1994). Other models of G protein-coupled receptors have placed the seven helices directly in a rhodopsin-like arrangement by interactive computer graphics techniques, and refined the models by molecular mechanical energy minimization and molecular dynamics simulations of the receptor with a ligand near a postulated binding site (HUTCHINS 1994; SYLTE et al. 1995; KRISTIANSEN and DAHL 1995).

Although the present three-dimensional models of G protein-coupled receptors for dopamine, serotonin and other neurotransmitters may have a fairly correct overall shape, it is premature to draw firm conclusions from such models about the fine structure of the receptors and detailed mechanisms of ligand binding. Up to now the main usefulness of such receptor models has been to interpret data from site-directed mutagenesis studies of drug–receptor interactions, and to design such experiments. Three-dimensional dopamine and serotonin receptor models have already proven to be useful for these purposes.

D. Molecular Modelling of Drug–Receptor Interactions

I. Electrostatic Fields Around Drug and Receptor Molecules

Some of the dopamine and serotonin receptor models which have been proposed have included both the seven membrane-spanning helices and the cytoplasmic and synaptic domains (DAHL et al. 1989, 1991b; EDVARDSEN et al. 1992; SYLTE et al. 1993; KRISTIANSEN and DAHL 1995). Although the cytoplasmic and synaptic parts of these models must be regarded as relatively crude estimates, calculation of the molecular surfaces and electrostatic potentials of the entire receptor models provided some interesting results. For all these models, the electrostatic field had a pronounced dipolar character, with the synaptic domain dominated by negative potentials and the cytoplasmic domain dominated by positive potentials. This is illustrated in Fig. 3 for our initial dopamine D2 receptor model. Interestingly, a similar dipolar character was seen in a recent 12-helical model of the presynaptic dopamine transporter (EDVARDSEN and DAHL 1994). These calculations confirm the "positive inside rule" for membrane proteins, which has been suggested by others from molecular biology experiments (NILSSON and VON HEIJNE 1990).

The dipolar character of the receptor models was not as apparent when judged only from the distribution of positively and negatively

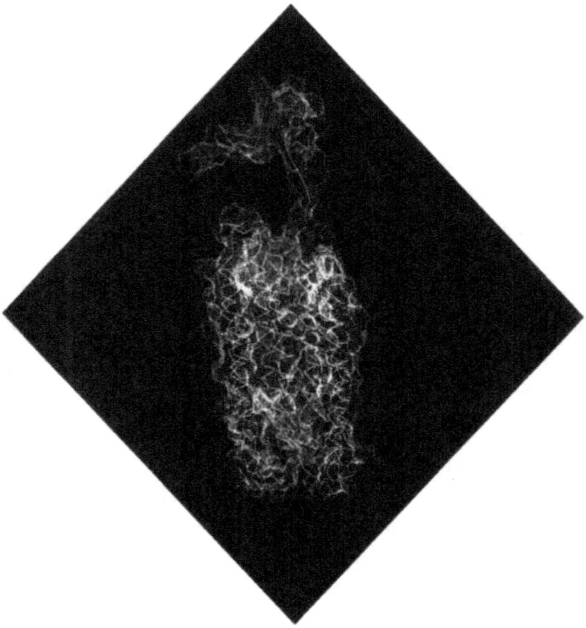

Fig. 3. Three-dimensional model of the rat dopamine D_2 receptor, viewed in the plane of the cell membrane. The *dots* show the molecular surface of the receptor, colour-coded according to electrostatic potentials (e, kcal/mol) 1.4 Å outside the surface: *blue*, e < –15; *white*, –15 < e < 15; *red*, e > 15. Reproduced from DAHL et al. (1991)

charged residues in the synaptic and cytoplasmic domains. The calculated electrostatic field around a protein depends both on the charges assigned to individual atoms and on the protein folding, such that positive and negative charges within a certain distance may partly neutralize each other. Although the folding of the synaptic and cytoplasmic segments in these receptor models is certainly not correct in every detail, the overall "negative outside – positive inside" character of the receptors may still be a real phenomenon.

The negative electrostatic field around most of the synaptic receptor domains suggests that electrostatic forces may attract protonated agonists and antagonists to the receptor. Molecular modelling studies demonstrated that sulphoxide metabolites of phenothiazine drugs (DAHL et al. 1992), and the *trans*-(Z)-isomers of thioxanthene drugs (SYLTE and DAHL 1991a,b), all of which are pharmacologically inactive, have strong negative electrostatic potentials around part of the molecules. Such negative electrostatic potentials, which reduce their electrostatic attraction to the D2 receptor, were not found for the pharmacologically active phenothiazine derivatives and *cis*-(E)-thioxanthenes. This may be one reason why the *trans*-(Z)-isomers of thioxanthene drugs are virtually devoid of D2 receptor binding affinity and antipsychotic activity.

II. Molecular Dynamics of Drug–Receptor Interactions

Many models of G protein-coupled receptors and drug interactions which have thus far been proposed were built and regarded as static structures, without considering different conformations of the ligand. However, proteins in their native state in cells have constantly changing geometries, with movements occurring on a femtosecond (10^{-15}s) time scale. The method of molecular dynamics simulations, which combines a molecular mechanics energy function with Newton's equations of motion for a molecular system, has been used both to study the naturally occurring internal movements in proteins and other biologically active molecules, and to refine three-dimensional molecular structures. Due to the addition of kinetic energy to the molecular system, a structure may move across conformational barriers and undergo substantial conformational changes during such simulations. For example, a seven-helical protein model may change from an initial circular arrangement of the helices into a more bacteriorhodopsin-like shape during 20–25 ps of molecular dynamics simulation in vacuo (JÄHNIG and EDHOLM 1992; EDVARDSEN et al. 1992).

Molecular dynamics simulations are always started from an energy-minimized set of atomic coordinates. Such simulations, even over a period of only a few picoseconds (10^{-12}s), involve a large number of computational steps, and simulations of macromolecules are facilitated by using a supercomputer.

It has previously been suggested that dopamine has an extended *trans* conformation when it interacts with a receptor (KOMISKEY et al. 1978). A series of conformationally restricted phenothiazine analogs, which were designed from this hypothesis, had no neuroleptic activity (GROL et al. 1981). However, dopamine, serotonin and all currently used antipsychotic drugs have flexible molecular structures, which may move rapidly between various conformations as the ligand approaches the receptor (EDVARDSEN and DAHL 1991, 1992; SYLTE and DAHL 1991a,b). Molecular dynamics simulations of ligand-receptor complexes have indicated that ligands may bind to G protein-coupled receptors by a "zipper" mechanism, which allows the ligand to adjust its conformation as it approaches and binds to the receptor (DAHL et al. 1991b; EDVARDSEN et al. 1992). It also appears from these simulations that the amine and hydroxyl groups of dopamine and serotonin may interact with more than one site in the receptor.

E. Conclusions

One of the aims of structure–activity relationship studies is to explain binding specificities of various ligands, and to make correct predictions about receptor-binding affinities of ligands which have not yet been examined in binding assays. Receptor modelling and molecular biology experiments have shown that several residues lining the central core of dopamine and serotonin receptors may interact with ligands and be of importance for the specificity of ligand recognition and binding. Molecular dynamics simulations of receptor models with an antagonist or a neurotransmitter have demonstrated that ligand interactions should be regarded as dynamic processes, which may be essential for understanding the molecular mechanisms.

Antipsychotic drug molecules and the neurotransmitters acetylcholine, dopamine and serotonin have flexible structures which may move rapidly between various conformations as the ligand approaches the receptor. Molecular dynamics simulations of ligand–receptor complexes suggest that neurotransmitter and antagonist drug molecules may bind to the receptors by "zipper" mechanisms, in which the ligand adjusts its conformation as it approaches and binds to the receptor.

High affinity drug–receptor binding requires complementarity between the drug and the receptor-binding pocket both in topology and in molecular electrostatic fields. Both the molecular electrostatic potentials and the three-dimensional structure of ligand molecules must therefore fulfil certain requirements for receptor recognition and binding. Neurotransmitters and antipsychotic drug molecules are protonated at physiological pH, which produces a positive electrostatic field around all or most of the molecule. Molecular modelling has indicated that G protein-coupled receptors have a bipolar charge distribution, with electrostatic fields being mainly negative in the synaptic domains and positive in the cytoplasmic parts. The electrostatic fields around synaptic receptor domains may guide the positively charged ligand molecules to their receptor binding sites, as also suggested for substrate–enzyme interactions (WADE et al. 1994).

It is possible that agonists and antagonists have different but overlapping binding sites in G protein-linked neurotransmitter receptors. Most receptor models place the conserved Asp residue in TMH 2 further down in the central core of the receptor than the conserved Asp residue in TMH 3. In such models the protonated neurotransmitter moves down the central core of the receptor, as its own positive electrostatic potentials are attracted to negative potentials created by aspartate residues in TMH 3 and TMH 2, to "dock" the transmitter in the receptor. This attraction between the transmitter and the receptor changes the conformation of the receptor protein, leading to G protein stimulation and subsequent intracellular biochemical changes. Antipsychotic drug molecules, which are protonated at physiological pH and therefore have a positive molecular electrostatic field, probably bind to a conserved Asp residue in TMH 3, and thereby prevent access of the neurotransmitter to its binding site further down in the central core of the receptor.

Biological experiments have produced substantial information about which receptor domains interact with G proteins. As the receptor models become more refined, a next step may be to build models of receptor–G protein complexes, use these models to suggest new experiments, and use the results of such experiments to refine the models of G protein–receptor complexes. Such computational-experimental approaches are likely to provide further solutions to the complex structural problems regarding the function of G protein-coupled receptors and the mechanisms of drug interactions.

References

Anfinsen CB (1973) Principles that govern the folding of protein chains. Science 181:223–230

Anfinsen CB, Haber E, Sela M, White FH (1961) The kinetics of formation of a native ribonclease during oxidation of the reduced polypeptide chain. Proc Natl Acad Sci USA 47:1309–1314

Asselin AA, Humber LG, Voith K, Metcalf G (1986) Drug design via pharmacophore identification. Dopaminergic activity of 3H-benz[e]indol-8-amines and their mode of interaction with the dopamine receptor. J Med Chem 29:648–654

Barbe J, Blanc A (1976) Chimie structurale – etude conformationelle en relation avec les activites biologiques d'une phénothiazine-2,10 substituée, la dimethothiazine. CR Acad Sci Paris 282 C:299–301

Barbe J, Blanc A, Hurwic J (1973) Physicochimie des diélectriques – etude conformationelle des phenothiazines N-substituées: application aux relations structure-activité. C R Acad Sci Paris 277 C:1071–1074

Carlsson A, Lindquist M (1963) Effect of chlorpromazine or haloperidol on formation of 3-methoxytyramine and normetanephrine in mouse brain. Acta Pharmacol Toxicol 20:140–144

Cox BA Henningsen RA, Spanoyannis A, Neve RL, Neve KA (1992) Contributions of conserved serine residues to the interactions of ligands with dopamine D2 receptors. J Neurochem 59:627–634

Dahl SG, Edvardsen Ø (1994) Molecular modeling of dopamine receptors. In: Niznik HY (ed) Dopamine receptors and transporters. Dekker, New York, pp 265–282

Dahl SG, Hough E, Hals PA (1986) Phenothiazine drugs and metabolites: molecular conformation and dopaminergic, alpha adrenergic and muscarinic cholinergic receptor binding. Biochem Pharmacol 35:1263–1269

Dahl SG, Edvardsen Ø, Heimstad E, Sylte I (1989) Three dimensional structure of the dopamine D2 receptor. In: Stefanis CN, Soldates CR, Rabavilas AD (eds) Psychiatry today. Excerpta Medica, Amsterdam, p 1889

Dahl SG, Edvardsen Ø, Sylte I (1991a) Molecular structure and dynamics of the dopamine D2 receptor and ligands. Clin Neuropharmacol 13 [Suppl 2]:41–42

Dahl SG, Edvardsen Ø, Sylte I (1991b) Molecular dynamics of dopamine at the D2 receptor. Proc Natl Acad Sci USA 88:8111–8115

Dahl SG, Kollman PA, Rao SN, Singh UC (1992) Structural changes by sulfoxidation of phenothiazine drugs. J Comput Aided Mol Design 6:207–222

Dixon RA, Kobilka BK, Strader DJ, Benovic JL, Dohlman HG, Frielle T, Bolanowski MA, Bennett CD, Rands E, Diehl RE, Mumford RA, Slater EE, Sigal IS, Caron MG, Lefkowitz RJ, Strader CD (1986) Cloning of the gene and cDNA for mammalian β-adrenergic receptor and homology with rhodopsin. Nature 321:75–79

Edvardsen Ø, Dahl SG (1991) Molecular structure and dynamics of serotonin. Mol Brain Res 9:31–37

Edvardsen Ø, Dahl SG (1992) Molecular dynamics and electrostatic potentials of dopamine. Mol Neuropharmacol 1:165–172

Edvardsen Ø, Dahl SG (1994) A putative model of the dopamine transporter. Mol Brain Res 27:265–274

Edvardsen Ø, Sylte I, Dahl SG (1992) Molecular dynamics of serotonin and ritanserin interacting with the 5-HT_2 receptor. Mol Brain Res 14:166–178

Grol CJ, Dijkstra D, Schunselaar W, Westerink BHC (1981) Synthesis and pharmacological evaluation of conformationally restricted phenothiazine analogues. J Med Chem 23:322–324

Harbert CA, Plattner JJ, Welch WM, Weissman A, Koe BK (1980) Neuroleptic activity of the 5-aryltetrahydro-gamma-carboline series. Conformational requirements for interaction with central dopamine receptors. Mol Pharmacol 17:38–42

Högberg T, Ramsby S, de Paulis T, Stensland B, Csoregh I, Wagner A (1986) Solid state conformations and antidopaminergic effects of remoxipride hydrochloride and a closely related salicylamide FLA 797, in relation to dopamine receptor models. Mol Pharmacol 30:345–351

Humber LG, Bruderlein FT, Voith K (1975) Neuroleptic agents of the benzocycloheptapyridoisoquinoline series. A hypothesis on their mode of interaction with the central dopamine receptor. Mol Pharmacol 11:833–840

Humber LG, Bruderlein FT, Philipp AH, Gotz M (1979) Mapping the dopamine receptor. 1. Features derived from modifications in ring E of the neuroleptic butaclamol. J Med Chem 22:761–767

Hutchins C (1994) Three-dimensional models of D1 and D2 dopamine receptors. Endocr J 2:7–23

Jähnig F, Edholm O (1992) Modeling of the structure of bacteriorhodopsin. A molecular dynamics study. J Mol Biol 226:837–850

Kaufman JJ, Manian AA (1972) Topological conformation similarities among antipsychotic drugs, narcotics and biogenic amines: a summary. Int J Quant Chem 6:375–381

Koch MH (1974) The conformation of neuroleptic drugs. Mol Pharmacol 10:425–437

Komiskey HL, Bossart JF, Miller DD, Patil PN (1978) Conformation of dopamine at the dopamine receptor. Proc Natl Acad Sci USA 75:2641–2643

Kozell LB, Machidas CA, Neve RL, Neve KA (1994) Chimeric D1/D2 dopamine receptors. Distinct determinants of selective efficacy, potency, and signal transduction. J Biol Chem 269:30299–30306

Kristiansen K, Dahl SG (1995) Molecular modeling of serotonin, ketanserin, ritanserin and their 5-HT2c receptor interactions (submitted for publication)

Liljefors T, Bogeso KP, Hyttel J, Wikstrom H, Svensson K, Carlsson A (1990) Pre- and postsynaptic dopaminergic activities of indolizidine and quinolizidine derivatives of 3-(3-hydroxyphenyl)-N-(n-propyl)piperidine (3-PPP). Further developments of a dopamine receptor model. J Med Chem 33:1015–1022

Manallack DT, Beart PM (1988) A three dimensional receptor model of the dopamine D2 receptor from computer graphic analyses of D2 agonists. J Pharm Pharmacol 40:422–428

McDowell JJH (1969) The crystal and molecular structure of chlorpromazine. Acta Crystallogr B 25:2175–2181

McDowell JJH (1974) The molecular structures of the phenothiazine derivatives, chlorpromazine, thiethylperazine and thioridazine, and a discussion of the mechanism of action. In: Bergman E, Pullman B (eds) Molecular and quantum pharmacology. Reidel, Dordrecht, pp 269–300

Nilsson I, Von Heijne G (1990) Fine-tuning the topology of a polytopic membrane protein: Role of positively and negatively charged amino acids. Cell 62:1135–1141

Olson GL, Cheung HC, Morgan KD, Blount JF, Todaro L, Berger L, Davidson AB, Boff E (1981) A dopamine receptor model and its application in the design of a new class of rigid pyrrolo[2,3-g]isoquinoline antipsychotics. J Med Chem 24:1026–1034

Philipp AH, Humber LG, Voith K (1979) Mapping the dopamine receptor 2. Features derived from modifications in the rings A/B region of the neuroleptic butaclamol. J Med Chem 22:768–773

Rognan D, Sokoloff P, Mann A, Martres MP, Schwartz JC, Costentin J, Wermuth CG (1990) Optically active benzamides as predictive tools for mapping the dopamine D2 receptor. Eur J Pharmacol 189:59–70

Savarese TM, Fraser CM (1992) In vitro mutagenesis and the search for structure-function relationships among G protein-coupled receptors. Biochem J 283:1–19

Schertler GFX, Villa C, Henderson R (1993) Projection structure of rhodopsin. Nature 362:770–772

Sylte I, Dahl SG (1991a) Three-dimensional structure and molecular dynamics of cis(Z)-and trans(E)-chlorprothixene. J Pharm Sci 80:735–740

Sylte I, Dahl SG (1991b) Molecular structure and dynamics of cis(Z)-and trans(E)-flupenthixol and clopenthixol. Pharm Res 8:462-70IS

Sylte I, Edvardsen, Dahl SG (1993) Molecular dynamics of the 5-HT_{1a} receptor and ligands. Protein Eng 6:691–700

Sylte I, Edvardsen Ø, Dahl SG (1995) Molecular modeling of the 5-HT1a receptor and UH-301 interactions (submitted for publication)

Tollenaere JP, Moereels H, Koch MHJ (1977) On the conformation of neuroleptic drugs in the three aggregation states and their resemblance to dopamine. Eur J Med Chem 12:199–211

Tota MR, Strader CD (1990) Characterization of the binding domain of the β-adrenergic receptor with the fluorescent antagonist carazolol. J Biol Chem 265:16891–16897

Viterbo D, Hansen LK, Hough E, Dahl SG (1984) The structures of dimethylaminopropyl phenothiazine drugs and their metabolites, IV. Methoxypromazine at 120 K. Acta Crystallogr C 42:889–892

Wade RC, Luty BA, Demchuk E, Madura JD, Davis ME, Briggs JM, McCammon JA (1994) Simulation of enzyme-substrate encounter with gated active sites. Struct Biol 1:65–69

Weber HP, Lybrand T, Singh U, Kollman P (1986) Analysis of the pharmacological properties of clozapine analogues using molecular electrostatic potential surfaces. J Mol Graphics 4:56–60

CHAPTER 3

Interaction of Antipsychotic Drugs with Dopamine Receptor Subtypes

DEBORAH HARTMAN, FREDERICK MONSMA, and OLIVIER CIVELLI

A. Introduction

Until the mid-twentieth century, patients suffering from psychiatric diseases such as schizophrenia were subjected to barbiturates, insulin coma, electro-convulsive therapy, and brain surgery, all of which provided little or no therapeutic benefit. The first major breakthrough in the treatment of psychiatric diseases occurred in the early 1950s with the discovery and introduction of chlorpromazine as the first effective neuroleptic (DELAY et al. 1952). Improved neuroleptic drugs were subsequently developed, and the search began for the common "site of action" of these drugs.

During the 1970s it was shown that the antipsychotic action of neuroleptics was correlated with their affinity to dopamine receptors in vitro (CREESE et al. 1976; SEEMAN et al. 1976), implicating overactivity of the central dopaminergic system in psychiatric diseases such as schizophrenia (CARLSSON 1988; SEEMAN 1987). In 1979, the dopamine receptors were found to be comprised of two distinct molecular entities which were referred to as D_1 and D_2 receptors (KEBABIAN and CALNE 1979). It was further shown that neuroleptics mostly act by specifically blocking the D_2 receptors, but that this blockade also results in the occurrence of serious deleterious side effects.

The recent discovery of several novel subtypes of dopamine receptors has given rise to the hope that characterization of the interactions of antipsychotic drugs with various receptor sites may lead to a better understanding of their relevant site of action. Enhanced antipsychotic activity with a minimized side effect profile might then be achieved by development of agents exhibiting a high selectivity for the relevant receptor site.

In this chapter we will examine the interaction of current antipsychotic drugs with the known subtypes of dopamine receptors in an attempt to point out the multiple sites of action for these compounds, and to indicate directions in which the future development of new and improved neuroleptics may be focused.

B. Molecular Biology of Dopamine Receptor Subtypes

Dopamine receptors have classically been divided into two subtypes, termed D_1 and D_2, which could be distinguished by their pharmacological characteristics as well as their effects on intracellular second messenger systems (KEBABIAN and CALNE 1979). However, in recent years, molecular cloning techniques have revealed the existence of five distinct subtypes of dopamine receptor, as well as two isoforms of the prototypical D_2 receptor which are generated by alternative mRNA splicing. Interestingly, these new subtypes of dopamine receptor appear to fall into two distinct families corresponding to the original division of distinct D_1 and D_2 receptor subtypes.

I. General Structural Features of Dopamine Receptors

The elucidation of the primary sequence of the dopamine receptors has revealed that they belong to the class of proteins known as the G protein-coupled receptor superfamily. In keeping with the basic structural features of this class of proteins, they exhibit seven highly hydrophobic stretches of approximately 24 amino acids which are believed to represent membrane-spanning domains. The proposed transmembrane topology provides an extracellular amino terminal (N-terminal) and an intracellular carboxyl terminal (C-terminal), with alternating intra- and extracellular loops connecting the transmembrane regions. Prior to the cloning of these receptors, studies had revealed that the receptor protein was glycosylated, and analysis of the primary amino acid sequence has revealed the existence of one or more consensus sites for N-linked glycosylation in the extracellular domains. In vitro mutagenesis studies have indicated that the ligand-binding domain of both agonists and antagonists probably involves these transmembrane domains, particularly the third and fifth transmembrane domains for agonist binding and including the seventh transmembrane domain for antagonist binding (see SAVARESE and FRASER 1992, for review). This concept is particularly important for understanding the pharmacological similarities and differences between dopamine receptor subtypes.

II. The D_1 Family of Dopamine Receptors

The D_1 family of dopamine receptors consists of two members, which have been referred to as D_1 and D_5 in the case of the human genes, and D_{1A} and D_{1B} in the case of the rat genes. These receptor genes were identified by various techniques using their homology with previously cloned G protein-coupled receptors as a means of isolation. The deduced amino acid sequences of the D_1 and D_5 receptors are highly homologous, particularly in the transmembrane domains, and share a common overall topology which includes a short third intracellular loop and long C-terminal cytoplasmic tail (see Fig. 1). At the level of nucleotide and amino acid sequence these are clearly distinct receptor proteins; however, on a pharmacological level with known ligands, these two sites are barely distinguishable. Moreover, both receptors have been shown to stimulate the activity of adenylyl cyclase.

```
Hum D1    M-------------------RTLNTSAMDGTGLVVERDFSVRILTACFLSLLILSTLLGNTLVCAAVIRF   51
Hum D5    MLPPGSNGTAYPGQFA--LYQQLAQGNAVGGSAGAPPLGPSQVVTACLLTLLIIWTLLGNVLVCAAIVRS   68
Hum D2L   M--DPLNLSWYDDDLERQ------NWSRPFNGSDGKADRPHYNYYATLLTLLIAVIVFGNVLVCMAVSRE   62
Hum D3    M--ASLS-----------QLSSHLNYTCGAENSTGASQARPHAYYALSYCALILAIVFGNGLVCMAVLKE   57
Hum D4.2  M--GNRSTADADGLLAGR------GPAAGASAGASAGLAGQGAAALVGGVLLIGAVLAGNSLVCVSVATE   62

Hum D1    RHLRSKVTNFFVISLAVSDLLVAVLVMPWKAVAEIAG-FWPFGSF-CNIWVAFDIMCSTASILNLCVISV  119
Hum D5    RHLRANMTNVFIVSLAVSDLFVALLVMPWKAVAEVAG-YWPFGAF-CDVWVAFDIMCSTASILNLCVISV  136
Hum D2L   KALQTT-TNYLIVSLAVADLLVATLVMPWVVYLEVVGE-WKFSRIHCDIFVTLDVMMCTASILNLCAISI  130
Hum D3    RALQTT-TNYLVVSLAVADLLVATLVMPWVVYLEVTGGVWNFSRICCDVFVTLDVMMCTASILNLCAISI  126
Hum D4.2  RALQTP-TNSFIVSLAAADLLLALLVLPLFVYSEVQGGAWLLSPRLCDALMAMDVMLCTASIFNLCAISV  131

Hum D1    DRYWAISSPFRYE---RKMTPKAAFILISVAWTLSVLISFIPVQLSWHKAKPTSPSDGN-ATSLAE----  182
Hum D5    DRYWAISRPFRYK---RKMTQRMALVMVGLAWTLSILISFIPVQLNWHRDQAASWGGLDLPNNLANWTPW  203
Hum D2L   DRYTAVAMPMLYN--TRYSSKRRVTVMISIVWVLSFTIS-CPLLFGLNN--AD-----------------  179
Hum D3    DRYTAVVMPVHYQHGTGQSSCRRVALMITAVWVLAFAVS-CPLLFGFNT--TG-----------------  177
Hum D4.2  DRFVAVAVPLRYN---RQGGSRRQLLLIGATWLLSAAVA-APVLCGLNDVRGR-----------------  181

Hum D1    ----------TIDNCDSSLSRTYAISSSVISFYIPVAIMIVTYTRIYRIAQKQIRRIAALERAAVHAKNCQ  242
Hum D5    EEDFWEPDVNAENCDSSLNRTYAISSSLISFYIPVAIMIVTYTRIYRIAQVQIRRISSLERAAEHAQSCR  273
Hum D2L   -----QNE-----CIIA-NPAFVVYSSIVSFYVPFIVTLLVYIKIYIVLR-RRRKRVNTKRS---SRAFR  233
Hum D3    -----DPTV----CSIS-NPDFVIYSSVVSFYLPFGVTVLVYARIYVVLKQRRRKRILTRQN---SQ---  231
Hum D4.2  -----DPAV----CRLE-DRDYVVYSSVCSFFLPCPLMLLLYWATFRGLQ----RWEVARRA-------K  229

Hum D1    TTTGNGKPVECSQPES------------------------------------------------------  259
Hum D5    SSAA------CA-PDT------------------------------------------------------  283
Hum D2L   AHLRAPLKGNCTHPEDMKLCTVIMKSNGSFPVNRRRVE--AARRAQELEMEMLSST-SPPERTRYSPIPP  300
Hum D3    -------------------CNSVRPG---FPQQTLSPD--PAH----LELK------------RYYSI--  259
Hum D4.2  LHGRAPRRPSGPGP------PSPTPPAPRLPQDPCGPD--CAPPAPGLPPDPCGSNCAPPDAVRAAALPP  291

Hum D1    -------------------------------------------------------------SFKMSFKRE  267
Hum D5    -------------------------------------------------------------SLRASIKKE  291
Hum D2L   SHHQLTLPDPSHHGLHSTPDSPAKPEKNGHAKDHPKIAKIFEIQTMPNGKTRTSLK--TMSRRKLSQQKE  368
Hum D3    -CQDTALGGPGFQERGGELKREEKTRNSLSPTIAPKLS--LEVRKLSNGRLSTSLKLGPLQPRGVPL-RE  324
Hum D4.2  --------------------QTPPQTRR--------------------------------RRRAKITGRE  309

Hum D1    TKVLKTLSVIMGVFVCCWLPFFILNCILPFCGS---GETQPF--CIDSNTFDVFVWFGWANSSLNPIIYA  332
Hum D5    TKVLKTLSVIMGVFVCCWLPFFILNCMVPFCSGHPEGPPAGFP-CVSETTFDVFVWFGWANSSLNPVIYA  360
Hum D2L   KKATQMLAIVLGVFIICWLPFFITHILNIHCD-----------CNIPPVLYSAFTWLGYVNSAVNPIIYT  427
Hum D3    KKATQMVAIVLGAFIVCWLPFFLTHVLNTHCQ----------TCHVSPELYSATTWLGYVNSALNPVIYT  384
Hum D4.2  RKAMRVLPVVVGAFLLCWTPFFVVHITQALCP----------ACSVPPRLVSAVTWLGYVNSALNPVIYT  369

Hum D1    -FNADFRKAFSTLLGCYRLCP-ATNNAIETVSINNNGAAMFSSHHEP-RGSISKECNLVYLIPHAV--GS  397
Hum D5    -FNADFQKVFAQLLGCSHFCS-RT--PVETVNISNE---LISYNQDI-VFHKEIAAAYIHMMPNAVTPGN  422
Hum D2L   TFNIEFRKAFLKILH--C.                                                    444
Hum D3    TFNIEFRKAFLKILS--C.                                                    401
Hum D4.2  VFNAEFRNVFRKALRACC.                                                    388

Hum D1    SEDLKKEEAAGIARPLE--KLSPALSVI------LDYDTDVSLEKIQPITQNGQHPT.             447
Hum D5    REVDNDEEEGPFDRMFQIYQTSPDGDPVAESVWELDCEGEISLDKITPFTPNGFH.               478
```

Fig. 1. Alignment of deduced amino acid sequences of cloned human dopamine receptors. Amino acids which are identical in at least three subtypes of dopamine receptor are *boxed*. *Heavy lines* indicate the proposed positions of putative transmembrane regions

1. The Dopamine D_1 Receptor

The D_1 dopamine receptor was initially cloned in 1990 from both rat and human genomic and cDNA libraries by four independent groups (DEARRY et al. 1990; MONSMA et al. 1990; SUNAHARA et al. 1990; ZHOU et al. 1990). The clones obtained from both species possessed an open reading frame of 446 amino acids, and exhibited 91% identity at the amino acid level. Both the human and rat genes lack introns in their coding regions, and the human gene has been localized to chromosome 5 (DEARRY et al. 1990; SUNAHARA et al. 1990; ZHOU et al. 1990). The D_1 receptor contains two potential extracellular N-linked glycosylation sites, one on the N-terminal tail, and the other on the second extracellular loop. There exists one consensus phosphorylation site for cAMP-dependent protein kinase in the third cytoplasmic loop, and a conserved cystine residue in the C-terminal tail which has been shown to serve as a site for palmitoylation (NG et al. 1994). The C-terminal tail also contains numerous serine and threonine residues which could serve as additional sites for regulatory phosphorylation.

Heterologous expression of the D_1 receptor clone, either transiently or stably, produced a receptor protein which exhibited the pharmacological and functional characteristics classically associated with the D_1 receptor found in tissue preparations (DEARRY et al. 1990; MONSMA et al. 1990; SUNAHARA et al. 1990; ZHOU et al. 1990). Saturable, high-affinity binding of the D_1 selective antagonists [^{3}H]-SCH 23390 or [^{125}I]-SCH 23982 to the recombinant receptor could be demonstrated, which showed the appropriate affinity and pharmacological specificity. Functionally, the recombinant receptors were shown to mediate activation of the enzyme adenylyl cyclase with a pharmacological profile identical to that described for endogenous receptor-expressing systems. From these data it was concluded that this DNA sequence in fact encoded the classical dopamine D_1 receptor.

This conclusion was further supported by the tissue distribution of the mRNA encoding this protein. Using both Northern blot and in situ hybridization analyses, D_1 receptor mRNA was shown to be predominantly localized to the striatum and nucleus accumbens, with lesser amounts seen in the olfactory tubercle and cortex (DEARRY et al. 1990; MONSMA et al. 1990; SUNAHARA et al. 1990; ZHOU et al. 1990), correlating well with data based on D_1 receptor binding activity (MANSOUR et al. 1991) and dopamine-stimulated adenylyl cyclase activity. Within the caudate putamen, about 50% of the medium-sized neurons appeared to express the D_1 receptor, as assessed by in situ hybridization (GERFEN et al. 1990; MEADOR-WOODRUFF et al. 1991). A small number of large, presumably cholinergic interneurons may also express low levels of D_1 receptor mRNA (LE MOINE et al. 1991). Current evidence indicates that many of the medium-sized neurons which express high levels of D_1 receptor mRNA also express substance P, and belong to the striatal-nigral projection system (GERFEN et al. 1990; LE MOINE et al. 1991).

2. The Dopamine D_5 Receptor

The second member of the D_1 receptor family, the D_5 or D_{1B} receptor, was cloned in 1991 from rat and human genomic DNA libraries by means of homology with the human D_1 or 5-HT1A receptors (GRANDY et al. 1991; JARVIE et al. 1993; SUNAHARA et al. 1991; TIBERI et al. 1991; WEINSHANK et al. 1991). This clone contained an intronless open reading frame of 477 deduced amino acids which was 50% homologous to the D_1 receptor overall, and up to 80% identical in the transmembrane regions (Fig. 1). As previously noted, the overall topology of the D_5 receptor was very similar to that proposed for the D_1 receptor, with a short third cytoplasmic loop and a long C-terminal cytoplasmic tail. Additionally, the D_5 receptor contains two potential N-linked glycosylation sites, one in the N-terminal and the other in the second extracellular loop; a protein-kinase A consensus phosphorylation site in the third cytoplasmic loop; and a conserved cystine residue in the C-terminal tail.

Transfection of the D_5 receptor clone into mammalian cells resulted in the expression of high-affinity, saturable [^{3}H]-SCH 23390 binding, with an affinity virtually identical to that seen with the D_1 receptor. Indeed, inhibition of [^{3}H]-SCH 23390 binding by a variety of dopaminergic agents reveals affinities which differ little from the D_1 receptor binding affinities. However, a notable exception is the affinity for the endogenous agonist dopamine, which binds to the D_5 receptor with a tenfold higher affinity than to the D_1 receptor (GRANDY et al. 1991; JARVIE et al. 1993; SUNAHARA et al. 1991; TIBERI et al. 1991; WEINSHANK et al. 1991). Stimulation of the recombinant D_5 receptor by dopamine or other D_1 agonists results in the activation of adenylyl cyclase; however, the maximal response achieved in cell lines expressing the D_5 receptor is consistently about 25% lower than that seen for D_1 receptors expressed at similar levels (JARVIE et al. 1993). While the molecular basis for these differences remains to be elucidated, it is possible that there will be functional consequences such as differential sensitivity to desensitization and downregulation.

Compared to the D_1 receptor, the regional distribution of the D_5 receptor is more restricted in the CNS, and it is generally expressed at much lower levels. By Northern blot analysis, the areas of highest expression in the human are the frontal and temporal cortex, with lower amounts found in the hippocampus, caudate, and brainstem (SUNAHARA et al. 1991; WEINSHANK et al. 1991). In situ hybridization in slices of rat brain indicated that D_5 receptor mRNA was localized to the hippocampus and the parafascicular nucleus of the thalamus (MEADOR-WOODRUFF et al. 1992).

III. The D_2 Family of Dopamine Receptors

The D_2 receptor family currently consists of four predominant proteins which are the products of three separate genes and one alternative splicing event. These are termed the D_2 (including the alternatively spliced D_{2L} and D_{2S}), D_3, and D_4 dopamine receptors. The grouping of these three genes into a single subfamily of dopamine receptors is based on sequence homologies, pharmacological profiles, and functional activities. These receptors exhibit approximately 40% amino acid homology overall and 56% within the transmembrane domains (Fig. 1); they all exhibit a "D_2-like" pharmacological profile, although their affinities for specific drugs may vary; and, with the possible exception of the D_3 receptor, they appear to couple to the inhibition of adenylyl cyclase through the Gi family of G proteins.

1. The Dopamine D_2 Receptor

The D_2 dopamine receptor was the first of the dopamine receptors to be cloned, through low-stringency hybridization with the beta-adrenergic receptor sequence (BUNZOW et al. 1988). The D_2 cDNA clone contained an open reading frame encoding 415 amino acids with 7 hydrophobic stretches. This sequence possessed three potential N-linked glycosylation sites in the putative extracellular N-terminal tail, and exhibited a relatively long third cytoplasmic loop and a short cytoplasmic C-terminal tail. Transfection of this clone into mouse fibroblasts resulted in the expression of high-affinity, saturable binding of the D_2 antagonist [^{3}H]-spiroperidol (BUNZOW et al. 1988), which showed a pharmacological profile indicative of binding to a D_2 dopamine receptor based on data obtained with rat striatum membranes (BUNZOW et al. 1988).

Soon after its identification, it was discovered that there are two isoforms of the D_2 receptor which are generated by alternative splicing of precursor mRNA (CHIO et al. 1990; DAL TOSO et al. 1989; EIDNE et al. 1989; GANDELMAN et al. 1991; GIROS et al. 1989; GRANDY et al. 1989; MACK et al. 1991b; MILLER et al. 1990; MONSMA et al. 1989; MONTMAYEUR and BORRELLI 1991; O'DOWD et al. 1990; O'MALLEY et al. 1990). This splicing event results in the inclusion (D_2long form or D_{2L}) or exclusion (D_2short form or D_{2S}) of an 87-base-pair exon, resulting in the addition of 29 amino acids to the third cytoplasmic loop as compared to the sequence originally described. In general, the long isoform is the predominant isoform expressed in the brain (DAL TOSO et al. 1989; GANDELMAN et al. 1991; GIROS et al. 1989; GRANDY et al. 1989; LE MOINE and BLOCH 1991; MACK et al. 1991a,b; MONSMA et al. 1989; NEVE et al. 1991; O'MALLEY et al. 1990; SNYDER et al. 1991). Based on in situ hybridization studies using probes which can distinguish between the splice variants, both isoforms appear to be coexpressed, and it is clear that the two isoforms do not represent presynaptic vs postsynaptic forms of D_2 (LE MOINE and BLOCH 1991).

Since the transmembrane regions of the D_{2L} and D_{2S} isoforms are identical, it is not surprising that the pharmacology of these isoforms is essentially identical. Moreover, D_{2L} and D_{2S} are virtually indistinguishable in their ability to inhibit adenylyl cyclase activity in some heterologous expression systems (LIU et al. 1992; SENOGLES 1994), as well as to stimulate PLC activation (LIU et al. 1992), mobilize intracellular calcium (HAYES et al. 1992), and activate Na^+/

H^+ exchange (NEVE et al. 1992). However, recent studies indicate that in fact these receptor isoforms can exhibit various signaling properties. For example, in the neuroblastoma/glioma hybrid NG108-15, LIU L-X et al. (1994) have reported significant differences between D_{2L} and D_{2S} signaling which appear to relate to their ability to couple to different classes of heterotrimeric G proteins in this cell line. In another study, Senogles has demonstrated that in GH4C1 cells the D_2 receptor isoforms couple preferentially to distinct forms of the Gi alpha subunit to inhibit adenylyl cyclase activation (SENOGLES 1994). While these are the first studies to demonstrate signaling differences between the D_2 receptor isoforms convincingly, it remains to be shown that such differences occur in vivo.

2. The Dopamine D_3 Receptor

The D_3 receptor was identified by virtue of its sequence homology with the D_2 receptor and other G protein-coupled receptors (SOKOLOFF et al. 1990). At the level of amino acid sequence, it is most homologous to the D_2 receptor, exhibiting 75% homology in the transmembrane domains. In addition, other structural features, such as a long third cytoplasmic loop and a short C-terminal domain, reinforce the similarity of the D_3 receptor to the previously cloned D_2 receptor, and the dissimilarity to D_1 and D_5 receptors (Fig. 1). Characterization of the gene for the D_3 receptor has revealed the existence of introns in the D_3 receptor gene (GIROS et al. 1990, 1991; MACK et al. 1991; PARK et al. 1995; SOKOLOFF et al. 1990) which give rise to alternate splicing events generating D_3 receptor isoforms, some of which encode truncated receptor proteins (GIROS et al. 1991; SNYDER et al. 1991). However, in the mouse two alternatively spliced forms of the D_3 receptor have been described which resemble the isoforms of the D_2 receptor in that they differ by the presence or absence of a 21-amino-acid segment in the third cytoplasmic loop (FISHBURN et al. 1993; PARK et al. 1995). Functional differences between these D_3 receptor isoforms remain to be investigated.

Interestingly, comparison of the murine and human amino acid sequences in the third cytoplasmic loop reveals that the murine D_3 receptor contains 46 additional amino acids which represent two short exons not yet described in the human (PARK et al. 1995). Since the gene structure for the human D_3 receptor is not known, it is not clear whether the published human cDNA sequence represents a short, alternatively spliced isoform, or whether these exonic sequences have been lost during human evolution. By analogy with the D_2 receptor, it is likely that such variation in the third cytoplasmic loop will have functional consequences, and thus further characterization of the human D_3 receptor gene is necessary for better understanding of the molecular nature of the D_3 receptor in the human.

When expressed in fibroblast cells, the D_3 receptor exhibits a pharmacology similar to that of the D_2 receptor, with most antagonists exhibiting a slightly higher affinity for the D_2 receptor (SOKOLOFF et al. 1990). Notable exceptions to this are the putative autoreceptor selective antagonists AJ76 and UH232. Agonist binding to the D_3 receptor also tends to exhibit higher affinity as compared to the D_2 receptor. However, agonist binding to the D_3 receptor in most cellular systems appears to be unaffected by the addition of guanine nucleotides, suggesting a lack of functional interaction with these signal-transducing complexes. Although the demonstration of second messenger responses to D_3 stimulation has been difficult, at least one group has been able to demonstrate inhibition of adenylyl cyclase activity and stimulation of extracellular acidification through D_3 receptors expressed in a Chinese hamster ovary (CHO) cell line, but this appears to require relatively high levels of expression (Chio et al. 1994b). Using the D_3 receptor transfected into various cell lines, other groups have shown D_3 receptor-mediated inhibition of calcium currents in differentiated NG108-15 cells (Seabrook et al. 1994); activation of c-fos and stimulation of [^{3}H]-thymidine incorporation in undifferentiated NG108-15 cells (Pilon et al. 1994), and inhibition of dopamine release in a mesencephalic cell line (Tang et al. 1994b).

The expression of D_3 receptor mRNA in the rat brain appears to be localized to regions receiving dopaminergic input from the A_{10} cell group including the nucleus accumbens, islands of Calleja, and the bed nucleus of the stria terminalis (Bouthenet et al. 1991; Sokoloff et al. 1990). D_3-receptor mRNA can also be found in other limbic brain areas including the olfactory tubercle, hippocampus, mammary nuclei, and hypothalamus. Small amounts of D_3-receptor message have also been observed in the caudate putamen and certain regions of the cerebral cortex. Importantly, the D_3 receptor is also expressed in dopaminergic cells in the substantia nigra, indicating it may function as a presynaptic autoreceptor, as suggested by pharmacological data. Overall, the pattern of D_3 receptor expression suggests that this dopamine receptor subtype may be well positioned to mediate the effects of dopamine on cognitive and emotional functions, and thus could prove to be a relevant target for antipsychotic drug therapy.

The characteristics of the D_3 receptor discussed above have also led to attempts to link mutations in the D_3 receptor gene with the etiology of schizophrenia (Crocq et al. 1992; Di Bella et al. 1994; Jonsson et al. 1993; Laurent et al. 1994; Mant et al. 1994; Morel 1993; Nanko et al. 1993b; Nimgaonkar et al. 1993; Nothen et al. 1993; Sabate et al. 1994; Wiese et al. 1993; Yang et al. 1993). Except for a restriction fragment length polymorphism observed for the D_3 gene which apparently is not associated with schizophrenia (Crocq et al. 1992; Di Bella et al. 1994; Jonsson et al. 1993; Laurent et al. 1994; Mant et al. 1994; Morel 1993; Nanko et al. 1993b; Nimgaonkar et al. 1993; Nothen et al. 1993; Sabate et al. 1994; Wiese et al. 1993; Yang et al. 1993), no significant mutations in the coding sequence have been noted. On the other hand, one group has reported a loss of D_3 receptor mRNA from parietal and motor cortex in postmortem brain samples from schizophrenic patients (Schmauss et al. 1993). However, since these samples have not been well controlled for medication history and other possible confounding variables, it will be necessary for such a finding to be replicated in a larger patient sample.

3. The Dopamine D_4 Receptor

The sequence of the D_2 receptor was also utilized to identify the third member of the D_2 receptor subfamily, the D_4 receptor. VAN TOL et al. (1991) screened a human neuroblastoma cell line cDNA library with a D_2 receptor probe at low stringency and isolated a partial cDNA clone which exhibited high homology to the D_2 receptor transmembrane regions 5, 6, and 7. This clone was then used to isolate the corresponding genomic clone from a human genomic library, which contained sequences corresponding to the N-terminal through the fourth transmembrane region, and contained several introns which interrupted the coding sequence in transmembrane regions 1, 3, 6, and the third cytoplasmic loop. These introns are in positions analogous to the introns found in both the D_2 and D_3 receptors (GIROS et al. 1990; GRANDY et al. 1989; MACK et al. 1991; O'MALLEY et al. 1990; PARK et al. 1995; SOKOLOFF et al. 1990; VAN TOL et al. 1991). The deduced amino acid sequence, based on the combined genomic and cDNA sequences, results in a 387-amino-acid open reading frame containing seven hydrophobic regions which correspond to the putative transmembrane regions found in the D_2 and D_3 receptors. Overall, the homology with the D_2 and D_3 receptors is 41% and 39%, respectively, and about 56% for both receptors within the transmembrane regions (Fig. 1). Likewise, the proposed transmembrane topology is very similar between the three receptor subtypes.

In order to analyze the pharmacological profile of the D_4 receptor, a hybrid gene/cDNA construct was utilized for transfection into COS-7 cells, since the full-length cDNA clone was never isolated, perhaps due to the extremely high G + C nucleotide content of the mRNA (VAN TOL et al. 1991). The recombinant receptor exhibited specific, saturable [^{3}H]spiperone binding with an affinity similar to that seen for the D_2 and D_3 receptors. Further characterization of the pharmacological profile revealed that the D_4 receptor generally bound agonists and antagonists with similar or somewhat lower affinity than did the D_2 receptor. A particularly notable exception to this profile was observed for the atypical antipsychotic clozapine and its congener octoclothepin, both of which bound to the D_4 receptor with an approximately tenfold higher affinity than to the D_2 receptor (VAN TOL et al. 1991, 1992). This observation is particularly interesting in that the affinity of clozapine for the D_4 receptor is close to the serum concentration achieved during successful clozapine therapy (SEEMAN 1992; see also Sec. C.II. below), leading to much interest in the D_4 receptor as a potential target for novel antipsychotic drug development. Subsequent studies, however, have reported a somewhat lower affinity and selectivity of clozapine for the D_4 receptor (CHIO et al. 1994a; LAHTI et al. 1993; MCHALE et al. 1994; MILLS et al. 1993) (see Table 2).

Heterologous expression of the D_4 receptor has also shown that this dopamine receptor couples to inhibition of adenylyl cyclase activity (CHIO et al. 1994a; MCHALE et al. 1994), stimulation of Na^+/H^+ exchange (CHIO et al. 1994a), and potentiation of stimulated arachidonic acid release (CHIO et al. 1994a), further reinforcing the similarity with the D_2 receptor.

Interestingly, the characterization of the sequence of the D_4 receptor in multiple individual human samples has revealed that the D_4 sequence is highly polymorphic. The most striking polymorphism involves a 48-base-pair sequence in the third cytoplasmic loop of the receptor (VAN TOL et al. 1992). In the original cDNA fragment isolated from a human neuroblastoma cell line, this sequence was repeated twice, but subsequently isolated genomic clones contained fourfold and sevenfold repeats (VAN TOL et al. 1992) with minor sequence variations. Further screening has led to the identification of individuals bearing up to ten repeats, represented by 18 different amino acid sequences (LICHTER et al. 1993). Although the functional significance of this sequence variation is not clear, one report indicated differential effects of sodium on antagonist binding depending on the number of repeat sequences (VAN TOL et al. 1992). However, more extensive characterization of the binding of dopamine to a variety of D_4 isoforms, and the effects of guanosine triphosphate (GTP) analogs on this binding, have revealed little influence of the polymorphic repeat sequence on D_4 receptor binding (ASGHARI et al. 1994). Nevertheless, other possible effects remain to be investigated, particularly with respect to receptor signaling.

The D_4 receptor has been implicated in the etiology of schizophrenia based on both its high affinity for clozapine and a recent report suggesting that D_4 receptor levels are increased in schizophrenic patients (SEEMAN et al. 1993). The discovery of extensive genetic polymorphism at the D_4 receptor locus lent further support to this hypothesis, and opened up the possibility that specific variants of the D_4 receptor might be associated with schizophrenic pathology. Numerous groups have subsequently conducted linkage studies examining D_4 receptor variants in schizophrenic patients, but the results thus far have found no evidence for linkage of D_4 receptor isoforms to genetic susceptibility for schizophrenia in the populations studied (BARR et al. 1993; CAMPION et al. 1994; COON et al. 1993; MACCIARDI et al. 1994; SHAIKH et al. 1994; SOMMER et al. 1993). On the other hand, a polymorphism involving the presence or absence of a 12-base-pair repeat in the N-terminal of the receptor has been shown to be associated with delusional disorders in an Italian population (CATALANO et al. 1993). However, the effect of such a change on the functional properties of the D_4 receptor has yet to be investigated.

C. Pharmacology of Neuroleptics at Recombinant Dopamine Receptors

The classical neuroleptics such as chlorpromazine and haloperidol have shown efficacy in alleviating acute "positive" psychotic symptoms (i.e., hallucinations and paranoia), and they also appear to reduce the number or frequency of psychotic episodes (COLE et al. 1966; DAVIS et al. 1980; KANE 1989). However, these drugs are associated with a high incidence of side effects (AYD 1961; KANE 1989), lack of efficacy in up to 30% of patients, and no benefit for the eventual outcome of the disease. Up to 40% of patients receiving these drugs experience neuroleptic-induced parkinsonism severe enough to require additional therapy, or to require complete withdrawal of neuroleptic treatment.

Most of the behavioral and neurophysiological effects induced by neuroleptics are traditionally believed to be mediated by blockade of D_2 receptors in the brain. Recent cloning and expression studies, however, have revealed that the classical neuroleptics bind not only to D_2 receptors, but also show high-affinity binding to D_3 and D_4 receptors. In the following section we will introduce both the classical and the newer neuroleptic drugs, and discuss their interactions with the known subtypes of dopamine receptors.

I. Traditional Neuroleptics and Related Compounds

Table 1 shows the binding affinities of a number of classical neuroleptic drugs and related compounds to recombinant dopamine receptors expressed in a wide range of heterologous cells including fibroblasts, an immortalized neuronal cell hybrid, embryonic kidney cells, and a glioma cell line. The chemical structures of these compounds are shown in Fig. 2.

Conventional antipsychotics can be divided into the two general categories of low potency, which includes chlorpromazine and thioridazine, and high potency, referring to haloperidol, trifluoperazine, pimozide, and fluphenazine. The former have less propensity for extrapyramidal side effects (EPS) but cause more sedative, autonomic, and anticholinergic effects. Studies with recombinant receptors have shown that chlorpromazine binds with relatively high affinity to all five dopamine receptor subtypes, and shows some preference for D_2 and D_3 receptors. Haloperidol, which is representative of compounds with higher potency, shows subnanomolar affinity for D_2 receptors; however, it also binds to D_3 and D_4 receptors with affinities in the nanomolar range. A complete pharmacological profile of haloperidol is shown in Table 3.

Other compounds which are no longer considered as potential neuroleptics but continue to play a major role in the characterization of dopamine receptor subtypes include (+)-butaclamol, which shows broad receptor specificity for all five DA receptor subtypes, but an about tenfold lower affinity for the D_4 receptor. Spiperone shows selectivity for the D_2-like receptors, with dissociation constants in the picomolar range, and an approximately 1000-fold lower affinity to the D_1-like receptors. Raclopride also shows strong selectivity for D_2 and D_3 receptors, and a 100- to 1000-fold lower affinity for D_4 and D_1 receptors, respectively. In contrast, SCH23390 binds with subnanomolar affinity to D_1 and D_5 receptors, but has negligible affinity for the D_2-like

Table 1. Pharmacological profiles of dopamine receptor subtypes

Drug		K_i(nM)	Species	Isoform	Cell line	Radioligand	Bmax	Reference
Chlorpromazine	D_2	1.1	human	long	Ltk-	[^{3}H]rac	2.5	MALMBERG et al. (1993)
		0.6	human	short	Ltk-	[^{3}H]rac	2.0	MALMBERG et al. (1993)
		1.3	rat	long	CHO	[^{3}H]YM	3.3	LAHTI et al. (1993)
		2.0	rat	long	CHO-K1	[^{3}H]spip	not given	LAWSON et al. (1994)
		2.8	rat	long	CHO-K1	[^{125}I]sulp	not given	SOKOLOFF et al. (1990)
	D_3	1.2	human		CHO	[^{3}H]rac	7.6	MALMBERG et al. (1993)
		6.1	rat		CHO-K1	[^{125}I]sulp	not given	SOKOLOFF et al. (1990)
	D_4	7.9	human	D4.2	HEK293	[^{3}H]spip	0.72	LAWSON et al. (1994)
		16	human	D4.2	COS-7	[^{3}H]YM	0.24	LAHTI et al. (1993)
		37	human	D4.2	COS-7	[^{3}H]spip	0.3	VAN TOL et al. (1991)
	D_1	16	human		BHK	[^{3}H]SCH	4.8	PEDERSEN et al. (1994)
		73	human		COS-7	[^{3}H]SCH	0.89	SUNAHARA et al. (1991)
	D_5	33	human		CHO	[^{3}H]SCH	2	PEDERSEN et al. (1994)
		133	human		COS-7	[^{3}H]SCH	0.41	SUNAHARA et al. (1991)
Haloperidol	D_2	0.67	human	long	Ltk-	[^{3}H]rac	2.5	MALMBERG et al. (1993)
		2.40	human	long	Ltk-	[^{3}H]dom	4	GRANDY et al. (1989)
		0.53	human	short	Ltk-	[^{3}H]rac	2.0	MALMBERG et al. (1993)
		2.30	human	short	CHO-K1	[^{125}I]sulp	0.7	MCALLISTER et al. (1993)
		0.45	rat	long	CHO-K1	[^{125}I]sulp	not given	SOKOLOFF et al. (1990)
		0.40	rat	long	CHO-K1	[^{3}H]spip	not given	LAWSON et al. (1994)
		0.90	rat	long	CHO	[^{3}H]YM	3.3	LAHTI et al. (1993)
		5.10	rat	long	Ltk-	[^{3}H]dom	not given	GRANDY et al. (1989)
		2.90	rat	short	C6 glioma	[^{125}I]epid	0.2–0.5	KOZELL et al. (1994)
	D_3	1.4	human		CHO-K1	[^{125}I]sulp	0.29	MCALLISTER et al. (1993)
		2.4	human		GH4C1	[^{125}I]sulp	0.09	SEABROOK et al. (1992)
		2.7	human		CHO	[^{3}H]rac	7.6	MALMBERG et al. (1993)
		9.8	rat		CHO-K1	[^{125}I]sulp	not given	SOKOLOFF et al. (1990)
		40.5	rat		Sf9	[^{125}I]NCQ298	5–15	BOUNDY et al. (1993)

	D_4	1.8	human	D4.2	CHO	[3H]spip	0.28	Chio et al. (1994)
		2.0	human	D4.2	HEK293	[3H]spip	0.72	Lawson et al. (1994)
		2.5	human	D4.2/4.4/4.7	COS-7	[3H]spip	0.3–1.0	Van Tol et al. (1992)
		2.9	human	D4.2	COS-7	[3H]YM	0.24	Lahti et al. (1993)
		5.1	human	D4.2	COS-7	[3H]spip	0.3	Van Tol et al. (1991)
		7.3	human	D4.2	Sf9	[3H]spip	5.0	Chabert et al. (1994)
	D_1	10	human		BHK	[3H]SCH	4.8	Pedersen et al. (1994)
		27	human		COS-7	[3H]SCH	0.89	Sunahara et al. (1991)
		110	rhesus macaque		C6 glioma	[3H]SCH	0.2–0.5	Kozell et al. (1994)
	D_5	27	human		CHO	[3H]SCH	2	Pedersen et al. (1994)
		48	human		COS-7	[3H]SCH	0.41	Sunahara et al. (1991)
		156	human		COS-7	[3H]SCH	4	Grandy et al. (1991)
Spiperone	D_2	0.13	human	long	Ltk-	[3H]dom	4	Grandy et al. (1989)
		0.05	rat	short	C6 glioma	[125I]epid	0.2–0.5	Kozell et al. (1994)
		0.07	rat	long	CHO-K1	[125I]sulp	not given	Sokoloff et al. (1990)
		0.28	rat	long	MN9D	[3H]spip	1.4	Tang et al. (1994)
		0.37	rat	long	CCL1.3	[3H]spip	1.1	Tang et al. (1994)
		0.35	rat	long	Ltk-	[3H]dom	not given	Grandy et al. (1989)
	D_3	0.32	human		CCL1.3	[3H]spip	0.65	Tang et al. (1994)
		0.71	human		MN9D	[3H]ship	0.55	Tang et al. (1994)
		0.61	rat		CHO-K1	[125I]sulp	not given	Sokoloff et al. (1990)
	D_4	0.05	human	D4.2	COS-7	[3H]spip	0.3	Van Tol et al. (1991)
		0.3	human	D4.2	Sf9	[3H]spip	5.0	Mills et al. (1993)
		0.4	human	D4.2	HEK293	[3H]spip	0.72	Lawson et al. (1994)
		1.4	rat	n.a.	MN9D	[3H]spip	0.1	Tang et al. (1994)
		4.0	rat	n.a.	CCL1.3	[3H]spip	0.12	Tang et al. (1994)
	D_1	179	human		BHK	[3H]SCH	4.8	Pedersen et al. (1994)
		220	human		COS-7	[3H]SCH	0.89	Sunahara et al. (1991)
		540	rhesus macaque		C6 glioma	[3H]SCH	0.2–0.5	Kozell et al. (1994)
	D_5	504	human		CHO	[3H]SCH	2	Pedersen et al. (1994)
		4500	human		COS-7	[3H]SCH	0.41	Sunahara et al. (1991)

Table 1. *Continued*

Drug		K_i(nM)	Species	Isoform	Cell line	Radioligand	Bmax	Reference
(+)-Butaclamol	D_2	0.9	human	long	Ltk-	[^{3}H]dom	4	GRANDY et al. (1989)
		0.5	rat	short	C6 glioma	[^{125}I]epid	0.2–0.5	KOZELL et al. (1994)
		0.7	rat	long	CCL1.3	[^{3}H]spip	1.1	TANG et al. (1994)
		1.2	rat	long	Ltk-	[^{3}H]dom	not given	GRANDY et al. (1989)
		2.6	rat	long	MN9D	[^{3}H]spip	1.4	TANG et al. (1994)
	D_3	4.1	human		MN9D	[^{3}H]spip	0.55	TANG et al. (1994)
		11.2	human		CCL1.3	[^{3}H]spip	0.65	TANG et al. (1994)
		4.6	rat		Sf9	[^{125}I]NCQ298	5–15	BOUNDY et al. (1993)
	D_4	40	human	D4.2	COS-7	[^{3}H]spip	0.3	VAN TOL et al. (1991)
		64	human	D4.2	CHO	[^{3}H]spip	0.28	CHIO et al. (1994)
		87	human	D4.2	HEK293	[^{3}H]spip	0.27	LAWSON et al. (1994)
		232	human	D4.2	Sf9	[^{3}H]spip	5.0	CHABERT et al. (1994)
		38	rat	n.a.	MN9D	[^{3}H]spip	0.1	TANG et al. (1994)
		51	rat	n.a.	CCL1.3	[^{3}H]spip	0.12	TANG et al. (1994)
	D_1	0.2	human		BHK	[^{3}H]SCH	4.8	PEDERSEN et al. (1994)
		3.0	human		COS-7	[^{3}H]SCH	0.41	SUNAHARA et al. (1991)
		2.8	rhesus macaque		C6 glioma	[^{3}H]SCH	0.2–0.5	KOZELL et al. (1994)
	D_5	0.3	human		CHO	[^{3}H]SCH	2	PEDERSEN et al. (1994)
		27	human		COS-7	[^{3}H]SCH	0.41	SUNAHARA et al. (1991)
		9.1	human		COS-7	[^{3}H]SCH	4	GRANDY et al. (1991)
Raclopride	D_2	2.3	human	long	Ltk-	[^{3}H]rac	2.5	MALMBERG et al. (1993)
		1.4	human	long	Ltk-	[^{3}H]rac	2.7	MOHELL et al. (1993)
		0.99	human	short	Ltk-	[^{3}H]rac	2.0	MALMBERG et al. (1993)
		1.8	rat	long	CHO-K1	[^{125}I]sulp	not given	SOKOLOFF et al. (1990)
		2.0	rat	long	CHO	[^{3}H]YM	3.3	LAHTI et al. (1993)
		1.0	rat	long	CHO-K1	[^{3}H]spip	not given	LAWSON et al. (1994)
	D_3	1.6	human		CHO-K1	[^{3}H]rac	3.57	MOHELL et al. (1993)
		1.8	human		CHO	[^{3}H]rac	7.6	MALMBERG et al. (1993)
		3.5	rat		CHO-K1	[^{125}I]sulp	not given	SOKOLOFF et al. (1990)

	D_4	237	human	D4.2	COS-7	[^{3}H]spip	0.3	VAN TOL et al. (1991)
		1344	human	D4.2	COS-7	[^{3}H]YM	0.24	LAHTI et al. (1993)
		1367	human	D4.2	Sf9	[^{3}H]spip	5.0	MILLS et al. (1993)
		1549	human	D4.2	HEK293	[^{3}H]spip	0.72	LAWSON et al. (1994)
		1600	human	D4.2/4.4/4.7	COS-7	[^{3}H]spip	0.3–1.0	VAN TOL et al. (1992)
		1650	human	D4.2	CHO	[^{3}H]spip	0.28	CHIO et al. (1994)
	D_1	18000						O'DOWD et al. (1994)
SCH23390	D_2	267	human	long	Ltk-	[^{3}H]rac	2.5	MALMBERG et al. (1993)
		560	human	short	CHO-K1	[^{125}I]sulp	0.7	MCALLISTER et al. (1993)
		284	human	short	Ltk-	[^{3}H]rac	2.0	MALMBERG et al. (1993)
		920	rat	short	C6 glioma	[^{125}I]epid	0.2–0.5	KOZELL et al. (1994)
	D_3	314	human		CHO	[^{3}H]rac	7.6	MALMBERG et al. (1993)
		370	human		CHO-K1	[^{125}I]sulp	0.29	MCALLISTER et al. (1993)
		334	rat		HEK293	[^{125}I]PIPAT	0.27	BURRIS et al. (1994)
	D_4	2564	human	D4.2	HEK293	[^{3}H]spip	0.72	LAWSON et al. (1994)
		3560	human	D4.2	COS-7	[^{3}H]spip	0.3	VAN TOL et al. (1991)
	D_1	0.50	human		BHK	[^{3}H]SCH	4.8	PEDERSEN et al. (1994)
		0.35	human		COS-7	[^{3}H]SCH	0.89	SUNAHARA et al. (1991)
		0.50	rhesus macaque		C6 glioma	[^{3}H]SCH	0.2–0.5	KOZELL et al. (1994)
	D_5	0.5/5.1	human		CHO	[^{3}H]SCH	2	PEDERSEN et al. (1994)
		0.3	human		COS-7	[^{3}H]SCH	0.41	SUNAHARA et al. (1991)
		0.6	human		COS-7	[^{3}H]SCH	4	GRANDY et al. (1991)
7-OH-DPAT	D_2	223	human	not given	CCL1.3	[^{3}H]spip	not given	MACKENZIE et al. (1994)
		61	rat	short	CHO	[^{125}I]sulp	0.9	LEVESQUE et al. (1992)
		3.6	rat	long	CHO	[^{3}H]7-OH-DPAT	0.24 (a)	GONZALES and SIBLEY (1995)
	D_3	7.1	human		CCL1.3	[^{3}H]spip	0.4	MACKENZIE et al. (1994)
		0.78	rat		CHO	[^{125}I]sulp	0.5	LEVESQUE et al. (1992)
		0.8	rat		HEK293	[^{125}I]PIPAT	0.27	BURRIS et al. (1994)
	D_4	650	human	D4.2	COS-7	[^{3}H]spip	not given	LEVESQUE et al. (1992)
	D_1	5300	human		CHO	[^{3}H]SCH	not given	LEVESQUE et al. (1992)

K_i, inhibition constant.
[^{3}H]dom, [^{3}H]domperidone; [^{125}I]epid, [^{125}I]epidipride; [^{3}H]rac, [^{3}H]raclopride; [^{3}H]spip, [^{3}H]spiperone; [^{125}I]sulp, [^{125}I]sulpiride; [^{3}H]YM, [^{3}H]YM-09151-2.

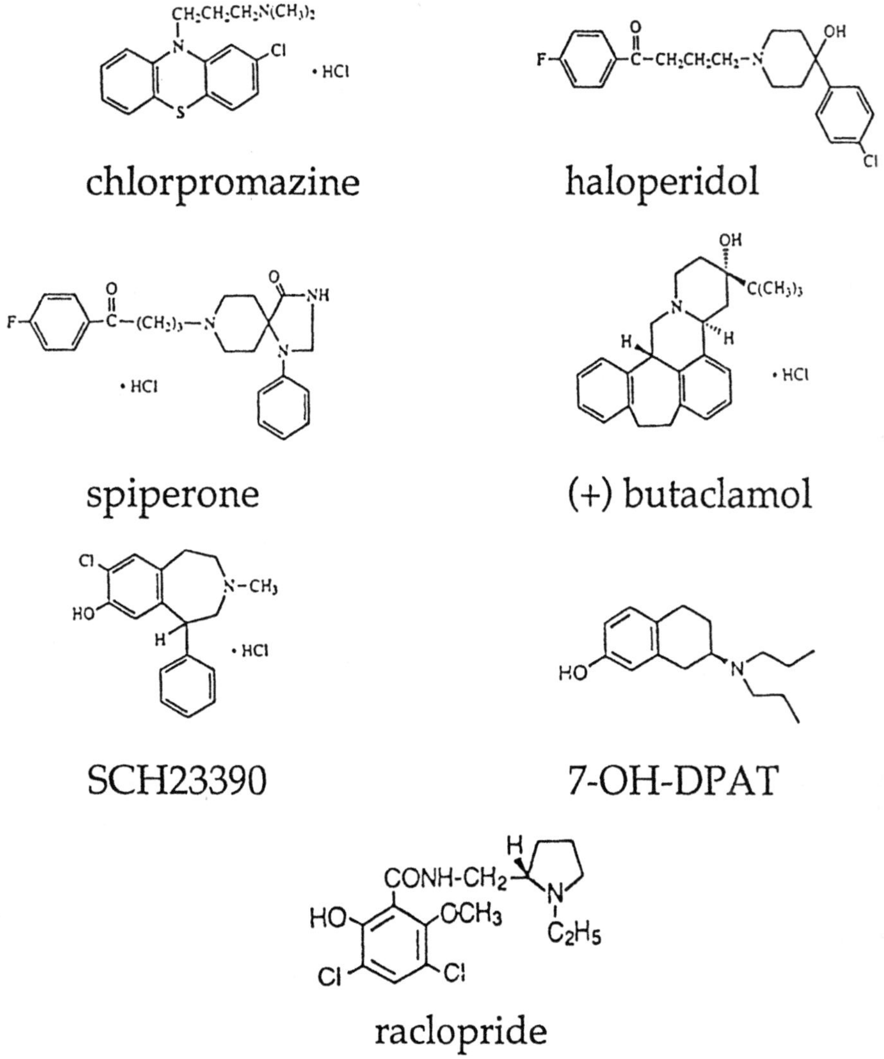

Fig. 2. Chemical structures of selected classical antipsychotics and related compounds

receptor family. Interestingly, the (+) enantiomers of the agonists N-propylnoraporphine and apomorphine, which are inactive at the D_2 receptor, show relatively high affinity binding to both isoforms of the D_4 receptor tested – the $D_{4.2}$, containing two repeat sequences, and the $D_{4.4}$, containing four repeat sequences (SEEMAN and VAN TOL 1993).

No significant pharmacological differences have been observed for the D_2 receptor isoforms, D_{2L} and D_{2S} (see GIROS et al. 1989 and Tables 1 and 2), or for the multiple D_4 receptor polymorphic variants (GRANDY et al. 1989). In both cases, however, these sequence variations may have consequences for receptor function since they are localized to the region which is directly involved in coupling to signal-transducing G proteins in the cell.

A relatively new compound, 7-OH-DPAT, deserves brief mention as a putative D_3 receptor-specific compound, although it has not been proposed as a potential neuroleptic drug since it acts as a dopamine receptor agonist. Initial studies of 7-OH-DPAT reported an approximately 100-fold selectivity for the rat D_3 receptor [inhibition constant (K_i) = 0.8nM] compared to the rat D_{2L} receptor (K_i = 61nM) expressed in CHO cells. In a second study, a 30-fold selectivity was reported for the D_3 over the D_2 receptor expressed in CCL1.3 fibroblast cells (MacKenzie et al. 1994). More recently, however, the dissociation constant of 7-OH-DPAT for the rat D_{2L} receptor, also expressed in CHO cells, was reported to be 3.6nM (Gonzalez and Sibley 1995), which would indicate only fourfold selectivity for the D_3 versus the D_2 receptor; however, in this study [^{3}H]7-OH-DPAT appeared to recognize only a subpopulation (about 15%) of the D_2 receptors labelled by [^{3}H]spiperone. Electrophysiological studies have recently shown that 7-OH-DPAT does behave as a D_2 receptor agonist in vivo, and does not discriminate between nigrostriatal and mesolimbic systems (Liu J-C et al. 1994). This suggests that the D_{3R} selectivity of this compound may not be sufficient for in vivo studies.

The locomotor or extrapyramidal side effects (EPS) of traditional neuroleptic drugs such as haloperidol are believed to be caused by blockade of D_2 receptors expressed at high levels in the nigrostriatal pathway, which controls voluntary movement. In the following sections, we discuss the search for new antipsychotics with decreased liability for EPS, and the discovery of clozapine, a new neuroleptic which has played a pivotal role in defining the concept of the "atypical" neuroleptic.

II. Clozapine and the Atypical Neuroleptics

In contrast to the somewhat disappointing clinical profiles of the typical neuroleptics, the dibenzazepine clozapine has been found to have therapeutic effects on both positive and negative symptoms of schizophrenia without producing EPS (Angst et al. 1971; Baldessarini and Frankenburg 1991; Stille and Hippius 1971) or elevation of prolactin levels (see Coward 1992). Clozapine differs from other antipsychotic agents in its pharmacological, biochemical, and clinical profiles, and has therefore been termed an "atypical neuroleptic." Clozapine has served as the prototype for development of a new class of neuroleptics; however, its own use is restricted because it has the potential to induce agranulocytosis, a potentially fatal immune condition, in approximately 3% of patients (Krupp and Barnes 1992).

Clozapine shows high-affinity binding in the nanomolar range to both the D_4 and D_2 receptors, somewhat lower affinity to the D_1 receptor, and submicromolar affinities to the D_3 and D_5 receptors (see Table 2). Clozapine binding to the D_4 receptor was initially reported to show a tenfold selectivity over the D_2 receptor (Van Tol et al. 1991). This was particularly exciting in light of other evidence showing that elevated D_4 receptor expression can be found in the CNS of schizophrenic patients, thereby implicating the D_4 recep-

Table 2. Summary of clozapine affinity at dopamine receptor subtypes

Receptor subtype	K_i (nM)	Species	Isoform	Cell line	Radioligand	Bmax	Reference
D_{2L}	60	human	long	Ltk-	[^{3}H]rac	2.5	Malmberg et al. (1993)
	138	human	long	Ltk-	[^{3}H]spip	4	Van Tol et al. (1993)
	56	rat	long	CHO-K1	[^{125}I]sulp	not given	Sokoloff et al. (1994)
	55	rat	long	CHO-K1	[^{3}H]spip	not given	Lawson et al. (1994)
	82	rat	long	CHO	[^{3}H]YM	3.3	Lahti et al. (1993)
	143	rat	long	MN9D	[^{3}H]spip	1.4	Tang et al. (1994)
	170	rat	long	CCL1.3	[^{3}H]spip	1.1	Tang et al. (1994)
D_{2S}	17	human	not given	CHO-K1	[^{3}H]spip	2.3	MacKenzie et al. (1994)
	35	human	short	Ltk-	[^{3}H]rac	2.0	Malmberg et al. (1993)
	140	rat	short	C6 glioma	[^{125}I]epid	0.2–0.5	Kozell et al. (1994)
D_3	55	human		GH4C1	[^{125}I]sulp	0.09	Seabrook et al. (1992)
	83	human		CHO	[^{3}H]rac	7.6	Malmberg et al. (1993)
	264	human		CHO-K1	[^{3}H]spip	1.9	MacKenzie et al. (1994)
	620	human		MN9D	[^{3}H]spip	0.55	Tang et al. (1994)
	480	human		CCL1.3	[^{3}H]spip	0.65	Tang et al. (1994)
	205	rat		Sf9	[^{125}I]NCQ298	5–15	Boundy et al. (1993)
	180	rat		CHO-K1	[^{125}I]sulp		Sokoloff et al. (1990)
	389	rat		HEK293	[^{125}I]PIPAT	0.27	Burris et al. (1994)
D_4	9	human	D4.2	COS-7	[^{3}H]spip	0.3	Van Tol et al. (1991)
	29	human	D4.2	COS-7	[^{3}H]YM	0.24	Lahti et al. (1993)
	22	human	D4.2	CHO	[^{3}H]spip	0.28	Chio et al. (1994)
	32	human	D4.2	HEK293	[^{3}H]spip	0.72	Lawson et al. (1994)
	54	human	D4.2	Sf9	[^{3}H]spip	5.0	Mills et al. (1993)
	35	rat	n.a.	CCL1.3	[^{3}H]spip	0.12	Tang et al. (1994)
	42	rat	n.a.	MN9D	[^{3}H]spip	0.1	Tang et al. (1994)

Human D_4 isoforms:							
	4.2	0 mM NaCl	D4.2	COS-7	[^{3}H]spip	0.3–1.0	Van Tol et al. (1992)
	25.3	120 mM NaCl	D4.2	COS-7	[^{3}H]spip	0.3–1.0	Van Tol et al. (1992)
	4	0 mM NaCl	D4.4	COS-7	[^{3}H]spip	0.3–1.0	Van Tol et al. (1992)
	28	120 mM NaCl	D4.4	COS-7	[^{3}H]spip	0.3–1.0	Van Tol et al. (1992)
	14	0 mM NaCl	D4.7	COS-7	[^{3}H]spip	0.3–1.0	Van Tol et al. (1992)
	21	120 mM NaCl	D4.7	COS-7	[^{3}H]spip	0.3–1.0	Van Tol et al. (1992)
D_1	18	human		BHK	[^{3}H]SCH	4.8	Pedersen et al. (1994)
	141	human		COS-7	[^{3}H]SCH	0.89	Sunahara et al. (1991)
	150	rhesus macaque		C6 glioma	[^{3}H]SCH	0.2–0.5	Kozell et al. (1994)
D_5	35	human		CHO	[^{3}H]SCH	2	Pedersen et al. (1994)
	250	human		COS-7	[^{3}H]SCH	0.41	Sunahara et al. (1991)
	406	human		COS-7	[^{3}H]SCH	4	Grandy et al. (1991)

For abbreviations, see Table 1.

tor in schizophrenia (SEEMAN et al. 1993). Additional studies, however, have revealed high-affinity binding of clozapine to the short form of the D_2 receptor (MALMBERG et al. 1993), and in fact clozapine binding to the long form of the D_2 receptor now also appears to be of higher affinity than originally thought (see Table 2 and references therein).

Such differences in clozapine-binding affinity to the dopamine receptor subtypes may be a consequence of different cell environments, as shown in Table 2. The tenfold selectivity of clozapine originally reported for D_4 receptors expressed in COS-7 cells was compared to D_2 receptors expressed in CHO cells (VAN TOL et al. 1992). It now appears, however, that the D_4 receptor expressed in COS-7 cells shows an unusually high affinity for clozapine (see Table 2). Subsequent studies in CHO cells have revealed selectivity ratios of only two- to threefold for the D_4 over the D_2 receptor. Furthermore, in an immortalized mesencephalic cell line, MN9D, clozapine selectivity for the D_4 over the D_2 receptor was found to be only threefold (TANG et al. 1994), and these same receptor clones expressed in mouse CCL1.3 fibroblast cells showed a fivefold selectivity in clozapine binding (TANG et al. 1994). Interestingly, clozapine shows a particularly high affinity for the D_3 receptor expressed in GH4C1 pituitary cells (SEABROOK et al. 1992). Other factors which may lead to the reported differences in affinity and selectivity are species-specific elements, as well as the choice of radioligand used to label the receptors. For example, D_2 receptors expressed in Ltk- cells showed an inhibition constant (K_i) for clozapine of 60 nM using [^{3}H]raclopride, and 138 nM using [^{3}H]spiperone (see Table 3).

Clozapine binding to D_2, D_3, and D_4 receptors has also been shown to be sensitive to sodium concentration. Table 2 shows the K_i values for three D_4 receptor isoforms in the presence and absence of 120 mM sodium chloride (taken from VAN TOL et al. 1992). In each case, clozapine affinity was reduced two- to sixfold in the presence of sodium chloride. MALMBERG et al. (1993) also showed that sodium chloride decreased clozapine affinity for both human D_{2S} and D_{2L} receptors about threefold; however, it had no effect on clozapine binding to human D_3 receptors. In a recent review article, Seeman showed an association between neuroleptic dissociation constants at the D_2 receptor and the free neuroleptic concentration in the plasma water of patients (SEEMAN 1992); in this analysis, only clozapine showed a lack of correlation due to its low affinity for the D_{2L} receptor (VAN TOL et al. 1991). In light of more recent data, however, the average clinical dose of clozapine in controlling schizophrenia can now also be shown to correlate with its affinity at the D_2 receptor (HOLLISTER 1995).

The recent "rediscovery" of clozapine as a neuroleptic drug without locomotor side effects has led to extensive studies examining the biochemical basis of clozapine action (COWARD 1992). Surprisingly, [^{3}H]clozapine labelling of postmortem human brain membranes from the striatum, amygdala, frontal cortex, and substantia nigra could not be displaced, even in part, by dopamine; instead, 50% of the binding in striatal membranes could be displaced by atropine and pirenzepine, confirming [^{3}H]clozapine binding to M1 muscarinic cholinergic receptors ex vivo (FLAMEZ et al. 1994). The complete pharmacological profile of clozapine (summarized in Table 3) suggests a broad molecular basis of action and is discussed in the next section, together with data from several new potential atypical neuroleptics currently under development.

III. New Antipsychotics

The concept of the "atypical neuroleptic" as defined by the functional properties of clozapine has led to great efforts to identify new antipsychotic drugs that discriminate between therapeutic antipsychotic effects and unwanted extrapyramidal side effects, and also show increased efficacy. Two major approaches to the development of atypical neuroleptics are (1) identification of safe compounds which show a broad pharmacological spectrum similar to

Table 3. Receptor affinities for atypical antipsychotics compared to haloperidol

Receptor	Clozapine	Olanzapine	Remoxipride	Risperidone	Haloperidol
D_1	**85**	**31**	>10000	**75**	**10**
D_{2L}	**60–150**	**11**	**125**	**1.5**	**0.5**
D_{2S}	**35**	**11**	**54**	**1.5**	**0.5**
D_3	**300**		969	**6.7**	**2**
D_4	**9–54**	**27**	3690		**2**
D_5	**35–400**				**27**
$5HT_{1A}$	875	>10000	>10000	**16**	1714
$5HT_{1B}$	1750	1980	>10000	1950	
$5HT_{1D}$	1060	760	8310	135	
$5HT_{2A}$	**8**	**5**	>10000	**0.6**	74
$5HT_{2C}$	**12**	**23**	>10000	**16**	5755
$5HT_3$	170	140	>10000	>10000	>10000
$5HT_6$	**4**	**2.5**	>5000	425	>5000
$5HT_7$	**6.3**	104	>5000	**1.39**	263
M_1	**1.8**	**1.9**	>3000	>3000	1700
M_2	**21**	**18**	>3000	>3000	2500
M_3	**13**	**25**	>3000	>3000	>3000
M_4	**12**	**13**	>3000	>3000	2700
M_5	**3.7**		>3000	>3000	1800
$alpha_1$	**7**	**19**	>10000	**2**	**46**
$alpha_2$	**8**	228	2904	**3**	360
$beta_1$	>10000	>10000	>10000	>10000	>10000
H_1	**6**	**7**	>10000	155	3630
GABA.A	>10000	>10000	>10000	>10000	>10000
sigma	>10000		60–180		5–50

Data from MOORE et al. (1993); ROTH et al. (1994); KÖHLER et al. (1990); WALKER et al. (1990); ASSIE et al. (1993); CANTON et al. (1990); WATLING et al. (1990).

clozapine, and (2) targeting preferential blockade of dopamine D_2-like receptors localized in the forebrain and limbic areas of the brain, which are believed to be associated with schizophrenia.

In contrast to D_2 receptors, both D_3 and D_4 receptors are expressed in relatively low abundance in the striatum. The localization of D_3 and D_4 receptor expression in the mesolimbic/mesocortical pathway (SOKOLOFF et al. 1990; VAN TOL et al. 1991), which controls emotional behavior and is implicated in affective and compulsive behaviors, suggests that neuroleptics selective for these receptor subtypes might have potent antipsychotic activity without producing locomotor side effects.

In animal models, an atypical profile can be illustrated by a significant separation, or high discrimination ratio, between the drugs doses causing EPS, and the doses acting on measures of antipsychotic activity such as blockade of apomorphine-enhanced locomotor activity and inhibition of conditioned avoidance response (CAR) in the rat, which are believed to be mediated by stimulation of DA receptors in the mesolimbic DA system (KELLY et al. 1975). The inhibition of CAR is widely accepted as a test predictive of the clinical antipsychotic potential of a compound (ARNDT 1982), while catalepsy in the rat, which is induced by D_2 receptor blockade in the striatum, is associated with the occurrence of EPS in the clinic (WORMS et al. 1983). Several recently identified compounds which show some properties of atypical neuroleptics are described in this section, and are shown in Fig. 3.

1. Olanzapine: A Second-Generation Clozapine-Like Compound

The broad pharmacological profile of clozapine has made it difficult to determine the precise nature of its mechanism of action. The search for safe, clozapine-like drugs has led to a compound called olanzapine which shows a similarly broad recognition of dopamine, 5-HT, adrenergic, histamine, and muscarinic receptors (Table 3). Interestingly, clozapine shows high affinity for both of the newly cloned 5-HT_6 and 5-HT_7 receptors, while olanzapine binds with high affinity only to the 5-HT_6 receptor. In contrast to clozapine, olanzapine was tolerated in humans with no sign of agranulocytosis (HOWARD and SEEGER 1993). Olanzapine and clozapine share selectivity for the M1 muscarinic receptor; however, this alone cannot explain their atypical profiles since combination of a typical dopamine receptor antagonist with the muscarinic antagonist, atropine, does not result in an atypical profile (BARTHOLINI 1980; SAYERS et al. 1976).

In animal behavior models, olanzapine does indeed show a separation of doses required to elicit inhibition of CAR (ED_{50} = 5.6 mg/kg po) from doses which induce catalepsy (ED_{50} = 23 mg/kg po) (TYE et al. 1992). Consistent with its atypical profile, olanzapine has been shown to produce increased levels of dopamine metabolites in rodents, demonstrating that it has dopamine receptor antagonist activity in vivo, and yet it shows very little effect on prolactin

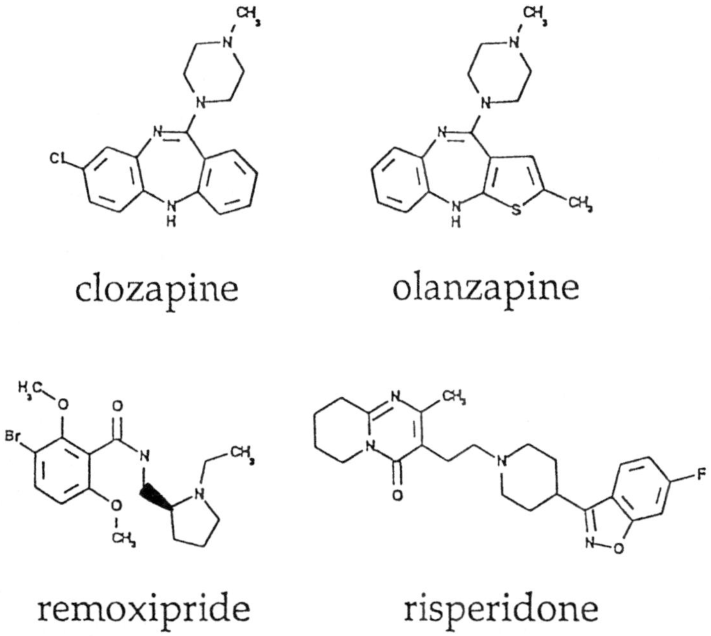

Fig. 3. Chemical structures of selected atypical antipsychotics

secretion in humans (BEASLEY et al. 1992). In an open-label study with ten schizophrenic patients, olanzapine has successfully shown antipsychotic action and a very low propensity to produce adverse side effects with respect to both EPS and prolactin changes (BEASLEY et al. 1992).

2. Remoxipride: A D_2 Receptor-Selective Substituted Benzamide

As shown in Table 3, remoxipride is a selective D_2 receptor antagonist with no affinity for 5-HT, muscarinic, adrenergic, or histamine receptors. In addition to the D_2 receptor, remoxipride also recognized the sigma binding site (KÖHLER et al. 1990). Surprisingly, this D_2-selective compound shows an atypical profile in animal behavior models. That is, (a) remoxipride blocks dopamine-induced hyperactivity at much lower doses than it induces catalepsy, (b) it is active in inhibition of CAR, and (c) it produces only a short, transient increase in prolactin levels (ÖGREN 1994).

What is the reason for this atypical profile? Several data suggest that remoxipride may recognize only a subpopulation of D_2 receptors. Binding data show a two- to threefold selectivity of remoxipride for the short vs the long form of the D_2 receptor (MALMBERG et al. 1993); interestingly, similar selectivity is also found for clozapine. Other data suggest that remoxipride binds preferentially to functionally coupled D_2 receptors. For example, repeated treatment with remoxipride did not produce the behavioral supersensitivity which is marked after haloperidol treatment (ÖGREN et al. 1990). In addition, remoxipride (a) does not increase the density of D_2 receptor binding in the striatum, (b) does not increase the levels of preproenkephalin mRNA in striatum, and (c) does not increase neurotensin concentrations in the caudate nucleus (KÖHLER et al. 1991; LEVANT et al. 1991).

Another possible explanation for the atypical profile of remoxipride that was put forward is that it may be acting on specifically localized D_2 receptors. This is suggested by the finding that remoxipride was more effective in blocking [^{3}H]raclopride binding in areas outside the striatum than within the striatum (KÖHLER et al. 1990, 1992). In vivo studies examining receptor occupancy revealed that even at high doses, remoxipride protected only 50% of the sulpiride or raclopride binding sites in the neostriatum from N-ethoxycarbonyl-2-ethoxy-1,2-dihydroquinolone (EEDQ) inactivation (ÖGREN 1994), suggesting that it occupies only a subpopulation of D_2 receptors. One exception was in the substantia nigra, where protection by remoxipride reached nearly 100%. Also interesting was the finding that remoxipride causes selective dopamine utilization in the dorsomedial part of the striatum, and induces c-fos expression only in the ventral striatum, or nucleus accumbens, an area which is specifically associated with antipsychotic effects (DEUTCH et al. 1992).

Since its introduction to the European market in 1990, it is estimated that more than 50000 patients have received remoxipride. Clinical trials involving over 2500 patients have shown that remoxipride (a) can improve both positive and negative symptoms of schizophrenia, (b) is at least as effective as haloperidol in both first-episode patients and severely psychotic patients, (c) can prevent relapse in chronic schizophrenics, and (d) was effective in approximately one-third of therapy-resistant patients tested (reviewed in LEWANDER 1994). In addition, side effects such as sedation, EPS, and neuroleptic-induced deficit syndrome are minimal. Unfortunately, however, remoxipride was withdrawn in February, 1994, due to a low risk of aplastic anemia (8 cases in 45000 patients), which can be fatal.

In conclusion, results obtained with remoxipride support the hypothesis that D_2 receptors may be responsible for the therapeutic effects of neuroleptics. The molecular mechanisms underlying the atypical profile of remoxipride are clearly different from that of clozapine, since remoxipride effects apparently cannot be related to blockade of 5-HT, muscarinic, or adrenergic receptors. The surprisingly low levels of EPS associated with this compound suggest that the atypical effects of remoxipride may be due to preferential effects on subpopulations of D_2 receptors organized in distinct but widespread neurochemical or functional compartments.

3. Risperidone: A D_2/5-HT_2 Receptor Antagonist

Several studies have suggested that the binding affinity of clozapine to 5-HT_2 receptors might contribute to its atypical profile (MELTZER 1992). For example, clinical evidence has shown that ritanserin, a selective $5\text{-}HT_2$ receptor antagonist, ameliorates negative symptoms of schizophrenia and can reduce the incidence of EPS (BERSANI et al. 1986). Risperidone has been developed as a combination D_2 plus $5\text{-}HT_2$ receptor antagonist. It shows subnanomolar affinity for the $5\text{-}HT_2$ receptor, with an about twofold lower affinity for the D_2 receptor; however, it also binds to the D_3, $5\text{-}HT_7$, alpha$_1$, alpha$_2$, and $5\text{-}HT_{1A}$ receptors with significant affinity (see Table 3). An important distinction between risperidone and clozapine is its total lack of anticholinergic activity.

In animal behavior experiments, risperidone also shows some separation between doses which produce antipsychotic effects (ED_{50} for inhibition of CAR = 1.3 mg/kg po) and doses which induce EPS (ED_{50} for induction of catalepsy = 4.3 mg/kg po) (MOORE et al. 1993). In clinical trials, risperidone has already shown efficacy in treating both the positive and negative symptoms of schizophrenia, and at low doses produces few adverse side effects (BORISON et al. 1992). These results suggest that compounds such as risperidone with combined specificities at multiple receptors may indeed have potential as atypical antipsychotic drugs.

D. Future Outlook and Hopes for Subtype-Specific Drugs

Schizophrenia today affects approximately 1% of the world population, and it has been estimated that care of schizophrenic patients costs $33 billion per year in the U.S. alone. While classical neuroleptics are clearly efficacious in alleviating many psychotic symptoms and reducing the frequency of psychotic episodes, the high incidence of side effects, lack of efficacy in certain patients, and the lack of effects on the eventual outcome of the disease clearly indicate the need for novel therapeutic agents. The development of improved neuroleptics will certainly require a better understanding and clearer diagnosis of psychiatric disease, which would be greatly facilitated by the identification of biochemical markers for disease subtypes. Perhaps the effectiveness of a drug such as clozapine is due to its broad pharmacological profile which can provide therapeutic benefit to the wide spectrum of different psychiatric diseases that are now referred to as schizophrenia.

Although the dopamine hypothesis of schizophrenia best explains our knowledge to date, a major remaining dilemma is that D_2 receptor blockade can be achieved almost immediately after administration of one or very few doses of a neuroleptic, while antipsychotic effects occur in patients only after days or weeks of treatment (SEEMAN 1987). In addition, it is generally true that the doses of a typical neuroleptic drug given to psychiatric patients are titrated to the highest acceptable level of adverse side effects. This correlation must be considered as an alternate explanation for the basis of the association between the clinical dose administered and the D_2 receptor drug affinities as described by SEEMAN (1992).

The recent identification of novel dopamine receptors which are clear targets for antipsychotic drugs has provided a potential for a better understanding of this complex disease. The ongoing development and testing of D_3 and D_4 receptor-selective drugs will provide new insight into their roles in psychoses, and holds the promise of some day revealing molecular mechanisms of psychiatric disease in humans.

References

Angst J, Bente D, Berner P et al. (1971) Das Klinische Wirkungsbild von Clozapine. Pharmacopsychiatry 4:200–211

Arndt J (1982) Pharmacological specificity of conditioned avoidance response inhibition in rats: inhibition by neuroleptics and correlation to dopamine receptor blockade. Acta Pharmacol Toxicol 51:321–329

Asghari V, Schoots O, van Kats S, Ohara K, Jovanovic V, Guan HC, Bunzow JR, Petronis A, Van TOl H (1994) Dopamine D_4 receptor repeat: analysis of different native and mutant forms of the human and rat genes. Mol Pharmacol 46:364–473

Assie M-B, Sleight AJ, Koek W (1993) Biphasic displacement of [3H]YM-09151-2 binding in the rat brain by thioridazine, risperidone, and clozapine, but not by other antipsychotics. Eur J Pharmacol 237:183–189

Ayd FR Jr (1961) A survey of drug-induced extrapyramidal reactions. J Am Med Assoc 175:1054–1060

Baldessarini RJ, Frankenburg FR (1991) Clozapine: a novel antipsychotic agent. N Engl J Med 324:646–754

Barr CL, Kennedy JL, Lichter JB, Van Tol H, Wetterberg L, Livak KJ, Kidd KK (1993) Alleles at the dopamine D_4 receptor locus do not contribute to the genetic susceptibility to schizophrenia in a large Swedish kindred. Am J Med Genet 48:218–22

Bartholini G (1980) Interactions of striatal dopaminergic, cholinergic, and GABA-ergic neurons: relation to extrapyramidal function. Trends Pharmacol Sci 1:138–140

Beasley CM, Montgomery S, Tye NC (1992) The behavioral pharmacology of olanzapine, a novel "atypical" antipsychotic agent. In: 2nd international conference on schizophrenia, Vancouver, Canada

Bersani G, Grispini A, Marini S, Pasini A, Valducci M, Ciani N (1986) Neuroleptic-induced extrapyramidal side effects: clinical perspectives with ritanserin, a new selective $5\text{-}HT_2$ receptor blocking agent. Curr Ther Res 40:492–499

Borison RL, Diamond BI, Pathiraja A, Meibach RC (1992) Clinical overview of risperidone. In: Meltzer HY (ed) Novel antipsychotic drugs. Raven New York, p 233

Boundy VA, Luedtke RR, Gallitano AL, Smith JE, Filtz TM, Kallen RG, Molinoff PB (1993) Expression and characterization of the rat D_3 dopamine receptor: pharmacologic properties and development of antibodies. J Pharmacol Exp Ther 264:1002–1010

Bouthenet ML, Souil E, Martres MP, Sokoloff P, Giros B, Schwartz JC (1991) Localization of dopamine D_3 receptor mRNA in the rat brain using in situ hybridization histochemistry: comparison with dopamine D_2 receptor mRNA. Brain Res 564:203–219

Bunzow JR, Van TH, Grandy DK, Albert P, Salon J, Christie M, Machida CA, Neve KA, Civelli O (1988) Cloning and expression of a rat D_2 dopamine receptor cDNA. Nature 336:783–787

Burris KD, Filtz TM, Chumpradit S, Kung M-P, Foulon C, Hensler JG, Kung HF, Molinoff PB (1994) Characterization of [^{125}I]-R-trans-7-OH-PIPAT binding to dopamine D_3 receptors in rat olfactory tubercle. J Pharmacol Exp Ther 268:935–942

Campion D, d'Amato T, Bastard C, Laurent C, Guedj F, Jay M, Dollfus S, Thibaut F, Petit M, Gorwood P et al. (1994) Genetic study of dopamine D_1, D_2, and D_4 receptors in schizophrenia. Psychiatry Res 51:215–230

Canton H, Verriele L, Colpaert FC (1990) Binding of typical and atypical antipsychotics to 5-HT_{1c} and 5-HT_2 sites: clozapine potently interacts with 5-HT_{1C} sites. Eur J Pharmacol 191:93–96

Carlsson A (1988) The current status of the dopamine hypothesis of schizophrenia. Neuropsychopharmacology 1:179–186

Catalano M, Nobile M, Novelli E, Nothen MM, Smeraldi E (1993) Distribution of a novel mutation in the first exon of the human dopamine D_4 receptor gene in psychotic patients. Biol Psychiatry 34:459–464

Chabert C, Cavegn C, Bernard A, Mills A (1994) Characterization of the functional activity of dopamine ligands at human recombinant dopamine D_4 receptors. J Neurochem 63:62–65

Chio CL, Hess GF, Graham RS, Huff RM (1990) A second molecular form of D_2 dopamine receptor in rat and bovine caudate nucleus. Nature 343:266–269

Chio CL, Drong RF, Riley DT, Gill GS, Slightom JL, Huff RM (1994a) D_4 dopamine receptor-mediated signaling events determined in transfected Chinese hamster ovary cells. J Biol Chem 269:11813–11819

Chio CL, Lajiness ME, Huff RM (1994b) Activation of heterologously expressed D_3 dopamine receptors: comparison with D_2 dopamine receptors. Mol Pharmacol 45:51–60

Cole JO, Goldberg SC, Klerman GL (1966) Phenothiazine treatment in acute schizophrenia. Arch Gen Psychiatry 10:246–261

Coon H, Byerley W, Holik J, Hoff M, Myles WM, Lannfelt L, Sokoloff P, Schwartz JC, Waldo M, Freedman R et al. (1993) Linkage analysis of schizophrenia with five dopamine receptor genes in nine pedigrees. AM J Hum Genet 52:327–334

Coward DM (1992) General pharmacology of clozapine. Br J Psychiatry 160:5–11

Creese I, Burt DR, Snyder SH (1976) Dopamine receptor binding predicts clinical and pharmacological potencies of antischizophrenic drugs. Science 192:481–483

Crocq MA, Mant R, Asherson P, Williams J, Hode Y, Mayerova A, Collier D, Lannfelt L, Sokoloff P, Schwartz JC et al. (1992) Association between schizophrenia and homozygosity at the dopamine D_3 receptor gene. J Med Genet 29:858–860

Dal Toso R, Sommer B, Ewert M, Herb A, Pritchett DB, Bach A, Shivers BD, Seeburg PH (1989) The dopamine D_2 receptor: two molecular forms generated by alternative splicing. EMBO J 8:4025–4034

Davis JM, Schaffer CB, Killian GA et al. (1980) Important issues in the drug treatment of schizophrenia. Schizophr Bull 6:70–87

Dearry A, Gingrich JA, Falardeau P, Fremeau RT Jr, Bates MD, Caron MG (1990) Molecular cloning and expression of the gene for a human D_1 dopamine receptor. Nature 347:72–76

Delay J, Deniker P, Harl J-M (1952) Traitement des états d'excitation et d'agitation par une méthode médicamenteuse dérivée de l'hibernothérapie. Ann Med Psychol 110(2):267–273

Deutch AY, Lee MC, Iadarola MJ (1992) Regionally specific effects of atypical neuroleptic drugs on striatal fos expression: the nucleus accumbens shell as a locus of antipsychotic action. Mol Cell Neurosci 3:332–341

Di Bella D, Catalano M, Strukel A, Nobile M, Novelli E, Smeraldi E (1994) Distribution of the MscI polymorphism of the dopamine D_3 receptor in an Italian psychotic population. Psychiatr Genet 4:39–42

Eidne KA, Taylor PL, Zabavnik J, Saunders PT, Inglis JD (1989) D_2 receptor, a missing exon. Nature 342

Fishburn CS, Belleli D, David C, Carmon S, Fuchs S (1993) A novel short isoform of the D_3 dopamine receptor generated by alternative splicing in the third cytoplasmic loop. J Biol Chem 268:5872–5878

Flamez A, De Backer J-B, Wilczak N, Vauquelin G, De Keyser J (1994) [^{3}H]Clozapine is not a suitable radioligand for the labelling of D_4 dopamine receptors in postmortem human brain. Neurosci Lett 175:17–20
Gandelman KY, Harmon S, Todd RD, O'Malley KL (1991) Analysis of the structure and expression of the human dopamine D_{2A} receptor gene. J Neurochem 56:1024–1029
Gerfen CR, Engber TM, Mahan LC, Susel Z, Chase TN, Monsma FJ Jr, Sibley DR (1990) D_1 and D_2 dopamine receptor-regulated gene expression of striatonigral and striatopallidal neurons. Science 250:1429–1432
Giros B, Sokoloff P, Martres MP, Riou JF, Emorine LJ, Schwartz JC (1989) Alternative splicing directs the expression of two D_2 dopamine receptor isoforms. Nature 342:923–926
Giros B, Martres MP, Sokoloff P, Schwartz JC (1990) Clonage du gène du récepteur dopaminergique D_3 humain et identification de son chromosome. CR Acad Sci [III] 311:501–508
Giros B, Martres MP, Pilon C, Sokoloff P, Schwartz JC (1991) Shorter variants of the D_3 dopamine receptor produced through various patterns of alternative splicing. Biochem Biophys Res Commun 176:1584–1592
Gonzalez AM, Sibley DR (1995) [^{3}H]7-OH-DPAT is capable of labeling dopamine D_2 as well as D_3 receptors. Eur J Pharmacol 272:R1–R3
Grandy DK, Marchionni MA, Makam H, Stofko RE, Alfano M, Frothingham L, Fischer JB, Burke-Howie KJ, Bunzow JR, Server AC, Civelli O (1989) Cloning of the cDNA and gene for a human D_2 dopamine receptor. Proc Natl Acad Sci USA 86:9762–9766
Grandy DK, Zhang Y, Bouvier C, Zhou Q-Y, Johnson RA, Allen L, Buck K, Bunzow JR, Salon J, Civelli O (1991) Multiple human D_5 dopamine receptor genes: a functional receptor and two pseudogenes. Proc Natl Acad Sci USA 88:9175–9179
Hayes G, Biden TJ, Selbie LA, Shine J (1992) Structural subtypes of the dopamine D_2 receptor are functionally distinct: expression of the cloned D_{2A} and D_{2B} subtypes in a heterologous cell line. Mol Endocrinol 6:920–926
Hollister LE (1995) Antipsychotic agents and lithium. In: Katzung BG (ed) Basic and clinical pharmacology. Prentice Hall, New York, p 432
Howard HR, Seeger TF (1993) Novel antipsychotics. Annu Rep Med Chem 28:39–47
Jarvie KR, Tiberi M, Silvia C, Gingrich JA, Caron MG (1993) Molecular cloning, stable expression and desensitization of the human dopamine D_{1B}/D_5 receptor. J Recept Res 13:573–590
Jonsson E, Lannfelt L, Sokoloff P, Schwartz JC, Sedvall G (1993) Lack of association between schizophrenia and alleles in the dopamine D_3 receptor gene. Acta Psychiatr Scand 87:345–349
Kane JM (1989) The current status of neuroleptic therapy. J Clin Psychiatry 50:322–328
Kebabian JW, Calne DB (1979) Multiple receptors for dopamine. Nature 277:93–96
Kelly PH, Seviour PW, Iversen SD (1975) Amphetamine and apomorphine responses in the rat following 6-OHDA lesions of the nucleus accumbens septi and corpus striatum. Brain Res 94:507–522
Köhler C, Hall H, Magnusson O, Lewander T, Gustafsson K (1990) Biochemical pharmacology of the atypical neuroleptic remoxipride. Acta Psychiatr Scand 82:27–36
Köhler C, Danielsson A, Magnusson O, Gustafsson K, Lewander T, Ogren SO, Ericson H (1991) Biochemical pharmacology of the atypical neuroleptic remoxipride. Schizophr Res 4:311–319
Köhler C, Radesäter AC, Karlsson-Boethius G, Bryske B, Widman M (1992) Regional distribution and in vivo binding of the atypical antipsychotic drug remoxipride. J Neural Transm 87:49–62
Kozell LB, Machida CA, Neve RL, Neve KA (1994) Chimeric D_1/D_2 dopamine receptors. J Biol Chem 269:30299–30306
Krupp P, Barnes P (1992) Clozapine-associated agranulocytosis: risk and aetiology. Br J Psychiatry 160:38–40

Lahti RA, Evans DL, Stratman NC, Figur LM (1993) Dopamine D4 versus D_2 receptor selectivity of dopamine receptor antagonists: possible therapeutic implications. Eur J Pharmacol 236:483–486

Laurent C, Savoye C, Samolyk D, Meloni R, Mallet J, Campion D, Martinez M, D'Amato T, Bastard C, Dollfus S (1994) Homozygosity at the dopamine D_3 receptor locus is not associated with schizophrenia. J Med Genet 31:260

Lawson CF, Mortimore RA, Schlachter SK, Smith MW (1994) Pharmacology of a human dopamine D_4 receptor expressed in HEK293 cells. Methods Find Exp Clin Pharmacol 16:303–307

Le Moine C, Bloch B (1991) Rat striatal and mesencephalic neurons contain the long isoform of the D_2 dopamine receptor mRNA. Brain Res Mol Brain Res 10:283–289

Le Moine C, Normand E, Bloch B (1991) Phenotypical characterization of the rat striatal neurons expressing the D_1 dopamine receptor gene. Proc Natl Acad Sci USA 88:4205–4209

Levant B, Bissette G, Widerlöv E, Nemeroff CB (1991) Alterations in regional brain neurotensin concentrations produced by atypical neuroleptic drugs. Regul Pept 32:193–201

Levesque D, Diaz J, Pilon C, Martres M-P, Giros B, Souil E, Schott D, Morgat J-L, Schwartz J-C, Sokoloff P (1992) Identification, characterization, and localization of the dopamine D_3 receptor in rat brain using [^{3}H]7-OH-DPAT. Proc Natl Acad Sci USA 89:8155–8159

Lewander T (1994) Overcoming the neuroleptic-induced deficit syndrome: clinical observations with remoxipride. Acta Psychiatr Sand 89 [Suppl 380]:64–67

Lichter JB, Barr CL, Kennedy JL, Van Tol H, Kidd KK, Livak KJ (1993) A hypervariable segment in the human dopamine receptor D_4 (DRD4) gene. Hum Mol Genet 2:767–773

Liu J-C, Cox RF, Greif GJ, Freedman JE, Waszczak BL (1994) The putative dopamine D_3 receptor agonist 7-OH-DPAT: lack of mesolimbic selectivity. Eur J Pharm 264:269–278

Liu L-X, Monsma FJ, Jr, Sibley DR, Chiodo LA (1994) D_{2S} and D_{2L} Receptors couple to K^+ currents in NG108-15 cells via different signal transduction pathways. Soc Neurosci Abstr 20:523 no 223.16

Liu YF, Civelli O, Grandy DK, Albert PR (1992) Differential sensitivity of the short and long human dopamine D_2 receptor subtypes to protein kinase C. J Neurochem 59:2311–2317

Macciardi F, Petronis A, Van Tol H, Marino C, Cavallini MC, Smeraldi E, Kennedy JL (1994) Analysis of the D_4 dopamine receptor gene variant in an Italian schizophrenia kindred. Arch Gen Psychiatry 51:288–293

Mack KJ, O'Malley KL, Todd RD (1991a) Differential expression of dopaminergic D_2 receptor messenger RNAs during development. Brain Res Dev Brain Res 59:249–251

Mack KJ, Todd RD, O'Malley KL (1991b) The mouse dopamine D_{2A} receptor gene: sequence homology with the rat and human genes and expression of alternative transcripts. J Neurochem 57:795–801

MacKenzie RG, VanLeeuwen D, Pugsley TA, Shih Y-H, Demattos S, Tang L, Todd RD, O'Malley KL (1994) Characterization of the human dopamine D_3 receptor expressed in transfected cell lines. Eur J Pharmacol 266:79–85

Malmberg A, Jackson DM, Eriksson A, Mohell N (1993) Unique binding characteristics of antipsychotic agents interacting with human dopamine D_{2A}, D_{2B}, and D_3 receptors. Mol Pharmacol 43:749–754

Mansour A, Meador WJ, Zhou QY, Civelli O, Akil H, Watson SJ (1991) A comparison of D_1 receptor binding and mRNA in rat brain using receptor autoradiographic and in situ hybridization techniqus. Neuroscience 45:359–371

Mant R, Williams J, Asherson P, Parfitt E, McGuffin P, Owen MJ (1994) Relationship between homozygosity at the dopamine D_3 receptor gene and schizophrenia. Am J Med Genet 54:21–26

McAllister G, Knowles MR, Patel S, Marwood R, Emms F, Seabrook GR, Granziano M, Borkowski D, Hey PJ, Freedman SB (1993) Characterization of a chimerich D_3/D_2 dopamine receptor expressed in CHO cells. FEBS Lett 324:81–86
McHale M, Coldwell MC, Herrity N, Boyfield I, Winn FM, Ball S, Cook T, Robinson JH, Gloger IS (1994) Expression and functional characterisation of a synthetic version of the human D_4 dopamine receptor in a stable human cell line. FEBS Lett 345:147–150
Meador-Woodruff JH, Mansour A, Civelli O, Watson SJ (1991) Distribution of D_2 dopamine receptor mRNA in the primate brain. Prog Neuropsychopharmacol Biol Psychiatr 15:885–893
Meador-Woodruff JH, Mansour A, Grandy DK, Damask SP, Civelli O, Watson SJ (1992) Distribution of D_5 dopamine receptor mRNA in rat brain. Neurosci Lett 145:209–212
Meltzer HY (1992) The importance of serotonin-dopamine interactions in the action of clozapine. Br J Psychiatry 160:22–29
Miller JC, Wang Y, Filer D (1990) Identification by sequence analysis of a second rat brain cDNA encoding the dopamine (D_2) receptor. Biochem Biophys Res Commun 166:109–112
Mills A, Allet B, Bernard A, Chabert C, Brandt E, Cavegn C, Chollet A, Kawashima E (1993) Expression and characterization of human D_4 dopamine receptors in baculovirus-infected insect cells. FEBS Lett 320:130–134
Mohell N, Sällemark M, Rosqvist S, Malmberg A, Högberg T, Jackson DM (1993) Binding characteristics of remoxipride and its metabolites do dopamine D_2 and D_3 receptors. Eur J Pharmacol 238:121–125
Monsma FJ Jr, McVittie LD, Gerfen CR, Mahan LC, Sibley DR (1989) Multiple D_2 dopamine receptors produced by alternative RNA splicing. Nature 342:926–929
Monsma FJ Jr, Mahan LC, McVittie LD, Gerfen CR, Sibley DR (1990) Molecular cloning and expression of a D_1 dopamine receptor linked to adenylyl cyclase activation. Proc Natl Acad Sci USA 87:6723–6727
Montamayeur JP, Borrelli E (1991) Transcription mediated by a cAMP-responsive promoter element is reduced upon activation of dopamine D_2 receptors. Proc Natl Acad Sci USA 88:3135–3139
Moore NA, Calligaro DO, Wong DT, Bymaster F, Tye NC (1993) The pharmacology of olanzapine and other new antipsychotic agents. Curr Opin Invest Drugs 2:281–293
Morel RT (1993) Association between schizophrenia and homozygostity at the dopamine D_3 receptor gene. J Med Genet 30:708–709
Nanko S, Hattori M, Ueki A, Ikeda K (1993a) Dopamine D_3 and D_4 receptor gene polymorphisms and Parkinson's disease. Lancet 342
Nanko S, Sasaki T, Fukuda R, Hattori M, Dai XY, Kazamatsuri H, Kuwata S, Juji T, Gill M (1993b) A study of the association between schizophrenia and the dopamine D_3 receptor gene. Hum Genet 92:336–338
Neve KA, Neve RL, Fidel S, Janowsky A, Higgins GA (1991) Increased abundance of alternatively spliced forms of D_2 dopamine receptor mRNA after denervation. Proc Natl Acad Sci USA 88:2802–2806
Neve KA, Kozlowski MR, Rosser MP (1992) Dopamine D_2 receptor stimulation of Na^+/H^+ exchange assessed by quantification of extracellular acidification. J Biol Chem 267:25748–25753
Ng GY, Mouillac B, George SR, Caron M, Dennis M, Bouvier M, O'Dowd BF (1994) Desensitization, phosphorylation and palmitoylation of the human dopamine D_1 receptor. Eur J Pharmacol 267:7–19
Nimgaonkar VL, Zhang XR, Caldwell JG, Ganguli R, Chakravarti A (1993) Association study of schizophrenia with dopamine D_3 receptor gene polymorphisms: probable effects of family history of schizophrenia? Am J Med Genet 48:214–217
Nothen MM, Cichon S, Propping P, Fimmers R, Schwab SG, Wildenauer DB (1993) Excess of homozygosity at the dopamine D_3 receptor gene in schizophrenia not confirmed. J Med Genet 30:708–709

O'Dowd BF, Nguyen T, Tirpak A, Jarvie KR, Israel Y, Seeman P, Niznik HB (1990) Cloning of two additional catecholamine receptors from rat brain. FEBS Lett 262:8–12

O'Dowd BF, Seeman P, George SR (1994) Dopamine receptors. In: Peroutka SJ (ed) Handbook of receptors and channels. CRC, Boca Raton p 95

O'Malley KL, Mack KJ, Gandelman KY, Todd RD (1990) Organization and expression of the rat D_{2A} receptor gene: identification of alternative transcripts and a variant donor splice site. Biochemistry 29:1367–1371

Ögren SO (1994) Potential mechanisms underlying the atypical antipsychotic profile of remoxipride. In: Palomo T, Archer T, Beninger R (eds) Strategies for studying brain disorders. Farrand, London, p 85

Ögren SO, Florvall L, Hall H, Magnusson O, Angeby-Möller K (1990) Neuropharmacological and behavioral properties of remoxipride in the rat. Acta Psychiatr Scand 82:21–26

Park B-H, fishburn S, Carmon S, Accili D, Fuchs S (1995) Structural organization of the murine D_3 dopamine receptor gene. J Neurochem 64:482–486

Pedersen UB, Norby B, Jensen AA, Schiodt M, Hansen A, Suhr-Jessen P, Scheideler M, Thastrup O, Andersen PH (1994) Characteristics of stably expressed human dopamine D_{1a} and D_{1b} receptors: atypical behavior of the dopamine D_{1b} receptor. Eur J Pharmacol 267:85–93

Pilon C, Lévesque D, Dimitriadou V, Griffon N, Martres M-P, Schwartz J-C, Sokoloff P (1994) Functional coupling of the human dopamine D_3 receptor in a transfected NG108-15 neuroblastoma-glioma hybrid cell line. Eur J Pharmacol/Mol Pharmacol 268:129–139

Roth BL, Craigo SC, Choudhary MS, Uluer A, Monsma FJ, Shen Y, Meltzer HY, Sibley DR (1994) Binding of typical and atypical antipsychotic agents to 5-hydroxytryptamine-6 and 5-hydroxytrypamine-7 recepters. Mol Pharm 268:1403–1410

Sabate O, Compion D, d'Amato T, Martres MP, Sokoloff P, Giros B, Leboyer M, Jay M, Guedj F, Thibaut F et al. (1994) Failure to find evidence for linkage or association between the dopamine D_3 receptor gene and schizophrenia. Am J Psychiatry 151:107–111

Savarese TM, Fraser CM (1992) In vitro mutagenesis and the search for structure-function relationships among G protein-coupled receptors. Biochem J 283:1–19

Sayers AC, Burki HR, Ruch W, Asper H (1976) Anticholinergic properties of antipsychotic drugs and their relation to extrapyramidal side effects. Psychopharmacol 51:15–22

Schmauss C, Haroutunian V, Davis KL, Davidson M (1993) Selective loss of dopamine D_3-type receptor mRNA expression in parietal and motor cortices of patients with chronic schizophrenia. Proc Natl Acad Sci USA 90:8942–8946

Seabrook GR, Patel S, Marwood R, Emms F, Knowles MR, Freedman S, MacAllister G (1992) Stable expression of human D_3 dopamine receptors in GH4C1 pituitary cells. FEBS Lett 312:123–126

Seabrook GR, Kemp JA, Freedman SB, Patel S, Sinclair HA, McAllister G (1994) Functional expression of human D_3 dopamine receptors in differentiated neuroblastoma x glioma NG108-15 cells. Br J Pharmacol 111:391–393

Seeman P (1987) Dopamine receptors and the dopamine hypothesis of schizophrenia. Synapse 1:133–152

Seeman P (1992) Dopamine receptor sequences: therapeutic levels of neuroleptics occupy D_2 receptors, clozapine occupies D_4. Neuropsychopharmacology 7:261–284

Seeman P, Van Tol HHM (1993) Dopamine D_4 receptors bind inactive (+)-aporphines, suggesting neuroleptic role. Sulpiride not stereoselective. Eur J Pharmacol 233:173–174

Seeman P, Lee T, Chan-Wong M, Wong K (1976) Antipsychotic drug doses and neuroleptic/dopamine receptors. Nature 261:717–719

Seeman P, Guan HC, Van Tol H (1993) Dopamine D_4 receptors elevated in schizophrenia. Nature 365:441–445

Senogles SE (1994) The D_2 dopamine receptor isoforms signal through distinct $G_{i\alpha}$ proteins to inhibit adenylyl cyclase. J Biol Chem 269:23120–23127
Shaikh S, Gill M, Owen M, Asherson P, McGuffin P, Nanko S, Murray RM, Collier DA (1994) Failure to find linkage between a functional polymorphism in the dopamine D_4 receptor gene and schizophrenia. Am J Med Genet 54:8–11
Snyder LA, Roberts JL, Sealfon SC (1991a) Alternative transcripts of the rat and human dopamine D_3 receptor. Biochem Biophys Res Commun 180:1031–1035
Snyder LA, Roberts JL, Sealfon SC (1991b) Distribution of dopamine D_2 receptor mRNA splice variants in the rat by solution hybridization/protection assay. Neurosci Lett 122:37–40
Sokoloff P, Giros B, Martres M-P, Bouthenet M-L, Schwartz J-C (1990) Molecular cloning and characterization of a novel dopamine receptor (D_3) as a target for neuroleptics. Nature 347:146–151
Sokoloff P, Andrieux M, Besançon R, Pilon C, Martres M-P, Giros B, Schwartz J-C (1992) Pharmacology of human dopamine D_3 receptor expressed in a mammalian cell line: comparison with the D_2 receptor. Eur J Pharmacol/Mol Pharmacol 225:331–337
Sommer SS, Lind TJ, Heston LL, Sobell JL (1993) Dopamine D_4 receptor variants in unrelated schizophrenic cases and controls. Am J Med Genet 48:90–93
Stille G, Hippius H (1971) Kritische Stellungnahme zum Begriff der Neuroleptika. Pharmacopsychiatry 4:182–191
Sunahara RK, Niznik HB, Weiner DM, Stormann TM, Brann MR, Kennedy JL, Gelernter JE, Rozmahel R, Yang Y, Israel Y, Seeman P, O'Dowd BF (1990) Human dopamine D_1 receptor encoded by an intronless gene on chromosome 5. Nature 347:80–83
Sunahara RK, Guan H-C, O'Dowd BF, Seeman P, Laurier LG, Ng G, George SR, Torchia J, Van Tol HHM, Niznik HB (1991) Cloning of the gene for a human dopamine D_5 receptor with higher affinity for dopamine than D_1 Nature 350:614–619
Tang L, Todd RD, Heller A, O'Malley KL (1994a) Pharmacological and functional characterization of D_2, D_3 and D_4 dopamine receptors in fibroblast and dopaminergic cell lines. J Pharmacol Exp Ther 268:495–502
Tang L, Todd RD, O'Malley KL (1994b) Dopamine D_2 and D_3 receptors inhibit dopamine release. J Pharmacol Exp Ther 270:475–479
Tiberi M, Jarvie KR, Silvia C, Falardeau P, Gingrich JA, Godinot N, Bertrand L, Yang-Feng TL, Fremeau RT Jr, Caron MG (1991) Cloning, molecular characterization, and chromosomal assignment of a gene encoding a second D_1 dopamine receptor subtype: differential expression pattern in rat brain compared with the D_{1A} receptor. Proc Nat Acad Sci USA 88:7491–7495
Tye NC, Moore NA, Calligaro D, Rees G, Sanger G, Beasley CM (1992) The preclinical pharmacology of olazapine, a novel "atypical" antipsychotic agent. In: 2nd internation conference on schizophrenia, Vancouver, Canada
Van Tol HHM, Bunzow JR, Guan H-C, Sunahara RK, Seeman P, Niznik HB, Civelli O (1991) Cloning of the gene for a human dopamine D_4 receptor with high affinity for the antipsychotic clozapine. Nature 350:610–614
Van Tol HHM, Wu CM, Guan H-C, Ohara K, Bunzow JR, Civelli O, Kennedy J, Seeman P, Niznik HB, Jovanovic V (1992) Multiple dopamine D_4 receptor variants in the human population. Nature 358:149–152
Walker JM, Bowen WD, Walker FO, Matsumoto RR, De Costa B, Rice KC (1990) Sigma receptors: biology and function. Pharmacol Rev 42:355–390
Watling KJ, Beer MS, Stanton JA, Newberry NR (1990) Interaction of the atypical neuroleptic clozapine with 5-HT_3 receptors in the cerebral cortex and superior cervical ganglion of the rat. Eur J Pharmacol 182:465–471
Weinshank RL, Adham N, Macchi M, Olsen MA, Branchek TA, Hartig PR (1991) Molecular cloning and characterization of a high affinity dopamine receptor ($D1\beta$) and its pseudogene. J Biol Chem 266:22427–22435

Wiese C, Lannfelt L, Kristbjarnarson H, Yang L, Zoega T, Sokoloff P, Ivarsson O, Schwartz JC, Moises HW, Helgason T (1993) No evidence of linkage between schizophrenia and D_3 dopamine receptor gene locus in Icelandic pedigrees. Psychiatry Res 46:69–78

Worms P, Broekkamp CLE, Lloyd K (1983) Behavioral effects of neuroleptics. In: Coyle JT, Enna SJ (eds) Neuroleptics: neurochemical, behavioral, and clinical perspectives. Raven, New York, p 93

Yang L, Li T, Wiese C, Lannfelt L, Sokoloff P, Xu CT, Zeng Z, Schwartz JC, Liu X, Moises HW (1993) No association between schizophrenia and homozygosity at the D_3 dopamine receptor gene. Am J Med Genet 48:83–86

Zhou Q-Y, Grandy DK, Thambi L, Kushner JA, Van Tol HHM, Cone R, Pribnow D, Salon J, Bunzow JR, Civelli O (1990) Cloning and expression of human and rat D_1 dopamine receptors. Nature 347:76–80

CHAPTER 4

Atypical Antipsychotic Drugs: Clinical and Preclinical Studies

S. HOSSEIN FATEMI, HERBERT Y. MELTZER, and BRYAN L. ROTH

A. Introduction

This chapter summarizes data which indicate that nondopaminergic systems may be essential for the activity of atypical antipsychotoic drugs. For the purposes of this chapter, an atypical antipsychotic drug is defined as a drug with demonstrated effectiveness in laboratory animal models of antipsychotic efficacy (e.g., blockade of amphetamine-induced hyperactivity or depolarization blockade of limbic dopamine neurons) but which induces catalepsy only at much higher doses and causes few or no extrapyramidal symptoms (EPS) in humans. It is probable that some of the compounds discussed here will not, in fact, prove to be antipsychotic in humans, despite their activity in many animal models of antipsychotic activity. This is most convincingly illustrated by 5-HT_3 antagonists such as zacopride which induce depolarization blockade of limbic dopamine (DA) neurons but do not induce catalepsy in rats and which are apparently devoid of clear-cut antipsychotic activity in humans (NEWCOMER et al. 1992).

A major limitation, then, of the classification of drugs as "typical" or "atypical" is that clinical information on many of them is scanty. The problem is exacerbated by the fact that clozapine, the prototypical antipsychotic drug, has a unique spectrum of efficacy, being useful for treatment-resistant schizophrenia (KANE et al. 1988), without causing tardive dyskinesia (LIEBERMAN and KOREEN 1993; MELTZER et al. 1989), and is effective in treating L-DOPA-induced psychosis and schizoaffective disorders in addition to causing few or no EPS. The only other drug currently approved for treatment of schizophrenia which may have a spectrum of efficacy similar to clozapine is risperidone. Risperidone, however, can induce EPS at high doses; additionally, it is not clear if it has the same degree of efficacy as clozapine (CHOUINARD et al. 1993; BORISON et al. 1992). Most of the animal studies cited used clozapine as the reference atypical antipsychotic drug, while only very few used more than one or two other putative atypicals. Several of the compounds reviewed have been used in only a few clinical studies; only a few have been tested in placebo-controlled, double-blind studies. Table 1 lists representative atypical antipsychotic drugs as well as their affinities for a variety of serotonin and DA receptors. As can be seen, the atypical antipsychotic drugs shown have relatively high affinities for serotonin compared to DA receptors.

The significance of this finding is summarized in the section on 5-HT receptors.

Despite these caveats, we will summarize information for many putative atypical antipsychotic drugs using the following format: First, we will consider evidence which implicates the specific neuronal system in the pathophysiology of schizophrenia (e.g., serotonergic neuronal system). Since much of this information is available from other sources, this will only be summarized here. Secondly, we will examine animal data which imply that putative atypical antipsychotic drugs may have effects on the systems examined. Thirdly, we will review preclinical and clinical findings regarding novel atypical antipsychotic drugs that may have specific actions on the neuronal systems in question.

B. Glutamate

L-glutamic acid (glutamate) is the major excitatory amino acid transmitter in the mammalian brain (ULAS and COTMAN 1993). Glutamate and some of its acid metabolites have been demonstrated to possess neurotoxic effects, presumably secondary to their effects on ionic movements across specific glutamate receptor channels which are abundantly present on dendrites (OLNEY 1969; OLNEY et al. 1971). It has recently been hypothesized that the excitotoxic effects of glutamate may be either increased or decreased depending on the chemical state of the nitric oxide group, leading to destruction or

Table 1. Binding of atypical antipsychotic drugs to serotonin receptors: comparison to D_4 and D_2 binding affinities

Drug	pK_i Values							
	5-HT_{2A}	5-HT_{2C}	5-HT_6	5-HT_7	D_2	D_4	$5HT_{2A}/D_2$	$5HT_{2A}/D_4$
Chlorotepine (+)	ND	ND	9.4	9.4	8.8		ND	ND
Zotepine	9.2	ND	9.0	8.8	9.0	8.23	0.2	0.97
Olanzepine	8.3	7.3	8.6	7.0	7.95	8.02	0.35	0.28
Clozapine	8.4	8.2	8.4	8.2	7.25	8.02	1.15	0.38
Rilapine	9.2	7.6	8.2	8.6	7.4	ND	2.2	ND
Fluperlapine	8.2	7.7	7.8	8.3	6.5	6.9	1.7	1.3
Tiosperone	10.2	8.0	7.6	9.2	9.3	7.9	0.9	2.3
Perlapine	7.9	ND	7.2	7.6	6.3	8.1	1.6	−0.2
Amperozide	7.9	5.9	7.2	6.3	6.3	ND	1.6	ND
Tenilapine	7.4	7.2	7.0	7.3	5.8	6.3	1.6	1.1
Risperidone	9.6	7.5	6.4	8.9	8.8	7.8	0.8	1.8
Melperone	7.1	5.9	5.9	6.3	6.7	6.6	0.4	0.5

Shown are pK_i values for selected atypical antipsychotic drugs for cloned 5-HT and dopamine receptors. Data are from ROTH et al. 1992; 1994 and ROTH et al., manuscript in preparation.
ND, not determined.

protection of neurons (LIPTON and ROSENBERG 1994). Several synthetic and naturally-occurring glutamate analogs which produce unique patterns of excitotoxicity have been indentified, including kainic acid, quisqualic acid, and *N*-methyl-D-aspartic acid (NMDA; see COTMAN and MONAGHAN 1987 for review).

I. Glutamate Receptors

Many receptor subtypes for glutamate exist, including both ion-channel receptors related to the nicotinic acetylcholine receptor superfamily and the metabotropic receptors. At least 19 distinct glutamate receptors exist in the mammalian central nervous system (DINGLEDINE and MCBAIN 1994; TIPS Receptor Nomenclature Supplement 1994). The ionotropic receptors consist of three major families: the NMDA, the α-amino-3-hydroxy-5-methyl-4-isoxazole propionic acid (AMPA), and the kainic acid receptors. The NMDA-receptor complex is tightly regulated with multiple binding sites, including those for (1) glutamate, (2) L-aspartic acid, (3) glycine agonist recognition sites, (4) a polyamine regulatory site (positive modulator), (5) a negative modulatory site for zinc, (6) a magnesium binding site, and (7) a phencyclidine (PCP) recognition site that can also bind ketamine, MK-801, and haloperidol (ULAS and COTMAN 1993; DINGLEDINE and MCBAIN 1994; KLECKNER and DINGLEDINE 1988). The NMDA receptor is unique with regard to the requirement of simultaneous binding of glutamate and glycine to achieve activation (KLECKNER and DINGLEDINE 1988). The AMPA site preferentially uses quisqualic acid as an agonist while the various kainic acid receptors are selective for kainic acid (TIPS Receptor Nomenclature Supplement 1994). There are at least five subtypes of NMDA receptors, four AMPA receptors, and five kainic acid receptors (TIPS Receptor Nomenclature Supplement 1994).

In addition to the ionotropic glutamate receptors there also exist at least six metabotropic receptors, two of which ($mGluR_1$, $mGluR_5$) utilize the inositol trisphosphate/diacylglycerol signal transduction pathway and four of which ($mGluR_2$, $mGluR_3$, $mGluR_4$ and $mGluR_6$) inhibit adenylate cyclase production (SCHOEPP and CONN 1993 and references cited therein). In addition to the neurotransmitter functions of glutamate, it is also incorporated into proteins; it is involved in fatty acid synthesis, the regulation of ammonia levels, and control of osmotic and anion balance; and it serves as a precursor for γ-aminobutyric acid (GABA) synthesis (DINGLEDINE and MCBAIN 1994).

II. Glutamate Hypothesis of Schizophrenia

The role of glutamate in schizophrenia has been supported by several lines of research which will only be summarized here (see DEUTSCH et al. 1989; KIM et al. 1980; RIEDERER et al. 1992). The original hypothesis stated that decreased glutamatergic neurotransmission played a role in schizophrenia (KIM et al.

1980) while a revised hypothesis was based on the observation that PCP can exacerbate schizophrenia and can cause a schizophrenia-like psychosis in control subjects (LUBY et al. 1959, 1962; ZUKIN and JAVITT 1991). Importantly, long-term administration of PCP can cause a psychosis in which both positive and negative symptoms appear (LUBY et al. 1959, 1962; ZUKIN and JAVITT 1991). However, some features of psychosis such as auditory hallucinations are rare. Additionally, many studies have reported a dysfunction of the glutamatergic system in schizophrenia, including low CSF levels of glutamate in schizophrenic patients (KIM et al. 1980); increased density of glutamate receptors in schizophrenia (NISHIKAWA et al. 1983); decreased kainic acid binding sites in the CA4 and CA3 regions of the hippocampus, dentate gyrus, and entorhinal cortex (KERWIN et al. 1990); decreased glutamate receptor gene expression in schizophrenia (HARRISON et al. 1991); and diminished glutamate release from synaptosomes isolated from postmortem brain tissue of schizophrenic patients (SHERMAN et al. 1991).

Dysfunction of the glutamate system is hyptothesized to alter dopaminergic neurotransmission as follows:

1. Hyperdopaminergic neurotransmission in mesocortical dopaminergic neurons may lead to hypoglutamatergic activity in corticofugal glutamate neurons (KORNHUBER and KORNHUBER 1986).
2. Hyperdopaminergic activity may be secondary to hypofunctioning NMDA receptors which are present on DA-containing nerve terminals where they may function as inhibitory heteroreceptors gating DA release (CARLSSON and CARLSSON 1990).
3. Hyperdopaminergic or hypoglutamatergic function can affect GABA-mediated inhibition of the excitatory thalamocortical tract (CARLSSON 1988).
4. Abnormal genetic regulation of NMDA receptors may lead to adult schizophrenic symptoms (ETIENNE and BAUDRY 1990).
5. Hyperdopaminergic activity may result from hyperinnervation of the frontal cortex as a result of abnormal early development of glutamatergic neurons (DEAKIN et al. 1989).

In addition to the human data cited above, many animal studies suggest that specific alterations of glutamate receptor function may be elicited by atypical antipsychotic drugs or that glutamate-active drugs may possess antipsychotic-like activity:

1. Intracerebroventricular infusion of the glycine agonists D-serine and D-alanine were reported to antagonize PCP- and MK-801-induced stereotypy or amphetamine-induced hyperactivity in rats (CONTRERAS 1990; HASHIMOTO et al. 1971; TRICKLEBANK et al. 1992).
2. The glycine antagonist R (+)HA966 antagonizes the PCP- and MK801-induced DA turnover in the nucleus accumbens, amygdala, and prefrontal cortex (HUTSON et al. 1991; BRISTOW et al. 1991).
3. Exogenous administration of glycine counteracts PCP-induced hyperactivity in mice, suggesting that glycine may be used in the treatment of schizophrenia (TOTH and LAJTHA 1986).

4. In some studies, acute administration of antipsychotic drugs may elevate glutamate levels in several brain areas (striatum, cortex) with drugs like clozapine having selective effects on cortical glutamate levels and typical antipsychotic drugs selectively increasing striatal glutamate levels (MOGHADDAM and BUNNEY 1990; PEHEK et al. 1991; BADGETT et al. 1993; YAMAMOTO and COOPERMAN 1994). The hypoglutamatergic hypothesis of schizophrenia suggests that increasing glutamate is beneficial; hence, clozapine and related drugs could be exerting their therapeutic effect via this route. However, very recent studies by YAMAMOTO and COOPERMAN (1994), who examined long-term drug treatment, found somewhat different results. These authors found that long-term clozapine treatment enhanced glutamate release only in the nucleus accumbens while having no effect on striatal or cortical glutamate release. The authors suggest that the differences between their results and those of BADGETT et al. (1993) may be accounted for by the fact that BADGETT et al. (1993) measured the tissue content of amino acids while YAMAMOTO and COOPERMAN (1994) quantified release via in vivo microdialysis. YAMAMOTO and COOPERMAN (1994) suggest that the ability of haloperidol to elevate striatal glutamate release is related to its ability to induce tardive dyskinesia via excitotoxic damage (GUNNE and ANDREN 1993). It is conceivable that the lack of induction of tardive dyskinesia due to clozapine is secondary to its inability to increase striatal levels of glutatmate. YAMAMOTO and COOPERMAN (1994) suggest that the ability of clozapine to enhance nucleus accumbens glutamate release is secondary to blockade of 5-HT_2 receptors, since the 5-HT_2 antagonist ketanserin blocks 5-HT's ability to inhibit glutamate release in the striatum (YAMAMOTO and COOPERMAN 1994; MAURA et al. 1988).
5. Clozapine inhibits [3H]-MK-801 binding with high affinity and has antiglutamatergic effects in vivo (LANG et al. 1992; LIDSKY et al. 1993). These results are difficult to reconcile with those in point 4 which suggest that clozapine actually enhances subcortical glutamate release. It has been suggested that the antiglutamatergic activity of clozapine may be the basis for a neuroprotective action (see MESHUL et al. 1992) in schizophrenia.

III. Glutamatergic Drugs

1. Glycine and Milacemide

In other studies, administration of the glutamate agonists glycine and milacemide yielded contradictory responses. In the first open-label study, WAZIRI (1988) reported favorable antipsychotic effects by exogenous glycine. Less dramatic, though positive, results were reported by COSTA et al. (1990) and pointed to a modest beneficial effect of glycine in the reduction of psychotic symptoms. In the glycine studies of POTKIN et al. (1992), the only significant improvement noted was in the clinical global impression (CGI). A discriminant function analysis also showed that glycine clearly improved emotional withdrawal, depression, hostility, and uncooperativeness with less improvement in mannerisms (Potkin et al. 1992).

Milacemide is an amino acetamide derivative which increases glycine concentrations after oral administration. Initial studies did not demonstrate any apparant therapeutic effects (Rossee et al. 1990). A double-blind crossover study found similar negative results of milacemide (Tamminga et al. 1992).

2. Umespirone

Umespirone, a putative atypical antipsychotic drug, has been reported to block the stereotypical behavior induced by the NMDA antagonist DL-2-amino-5-phosphonovaleric acid (AP5) selectively. Thus, in one study umespirone blocked AP5-induced sniffing, a potential glutamate-mediated effect (Schmidt et al. 1991). Umesprione also appears to bind to the high-affinity state of the sigma receptor (Itzhak et al. 1990). Clozapine as well can block NMDA-mediated responses (Rao et al. 1991), so it is conceivable that umespirone could possess clozapine-like atypical antipsychotic activity.

3. Others

Recent discovery of several glycine site partial agonists with anticonvulsant and neuroprotective properties have stimulated the development of potential atypical antipsychotic drugs. Two such agents are (+)HA966 and L687414 (Singh et al. 1990; Kemp and Leeson 1993), both of which are low-efficacy partial glycine antagonists that have good CNS-penetrating abilities (Singh et al. 1990; Saywell et al. 1991). In rodent models, (+)HA966 and L687414 are not PCP-like and have better separation between their anticonvulsant and sedative effects than MNDA ion channel blockers (Singh et al. 1990; Kemp and Leeson 1993). R(+)HA966 blocks the selective activation of mesolimbic dopaminergic systems by amphetamine and amphetamine-induced hyperactivity (Hutson et al. 1991). This property contrasts with the PCP-induced activation of the mesolimbic dopamine system. The development of glutamate agonists which may act at various binding sites of the NMDA receptor complex (i.e., the polyamines spermine or spermidine) or affect non-NMDA sites could be of clinical value in treating schizophrenia. Table 2 summarizes findings obtained with glutamatergic agents.

C. γ-Aminobutyric Acid (GABA)

I. GABA Receptors

γ-Aminobutyric acid (GABA) is the major inhibitory neurotransmitter in the mammalian central nervous system. Two classes of GABA receptors exist: $GABA_A$ and $GABA_B$. $GABA_A$ receptors are heterooligomeric structures which contain GABA-gated anionic channels (SCHOFIELD et al. 1987). Benzodiazepines act as positive modulators of GABA actions at the $GABA_A$ receptors (COSTA 1991 for review). There is a high degree of structural microheterogeneity present in the expression of GABA receptors identified in various regions of rat brain (OLSEN and TOBIN 1990). Four subunits (α, β, γ, and δ) exist, each of which has multiple subtypes (six α subunits, four β subunits, three γ, and a single δ subunit) so that multiple combinations of subunits are possible. The α subunit determines molecular recognition properties while the γ subunit appears to be absolutely required for benzodiazepine binding (SIEGEL et al. 1990; TIPS RECEPTOR NOMENCLATURE SUPPLEMENT 1994). Interestingly, activation of $GABA_A$ receptors by isoguvacine, 4,5,6,7-tetrahydroisoxazolo-5,4C-pyridin-3-ol (THIP) or muscimol can produce psychosis in humans (HOEHN-SARIC 1983). Additionally, reports indicate that GABA uptake inhibitors and benzodiazepines can, in some cases, worsen schizophrenic symptoms, presumably via activation of $GABA_A$ receptors (SEDMAN et al. 1990; MELDRUM 1982); typically, however, benzodiazepines are used to augment antipsychotic activity. γ-Vinyl-GABA, a GABA-mimetic, has been tested as a potential treatment for tardive dyskinesia with only mixed results, though it does appear to have anticonvulsant properties (FONNUM 1987).

$GABA_B$ receptors are involved in inhibiting excitatory synaptic transmission between neurons and by coupling to G proteins via inhibition of adenylate cyclase activity. $GABA_B$ receptors exist in multiple subtypes as well. GABA-ergic receptors (both A and B) have been localized to neocortex, hippocampus, basal ganglia, thalamus, hypothalamus, amygdala, cerebellum, medulla, spinal cord, and several peripheral tissues (BONANNO and RAITERI

Table 2. Effect of glutamatergic, adrenergic, and cholecystokinergic drugs on various clinical and preclinical measures of antipsychotic activity

Drug	Depol Block		Inh. DA response	Catalepsy	Inh. CAR	EPS	Anti-psychotic activity
	A9	A10					
Glycine	?	?	?	–	?	–	+ (Negative symptoms)
Milacemide	?	?	?	–	?	–	–
Umespirone	?	?	?	–	?	?	?
R(+)-HA966	?	?	+	–	?	?	?
Prazosin + Haloperidol	+	–	+	?	?	?	+
LY26291	+	–	?	–	?	?	?

Inh. DA response, inhibition of responses induced by apomorphine, amphetamine, or other dopaminergic agents; Depol Block, the ability of a drug to induce depolarization blockade after chronic administration; Inh. CAR, the ability of a drug to inhibit conditioned avoidance response; EPS, the ability of a drug to induce extrapyramidal symptoms in humans.

1993). GABA-ergic neurons are ubiquitously distributed throughout the central nervous system where they function as interneurons. In this regard, it is interesting to note that 5-HT_{2A} receptors (which may be important for the actions of clozapine) are localized on GABA-ergic interneurons (MORILAK et al. 1993; GARLOW et al. 1993).

For many years it has been known that typical neuroleptics alter GABA receptor functioning. In the substantia nigra, for example, long-term neuroleptic administration has been shown to increase [3H]-GABA and [3H]-muscimol binding (GALE 1980; SEE et al. 1989) and to decrease GABA turnover (MARCO et al. 1976; MAO et al. 1977). Since the major efferent pathways from the striatum are GABA-ergic, neuroleptic-induced alterations of GABA function are thought to be involved in the etiology of tardive dyskinesia (GUNNE and HAGGSTROM 1983). Few studies have examined the effects of atypical antipsychotic drugs on GABA-ergic functioning.

SEE et al. (1990) found that long-term clozapine or raclopride administration had no effect on [3H]-muscimol binding in the substantia nigra as measured by quantitative receptor autoradiography, though haloperidol did increase [3H]-muscimol binding in agreement with other studies. These results are consistent with the notion that clozapine and raclopride have a low propensity to induce tardive dyskinesia. Clozapine was reported to increase GABA release in the globus pallidus and ventral striatum in rats using microdialysis while haloperidol decreased GABA release in these two areas (DREW et al. 1990). However, YAMAMOTO and COOPERMAN (1994) reported that haloperidol and clozapine increased GABA efflux in the ventral pallidum.

II. GABA Hypothesis of Schizophrenia and Clinical Studies of GABA-ergic Drugs

VANKAMMEN (1977; VANKAMMEN and GELERNTER 1987) proposed the GABA hypothesis of schizophrenia which posits that decreased GABA inhibition of DA neurotransmission is responsible for the genesis of schizophrenic symptoms. Additional support was provided by REYNOLDS et al. (1990) and SIMPSON et al. (1989), who reported lower-than-normal GABA uptake sites in the hippocampus and subcortical brain regions of schizophrenic patients. The reduction in GABA uptake sites in the hippocampus correlated with an increase in DA in the amygdala, and both changes were preferentially found in the left hemisphere. Clinical trials of GABA-ergic medications have been disappointing with a general lack of positive results and some reports indicating a worsening of psychotic symptoms. An extensive review by WOLKOWITZ and PICKAR (1991) on the role of benzodiazepines in the treatment of schizophrenia indicated that the response to benzodiazepines alone was highly variable, with some improvement in 30%–50% of the patients. Moreover, the potential utility of benzodiazepines became more evident when they were used as adjunctive treatments in the acute management of excited schizophrenic patients. A recent double-blind, placebo-controlled trial of alprazolam augmentation of antipsychotic drug actions, however, suggested that about 50% of the patients improved with the treatment, though no characteristics of responders versus nonresponders were evident (WOLKOWITZ et al. 1992). Thus, although augmenting GABA-ergic function may be useful in the treatment of schizophrenia, the evidence suggests that such agents possess little atypical antipsychotic activity. It is conceivable that the relative inability of clozapine to cause tardive dyskinesia is secondary to specific effects on GABA-ergic systems, perhaps mediated via 5-HT_{2A} receptor blocking activity.

D. Acetylcholine

I. Acetylcholine Receptors

Acetylcholine receptors are divided into nicotinic and muscarinic classes based on classical pharmacologic criteria (TIPS RECEPTOR NOMENCLATURE SUPPLEMENT 1994). Nicotinic receptors belong to the same family as the GABA receptors in which four subunits (α, β, γ, and δ) exist in muscle-type (α, β, γ, and δ) or in neuronal-type (α and β only) configurations. Several varieties of the α and β subunits have been identified by molecular cloning technologies. Muscarinic receptors belong to the G-protein superfamily of receptors; at least five subtypes of muscarinic receptors exist (TIPS Receptor Nomenclature Supplement 1994). Activation of the muscarinic receptor modulates learning, memory, attention, mood, nociception, motor function, emotion, sleep, and neuroendocrine functions. Each of the five subtypes of muscarinic receptors shows a distinct regional brain distribution and receptor pharmacology.

II. Muscarinic Hyperactivity in Schizophrenia?

Early studies by several investigators attributed the psychotic symptoms of schizophrenia to dopaminergic-cholinergic imbalances (FRIEDHOFF and ALPERT 1973; JANOWSKI et al. 1973; SINGH and KAY 1985). TANDON and GREDEN (1989), however, have suggested a more prominent role for muscarinic cholinergic receptor hyperactivity in the pathogenesis of negative symptoms of schizophrenia. This thesis is based on the following types of observations: first, infusion of physostigmine in normal controls leads to a behavioral syndrome similar to negative symptoms in schizophrenia (SINGH and KAY 1985 for review). Second, findings of reduced pain perception, hypersalivation, increased water intake, relative hypothermia, and pupillary abnormalities were reported in unmedicated schizophrenics by KRAEPELIN (1919) and are consistent with a muscarinic receptor hyperactive state. Third, anticholinergic agents have occasionally been reported to be effective in treating negative symptoms of schizophrenia, though it is not clear if this was merely due to a reduction in the EPS of neuroleptics (TANDON et al. 1988; TANDON and Greden 1987; Goff et al. 1994). Fourth, some antipsychotic drugs with high anticholinergic activity (e.g., clozapine, fluperlapine, zotepine) are atypical and may reduce the magnitude of negative symptoms (Bolden et al. 1992; Meltzer et al. 1989). Fifth, older literature suggests that cholinomimetic agents exacerbate negative symptoms (Rowntree et al. 1950; Modestin et al. 1973). Sixth, a correlation exists comparing increased cholinergic activity and reduced REM-latencies and reduced slow-wave sleep in patients with negative symptoms (Ganguli et al. 1987; Tandon et al. 1992).

Based on these findings, one might predict that the degree of atypicality of an antipsychotic drug would be related to its anticholinergic activity. Richelson's group recently tested this hypothesis by assessing the affinities of a number of typical and atypical antipsychotic drugs at the five cloned human muscarinic receptors (Bolden et al. 1992). In general, no clear pattern emerged from this study. Thus, for instance, clozapine had high affinities for all five muscarinic receptors while melperone and risperidone (Leysen et al. 1993; both putative atypical antipsychotic drugs) had very low affinities. Additionally, thioridazine, which is not an atypical antipsychotic, had a generally higher affinity for all five muscarinic receptors than clozapine. This issue is complicated by the possibility that antipsychotic drugs may enhance cholinergic activity by indirect mechanisms. Thus, two studies show that the ability of clozapine to increase DA turnover in the rat striatum can be blocked by pretreatment with the anticholinergic scopolamine (Rivest and Marsden 1991; Meltzer et al. 1994). Taken together, these results cast doubt on the hypothesis that anticholinergic activity per se is essential for an atypical antipsychotic drug.

Whole animal studies have done little more to clarify the role of cholinergic blockade in the action of atypical antispychotic drugs like clozapine. Thus, some studies have suggested that long-term administration of anticholinergic drugs with haloperidol will change the pattern of inactivation of DA neurons to a selective inactivation of mesolimbic neurons (Chiodo and Bunney 1985) while more recent workers were unable to replicate these findings (Gardner et al. 1993). Other animal studies have shown that simple addition of an anticholinergic drug to haloperidol does not mimic clozapine's effects precisely (Coward et al. 1989).

Human studies are equally ambiguous. Tandon et al. (1988) reported on the acute effects of trihexiphenidyl, an anticholinergic agent, on positive and negative symptoms in schizophrenic patients. Trihexiphenidyl produced a significant increase in positive symptoms while reducing the negative symptoms. In contrast, Goff et al. (1994) reported minimal or no effects by anticholinergic agents on negative or psychotic symptoms in a group of relatively stable schizophrenic patients who had been neuroleptic-free for 8 weeks in a double-blind, placebo-controlled study. Additionally, anticholinergic medications do not prevent the development of tardive dyskinesia (Gerlach 1977). Thus the ability of anticholinergic drugs to function as atypical antipsychotics is currently unclear.

E. Norepinephrine

Norepinephrine (NE) binds to two major classes of receptors: α and β. All of the α and β receptor subtypes are G protein-linked. To date, four subtypes of α_1-adrenoceptors, three subtypes of α_2-adrenoceptors, and three subtypes of β-adrenoceptors have been cloned. The α_1-adrenoceptors all appear to be coupled to phosphoinositide hydrolysis while the α_2-adrneoceptors all inhibit cyclic adenosine monophosphate (cAMP) production. β-adrenoceptors all activate adenylate cyclase production.

Historically, the first reference to the involvement of NE in the pathophysiology of schizophrenia was made by STEIN and WISE (1971). These investigators observed a reduction in the activity of NE-synthesizing enzymes in the brainstems of schizophrenic patients. Subsequent studies could not confirm these findings, though later reports demonstrated an increase in NE input to the limbic forebrain in schizophrenics and established a relationship between NE levels and the intensity of schizophrenic symptoms and drug response (VANKAMMEN et al. 1990). Additionally, BREIER et al. (1994) recently demonstrated that clozapine's ability to elevate plasma NE levels was directly correlated with its ability to improve positive symptoms (as measured by the BPRS). BREIER et al. (1994) suggest that the superior efficacy and atypical features of clozapine correlate with peripheral noradrenergic stimulation, perhaps via α_1-adrenoceptor blockade. However, these observations may simply reflect the correlation of clinical response to plasma clozapine levels (HASAGAWA et al. 1993).

CSF NE levels have been reported to be increased in patients with schizophrenia. Various investigators have reported increased NE plasma levels in patients with schizophrenia (VANKAMMEN et al. 1989). Varying levels of α_2-receptor binding in brains of schizophrenic patients have been reported (VANKAMMEN and KELLY 1991). Several authors have postulated that noradrenergic systems may be dysregulated during stages of acute psychosis (VANKAMMEN 1991). The same studies point to suppression of central noradrenergic neurons during antipsychotic drug treatment (GOMES et al. 1980). Upon discontinuation of neuroleptic therapy, the rate of relapse correlates with activation of noradrenergic systems and a worsening of psychotic symptoms (GOMES et al. 1980; VANKAMMEN and KELLY 1991; VANKAMMEN 1991).

Clinical studies employing various noradrenergic agonists and antagonists reveal several findings of interest. First, β-adrenoceptor antagonists such as propranolol are ineffective alone but are clearly beneficial for akathisia. Secondly, α_1-adrenoceptor antagonists are generally devoid of antipsychotic activity as demonstrated by negative studies using prazosin (HOMMER et al. 1984). Thirdly, α_2-adrenoceptor agonists such as clonidine may have moderate therapeutic effects while α_2-adrenoceptor antagonists (e.g., yohimbine) generally worsen psychotic symptoms in acutely ill schizophrenics (FREEDMAN et al. 1982; HOLMBERG and GERSHON 1961). Yohimbine, an α_2-adrenoceptor antagonist, enhances noradrenergic tone (by inhibiting autoreceptor activity), and may mimic the symptoms of sensory overload and disruption of sensory gating which occur in patients with schizophrenia (STEVENS et al. 1993).

I. α_1-Adrenergic Receptor Involvement in Atypical Antipsychotic Drug Actions

In terms of atypical antipsychotic drugs, clozapine has potent α_1-adrenoceptor receptor blocking activity (COHEN and LIPINSKI 1986), while risperidone and sertindole have more modest α_1-adrenoceptor receptor blocking activity (LEYSEN et al. 1993; HOLLISTER 1994). Umespirone, a potential atypical antipsychotic, has a relatively high affinity for the α_1-adrenoceptor (ITZHAK et al. 1990). Other putative atypical antipsychotic drugs (e.g., melperone, MDL100907) have minimal α_1-adrenoceptor antagonsit activity (SORENSEN et al. 1993; LEYSEN et al. 1993).

BALDESSARINI et al. (1992) have suggested that the α_1-adrenoceptor antagonistic effects of clozapine are essential for its atypical nature. Importantly, long-term clozapine treatment induces a supersensitivity of α_1-adrenoceptors, similar to the effects of prazosin and reserpine (COHEN and LIPINSKI 1986). Additionally, these authors reported that coadministration of an α_1-adrenoceptor antagonist (prazosin) with haloperidol led to a pattern of inactivation of rat A9 and A10 DA neurons which was indistinguishable from that induced by clozapine (BALDESSARINI et al. 1992). Finally, recent studies by RAO et al. (1991) suggest that the reason clozapine is able to block the NMDA-mediated responses in vivo is due to α_1-adrenoceptor blockade via a complicated interaction. If true, this would tie together the putative antiglutamate and antinoradrenergic actions of clozapine.

F. Cholecystokinin (CCK)

The existence of cholecystokinin (CCK) in the CNS was first described by VANDERHAEGHEN et al. (1975). In at least 40% of the ventral mesencephalic neurons, DA and CCK are colocalized. Based on electrophysiological evidence, DA and CCK appear to interact at the synaptic level in the nucleus accumbens (see CRAWLEY 1991 for review). Two subtypes of CCK receptors have been identified: CCKA and CCKB. Both receptors are coupled to phosphoinositide hydrolysis, with the CCKA receptor predominating in the periphery and the CCKB receptor found throughout the brain. Several CCKA and CCKB receptor antagnosists have been developed, including devazepide; A65186 and lorgluride (CCKA-selective); and L365260, CI988, and LY262684 (CCKB-selective).

Involvement of CCK-containing dopaminergic pathways in patients with schizophrenia and other psychiatric disorders such as Parkinson's disease and drug addiction have been suggested (CRAWLEY 1991). ROBERTS et al. (1983) showed decreased CCK-like immunoreactivity in the temporal cortex of schizophrenics while GARVER et al. (1990) reported reduced basal levels of CSF CCK in drug-free schizophrenics. In contrast, other studies show either no difference (GJERRIS et al. 1984) or increased CCK (GERNER et al. 1985) in the CSF of schizophrenic patients. More recently, SUZUKI et al. (1993) showed significant increases of CSF CCK in patients. The reason for these disparate results is not known, but could represent differences in patient selection.

I. CCK-ergic Drugs

1. LY262691

RASMUSSEN et al. (1991b) used the selective CCKB antagonist LY262691 to study the effects of chronic treatment on A9 and A10 dopaminergic neurons and found inactivation of both neuronal types. The CCKA-selective drug LY219057 had no effect on the spontaneous activity of A10 or A9 cells, indicating that the effect of LY262691 is specific for the CCKB receptor. LY262691 also failed to induce catalepsy in rats (RASMUSSEN et al. 1993). Although these results are promising, LY262691 has not yet been evaluated in humans with schizophrenia.

2. Caerulin

Based on human studies of altered CCK levels in schizophrenia, initial open trials suggested antipsychotic activity following systemic administration of the CCK analogue caerulin (ALBUS 1988; STOESSL 1989). However, controlled, double-blind crossover studies showed no significant effect with systemically administered caerulin (CRAWLEY 1991). These negative results have been explained by a lack of peptide entry into the brain.

3. Others

Further studies with CCK8 and proglumide (a weak CCK antagonist) showed no antipsychotic effects (INNIS et al. 1986; MONTGOMERY and GREEN 1988). The use of CCK antagonists is to inhibit the potent excitatory effects of CCK on dopaminergic cells. Previous studies indicated that CCK may facilitate DA-induced behavior in the medial part of the posterior nucleus accumbens. Thus, both CCKB receptor antagonists and agonists as well as CCKA agonists may act as atypical antipsychotic drugs.

G. Neurotensin

Neurotensin (NT) is an endogenous tridecapeptide present in the central and peripheral nervous systems with the highest concentrations in the hypothalamus, substantia nigra, periaqueductal gray matter, and the limbic system (KOBAYASHI et al. 1977). The NT receptor has been cloned and is in the G protein-coupled supergene family. This receptor is coupled to phosphoinositide hydrolysis and may also regulate adenylate cyclase activation.

Direct administration of NT into the CNS causes effects similar to the antipsychotic actions of chlorpromazine and haloperidol: (1) potentiation of barbiturate- and ethanol-induced narcosis; (2) inhibition of conditioned avoidance; (3) decrease of locomotor hyperactivity induced by methylphenidate, cocaine, and *d*-amphetamine; (4) inhibition of intracranial self-administration; and (5) increase in the metabolic activity of presynaptic DA neurons (see NEMEROFF et al. 1992 for recent review).

I. Neurotensin and Schizophrenia

WIDERLOV et al. (1982) reported decreased CSF NT concentrations in some patients with schizophrenia compared to controls; treatment with neuroleptics was associated with a normalization of NT levels. Two additional studies (LINDSTROM et al. 1988; NEMEROFF et al. 1989) also reported decreased NT levels in the CSF of drug-free schizophrenic patients. GARVER et al. (1991) reported decreased CSF NT levels in a subgroup of psychotic women whose NT concentrations increased following treatment with haloperidol. All of these results implied a deficit of NT in schizophrenia which was normalized by antipsychotic drug treatment.

II. Effects of Atypical Antipsychotic Drugs on Neurotensin Systems

NEMEROFF et al. (1992) studied the effects of several atypical antipsychotic drugs on CNS NT levels. Rimcazole and sulpiride increased caudate nucleus NT concentrations. S (+)-N-n-propylnorapomorphine, a direct active DA agonist, increased NT levels in the nucleus accumbens, caudate nucleus, and substantia nigra (NEMEROFF et al. 1991). CI-943 (8-ethyl-7,8-dihydro-1,3,5-trimethyl-1-H-imidazo [1,2-C]-pyrazolo-[3,4-e]-pyrimidine) also increased NT in the nucleus accumbens and caudate nucleus, hypothalamus, and substantia nigra (LEVANT et al. 1991). Both acute and long-term administration of the putative atypical antipsychotic durg BMY 14802, a sigma receptor antagonist, caused a dose-dependent increase in NT concentrations in the nucleus accumbens as well as anterior and posterior caudate nucleus while decreasing NT levels in the frontal cortex (NEMEROFF et al. 1992). Finally, LEVANT et al. (1992) found increased expression of proNT mRNA in the nucleus accumbens and dorsolateraol caudate nucleus following treatment with BMY 14802.

MERCHANT and DORSA (1993) have studied the mechanisms by which atypical antipsychotic drugs regulate NT mRNA levels. These authors reported that atypical antipsychotic drugs increased expression of NT mRNA in the shell of the nucleus accumbens (a region of brain activated by the limbic system) while typical antipsychotic drugs increased NT mRNA in the dorsolateral striatum (MERCHANT and DORSA 1993). The haloperidol-induced increase in NT mRNA in the striatum was abolished by antisense oligonucleotides to c-*fos*, indicating that c-*fos* may function as a *trans*-acting factor to increase NT by haloperidol (MERCHANT 1994). Thus, these studies appear to indicate that atypical antipsychotic drugs like clozapine selectively increase NT mRNA in the shell of the nucleus accumbens via a c-*fos*-independent manner. BOLDEN-WATSON et al. (1993) have shown, in a similar manner, that haloperidol but not clozapine increases NT mRNA in the rat substantia nigra.

A report by CUSACK et al. (1993) demonstrated development of novel NT mimetics and investigated their effects on murine neuroblastoma cell lines which endogenously express NT receptors. Both drugs antagonized the NT-stimulated production of cyclic guanosine monophosphate (cGMP) with Ki values in the low micromolar range. One of these compounds (mimic 2) binds to a high-affinity site when acting as an agonist and a low-affinity site when acting as an antagonist. Thus, these compounds appear to be partial agonists. Future studies with other analogues will be necessary to further evaluate the potenial for NT as an atypical antipsychotic drug.

H. Sigma Receptors

I. Sigma Receptors

MARTIN et al. (1976) were the first to use the term "sigma receptor" when they showed that certain behavioral effects of SKF 10047 (*N*-allyl-normetazocine) and other benzomorphans could not be classified by existing pharmacology. Interestingly, cyclazocine, another benzomorphan, has clear psychotomimetic activity in humans (KEATS and TOLFORD 1964), suggesting a possible role for benzomorphan antagonists in the treatment of schizophrenia. CLAY and BROUGHAM (1975) first reported on interaction of antipsychotic drugs with opiate receptors showing displacement of 3H-naloxone binding by haloperidol in a competitive, sodium-dependent manner. VAUPEL (1983) classified the nonopiate nature of haloperidol-displaceable naloxone binding in studies which failed to demonstrate reversal of (+/−)SKF 10047-induced canine delirium by opiate receptor antagonists. Additonally, prior studies by ZUKIN et al. (1986) showed that an *N*-allyl-normetazocine-sensitive binding site could be labelled with both 3H-phencyclidine and 3H-cyclazocine (see ZUKIN et al. 1986 and references cited therein). Although the PCP site was later shown to represent the glutamate receptor, the nonopioid nature of the site labelled by SKF10047 (which has nanomolar affinity for haloperidol) led to the classification of this receptor as the sigma site.

Additional reports further confirmed the existence of the sigma binding site and suggested that haloperidol's antipsychotic actions may be partially due to its sigma receptor blockade (LARGENT et al. 1987, 1988). These findings led to the hypothesis that the antipsychotic activities of haloperidol could be separated from its "neuroleptic" activities by preparing compounds with high specificity for the sigma receptor (Largent et al. 1988). Interestingly, some sigma binding sites are colocalized with dopaminergic neurons, and many sigma ligands affect dopamine neuronal activity (Weber et al. 1986; Wachtel and White 1988). Taken together, these reports have encouraged a role for drugs which act as sigma receptor ligands in the treatment of schizophrenia.

II. Preclinical Studies

In preclinical studies, sigma ligands have been shown to have many properties expected of atypical antipsychotic drugs. Thus, Ujike et al. (1992) showed that BMY14802 blocked methamphetamine-induced behavioral sensitization (a model for acute psychosis) without inducing catalepsy. BMY14802, when given long-term, selectively inactivates the mesolimbic (A9) DA neurons without affecting nigrostriatal (A10) DA neurons (Wachtel and White 1988). Studies with the relatively selective combined 5-HT_{2A}/sigma antagonist, DUP 734, showed a number of properties consistent with an atypical antipsychotic drug profile (Cook et al. 1992; Tam et al. 1992). Finally, SKF10047 (a sigma agonist) stimulates mesolimbic dopamine neurons, while rimcazole (a sigma antagonist) blocks SKF10047's activity without having any effect on basal activity (Ceci et al. 1988). Taken together, these results all suggest that sigma "antagonists" should possess atypical antipsychotic drug activity.

Other studies have cast doubt on the nature and localization of sigma receptors. Knight et al. (1991) and McCann et al. (1989) showed that sigma binding sites may be intracellular! Additionally, a number of diverse pharmacologic agents showed affinities for the sigma binding sites, many of which lack antipsychotic antivity (Walker et al. 1990), suggesting that the sigma site may represent a form of microsomal flavin-containing monooxygenase or cytochrome P450-related enzyme (Tricklebank et al. 1992 for review).

Walker et al. (1990) suggest that the role of sigma receptors in psychotic illness remains inconclusive for the following reasons. The positive evidence consists of the following points: (1) a high concentration of sigma sites exist in limbic brain areas, (2) there is a moderate-to-high affinity for haloperidol and certain other antipsychotic drugs, and (3) certain atypical antipsychotics bind to sigma sites. The negative evidence for involvement of sigma sites in the actions of atypical antipsychotic drugs is as follows: (1) high concentrations of sigma sites also exist in motor areas, (2) sigma receptors have a weak affinity for rimcazole and many other antipsychotic drugs, (3) there is no proven connection to the putative actions of BMY 14802, and (4) there is no evidence for a role in the mediation of the psychotomimetic effects in humans.

III. Specific Agents

Multiple sigma ligands have been used in human and animal studies which include the following:

1. *Rimcazole (BW234U)* does not induce catalepsy in rats but does antagonize hyperactivity and other responses (climbing behavior and aggression) induced by DA agonists (FERRIS et al. 1986), suggesting an atypical antipsychotic profile. Rimcazole was used in four open clinical trials in which an overall improvement of 50% of the patients was noted (TAYLOR and SCHLEMMER 1992). Double-blind studies failed to demonstrate antipsychotic efficacy (see TAYLOR and SCHLEMMER 1992).
2. *Cinuperone* is known to antagonize apomorphine-induced climbing behavior in mice at doses which do not induce catalepsy (HOCK et al. 1985). Cinuperone's clinical use was terminated due to orthostasis (TAYLOR and SCHLEMMER 1992).
3. *Tiosperone*, which also has a high $5\text{-}HT_{2A}/D_2$ ratio (MELTZER et al. 1989) and also binds to $5\text{-}HT_{2C}$, $5\text{-}HT_6$, and $5\text{-}HT_7$ receptors (ROTH et al. 1992, 1994), was developed by Bristol-Myers as a potential antipsychotic drug (TAYLOR and SCHLEMMER 1992) because of its atypical features in rats (WHITE and WANG 1986). Although some clinical trials indicated efficacy (MOORE et al. 1987), the dose used may have been excessive so that EPS occurred (MELTZER, personal communication).
4. *BMY13980* was also selected for development (NEW et al. 1988) because of its D_2 and sigma receptor binding activity. BMY13980 blocked conditioned avoidance and apomorphine-induced stereotypy (TAYLOR and SCHLEMMER 1992) without causing catalepsy. BMY13980 also selectively inactivated mesolimbic but not nigrostriatal DA neurons after long-term administration (WHITE and WANG 1986). Because of animal toxicology data (adrenal hyperplasia; TAYLOR and SCHLEMMER 1992), BMY13980 was dropped from further clinical testing.
5. *BMY 14802* has a high affinity for sigma sites in vitro and in vivo and a high $5\text{-}HT_{1A}$ receptor affinity (YEVICH et al. 1992). It has been used extensively in animal studies where it displays many of the attributes of an atypical antipsychotic drug (TAYLOR et al. 1991). BMY14802 has many clozapine-like properties including specific effects in the conditioned-avoidance paradigm (TAYLOR et al. 1991), selective inhibition of A9 DA neurons (WACHTEL and WHITE 1988), and attenuation of stereotypic behaviors in amphetamine-treated macaques (TAYLOR and SCHLEMMER 1992). A recent multicenter safety and efficacy study with BMY 14802 showed no significant improvement in psychiatric symptoms (as measured by the BPRS and CGI) as well as no significant EPS (GEWIRTZ et al. 1994).
6. *Remoxipride* is a substituted benzamide with low affinity for D_2 receptors as well as sigma receptors. It is effective in schizophrenia and has few EPS. Remoxipride induces catalepsy in rats only at high doses (ÖGREN et al. 1993). Clinically, remoxipride has a low potential to induce EPS in humans (CHOUINARD 1990). Remoxipride has now been withdrawn secondary to induction of aplastic anemia.
7. *NPC 16377* is a potent and highly selective sigma ligand with atypical antipsychotic features and neuroprotective properties (CLISSOLD et al.

1993). It has not yet been tested in humans. NPC16377 reverses amphetamine-induced hyperactivity and conditioned avoidance responses without reducing escape behavior and without producing EPS (CLISSOLD et al. 1993).

8. *Umespirone* has a relatively high affinity for the sigma site and has atypical antipsychotic-like activity in rats and mice (ITZHAK et al. 1990). As previously mentioned, umespirone also blocks glutamate receptor-mediated activity in mice (SCHMIDT et al. 1991) and shows α_1-adrenoceptor blocking activity in vitro.
9. *DUP 734* has a number of properties in rats and mice which suggest it may have atypical antipsychotic drug activity including: (1) no induction of catalepsy or antagonism of avoidance behavior, (2) blockade of mescaline- and 5-hydroxytryptophan-induced behaviors, (3) blockade of the effects of the sigma agonist ((+)-3PPP) on substantia nigra electrophysiological behavior, and (4) blockade of rotational behavior induced by SKF 10047 in lesioned rats (COOK et al. 1992; TAM et al. 1992). Thus, DUP734 has many properties consistent with atypical antipsychotic drug activity.

Clinical trials of sigma ligands have failed to provide convincing evidence of efficacy. This is particulary clear with BMY14802, which was absolutely devoid of clinical efficacy in a recently published trial (GEWIRTZ et al. 1994). With the exception of remoxipride, whose D_2 antagonism may account for its efficacy as an antipsychotic, none of the compounds tested have gone on to further use. Thus, at present, the outlook for sigma receptors as a target of antipsychotic drug development is not sanguine. Table 3 summarizes data obtained with sigma agents.

I. Opioids

Endogenous opioid peptides (endorphins, dynorphins, enkephalins) have been implicated in the modulation of pain, stress, addiction, sexuality, and motivation (NABER 1993). Moreover, recent reports indicate that dysfunction of opioidergic systems may be responsible for some psychiatric disorders. At least three families of opiate receptors have been cloned: μ, δ, and κ. The μ receptor has at least two subtypes; additionally, there is evidence that μ and δ receptors may allosterically interact within the plane of the membrane (TRAYNOR and ELLIOTT 1993). The δ site prefers enkephalin peptides, the μ site binds endorphins, and the κ site has a high affinity for dynorphins. κ receptors also exist in multiple states (or sites). All of the thus far cloned opiate receptors are members of the G protein supergene family.

Endogenous opioid peptides were initially reported to be elevated in the CSF of schizophrenic patients (TERENIUS et al. 1976), and increased after haloperidol treatment (RIMON et al. 1983). Further studies have not revealed convincing and consistent alterations in schizophrenic subjects (NABER 1993). Measurements of plasma levels of opiates have given inconsistent results, as have measurements in postmortem brain tissues (NABER 1993). Clozapine has been reported to decrease the levels of enkephalin mRNA following long-term drug treatment while haloperidol had no effect (MERCUGLIANO and CHESSELET 1992).

Treatment paradigms using various opioid agonists or antagonists have been inconclusive (NABER 1993). Some open trials with β-endorphin reported efficacy in schizophrenia (KLINE et al. 1977) but double-blind studies did not (GERNER et al. 1980; PICKAR et al. 1981; BERGER et al. 1980). Likewise, treatments with des-tyrosyl-γ-endorphin failed to show clinical benefits (TAMMINGA et al. 1981; MELTZER et al. 1982). Similarly, double-blind studies of FK 33-824 (a metenkephalin analogue which binds to μ and δ receptors) showed no effect (JUNGKUNZ et al. 1983). Studies with naloxone, an opioid antagonist, have likewise been inconclusive with early open studies showing an effect (WATSON et al. 1987) and a multicenter trial showing no effect (PICKAR et al. 1989). Some authors, however, have noted improvement in tardive dyskinesia with naloxone treatment (BLUM et al. 1987; LINDEMAYER et al. 1988). Thus, although the opioid neuronal system is localized in many brain areas thought to be essential for atypical antipsychotic drug actions, there is no convincing evidence to suggest that drugs which act on this system may be effective treatments for schizophrenia.

J. Serotonin

Crude preparations of serotonin (5-hydroxytryptamine; 5-HT) were used for pharmacologic purposes as early as 1868 and 5-HT was chemically identified in

Table 3. Effect of various sigma compounds on clinical and preclinical measures of antipsychotic activity

Drug	Depol block		Inh. DA response	Catalepsy	Inh. CAR	EPS	Antipsychotic activity
	A9	A10					
BMY14802	+	?	+	–	+	–	–
Rimcazole	?	?	+	–	?	–	–
Cinuperone	?	?	+	–	?	?	?
Tiospirone	+	–	?	Less	?	+/–	+
BMY13980	?	?	+	–	+	?	?
Remoxipride	+	–	+	–	+	+/–	+
NPC 16377	?	?	+	–	+	?	?
DUP734	?	?	+	–	+	?	?

Inh. DA response, inhibition of responses induced by apomorphine, amphetamine, or other dopaminergic agents; Depol Block, the ability of a drug to induce depolarization blockade after chronic administration; Inh. CAR, the ability of a drug to inhibit conditioned avoidance response; EPS, the ability of a drug to induce extrapyramidal symptoms in humans.

1948 by RAPPORT et al. The first indications that 5-HT might be involved in the pathophysiology of schizophrenia came with the discovery that certain ergots (e.g., lysergic acid diethylamide) and tryptamine derivatives (e.g., *N,N′*-dimethyltryptamine, bufotenine, psilocybin) were hallucinogenic and induced many of the prototypical symptoms of schizophrenia (WOOLEY and SHAW 1954). Furthermore, some investigators have suggested that endogenous hallucinogens might be produced in schizophrenia, though these findings were unable to be replicated at a later date (BLEICH et al. 1988; GILLAN et al. 1976).

I. 5-HT Receptors and Schizophrenia

At least 14 separate 5-HT receptors existing in seven major families have been identified by molecular cloning techniques: 5-HT_1, 5-HT_2, 5-HT_3, 5-HT_4, 5-HT_5, 5-HT_6, and 5-HT_7. The 5-HT_1 family has at least five members (5-HT_{1A} to 5-HT_{1F}), the 5-HT_2 has three members (5-HT_A, 5-HT_B, 5-HT_C) and the 5-HT_5 has two members. The 5-HT_3 receptor belongs to the GABA receptor supergene family while all of the other thus far cloned 5-HT receptors are members of the seven-helix, G protein-coupled supergene family (TIPS RECEPTOR NOMENCLATURE SUPPLEMENT 1994).

The evidence implicating 5-HT in the pathogenesis of schizophrenia has come from many sources which will be summarized only briefly here; recent reviews may be found elsewhere (ROTH and MELTZER 1995; ROTH 1994). Firstly, 5-HT agonists like LSD and *N,N′*-dimethyltryptamine are hallucinogenic in humans and their ability to induce hallucinations correlates well with their ability to bind to the 5-HT_{2A} receptor (GLENNON 1987; GLENNON et al. 1984). Although the psychosis induced by hallucinogens shares some characteristics with schizophrenia beyond visual hallucinations (e.g., conceptual disorganization), major differences have been found, especially the failure of hallucinogens to induce auditory hallucinations and negative symptoms (SZARA 1967; FISCHMAN 1983; HOLLISTER 1962). Secondly, alterations in 5-HT receptors, especially 5-HT_{1A} and 5-HT_{2A} subtypes, have been reported in schizophrenia. In general, investigators have noted decreased numbers of 5-HT_{2A} receptors (ARORA and MELTZER 1991; BENNETT et al. 1979; MITA et al. 1986) and increased numbers of 5-HT_{1A} receptors (HASHIMOTO et al. 1991), though conflicting results have been obtained (JOYCE et al. 1993; REYNOLDS et al. 1983; WHITAKER et al. 1981). Additionally, increased numbers of 5-HT uptake sites have been measured in brains from schizophrenic patients (JOYCE et al. 1993). Thirdly, alterations in 5-HT metabolism have been measured in schizophrenia. Plasma or platelet 5-HT concentrations are not consistently abnormal. CSF 5-HT appears to be strongly negatively correlated with ventricular enlargement (BLEICH et al. 1988; POTKIN et al. 1983). Fourthly, alterations in serotonergic neuroendocrine challenge studies have been measured in schizophrenia. In general, decreases in 5-HT_2-mediated responses have been seen in unmedicated patients (see MELTZER et al. 1993; ROTH and MELTZER 1995, and ROTH 1994 for review). For these studies, 5-HT receptor active agents are given and the neuroendocrine measurements (e.g., growth hormone, cortisol, prolactin) are quantified. Additionally, atypical antipsychotic drugs like clozapine appear to blunt the responses to challenge agents such as m-chlorophenylpiperzine (mCPP), MK-212, and cimetidine (an indirect 5-HT agonist) (see ROTH and MELTZER 1995). Fifthly, atypical antipsychotic drugs like clozapine and risperidone appear to occupy 5-HT_{2A} receptors in vivo (GOYER et al. 1993; NYBERG et al. 1995). Sixthly, many atypical antipsychotic drugs possess a high affinity for various 5-HT receptors (MELTZER et al. 1989; ROTH et al. 1992, 1994).

The prototypical atypical antipsychotic drug is clozapine (MELTZER 1989a,b; KANE et al. 1988). Clozapine was first shown to be a 5-HT receptor antagonist in the early 1980s by FINK et al. (1984). Since then, clozapine has been shown to have a high affinity for at least four separate 5-HT receptors including 5-HT_{2A}, 5-HT_{2C}, 5-HT_6, and 5-HT_7 (CANTON et al. 1990; MELTZER et al. 1989; ROTH et al. 1992, 1994). Additionally, clozapine has a modest affinity for the 5-HT_3 receptor, though in vivo it does appear to blunt 5-HT_3 receptor responses, at least in the medial prefrontal cortex (EDWARDS et al. 1991). Thus, for instance, EDWARDS et al. (1991) showed that clozapine blocked the stimulation of phosphoinositde hydrolysis induced by 2-methyl-5HT (EDWARDS et al. 1991). Clozapine appears to occupy 5-HT_{2A} receptors in vivo as well in both laboratory animals and humans (GOYER et al. 1993; STOCKMEIER et al. 1993). Acute (24-h) and long-term (7- to 10-day) clozapine treatment has been demonstrated to downregulate both 5-HT_{2A} and 5-HT_{2C} receptors (MATSUBARA and MELTZER 1989; KUOPPAMAKI et al. 1993a), though the exact mechanism responsible for this downregulation remains unknown. Interestingly, the major metabolite of clozapine, *N*-desmethylclozapine, is a potent 5-HT_{2C} antagonist (ROTH and CIARANELLO 1991; KUOPPAMAKI et al. 1993b).

MELTZER et al. (1989) were the first to convincingly demonstrate a correlation between the atypical nature of a large number of drugs and their ability to bind to 5-HT_{2A} receptors, although ALTAR et al. (1988) had earlier suggested that 5-HT_2/D_2 ratios could be important for clozapine's actions. In a landmark study, MELTZER et al. (1989) showed that most putative atypical antipsychotic drugs could be correctly classified based on their 5-HT_{2A}/D_2 affinity ratios. This work was followed by studies which demonstrated that in vivo, most putative atypical antipsychotic drugs tested displayed greater occupancies of 5-HT_{2A} versus D_2 receptors (STOCKMEIER et al. 1993). These results have been replicated by others (MATSUBARA et al. 1993).

A large number of putative atypical antipsychotic drugs have been developed and tested in the treatment of schizophrenia. These fall into several categories: (1) relatively selective 5-$HT_{2A/2C}$ antagonists (e.g., mianserin, ICI169369, MDL100907), (2) drugs with high 5-HT_{2A}/D_2 affinity ratios (e.g., clozapine, risperidone, olanzapine, amperozide, and others), (3) 5-HT_3 antagonists (e.g., zacopride and ondansetron), (4) nonselective 5-HT receptor antagonists (e.g., methysergide), (5) 5-HT reuptake inhibitors (e.g., fluoxetine). These drugs will be reviewed separately.

II. Selective 5-$HT_{2A/2C}$ Antagonists

1. Ritanserin

Ritanserin was the first reasonably selective 5-$HT_{2A/2C}$ antagonist to be used in the treatment of schizophrenia. Preclinical studies suggested that it might have atypical antipsychotic properties (Niemegeers et al. 1990; Ugedo et al. 1989) while clinical testing showed that it reduced EPS and, possibly, negative symptoms in patients with schizophrenia (Bersani et al. 1991; Strauss and Klieser 1991). A recent placebo-controlled trial using 33 patients demonstrated that ritanserin decreased negative symptoms when added to stable neuroleptic treatment (Duinkerke et al. 1993). Weisel et al. (1994) recently reported that ritanserin was effective in decreasing positive and negative symptoms in a group of schizophrenic patients in acute exacerbation with no increase in EPS. This was an open study which requires replication under controlled conditions. If verified, it raises the issue of whether 5-$HT_{2A/2C}$ receptor blockade per se is sufficient for antipsychotic activity.

2. ICI169369 and MDL100907

These agents have both been shown by preclinical studies to induce depolarization blockade of mesolimbic but not mesocortical DA neurons (Saller et al. 1990; Sorensen et al. 1993). MDL100907 (R (+)-alpha-(2,3-dimethoxyphenyl)-1-[2-(4-fluorophenylethyl)]-4-piperidine-methanol) has selectivity for the 5-HT_{2A} receptor while ICI169369 is a relatively nonselective 5-$HT_{2A/2C}$ antagonist. ICI169369 was evaluated for its antipsychotic potency and ability to decrease neuroleptic-induced EPS. The results have not been reported. MDL100907 blocks amphetamine-stimulated locomotion in mice without blocking climbing behavior (Sorensen et al. 1993).

3. Mianserin

Mianserin is a 5-$HT_{2A/2C}$ antagonist which downregulates 5-HT_{2A} receptors after acute and long-term treatment, apparently by posttranslational mechanisms (Roth and Ciaranello 1991). One study indicated that mianserin might improve both positive and negative symptoms when added to typical neuroleptic medications (Rogue and Rogue 1992).

III. Mixed 5-HT_2/D_2 Antagonists

A vast number of mixed 5-HT_2/D_2 antagonists have been developed and tested both clinically and preclinically (see Meltzer et al. 1989; Roth et al. 1994 for examples).

1. Clozapine

The prototypical compound is clozapine and was first noticed to be devoid of EPS several years ago (see MATZ et al. 1974). ALTAR et al. (1988), RASMUSSEN and AGHAJANIAN (1988), MELTZER et al. (1989), and others (NIEMEGEERS et al. 1990; SEEMAN 1990) have all proposed that the low incidence of EPS seen with clozapine is due to its relatively high occupancy of 5-HT_{2A} receptors relative to D_2 receptors. Positron emission tomography (PET) studies have indicated that clozapine preferentially occupies 5-HT_2 versus D_2 receptors in humans (GOYER et al. 1993). The evidence for clozapine's atypical nature is reviewed in other chapters in this book.

2. Risperidone

Risperidone is the first approved drug which was specifically designed to have a relatively high 5-HT_{2A}/D_2 ratio (MEERT et al. 1989; MESOTTEN et al. 1989; CHOUINARD et al. 1993; BORISON et al. 1992). Risperidone also has a high affinity for α_1-adrenoceptors and 5-HT_7 receptors, moderate affinities for 5-HT_{2C} and 5-HT_6 receptors, and low affinities for other neurotransmitter receptors (LEYSEN et al. 1993; ROTH et al. 1992, 1994), including sigma receptors. In animal models, risperidone behaves as a classical atypical antipsychotic drug by inhibiting conditioned avoidance responses and showing a relative lack of induction of catalepsy (MEERT et al. 1989; NIEMEGEERS et al. 1990; LEYSEN et al. 1993). In humans, risperidone shows dose-dependent effects. At relatively low doses (4–8 mg/day) risperidone is an effective antipsychotic drug which produces few EPS, while at higher doses, risperidone induces EPS similarly to haloperidol. At 6 mg/day, risperidone is superior to 20 mg/day haloperidol for reduction of both positive and negative symptoms while this superiority is lost at higher doses (CHOUINARD et al. 1993). The Canadian multicenter trial also showed that risperidone significantly decreased symptoms of tardive dyskinesia (CHOUINARD et al. 1993; MARDER and MEIBACH 1994). Taken together, these results indicate that risperidone has atypical features which are dose-dependent; these results support the hypothesis that 5-HT_{2A} antagonism is an essential feature of atypical antipsychotic drugs (MELTZER et al. 1989).

3. Melperone

Melperone has been shown clinically in at least two studies (HARNRYD et al. 1989; CHRISTENSSON 1989) to reduce psychotic symptoms and have a lower incidence of EPS. Melperone is also effective in treating some neuroleptic-resistant schizophrenic patients (MELTZER, unpublished observation) and appears to have a very weak ability to induce tardive dyskinesia, even in elderly patients who are vulnerable to developing dyskinesias (CHRISTENSSON 1989). Melperone is relatively selective for 5-HT_{2A} receptors, having little affinity for other 5-HT receptors (ROTH et al. 1992, 1994). It may also have a higher affinity for the D_4 receptor than for the D_2 (LAHTI et al. 1993; ROTH et al., manuscript in preparation).

4. Olanzapine

Olanzapine is a heterocyclic compound which binds to 5-HT_{2A} and 5-HT_6 receptors with high affinity and which, like clozapine, has high anticholinergic activity (ROTH et al. 1994). Olanzapine behaves as a classical atypical antipsychotic (MOORE et al. 1992). In a double-blind, placebo-controlled trial, olanzapine was found to be more effective than haloperidol in decreasing BPRS total and negative symptom scores (BEASLEY et al., submitted).

5. Amperozide

Amperozide is a diphenylbutylperazine which in animals has atypical antipsychotic activity. Amperozide has a high affinity for 5-HT_{2A} receptors and such low D_2 binding affinity that it would not be expected to occupy D_2 sites at clinically effective doses (MELTZER and STOCKMAIER 1992; ROTH et al. 1994). Acute administration of amperozide enhances rat cortical and striatal dopamine efflux (PEHEK et al. 1993). The effects on corticolimbic DA release could be secondary to 5-HT_{2A} receptor blockade. Amperozide also increases striatal DA release via a carrier-mediated process (YAMAMOTO and MELTZER 1992). In humans, one open-label study indicated that amperozide had a low incidence of EPS and was effective at reducing psychotic symtoms (AXELSSON et al. 1991). In another multicenter trial, amperozide also showed marked antipsychotic effects only in female schizophrenics while being relatively ineffective in male schizophrenics, compared to haloperidol (MELTZER et al., in preparation). Amperozide does not cause catalepsy in rats (CHRISTENSSON and BJORK 1990). Since ritanserin (a 5-$HT_{2A/2C}$ antagonist) enhances DA cell firing (UGEDO et al. 1989), it is tempting to speculate that the effects of amperozide are secondary to 5-HT_{2A} receptor blocking activities.

6. Fluperlapine

Fluperlapine is a heterocyclic compound similar in structure to clozapine with a clozapine-like efficacy and EPS profile in schizophrenic patients (WOGGON et al. 1984, 1986). Fluperlapine does not cause catalepsy in rats (GUDELSKY et al. 1987). Fluperlapine, like clozapine, has a high affinity for 5-HT_{2A}, 5-HT_{2C}, 5-HT_6, and 5-HT_7 receptors (ROTH et al. 1992, 1994). Because of several cases of agranulocytosis, fluperlapine was withdrawn from further study. RMI-81,582 is structurally similar to fluperlapine and may have antipsychotic activity based on its high in vivo 5-HT_{2A}/D_2 occupancy ratio (STOCKMEIER et al. 1993).

7. Tiosperone

Tiosperone has atypical features in animal studies and binds with high affinity to 5-HT_{2A}, 5-HT_{2C}, and 5-HT_7 receptors in addition to its aforementioned sigma receptor affinity (MELTZER et al. 1989; ROTH et al. 1992, 1994; TAYLOR and SCHLEMMER 1992). One double-blind study suggested equivalency with haloperidol while producing fewer EPS (MOORE et al. 1987).

8. Zotepine

Zotepine was shown in a double-blind study to be superior to haloperidol in patients with predominantly negative symptoms (BARNAS et al. 1992). Zotepine, like clozapine, has a high affinity for 5-HT_{2A}, 5-HT_{2C}, 5-HT_6, and 5-HT_7 receptors (MELTZER et al. 1989; ROTH et al. 1992, 1994). Zotepine also produces a low incidence of EPS in humans.

9. Others

Perlapine, rilapine, CP-88059, seroquel (ICI 204,676), and HP-370 have all been classified as atypical based on animal studies (see MELTZER et al. 1989). CP-88059 and seroquel are in phase II and/or phase III clinical studies. Sertindole has a relatively high 5-HT_{2A}/D_2 ratio and is in phase III clinical trials; in rats sertindole has atypical features (HOLLISTER 1994). Several pyridobenzodiazepines which have potential atypical antipsychotic actions and high 5-HT_{2A}/D_2 ratios have been synthesized by LIEGEOIS et al. (1993).

IV. 5-HT_3 Antagonists

Zacopride, zatosetron, and ondansetron have been shown in preclinical studies to inactivate mesolimbic versus nigrostriatal DA neurons preferentially and not to induce catalepsy in rats (COSTALL et al. 1990, 1991; RASMUSSEN et al. 1991a). MDL73147EF has been shown to facilitate latent inhibition (a potential behavioral marker of atypical antipsychotic drug actions) in the rat. Open-label trials of zacopride and ondansetron disclosed no therapeutic efficacy for the positive symptoms of schizophrenia, although EPS may have been significantly decreased with zacopride (COSTALL 1993; NEWCOMER et al. 1992). Clozapine has a relatively low affinity for the 5-HT_3 receptor, although in rats this drug has been shown to block certain biochemical effects of 5-HT_3 antagonists in vivo (EDWARDS et al. 1991).

V. Nonselective 5-HT Receptor Antagonists

Methysergide, a non selective 5-HT receptor antagonist, was shown in one open-label trial to diminish psychotic symptoms in a single individual with schizophrenia (MENDELS 1967). Another study indicated that methysergide worsened psychosis (SKORZEWSAK and LAL 1989). Methysergide is a partial DA agonist which binds to many 5-HT receptors. mCPP is a mixed agonist-antagonist which has been reported to worsen psychotic symptoms in schizophrenic individuals, presumably via its 5-$HT_{2A/2C}$ agonist actions (KRYSTAL et al. 1993). Fenfluramine induces 5-HT release and has been shown to worsen psychotic symptoms (MARSHALL et al. 1989; KOLAKOWSKA et al. 1987). Buspirone is a selective 5-HT_{1A} partial agonist which has had mixed effects on positive and negative symptoms and which may improve symptoms of tardive dyskinesia (BRODY et al. 1990; GOFF et al. 1991; MOSS et al. 1993). Table 4 summarizes data obtained with serotonergic agents.

VI. 5-HT Reuptake Inhibitors

An open 6-week trial with fluoxetine showed that positive and negative symptoms were decreased in nine neuroleptic-treated patients with schizophrenia (GOFF et al. 1990). These results are interesting in light of studies by JOYCE et al. (1993) who showed an increased density of 5-HT transporters in the striatum of schizophrenics patients in postmortem studies. The JOYCE et al. (1993) findings have recently been called into question by the observation of STOCKMEIER (1994) that age-related changes were not taken into account. According to STOCKMEIER (1994), the schizophrenic patients differed significantly from the controls in age and these differences were large enough to cause changes considered to be due possibly to age alone.

K. Conclusions

Taken together all of these findings support the hypothesis that multiple effector systems are involved in the efficacy of atypical antipsychotic drugs like clozapine. It is unlikely that developing highly selective drugs for subtypes of DA receptors (e.g., D_4, D_3) is as likely to be effective as is an approach which targets multiple receptors simultaneously (e.g., $5\text{-}HT_{2A}/D_4$ and or $5\text{-}HT_6/D_2$). It is also possible that atypical antipsychotic drugs which differ in mechanism of

Table 4. Effects of various serotonergic compounds on clinical and preclinical measures of antipsychotic activity

Drug	Depol block		Inh. DA response	Catalepsy	Inh. CAR	EPS	Antipsychotic activity
	A9	A10					
5-HT2A/2C antagonists							
Ritanserin	+	–	+	–	+	–	+
MDL 100907	+	–	+	–	+	?	?
5-HT2A/D2 antagonists							
Clozapine	+	–	+	–	+	–	+
Risperidone	–	–	+	Less	+	+/– (Dose-dependent)	+
Melperone	?	?	+	–	?	–	+
Olanzapine	+	–	+	–	+	–	+
Amperozide	?	?	+	–	?	–	+/–
Fluperlapine	?	?	?	–	?	–	+
Ziprasidone	+	–	+	–	+	–	+
Sertindole	+	–	+	–	+	–	+
Seroquel	+	–	+	–	+	–	+

Inh. DA response, inhibition of responses induced by apomorphine, amphetamine, or other dopaminergic agents; Depol Block, the ability of a drug to induce depolarization blockade after chronic administration; Inh. CAR, the ability of a drug to inhibit conditioned avoidance response; EPS, the ability of a drug to induce extrapyramidal symptoms in humans.

action and clinical profile will be developed based on such a scheme (e.g., effectiveness for neuroleptic-resistant schizophrenia vs low EPS profile). At the present time, the scheme for developing effective atypical antipsychotic drugs with the most success has been that combining 5-HT_{2A}- and D_2-blocking activities. Thus, compounds like risperidone and olanzapine clearly show atypical antipsychotic properties, though none are, as yet, equivalent in efficacy to clozapine. With the discovery of many additional 5-HT receptors (e.g., 5-HT_6 and 5-HT_7) as potential targets for atypical antipsychotic drugs, the future outlook is favorable for the development of newer and safer drugs for the treatment of schizophrenia.

References

Albus M (1988) Cholecystokinin. Prog Neuropsychopharmacol Biol Psychiatry 12:S5–S21

Altar CA, Boyer WC, Wasley A, Liebman JM, Wood PL, Gerhardt SG (1988) Dopamine neurochemical profile of atypical antipsychotics resemble that of D-1 antagonists. Naunyn-Schmiedebergs Arch Pharmacol 338:162–168

Arora RC, Meltzer HY (1991) Serotonin2 (5-HT_2) receptor binding in the frontal cortex of schizophrenic patients. J Neurotransm 85:19–29

Axelsson R, Nilsson A, Christensson E, Bjork A (1991) Effects of amperozide in schizophrenia: an open study of a potent 5-HT2 receptor antagonist. Psychopharmacology (Berl) 104:287–292

Badgett M, Wrona CT, Newcomer JW, Csernansky JG (1993) Subcortical excitatory amino acid levels after acute and chronic administration of typical and atypical neuroleptics. Eur J Pharmacol 230:245–250

Baldessarini RJ, Huston-Lyons D, Campbell A, Marsh E, Cohen BM (1992) Do central antiadrenergic actions contribute to the atypical properties of clozapine? Br J Psychiatry 160:12–16

Barnas C, Stuppack CH, Miller C, Harig C, Sperner-Unterweger B, Fleischhacker WW (1992) Zotepine in the treatment of schizophrenic patients with prevailingly negative symptoms. A double-blind trial vs haloperidol. Int Clin Psychopharmacol 7:23–27

Bennett JP, Enna SJ, Bylund DB, Gillin JC, Wyatt RC, Snyder SH (1979) Neurotransmitter receptors in the frontal cortex of schizophrenics. Arch Gen Psychiatry 36:927–934

Berger PA, Watson SJ, Akil H, Ellicot GR, Rubin RT, Pfferbaum A, Davis KL, Barchas JD LI CH (1980) β-endorphin and schizophrenia. Arch Gen Psych 37:635–640

Bersani G, Pozzi F, Marini S, Grispini A, Pasini A, Ciani N (1991) 5-HT2 receptor antagonism in dysthymic disorder: a double-blind placebo-controlled study with ritanserin. Acta Psychiatr Scand 83:244–248

Bleich A, Brown SL, Kahn R, Van Praag HM (1988) The role of serotonin in schizophrenia. Schizophr Bull 14:297–315

Blum K, Nisipeanu PF, Roberts E (1987) Naloxone in tardive dyskinesia. Psychopharmacology 93:538

Bolden C, Cusack B, Richelson E (1992) Antagonism by anitmuscarinic and neuroleptic compounds at the five cloned human muscarinic cholinergic receptors expressed in chinese hamster ovary cells. J Pharmacol Exp Ther 260:576–580

Bolden-Watson C, Watson MA, Murray KD, Isackson PJ, Richelson E (1993) Haloperidol but not clozapine increases neurotensin receptor mRNA levels in rat substantia nigra. J Neurochem 61:1141–1143

Bonanno G, Raiteri M (1993) Multiple $GABA_B$ receptors. Trends Pharmacol Sci 14:259–261

Borison RL, Pathraja AP, Diamond BI, Meibach RC (1992) Risperidone: clinical safety and efficacy in schizophrenia. Schizophr Res 28:213–218
Breier A, Buchanan RW, Waltrip RW, Listwak S, Holmes C, Goldstein DS (1994) The effect of clozapine on plasma norepinephrine: relationship to clinical efficacy. Neuropsychopharmalogy 10:1–7
Bristow LS, Barcutt LJ, Hutson PH, Torn L, Tricklebank MD (1991) R(+)-HA-966, a glycine/NMDA receptor antagonist, attenuates PCP and amphetamine-induced hyperactivity in rodents. Br J Pharmacol 102:68
Brody D, Adler LA, Kuo T, Aubust B, Rotrosen J (1990) Effects of buspirone in seven schizophrenic subjects. J Clin Psychopharmacol 10:68–69
Canton H, Verriele L, Colpaert FC (1990) Binding of typical and atypical antipsychotics to 5-HT1c and 5-HT2 sites: clozapine potently interacts with 5-HT1c sites. Eur J Pharmacol 191:93–96
Carlsson A (1988) The current status of the dopamine hypothesis of schizophrenia. Neuropsychopharmacology 1:179–186
Carlsson M, Carlsson A (1990) Systems within the basal ganglia: implications for schizophrenia and Parkinson's disease. Trends Neuro Sci 13:272–276
Ceci A, Smith M, French ED (1988) Activation of the A10 mesolimbic system by the sigma-receptor agonist (+)SKF 10,047 can be blocked by rimcazole, a novel putative antipsychotic. Eur J Pharmacol 140:121–122
Chiodo LA, Bunney SB (1985) Possible mechanisms by which repeated clozapine administration differentially affects the activity of two subpopulations of midbrain dopamine neurons. J Neurosci 5:2329–2344
Chouinard G (1990) A placebo-controlled clinical trial of remoxapride and chlorpromazine in newly admitted schizophrenic patients with acute exacerbation. Acta Psychiatr Scand 83 [Suppl 358]:111–119
Chouinard G, Jones B, Remington G, Bloom D, Addington D, Macewan GW, Labelle A, Beuclair L, Arnott W (1993) A Canadian multicenter placebo-controlled study of fixed doses of risperidone and haloperidol in the treatment of chronic schizophrenia. J Clin Psychopharmacol 13:25–40
Christensson EG (1989) Pharmacological data of the atypical neuroleptic melperone. Acta Psychiatr Scand [Suppl] 352:7–15
Christensson E, Bjork A (1990) Amperozide: a new pharmacological approach in the treatment of schizophrenia. Pharmacol Toxicol 66:5–7
Clay GA, Brougham LR (1975) Haloperidol binding to an opiate receptor. Biochem Pharmacol 24:1363–1367
Clissold DB, Pontecorvo MJ, Jones BE, Abreu ME, Karbon EW, Erickson RH, Natalie KJ, Borosky S, Hartman T, Mansbach RS, Balster RL, Ferkany JW, Enna SJ (1993) NPC16377, a potent and selective sigma ligand. II. Behavioral and neuroprotective profile. J Pharmacol Exp Ther 265:876–886
Cohen BM, Lipinski JF (1986) In vivo potencies of antipsychotic drugs in blocking alpha1 noradrenergic and dopamine D_2 receptors: implications for drug mechanisms of action. Life Sci 39:2571–2586
Contreras PC (1990) D-serine antagonized phencyclidine- and MK-801-induced stereotyped behavior and ataxia. Neuropharmacology 292:291–293
Cook L, Tam SW, Rohrback KW (1992) DuP 734 [1-(cyclopropylmethyl)-4-(2′(4″-fluorophenyl)-2′-oxyethyl)piperidien HBr], a potential antipsychotic agent: preclinical behavioral effects. J Pharmacol Exp Ther 263:1159–1166
Costa E (1991) The allosteric modulation of GABAA receptors: seventeen years of research. Neuropsychopharmacology 4:225–235
Costa J, Khaled E, Sramek J, Bunney WE, Potkin S (1990) An open trial of glycine as an adjunct to neuroleptics in chronic treatment-resistant schizophrenics. J Clin Psychopharmacol 10:71–72
Costall B (1993) The breadth of action of the 5-HT_3 receptor antagonists. Int Clin Psychopharmacol [Suppl] 2:3–9
Costall B, Naylor RJ, Tyers MB (1990) The psychopharmacology of 5-HT3 receptors. Pharmacol Ther 47:181–202

Costall B, Domeney AM, Kelly ME, Naylor RJ, Meltzer HY (1993) 5-HT_3 receptor antagonists: preclinical and clinical considerations. In: Racagni IG, Brunelleo N, Fukuda T (eds) Biological psychiatry, vol 1. Elsevier, Amsterdam, pp 538–540

Cotman CW, Monaghan DT (1987) Chemistry and anatomy of excitatory amino acid systems. In: Meltzer HY (ed) Psychopharmacology, the third generation of progress. Raven, New York, pp 197–210

Coward DM, Imperato A, Urwyler S, White TG (1989) Biochemical and behavioral properties of clozapine. Psychopharmacology (Berl) 99:S6–S12

Crawley JN (1991) Cholecystokinin-dopamine interactions. Trends Pharmacol Sci 12:232–236

Cusack B, Richelson E, Pang Y-P, Zaidi J, Kozikowski AP (1993) Pharmacological studies on novel neurotensin mimetics: discovery of a pharmacologically unique agent exhibiting concentration-dependent dual effects as antagonist and agonist. J Pharmacol Exp Ther 44:1036–1040

Deakin JFW, Slater P, Simpson MDC, Gilchrist AC, Skan WJ, Royston MC, Reynolds GP, Cross AJ (1989) Frontal cortical and left temporal glutaminergic dysfunction in schizophrenia. J Neurochem 52:1781–1786

Deutsch SI, Weizman A, Goldman ME, Morihisa JM (1988) The sigma receptor: novel site implicated in psychosis and antipsychotic drug efficacy. Clin Neuropharmacol 11:105–119

Deutsch SI, Mastropaolo J, Schwartz BL, Rosse RB, Morihisa JM (1989) A "glutamergic hypothesis" of schizophrenia. Rationale for pharmacotherapy with glycine. Clin Neuropharmacol 12:1–13

Dingledine R, McBain CJ (1994) Excitatory amino acid transmitters. In. Siegel GJ et al. (eds) Basic neurochemistry: molecular, cellular and medical aspects. Raven, New York, pp 367–387

Drew KL, O'Conner WT, Kehr J, Ungersted U (1990) Regional specific effects of clozapine and haloperidol on GABA and dopamine release in rat basal ganglia. Eur J Pharmacol 187:385-397

Duinkerke SJ, Botter PA, Jansen AAI, Van Dongen PAM, Van Haaften AJ, Boom AJ, Van Laarhoven JHM, Busard HLSM (1993) Ritanserin, a selecteive 5-HT2c/1c antagonist, and negative symptoms in schizophrenia: a placebo-controlled double-blind trial. Br J Psychiatry 163:451–455

Edwards E, Ashby CR, Wang RY (1991) The effect of typical and atypical antipsychotic drugs on the stimulation of phosphoinositide hydrolysis produced by the 5-HT_3 receptor agonist 2-methyl-serotonin. Brain Res 545:276–278

Etienne P, Baudry M (1990) Role of excitatory amino acid neurotransmission in synaptic plasticity and pathology. An integrative hypothesis concerning the pathogenesis and evolutionary advantages of schizophrenia-related genes. J Neurotransm 29:39–48

Ferris RM, Tang FLM, Chang K-J, Russell A (1986) Evidence that the potential antipsychotic agent rimcazole (BW 234U) is a specific competitive antagonist of sigma sites in brain. Life Sci 38:2329–2337

Fink H, Morgenstern R, Oelssner W (1984) Clozapine – a serotonin antagonist? Pharmacol Biochem Behav 20:513–517

Fischman LG (1983) Dreams, hallucinogenic drug states, and schizophrenia: a psychological and biological comparison. Schizophr Bull 9:73–94

Fleischhacker WW, Miller CH, Schett P, Barnas C, Ehrmann H (1991) The Hillside Akathisia Scale: a reliability comparison of the English and German versions. Psychopharmacology (Berl) 105:141–144

Fonnum F (1987) Biochemistry, anatomy and pharmacology of GABA neurons. In: Meltzer HY (ed) Psychopharmacology, the third generation of progress. Raven, New York, pp 173–182

Freedman R et al. (1982) Clonidine treatment of schizophrenia. Double blind comparison to placebo and neuroleptic drugs. Acta Psychiatr Scand 65:35–45

Freeman AS, Chiodo LA (1988) Electrophysiologic effects of cholecystokinin octapeptide on identified rat nigrostriatal dopaminergic neruons. Brain Res 439:266

Friedhoff AJ, Alpert M (1973) A dopaminergic-cholinergic mechanism in the production of psychotic symptoms. Biol Psychiatry 6:165–169
Gale K (1980) Chronic blockade of dopamine receptors by anti-schizophrenic drugs enhances GABA binding in substantia nigra. Nature 283:569–570
Ganguli R, Reynolds CF, Kupfer DJ (1987) Electroencephalographic sleep in young never-medicated schizophrenics. Arch Gen Psychiatry 44:36–44
Gardner EL, Walker LS, Paredes W (1993) Clozapine's functional mesolimbic selectivity is not duplicated by the addition of anticholinergic action to haloperidol: a brain simulation study in the rat. Psychopharmacology (Berl) 110:119–124
Garlow S, Morilak D, Dean RS, Roth BL, Ciaranello RD (1993) Production and characterization of an antibody for the 5-HT_2 receptor which labels a subpopulation of rat forebrain neurons. Brain Research 615:113–120
Garver DL, Beinfeld MC, Yao JK (1990) Cholecystokinin, dopamine and schizophrenia. Psychopharmacol Bull 26:377–380
Garver DL, Bissette G, Yao JK, Nemeroff CB (1991) CSF neurotensin concentration in psychoses: relationship to symptoms and drug response. Am J Psychiatry 148:484–488
Gerlach J (1977) The relationship between parkinsonism and tardive dyskinesia. Am J Psychiatry 143:781–784
Gerner RH, Catlin DH, Gorelick DA, Hui Kk li CH (1980) β-endorphin. Intravenous infusion causes behavioral change in psychiatric patients. Arch Gen Psych 37:642–647
Gerner RH, VanKammen DP, Ninan PT (1985) Cerebrospinal fluid cholecystokinin, bombesin and somatostatin in schizophrenia and normals. Prog Neuropsychopharmacol Biol Psychiatry 9:73–82
Gewirtz GR, Gorman JM, Volavka J, Macaluso J, Gribkoff G, Taylor DP, Borison R (1994) BMY14802, a sigma receptor ligand for the treatment of schizophrenia. Neuropsychopharmacology 10:37–40
Gillan JC, Kaplan J, Stillman R, Wyatt RJ (1976) The psychedelic models of schizophrenia: the case of N,N′-dimethyltryptamine. Am J Psychiatry 133:203–208
Gjerris A, Rafaelson OJ, Vendsborg P, Fahrenkrug J, Rehfeld JF (1984) Vasoactive intestinal peptide decreased in cerebrospinal fluid (CSF) in atypical depression. J Affective Disord 7:325–337
Glennon RA (1987) Central serotonin receptors as targets for drug research. J Med Chem 30:1–12
Glennon RA, Titler M, Mckenney JD (1984) Evidence for 5-HT2 involvement in the mechanism of action of hallucinogenic agents. Life Sci 35:2505–2511
Goff DC, Brotrian AW, Aiates M, Mccormack S (1990) Trial of fluoxetine added to neuroleptics for treatment resistent schizophrenic patients. Am J Psychiatry 14:492–493
Goff DC, Midha KK, Brotman AW, Mccormick S, Waites M, Amico ET (1991) An open trial of buspirone added to neuroleptics in schizophrenic patients. J Clin Psychopharmacol 11:193–197
Goff DC, Amico E, Dreyfuss D, Ciraulo D (1994) A placebo-controlled trial of trihexyphenidyl in unmedicated patients with schizophrenia. Am J Psychiatry 151:429–431
Gomes UCR, Shanley BC, Potgieter L et al. (1980) Noradrenergic overactivity in chronic schizophrenia: evidence based on cerebrospinal fluid noradrenaline and cyclic nucleotide concentrations. Br J Psychiatry 137:346–351
Govoni S, Hong JS, Yang HYT, Costa E (1980) Increase of neurotensin content elicited by neuroleptics in nucleus accumbens. J Pharmacol Exp Ther 215:413–417
Goyer PF, Berridge MS, Semple WE, Morris ED, Wong DF, Schulz SC, Miraldi F, Meltzer HY (1993) Dopamine-2 and serotonin-2 receptor indices in clozapine treated schizophrenic patients. Schizophr Res 9:199
Gudelsky GA, Koenig JI, Simonovic M, Koyama T, Ohmori T, Meltzer HY (1987) Differential effects of haloperidol, clozapine and fluperlapine on tuberoinfun-

dibular dopamine neurons and prolactin secretion in the rat. J Neural Transm 68:227–240
Gunne LM, Andren PE (1993) An animal model for coexisting tardive dyskinesia and tardive parkinsonism: a glutamate hypothesis for tardive dyskinesia. Clin Neuropharmacol 16:90–95
Gunne LM, Haggstrom JE (1983) Reduction of nigral glutamic acid decarboxylase in rats with neuroleptic induced dyskinesia. Psychopharmacology (Berl) 81:191–194
Harnryd C, Bjerkenstadt L, Gullberg B (1989) A clinical comparison of melperone and placebo in schizophrenic women on a milieu therapeutic ward. Acta Psychiatr Scand [Suppl] 352:40–47
Harrison PJ, Mclaughlin D, Kerwin RW (1991) Decreased hippocampal expression of a glutamate receptor gene in schizophrenia. Lancet 337:450–452
Hasegawa M, Gutierrez-Esteinou R, Way L, Meltzer HY (1993) Relationship between clinical efficacy and clozapine plasma concentrations in schizophrenia: effect of smoking. J Clin Psychopharmacol 13:383–390
Hashimoto A, Nishikawa T, Oko T, Takahashi K (1971) D-alanine inhibits methampehtamine-induced hyperactivity in rats. Eur J Pharmacol 202:105
Hashimoto T, Nishino N, Nakai H, Tanaka C (1991) Increase in serotonin 5-HT1A receptors in prefrontal and temporal cortices of brains from patients with chronic schizophrenia. Life Sci. 48:355–363
Hock FJ, Kruse H, Gerhards HJ, Konz E (1985) Pharmacological effects of HR 375: a new potential antipsychotic agent. Drug Development Res 16:301–311
Hoehn-Saric R (1983) The effect of THIP on chronic anxiety. Psychopharmacology (Berl) 80:338–391
Hollister LE (1962) Drug-induced psychoses and schizophrenic reactions, a critical comparison. Ann NY Acad Sci 96:80–88
Hollister LE (1994) New psychotherapeutic drugs. J Clin Psychopharmacol 14:50–63
Holmberg G, Gershon S (1961) Autonomic and psychic effects of yohimbine hydrochloride. Psychopharmacology (Berl) 2:93–106
Hommer DW, Zahn TP, Pickar D et al. (1984) Prazosin, a specific alpha1-noradrenergic receptor antagonist has no effect on symptoms but increases autonomic arousal in schizophrenic patients. Psychiatry Res 11:193–204
Hutson PH, Thorn L, Bristow LS, Tricklebank MD (1991) R (+)-HA-966, a glycine/NMDA receptor antagonist, blocks increased mesolimbic dopamine metabolism induced by psychostimulant drugs. Br J Pharmacol 102:67
Innis RB, Bunney BS, Charney DS, Price LH, Glazer WM, Sternberg DE, Rubin AL, Heninger GR (1986) Does the cholecyctokinin antagonist proglumide possess antipsychotic activity? Psychiatry Res 18:1–7
Itzhak Y, Ruhland M, Krahling H (1990) Binding of umespirone to the sigma receptor: evidence for multiple affinity states. Neuropharmacology 29:181–184
Janowski DS, El-Yousef MK, Davis JM, Sekerke HJ (1973) Antagonistic effects of physostigmine and methylphenidate in man. Am J Psychiatry 130:1370–1376
Joyce JN, Shane A, Lexow N, Winokur A, Casanova MF, Kleinman JE (1993) Serotonin uptake sites and serotonin receptors are altered in the limbic system of schizophrenics. Neuropsychopharmacology 8:315–336
Jungkunz G, Nedopil N, Ruther E (1983) Acute effects of the synthetic analogue of methionine enkephalin FK 33-824 on depressive symptoms. Pharmacopsychiatry 16:90–92
Kane J, Honigfield G, Singer J, Meltzer HY and the Clozaril Collaborative Study Group (1988) Clozapine for the treatment-resistant schizophrenic. Arch Gen Psychiatry 45:789–796
Keats AS, Telford J (1964) Narcotic antagonists as analgesics. In: Gould RF (ed) Clinical aspects, molecular modification in drug design: advances in chemistry. American Chemical Society, Washington, pp 170–176
Kemp JA, Leeson PD (1993) The glycine site of the NMDA receptor-five years on. Trends Pharmacol Sci 14:20–25

Kerwin R, Patel S, Meldrum B (1990) Quantitative autoradiographic analysis of glutamate binding sites in the hippocampal formation in normal and schizophrenic brain post mortem. Neuroscience 39:25–32

Kim JS, Kornhuber HH, Schmid-Burgk W, Holzmuller B (1980) Low cerebrospinal fluid glutamate in schizophrenic patients and a new hypothesis on schizophrenia. Neurosci Lett 20:379–382

Kleckner NW, Dingledine R (1988) Requirement for glycine in activation of NMDA receptors expressed in Xenopus oocytes. Science 241:835–837

Kline NN, Li CH, Lehman HE, Lathja A, Laski E, Cooper T (1977) β-endorphin-induced changes in schizophrenic and depressed patients. Arch Gen Psych 34:1111–1113

Knight AR, Wahle A, Wong EHF, Middlemiss JC (1991) The subcellular distribution and pharmacology of the sigma recognition site in the guinea-pig brain and liver. Mol Neuropharmacol 1:71–73

Kobayashi RM, Brown M, Vale W (1977) Regional distribution of neurotensin and somatostatin in rat brain. Brain Res 126:584–588

Kolakowska T, Gadhvi H, Molyneux S (1987) An open clinical trial of fenfluramine in chronic schizophrenia: a pilot study. Int Clin Psychopharmacol 2:83–88

Kornhuber J, Kornhuber ME (1986) Presynaptic dopaminergic modulation of cortical input to the striatum. Life Sci 39:669–674

Kraepelin E (1919) Dementia Praecox and Paraphrenia. Kreiger, New York

Krystal JH, Seibyl JP, Price LH, Woods SW, Heninger GR, Aghajanian GK, Charney DS (1993) m-Chlorophenylpiperazine (MCPP) effects in neuroleptic-free schizophrenic patients. Arch Gen Psychiatry 50:624–635

Kuoppamaki M, Seppala T, Syvalahti E, Hietala J (1993a) Chronic clozapine treatment decreases 5-hydroxytryptamine 1c receptor density in the rat choroid plexus: comparison with haloperidol. J Pharmacol Exp Ther 264:1262–1267

Kuoppamaki M, Syvalahti E, Hietala J (1993b) Clozapine and N-desmethylclozapine are potent 5-HT1C receptor antagonists. Eur J Pharmacol 245:179–182

Lahti RA, Evans DL, Stratman NC, Figur LM (1993) Dopamine D_4 versus D_2 receptor selectivity of dopamine receptor antagonists: possible therapeutic implications. Eur J Pharmacol 236:483–486

Lang A, Vasar E, Soosaar A, Harro J (1992) The involvement of sigma and phencyclidine receptors in the action of antipsychotic drugs. Pharmacol Toxicol 71:132–138

Largent BL, Wikstrom H, Gundlach AL, Snyder SH (1987) Structural determinants of sigma receptor affinity. Mol Pharmacol 32:772–784

Largent BL, Wikstrom H, Snowman AM, Snyder, SH (1988) Novel antipsychotic drugs share high affinity of sigma receptors. Eur J Pharmacol 155:345–347

Levant B, Bissette G, Davis MD, Heffner TG, Nemeroff CB (1991) Effects of Cl-943, a potential antipsychotic drug, and haloperidol on regional brain neurotensin concentrations. Synapse 9:225–230

Levant B, Merchant KM, Dorsa DM, Nemeroff CB (1992) BMY 14802, a potential antipsychotic drug, increases expression of proneurotensin mRNA in the rat striatum. Mol Brain Res 12:279–284

Leysen JE, Janssen PMF, Schotte A, Luyten WHML, Megens AAHP (1993) Interaction of antipsychotic drugs with neurotransmitter receptor sites in vitro and in vivo in relation to pharmacological and clinical effects: role of 5HT2 receptors. Psychopharmcrology (Berl) 112:S40–S54

Lidsky TI, Yablonsky-Alter E, Zuck L, Banerjee SP (1993) Anti-glutamatergic effects of clozapine. Neurosci Lett 163:155–158

Lieberman JA, Koreen AR (1993) Neurochemistry and neuroendocrinology of schizophrenia: a selective review. Schizophr Bull 19:371–429

Liegeois JF, Bruhwyler J, Damas J, Nguyen TP, Chleide EM, Mercier MG, Rogister FA, Delarge JE (1993) New pyridobenzodiazepines as potential antipsychotics: synthesis and neurochemical study. J Med Chem 36:2107–2114

Lindenmayer J-P, Gardner E, Goldberg E, Opler LA, Kay SR, van Praag HM, Weiner M, Zukin S (1988) High dose naloxone in tardive dyskinesia. Psychiatry Res 26:19–28

Lindstrom LH, Widerlov E, Bissette G, Nemeroff CB (1988) Reduced CSF neurotensin concentration in drug-freee schizophrenic patients. Schizophr Res 1:55–59

Lipton SA, Rosenberg PA (1994) Excitatory amino acids as a final common pathway in neurologic disorders. N Engl J Med 330:613–622

Luby ED, Cohen BD, Rosenbaum F et al. (1959) Study of a new schizophreniomimetic drug Sernyl. AMA Arch Neurol Psychiatry 81:363–369

Luby ED, Gottlieb JS, Cohen BD, Rosenbaum G, Domino EF (1962) Model psychosis and schizophrenia. Am J Psychiatry 119:61–65

Mao CC, Cheney DL, Marco E, Revuelta A, Costa E (1977) Turnover times of gamma-aminobutyric acid and acetylcholine in nucleus caudatus, nucleus accumbens, globus pallidus and substantia nigra: effects of repeated adminstration of haloperidol. Brain Res 132:375–379

Marco E, Mao CC, Cheney DL, Revuelta A, Costa E (1976) The effects of antipsychotics on the turnover rate of GABA and acetylcholine in rat brain nuclei. Nature 264:363–365

Marder S, Meibach R (1994) Risperidone in the treatment of schizophrenia. Am J Psychiatry 151:825–835

Marshall BD, Glynn SM, Midha KK, Hubbard JW, Bowen LL, Banzett L, Mintz J, Liberman RP (1989) Adverse effects of fenfluramine in treatment refractory schizophrenia. J. Clin. Psychopharmacol 9:110–115

Martin WR, Eades CG, Thompson JA, Huppler RE, Gilbert PE (1976) The effects of morphine and morphine-like drugs in the non-dependent and morphine-dependent chronic spinal dog. J Pharmacol Exp Ther 197:517–532

Matsubara S, Meltzer HY (1989) Effect of typical and atypical antipsychotic drugs on 5-HT2 receptor density in rat cerebral cortex. Life Sci 45:1397–1406

Matsubara S, Matsubara R, Kusumi I, Koyama T, Yamashita I (1993) Dopamine D1, D2 and serotonin2 receptor occupancy by typical and atypical antipsychotic drugs in vivo. J Pharmacol Exp Ther 265:498–508

Matz R, Rick W, Oh D et al. (1974) Clozapine-a potential antipsychotic agent without extrapyramidal manifestations. Curr Ther Res 16:687–695

Maura G, Roccatagliata R, Ulivi M, Raiteri M (1988) Serotonin glutamate interaction in rat cerebellum: involvement of 5-HT_1 and 5-HT_2 receptors. European Journal of Pharmacology 3:31–38

McCann DJ, Rabin RA, Rens-Domiano S, Winter JC (1989) Phencyclidine/SKF 10,047 binding sites: evaluation of function. Pharmacol Biochem Behav 32:87–94

Meert TF, Dehaes P, Janssen PA (1989) Risperidone (R 64766), a potent and complete LSD antagonist in drug discrimination by rats. Psychopharmacology (Berl) 97:206–212

Meldrum B (1982) GABAA and acute psychosis. Psychiatr Med 12:1–5

Meltzer HY, Stockmeier CA (1992) Occupancy of D_2 receptors. Arch Gen Psychiatry 49:588–589

Meltzer, HY, Busch DA, Tricou BJ, Robertson A (1982) Effect of (des-tyr)-γ-endorphin in schizophrenia. Psychiatry Research 6:313–326

Meltzer HY, Matsubara S, Lee J-C (1989) Classification of typical and atypical antipsychotic drugs on the basis of dopamine D-1, D-2 and serotonin2 pK_i values. J Pharmacol Exp Ther 251:238–246

Meltzer HY, Maes M, Lee MA (1993) The cimetidine-induced increase in prolactin secretion in schizophrenia: effect of clozapine. Psychopharmacology 112:S95–S104

Meltzer HY, Chai BL, Thompson PA, Yamamoto BK (1994) Effect of scopolamine on the efflux of dopamine and its metabolites following clozapine, haloperidol or thioridazine. Journal of Pharmacology and Experimental Therapeutics 268:1452–1461

Mendels J (1967) The effect of methysergide (a serotonine agent) on schizophrenia: a preliminary report. Br. J. Psychiatry 124:157–160
Merchant KM (1994) c-fos antisense oligonucleotide specifically attenuates haloperidol-induced increases in neurotensin/neuromedin N mRNA expression in rat dorsal striatum. Mol Cell Neurosci 5:336–344
Merchant KM, Dorsa DM (1993) Differential induction of neurotensin and c-fos gene expression by typical verses atypical antipsychotics. Proc Natl Acad Sci USA 90:3447–3451
Mercugliano M, Chesselet MF (1992) Clozapine decreases enkephalin mRNA in rat striatum. Neurosci. Lett. 136:10–14
Meshul CK, Janowsky A, Casey DR, Stallbaumer RK, Taylor B (1992) Effect of haloperidol and clozapine on the density of "perforated" synapses in caudate, nucleus accumbens and medial prefrontal cortex. Psychopharmacology (Berl) 106:45–52
Mesotten F, Suy F, Pietquin M, Burton P, Heylen S, Gelders Y (1989) Therapeutic effects and safety of increasing doses of risperidone (R 64766) in psychotic patients. Psychopharmacology 99:445–449
Mita T, Hanada S, Nishino N, Kuno T, Nakai H, Yamadori T, Mizol Y, Tanaka C (1986) Decreased serotoninS2 and increased dopamine D2 receptors in chronic schizophrenics. Biol Psychiatry 21:1407–1414
Modestin J, Schwartz RB, Hunger J (1973) Zur Frage der Beeinflussung schizophrener Symptome durch Physostigmin. Pharmakopsychiatr Neuropsychopharmakol 9:300–304
Moghaddam B, Bunney BS (1990) Acute effects of typical and atypical antipsychotic drugs on the release of dopamine from prefrontal cortex, nucleus accumbens, and striatum of the rat: an in vivo microdialysis study. J. Neurochem. 54:1755–1760
Montgomery SA, Green M (1988) The use of cholecystokinin in schizophrenia: a review. Psychol Med 18:593–603
Moore NA, Tye NC, Axton MS, Risius, FC (1992) The behavioral pharmacology of olanzapine, a novel "atypical" antipsychotic agent. J Pharmacol Exp Ther 262:545–551
Moore NC, Meyendorff E, Yeragani V, Lewitt PA, Gershon S (1987) Tiaspirone in schizophrenia. J Clin Psychiatry 7:98–101
Moran PM, Moser PC (1992) MDL 73,147EF, a 5-HT3 antagonist, facilitates latent inhibition in the rat. Pharmacol. Biochem Behav 42:519–522
Morilak DA, Garlow SJ, Ciaranello RD (1993) Immunocytochemical localization and description of neurons expressing serotonin2 receptors in the rat brain. Neuroscience 54:701–717
Moss LE, Neppe VM, Drevets WC (1993) Buspirone in the treatment of tardive dyskinesia. J Clin Psychopharmacol 13:204–208
Naber D (1993) Opioids in the etiology and treatment of psychiatric disorders. In: Herz A (ed) Opioids III. Springer, Berlin Heidelberg New York, pp 781–801
Nemeroff CB, Bissette G, Widerlov E, Beckman HH, Gerner R, Manberg PJ, Lindstrom L, Prang AJ, Gattaz W (1989) Neurotensin-like immunoreactivity in cerebrospinal fluid of patients with schizophrenia, depression, anorexia nervosa-bulemia and premenstruel-syndrome. J Neuropsychiatry Clin Neurosci 1:16–20
Nemeroff CB, Kilts CD, Levant B, Bissette G, Campbell A, Baldessarini RJ (1991) Effects of the isomers of N-n-propylnorapomorphine and haloperidol on regional concentrations of neurotensin in rat brain. Neuropsychopharmacology 4:27–33
Nemeroff CB, Levant B, Myers B, Bissett B (1992) Neurotensin, antipsychotic drugs and schizophrenia. Ann NY Acad Sci 146–156
New JP, Yevich JP, Temple DL, New KB, Gross SM, Schlemmer RF, Eison MS, Taylor DP, Riblet LA (1988) Atypical antipsychotic agents: patterns of activity in a series of 3-substituted 2-pyridinyl-1-piperazine derivatives. J Med Chem 31:618–624
Newcomer JW, Faustman WO, Zipursky RB, Csernansky JG (1992) Zacopride in schizophrenia: a single-blind serotonin type 3 antagonist trial. Arch Gen Psychiatry 49:751–752

Niemegeers CJ, Awouters F, Janssen PA (1990) Serotonin antagonism involved in the antipsychotic effect. Confirmation with ritanserin and risperidone. Encephale 16:147–151
Nishikawa T, Takashima M, Toru M (1983) Increased [3H]-kainic acid binding in the prefrontal cortex in schizophrenia. Neurosci Lett 40:245–250
Nordstrom AL, Farde L, Haldin C (1993) High 5-HT2 receptor occupancy in clozapine treated patients demonstrated by PET. Psychopharmacology (Berl) 110:365–367
Ogren SO, Lundstrom J, Nilsson LB (1993) Concentrations of remoxipride and its phenolic metabolites in rat brain and plasma. Relationship to extrapyramidal side effects and atypical antipsychotic profile. J Neural Transm 94(3):199–216
Olney JW (1969) Brain lesions, obesity, and other disturbances in mice treated with monosodium glutamate. Science 164:719–721
Olney JW, Ho OL, Rhee V (1971) Cytotoxic effects of acidic and sulphur containing amino acids on the infant mouse central nervous system. Exp. Brain Res 14:61–76
Olsen RW, Tobin AJ (1990) Molecular biology of GABAA receptors. FASEB J 4:1469–1480
Pehek EA, Yamamoto BK, Meltzer HY (1991) The effects of clozapine on dopamine, 5-HT and glutamate release in the rat medial prefrontal cortex. Schizophr Res 4:323
Pehek EA, Meltzer HY, Yamamoto BK (1993) The atypical antipsychotic drug amperozide enhances rat cortical and striatal dopamine efflux. Eur J Pharmacol 240:107–109
Pickar D, Davis GC, Schulz SC, Extein I, Wagner R, Naber D, Gold PW, van Kammen DP, Goodwin FK, Wyatt RJ, Li CH, Bunney WE (1981) Behavioral and biological effects of acute β-endorphin injection in schizophrenic and depressed patients. Am J Psych 138:160–166
Pickar D, Bunney WE, Douillet P, Sethi BB, Sharma M, Vartanian ME, Lideman RP, Naber D, Leibl K, Yamashita I, Koyama T, Verhoeve WMA, Vartanian F, Morozov PV, Khac TN (1989) Repeated naloxone administration in schizophrenia: a phase II World Health Organization study. Biol Psych 25:440–448
Potkin SG, Weinberger DR, Linoila M, Wyatt RJ (1983) Low CSF 5-hydroxyindoleacetic acid in schizophrenic patients with enlarged cerebral ventricles. Am. J. Psychiatry 140:21–25
Potkin SG, Costa J, Roy S, Sramek J, Jin Y, Gulasekaram B (1992) Glycine in the treatment of schizophrenia: theory and practical results. In: Meltzer HY (ed) Novel antipsychotic drugs. Raven, New York, pp 179–188
Rao TS, Contreras PC, Cler JA, Emmett Mr, Mick SJ, Iyengar S, Woods PL (1991) Clozapine attenuates *N*-methyl-D-asparte receptor complex-mediated responses in vivo: tentative evidence for a functional modulation by a noradrenergic mechanism. Neuropharmacology 30:557–565
Rapport MM, Green AA, Page IH (1948) Crystalline serotonin. Science 108:329
Rasmussen K, Aghajanian GK (1988) Potency of antipsychotics in reversing the effects of a hallucinogenic drug on locus coeruleus neurons correlates with 5-HT2 binding affinity. Neuropsychopharmacology 1:101–107
Rasmussen K, Stockton ME, Czachura JF (1991a) The 5-HT3 receptor antagonist zatosetron decreases the number of spontaneously active A10 dopamine neurons. Eur J Pharmacol 205:113–116
Rasmussen K, Stockton ME, Czachura JF, Howbert JJ (1991b) Cholecystokinin (CCK) and schizophrenia: the selective CCKB antagonist LY262691 decreases midbrain dopamine unit activity. Eur J Pharmacol 209:135–138
Rasmussen K, Czachura JF, Stockton ME, Howbert JJ (1993) Electrophysiological effects of diphenylpyrazolidoinone cholecystokinin-B and cholecystokinin-A antagonists on midbrain dopamine neurons. J Pharmacol Exp Ther 264:480–488
Reynolds GP, Rossor M, Iversen LL (1983) Preliminary studies of human cortical 5-HT2 receptors and their involvement in schizophrenia and neuroleptic drug action. J Neural Transm 18 [Suppl]:273–277

Reynolds GP, Czudek C, Andrews HB (1990) Deficit and hemispheric asymmetry of GABA uptake sites in hippocampus in schizophrenia. Biol Psychiatry 27:1038–1044

Riederer P, Lange KW, Kornhuber J, Danielczyk W (1992) Glutamatergic-dopaminergic balance in the brain. Its importance in motor disorders and schizophrenia (Review). Arzneimittelforschung 42:265–268

Rimon R, Terenius L, Averbuch I, Belmaker RH (1983) High-dose haloperidol increases CSF opioid activity in patients with chronic schizophrenia. Pharmacopsychiatry 16:9–12

Rivest R, Marsden CA (1991) Muscarinic antagonists attenuate the increase in accumbens and striatum dopamine metabolism produced by clozapine but not by haloperidol. Br J Pharmacol 104:234–238

Roberts GW, Ferrier N, Lee Y, Crow TJ, Johnstone EC, Owens DGC, Baraese-Hamilton AJ, Mcgregor G, O'shaughnessey D, Polak JM, Bloom SR (1983) Peptides, the limbic lobe and schizophrenia. Brain Res 288:199–211

Rogue A, Rogue P (1992) Mianserin in the management of schizophrenia. Schizophrenia 1992: an international conference. Vancouver, BC, p 135

Rossee RB, Schwartz BL, Leighton RE, Davis RE, Deutsch SI (1990) An open-label trial of milacemide in schizophrenia: an NMDA intervention strategy. Clin Neuropharmacol 13:348–354

Roth BL (1994) Multiple serotonin receptors: clinical and experimental aspects. Annul Clin Psychiatry 6:67–78

Roth BL, Ciaranello RD (1991) Mianserin decreases 5-HT2 radioligand binding without altering 5-HT_2 receptor mRNA levels. Eur J Pharmacol 207:169–172

Roth BL, Meltzer HY (1995) The role of serotonin in schizophrenia. In: Bloom F, Kupfer DJ (eds) Psychopharmacology: a 4th generation of progress. Raven, New York 1215–1228

Roth BL, Ciaranello RD, Meltzer HY (1992) Binding of typical and atypical antipsychotic agents with transiently expressed 5-HT1c receptors. J Pharmacol Exp Ther 260:1361–1365

Roth BL, Craigo SC, Choudhary MS, Uluer A, Monsma FJ, Shen Y, Meltzer HY, Sibley DR (1994) Binding of typical and atypical antipsychotic agents to 5-hydroxytryptamine6 (5-HT6) and 5-hydroxytryptamine7 (5-HT7) receptors. J Pharmacol Exp Ther 268:1403–1410

Rowntree DW, Nevin S, Wilson A (1950) The effects of diisopropylfluorophosphate in schizophrenia and manic-depressive psychosis. J Neurol Neurosurg Psychiatry 13:47–62

Saller CF, Czupryna MJ, Salama AI (1990) 5-HT2 receptor blockade by ICI 169369 and other 5-HT2 antagonists modulates the effects of D-2 dopamine receptor blockade. J Pharmacol Exp Ther 253:1162–1170

Saywell K, Singh L, Oles RJ, Vass C, Leeson PD, Williams BJ, Tricklebank MD (1991) The anticonvulsant properties in the mouse of the glycine/NMDA receptor antagonist, L-687414. Br J Pharmacol 102:66

Schmidt WJ, Krahling H, Ruhland M (1991) Antagonism of AP-5-induced sniffing stereotypy links umespirone to atypical antipsychotics. Life Sci 48:499–505

Schoepp DD, Conn PJ (1993) Metabotropic glutamate receptors in brain function and pathology. Trends Pharmacol Sci 14:13–20

Schofield PR, Darlison MG, Fujita N, Burt DR, Stephenson FA, Rodriguez H, Rhee LM, Ramachandron J, Reale V, Glencorse IA, Seeburg PH, Barnard EA (1987) Sequence and functional expression of the GABAA receptor shows a ligand-gated receptor super family. Nature 328:221–227

Sedman AJ, Gimlet GP, Sayed AJ, Posvar EL (1990) Initial human safety and tolerance study of a GABA uptake inhibitor, CI966: potential role of GABA as a mediator of schizophrenia and anemia. Drug Dev Res 21:235–242

See RE, Aravagiri M, Ellison GD (1989) Chronic neuroleptic treatment in rats produces persistant changes in GABAa and dopamine D-2 but not dopamine D-1 receptors. Life Sci 44:229–236

See RE, Toga AW, Ellison G (1990) Autoradiographic analysis of regional alteration in brain receptors following chronic administration and withdrawl of typical and atyipcal neuroletpics in rats. J Neural Transm 82:93–109

Seeman P (1990) Atypical neuroleptics: role of multiple receptors, endogenous dopamine, and receptor linkage. Acta Psychiatr Scand 82 [Suppl 358]:14–20

Sherman AD, Hegwood TS, Baruah S, Waziri R (1991) Deficient NMDA-mediated glutamate release from synaptosomes of schizophrenics. Biol Psychiatry 30:1191–1198

Siegel E, Baur R, Trube G, Mohler H, Malherbe P (1990) The effect of subunit composition of rat brain GABAA receptors on channel function. Neuron 5:703–711

Simpson MDC, Royston MC, Slater P, Deakin JFW, Skam WJ (1989) Reduced GABA uptake sites in the temporal lobe in schizophrenia. Neurosci Lett 107:211–215

Singh L, Menzies R, Tricklebank MD (1990) The discriminative stimulus properties of (+)-HA-966, an antagonist at the glycine/NMDA receptor. Eur J Pharmacol 186:129–132

Singh L, Donald E, Foster AC, Hudon PH, Iverson LL, Iverson SD, Kemp JA, Leeson PD, Marshall GR, Oles RJ, Priestley T, Thorn L, Tricklebank MD, Vass CA, Williams BJ (1990) Enantionmers of HA-966 (3-amino-1-hydroxypyrrolid-2-one) exhibit distinct central nervous system effects: (+)-HA-966 is a selective glycine/NMDA receptor antagonist, but (−) HA-966 is a potent γ-butyrolactone-like sedative. Proc Natl Acad Sci USA 87:347–351

Singh MM, Kay SR (1985) Pharmacology of central cholinergic mechanisms and schizophrenic disorders. In: Singh MM, Warburton DM, Lal H (eds) Central cholinergic mechanisms and adaptive functions. Plenum, New York, pp 247–308

Skorzewsak A, Lal S (1989) Methysergide-induced psychosis: case report with long-term follow-up. Neuropsychobiology 22:125–127

Sorensen SM, Kehne JH, Fadayel EM, Humphreys TM, Ketteler HJ, Sullivan C, Taylor VL, Schmidt CJ (1993) Characterization of the 5-HT2 antagonist MDL 100907 as a putative atypical antipsychotic: behavioral, electrophysiological and neurochemical studies. J Pharmacol Exp Ther 266:684–691

Stein L, Wise CD (1971) Possible etiology of schizophrenia: progressive damage to the noradrenergic reward system by 5-hydroxydopamine. Science 171:1032–1036

Stevens KE, Meltzer J, Rose GM (1993) Disruption of sensory gating by the $\alpha 2$ selective noradrenergic antagonist yohimbine. Biol Psychiatry 33:130–132

Stockmeier CA (1994) On age and serotonin receptor binding in the human brain. Neuropsychopharmacology 11:143–144

Stockmeier CA, Dicarlo JJ, Zhang Y, Thompson P, Meltzer HY (1993) Characterization of typical and atypical antipsychotic drugs based on in vivo occupancy of serotonin2 and domapine2 receptors. J Pharmacol Exp Ther 266:1374–1384

Stoessl AJ (1989) Peptide-dopamine interactions in the central nervous system: implications for neuropsychiatric disorders. J Psychopharmacol 3:99–120

Strauss WH, Klieser E (1991) Psychotropic effects of ritanserin, a selective S2 antagonist: an open study. Eur Neuropsychopharmacol 1:101–105

Suzuki T, Moroji T, Hori T, Baba A, Kaniai N, Koizumi J (1993) Autoradiographic localization of CCK-8 binding sites in the rat brain: effects of chronic methamphetamine administration on these sites. Biol Psychiatry 34:781–790

Szara S (1970) DMT (N,N-dimethyltryptamine) and homologues: clinical and pharmacological considerations. In: Efron DH (ed) Psychotomimetic drugs. Raven, New York, pp 275–286

Tam SW, Steinfels GF, Gilligan PJ, Schmidt WK, Cook L (1992) DuP734, a sigma and 5-hydroxytryptamine2 receptor antagonist: receptor-binding, electrophysiological and neuropharmacological profiles. J Pharmacol Exp Ther 263:1167–1174

Tamminga CA, Tighe PJ, Chase TN, Defriates EG, Schaffer MH (1981) Des-tyrosine-gamma-endorphin administration in chronic schizophrenia. Arch Gen Psych 38:167–168

Tamminga CA, Cascella N, Fakouhi TD, Herting RL (1992) Enhancement of NMDA-mediated transmission in schizophrenia: effects of milacemide. In: Meltzer HY (ed) Novel antipsychotic drugs. Raven, New York, pp 171–177
Tandon R, Greden JF (1987) Trihexyphenidyl treatment of negative schizophrenic symptoms. Acta Psychiatr Scand 76:732
Tandon R, Greden JF (1989) Cholinergic hyperactivity and negative schizophrenic symptoms. Arch Gen Psychiatry 46:745–753
Tandon R, Greden JF, Silk KR (1988) Treatment of negative schizophrenic symptoms with trihexyphenidyl. J Clin Psychiatry 8:212–215
Tandon R, Shipley JE, Taylor S, Greden JF, Eiser A, Dequardo J, Goodson J (1992) Electroencephalographic sleep abnormalities in schizophrenia: relationship to positive/negative symptoms and prior neuroleptic treatment. Arch Gen Psychiatry 49:185–194
Taylor DP, Schlemmer RF (1992) Sigma "antagonists" potential antipsychotics? In: Meltzer HY (ed) Novel antipsychotic drugs. Raven, New York, pp 189–201
Taylor DP, Eison MS, Moon SN, Yocca FD (1991) BMY14802: a potential antipsychotic with selective affinity for sigma-binding sites. In: Tamminga, CA, Schulz SC (eds) Advances in neuropsychiatry and psychopharmacology. Raven, New York, pp 307–315
Terenius L, Wahlstrom A, Lindstrom L, Widerlov E (1976) Increased CSF levels of endorphins in chronic psychosis. Neurosci Lett 3:157–162
TIPS Receptor Nomenclature Supplement (1994) Excitatory amino acid receptors (ionotropic). In: 1994 Receptor and ion channel nomenclature supplement. Trends Pharmacol Sci [Suppl]:19
Toth E, Lajtha A (1986) Antagonism of phencyclidine-induced hyperactivity by glycine in mice. Neurochem Res 11:393–400
Traynor JR, Elliott J (1993) Delta-opioid receptor subtypes and cross-talk with mu receptors. Trends Pharmacol Sci 14:84–86
Tricklebank MD, Bristow LJ, Hutson PH (1992) Alternative appraoches to the discovery of novel antipsychtoic agents. Prog Drug Res 38:299–336
Ugedo L, Grenhoff J, Svensson TH (1989) Ritanserin, a 5-HT2 receptor antagonist, activates midbrain dopamine neurons by blocking serotonergic inhibition. Psychopharmacology (Berl) 98:45–50
Ujike H, Kanzaki A, Okummura K, Akiyama K, Otsuki S (1992) Sigma antagonist BMY 14802 prevents methamphetamine-induced sensitization. Life Sci 50:PL129–PL134
Ulas J, Cotman CW (1993) Excitatory amino acid receptors in schizophrenia. Schizophr Bull 19:105–117
Vanderhaeghen JJ, Signeau JC, Gepts W (1975) New peptide in the ventral CNS reacting with antigastrin antibodies. Nature 257:604–605
VanKammen DP (1977) Gamma-aminobutyric acid (GABA) and the dopamine hypothesis of schizophrenia. Am J Psychiatry 134:138–143
VanKammen DP (1991) The biochemical basis of relapse and response in schizophrenia: review and hypothesis. Psychol Med 21:881–895
VanKammen DP, Gelernter J (1987) Biochemical instability in schizophrenia. II. The serotonin and gamma-aminobutyric acid systems. In: Meltzer HY (ed) Psychopharmacology: the third generation of progress. Raven, New York, pp 753–758
VanKammen DP, Kelly M (1991) Dopamine and norepinephrine activity in schizophrenia: an integrative perspective. Schizophr Res 4:173–191
VanKammen DP, Peters J, VanKammen WB, Nugent A, Goetz KL, Yao J, Linnoila M (1989) CSF norepinephrine in schizophrenia is elevated prior to relapse after haloperidol withdrawal. Biol Psychiatry 26:176–188
VanKammen DP, Peters J, Yao J, VanKammen WB, Neylen T, Shaw D, Linnoila M (1990) Norepinephrine in acute exacerbations of chronic schizophrenia. Arch Gen Psychiatry 47:161–168
Vaupel DP (1983) Naltrexone fails to antagonize the sigma effects of PCP and SKF-10047 in the dog. Eur J Pharmacol 92:269–274

Wachtel SR, White FJ (1988) Electrophysiological effects of BMY 14802, a new potential antipsychotic drug, on midbrain dopamine neurons in the rat: acute and chronic studies. J Pharmacol Exp Ther 244:410–416

Walker JM, Bowen WD, Walker FO, Matsumoto RR, Decosta B, Rice KC (1990) Sigma receptors: biology and function. Pharmacol Rev 42:355–402

Watson M, Roeske WR,Yamamura HI (1987) Cholinergic receptor heterogeneity. In: Meltzer HY (ed) Psychopharmacology: the third generation of progress. Raven, New York, pp 241–248

Waziri R (1988) Glycine therapy of schizophrenia. Biol Psychiatry 23:210–211

Weber E, Sonders M, Quarum M, Mclean S, Pou S, Keana JFW (1986) 1,3-Di(2-[5-3H]tolyl)guanidine: a selective ligand that labels sigma-type receptors for psychotomimetic opiates and antipsychotic drugs. Proc Natl Acad Sci USA 83:8784–8788

Whitaker PM, Crow TJ, Ferrer IN (1981) Tritiated LSD binding in frontal cortex in schizophrenia Arch Gen Psychiatry 38:278–280

White FJ, Wang RY (1986) Effects of tiaspirone (BMY-13859) and a chemical congener (BMY-13980) on A9 and A10 dopamine neurons in the rat. Neuropharmacology 25:995–1001

Widerlov E, Lindstrom LH, Besev G, Manberg P, Nemeroff CB, Breese GR, Kizer JS, Prange AJ (1982) Subnormal CSF levels of neurotensin in a subgroup of schizophrenic patients. Am J Psychiatry 139:1122–1126

Wiesel F-A, Nordström A-L, Farde L, Eriksson B (1994) An open clinical and biochemical study of ritanserin in acute patients with schizophrenia. Psychopharmacology 114:31–38

Woggon B, Heinreich K, Kufferele B, Miller-Oberlinghauser, Poldinger W, Ruther E, Scheid HW (1984) Results of a multicenter AMPD study with fluperlapine in schizophrenic patients. Arzneimittelforschung 34:122–124

Woggon B, Beckman H, Heinrich K, Linden M, Krebs E, Kufferle B, Pflug B, Ruther E, Scheid HY (1986) Clinically relevant differences in the therapeutic profile of fluperlapine compaired to haloperidol in the treatment of schizophrenic psychosis: results of a multicenter double-blind study. Pharmacopsychiatry 19:204–205

Wolkowitz OM, Pickar D (1991) Benzodiazepines in the treatment of schizophrenia: a review and reappraisal. Am J Psychiatry 148:714–726

Wolkowitz OM, Turetsky N, Reus VI, Hargreaves WA (1992) Benzodiazepine augmentation of neuroleptics in treatment-resistant schizophrenia. Psychopharmacol Bull 28:291–295

Wooley DW, Shaw E (1954) A biochemical and pharmacological suggestion about certain mental disorders. Proc Natl Acad Sci USA 40:228–231

Yamamoto B, Cooperman MA (1994) Differential effects of chronic antipsychotic drug treatment on extracellular glutamate and dopamine concentratons. J Neurosci 14 (in press)

Yamamoto BK, Meltzer HY (1992) The effect of the atypical antipsychotic drug amperozide, on carrier-mediated striatal dopmaine release measured in vivo. J Pharmacol Exp Ther 263:180–185

Yevich JP, New JS, Lobeck WG, Dextraze P, Bernstein E, Taylor DP, Yocca FD, Eison MS, Temple DL (1992) Synthesis and biological characterization of alpha-(4-fluorophenyl)-4-(5-fluoro-2-pyrimidinyl)-1-piperazinebutanol and analogues as potential atypical antipsychotic drugs. J Med Chem 35:4516–4525

Zukin SR, Javitt DC (1991) The brain NMDA receptor, psychotomimetic drug effects, and schizophrenia. In: Tasman A, Goldfinger SM (eds) Review of psychiatry. American Psychiatric, Washington, pp 480–498

Zukin SR, Tempel A, Gerdner EL, Zukin RS (1986) Interaction of [3H]-(−)SKF-10047 with brain sigma receptors: characterization and autoradiographic visualization. J Neurosci 46:1032–1041

CHAPTER 5

Sites and Mechanisms of Action of Antipsychotic Drugs as Revealed by Immediate-Early Gene Expression

ARIEL Y. DEUTCH

A. Introduction

In the forty-odd years since the introduction of chlorpromazine as an antipsychotic drug, dozens of drugs have been introduced as effective agents for the treatment of schizophrenia. These antipsychotic drugs (APDs), which have become the primary treatment for schizophrenia, have substantially reduced symptomatology and improved the quality of life for untold numbers of patients. However, APDs represent at best a moderately effective treatment for schizophrenia. They do not cure patients, and in virtually all patients there are residual deficits in function, particularly within the social domain and with respect to cognition. Moreover, substantial numbers of patients do not show any significant improvement, due to lack of treatment response or because unacceptable side effects of the APDs lead to discontinuation of treatment. This lack of therapeutic response or poor response in a relatively large percentage of patients has resulted in continued efforts to uncover new APDs that are effective in a broader group of patients, are more effective in treating certain symptom constellations, and have markedly reduced or absent side effects. Despite continued investigations into the mechanisms of action of APDs, the development of new "atypical" pharmacological treatments has been painfully slow.

Most hypotheses addressing the mechanisms of action of APDs have focused on central dopamine (DA) systems. The emphasis on DA systems is reasonable, since all the known antipsychotic drugs (defined on the basis of double-blind studies) are antagonists at the D_2 DA receptor (see Chap. 3 in this volume by CIVELLI), and since there is a very good correlation between affinity for the D_2 receptor and the doses of various APDs that are commonly used in clinical practice. Thus, despite the fact that there are no consistent changes in various indices of DA function in postmortem samples from schizophrenic patients (see GOLDSTEIN and DEUTCH 1992), a dopaminergic perspective dominates research into the mechanisms of action of APDs.

The focus on central DA systems in studies of schizophrenia and pharmacological approaches to the treatment of schizophrenia resulted in investigators initially studying those areas of the brain that receive a dense DA innervation. Early studies of DA systems were hampered by a relative lack of sensitivity in methods used to measure DA, thus leading to studies being restricted to the caudate nucleus and putamen, since the striatal DA innervation is the most dense of the forebrain DA projections. While studies of striatal DA function have been enormously influential, such studies have arguably been of greater significance in defining the role of DA in Parkinson's disease rather than schizophrenia. These studies do have relevance for attempts to understand the mechanisms of action of APDs, since virtually all APDs have parkinsonian (extrapyramidal) side effects.

The initial focus on neostriatal dopaminergic function in most studies of schizophrenia and APDs also seemed appropriate because there was a pervasive belief that the therapeutic properties of APDs might be inextricably linked to their side effects, i.e., that a drug lacking extrapyramidal side effects (EPS) could not be an effective antipsychotic agent. This contention was not overturned until the (delayed) acceptance of clozapine as an effective APD, when it became convincingly clear that the therapeutic actions of APDs could be dissociated from extrapyramidal side effects (EPS; see Chaps. 11–13 in this volume).

Over the quarter century since the DA hypothesis of schizophrenia was formulated, our knowledge of chemical neuroanatomy has expanded considerably. We now know that there are several specific DA innervations of both cortical regions and subcortical sites. These anatomical studies have been paralleled by the elucidation of multiple subtypes of DA receptors, and the realization that many APDs exhibit high affinities for certain nondopaminergic as well as dopaminergic receptors. These anatomical and pharmacological observations have led to the realization that multidisciplinary approaches permitting the definition of the receptors with which APDs interact and the delineation of the precise sites of action are probably the best strategy for defining the mechanisms of action of APDs. Such a melding of pharmacological and anatomical techniques also fits well with a growing disenchantment with the major emphasis of preclinical studies of APDs being directed to DA systems (see Chap. 4, this volume). There has been a gradual shift to a more extended view of dopamine systems that emphasizes both their afferent regulation and the ability of DA to regulate extended postsynaptic projection systems.

B. Immediate-Early Gene Expression as a Method to Assess the Sites and Mechanisms of Action of Antipsychotic Drugs (APDs)

One approach that permits the regionally-specific effects of drugs to be examined with both excellent anatomical and pharmacological resolution is to follow the induction of an immediate-early gene in response to a drug challenge. Immediate-early genes are transcriptional regulators that are rapidly induced and rapidly subside following some acute activation of a neuron. The reader is referred to recent reviews concerning the actions of immediate-early genes for a general overview of the subject matter (HE and ROSENFELD 1991; MORGAN and CURRAN 1991, 1995). The ability of various pharmacological treatments to modify immediate-early gene induction is now well established, and appears to reflect a metabolic activation of the neuron (DRAGUNOW and FAULL 1989; HUNT et al. 1987; SAGAR et al. 1988). c-*fos* expression is limited to neurons under physiological conditions, but after trauma Fos may be expressed in non-neuronal cells, including glia (MUGNAINI et al. 1989). In contrast, immunohistochemical studies of the protein encoded for by a related immediate-early gene, Fos B, revealed that this protein is not expressed in glia and ependymal cells after nonspecific damage such as ischemia (DRAGUNOW 1990). It is important to note that although the induction of Fos frequently correlates with neuronal depolarization, there are situations in which c-*fos* is induced in neurons that are not depolarized (BULLITT 1990). Thus, immediate-early gene induction appears to be a convenient marker for neurons that are metabolically activated, rather than an invariant marker of neurons that are depolarized. There are several methodological concerns that must be met for studies of immediate-early gene induction to be meaningful; these have been detailed in several publications (BULLITT 1990; DEUTCH and DUMAN 1995; DEUTCH et al. 1991a; DRAGUNOW and FAULL 1989).

Assessment of immediate-early gene expression is now an established method for determining the sites of action of both drug and environmental challenges. In particular, the use of immunohistochemical or in situ hybridization histochemical methods permits the detection of single cells responding to a challenge, while immunoblot or Northern blot analyses are readily applicable to studies of the pharmacological substrates of a given treatment. In addition, immediate-early genes are transcriptional regulators, and as such may offer important clues to the long-term modifications induced by APDs that occur through changes in gene expression.

Most studies of immediate-early gene induction in response to APDs have focused on expression of c-*fos* or its protein product Fos; relatively few studies have examined other members of the same or different immediate-early gene families. Our review of the effects of APDs on immediate-early gene expression will therefore by necessity be narrowly focused on the changes in brain function revealed by expression of Fos. Another factor has further narrowed our review: there are very few studies of the effects of APDs in areas other than the basal ganglia and allied cortical and limbic areas. As with all reviews, this one will be rapidly outdated. We can only hope that this review becomes outdated because the mechanisms of action of typical and atypical APDs are conclusively determined.

C. Effects of APDs on Immediate-Early Gene Induction in the Striatal Complex

The striatal complex was the first area in which immediate-early responses to APDs were examined. There is complete agreement between investigators that acute administration of virtually all D_2 DA antagonists, including APDs, markedly increases c-*fos* mRNA levels or levels of Fos protein in the striatal complex of the rat in a regionally-specific manner. Typical APDs, but not clozapine, induce expression in the dorsal striatum (caudatoputamen), whereas all APDs (including clozapine) increase Fos expression in the shell compartment of the nucleus accumbens. D_2 antagonists also increase the expression of several other immediate-early genes. Despite these consistent findings, the precise mechanisms through which APDs increase immediate-early gene expression remain unclear.

I. Effects of APDs on Regionally Specific Striatal Immediate-Early Gene Expression

Early studies of the ability of D_2 antagonists to elicit immediate-early gene expression focused on the striatal complex. These studies revealed that acute administration of D_2 antagonists increases expression of c-*fos* mRNA and its protein product Fos in a temporally- and regionally-specific manner. D_2 antagonists typically increased Fos throughout the striatal complex, involving both dorsal striatal (caudatoputamen) and ventral striatal (nucleus accumbens, islands of Calleja) sites. As more attention focused on the effects of D_2 antagonists on Fos expression in the striatum, several different APDs with different clinical profiles were tested. These studies revealed that D_2 antagonists with extrapyramidal side effect (EPS) liability strongly induced Fos expression in neurons of the dorsal striatum; clozapine and several putative atypical APDs either did not induce Fos in the dorsal striatum or did so only upon administration of high doses. In contrast, all clinically effective APDs increased Fos expression in the shell of the nucleus accumbens (NAS).

1. Dorsal Striatum

The APD-elicited increase in expression of c-*fos* mRNA or Fos protein in the dorsal striatum is not homogeneous. Thus, administration of APDs results in a more marked increase in the numbers of neurons in which the c-*fos* transcript can be detected or in which Fos-like immunoreactivity (-li) is observed in the lateral caudatoputamen (CP). There is a somewhat less marked effect of APDs on the numbers of neurons expressing Fos in the medial striatum. Interestingly, APDs do not markedly enhance the numbers of neurons expressing Fos or the c-*fos* transcript in the central aspects of the CP. This regional pattern of striatal Fos induction in response to haloperidol challenge has been repeatedly reported (Deutch et al. 1992; Dilts et al. 1993;

MacGibbon et al. 1994; Merchant et al. 1994; Nguyen et al. 1992; Robertson and Fibiger 1992), although one report suggests that moderate doses of haloperidol result in a homogeneous distribution of Fos-li cells (Dragunow et al. 1990).

The pattern of striatal Fos expression noted above is seen after administration of virtually all APDs that have been examined. The major exception is the atypical APD clozapine, administration of which does not significantly induce Fos in the CP, at least at doses other than those approaching lethality (Deutch et al. 1992; MacGibbon et al. 1994; Robertson et al. 1994a; Sebens et al. 1995). Other compounds that are putative atypical APDs, including remoxipride, induce Fos expression in the CP after high but not low doses (Deutch et al. 1992). Still other putative atypical APDs, such as risperidone, do not induce Fos in the dorsolateral CP at relatively low doses (e.g., 0.25 mg/kg) but show a steep dose–response curve, with a significant induction at doses as low as 1.25 mg/kg (personal observations).

The pattern of Fos induction noted above is seen in the precommissural CP in adult rodents treated with moderate-to-high doses of most APDs (Deutch et al. 1992; Dilts et al. 1993; MacGibbon et al. 1994; Merchant and Dorsa 1994). However, very low doses of APDs elicit a more widespread effect throughout the striatum. For example, acute administration of low doses of haloperidol (0.25 mg/kg) have been reported to increase the numbers of neurons expressing Fos-li across the mediolateral extent of the CP (Elibol et al. 1994) of the precommissural striatum. Higher doses (1.0–5.0 mg/kg) appear to increase Fos in the more restricted (dorsolateral CP > medial CP > central CP) pattern. In addition to the mediolateral differences in haloperidol-elicited Fos expression in the CP, there also appears to be a rostrocaudal gradient, with more rostral CP areas more strongly impacted by APD challenge (MacGibbon et al. 1994; Robertson and Fibiger 1992).

The compartmental organization of the striatum is well known, with histochemically distinct compartments of the CP showing difference in both afferent and efferent projections as well as the density of neurochemically distinct intrinsic neurons (Fuxe et al. 1979; Gerfen 1992; Graybiel 1984). Several authors have commented that administration of haloperidol leads to a somewhat patchy appearance of Fos-li neurons in the CP (Dilts et al. 1993; MacGibbon et al. 1994; Nguyen et al. 1992). However, the "patchiness" of the distribution of Fos-li neurons seen after APD challenge does not correspond to either patch (striosomal) or matrix (diffuse) compartments of the striatum: haloperidol increases the numbers of Fos-li neurons in both the patch and matrix compartments of the CP (Deutch et al. 1992).

The pattern of dorsal striatal Fos expression observed in response to moderate doses of haloperidol has been extensively characterized in the rat. There are few data concerning the effects of APDs on immediate-early gene expression in other species. Among these are the recent finding that acute haloperidol administration results in a pattern of Fos expression in a non-human primate that is comparable to that observed in the rat, i.e., a very strong induction of the immediate-early gene in the dorsolateral putamen and caudate nucleus, with a less marked effect in the medial (periventricular) caudate nucleus and ventromedial putamen (Deutch et al. 1995a).

2. Ventral Striatum

Typical APDs increase the numbers of Fos-li neurons or neurons expressing c-*fos* mRNA in the dorsolateral CP; clozapine differs sharply from other APDs by not significantly increasing Fos expression in the dorsolateral CP. However, all APDs that have been examined increase Fos expression in the nucleus accumbens (NAS), specifically in the shell compartment.

Over the past 5 years it has become clear that there are distinct compartments in the NAS that can be differentiated on the basis of efferent projections and afferent projections, pharmacology, and function (Brog et al. 1993; Deutch and Cameron 1992; Deutch et al. 1993; Heimer et al. 1991; Zahm and Brog 1992; Zahm and Heimer 1990). Current classification indicates the presence of several NAS compartments, including the shell, core, septal pole, and rostral pole regions (see Brog et al. 1993; Deutch et al. 1993).

The functional attributes of these compartments appears to reflect in large part the underlying connectivity of the compartments. For example, the core of the NAS projects to classical extrapyramidal sites, and the functional attributes of this accumbal compartment mirror to a considerable degree those of the contiguous CP; the shell of the NAS has connections with different sites, and the functional attributes of this compartment are more "limbic"-like (see Deutch et al. 1993).

The effects of haloperidol and raclopride are seen throughout the NAS. These typical APDs result in an increase in the number of Fos-li in the shell compartment, but they particularly target the more dorsal aspects of the shell; typical APDs such as haloperidol also strongly increase Fos expression in the NAS core (Deutch et al. 1992; Nguyen et al. 1992; Robertson and Fibiger 1992). Clozapine results in a different pattern of Fos induction in the NAS: Fos-li neurons are more diffusely distributed across the dorsoventral domain of the shell, and the number of Fos-li cells is markedly increased in the septal pole (Deutch et al. 1992; Robertson and Fibiger 1992). However, clozapine differs most strikingly from typical APDs by not increasing the numbers of Fos-li neurons in the core compartment: Fos-li cells are observed at the border of the shell and core (the so-called shore), but there is not a significant increase in the number of Fos-li cells in the core relative to vehicle-treated animals (Deutch et al. 1992). Within the NAS, there is also the rostrocaudal gradient of Fos expression reported for the more dorsal CP (MacGibbon et al. 1994; Robertson and Fibiger 1992; Robertson et al. 1994a).

Clozapine differs from typical APDs by substantially increasing the numbers of Fos-li neurons in the islands of Calleja, whereas typical APDs less markedly (albeit significantly) enhance Fos expression in the islands of Calleja (MacGibbon et al. 1994). However, the number of Fos-li cells in the olfactory tubercle is marginally increased by both typical APDs and the atypical APD clozapine (Dilts et al. 1994; Robertson and Fibiger 1992).

II. Striatal Immediate-Early Gene Induction: Fos, Fos-Related Antigens, and Others

As noted above, most studies of the effects of APDs on striatal immediate-early gene expression have focused on c-*fos*. Relatively few have focused on a group of related proteins that share a conserved peptide sequence, the Fos-related antigens (Fras), and still fewer have addressed the issue of APD-elicited changes in other immediate-early gene family members. The initial focus on c-*fos* stemmed from the early studies promulgating expression of Fos as a marker of cells activated by a particular treatment (Dragunow and Faull 1989; Hunt et al. 1987; Sagar et al. 1988). As will be evident, the early decision to focus on c-*fos* was fortuitous, allowing clear discernment of the differences between typical and atypical APDs in the ability to induce the expression of immediate-early genes in a regionally-specific manner.

1. Fos Versus Fras

The ease and relative rapidity with which immediate-early gene expression can be assessed using immunohistochemical methods has led to this approach being the primary method for analysis of immediate-early gene expression in response to various challenges, including APDs. However, all immunohistochemical methods are critically dependent on the specificity of antisera. Many, if not most, antisera that are used in studies of Fos expression are generated against the M-peptide, a sequence common to Fos and Fras, which has proven to be a useful immunogen and allows the use of these antibodies in a wide variety of species. However, the use of antibodies that recognized Fos and Fras has led to some difficulties. Primary among these is the fact that the ability to determine a specific effect is hampered by the difference in the time courses of Fos and Fra expression. Fos is rapidly induced and rapidly subsides, whereas Fra proteins show a relatively long duration of expression following a single challenge (Dilts et al. 1993; Dragunow et al. 1990; MacGibbon et al. 1994).

The difference in time courses between Fos and Fras has clouded the issue of clozapine effects on the lateral CP. In situ hybridization and Northern blot studies have revealed that clozapine does not increase c-*fos* expression in the lateral CP. However, there are no comparable data for various Fra transcripts; in fact, many of the Fras have yet to be cloned. Immunohistochemical studies by MacGibbon et al. (1994) have suggested that clozapine induces Fras but not Fos in the CP. However, immunohistochemical studies with antibodies that recognize both Fos and Fras have not detected an increase in the number of striatal immunoreactive neurons in response to clozapine challenge, using comparable or higher doses of clozapine in animals killed at the same or longer posttreatment interval (Deutch et al. 1992; Robertson and Fibiger 1992). Studies examining specific Fra mRNAs will be required to resolve the different conclusions reached concerning dorsal striatal Fra expression in response to APDs.

Immunohistochemical studies of the ventral striatum have uniformly indicated that APDs increase expression of Fos and Fras in the NAS shell (Deutch et al. 1992; Dilts et al. 1993; MacGibbon et al. 1994; Nguyen et al. 1992; Robertson and Fibiger 1992). Strikingly, Merchant and Dorsa (1993) found that while haloperidol markedly increased expression of c-*fos* mRNA in the dorsolateral striatum, there was little induction in the nucleus accumbens. We have observed a weak but significant effect of haloperidol relative to vehicle injections on c-*fos* mRNA in the NAS, particularly the septal pole region and shell of the NAS (unpublished observations); administration of vehicle also increased c-*fos* expression relative to noninjected controls. MacGibbon et al. (1994) state that they also found an increase in c-*fos* mRNA in the NAS. The ability of vehicle injections to induce c-*fos* expression presumably results in the inability to demonstrate a strong APD-elicited induction of c-*fos* mRNA in the NAS, although a small but significant effect can be realized. While there is a significant effect of haloperidol and clozapine on c-*fos* mRNA levels in the NAS, immunohistochemical studies (using both Fos-selective antibodies and antibodies that recognize both Fos and Fras) (Dilts et al. 1993), and immunoblot studies to reveal Fos and Fras (personal observations) indicate that the striking APD-elicited increase in the numbers of Fos-li neurons in the nucleus accumbens primarily reflects induction of several Fras.

The importance of defining the precise species of proteins that are expressed lies in the idea that varying combinations of Fos or Fras may differentially regulate transcription of different genes. The transcriptional activity of the Fos and Jun families of immediate-early genes depends on the formation of dimers (Vogt and Bos 1990). Members of the Fos family, including Fos and various Fras, can form heteromers with Jun; the affinity with which these heteromers bind to DNA, and thereby regulate transcription of target genes, appears to depend on the protein combination (Morgan and Curran 1991; Nakabeppu et al. 1988; Smeal et al. 1989; Zerial et al. 1989). In addition, some immediate-early gene proteins can form heteromers with other DNA binding proteins. A recent paper by Hope et al. (1994), in a study examining the relative specificity of acute and chronic electroconvulsive shock for Fos and Fras, found data consistent with the concept of differential binding of the AP-1 complex for promotor sequences dependent on its dimer constituents: the different composition of the chronic AP-1 complex led to a more avid binding to the calcium response element than the acute complex.

2. Other Immediate-Early Genes

In a few cases investigators have examined the responses of immediate-early genes other than c-*fos* to APD challenges. These include other members of the Fos family (e.g., Fos B), members of the Jun family (including c-*jun* and Jun B), and such immediate early genes as zif268 (also known as Krox24, NGFI-A, TIS8, and ERG-1). The specific immediate-early genes induced by APDs are of importance since different combinations of the gene products differentially regulate transcription (CURRAN and MORGAN 1991; NAKABEPPU et al. 1988; SMEAL et al. 1989; ZERIAL et al. 1989).

The effects of haloperidol and clozapine on immediate-early gene expression in the lateral CP and NAS are shown in Table 1. What is apparent is that among the immediate-early genes examined to date, none has shown the pattern of c-*fos*, with induction in the CP by typical APDs and NAS by all APDs.

It is reasonable to question the significance of the observations of regionally selective immediate-early gene expression if only one of these genes is induced in a manner that fits with current notions of the actions of APDs. The importance of the c-*fos* data resides in the target gene that is being regulated

Table 1. Effects of haloperidol and clozapine on immediate-early gene expression in the lateral CP and NAS

	CP		NAS	
	HPD	HPD + CLZ	HPD	HPD + CLZ
c-*Fos*	+++			+++
Fras		+++/---		+++/---
Fos B	+++		+++	
c-*Jun*		---		---
Jun B		+++	+++	
Jun D		---		---
Zif268		+++		+++

CP, caudatoputamen; NAS, nucleus accumbens.
Data are derived from reports of DEUTCH et al. (1992), DILTS et al. (1993), MACGIBBON et al. (1994), NGUYEN et al. (1992), ROGUE and VINCENDON (1992). There is broad agreement between investigators concerning the effects of APDs on immediate-early genes. An exception is the effects of haloperidol on c-*jun*: ROGUE and VINCENDON (1992) reported that haloperidol increased c-*jun* mRNA in the striatum, whereas MACGIBBON et al. (1994) did not observe an increase in either c-*jun* mRNA or protein in the lateral striatum. The other discrepancy is the induction of Fras in the lateral striatum by clozapine (see text); accordingly, we have marked the appropriate spaces as +++/--- to reflect the disagreement.

by c-*fos*. Identification of the targets of APD-elicited c-*fos* actiation will open new avenues to the mechanisms underlying the therapeutic actions of APDs.

III. Mechanisms of Antipsychotic Drug-Elicited Striatal Fos Expression

Although the effects of APDs on striatal immediate-early gene expression have now been extensively investigated, the receptor mechanisms underlying the ability of APDs to enhance Fos expression remain unclear. A number of receptors have been suggested to play a role in the APD-induced expression of Fos in striatal neurons, including DA, excitatory amino acid, muscarinic cholinergic, and adenosine receptors. It appears that D_2 receptors play a critical role, but the degree to which this reflects actions on the postsynaptic neuron that expresses Fos or a presynaptic receptor to evoke DA release have not been conclusively established. It is probable that there is a complex interplay of receptor mechanisms involved, with a number of necessary but not sufficient conditions to induce Fos expression in striatal DA neurons. Moreover, many of the receptor antagonists that modify APD-elicited Fos expression appear to do so by impacting on postreceptor transduction mechanisms that depend on the primary drive eliciting Fos expression.

1. Dopamine Receptors and APD-Elicited Increases in Striatal Fos

The observation that acute administration of different APDs induces Fos in the striatal complex suggests that interaction with the DA D_2 receptor is a critical factor. This suggestion is bolstered by the observation that relatively selective D_2 antagonists, such as raclopride and sulpiride (Robertson and Fibiger 1992; Robertson et al. 1994a), increase the number of Fos-li neurons in the lateral striatum. However, several findings indicate that the mechanisms subserving APD-induced increases in striatal Fos expression are not as clear as expected.

a) D_2 Dopamine Receptors

The importance of the D_2 autoreceptor in the APD-elicited increase in striatal Fos expression is illustrated by recent findings with inbred mouse strains that differ in cataleptic response to haloperidol. Hitzemann and colleagues have generated mice that differ markedly in the cataleptic response to acute haloperidol challenge (Kanes et al. 1993; Patel et al. 1995; Qian et al. 1993). The mice that show a high cataleptic response have a markedly increased density of D_2 receptors in the substantia nigra (Aq DA cell body region), with a significant (20%–30%) increase in number in the lateral and caudal striatum. In contrast, there is no difference in striatal D_1 receptor density across the two lines of mice (Kanes et al. 1993; Qian et al. 1993). Relative to those mice that show little cataleptic response to haloperidol, the line of mice that responds strongly to haloperidol (and which exhibits an elevated D_2 receptor density) shows a dramatic increase in the number of Fos-li cells in the lateral striatum (but not nucleus accumbens) after haloperidol challenge (PATEL et al. 1995). These observations indicate that the increased number of D_2 receptors enhances the striatal Fos response to APD challenge, and suggests that DA autoreceptors may be the primary determinant. Also consistent with this proposal is the observation that pretreatment of rats with D_2 receptor agonists attenuates the Fos response to haloperidol.

Accordingly, the increase in Fos expression elicited by APDs should be due to release of DA which in turn interacts with postsynaptic DA receptors. Consistent with this speculation is the observation that relatively high doses of the indirect DA agonists cocaine and amphetamine increase striatal Fos expression (DAUNAIS and MCGINTY 1994; GRAYBIEL et al. 1993; YOUNG et al. 1991). Cocaine and amphetamine elevate extracellular DA levels by blocking the DA transporter, but also increase levels of norepinephrine and serotonin. This raises the possibility that the mechanism subserving amphetamine- or cocaine-induced striatal Fos expression may differ from that underlying the Fos response to APDs. Moreover, the pattern of striatal Fos expression observed in response to psychostimulants differs between different psychostimulants as well as from that seen after APD challenge (JOHANSSON et al. 1994). However, the specific DA reuptake blocker GBR-12909 also increases Fos expression in the striatum (IADAROLA et al. 1993), suggesting that increased extracellular DA levels represent the proximate cause of striatal Fos induction.

There have been several studies aimed at determining the specific postsynaptic DA receptor that is critical for APD-elicited Fos induction. Early observations indicated that pretreatment of animals with D_2-like agonists prevents haloperidol-elicited striatal Fos induction (MILLER 1990; ROGUE and VINCENDON 1992), thereby suggesting that D_2 receptors on striatal medium spiny neurons are the critical determinant. However, several studies have reported that acute administration of D_2 agonists such as quinpirole and quinelorane, which target $DA_{2/3/4}$ receptors, do not increase Fos expression (LAHOSTE et al. 1993; MILLER 1990; ROGUE and VINCENDON 1992; WIRTSCHAFTER and ASIN 1994). In contrast to these findings, DRAGUNOW et al. (1990) reported that the D_2-like agonist YM 09151-2 (emonapride) increases expression of Fos and Fras in the striatum. The reason for this discrepancy is not clear, but may involve the moderate affinity that YM 09151-2 displays for the 5-HT_{1a} receptor, since serotonergic agonsits increase striatal DA levels (BENLOUCIF and GALLOWAY 1991; BENLOUCIF et al. 1993; RASMUSSEN et al. 1994).

b) D_1 Dopamine Receptors

The involvement of the D_1 receptor in APD-evoked Fos expression is unclear. Several reports have indicated that acute administration of the D_1 agonist SKF

38393 does not evoke Fos expression in the striatum of the normal rat (COLE et al. 1992; H.A. ROBERTSON et al. 1989, 1991; G.S. ROBERTSON et al. 1992). There has been a report in mice that SKF 38393 does induce c-*fos* in the striatum, but only in its periventricular and caudal regions (ARNAULD et al. 1993). This observation emphasizes the heterogeneity of the striatum and the striatal response to drug challenges. However, the pattern of c-*fos* mRNA expressing cells in the study of ARNAULD and colleagues (1993) is clearly different from that elicited by neuroleptic challenge.

The differences between the effects of D_1 agonists on striatal Fos expression and those of neuroleptics may be due to the fact that SKF 38393 is a partial agonist at the D_1 site. WIRTSCHAFTER and ASIN (1994) have recently demonstrated that the full D_1 agonist A-77636 increases Fos expression in the striatum. However, the pattern of Fos expression in the striatum in response to A-77636 differed from that observed with neuroleptic drugs, with induction of Fos in the medial but not lateral striatum. This finding roughly parallels the data of ARNAULD et al. (1993), who indicated that SKF 38393 increased Fos in the medial periventricular striatum.

In addition to the differences in the intra-striatal pattern of Fos expression observed with D_1 agonists, the observation that pretreatment of rats with the D_1 antagonist SCH-23390 does not prevent the haloperidol-elicited increase in the numbers of striatal Fos-li neurons (DRAGUNOW et al. 1990) argues against a crucial involvement of the D_1 receptor in subserving the Fos response to APD challenge.

c) Concurrent D_2/D_1 Receptor Occupancy and Striatal Fos Expression

Available data suggest that neither the D_2 nor D_1 postsynaptic receptor alone is critical for the APD-elicited Fos response in striatal neurons. It is therefore likely that occupancy of both receptors is necessary. This speculation is bolstered by the observation that high (5.0 mg/kg) but not low (0.25 mg/kg) doses of the mixed D_2–D_1 agonist apomorphine increases striatal Fos expression (DILTS et al. 1993; MANDEL et al. 1992); the fact that apomorphine exhibits a much higher affinity for the D_2 receptor than the D_1 site may explain the high doses of apomorphine required to elicit striatal Fos expression.

The hypothesis that concurrent D_2 and D_1 occupancy is required to elicit striatal Fos expression is consistent with several reports. LAHOSTE et al. (1993) directly addressed this possiblity by examining the effects of administration of D_2 and D_1 agonists alone or concurrently, and concluded that striatal Fos expression reflects synergism at D_1 and D_2 sites. Consistent with this speculation are the data of WIRTSCHAFTER and ASIN (1994), who reported that doses of quinpirole that failed to alter Fos expression significantly potentiated the effects of the D_1 agonist A-77636. Interestingly, WIRTSCHAFTER and ASIN reported that coadministration of quinpirole with A-7736 resulted in Fos expression in the lateral striatum, while A-7736 alone elicited Fos expression in the medial striatum. The difference in regional effects of the D_1 agonist and the combined D_1–D_2 treatment again suggests that neither D_1 nor D_2 agonists alone are sufficient to mimic the effects of acute neuroleptics. Thus, the best explanation for the postsynaptic actions mediating APD-elicited Fos expression may be concurrent D_2 and D_1 occupancy, with neither alone being a sufficient condition in the intact animal.

It is important to recognize that this statement applies to the intact animal, and probably is not applicable to other situations. Animals with 6-OHDA lesions of the striatal DA innervation exhibit an increased number of striatal Fos-li neurons in response to acute neuroleptic challenge (Beretta et al. 1992; Jian et al. 1993; Robertson and Staines 1994; Robertson et al. 1989). This has been a fruitful area of research, particularly for studies aimed at defining the phenotype of striatal neurons that express Fos in response to APD challenge. However, the many compensatory responses that occur in response to striatal DA denervation confound clear interpretation of the mechanisms subserving the striatal Fos response to acute APD challenge. Moreover, chronic treatment with APDs have been reported to alter the response to acute APD challenge (see following), and hence the generalizability of the data derived from studies of DA-denervated rats is not clear.

2. Involvement of Excitatory Amino Acid Receptors in Neuroleptic-Elicited Striatal Fos Expression

The role of glutamatergic-dopaminergic interactions in regulation of striatal function has been a topic of considerable interest in basic and clinical neuroscience (see Deutch 1992, 1993). Glutamate has been reported to evoke DA release in the striatum through an impulse-independent mechanism (Chesselet 1984; see Deutch 1992), and APDs have recently been shown to regulate expression of different glutamate receptor subunits (Fitzgerald et al. 1995).

The haloperidol-elicited increase in striatal Fos expression is markedly attenuated by pretreatment with high doses of the noncompetitive *N*-methyl-D-aspartate (NMDA) receptor antagonist MK-801 (Dragunow et al. 1990; Ziolowska and Hollt 1993). The involvement of an NMDA receptor is consistent with an emerging body of data indicating that glutamate release evoked from corticostriatal neurons can increase Fos expression in striatal neurons (Cenci and Bjorklund 1993). Thus, stimulation of the corticostriatal neurons increases expression of Fos and JunB in striatal medium spiny neurons (Beretta et al. 1994; Fu and Beckstead 1994; Parthasarathy et al. 1994), operating through an NMDA receptor (Aronin et al. 1991).

Although the ability of glutamate or related excitatory amino acids to evoke DA release offers a possible mechanism to account for the effects of excitatory amino acids on Fos expression, several findings suggest that the effects of cortical stimulation on Fos expression do not directly involve altered extracellular DA levels. The ability of glutamate to evoke DA release in the striatum has been called into question, with the suggestion made that DA stimulates DA synthesis but not release in the caudatoputamen (Keefe et al. 1993; see Deutch 1993). More directly, lesions of the striatal DA innervation do not attenuate the effects of electrical stimulation of the cortex on striatal Fos expression (Fu and Beckstead 1994). Thus, the Fos induction attributable to activation of corticostriatal neurons does not necessarily involve augmented DA release.

Conversely, pharmacological data argue against neuroleptics evoking glutamate release and thereby increasing Fos expression. Acute haloperidol administration does not increase levels of glutamate or aspartate in the striatum (Daly and Moghaddam 1993), although chronic haloperidol does (Yamamoto and Cooperman 1994). Similarly, neither D_2 nor D_1 DA agonists alter basal levels of glutamate in the striatum (Yamamoto and Davy 1992).

These data suggest that although the NMDA receptor antagonist MK-801 blocks the neuroleptic-elicited increase in striatal Fos expression, it is unlikely that haloperidol and other neuroleptic drugs induce Fos through a change in striatal excitatory amino acid levels. Instead, these data suggest that changes in NMDA receptor occupancy on medium spiny neurons alter the effects of dopamine though changes in postreceptor transduction interactions.

3. Cholinergic and Adenosine Receptors

a) Muscarinic Cholinergic Receptors

The regulation of striatal medium spiny neurons by both DA afferents and striatal cholinergic neurons has long been exploited as a means of treamtent of Parkinson's disease: anticholinergic drugs have been a mainstay of the treatment of Parkinson's disease early in the course of the disease. Accordingly, it might be expected that manipulation of muscarinic cholinergic receptors would influence neuroleptic-evoked Fos expression in the striatum.

Guo et al. (1992) reported that pretreatment of rats with high doses of scopolamine substantially reduced haloperidol-elicited Fos induction in the striatum, but not in the nucleus accumbens. However, Bernard et al. (1993) have recently indicated that scopolamine itself induces Fos expression in the striatum. Unfortunately, Guo and colleagues did not report any data concerning the effects of scopolamine alone. The suggestion that scopolamine alone increases Fos but attenuates haloperidol-elicited Fos raises the interesting possibility that the muscarinic cholinergic antagonist and haloperidol together operate through different transduction mechanisms to counteract the individual positive effects of each drug individually.

Available data suggest that the ability of scopolamine to block the Fos response to haloperidol is not due to regulation of DA release in the striatum. Meltzer et al. (1994) have reported that scopolamine, although without intrinsic activity, attenuates the clozapine-elicited increase in striatal DA levels, but does not attenuate haloperidol-induced increases in striatal extracellular

DA levels. These data suggest that the ability of scopolamine to attenuate neuroleptic-elicited Fos expression in the striatum occurs at the level of the medium spiny neuron, and thus involves the transduction mechanisms subserving transmitter-evoked Fos expression.

b) *Adenosine A_2 Receptors*

Adenosine A_2 receptors are present on striatal medium spiny neurons that express the D_2 DA receptor (FINK et al. 1992). The colocalization of these receptors suggests that adenosine ligands may alter the striatal Fos response to neuroleptic challenge. This is the case: adenosine A_2 antagonists reduce the number of Fos-li striatal neurons by about 50% (BOEGMAN and VINCENT 1995). The finding that adenosine blocks the effects of haloperidol again suggests that key transduction mechanisms, rather than alterations in DA release, subserve the ability of several antagonists, including glutamatergic cholinergic, and adenosine receptor agents, to block neuroleptic-elicited striatal Fos response.

IV. Acute Versus Chronic Effects of APDs on Striatal Fos Expression

Most studies of the effects of APDs on striatal Fos expression have used acute administration paradigms. There have been very few studies assessing the effects of long-term administration of APDs.

There are two major reasons for the lack of attention devoted to studies of long-term treatment with APDs and immediate-early gene expression. The first is that there is an increasing awareness that although antipsychotic effects are not *fully* manifested until several weeks of treatment have elapsed, some degree of specific antipsychotic effects is realized rapidly. For example, STERN et al. (1993) reported that in 72% of cases one could correctly classify overall treatment outcome by the third day of APD administration, using changes in BPRS scores as an index of improvement. In addition to the decrease in total BPRS, subscales measuring psychosis, tension, and anergia were decreased. Approximately half of the eventual improvement was observed by day 3. STERN and colleagues suggest that the long prestudy hospitalization and drug-free period in this study argue against milieu determining response, and that the lack of response of the hostility/suspiciousness subscale argues against sedative properties of APDs contributing to observed effects. Thus, there is a rapid improvement that occurs on initiation of APD therapy; the delay until maximal effects are reached may simply reflect a lag period between the initiation of a long-term change in certain neurons, marked by an acute Fos response, and its peak effect.

Perhaps more important is the fact that Fos is a transcriptional regulator that is rapidly induced and rapidly subsides. The significance of Fos is that this and other immediate-early genes cooperatively regulate transcription of several target genes that have appropriate upstream response elements. Thus, the decision to study the effects of acute APD adminstration is appropriate since the IEG signifies a long-lasting genomic change in a neuron.

Nonetheless, there are compelling reasons to undertake studies of the chronic effects of APDs on immediate-early gene expression. A major reason is that the protein products of immediate-early genes autoregulate their transcription (Schonthal et al. 1989). For example, c-Fos and Fos B repress the c-*fos* promoter (Nakabeppu and Nathans 1991; Sassone-Corsi et al. 1988; Schonthal et al. 1989), but a truncated form of Fos B, termed Δ Fos B, can inhibit this repression and Fos/Jun and Jun/Jun transcriptional activity (Nakabeppu and Nathans 1991). Recently, Robertson et al. (1994b) suggested that the long-lasting increase in the number of Fos-li neurons observed in response to dopaminergic denervation (Dragunow et al. 1991) is primarily attributable to the enhanced expression of Δ Fos B. Since data are accruing that suggest that different functional AP-1 complexes probably dictate different regulation of target genes, it will be important to determine the degree to which Fos, Fras, and other immediate-early genes are differentially regulated by chronic APDs. In particular, such studies in the striatum may shed light on the mechanisms involved in tardive dyskinesia, a condition that emerges only after chronic treatment with APDs.

There have been few reports that document effects of chronic treatment with APDs on expression of immediate-early genes. Surprisingly, Miller (1990) reported that acute haloperidol challenge in animals treated with haloperidol for 12 days prior to testing resulted in an increase in c-*fos* mRNA in the stiatum. However, Sebens et al. (1995) have recently reported that in many but not all forebrain regions, chronic administration of haloperidol and clozapine results in a decrease in the Fos response to acute APD challenge. In rats receiving chronic treatment with haloperidol, the regions that showed a decreased number of Fos-li neurons in response to acute haloperidol challenge included the lateral striatum, nucleus accumbens, and central nucleus of the amygdala. In contrast, there was no apparent tolerance to haloperidol in the medial striatum and lateral septum. Conversely, there was a significant tolerance to chronic clozapine treatment in the lateral septum. The use of a "pan" Fos antibody in the studies of Sebens and associates (1995) precludes further analysis of the precise immediate-early genes that may be differentially regulated by APDs.

V. What Is the Transcriptional Target of APD-Elicited Striatal Fos Expression?

Studies of immediate-early gene responses to APD challenges have provided very useful information concerning the sites of action of various APDs, and also offered insights into the receptor mechanisms subserving the actions of APDs. The next set of questions to be addressed is equally as important: what target genes are being regulated by immediate-early genes such as c-*fos* that are induced by APDs?

Unfortunately, there is no direct method for extrapolating from the immediate-early gene to the target gene. If the target gene has a consensus AP-1 site, it is a potential target for c-*fos*. Unfortunately, this does not narrow the field of potential targets significantly. And one cannot deduce from the composition of the AP-1 complex the identity of a target gene. Accordingly, we are now limited to making educated guesses and then following these guesses with critical experiments to ascertain if a specific target gene is being regulated.

1. Neurotensin

The tridecapeptide neurotensin (NT) is colocalized with DA in A10 dopaminergic neurons (HOKFELT et al. 1984; DEUTCH and BEAN 1995). In addition, a large body of data indicates that changes in dopaminergic transmission can alter NT release and synthesis; the regulation of NT by DA occurs in the striatal complex as well as other forebrain regions (DEUTCH and ZAHM 1992; NEMEROFF and CAIN 1985).

Although NT-li axon terminals are abundant in the striatum, immunohistochemical studies have revealed only rare NT-li perikarya (JENNES et al. 1982; UHL et al. 1977). However, acute reserpine administration and administration of DA receptor antagonists, including raclopride and other D_2 selective antagonists, markedly elevate striatal NT concentrations and the number of demonstrable NT-li perikarya in the caudatoputamen and ventral striatum (BEAN et al. 1989; EGGERMAN and ZAHM 1988; FREY et al. 1986). The ability to detect NT-li perikarya in the striatum appears to reflect two different processes: (1) a DA-evoked decrease in the release of NT from striatopallidal neurons and the accumulation of peptide in the soma and (2) an increase in neurotensin/neuromedin N (NT/M) transcription and synthesis (see DEUTCH and ZAHM 1992); the two mechanisms appear largely (but not exclusively) to affect different populations of striatal NT neurons (DEUTCH and ZAHM 1992).

The ability of neuroleptics to increase markedly the number of NT-li neurons and NT concentrations in the striatum is seen after administration of most but not all APDs: clozapine does not significantly increase NT concentrations in the lateral striatum (FREY et al. 1986). In situ hybridization studies have revealed that administration of APDs markedly increases expression of NT/N mRNA in the dorsolateral striatum and shell of the nucleus accumbens, the areas in which APDs markedly enhance Fos expression (AUGOOD et al. 1991; MERCHANT et al. 1992, 1994; MERCHANT and DORSA 1993). These observations suggest that Fos may be a transcriptional regulator of the NT/N gene, a speculation consistent with the presence of a consensus AP-1 site in the upstream promoter region of the NT/N gene (DOBNER et al. 1992).

More recent anatomical studies have suggested that c-*fos* is expressed in some but not all NT cells in the striatum after acute haloperidol challenge. Merchant and Miller (1994) reported that in haloperidol-treated rats, c-*fos* mRNA was colocalized with NT in about 75% of the NT cells in the dorsolateral striatum. Complementary studies showed that injections of c-*fos* antisense oligonucleotides into the striatum reduced by about one-half the number of cells expressing NT/N mRNA in the dorsolateral striatum, although the magnitude of the increase in preproenkephalin mRNA-containing cells was not altered (Merchant 1994). These data suggest that NT may be a target gene for c-*fos*; confirmatory studies designed to test this hypothesis more conclusively are in progress.

Although the studies of Merchant and colleagues find that DA antagonists elicit c-*fos* expression in a clear majority of NT/N mRNA-expressing striatal neurons, data from immunohistochemical studies have not led to the same conclusion. Senger et al. (1993) found that Fos-li and NT-li were colocalized in a minority of striatal cells after administration of the D_2 antagonist eticlopride. However, the use of immunohistochemistry to visualize NT-li cells after D_2 antagonism requires that one wait for a relatively extended period of time, while Fos protein expression is increased and subsides on a more rapid time scale. In addition to the difficulties in matching these competing events, the striatal region in which colocalized Fos- and NT-li cells were reported by Senger et al. (1993) differed from that analyzed by Merchant and colleagues. This raises the possibility that there are two populations of striatal NT-containing neurons, and that only in one of these populations do neurons colocalize Fos and NT in response to acute D_2 receptor antagonist treatment. This is consistent with previous findings indicating two functionally distinct populations of striatal NT cells (see Deutch and Zahm 1992).

2. Enkephalin

The opioid peptides [Met^5] and [Leu^5]enkephalin are products of the preproenkephalin (PPE) gene. The D_2 receptor is localized to striatopallidal neurons that express PPE mRNA (Le Moine et al. 1990). Dopamine D_2 receptor antagonists, including haloperidol, have been shown to increase PPE gene expression in the striatum (Chen et al. 1993; Jaber et al. 1994; Soghomonian 1994); similarly, surgical denervation of the striatal DA innervation increases PPE expression (Soghomonian 1994; Voorn et al. 1987). Moreover, studies in rats sustaining striatal DA lesions suggest that haloperidol-elicited Fos expression occurs primarily in striatopallidal enkephalin cells (Robertson et al. 1992). These data suggest that PPE may be a target of Fos regulation.

However, there is a temporal dissociation between haloperidol-elicited PPE and Fos expression(Konradi et al. 1993), suggesting that D_2 antagonist-mediated expression of PPE may not be regulated by Fos. Konradi et al. found that this was indeed the case: gel shift assays revealed that Fos protein did not interact with the PPE enhancer, but the cAMP-response element binding protein (CREB) did.

3. Glutamic Acid Decarboxylase

Medium spiny neurons comprise about 95% of striatal cells. These GABAergic neurons are regulated by DA. Two forms of the biosynthetic enzyme for GABA, glutamic acid decarboxylase (GAD), are encoded for by distinct genes, giving rise to proteins with different masses, cofactor sensitivities, intracellular distributions, and susceptibility to regulation (ERLANDER et al. 1991; KAUFMAN et al. 1991). The larger form, GAD_{67}, has been shown to be regulated by several treatments, including striatal DA denervation (CHESSELET et al. 1993; SOGHOMONIAN 1994).

Fos is expressed in GABAergic neurons in the striatum, and GAD thus represents a potential target for Fos. We have recently observed that in rats sacrificed 4h after a single injection of haloperidol there is an increase in striatal GAD_{67} mRNA levels (unpublished observations). These data indicate that the temporal characteristics of changes in striatal GAD_{67} expression are consistent with this gene being a target for Fos, and warrant further investigation of GAD.

D. Preferential Induction of Fos in the Prefrontal Cortex (PFC) by Clozapine

Studies of the effects of APDs on Fos expression in the striatum led to two key findings that have now been replicated by several groups. Typical APDs but not clozapine induce Fos expression in the dorsolateral striatum. In contrast, all APDs, notably including clozapine, induce Fos expression in the shell of the nucleus accumbens. These observations leave open the question of the site(s) at which clozapine exerts its effects in neuroleptic-resistant patients and in patients with prominent negative symptoms. A substantial number of schizophrenic patients do not respond to conventional APDs. However, in many cases patients who are resistant to conventional APDs respond to clozapine treatment. Moreover, clozapine is more effective in reducing negative symptomatology than are conventional APDs. Since all APDs act in the shell of the nucleus accumbens, it is reasonable to look to other CNS sites as the regions in which the unique therapeutic effects of clozapine are manifested. One such area is the prefrontal cortex.

Current views of striatal neurons emphasize different striatal sectors as part of extended cortico-striatal-pallido-fugal systems that are (in large part) spatially segregated (ALEXANDER et al. 1990; DEUTCH et al. 1993; SWERDLOW and KOOB 1987). ROBERTSON and FIBIGER (1992) reported that clozapine differed from the typical APDs haloperidol and raclopride by increasing the number of Fos-li neurons in the medial prefrontal cortex (PFC) of the rat. In anatomical studies of the afferents of the nucleus accumbens, we observed that cortical afferents to the shell compartment were derived from the infralimbic and ventral prelimbic parts of the PFC, but that these cortical areas did not prominently innervate the accumbal core (Brog et al. 1993). Most other cortical regions were found to innervate both the shell and core compartments of the nucleus accumbens (Brog et al. 1993). These anatomical data, in conjunction with physiological data emphasizing dysfunction of specific frontal cortical regions in schizophrenia (Berman and Weinberger 1990; Weinberger 1987), led us to examine the effects of clozapine and other APDs on Fos expression in the PFC, with particular emphasis on the ventral portions of the medial PFC.

I. Regional Effects of APDs on Fos Expression in the PFC

1. Effects of APDs on Fos Expression in the Medial PFC

In order to assess regionally the effects of clozapine in the PFC, we first replicated the finding of Robertson and Fibiger (1992) that clozapine but not haloperidol increased the number of PFC neurons expressing Fos (Deutch and Duman 1995). Our observations are consistent with recent reports indicating that clozapine increases the number of Fos-li neurons in the medial PFC (Merchant et al. 1995; Robertson and Fibiger 1992; Robertson et al. 1994a). In contrast, MacGibbon et al. (1994) did not observe a significant effect of clozapine on Fos expression in the PFC; there is no satisfactory explanation for this discrepancy.

In our evaluation of the distribution of Fos-li neurons, we observed that clozapine significantly increased the number of Fos-li cells in the deep, but not superficial, layers of the PFC (Deutch and Duman 1995). Moreover, clozapine elicited significant increases in the numbers of Fos-li neurons only in certain cytoarchitectonically defined subdivisions of the PFC. An increase in Fos-li neurons was seen in the ventral aspects of the medial PFC (the infralimbic and prelimbic cortices), but not in the more dorsally situated medial precentral ("shoulder") cortex. The clozapine-elicited increase in Fos expression was primarily manifested in PFC pyramidal neurons; in addition, a subset of the calbindin-containing interneurons in the PFC expressed Fos-like immunoreactivity in response to clozapine challenge (Deutch and Duman 1995).

The ability of clozapine but not haloperidol or raclopride to increase the number of Fos-li neurons in the PFC suggested that the therapeutic actions of clozapine in neuroleptic nonresponders might be due to unique actions of clozapine in the PFC. We therefore examined the effects of haloperidol and clozapine on Fos expression in the PFC using immunoblot analyses (Deutch and Duman 1995). This approach allowed us to determine if the clozapine-elicited increase in the number of PFC Fos-li neurons reflected an increase in Fos protein itself, or alternatively if the response was primarily attributable to Fos-related antigens.

Immunoblot studies revealed that the clozapine-elicited increase in the number of Fos-li neurons in the PFC primarily represents an induction of Fos protein. Thus, an increase in Fos protein (55 kD) was observed in the microdissected samples of the infralimbic and prelimbic cortex, but not in the "shoulder" cortex (DEUTCH and DUMAN 1995). In situ hybridization histochemistry has also revealed that c-*fos* mRNA is increased in the PFC in response to clozapine (MERCHANT et al. 1995). In addition to the induction of Fos, clozapine also increased expression of several Fras. These Fras share with Fos the ability to form dimers with Jun, and thereby alter transcription in certain genes (COHEN et al. 1989; FRANZA et al. 1987; NISHINA et al. 1990). Interestingly, a Fra of approximately 39 kD was specifically increased by clozapine but not haloperidol in the infralimbic but not prelimbic or medial precentral cortices.

2. Correlations Between PFC Expression and Clinical Status

Several correlations between the specificity of clozapine effects on PFC Fos expression and the clinial actions of clozapine can be drawn. The therapeutic response to clozapine is not observed in patients with prefrontal cortical sulcal atrophy (FRIEDMAN et al. 1991; HONER et al. 1995), suggesting that the PFC is a primary site at which clozapine acts in neuroleptic nonresponders; this is consistent with the observation that clozapine selectively induces Fos expression in the PFC. In addition, neuropsychological evaluations and in vivo imaging studies have led to the conclusion that PFC dysfunction correlates with negative symptoms (BERMAN and WEINBERGER 1990; BROWN and WHITE 1991; see DAVIS et al. 1992 and DEUTCH 1992). These observations suggest that the ability of clozapine to target negative symptoms and treat neuroleptic nonresponders may be reflected by the ability of clozapine to induce Fos expression in the PFC, and lend credence to the supposition that the unique therapeutic actions of clozapine may be exerted in the medial PFC.

Our data point to the ventral aspects of the medial PFC of the rat as a critical site of action of clozapine. The PFC of the rat is comprised of several distinct cytoarchitectonic regions (KRETTEK and PRICE 1978) that become spatially segregated in primate species. It would be desirable to extrapolate from our findings to the relevant cortical region in humans, if for no other reason than to target the region(s) in imaging and postmortem studies. Unfortunately, it is difficult (if not impossible) to identify unambiguously the homologous cortical regions of rodent frontal lobes and those of primates, including man. Homologous regions in different species can be identified on the basis of cytoarchitectonics, hodology, or functional attributes. There is relatively good agreement that the infralimbic cortex of the rat (area 25) probably corresponds to the medial orbital cortex in primates and possibly to the ventral aspects of the anterior cingulate region (PREUSS 1995). However, there is some disagreement as to the homology between the primate frontal cortices and the rodent prelimbic cortex (area 32). Although some have suggested that the prelimbic cortex may correspond to the dorsolateral PFC (area 46) of the primate, cytoarchitectonic considerations and issues of connectivity argue against this interpretation. Indeed, these factors have led Preuss (1995) to argue cogently that there is no rodent homologue of the dorsolateral PFC of primates. However, functional considerations (such as the deficits in delayed response tasks) do suggest that the prelimbic cortex of the rat may correspond to the dorsolateral frontal cortex of humans and other primates. What does seem relatively clear is that clozapine appears to exert several unique effects (ranging from changes in extracellular DA concentration to Fos expression) in the medial PFC of the rat, particularly in the more ventral regions. The unique responses to clozapine observed in the rat are matched by similarly unique responses to clozapine in humans (Baldessarini and Frankenburg 1991; Kane et al. 1988). It is unlikely that the unique effects of clozapine in the rat and in humans are unrelated, and it can be argued that several frontal cortical regions in humans be considered to mediate the response to clozapine; these would include the ventromedial orbital cortex, the ventral anterior pole of the cingulate cortex, and the dorsolateral PFC.

The regional specificity of clozapine-induced Fos expression within the medial PFC of the rat also has certain parallels to another set of clinical findings. The medial precentral (shoulder) cortex, in which clozapine does not induce Fos expression, appears to be the rodent homologue of the frontal eye field, supplementary motor, and premotor areas (Passingham et al. 1988). Schizophrenic patients exhibit smooth eye tracking deficits (Levy et al. 1993; Lieberman et al. 1993). These pursuit deficits are not improved by APDs, including clozapine (Levy et al. 1993); indeed, Friedman et al. (1993) suggested that clozapine at higher doses may actually impair performance. The fact that clozapine does not alter the smooth eye tracking deficit is consistent with the fact that clozapine does not induce Fos in the medial precentral cortex (frontal eye field), since this region subserves eye tracking. Moreover, there is a correlation between performance on smooth eye tracking performance and performance on the Wisconsin Card Sort Test, which schizophrenic patients cannot perform to criterion (Littman et al. 1991; Rosse et al. 1993).

II. Receptor Mechanisms of Clozapine-Elicited Increase in PFC Fos Expression

1. D_1 and D_2 Dopamine (DA) Receptors and Clozapine-Elicited Fos Expression

The response of the PFC to clozapine but not haloperidol and raclopride administration suggests that the frontal cortical regions in which Fos is induced may be the sites of action of clozapine in neuroleptic-resistant schizophrenics or in patients with negative symptoms that do not respond well to conventional neuroleptics. We reasoned that the ability to define the receptor mechanism(s) through which clozapine induced Fos expression in the PFC might prove useful in defining a mechanism of action of atypical APDs. Accordingly, we used immunoblot methods to examine the response of the PFC to several APDs that are D_2 receptor antagonists. Neither raclopride, sulpiride, or remoxipride increased Fos expression in the PFC (DEUTCH and DUMAN 1995). These three APDs have different affinities for D_2-like receptors: raclopride has high affinities for D_2 and D_3 receptors, while sulpiride has high affinities for D_2, D_3, and D_4 receptors (MALMBERG 1993; SEEMAN and VAN TOL 1993; SOKOLOFF et al. 1992); remoxipride has a moderate affinity for the D_2 receptor, but four active metabolites of remoxipride that are present in plasma all show affinities for D_2 and D_3 receptors in the low nanomolar range (MOHELL et al. 1993). Thus, the inability of raclopride, sulpiride, and remoxipride to induce PFC Fos expression suggests that simple antagonism at any of the known D_2-like receptors does not account for the unique actions of clozapine. This conclusions is bolstered by a recent report examining a number of known and putative APDs, which revealed that several D_2-like antagonists did not induce Fos in the PFC (ROBERTSON et al. 1994a; MERCHANT et al. 1995).

While there is broad agreement over the effects of APDs on Fos expression in the PFC, our observations (DEUTCH and DUMAN 1995) and those of ROBERTSON et al. (1994a) and MERCHANT et al. (1995) diverge with respect to the actions of remoxipride. Remoxipride, as noted earlier, is D_2-like antagonist, which in preclinical behavioral models resembles an atypical APD (ÖGREN et al. 1990; see chapter by ÖGREN, this volume). Clinial trials indicate that treatment with remoxipride results in a good therapeutic response with significantly lower incidence of EPS than thioridazine or haloperidol (MENLEWICZ et al. 1990; PFLUG et al. 1990). Interestingly, preliminary data suggest that remoxipride may treat neuroleptic-resistant patients (CONLEY et al. 1994).

ROBERTSON and associates (1994a) reported that remoxipride (1.5–3.0 mg/kg) increased the numbers of Fos-li neurons in the PFC, in contrast to our negative finding with this substituted benzamide (DEUTCH and DUMAN 1995). In fact, we observed a consistent trend toward a decrease in Fos expression in the PFC after remoxipride (2.5 mg/kg) challenge. We assessed expression of Fos protein by immunoblots in the infralimbic and prelimbic regions of the PFC, whereas ROBERTSON et al. (1994a) examined the numbers of Fos-li neurons (thereby assessing Fos and Fras) in the PFC. It is possible that there is a unique effect of remoxipride in the more dorsal PFC (medial precentral cortex), but in preliminary immunohistochemical studies we have not observed an effect of remoxipride in this region. Recently, MERCHANT et al. (1995) found that remoxipride increased c-*fos* mRNA expression in the PFC, particularly in the infralimbic cortex. However, MERCHANT and colleagues noted that only low doses (0.3 mg/kg) but not higher doses (≥0.6 mg/kg) of remoxipride significantly induced c-*fos* expression. Interestingly, MERCHANT et al. noted a trend toward a decrease in PFC expression of c-*fos* at 2.5 mg/kg, comparable to the dose we tested. These data suggest that remoxipride may have an unusual dose-related effect on PFC Fos expression.

ROBERTSON et al. (1994a) also tested several other putative atypical APDs, and found that several of these increased the numbers of Fos-li neurons in the PFC; in most cases these compounds remain to be proven as atypical APDs (on the basis of lack of EPS or improving the clinical status of neuroleptic-resistant patients) in double-blind trials. Interestingly, ROBERTSON and colleagues identified certain compounds known to be typical APDs as increasing the numbers of Fos-li neurons in both the PFC and lateral striatum. These observations await independent confirmation, but are potentially very important since they would reduce the predictive value of following Fos expression in the PFC as an index of atypical APD actions.

Clozapine shows significant affinities for both D_2-like and D_1 DA receptors. The affinity of clozapine for the D_1 receptor under in vivo conditions is approximately the same as for the D_2 receptor (ANDERSSON 1989), and in vitro determinations indicate a somewhat higher affinity of clozapine for the D_1 site. We therefore examined the effect of the D_1 receptor antagonist SCH-23390 on PFC Fos expression: no significant change in Fos protein levels in the PFC was observed (DEUTCH and DUMAN 1995). We also pretreated rats with SCH-23390 prior to clozapine challenge; this pretreatment did not modify the response to clozapine. Since SCH-23390 is an antagonist at D_1 and D_5 (D_{1b} in the rodent) receptors, our data suggest that the actions of clozapine in the PFC do not critically depend on simple D_1 antagonism.

2. Nondopaminergic Receptors and the PFC Fos Response to Clozapine

Increasing attention has been devoted to the possible role of nondopaminergic receptors SKF subserving the actions of APDs, particularly clozapine (see chapter by ROTH, this volume). In an attempt to determine further the receptor mechanisms that subserve the actions of clozapine in the PFC, we administered drugs that shared with clozapine affinities for receptors other than the DA receptors.

Clozapine has a high affinity for serotonergic receptors, and in particular the affinity of clozapine for 5-HT_{2a} receptors has been suggested to correlate with the atypical APD profile (MATSUBARA et al. 1993; MELTZER et al. 1989). We therefore administered the mixed 5-$HT_{2a/2c}$ antagonist ritanserin in an effort to mimic the actions of clozapine. Ritanserin was without effect (DEUTCH and DUMAN 1995), suggesting that antagonism at the 5-HT_{2a} or 5-HT_{2c} receptors is not crucial for the actions of clozapine. MELTZER and associates (1989) noted that there is a relatively strong correlation between the ratio of affinities of several putative atypical APDs for 5-HT_{2a} : D_2 receptors and low EPS liability; this observation led to the influential hypothesis that the atypical APD profile is determined by concurrent high 5-HT_{2a} occupancy and relatively low D_2 receptor antagonism (MELTZER 1994; MELTZER et al. 1989). Two approaches were used to test the effects of simultaneous antagonism at the 5-HT_{2a} and D_2 receptors. The first approach was to administer concurrently ritanserin (3.0 mg/kg) and (±) sulpiride (100 mg). We also examined the effects of the putative atypical APD risperidone, an antagonist at both 5-HT_2 and D_2 receptors, on PFC Fos expression; risperidone is an APD that lacks EPS liability at low doses (CLAUS et al. 1992). Neither treatment increased Fos in the PFC (DEUTCH and DUMAN 1995), suggesting that the unique action of clozapine in the PFC does not occur through simultaneous 5-HT_2:D_2 receptor antagonism. Consistent with our findings, ROBERTSON et al. (1994a) have found that risperidone does not increase the number of Fos-li neurons in the PFC.

We also examined the effects of the α_1-noradrenergic antagonist prazosin and the muscarinic cholinergic antagonist scopolamine on PFC Fos expression, since clozapine exhibits significant affinities for these receptors. Neither prazosin nor scopolamine increase Fos expression in the PFC.

3. Mechanisms of Clozapine-Elicited Increases in PFC Fos Expression

a) Possible Role of a Novel DA Receptor in Mediating Clozapine-Elicited Effects

The mechanisms that contribute to the unique ability of clozapine among clinically effective APDs to induce Fos expression in the PFC remain unclear. The most likely receptors through which clozapine acts do not appear to contribute to the unique actions of the atypical APD. One possibility that might account for the clozapine-elicited increase in Fos expression was suggested by the observation that clozapine increases extracellular DA levels to a significantly greater degree than do other APDs (MOGHADDAM and BUNNEY 1990; see chapter in this volume by KALIVAS). Moreover, the ability of clozapine to increase extracellular DA levels in the PFC is not dampened by chronic administration of the drug, but actually augmented (MOGHADDAM 1994; MOGHADDAM and BUNNEY 1990; YAMAMOTO et al. 1994; YOUNGREN et al. 1994).

We therefore examined the effects of the mixed D_1-D_2 agonist apomorphine and the D_2-like agonist quinpirole on Fos expression in the PFC. Both DA agonists markedly increased expression of Fos and Fras (as shown by immunoblots) in the PFC, with quinpirole exerting a more pronounced effect (DEUTCH and DUMAN 1995). The effects of both agonists were observed at 2 h after administration, but had markedly decreased by 4 h after treatment.

In contrast to our observations, LAHOSTE et al. (1994) have reported preliminary data indicating that high doses of the D_2-like agonist quinpirole or the D_1 agonist SKF 38393 failed to induce Fos expression in the frontal and parietal cortices; in contrast, combined administration increased the number of Fos-li in the frontal and parietal cortices. After treatment of rats for 5 days with reserpine to break down D_1–D_2 synergism, LAHOSTE et al. (1994) found an increase in cortical Fos to either agonist alone.

The reason for the discrepancy between our data and those of LAHOSTE and MARSHALL is not clear. One possibility is that we used lower doses (0.2–1.0 mg/kg) of quinpirole, whereas LAHOSTE and MARSHALL used very high doses of the agonist (3.0 mg/kg), which may interact with other receptors (e.g., the H_2 histamine site). Another possible explanation is that we examined Fos expression in the infralimbic and prelimbic, but not shoulder, cortices in our immunoblot studies, whereas they measured the numbers of Fos-li in the frontal cortex.

The observation that DA agonists increased Fos is consistent with the ability of clozapine to sharply increase extracellular DA levels in the PFC. However, as discussed above, rats pretreated with the D1 agonist SCH-23390 prior to clozapine challenge still respond to the atypical APD. We therefore elected to examine the effects of pretreatment with a high dose of sulpiride, an antagonist at $D_{2/3/4}$ receptors. Sulpiride pretreatment did not reduce the Fos response to clozapine. Thus, neither D_1-like nor D_2-like antagonists blocked the clozapine-elicited increase in PFC Fos expression, despite the fact that DA agonists increased Fos.

We interpret these data to suggest the presence of a novel DA receptor in the PFC that does not correspond to the cloned D_1–D_5 receptors. This suggestion is consistent with electrophysiological data that indicate that the inhibitory actions of DA on prefrontal cortical pyramidal cells exhibits a distinctly anomalous pharmacology (SESACK and BUNNEY 1989; THIERRY et al. 1986), quite distinct from that of the known DA receptors. Moreover, there are two peripheral DA receptors in the rat, one in the kidney and another in brown adipose tissue (HUO et al. 1991; NISOLI et al. 1992); the pharmacology of these peripheral DA receptors does not correspond to the D_1–D_5 receptors nor do these sites express D_2 or D_1 transcripts. Interestingly, the pharmacological profiles of the three anomalous DA receptors in the PFC, renal medulla, and brown adipose tissue, are all distinct. These observations suggest that there are additional uncharacterized DA receptors.

The nature of the D_2-like DA receptor that we posit to mediate the response to DA is probably quite different than that of known D_2 isoforms. First, the known D_2-like receptors are negatively linked (or do not regulate) cyclase, and as such would not be expected to enhance Fos expression. A novel D_2-like receptor would resemble the known D_2 isoforms in the general sense of displaying significant affinities for certain D_2-like agonists and antagonists, but might be expected to have different transduction mechanisms from the known D_2 receptors. One possibility is a D_2-like receptor coupled to phosphoinositide turnover. A second feature of a putative D_2 receptor that mediates clozapine effects in the PFC would be its localization. Since DA agonists enhance Fos expression, and since clozapine enhances extracellular DA levels, it is reasonable to speculate that the receptor might be a release-modulating autoreceptor that differs from known DA autoreceptors or an impulse-modulating autoreceptor on a restricted population of mesocortical DA neurons.

b) *Targeting of Multiple Receptors*

One of the current hypotheses advanced to explain the actions of clozapine is based on the relatively high ratio of 5-HT_{2a} : D_2 receptor antagonism. However, based on our experiments (Deutch and Duman 1995) and those others (Robertson et al. 1994a) it appears unlikely that concurrent antagonism of serotonergic and dopaminergic receptors is the determinant of the clozapine-elicited increases in Fos in the PFC: neither co-administration of selective serotonergic and dopaminergic receptor antagonists nor administration of risperidone mimicked the effect of clozapine.

Another possibility is that multiple receptors must be occupied. To those pharmacologists interested in uncovering the mechanisms of action of clozapine, to state simply that clozapine is a "dirty" drug due to its interaction with multiple receptors represents a gross understatment. Clozapine targets a large number of different receptors, and it is therefore exceedingly difficult (if not impossible) to arrive at the correct combination of receptor antagonists to test. Furthermore, the issue of the relative affinities of the drugs for these receptors renders any interpretation of a multireceptor drug combination suspect.

These caveats in mind, two cases may nonetheless be instructive. The first is that we have found that administration of loxapine (a dibenzodiazepine that is a structural analog of clozapine) does not induce Fos expression in the PFC. Although loxapine has comparable affinities to those displayed by clozapine for many receptors, it is a typical APD, with considerable EPS liability. Another drug that also targets many receptors is risperidone. Usually discussed in terms of its affinities for 5-HT_{2a} and D_2 receptors, risperidone exhibits significant affinities for several other receptors (Leysen et al. 1988). Despite the fact that these two drugs display high-to-moderate affinities for a large number of receptors, and therefore somewhat resemble clozapine in the lack of receptor selectivity, neither loxapine nor risperidone (which have different clinical profiles) induces Fos in the PFC.

c) *Transsynaptic Events as a Possible Determinant of Clozapine-Induced Changes*

One final possibility to be entertained is that the unique effect of clozapine on PFC Fos expression is not due to actions in the PFC, but rather to transsynaptic activation of cortical neurons. We and others have examined the effects of APDs on Fos expression using systemic administration of these drugs. The decision of investigators to study systemic administration of APDs is consistent with the clinical use of these agents, and also reflects the difficulty of interpreting changes in immediate-early genes after direct intraparenchymal administration of drugs, since certain immediate-early genes are induced by trauma. Given that clozapine is administered systemically, it clearly acts in several brain regions, some of which project to the PFC and regulate the activity of cortical neurons. Among these sites may be the medial thalamus, basolateral amygdala, entorhinal cortex, and hippocampus, all of which provide glutamatergic inputs to the PFC, and the basal forebrain and laterodorsal tegmental nucleus, which provide cholinergic and peptidergic projections to the PFC. The effects of lesions of these sites on subsequent responses to APDs have not been systematically explored (see chapter by Bardgett, this volume). In view of the frequent reports of neuronal loss in schizophrenia, it would be of considerable interest to determine if neuronal loss in certain areas alters the therapeutic response to APD treatment.

One other reason to suspect that transsynaptic events may be a critical determinant of the effects of clozapine in the PFC rests on the paradoxical observation that D_2 agonists increase Fos expression in the PFC. The D_2 receptor is negatively linked to adenyl cyclase; data from other systems suggests that activation of cyclase enhances Fos expression (Sheng et al. 1990; Weih et al. 1990). Thus, the suggestion that transsynaptic induction of Fos warrants investigation fits well with the observation that excitatory amino acid transmitters operating at NMDA receptors strongly induce Fos expression (Lerea and McNamara 1993; Morgan and Curran 1986; Sonnenberg et al. 1989; Vaccarino et al. 1992).

Unfortunately, we do not know if clozapine or other APDs alter Fos in afferents to the PFC. The exception to this statement is the medial thalamus, where the thalamic paraventricular nucleus (PV) responds to clozapine by sharply increasing Fos expression, but in which most APDs do not alter expression of the immediate-early gene (Deutch et al. 1995b; see below). Our preliminary data indicate that lesions of the PV do not alter the ability of clozapine to elicit a Fos response in the PFC. However, we did not examine the effects of lesions of the mediodorsal thalamus; a recent report by Erdtsieck et al. (1995) indicates that infusion of biccuculine into the mediodorsal thalamic nucleus to disinhibit thalamic projections to the PFC results in a rapid increase in c-*fos* expression in the PFC.

Obviously, there are many sources of afferents to the PFC that should be explored in the context of modification of the Fos response to clozapine, and in general as determinants of responsiveness to APDs (see Bardgett, this volume).

III. Transcriptional Targets of Clozapine's Actions in the PFC

There are essentially no studies that have addressed the transcriptional targets of immediate-early genes in the PFC. This may in part stem from the apparent selective effect of clozapine in the PFC. There are obviously a large number of potential targets. Anatomical data indicate that clozapine-elicited Fos expression in the PFC is predominantly observed in pyramidal cells (DEUTCH and DUMAN 1995). This observation, unfortunately, does little to narrow down the potential targets, although DA receptors are probably reasonable candidates. However, D_2 and D_1 receptors are present on both pyramidal cells and interneurons. We found that Fos is expressed in a small subset of interneurons in the PFC (DEUTCH and DUMAN 1995), particularly in a subpopulation of calbindin-containing cells. This observation becomes of interest since we have observed that clozapine (but not haloperidol) significantly increases levels of the GluR1 AMPA receptor subunit in the PFC (FITZGERALD et al. 1995); the GluR1 subunit is expressed on some prefrontal cortical calbindin cells (MARTIN et al. 1993). Future studies will examine if Fos-li is colocalized with this AMPA subunit in the PFC.

E. Effects of APDs on Immediate-Early Gene Expression in Other CNS Sites

Studies of the effects of APDs on immediate-early gene expression in the CNS have to date focused primarily on the basal ganglia and allied cortical regions. In view of the reasonably well-characterized DA innervations of these regions, and current concepts of the pathophysiology of schizophrenia, the striatal complex and PFC are excellent starting points. As noted earlier, however, the lack of clearly defined alterations in striatal and cortical DA systems in schizophrenia has led to a resurgence of interest in other regions of the brain as potential sites of dysfunction and thus of the therapeutic effects of APDs.

I. Thalamic Paraventricular Nucleus

The effect of clozapine in the PFC prompted us to examine the effects of haloperidol and clozapine in the mediodorsal thalamic nucleus (MD), a major source of afferents to the PFC; indeed, the PFC has been defined on the basis of receiving MD projections (KRETTECK and PRICE 1977; LEONARD 1969). In our initial immunohistochemical study, we were surprised to note that neither haloperidol or clozapine changed the number of Fos-li neurons in the MD. However, there was a striking clozapine-elicited increase in the thalamic paraventricular nucleus (PV), a midline nucleus situated between the bilateral mediodorsal nuclei (DEUTCH et al. 1995b). Although clozapine markedly enhanced Fos expression in the PV, haloperidol had no effect at comparable doses.

Subsequent immunoblot experiments revealed similarities between the PV and the PFC: with the exception of clozapine, most APDs, including raclopride, sulpiride, and remoxipride, had no effect on PV Fos expression. Similarly, there was no effect of combined antagonism at serotonergic and dopaminergic receptors (concurrent ritanserin and sulpiride administration or risperidone treatment), and no effect of alpha adrenergic or muscarinic cholinergic antagonists (Deutch et al. 1995b).

However, haloperidol at very high doses (5.0 mg/kg) induced Fos protein in the PV, albeit not to the same degree as 15 or 30 mg/kg clozapine. In addition, loxapine administration resulted in a small but significant increase in Fos expression in the PV.

In attempting to define the mechanism subserving the actions of clozapine in the PV, we administered apomorphine and quinpirole; both DA agonists increased Fos expression. However, recalling the pharmacology in the PFC, pretreatment of rats with either the D_1-like antagonist SCH-23390 or the D_2-like antagonist (±) sulpiride did not reduce reduce the clozapine-elicited increase in Fos expression in the PV.

The PV has a DA innervation that originates in the ventral tegmental area and hypothalamus (Groenewegen 1988; Öngür et al. 1995; Otake and Ruggiero 1995; Takada et al. 1990). Moreover, PV neurons project to forebrain DA regions, including the ventral aspects of the PFC, the shell of the NAS, the basolateral and central nuclei of the amygdala, and the hippocampus and entorhinal cortex (Berendse and Groenewegen 1990, 1991; Christie et al. 1987; Turner and Herkenham 1991).

These anatomical features suggest that the PV be considered as a novel site of action of APDs, in particular atypical APDs (Deutch et al. 1995b). It is interesting to note that the paraventricular thalamus is one of the few areas in the brain in which gliosis has been reported in schizophrenia (Stevens 1982; Nieto and Escobar 1972). One must also consider, however, the possibility that the PV is a site of action for clozapine's untoward effects, rather than the therapeutic effects of clozapine. Clozapine has a number of side effects other than agranulocytosis (Alvir and Lieberman 1994), including a high incidence of weight gain (Umbricht et al. 1994) and an increased incidence of seizures (Devinsky and Pacia 1994). The midline thalamic nuclei, including the PV, are part of the reticular activating system, and have long been associated with seizure activity. However, the observation that high doses of haloperidol also induce Fos in the PV suggests that the PV is not a site at which clozapine exerts its proconvulsant properties, since APDs with pronounced EPS liability appear to be the least likely to result in seizures (Itil and Soldatos 1980). Furthermore, most seizures associated with clozapine treatment are grand-mal seizures, whereas stimulation of the medial thalamus elicits different forms of seizure discharge.

In summary, the pharmacological characterization of the Fos response in the PV to APDs is strikingly similar to that of the PFC. Interestingly, the only DA receptor that has been localized in the PV is the D_3 receptor (Mansour and Watson 1995). This contrasts with the PFC, where D_3 receptor transcripts are not present in detectable levels, but where D_2 and D_1 receptors are expressed on neurons (Mansour and Watson 1995). Thus, the parallels between the PFC and PV responses to APDs, coupled with the different receptor populations present, again suggest that a novel DA receptor may subserve the effects of clozapine in these regions.

II. Lateral Septal Nucleus

Several authors have commented on the ability of haloperidol and other neuroleptics to induce Fos expression in the lateral septum (DILTS et al. 1993; MACGIBBON et al. 1994; ROBERTSON and FIBIGER 1992; ROBERTSON et al. 1994a; SEBENS et al. 1995). Indeed, ROBERTSON and colleagues (1994a) note that among several APDs they have tested, only risperidone failed to increase the numbers of Fos-li in the lateral septum.

Clozapine potently increases Fos in the lateral septum (ROBERTSON et al. 1994a; SEBENS et al. 1995). Although both haloperidol and clozapine increase Fos and Fras in the lateral septum (DILTS et al. 1993; MACGIBBON et al. 1994), clozapine also selectively induces expression of Jun B and zif268 in this region.

The ability of most APDs to induce immediate-early genes in the lateral septum suggests that this region may be an important site of action for the therapeutic effects of APDs. Although the functional significance of septal Fos induction is not clear, the septum figured prominently in early speculations concerning neural dysfunction in schizophrenia (GAREY et al. 1974; HEATH et al. 1970).

III. Other CNS Regions

DILTS et al. (1993) suggested that haloperidol increased expression of Fos and Fras in several regions, including the cingulate and pyriform cortex; both of these areas receive inputs from midbrain DA neurons. Somewhat surprisingly, MACGIBBON et al. (1994) indicated that neither haloperidol nor clozapine altered Fos expression in the hippocampus. In an initial examination of the effects of haloperidol in the primate, we observed a significant increase in the number of Fos-li neurons in the entorhinal, but not inferotemporal, cortex (DEUTCH et al. 1995a).

SEBENS et al. (1995) have recently reported that both haloperidol and clozapine increase the numbers of Fos-li neurons in the central nucleus of the amygdala. This region receives a DA innervation from the A10 (FALLON et al. 1978) and A8 (DEUTCH et al. 1988) DA cell groups, and alterations in the DA systems of the amygdala have been implicated in schizophrenia (see GOLDSTEIN and DEUTCH 1992).

MACGIBBON and coworkers noted an inconsistent effect of clozapine in the ventral tegmental area (VTA), but not in the substantia nigra; it remains to be determined if the effect of clozapine in the VTA can be consistently induced, and if any effect is manifested in DA neurons (as opposed to nondopaminergic cells) of the VTA. In studies of the effects of stress on midbrain DA neurons, we found that only a selected population of midbrain DA cells appears to have the capacity to express Fos (Deutch et al. 1991a).

There have been only two reports of which we are aware concerning the effects of APDs on neurons in the hypothalamus and pituitary. Sebens et al. (1995) reported that clozapine but not haloperidol markedly increased the number of Fos-li neurons in the hypothalamic paraventricular and supraoptic nuclei. This observation is interesting in light of the syndrome of polydipsia and intermittent hyponatremia in some psychotic patients (Vieweg et al. 1989), which is improved on clozapine but not neuroleptic treatment (Leadbetter and Shutty 1994).

In a study focused on determining the effects of estrogens on c-*fos* and c-*myc* expression in the anterior pituitary, Chernavsky et al. (1993) reported that haloperidol increased c-*myc* expression in the adenohypophysis of animals previously treated with estrogen, and that the estrogen-elicited increase in c-*fos* and c-*myc* expression has blocked by the DA agonist bromocriptine.

F. Conclusions

Over the past 5 years there has been a remarkable interest in immediate-early gene expression as an index of functional activation of neurons. Neuroscience research has long sought measures to reflect neuronal activity. Until the first studies suggesting that c-*fos* could be used to monitor activity of neurons (Hunt et al. 1987; Sagar et al. 1988; Dragunow and Faull 1989), the primary methods used were 2-deoxyglucose (2-DG) accumulation and cytochrome oxidase histochemistry. The 2-DG method has proven very useful, but primarily offers a picture of activity in nerve terminals, and thus provides a complementary view to that revealed by studies of Fos expression, which offer single-cell resolution. Cytochrome oxidase histochemistry has also been very useful; in expert hands it can be used to gain some information at the single-cell level, but it is better suited for regional studies. Moreover, the dynamic range of cytochrome oxidase appears to be smaller than that obtained with Fos or 2-DG measurements.

The ability to discern certain populations of neurons that are activated by pharmacological treatments has opened windows to the sites in the CNS where the therapeutic or side effects of APDs are manifested. Studies of immediate-early gene expression have also allowed us to probe the receptor mechanisms underlying the regionally-specific actions of APDs, and have started to reveal receptor mechanisms, as yet poorly understood, that differentiate the atypical APD clozapine from typical APDs. Finally, studies that have charted the responses of neurons by examining c-*fos* have proven to be a very useful screening method for defining new APDs and predicting their clinical profile as typical or atypical.

I. Methodological Issues

The ability to follow immediate-early gene expression has provided an easy method for assessing functional activation of neurons. Unfortunately, there have been increasingly frequent contradictions in the literature concerning the effects of APDs on immediate-early gene expression. The rapid and dramatic responses of immediate-early genes to pharmacological and environmental perturbations render this method both exquisitely sensitive and difficult to control. Accordingly, there are many variables to which attention must be devoted if studies of immediate-early gene expression are to be useful.

Issues such as the day–light cycle, time of sacrifice, interruption of animals by support staff for purposes of animal care in the same room, and control of extraneous signals (such as paging announcements) make it difficult to obtain reproducible and reliable data without instituting measures designed to overcome specifically these intrusions. Key issues such as vehicle administration, comparability of the stress of handling during injections, and comparability of the test agents for discomfort on administration all must be addressed. These issues are particularly important in many key regions of interest, including the PFC, NAS shell and septal pole region, lateral septum, and amygdala, since these areas have extensive autonomic connections. The temporal interval between treatment and sacrifice must be determined empirically for a given region of the brain and challenge condition, and the interval between anesthetic administration and perfusion be kept brief. Finally, it is critical that on any given day all groups in an experiment be represented, both control and treatment groups. This generally necessitates repeating experiments in relatively small groups of animals several times.

Issues concerning immunohistochemical detection also must be addressed. These include the requirement that the same antibody batches and solutions be used on any given day to treat tissue from different animals from the various (control and experimental) groups. Immunoperoxidase methods require that one standardize the development of the reaction product for vehicle-injected animals and that the same time of development be rigidly followed for the experimental animal samples. Sampling for quantitation must be consistent and reliable; given the heterogeneity present in such areas as the striatum, much less an area as heterogeneous as the PFC, it is imperative that investigators consistently measure the numbers of Fos-li neurons in the same designated region.

Unfortunately, the seeming simplicity of the method appears designed to tempt one to circumvent or neglect critical issues. This has led to discrepant findings in certain cases. One of the current uses of Fos immunohistochemistry is as a screen for drugs in development as APDs. Accordingly, issues of reliability emerge. As investigators become more aware of the key issues, discrepant findings will be less likely to reflect lack of attention to detail and more likely to lead us to key factors that govern responsiveness to APDs.

II. Validity of Fos Expression as a Model of the Actions of APDs

The reliability of the changes in immediate-early gene expression are of considerable concern because of the decision of several pharmaceutical firms to examine potential APDs by following Fos expression. The use of Fos expression as a model of the actions of typical and atypical APDs clearly has face validity, but in addition appears to have predictive validity and has good construct validity (see GEYER and MARKOU 1995). Using the specific regional expression of Fos as a marker for typical and atypical APDs, with induction of Fos in the shell of the accumbens marking antipsychotic efficacy, induction of Fos in the dorsolateral striatum serving as an index of EPS liability, and Fos induction in the PFC coupled with lack of caudatoputamen expression as an index of atypical APD, all known APDs (defined on the basis of double-blind studies) are revealed by the accumbal pattern of expression. Similarly, the correlation between induction of Fos in the dorsolateral striatum and EPS liability appears excellent.

The issue with atypical APDs is obviously clouded by the fact that only one atypical APD, clozapine, is available (see DEUTCH 1995); there are putative atypical APDs, but to date none have been shown to fulfill the criteria of efficacy in schizophrenic patients, including treatment-resistant patients, lack of EPS, and efficacy in the treatment of negative symptoms. It is possible that there are several types of atypical APDs, with different properties with regard to neuroleptic nonresponders, EPS liability, and negative symptoms (see DEUTCH et al. 1991b; see chapter by Csernansky in this volume). Until recently, only clozapine was reported to induce Fos in the PFC. However, ROBERTSON et al. (1994b) have recently suggested that some putative APDs induce Fos in the PFC.

One other drug offers a test of the construct validity of Fos expression as a marker for antipsychotic efficacy. Reserpine, which is an effective APD but has been abandoned because of side effects, induces Fos in the striatum (COLE and DIFIGLIA 1994); the mechanism of action of reserpine appears to involve a D_2 mechanism, since quinpirole but not SCH 23390 pretreatment blocks the induction of Fos (COLE and DIFIGLIA 1994).

III. Future Directions

We are beginning to understand the receptor mechanisms through which APDs induce Fos in various brain regions. In the striatum a D_2 receptor appears to be a necessary but not sufficient condition. Current data suggest that future attention be devoted to examination of transduction mechanisms rather than receptors as a key factor, since many treatments that modify APD-elicited Fos expression do not alter extracellular DA levels.

Our knowledge of the receptor mechanisms that underlie the clozapine-elicited induction of Fos in the PFC is poor. Current data point to a putative DA D_2-like receptor that does not correspond to one of the five cloned forms. There are peripheral DA receptors that have been pharmacologically demonstrated but not yet cloned, and there are central DA receptors that appear to alter phosphoinositide turnover and regulate calcium currents that have not been cloned. It appears likely that additional DA receptors may be discovered in the near future, one of which may provide the key understanding of how atypical APDs act.

We are now poised to exploit the immediate-early gene data by determining the target genes that are regulated by APDs. There is no a priori method to assess these target genes, however. Thus, additional data will be required to direct our attention to specific genes. The anatomical resolution afforded by immunohistochemical and in situ hybridization histochemical methods for the detection of immediate-early gene expression may help guide the decision to focus on a particular gene, since a specific cellular phenotype can be identified as critical. These anatomical data can then be used in conjunction with pharmacological and molecular methods to investigate the involvement of a potential target gene. However, even if a particular target gene is regulated by Fos or another immediate-early gene, it will still be necessary to demonstrate regulation of the target by APD treatments.

Most studies of the actions of APDs on immediate-early genes have focused on the striatal complex and the PFC. However, the myriad actions of APDs and the different clinical profiles of these drugs suggest that attention be devoted to other regions as well. Postmortem studies of schizophrenia have revealed (unfortunately, in an inconsistent way) changes in neuronal number or density in a large number of brain regions. It seems likely that extended circuits will hold the key to APD actions, and thus that researchers interested in the mechanisms of action of APDs expand their neuroanatomical horizons.

Acknowledgements. I am indebted to Drs. Kalpana Merchant and D. Scott Zahm for extended discussions on the effects of antipsychotic drugs on immediate-early gene and peptide expression, and for sharing with me their unpublished observations on the effects of antipsychotic drugs on Fos expression. This work was supported in part by grants MH-45124 and MH-25642 from the National Institute of Mental Health, by the National Parkinson Foundation Center for Excellence at Yale University, and by the Veterans Administration National Center for Schizophrenia Research and National Center for Post-Traumatic Stress Disorder at the West Haven VA Medical Center.

References

Alexander GE, Crutcher MD, DeLong MR (1990) Basal ganglia-thalamocortical circuits: parallel substrates for motor, oculomotor, "prefrontal" and "limbic" functions. Prog Brain Res 85:119–146

Alvir JAJ, Lieberman JA (1994) Agranulocytosis: incidence and risk factors. J Clin Psychiatry 55 [Suppl B]:137–138

Andersen PH (1988) Comparison of the pharmacological characteristics of [^{3}H]raclopride and [^{3}H]SCH 23390 binding to dopamine receptors in vivo in mouse brain. Eur J Pharmacol 146:113–120
Arnauld E, Arsaut J, Demotes-Mainard J (1993) Functional heterogeneity of the caudate-putamen as revealed by c-fos induction in response to D1 receptor activation. Mol Brain Res 18:339–342
Aronin N, Chase K, Sagar SM, Sharp FR, Difiglia M (1991) *N*-Methyl-D-aspartate receptor activation in the neostriatum increases c-fos and fos-related antigens selectively in medium-sized neurons. Neuroscience 44:409–420
Augood SJ, Kiyama H, Faull RLM, Emson PC (1991) Differential effects of acute dopaminergic D1 and D2 receptor antagonists on proneurotensin mRNA expression in rat striatum. Mol Brain Res 9:341–346
Baldessarini RJ, Frankenburg FR (1991) Clozapine: a novel antipsychotic drug. N Engl J Med 324:746–754
Bean AJ, During MJ, Deutch AY, Roth RH (1989) The effects of dopamine depletion on striatal neurotensin: biochemical and immunohistochemical studies. J Neurosci 9:4430–4438
Benloucif S, Galloway MP (1991) Facilitation of dopamine release in vivo by serotonin agonists: studies with microdialysis. Eur J Pharmacol 200:1–18
Benloucif S, Keegan MJ, Galloway MP (1993) Serotonin-facilitated dopamine release in vivo: pharmacological characterization. J Pharmacol Exp Ther 265:373–377
Berendse HW, Groenewegen HJ (1990) Organization of the thalamostriatal projections in the rat, with special emphasis on the ventral striatum. J Comp Neurol 299:187–228
Berendse HW, Groenewegen JH (1991) Restricted cortical termination fields of the midline and intralaminar nuclei in the rat. Neuroscience 42:73–102
Beretta S, Robertson HA, Graybiel AM (1992) Dopamine and glutamate agonists stimulate neuron-specific expression of Fos-like protein in the striatum. J Neurophysiol 68:767–777
Berretta S, Parthasarathy HB, Graybiel AM (1994) Differential induction of immediate early genes (IEGs) by cortical activation in striatal neuronal subpopulations. Soc Neurosci Abstr 20:406.2
Berman KF, Weinberger DR (1990) The prefrontal cortex in schizophrenia and other neuropsychiatric diseases: in vivo physiological correlates of cogitive deficits. Prog Brain Res 85:521–538
Bernard V, Dumartin B, Lamy E, Black B (1993) Fos immunoreactivity after stimulation or inhibition of muscarinic receptors indicates anatomical specificity for cholinergic control of striatal efferent neurons and cortical neurons in the rat. Eur J Neurosci 5:1218–1225
Boegman RJ, Vincent SR (1995) Involvement of adenosine and glutamate receptors in the induction of c-fos in the striatum by haloperidol. J Neurosci (submitted)
Brog JS, Zahm DS (1995) Morphology and Fos immunoreactivity reveal two subpopulations of striatal neurotensin neurons following acute 6-hydroxydopamine lesions and reserpine administration. Neuroscience 65:71–86
Brog JS, Salyapongse A, Deutch AY, Zahm DS (1993) The afferent innervation of the core and shell of the accumbens part of the rat ventral striatum: immunohistochemical detection of retrogradely transported Fluoro-gold. J Comp Neurol 338:255–273
Brown KW, White T (1991) The association among negative symptoms, movement disorders, and frontal lobe psychological deficits in schizophrenic patients. Biol Psychiatry 30:1182–1190
Bullitt E (1990) Expression of c-*fos*-like protein as a marker for neuronal activity following noxious stimulation in the rat. J Comp Neurol 296:517–530
Canton H, Verriele L, Colpaert FC (1990) Binding of typical and atypical antipsychotics to 5-HT_{1c} and 5-HT_2 sites: clozapine potently interacts with 5-HT_{1c} sites. Eur J Pharmacol 191:93–96

Cenci MA, Bjorklund A (1993) Transection of corticostriatal afferents reduces amphetamine and apomorphine-induced striatal Fos expression and turning behavior in unilaterally 6-hydroxydopamine-lesioned rats. Eur J Neurosci 5:1062–1070

Chen JF, Aloyo VJ, Weiss B (1993) Continous treatment with the D_2 dopamine receptor antagonist quinpirole decreases D_2 dopamine receptors, D_2 dopamine receptor messenger RNA and proenkephalin messenger RNA, and increases mu opioid receptors in mouse striatum. Neuroscience 54:669–680

Chernavsky AC, Valerani AV, Burdman JA (1993) Haloperidol and oestrogens induce c-myc and c-fos expression in the anterior pituitary gland of the rat. Neurol Res 15:339–343

Chesselet MF (1984) Presynaptic regulation of neurotransmitter release in the brain: facts and hypotheses. Neuroscience 12:347–375

Chesselet MF, Mercugliano M, Soghomonian JJ, Salin P, Qin Y, Gonzales C (1993) Regulation of glutamic acid decarboxylase gene expression in efferent neurons of the basal ganglia. Prog Brain Res 99:143–154

Christie MJ, Summers RJ, Stephenson JA, Cook CJ, Beart PM (1987) Excitatory amino acid projections to the nucleus accumbens septi in the rat: a retrograde transport study utilizing D[^{3}H]aspartate and [^{3}H]GABA. Neuroscience 22:425–439

Claus A, Bollen J, De Cuyper H, Eneman M, Malfroid M, Peuskens J, Heylen S (1992) Risperidone versus haloperidol in the treatment of chronic schizophrenic inpatients: a multicenter double-blind comparative study. Acta Psychiatr Scand 85:295–305

Cohen BM, Lipinski JF (1986) In vivo potencies of antipychotic drugs in blocking alpha 1 noradrenergic and dopamine D2 receptors: implications for mechanisms of action. Life Sci 39:2571–2580

Cohen DR, Ferreira PC, Gentz R, Franza BR Jr, Curran T (1989) The product of a fos-related gene, fra-1, binds cooperatively to the AP-1 site with Jun: transcription factor AP-1 is comprised of multiple protein complexes. Genes Develop 3:173–184

Cole AJ, Bhat RV, Patt C, Worley PF, Baraban JM (1992) D_1 dopamine receptor activation of multiple transcription factor genes in rat striatum. J Neurochem 58:1420–1426

Cole DG, DiFiglia M (1994) Reserpine increases Fos activity in the rat basal ganglia via a quinpirole-sensitive mechanism. Neuroscience 60:115–123

Conley R, Tamminga CA, Nguyen JA (1994) Clinical actions of remoxipride. Arch Gen Psychiatry 51:1001

Daly DA, Moghaddam B (1993) Actions of clozapine and haloperidol on the extracellular levels of excitatory amino acids in the prefrontal cortex and striatum of conscious rats. Neurosci Lett 152:61–64

Daunais JB, McGinty JF (1994) Acute and chronic cocaine administration differentially alters striatal opioid and nuclear transcription factor mRNAs. Synapse 18:35–45

Deutch AY (1992) The regulation of subcortical dopamine systems by the prefrontal cortex: interactions of central dopamine systems and the pathogenesis of schizophrenia. J Neur Trans [Suppl] 36:61–89

Deutch AY (1993) Prefrontal cortical dopamine and the elaboration of functional corticostriatal circuits: implications for schizophrenia and Parkinson's disease. J Neural Transm 91:197–221

Deutch AY (1995) Mechanisms of action of clozapine in the treatment of neuroleptic-resistant and neuroleptic-intolerant schizophrenia. Eur Psychiatr 10:39–46

Deutch AY, Bean AJ (1995) Colocalization in dopamine neurons. In: Bloom FE, Kupfer DJ (eds) Psychopharmacology: the fourth generation of progress. Raven, New York, pp 197–206

Deutch AY, Cameron DS (1992) Pharmacological characterization of dopamine systems in the nucleus accumbens core and shell. Neuroscience 46:49–56

Deutch AY, Duman RS (1995) The effects of antipsychotic drugs on prefrontal cortical Fos expression: cellular localization and pharmacological characterization. Neuroscience (in press)

Deutch AY, Zahm DS (1992) The current status of neurotensin-dopamine interactions: issues and speculations. Ann NY Acad Sci 668:232–252

Deutch AY, Goldstein M, Baldino F Jr, Roth RH (1988) The telencephalic projections of the A8 dopamine cell group. Ann NY Acad Sci 537:27–50

Deutch AY, Lee MC, Gillham MH, Cameron D, Goldstein M, Iadorola MJ (1991a) Stress selectively increases Fos protein in dopamine neurons innervating the prefrontal cortex. Cerebr Cortex 1:273–292

Deutch AY, Moghaddam B, Innis R, Krystal JH, Aghajanian GK, Bunney BS, Charney DS (1991b) Mechanisms of action of atypical antipsychotic drugs: implications for novel therapeutic strategies for schizophrenia. Schizophr Res 4:121–156

Deutch AY, Lee MC, Iadorola MJ (1992) Regionally specific effects of atypical antipsychotic drugs on striatal Fos expression: the nucleus accumbens shell as a locus of antipsychotic action. Mol Cell Neurosci 3:332–341

Deutch AY, Zahm DS, Bourdelais AJ (1993) The nucleus accumbens core and shell: delineation of corticostriatal circuits and their functional attributes. In: Kalivas PW, Barnes CD (eds) Limbic motor circuits and neuropsychiatry. CRC, Boca Raton, p 45

Deutch AY, Lewis DA, Redmond DE Jr, Roth RH (1995a) The effects of D_2 dopamine antagonists on Fos protein expression in the striatal complex of the non-human primate *Circopithecus aethiops*. Synapse (in press)

Deutch AY, Öngür D, Duman RS (1995b) Fos induction by antipsychotic drugs in the medial thalamus: a novel locus of antipsychotic drug action. Neuroscience 66:337–346

Devinsky O, Pacia SV (1994) Seizures during clozapine therapy. J Clin Psychiatry 55 [Suppl B]:153–156

Dilts RP Jr, Helton TE, McGinty JF (1993) Selective induction of Fos and FRA immunoreactivity within the mesolimbic and mesostriatal dopamine terminal fields. Synapse 13:251–263

Dobner PR, Kislauskis E, Bullock BP (1992) Cooperative regulation of neurotensin/neuromedin N gene expression in PC12 cells involves AP-1 transcription factors. Ann NY Acad Sci 668:17–29

Dragunow M (1990) Presence and induction of Fos B-like immunoreactivity in neural, but not non-neural, cells in adult rat brain. Brain Res 533:324–328

Dragunow M, Faull R (1989) The use of c-*fos* as a metabolic marker in neuronal pathway tracing. J Neurosci Methods 29:261–265

Dragunow M, Robertson GS, Faull RLM, Robertson HA, Jansen K (1990) D2 dopamine receptor antagonists induce Fos and related proteins in striatal neurons. Neuroscience 37:287–294

Dragunow M, Leah JD, Faull RLM (1991) Prolonged and selective induction of Fos-related antigen(s) in striatal neurons after 6-hydroxydopamine lseions of the rat substantia nigra pars compacta. Mol Brain Res 10:355–358

Eggerman KW, Zahm DS (1988) Numbers of neurotensin-immunoreactive neurons selectively increased in rat ventral striatum following acute haloperidol administration. Neuropeptides 11:125–132

Elibol B, Moratalla R, Hiroi N, Graybiel AM (1994) Low vs. high dose and acute vs. subchronic haloperidol treatment induces distinct patterns of immediate-early gene expression in the striatum. Soc Neurosci Abstr 20:990

Erdtsieck EBHW, Feenstra MGP, Botterblon MHA, van Uum HFM, Sluiter AA, Heinsbroek RPW (1995) c-Fos expression in the rat brain after pharmacological stimulation of the rat "medio-dorsal" thalamus by means of microdialysis. Neuroscience 66:115–131

Erlander MG, Tillakaratne NKT, Feldblum S, Patel N, Tobin AJ (1991) Two genes encode distinct glutamate decarboxylases. Neuron 7:91–100

Fallon JH, Koziell DA, Moore RY (1978) Catecholamine innervation of the basal forebrain II. Amygdala, suprarhinal cortex and entorhinal cortex. J Comp Neurol 180:509–532

Fink JS, Weaver DR, Rivkees SA, Peterfreund RA, Pollack AE, Adler EM, Reppert SM (1992) Molecular cloning of the rat A_2 adenosine receptor: selective co-expression with D_2 dopaminee receptors in rat striatum. Mol Brain Res 14:186–195

Fitzgerald LW, Deutch AY, Gasic G, Heinemann SF, Nestler EJ (1995) Regulation of cortical and subcortical glutamate receptor subunits by chronic antipsychotic drugs. J Neurosci 15:2453–2461

Franza BR Jr, Sambucetti LC, Cohen DR, Curran T (1987) Analysis of Fos protein complexes and Fos-related antigens by high-resolution two-dimensional gel electrophoresis. Oncogene 1:213–221

Frey P, Fuxe K, Eneroth P, Agnati LF (1986) Effects of acute and long-term treatment with neuroeptics on reginal telencephalic neurotensin levels in the male rat. Neurochem Int 8:429–434

Friedman L, Knutson L, Shurell M, Meltzer HY (1991) Prefrontal sulcal prominence is inversely related to response to clozapine in schizophrenia. Biol Psychiatry 29:865–877

Friedman L, Jesberger JA, Meltzer HY (1992) Effect of typical antispychotic medications and clozapine on smooth pursuit performance in patients with schizophrenia. Psychiatry Res 41:25–36

Fu L, Beckstead RM (1994) Cortical stimulation induces fos expression in striatal neurons. Neuroscience 46:329–334

Fuxe K, Fredholm B, Agnati LF, Corrodi H (1979) Dopamine receptors and ergot drugs. Evidence that an ergolene derivative is a differential agonist at subcortical limbic dopamine receptors. Brain Res 146:295–311

Garey RE, Heath RG, Harper JW (1974) Focal electroencephalographic changes induced by anti-septal antibodies. Biol Psychiatry 8:75–88

Gerfen CR (1992) The neostriatal mosaic. Annu Rev Neurosci 15:285–320

Geyer MA, Markou A (1995) Animal models of psychiatric disorders. In: Bloom FE, Kupfer DJ (eds) Psychopharmacology: the fourth generation of progress. Raven, New York, pp 631–642

Goldstein M, Deutch AY (1992) Dopaminergic mechanisms in the pathogenesis of schizophrenia. FASEB J 6:2413–2421

Graybiel AM (1984) Correspondence between the dopamine islands and striosomes of the mammalian striatum. Neuroscience 13:1157–1187

Graybiel AM, Moratalla R, Robertson HA (1993) Amphetamine and cocaine induce drug-specific activation of the c-*fos* gene in striosome-matrix compartments and limbic subdivisions of the striatum. Proc Natl Acad Sci USA 87:6912–6916

Groenewegen HJ (1988) Organization of the afferent connections of the mediodorsal thalamic nucleus in the rat, related to the mediodorsal-prefrontal topography. Neuroscience 24:379–431

Guo N, Robertson GS, Fibiger HC (1992) Scopalamine attentuates haloperidol-induced c-fos expression in the striatum. Brain Res 588:164–167

He X, Rosenfeld MG (1991) Mechanisms of complex transcriptional regulation: implications for brain development. Neuron 7:183–196

Heath RG, Guschwan AF, Coffey JW (1970) Relation of taraxein to schizophrenia. Dis Nerv Syst 31:391–395

Heimer L, Zahm DS, Churchill L, Kalivas PW, Wohltmann C (1991) Specificity in the projection patterns of the accumbal core and shell. Neuroscience 41:89–126

Hokfelt T, Everitt BJ, Theodorsson-Norheim E, Goldstein M (1984) Occurrence of neurotensin-like immunoreactivity in subpopulations of hypothalamic, mesencephalic, and medullary catecholamine neurons. J Comp Neurol 222:543–559

Honer WG, Simth GN, Lapointe JS, MacEwen GW, Kopala L, Altman S (1995) Regional cortical anatomy and clozapine response in refractory schizophrenia. neuropsychopharmacology

Hope BT, Nye HE, Kelz MB, Self DW, Iadarola MJ, Nakabeppu Y, Duman RS, Nestler EJ (1994) Induction of a long-lasting AP-1 complex composed of altered Fos-like proteins in brain by chronic cocaine and other chronic treatments. Neuron 13:1235–1244

Huo T, Ye MW, Healy DP (1991) Characterization of a dopamine receptor (DA_{2k}) in the kidney inner medulla. Proc Natl Acad Sci USA 88:3170–3174

Hunt SP, Pini A, Evan G (1987) Induction of c-*fos*-like protein in spinal cord neurons following sensory stimulation. Nature 328:632–634

Iadarola MJ, Chuang EJ, Yeung C-L, Hoo Y, Silverthorn M, Gu J, Draisci G (1993) Induction and suppression of protooncogene in rat striatum after single or multiple treatments with cocaine or GBR 12909. In: Grzanna R, Brown RM (eds) Activation of immediate-early genes by drugs of abuse. NIDA Res Monogr 125:181–211

Itil TM, Soldatos C (1980) Epileptogenic side effects of psychotropic drugs. J Am Med Assoc 244:1460–1463

Jaber M, Tison F, Fournier MC, Bloch B (1994) Differential influence of haloperidol and sulpiride on dopamine receptors and peptide mRNA levels in the rat striatum and pituitary. Mol Brain Res 23:14–20

Jenkins R, O'Shea R, Thomas KL, Hunt SP (1993) c-jun Expression in substantia nigra neurons following striatal 6-hydroxydopamine lesions in the rat. Neuroscience 53:447–455

Jennes L, Stumpf WE, Kalivas PW (1982) Neurotensin: topographic distribution in rat brain by immunohistochemistry. J Comp Neurol 210:211–224

Jian M, Staines WA, Iadarola MJ, Robertson GS (1993) Destruction of the nigrostriatal pathway increases Fos-like immunoreactivity predominantly in striatopallidal neurons. Mol Brain Res 19:156–160

Johansson B, Lindstrom K, Fredholm BB (1994) Differences in the regional and cellular localization of c-*fos* messenger RNA induced by amphetamine, cocaine and caffeine in the rat. Neuroscience 59:837–849

Kane JM, Honigfeld G, Singer J, Meltzer H, and the Clozaril Collaborative Study Group (1988) Clozapine for the treatment-resistant schizophrenic. Arch Gen Psychiatry 45:789–796

Kanes SJ, Hitzemann BA, Hitzemann RJ (1993) On the relationship between D2 receptor density and neuroleptic-induced catalepsy among eight inbred strains of mice. J Pharmacol Exp Ther 267:538–547

Kaufman DL, Houser CR, Tobin AJ (1991) Two forms of the γ-aminobutyric acid synthetic enzyme glutamate decarboxylase have distinct intraneuronal distributions and co-factor interactions. J Neurochem 56:720–723

Keefe KA, Zigmond MJ, Abercrombie ED (1993) In vivo regulation of extracellular dopamine in the neostriatum: influence of impulse activity and local excitatory amino acids. J Neural Transm (Gen) 91:223–240

Konradi C, Kobierski LA, Nguyen TV, Heckers S, Hyman SE (1993) The cAMP-response element binding protein interacts, but Fos protein does not interact, with the proenkephalin enhancer in rat striatum. Proc Natl Acad Sci USA 90:7005–7009

Krettek JE, Price JL (1977) The cortical projections of the mediodorsal nucleus and adjacent thalamic nuclei in the rat. J Comp Neurol 171:157–191

LaHoste GJ, Yu J, Marshall JF (1993) Striatal Fos expression is indicative of dopamine D1/D2 synergism and receptor supersensitivity. Proc Natl Acad Sci USA 90:7451

LaHoste GJ, Ruskin DN, Marshall JF (1994) Cortical Fos expression following dopaminergic stimulation: D1/D2 synergism and its breakdown. Soc Neurosci Abstr 20:406.3

Leadbetter RA, Shutty MS (1994) Differential effects of neuroleptics and clozapine on polydipsia and intermittent hyponatremia. J Clin Psychiatry 55 [Suppl B]:110–113

Le Moine C, Normand E, Guitteny AF, Fouque B, Teoule R, Bloch B (1990) Dopamine receptor gene expression by enkephalin neurons in rat forebrain. Proc Natl Acad Sci USA 87:230–234

Leonard CM (1969) The prefrontal cortex of the rat. I. Cortical projections of the mediodorsal nucleus. II. Efferent connections. Brain Res 12:321–343
Lerea LS, McNamara JO (1993) Ionotrophic glutamate receptor subtypes activate c-*fos* transcription by distinct calcium-requiring intracellular signalling pathways. Neuron 10:31–41
Levy DL, Holzman PS, Matthysse S, Mendell NR (1993) Eye tracking dysfunction and schizophrenia: a critical perspective. Schizophr Bull 19:461–536
Leysen JE, Gommeren W, Eeens A, de Courcelles DC, Stoof JC, Jannsen PAJ (1988) Biochemical profile of risperidone, a new antipsychotic. J Pharmacol Exp Ther 247:661–670
Lieberman JA, Jody D, Alvir JMJ, Ashtari M, Levy DL, Bogerts B, Degreef G, Mayerhoff DI, Cooper T (1993) Brain morphology, dopamine, and eye-tracking abnormalities in first-episode schizophrenia. Arch Gen Psychiatry 50:357–368
Littman RE, Hommer DW, Clem T, Ornsteen ML, Ollo C, Pickar D (1991) Correlation of Wisconsin Card Sorting Test performance with eye tracking in schizophrenia. Am J Psychiatry 148:1580–1582
MacGibbon GA, Lawlor PA, Bravo R, Dragunow M (1994) Clozapine and haloperidol produce a different pattern of immediate early gene expression in rat caudate-putamen, nucleus accumbens, lateral septum, and islands of Calleja. Mol Brain Res 23:21–32
Malmberg A, Jackson DM, Eriksson A, Mohell N (1993) Unique binding characteristics of antipsychotic agents interacting with human D_{2A}, D_{2B}, and D_3 receptors. J Pharmacol Exp Ther 43:749–754
Mandel RJ, Wictorin K, Cenci MA, Bjorklund A (1992) Fos expression in intrastriatal striatal grafts: regulation by host dopaminergic afferents. Brain Res 583:207–215
Mansour A, Watson SJ (1995) Dopamine receptor expression in the CNS. In: Bloom F, Kupfer D (eds) Psychopharmacology: the fourth generation of progress. Raven, New York, pp 207–219
Martin LJ, Blackstone CD, Levey AI, Huganir RL, Price DL (1993) AMPA glutamate receptor subunits are differentially distributed in rat brain. Neuroscience 53:327–358
Matsubara S, Matsubara R, Kusumi L, Koyama T, Yamashita I (1993) Dopamine D_1, D_2 and serotonin2 receptor occupation by typical and atypical antipsychotic drugs in vivo. J Pharmacol Exp Ther 265:498–508
Meltzer HY (1994) An overview of the mechanism of action of clozapine. J Clin Psychiatry 55 [Suppl B]:47–52
Meltzer HY (1995) Atypical antipsychotic drugs. In: Bloom FE, Kupfer DJ (eds) Psychopharmacology: the fourth generation of progress. Raven, New York, pp 1277–1286
Meltzer HY, Matsubara S, Lee J-C (1989) Classification of typical and atypical antispychotic durgs on the basis of D-1, D-2, and serotonin2 pKi values. J Pharmacol Exp Ther 251:238–246
Meltzer HY, Chai BL, Thompson PA, Yamamoto BK (1994) Effect of scopalamine on the efflux of DA and its metabolites after clozapine, haloperidol, or thioridazine. J Pharmacol Exp Ther 268:1452–1461
Menlewicz J, Bleeker E, Cosyns P, Deleu G, Lostra F, Masson A, Mertens C, Parent M, Peuskens J, Suy E, de Wile J, Wilmotte J, Norgard J (1990) A double-blind comparative study of remoxipride and haloperidol in schizophrenic and schizophreniform disorders. Acta Psychiatr Scand 82 [Suppl 358]:138–141
Merchant KM (1994) c-*fos* antisense oligonucleotide specifically attentuates haloperidol-induced increases in neurotensin/neuromedin N mRNA expression in rat dorsal striatum. Mol Cell Neurosci 5:336–344
Merchant KM, Dorsa DM (1994) Differential induction of neurotensin and c-fos gene expression by typical versus atypical antipsychotic drugs. PNAS 90:3447–3451
Merchant KM, Miller MA (1994) Coexpression of neurotensin and c-fos mRNAs in rat neostriatal neurons following acute haloperidol. Mol Brain Res 23:271–277

Merchant KM, Dobie DJ, Dorsa DM (1992) Expression of the proneurotensin gene in the rat brain and its regulation by antipsychotic drugs. Ann NY Acad Sci 668:54–69

Merchant KM, Hanson GR, Dorsa DM (1994) Induction of neurotensin and c-fos mRNA in distinct subregions of rat neostriatum after acute methamphetamine: comparison with acute haloperidol effects. J Pharmacol Exp Ther 269:806–812

Merchant KM, Figur LM, Evans DL (1995) Induction of c-*fos* mRNA in rat medial prefrontal cortex by antipsychotic drugs: role of dopamine D_2 and D_3 receptors. Cerebr Cortex (submitted)

Miller JC (1990) Induction of c-fos mRNA expression in rat striatum by neuroleptic drugs. J Neurochem 54:1453–1455

Moghaddam B (1994) Preferential activation of cortical dopamine neurotransmission by clozapine: functional significance. J Clin Psychiatry 55 [Suppl B]:27–29

Moghaddam B, Bunney BS (1990) Acute effect of typical and atypical antipsychotic drugs on the release of dopamine from the prefrontal cortex, nucleus accumbens, and striatum of the rat: an in vivo microdialysis study. J Neurochem 54:1755–1760

Mohell N, Sallemark M, Rosqvist S, Malmberg A, Hogberg T, Jackson DM (1993) Binding characteristics of remoxipride and its metabolites to dopamine D_2 and D_3 receptors. Eur J Pharmacol 238:121–125

Morgan JI, Curran T (1986) Role of ion flux in the control of c-*fos* expression. Nature 322:552–555

Morgan JI, Curran T (1991) Stimulus-transcription coupling in the nervous system: involvement of the inducible proto-oncogenes *fos* and *jun*. Annu Rev Neurosci 14:421–451

Morgan JI, Curran T (1995) Proto-oncogenes: beyond second messengers. In: Bloom FE, Kupfer DJ (eds) Psychopharmacology: the fourth generation of progress. Raven, New York, pp 631–642

Mugnaini E, Berrebi AS, Morgan JI, Curran T (1989) Fos-like immoreactivity induced by seizure in mice is specifically associated with euchromatin in neurons. Eur J Neurosci 1:46–52

Nakabeppu Y, Nathans D (1991) A naturally occuring trancated form of FosB than inhibits Fos/Jun transcriptional activity. Cell 64:751–759

Nakabeppu Y, Ryder K, Nathans D (1988) DNA binding activities of three murine Jun proteins: stimulation by Fos. Cell 55:907–915

Nemeroff CB, Cain ST (1985) Neurontensin-DA interactions in the CNS. Trends Pharmacol Sci 6:201–205

Nguyen TV, Kasofsky B, Birnbaum R, Cohen BM, Hyman SE (1992) Differential expression of c-fos and zif268 in rat striatum following haloperidol, clozapine, and amphetamine. Proc Natl Acad Sci USA 89:4270–4274

Nieto D, Escobar A (1972) Major psychoses. In: Minkler J (ed) Pathology of the nervous system, vol 3. New York, McGraw-Hill, pp 2654–2665

Nishina H, Sato H, Suzuki T, Sato M, Iba H (1990) Isolation and characterization of fra-2, an additional member of the fos gene family. Proc Natl Acad Sci USA 87:3619–3623

Nisoli E, Tonello C, Memo M, Carruba MO (1992) Biochemical and functional identification of a novel dopamine receptor subtype in rat brown adipose tissue. Its role in modulating sympathetic stimulation-induced thermogenesis. J Pharmacol Exp Ther 263:823–829

Ögren S-O, Florvall L, Hall H, Magnusson O, Ängeby-Möller K (1990) Neuropharmacological and behavioral properties of remoxipride in the rat. Acta Psychiatr Scand 82 [Suppl 358]:21–26

Öngür D, Cameron DS, Goldstein M, Deutch AY (1995) Anatomical and pharmacological characterization of the dopamine innervation of the thalamic paraventricular nucleus (submitted)

Otake K, Ruggiero DA (1995) Monoamines and nitric oxide are employed by afferents engaged in midline thalamic regulation. J Neurosci 15:1891–1911

Parthasarathy HB, Beretta S, Graybiel AM (1994) Cortical stimulation induces selective patterns of Jun B expression in the striatum. Soc Neurosci Abstr 20:406.1

Passingham RE, Myers C, Rawlins N, Lightfoot V, Fearn S (1988) Premotor cortex in the rat. Behav Neurosci 102:101–109

Patel N, Hitzemann B, Hitzemann R (1995) Genetics and the haloperidol-induced increase of Fos in the mouse striatum. Soc Neurosci Abstr 16:807.6

Pflug B, Bartels M, Bauer H, Bunse J, Gallhofer B, Haas S, Kanzow WT, Klieser E, Stein D, Weiselmann G (1990) A double-blind multicentre study comparing remoxipride, controlled release formulation, with haloperidol in schizophrenia. Acta Psychiatr Scand 82 [Suppl 358]:142–146

Preuss TM (1995) Do rats have prefrontal cortex? The Rose-Woolsey-Akert program reconsidered. J Cogn Neurosci 7:1–24

Qian Y, Hitzemann B, Yount GL, White JD, Hitzemann R (1993) D_1 and D_2 dopamine receptor turnover and D_2 messenger RNA levels in the neuroleptic-responsive and the neuroleptic nonresponsive lines of mice. J Pharmacol Exp Ther 267:1582–1590

Quinn JP, Takimoto M, Iadarola MJ, Holbrook N, Levens D (1989) Distinct factors bind the AP-1 consensus sites in gibbon ape leukemia virum and simian virus 40 enhancers. J Virol 63:1737–1742

Rasmussen AM, Goldstein LE, Deutch AY, Bunney BS, Roth RH (1994) 5-HT_{1a} agonist+8-OH-DPAT modulates basal and stress-induced changes in medial prefrontal cortical dopamine. Synapse 18:218–224

Robertson GS, Fibiger HC (1992) Neuroleptics increase c-fos expression in the forebrain: contrasting effects of haloperidol and clozapine. Neuroscience 46:315–328

Robertson GS, Staines HC (1994) D_1 dopamine receptor agonist-induced Fos-like immunoreactivity occurs in basal forebrain and mesopontine tegmentum cholinergic neurons and striatal neurons immunoreactive for neuropeptide Y. Neuroscience 59:375–387

Robertson GS, Vincent SR, Fibiger HC (1992) D_1 and D_2 dopamine receptors differentially regulate c-fos expression in striatonigral and striatopallidal neurons. Neuroscience 49:285–296

Robertson GS, Matsumura H, Fibiger HC (1994a) Induction patterns of Fos-like immunoreactivity in the forebrain as predictors of atypical antipsychotic activity. J Pharmacol Exp Ther 271:1058–1066

Robertson GS, Doucet J-P, Hope BT, Nestler EJ, Nakabeppu Y, Iadarola MJ, Wigle N, St-Jean M (1994b) Truncated FosB is responsible for the long lasting increase in striatal Fos-like immunoreactivity produced by dopaminergic denervation. Soc Neurosci Abstr 20:406.4

Robertson HA, Peterson MR, Murphy K, Robertson GS (1989) D_1-dopamine receptor agonists selectively activate c-*fos* independent of rotational behaviour. Brain Res 503:346–349

Robertson HA, Paul ML, Moratalla R, Graybiel AM (1991) Expression of the immediate early gene c-fos in basal ganglia: induction by dopaminergic drugs. Can J Neurol Sci 18:380–383

Rosse RB, Schwartz BL, Yim SY, Deutsch SI (1993) Correlation between antisaccade and Wisconsin Card Sorting Test performance in schizophrenia. Am J Psychiatry 150:333–335

Rogue P, Vincendon G (1992) Dopamine D_2 receptor antagonists induce immediate early genes in the rat striatum. Brain Res Bull 29:469–472

Sagar SM, Sharp FR, Curran T (1988) Expression of c-*fos* protein in brain: metabolic mapping at the cellular level. Science 240:1328–1331

Sassone-Corsi P, Sisson JC, Verma I (1988) Transcriptional autoregulation of the protooncogene fos. Nature 334:314–319

Schonthal A, Buscher M, Angel P, Rahmsdorf HJ, Ponta H, Hattori K, Chiu R, Karin M, Herrlich P (1989) The Fos and Jun/AP-1 proteins are involved in the downregulation of Fos transcription. Oncogene 4:629–636

Sebens JB, Koch T, Ter Horst GJ, Korf J (1995) Differential Fos-protein induction in rat forebrain regions after acute and long-term haloperidol and clozapine treatment. Eur J Pharmacol 273:175–182

Senger B, Brog JS, Zahm DS (1993) Subsets of neurotensin-immunoreactive neurons in rat striatal complex following antagonism of the dopamine D_2 receptor: an immunohistochemical double-label study using antibodies against Fos. Neuroscience 57:649–660

Seeman P, Van Tol HHM (1993) Dopamine D_4 receptors bind inactive (+)-aporophines, suggesting neuroleptic role. Sulpiride not stereoselective. Eur J Pharmacol 233:173–174

Sesack SR, Bunney BS (1989) Pharmacological characterization of the receptor mediating electrophysiological responses to dopamine in the rat medial prefrontal cortex: a microiontophoretic study. J Pharmacol Exp Ther 248:1323–1333

Sheng M, McFadden G, Greenberg M (1990) Membrane depolarization and calcium induce c-*fos* transcription via phosphorylation of transcription factor CREB. Neuron 4:571–582

Smeal T, Angel P, Meek J, Karin M (1989) Different requirements for formation of Jun:Jun and Jun:Fos complexes. Genes Dev 3:2091–2100

Soghomonian JJ (1994) Differential regulation of glutamate decarboxylase and preproenkephalin mRNA levels in the rat striatum. Brain Res 640:146–154

Sokoloff P, Andrieux M. Besancon R, Pilon C, Martres M-P, Giros B, Schwartz J-C (1992) Pharmacology of human dopamine D_3 receptor expressed in a mammalian cell line: comparison with D_2 receptor. Eur J Pharmacol (Mol Pharm) 225:331–337

Sonnenberg JL, Mitchelmore C, MacGregor-Leon PF, Hempstead J, Morgan JI, Curran T (1989) Glutamate receptor agonists increase expression of Fos, Fra, and AP-1 DNA binding activity in mammalian brain. J Neurosci Res 24:72–80

Stern RG, Kahn RS, Harvey PD, Amin F, Apter SH, Hirschowitz J (1993) Early response to haloperidol treatment in chronic schizophrenia. Schizophr Res 10:165–171

Stevens JR (1982) Neuropathology of schizophrenia. Arch Gen Psychiatry 39:1131–1139

Swerdlow NR, Koob GF (1987) Dopamine, schizophrenia, mania, and depression: toward a unified hypothesis of cortico-striato-pallido-thalamic function. Behav Brain Sci 10:197–245

Takada M, Campbell KJ, Moriizumi T, Hattori T (1990) On the origin of the dopaminergic innervation of the paraventricular thalamic nucleus. Neurosci Lett 115:33–36

Thierry AM, Le Douarin C, Penit J, Ferron A, Glowinski J (1986) Variation in the ability of neuroleptics to block the inhibitory influence of dopaminergic neurons on the activity of cells in the rat prefrontal cortex. Brain Res Bull 16:155–160

Turner BH, Herkenham M (1991) Thalamoamygdaloid projections in the rat: a test of the amygdala's role in sensory processing. J Comp Neurol 313:295–325

Uhl GR, Kuhar M, Snyder SH (1977) Neurotensin: immunohistochemical localization in rat central nervous system. Proc Natl Acad Sci USA 74:4059–4063

Umbricht DSG, Pollack S, Kane JM (1994) Clozapine and weight gain. J Clin Psychiatry 55 [Suppl B]:157–160

Vaccarino FM, Hayward MD, Nestler EJ, Duman RS, Tallman JF (1992) Differential induction of immediate-early genes by excitatory amino acid receptor types in primary cultures of cortical and striatal neurons. Mol Brain Res 12:233–241

Vieweg VWR, Godleski LS, Pulliam WR, Schofield WP, Saathoff GB, Hundley PL, Yank GR (1989) Development of water dysregulation during Arieti's third stage of schizophrenia? Biol Psychiatry 26:775–780

Vogt PK, Bos TJ (1990) Jun: oncogene and transcription factor. Adv Cancer Res 55:1–35

Voorn P, Roest G, Groenewegen HJ (1987) Increase of enkephalin and decrease of substance P immunoreactivity in the dorsal and ventral striatum of the rat after midbrain 6-hydroxydopamine lesions. Brain Res 412:391–396

Weih F, Stewart AF, Boshart M, Nitsch D, Schutz G (1990) In vivo monitoring of a cAMP-stimulated DNA-binding activity. Genes Dev 4:1437–1449
Weinberger DR (1987) Implications of normal brain development for the pathogenesis of schizophrenia. Arch Gen Psychiat 44:660–669
Wirtschafter D, Asin KE (1994) Interactive effects of stimulation of D_1 and D_2 dopamine receptors on fos-like immunoreactivity in the normosensitive rat striatum. Brain Res Bull 35:85–91
Yamamoto BK, Cooperman MA (1994) Differential effects of chronic antipsychotic drug treatment on extracellular glutamate and dopamine concentrations. J Neurosci 14:4159–4166
Yamamoto BK, Davy S (1992) Dopaminergic modulation of glutamate release in striatum as measured by microdialysis. J Neurochem 58:1736–1742
Yamamoto BK, Pehek EA, Meltzer HY (1994) Brain region effects of clozapine on amino acid and monoamine transmission. J Clin Psychiatry 55 [Suppl B]:8–14
Young ST, Porrino LJ, Iadarola MJ (1991) Cocaine induces striatal c-Fos-immunoreactive proteins via dopaminergic D_1receptors. Proc Natl Acad Sci USA 88:1291–1295
Youngren KD, Moghaddam B, Bunney BS, Roth RH (1994) Preferential activation of dopamine overflow in prefrontal cortex produced by chronic clozapine treatment. Neurosci Lett 165:41–44
Zahm DS, Brog JS (1992) On the significance of subterritories in the "accumbens" part of the rat ventral striatum. Neuroscience 50:751–767
Zahm DS, Heimer L (1990) Two transpallidal pathways originating in rat nucleus accumbens. J Comp Neurol 302:437–446
Zerial M, Toschi L, Ryseck R-P, Schuermann M, Muller R, Bravo R (1989) The product of a novel growth factor activated gene, fos B, interacts with JUN proteins enhancing their DNA binding activity. EMBO J 8:805–813
Ziolkowska B, Hollt V (1993) The NMDA receptor antagonist MK-801 markedly reduces the induction of c-fos gene by haloperidol in the mouse striatum. Neurosci Lett 156:39–42.

CHAPTER 6

Basic Neurophysiology of Antipsychotic Drug Action

Patricio O'Donnell and Anthony A. Grace

A. Introduction

The clinical actions of antipsychotic drugs (APDs) have typically been attributed to their effects on dopaminergic systems in the brain. Despite the controversy regarding whether or not blockade of dopamine (DA) receptors is both necessary and sufficient for APDs to exert their therapeutic actions, it is evident that the mesolimbic and/or mesocortical DA systems play some role in the clinical efficacy of these drugs. Similarly, the development of motor side effects following long-term treatment with classical neuroleptics appears to be contingent on their actions on the motor-related nigrostriatal DA projection. The focus of this chapter will be on reviewing studies of the physiology of DA systems as it relates to APD action, with particular emphasis on DA cell firing, how it is generated, how it is controlled, and how long-term treatment with APDs can lead to its inactivation, a condition known as depolarization block. In addition, this chapter will encompass recently described actions of APDs as they involve the modulation of information processing at a network level within structures receiving dopaminergic input (i.e., the striatum and nucleus accumbens). Briefly, it is now known that the ability of DA to modulate electrical coupling between neurons in these structures can be modified by long-term treatment with APDs. Thus, the actions of APDs within the basal ganglia appear to be rather complex, and several levels of analyses must be synthesized in order to gain a more comprehensive perspective of their actions.

B. The Dopamine Hypothesis of Schizophrenia

I. Antipsychotic Drugs as D_2 Blockers

The hypothesis that schizophrenia is due to a hyperdopaminergic state was developed as a consequence of the proposal that neuroleptics are DA antagonists (Carlsson and Lindqvist 1963). This theory gained considerable strength when Ian Creese and colleagues reported that the binding affinities of neuroleptics for DA receptors were highly correlated with their clinical potency (Creese et al. 1976). This correlation with clinical potency was later refined to a specific subtype of DA receptor, i.e., the D_2 receptor (KEBABIAN and CALNE 1979; SEEMAN 1987).

Additional evidence for the involvement of the DA system in psychosis was derived from the use of DA agonists or the DA precursor L-DOPA. A number of patients with Parkinson's disease treated with L-DOPA exhibit psychotic episodes as side effects (CUMMINGS 1992; JENKINS and GROH 1970), and schizophrenics receiving L-DOPA or the DA agonist amphetamine often show a worsening of their condition (ANGRIST et al. 1974). In addition, amphetamine abuse in otherwise normal individuals can induce a state resembling schizophrenia (ANGRIST et al. 1971, 1974; SNYDER 1973); thus, amphetamine psychosis has served as a pharmacological model of schizophrenia (ANGRIST and VAN KAMMEN 1984; SNYDER 1972). These results gave substantial support to the contention that schizophrenic symptoms (or at least the positive symptoms) result from a hyperactive DA system (SNYDER 1972, 1973).

II. Shortcomings of the Dopamine Hypothesis

Despite the early preponderance of evidence implicating an overactive DA system in schizophrenia, subsequent studies introduced substantial doubt into this simple model of the genesis of this disorder. Thus, although APDs produce maximal blockade of DA receptors within minutes of their acute administration (SEDVALL et al. 1986), their clinical actions do not become apparent until the patients have received APDs for weeks (PICKAR et al. 1984). Therefore, the mechanism of action of APDs appears to depend on delayed changes in the system rather than simply a direct blockade of DA receptors. Furthermore, DA systems are capable of exhibiting a remarkable degree of plasticity (HOLLERMAN and GRACE 1990; ZIGMOND et al. 1990), in that symptoms of Parkinson's disease appear only when more than 70% of the DA cells are destroyed by the disease (RIEDERER and WUKETICH 1976). Before that point is reached, the surviving cells are capable of compensating efficiently for this massive loss of DA stimulation. Likewise, one might predict that the DA system in the schizophrenic brain would have the capacity to compensate for pathological alterations in DA levels as well as for the blockade of DA receptors by APDs.

Despite the pharmacological data, substantial evidence for an increase in DA levels in the schizophrenic brain is still lacking (CARPENTER and BUCHANAN 1994; GRACE 1991a). Indeed, many patients exhibit alterations in the opposite direction from what may be predicted, such as low levels of the DA metabolite homovanillic acid (HVA) in the CSF (HERITCH 1990; POST et al. 1975; VAN KAMMEN et al. 1986), particularly in those patients that have a predominance of negative symptoms (VAN KAMMEN et al. 1986). Furthermore, low doses of the indirect DA agonist amphetamine have been reported to improve negative schizophrenic symptoms in a population of schizophrenics (ANGRIST and VAN KAMMEN 1984; GOLDBERG et al. 1991).

These findings led some investigators to speculate that, at least in some patients, there may be a decrease rather than an increase in DA activity in the schizophrenic brain (WEINBERGER 1987; WYATT 1986). Thus, it may be possible that a concidence of high and low DA activity is present in a region-specific manner in schizophrenia (DAVIS et al. 1991; GRACE 1991a). This hypodopaminergic state is proposed to be dependent on a dysfunction of the prefrontal cortex (CSERNANSKY et al. 1991; DAVIS et al. 1991; GRACE 1991a). Indeed, hypofrontality has been observed using regional cerebral blood flow assessments (BERMAN et al. 1986; BUCHSBAUM et al. 1982; FRISTON 1992; LIDDLE et al. 1992a,b; TAMMINGA et al. 1992) or by the Wisconsin Card Sorting Test, which is sensitive to cognitive processes attributed to the prefrontal cortex (FEY 1951; LIDDLE and MORRIS 1991). Recent models propose that, as a consequence of this hypofrontality, a hypodopaminergic state arises, with the result that the subsequent compensatory mechanisms it activates could in turn result in an increase in spike-dependent DA release (GRACE 1991a, see below) or in abnormally high DA release triggered in response to stress (CSERNANSKY et al. 1991). Since negative symptoms in schizophrenics resemble in many ways a hypofrontality syndrome, it has been proposed that a hypofrontality-dependent reduction in basal dopaminergic activity is responsible for the appearance of negative symptoms (CARPENTER and BUCHANAN 1994; DAVIS et al. 1991; GRACE 1991a; VAN KAMMEN et al. 1986), whereas the coexisting hyperreactivity of the mesolimbic DA system may be held responsible for the development of positive symptoms (GRACE 1991a).

C. Acute Physiology of Antipsychotic Drug Action

I. Dopamine Cell Identification and Physiology

The physiological properties of DA cells have been extensively studied over the past 20 years. Initial insight into the regulation of their activity was gained from extracellular single unit recordings (BUNNEY and GRACE 1978; BUNNEY et al. 1973; CHIODO and BUNNEY 1983; GRACE and BUNNEY 1983a–c; KAMATA and REBEC 1984) and, more recently, a more detailed knowledge of the function of these cells was acquired using the more sophisticated technique of intracellular recordings, either from brain slices in vitro (GRACE 1990, 1991b; GRACE and ONN 1989; MERCURI et al. 1992; SHEPARD and BUNNEY 1991; YUNG et al. 1991) or in vivo from intact, anesthetized rats (GRACE and BUNNEY 1983a,b, 1984a, 1986).

When lowering electrodes through the brain in pharmacological studies, it is important to ascertain the neurochemical identity of the cell to be studied. This was initially accomplished by injecting the DA precursor L-DOPA by iontophoresis in the vicinity of a putative DA cell during electrophysiological recordings. Subsequent histochemical examination revealed cells with intense fluorescence surrounding the site of the electrode tip (Bunney et al. 1973). Subsequent studies using intracellular recordings allowed for the direct injection of L-DOPA into single putative DA cells (Grace and Bunney 1980, 1983a). Direct identification of DA cells was also obtained with in vitro intracellular recordings by combining electrophysiological characterization of DA cells (see below) with labeling for tyrosine hydroxylase (Grace and Onn 1989).

Dopamine-containing neurons located in the substantia nigra or the ventral tegmental area (VTA) exist in two activity states in the anesthetized rat: either hyperpolarized and nonfiring, or firing spontaneously with a slow irregular pattern (Grace and Bunney 1984a; Grace 1987). When recorded extracellularly, these cells exhibit a characteristic waveform that has a marked break between the initial segment spike and the somatodendritic component (Bunney et al. 1973; Grace 1987; Grace and Bunney 1980, 1983a) that allows for a reliable electrophysiological identification of these cells.

The physiology of DA cell membranes can also be examined in the absence of the influence of many of the feedback and afferent inputs by performing recordings in rat brain slices containing the substantia nigra, pars compacta (SNc) and the VTA. As observed in vivo, most of the DA cells in this preparation fire spontaneously, with their spike activity generated by an endogenous depolarizing current that precedes action potential discharge, thus resulting in a pacemaker-like regular firing pattern (Fig. 1; Grace and Onn 1989; Johnson and North 1992; Mercuri et al. 1992; Wang and French 1993a,b; Wu et al. 1994).

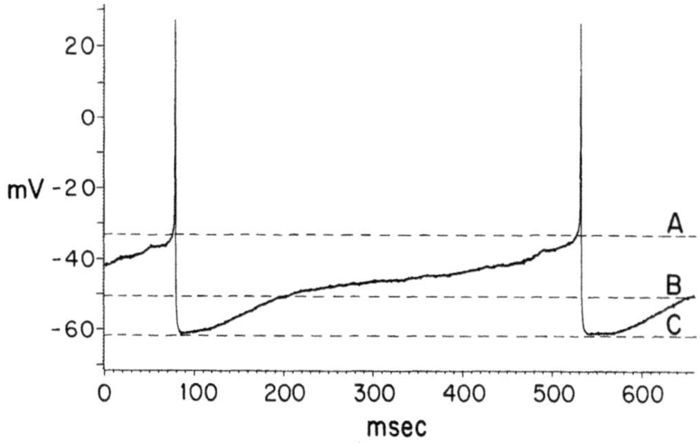

Fig. 1. Spontaneous changes in membrane potential underlying spike activity in dopamine (DA) cells recorded in vitro. A voltage-dependent slow depolarization brings the membrane potential from its resting potential (*dashed line B*) to its unusually high spike threshold (*dashed line A*). The action potential is followed by a marked afterhyperpolarization (*dashed line C*) that hyperpolarizes the neuron below its resting potential. The subsequent return of the membrane potential to baseline then merges with the onset of the slow depolarization for the next spike (from Grace and Onn 1989, with permission)

II. Acute Actions of Antipsychotic Drugs on Dopamine Cell Physiology

Studies on the physiology of the dopamine system in animals have repeatedly found changes in DA cell activity following treatment with APDs. Acute administration of APDs affects DA cell physiology in several ways: (1) by inducing an increase in firing rate, (2) by inducing an increase in the number of cells firing, and (3) by increasing the degree of burst firing (BUNNEY and GRACE 1978; BUNNEY et al. 1973; GRACE and BUNNEY 1984a,b; IWATSUBO and CLOUET 1977; MILLER et al. 1981; PUCAK and GRACE 1994; TUNG et al. 1991).

1. Acute Antipsychotic Drugs Increase Firing Rate

Acute administration of APDs induces an increase in DA cell firing rate (BUNNEY and GRACE 1978; BUNNEY et al. 1973; GRACE and BUNNEY 1984a,b; IWATSUBO and CLOUET 1977; MILLER et al. 1981; PUCAK and GRACE 1994; TUNG et al. 1991). Since this excitatory action may be relevant to the delayed blockade of activity known as depolarization block, we will review here the factors controlling firing rate, which are potential targets of APD action.

DA neurons recorded from rat midbrain slices exhibit a highly regular pattern of firing (GRACE 1988; GRACE and ONN 1989; MUELLER and BRODIE 1989) and a comparatively depolarized spike threshold (Fig. 1; GRACE and ONN 1989). Spike firing is dependent on (Fig. 1): (a) a voltage-dependent, tetrodotoxin (TTX)-sensitive, slow, pacemaker-like depolarization preceding spikes that serves to bridge the membrane potential from resting levels to this depolarized spike threshold (GRACE 1990, 1991b; GRACE and ONN 1989); (b) an initial segment spike (GRACE and ONN 1989) that triggers (c) high-threshold dendritic calcium spikes (CHIODO and KAPATOS 1992; GRACE and ONN 1989; HOUNSGAARD et al. 1992); these are followed by (d) a calcium-dependent large-amplitude afterhyperpolarization (Fig. 1; GRACE 1991b; GRACE and ONN 1989; SHEPARD and BUNNEY 1991; WALSH et al. 1991) which, as a rebound, triggers (e) a low-threshold calcium spike (GRACE and ONN 1989; WALSH et al. 1991; YUNG et al. 1991). In addition, the spike activity is further modulated by prominent passive membrane conductances: (a) a delayed repolarization (GRACE and ONN 1989; HARRIS and GREENFIELD 1991) and (b) instantaneous and time-dependent anomalous rectification (GRACE 1991b; GRACE and ONN 1989; HARRIS and GREENFIELD 1991; MUELLER and BRODIE 1989; WALSH et al. 1991; WANG and FRENCH 1993b). These cells can be found either tonically active or silent, and transitions between these states are likely to occur, since silent neurons can be induced to fire action potentials by drug treatment (GRACE 1991b; GRACE and BUNNEY 1984b).

Physiological properties of DA neurons recorded intracellularly in vivo are generally consistent with those observed in the in vitro preparation. Thus, in vivo studies also show a slow firing pattern, apparently under the control of: (a) a slow depolarizing potential that brings the cell membrane potential to spike threshold (GRACE and BUNNEY 1984b); (b) an initial segment spike, which is probably a low-threshold sodium spike triggered at the axon hillock (GRACE and BUNNEY 1983b); (c) somatodendritic spikes; and (d) a long-duration afterhyperpolarization (GRACE and BUNNEY 1983b, 1984b), similar to that shown using in vitro recordings.

DA cell firing is also under the influence of autoreceptors located in the somatodendritic region of DA cells (BUNNEY et al. 1973; GROVES et al. 1975). These receptors were traditionally considered to be of the D_2 type (BOWERY et al. 1994; BRODIE and DUNWIDDIE 1987), but recent work suggests that they may be of the D_3 subtype (ACKERMAN et al. 1993) or a D_2/D_3 combination (SCHWARTZ et al. 1992). Somatodendritic autoreceptors can be activated by the release of dendritic DA (GROVES et al. 1975; KALIVAS and DUFFY 1991; SANTIAGO and WESTERINK 1991, 1992; ZHANG et al. 1994), and their activation leads to a decrease in DA cell firing (BUNNEY et al. 1973; ACKERMAN et al. 1993; BOWERY et al. 1994; BRODIE and DUNWIDDIE 1987; GRACE and BUNNEY 1983a; GRACE and ONN 1989; JOHNSON and NORTH 1992; KELLAND et al. 1990; WANG and FRENCH 1993b) and to a membrane hyperpolarization (BOWERY et al. 1994; JOHNSON and NORTH 1992; WANG and FRENCH 1993b). This activation of somatodendritic autoreceptors may inhibit DA cell activity via the activation of a G protein-mediated pathway (INNIS and AGHAJANIAN 1987; LIU et al. 1994), resulting in an increase in K^+ conductances (CHIODO 1992; LACEY et al. 1987). These effects are different from those attributed to presynaptic autoreceptors located on DA terminals in the accumbens and striatum that control DA synthesis and release (TEPPER et al. 1984). The sensitivity of the somatodendritic receptors is changed by long-term administration of DA agonists (KAMATA and REBEC 1984; KLAWANS and MARGOLIN 1975) and in the cells surviving DA depletion by 6-hydroxydopamine (PUCAK and GRACE 1991); thus, the compensatory mechanisms in which they participate can be adjusted to the condition created as a result of continuous receptor blockade or stimulation.

The effects produced by acute administration of classical versus atypical APDs exhibit regional differences. Acute administration of haloperidol (HAL) induces an increase in firing rate in the DA cells located in both the SNc and the VTA (BUNNEY et al. 1973), whereas the atypical APD clozapine (CLZ) only induces changes in DA cells located in the VTA (BUNNEY and AGHAJANIAN 1975; DEUTCH et al. 1991). Similarly, acute treatment with HAL or CLZ can also induce other neurochemical changes in the striatal complex with similar regional selectivity. Both drugs may induce the expression of immediate early genes in the accumbens, whereas only HAL induces immediate early gene expression in the dorsal striatum (DEUTCH et al. 1992; MACGIBBON et al. 1994; NGUYEN et al. 1992; ROBERTSON and FIBIGER 1992) (see Chap. 4); similar patterns of response were described for the expression of neurotensin mRNA (MERCHANT and DORSA 1993), basal DA levels (ICHIKAWA and MELTZER 1992), DA turnover (CSERNANSKY et al. 1993), and, dihydroxyphenylacetic acid (DOPAC) levels (KAROUM et al. 1994).

Although their role in relation to clinical APD actions is less clear, other neurotransmitters and modulators can also affect DA cell activity. DA cell firing can be increased by intranigral or intra-VTA infusion of acetylcholine (LICHTENSTEIGER et al. 1976), neurotensin (MERCURI et al. 1993; SHI and BUNNEY 1992), cholecystokinin (BRODIE and DUNWIDDIE 1987; FREEMAN and CHIODO 1987; HOMMER et al. 1985), phencyclidine (FRENCH 1994), morphine (GIFFORD and WANG 1994; HENRY et al. 1992), and brain-derived neurotrophic factor (SHEN et al. 1994).

2. Acute Antipsychotic Drug Administration Increases Dopamine Cell Burst Firing

The changes induced by APDs on the DA cell firing pattern can be evaluated by quantifying the degree of burst firing. On occasion, DA neuron spike discharge occurs in a burst pattern, with each burst of spikes comprised of a series of 3–10 action potentials of decreasing amplitude and increasing duration (Fig. 2; BUNNEY et al. 1973; GRACE and BUNNEY 1983a). Acute administration of HAL induces an increase in DA cell firing rate and in the proportion of spikes the cells fire in a bursting pattern (BUNNEY and GRACE 1978; BUNNEY et al. 1973; GRACE and BUNNEY 1984a,b; IWATSUBO and CLOUET 1977; MILLER et al. 1981; TUNG et al. 1991). These bursts can also be elicited by glutamate administration (GRACE and BUNNEY 1984a) or by the activation of NMDA receptors (CHERGUI et al. 1993, 1994; OVERTON and CLARK 1992; JOHNSON et al. 1992; SEUTIN et al. 1993), and firing of these cells can be regularized by administration of *N*-methyl-D-aspartate (NMDA) receptor antagonists (CHERGUI et al. 1993). Furthermore, the stimulation of glutamatergic projections from the subthalamic nucleus also induces burst firing (CHERGUI et al. 1994; SMITH and GRACE 1992), presumably via activation of NMDA receptors (CHERGUI et al. 1994). Conversely, administration of the gamma-aminobutyric acid (GABA)-B receptor agonist baclofen causes burst-firing neurons to enter a non-bursting mode (GRACE and BUNNEY 1980; ENGBERG et al. 1993). Therefore, it appears that DA cell burst firing is the result of a delicate balance between excitatory and inhibitory inputs to these cells, with glutamatergic afferents possibly playing a significant role in this pattern of activity.

However, unlike the DA cells recorded in the intact preparation, DA cells recorded in vitro have not been observed to discharge spikes in a burst pattern (GRACE 1988; GRACE and ONN 1989; HARRIS and GREENFIELD 1991; MERCURI et al. 1992; SANGHERA et al. 1984; WANG and FRENCH 1993a; WU et al. 1994; YUNG et al. 1991), even when NMDA is applied to the slices (MERCURI et al. 1992; SEUTIN et al. 1990; WANG and FRENCH 1993b; but see SEUTIN et al. 1993). Approximately 40% of the DA cells recorded in vivo exhibit spontaneous activity with some degree of burst firing that can be observed as a group of spikes arising from a membrane depolarizing event, with the spikes exhibiting

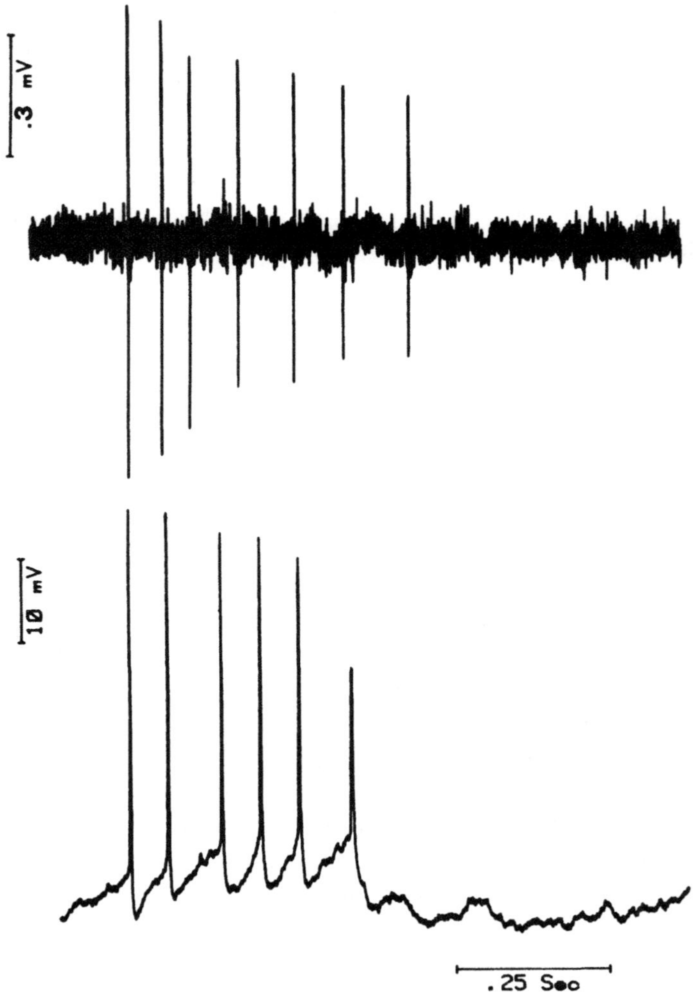

Fig. 2. DA cells recorded in vivo exhibit spontaneously occurring bursts of action potentials. ***Top*** Extracellular recording of a DA cell discharging spikes in bursts, with the spikes exhibiting a progressive decrease in amplitude and increase in duration as the burst progresses. ***Bottom*** In vivo intracellular recordings of DA cells also revealed the presence of spontaneous burst firing. The spikes within the burst also exhibited progressive increases in duration and decreases in amplitude (from GRACE and BUNNEY 1984a, with permission)

a progressive decrease in amplitude and increase in duration within the burst (Fig. 2; GRACE and BUNNEY 1983a, 1984a).

A number of transmitters, including DA, can modulate DA cell firing and bursting. The control of burst firing is likely to have a significant impact on how DA cells affect their postsynaptic targets, given the increased efficiency of DA release that results from a burst of action potentials when compared to single spikes (GONON 1988; GRACE and BUNNEY 1984a; NISSBRANDT et al.

1994). This is particularly relevant to APD actions since some of the acute and many of the chronic effects depend on changes in inputs to DA cells occurring by way of the striatonigral projection (Bunney and Grace 1978; Pucak and Grace 1991, 1994). Furthermore, local infusion of glutamate as well as direct application of glutamate agonists to DA cells in vivo by means of microiontophoresis increases their firing rate and causes them to fire in bursts (Chergui et al. 1993, 1994; Grace and Bunney 1984a; Suaud-Chagny et al. 1992). This glutamatergic innervation appears to depend at least in part on afferents from the subthalamic nucleus, since subthalamic nucleus stimulation is able to induce burst firing in SNc neurons, and lesions of this structure cause a regularization of DA cell discharge (Smith and Grace 1992). The glutamatergic innervation of the VTA, if it exists, is not as clear. Another input that may have an impact on DA cell firing and burst discharge is the feedback to DA cells mediated by afferents arising from their target neurons in the striatum or nucleus accumbens (Iwatsubo and Clouet 1977). These projections originate from GABAergic medium spiny neurons, and may in turn inhibit cell firing in the SNc and VTA. Although it has been shown that cortical stimulation will evoke a burst of spikes in a small number of DA neurons (Gariano and Groves 1988), it is not clear whether this is a direct effect or is dependent on activation of neurons in the nucleus accumbens (Harden and Grace 1994).

3. Acute Antipsychotic Drug Treatment Increases the Number of Dopamine Cells Firing Spontaneously

Early studies of APD action yielded evidence that a substantial proportion of the DA cells in intact rats do not discharge spikes spontaneously, but may be activated by APD treatments. However, quantifying this phenomenon during drug treatment would have been impossible using the standard electrophysiological measures present at that time. Therefore, a method of assessing the relative number of active cells and its alteration in response to different treatments was developed (Bunney and Grace 1978). This technique is based on counting the number of active cells that could be observed firing spontaneously during the passage of a recording electrode through the DA cell region in a predetermined pattern. This resulted in a measure of the number of cells firing per electrode track that was found to be highly reproducible between rats. With this method, acute administration of haloperidol (HAL) was found to increase the number of cells firing per electrode track (Bunney and Grace 1978).

D. Physiology of Chronic Antipsychotic Drug Treatment

I. Tolerance to Antipsychotic Drug Action

Repeated treatment with APDs has been found to produce unique responses with respect to behavioral, neurochemical, and physiological measures when compared to the effects occurring with acute APD administration. Acute administration of HAL induces a dose-dependent decrease in motor activity and an increase in striatal DA levels (IMPERATO and DICHIARA 1985; ZETTERSTRÖM et al. 1985). However, its antipsychotic action and parkinsonian side effects in schizophrenic patients require repeated treatment, suggesting that repeated administration has qualitatively distinct effects on the DA system. Furthermore, studies have shown that repeated APD administration may result in time-dependent changes in a number of DA parameters. For example, there is a sensitization to the increase in DA levels after repeated HAL administration (CSERNANSKY et al. 1990). On the other hand, following the initial increase in DA turnover caused by autoreceptor blockade, there is a time-dependent decrease in DA metabolites. Despite these initial alterations in DA system dynamics, there is a remarkable absence of clinical tolerance with repeated APDs. Indeed, the clinical efficacy of APDs improves with time, resulting in a reversal of symptoms after a few weeks of treatment.

II. Chronic Treatment with Antipsychotic Drugs and Dopamine Cell Depolarization Block

Although acute administration of APDs induces an increase in the firing rate of DA cells, long-term APD treatment results in a substantially different response. In contrast to its acute actions, repeated administration of HAL for 3 or more weeks results in a significant reduction in the number of spontaneously active DA cells recorded (BUNNEY and GRACE 1978; CHIODO and BUNNEY 1983; SKARSFELDT 1994; WACHTEL and WHITE 1992; WHITE and WANG 1983). Administration of other typical APDs (e.g., sulpiride or remoxipride) for 21 or more days also results in a decrease in the number of DA cells per track (SKARSFELDT 1993). Thus, it appears that a majority of DA cells are rendered completely silent by these drugs. Furthermore, this state appears to be different from that present in quiescent cells of untreated animals, since neurotransmitters that would normally activate silent DA cells do not activate the cells rendered silent by chronic APD treatment (Fig. 3A,B; BUNNEY and GRACE 1978). This blockade is not the result of an increased activity of an inhibitory GABAergic projection to these cells, since spike firing can be restored in these "silent" cells by the direct application of GABA to those nonfiring cells by means of microiontophoresis (Fig. 3C; BUNNEY and GRACE 1978). Furthermore, this is not the result of a change in cellular response to GABA, since an increase in the amount of GABA iontophoresed beyond that needed for the reversal of depolarization block results in an inhibition in firing rate (BUNNEY and GRACE 1978). Using in vivo intracellular recordings, the silent DA cells in chronic HAL-treated animals were found to be depolarized to a point at which their spike generating zone was inactivated. Their activity could be restored if the membrane was hyperpolarized by current injection or by administration of systemic DA agonists (GRACE and BUNNEY 1986). Thus,

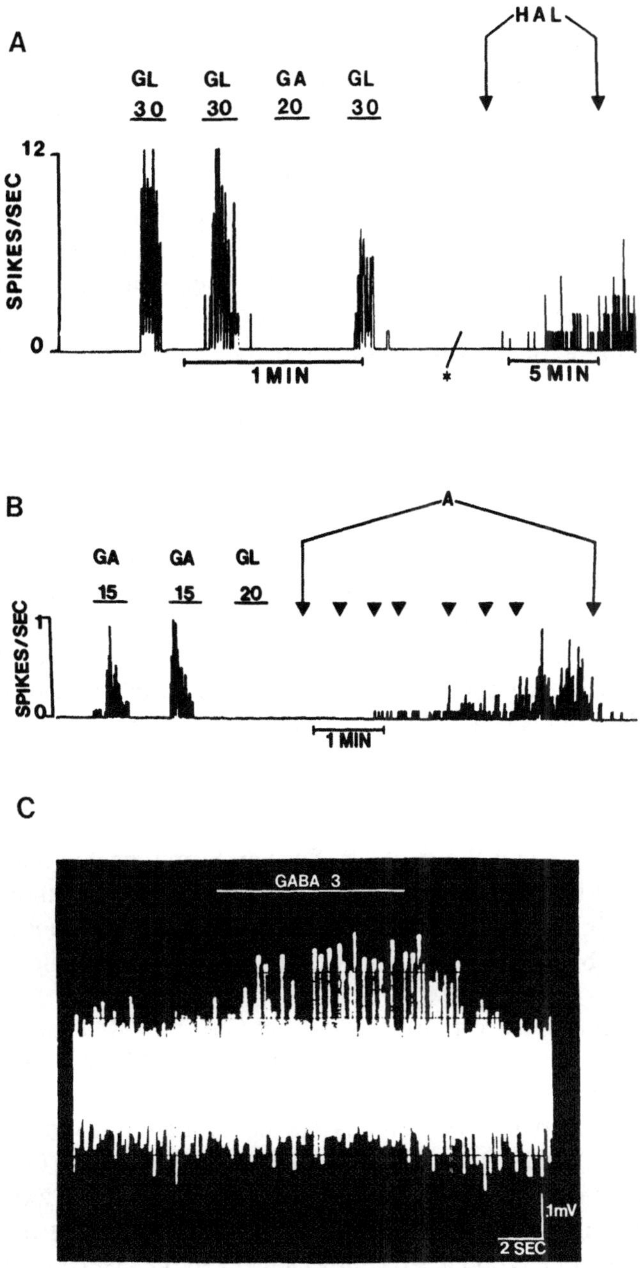

Fig. 3A–C. Depolarization block in DA neurons that occurs after repeated administration of haloperidol (*HAL*) can be reversed by application of the inhibitory neurotransmitter gamma-aminobutyric acid acid. **A** In DA cells recorded from control animals, iontophoretic application of glutamate (GL) exhibits its typical excitatory effect on the cell to induce spike discharge, whereas GABA (GA) fails to activate spike firing. Application of HAL causes the cell to begin firing spontaneously. **B** Iontophoretic application of GABA restores firing in a DA cell in depolarization block, whereas application of the excitatory neurotransmitter glutamate fails to induce cell activation. Furthermore, repeated application of apomorphine (*A*) causes the cell to begin firing spontaneously. **C** Trace from an extracellular recording of a DA cell in depolarization block that is activated by iontophoretic application of GABA (from Bunney and Grace 1978, with permission)

this silent state of DA cells occurs as a result of excessive depolarization of their membranes, and was initially named "depolarization inactivation" (Bunney and Grace 1978). Since further experiments involving intracellular recording of DA cells provided evidence that this state was indeed dependent on an excessive membrane depolarization that prevents spikes from being triggered, in a manner similar to what had been described in the spinal cord (Zieglgansberger and Puil 1983), it was later identified as "depolarization block" (Grace 1992; Grace and Bunney 1986).

Long-term treatment with classical and atypical APDs induce depolarization block of DA cells in a regionally selective manner. On the one hand, DA cells in the substantia nigra pars compacta that project primarily to the motor aspect of the striatal complex (i.e., the dorsal striatum) enter depolarization block after 3 or more weeks of treatment with the classic neuroleptic HAL, but not with the atypical APD clozapine (CLZ) (Chiodo and Bunney 1983; White and Wang 1983). On the other hand, DA cells located in the limbic-related VTA, which projects to the prefrontal cortex and nucleus accumbens, enter depolarization block following chronic treatment with either HAL or CLZ (Chiodo and Bunney 1983; White and Wang 1983). Thus, it is now recognized that drugs that induce depolarization block in DA cells that project to the accumbens (i.e., VTA cells) also are effective antipsychotic drugs (e.g., HAL and CLZ), whereas drugs that induce depolarization block of the substantia nigra DA cells projecting to the dorsal striatum will also lead to the development of extrapyramidal symptoms (e.g., HAL) (Chiodo and Bunney 1983; Grace 1992; White and Wang 1983). Furthermore, repeated administration of low doses of the D_2 antagonist metoclopramide, which is known to exert motor effects in the absence of significant antipsychotic actions, results in depolarization block of SNc DA cells, but not VTA DA cells (White and Wang 1983).

Since its initial observation (Bunney and Grace 1978), a large number of groups have confirmed that repeated treatment with APDs will cause depolarization block of DA cell firing in rats (Chiodo and Bunney 1983, 1985; Goldstein and Litwin 1988; Goldstein et al. 1989; Grace and Bunney 1986; Henry et al. 1992; Jiang et al. 1988; Kabzinski et al. 1987; Meltzer et al. 1989; Sorensen et al. 1989; Skarsfeldt 1988, 1993, 1994; Szewczak et al. 1990; Todorova and Dimpfel 1994; Wachtel and White 1992; White and Wang 1983). However, one group has recently challenged the assertion that DA cell depolarization block can occur in animals in the absence of anesthesia, and instead suggested that depolarization block may be a consequence of the additive excitatory interaction between the anesthetic and the APD (Mereu et al. 1994). Although it is well established that anesthesia will increase the firing rate of DA cells (Bunney et al. 1973), it is also an established fact that anesthesia will blunt the excitatory actions of APDs on DA cells (Bunney et al. 1973; Mereu et al. 1983, 1984). Therefore, one would predict that nonanesthetized rats would be more likely to show depolarization block than the anesthetized rats. This is consistent with earlier studies showing the presence of APD-induced depolarization block in nonanesthetized rats (BUNNEY and GRACE 1978). Furthermore, additional studies have provided biochemical (ABERCROMBIE et al. 1989; BLAHA and LANE 1987; CHEN et al. 1991; DEUTCH and ROTH 1988; FINLAY et al. 1987; LANE and BLAHA 1987) and behavioral (DOHERTY and GRATTON 1991; ROMPRÉ and WISE 1989) correlates of depolarization block in the nonanesthetized animal. One possibility that may account for the difficulties experienced by MEREU et al. (1994) may be related to our earlier findings that rats exposed to extreme stressors fail to show depolarization block (GRACE and BUNNEY, unpublished observations), which may play a substantial role when using paralyzed, nonanesthetized rats (MEREU et al. 1994). Thus, on the basis of overwhelming evidence contradicting the MEREU et al. (1994) report, APD-induced depolarization block occurs independently of the presence of anesthesia.

III. Analysis of Mechanisms Contributing to Depolarization Block: Compensatory Systems Involved in the Recovery of Function After Dopamine Lesions

The cellular mechanisms which lead to the development of depolarization block are difficult to study electrophysiologically due to the requirement for repeated drug treatment. However, a new model of depolarization block has been developed, which is also based on compensatory changes induced in the DA system secondary to a decreased DA receptor stimulation; however, in this case the compensatory changes are first induced by producing a partial lesion of the DA innervation of the striatum. In this model, rats with a 6-hydroxydopamine-induced partial lesion of the DA system that receive a dose of APD that is too small to affect controls exhibited a profound akinesia. Electrophysiological recordings of the residual DA neurons in these lesioned rats revealed that a similar dose of APD also resulted in acute depolarization block of the nonlesioned DA cells. Extremely high doses of APD failed to induce depolarization block of DA cells in intact rats; however, in animals that had been partially depleted of DA with 6-hydroxydopamine 1 month prior to recording, depolarization block was induced in the surviving DA cells by the acute administration of only 0.1 mg/kg of HAL i.v. (Fig. 4; HOLLERMAN et al. 1992; HOLLERMAN and GRACE 1989). Other manipulations were also found to be capable of inducing DA cell depolarization block acutely: (1) Local or systemic administration of morphine, which increases the firing rate of VTA cells, combined with a D_2 antagonist results in the cells entering depolarization block (HENRY et al. 1992); and (2) iontophoresis of glutamate or the glutamate agonists AMPA and NMDA at increasing concentrations onto nigral DA cells yield progressively larger increases in firing rate, with the higher doses resulting in a large increase of firing giving way to a drastic reduction in firing (OVERTON and CLARK 1992; WU et al. 1994) or depolarization block (GRACE and BUNNEY 1986). Thus, activation of glutamate receptors at a sufficient level

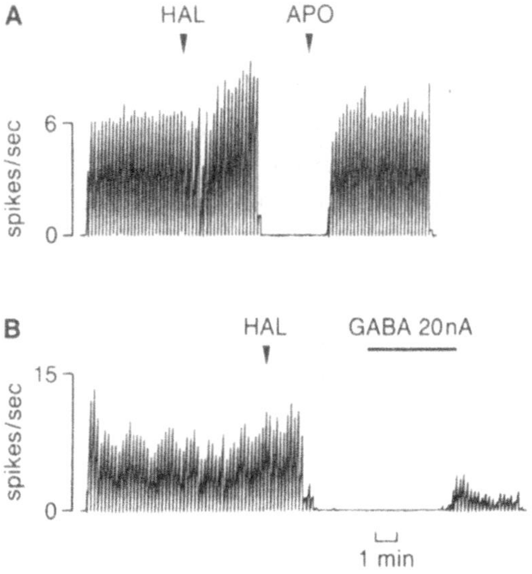

Fig. 4A,B. Although extremely high doses of antipsychotic drugs (APDs) fail to induce DA cell depolarization block in intact rats, DA cell depolarization block can be induced acutely in rats with partial DA lesions by administering comparatively low doses of HAL. **A** Systemic administration of HAL causes a progressive increase in firing rate of a spontaneously firing DA neuron until spike activity ceases secondary to the onset of depolarization block. Subsequent administration of apomorphine (*APO*) results in the restoration of spontaneous spike firing. **B** Following acute induction of depolarization block by HAL, iontophoretic application of GABA (*bar*) reverses the depolarization block to allow the cell to resume spontaneous spike firing. Full-amplitude spikes were observed in this case after cessation of GABA application, presumably due to a potent GABA-induced hyperpolarization that was sufficient to reverse depolarization block and inhibit spontaneous spike discharge (from HOLLERMAN and GRACE 1989, with permission)

to cause an excessive depolarization of these cells can cause depolarization blockade of their firing.

Evidence suggests that the striatonigral feedback pathway is involved in the development of depolarization block. Since APDs are primarily DA antagonists and their clinical efficacy appears to be related to their ability to block D_2 receptors (SEEMAN 1987), the sites potentially responsible for inducing depolarization block may be either the somatodendritic autoreceptors in the substantia nigra/VTA or the postsynaptic DA receptors in the striatum and nucleus accumbens. Unilateral striatal lesions prior to chronic HAL administration prevent the appearance of depolarization block in the substantia nigra ipsilateral to the lesion (BUNNEY and GRACE 1978). Furthermore, transections of striatonigral fibers reverse depolarization block in rats chronically treated with APDs (CHIODO and BUNNEY 1983). Finally, systemic administration of the GABA-B agonist baclofen, which blocks GABA release from striatonigral GABAergic fibers (NISENBAUM et al. 1993), will also reverse

HAL-induced depolarization block (GRACE and BUNNEY 1980). Thus, the integrity of the striatonigral pathway is required for both the onset and maintenance of depolarization block.

E. Dual Mode of Dopamine Release: Tonic Versus Phasic

The simplistic view of schizophrenic symptoms as arising from a hyperactive DA system has been challenged by a number of observations, as reviewed above. Furthermore, despite intensive investigations, studies to date have failed to localize a pathological condition within the DA system of schizophrenics. On the other hand, there is increasing evidence from metabolic and imaging studies of schizophrenics that suggests the involvement of structures such as the prefrontal cortex, temporal lobe, entorhinal cortex, and amygdala. There is substantial evidence that pathological changes in the prefrontal cortex may contribute to this disorder, including: (a) a reduction in regional cerebral blood flow in the dorsolateral aspect of the prefrontal cortex of schizophrenics (BERMAN et al. 1986; BUCHSBAUM et al. 1982; FRISTON 1992; LIDDLE et al. 1992a,b; WEINBERGER et al. 1992); (b) poor performance in neuropsychological tests sensitive to prefrontal cortical function (LIDDLE and MORRIS 1991; RAINE et al. 1992); (c) structural alterations observed with magnetic resonance imaging (MRI) (RAINE et al. 1992); (d) a reduction in the release of GABA and glutamate in tissue obtained from the prefrontal cortex of schizophrenics (SHERMAN et al. 1991), and (e) alterations in phosphomonoesters in the frontal cortical fibers (PETTEGREW et al. 1991). On the other hand, several studies have provided evidence for anatomical disruptions occurring in the amygdala-entorhinal complex (ARNOLD et al. 1991) and in the hippocampus (FALKAI and BOGERTS 1986; KOVELMAN and SCHEIBEL 1984) in this disorder. A potential involvement of these structures is further supported by MRI studies of schizophrenics in which a reduction in the size of the left temporal lobe has been reported (BARTA et al. 1990; BOGERTS et al. 1993; GUR and PEARLSON 1993; KAWASAKI et al. 1993; SHENTON et al. 1992; SUDDATH et al. 1990; WADDINGTON 1993) and by positron emission tomography (PET) studies revealing the presence of diminished metabolic activity in the hippocampus of schizophrenics (TAMMINGA et al. 1992). A potential involvement of these cortical projections in schizophrenia is supported by studies showing that the indirect glutamate NMDA antagonist phencyclidine can mimic psychosis, and in schizophrenics it exacerbates symptoms (LAHTI et al. 1993) in a manner that is specific for the symptom features of the individual patient (JAVITT and ZUKIN 1991). Furthermore, there is evidence that the cortical glutamatergic system may regulate subcortical DA neurons. Thus, in animal studies it has been shown that a prefrontal cortical lesion can exacerbate amphetamine-induced behaviors (IVERSEN et al. 1971; JASKIW et al. 1990).

Although substantial behavioral, biochemical, and anatomical evidence exists that supports a glutamate–DA interaction, the cellular mechanism underlying this phenomenon is unclear. On one hand, empirical data shows that the prefrontal cortex functionally suppresses subcortical DA function. On the other hand, several studies have provided evidence for a glutamate-dependent DA release in subcortical structures. Thus, there is a considerable amount of evidence suggesting that DA release can be induced in the striatum by administration of excitatory amino acids (cf. GRACE 1991a). For example, administration of glutamate agonists into the striatal complex, as well as stimulation of cortical afferents, can elicit DA release, as shown using brain slices (ASENCIO et al. 1991; CAI et al. 1991; GIORGUIEFF et al. 1977; JONES et al. 1987; KREBS et al. 1991a,b; ROBERTS and ANDERSON 1979), synaptosomes (JOHNSON and JENG 1991; WANG 1991), in vivo assessments using a push-pull cannula (CHÉRAMY et al. 1986; LEVIEL et al. 1990; NIEOULLON et al. 1978), or, more recently, with in vivo microdialysis (CARROZZA et al. 1991, 1992; CARTER et al. 1988; IMPERATO et al. 1990; MARTÍNEZ-FONG et al. 1992; MOGHADDAM et al. 1990; SHIMIZU et al. 1990; YOUNGREN et al. 1993). This release appears to be independent of DA cell firing, since it cannot be prevented by the sodium channel blocker tetrodotoxin (CARTER et al. 1988; CHÉRAMY et al. 1986; CLOW and JHAMANDAS 1989; GIORGUIEFF et al. 1977; KREBS et al. 1991b; ROBERTS and ANDERSON 1979; but see KEEFE et al. 1993). Furthermore, activation of the prefrontal cortex by local glutamate injections increases the DA oxidation peak in voltammograms recorded from the nucleus accumbens, whereas infusion of lidocaine in the prefrontal cortex reversibly reduces this peak (MURASE et al. 1993). Indeed, recent studies by GONZALEZ BURGOS et al. (1995) in other systems have demonstrated a biophysical basis for separate spike-dependent and spike-independent modes of neurotransmitter release. An anatomical substrate for such DA-glutamate interactions may indeed be present. Thus, although no axo-axonic synapses have been reported in the striatal complex, electron microscopic studies have shown that terminals labeled with tyrosine hydroxylase antibodies often occur in close apposition with corticostriatal terminals (BOUYER et al. 1984; SESACK and PICKEL 1990, 1992).

How can the apparent conflict between prefrontal cortical activation of subcortical DA-mediated behaviors be reconciled with evidence for glutamate-mediated DA release? These interactions led to the formulation of the hypothesis of a dual control of DA release in the striatum (GRACE 1991a). Most behavioral responses observed with DA appear to be dependent on phasic, spike-dependent release (Fig. 5) that allows for a transient rise in synaptic DA which is rapidly removed from the synaptic cleft by high-speed reuptake mechanisms (IVERSEN 1975). In addition to this phasic, spike-dependent release of DA by the terminals in the striatal complex, there appears to be a tonic, spike-independent DA release that may be driven by the presynaptic actions of glutamatergic cortical inputs onto DA terminals within the striatum (Fig. 6; GRACE 1991a). This constant glutamatergic input is proposed to underlie steady-state, very low levels of extracellular DA in the striatum.

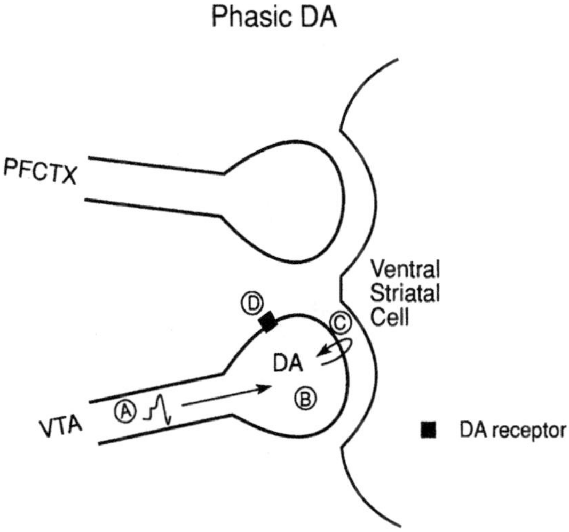

Fig. 5. The phasic component of DA release. The phasic or transient component is proposed to be the DA release caused by spike activity in DA neurons (*A*). In the DA terminal in the striatal complex (*B*), the action potential induces DA release onto the synaptic cleft, where it acts upon DA receptors on postsynaptic cells before it is rapidly cleared from the synaptic space by reuptake into the DA terminals (*C*). Because the phasically released DA does not escape from the synaptic cleft, it fails to stimulate the presynaptic DA autoreceptors located in the extrasynaptic space (*D*) (from GRACE 1992, with permission). *PFCTX*, prefrontal cortex; *VTA*, ventral tegmental area

The tonic nature of the glutamate-dependent DA release is further supported by the ability of glutamate antagonists to attenuate the increase in extracellular DA levels induced by DA uptake blockers (MOGHADDAM and BOLINAO 1994). Although this steady-state DA concentration exists at only a small fraction of the concentration of synaptic DA, the levels measured by microdialysis techniques are well within the sensitivity range of DA autoreceptors on DA terminals (cf. O'DONNELL and GRACE 1994). Thus, small glutamate-induced increases in extracellular DA should act to down-regulate the much larger spike-dependent DA release occurring within the synapse. As a consequence, the amount of DA to be phasically released may be an inverse function of the basal level of tonic DA present in the extrasynaptic space.

Such an interaction could account for the apparent glutamate-DA dysfunction observed in schizophrenics (GRACE 1991a). In this model, a pathological decrease in the activity of prefrontal cortical projections to the accumbens leads to a reduction in tonic, glutamate-dependent DA release, which in turn would produce a decrease in the basal extracellular DA levels in the striatum. The resultant decrease in DA terminal autoreceptor stimulation would thus produce an abnormally large enhancement of spike-dependent DA. As a result of this up-regulation, stimulus-dependent spike activity in the nigrostriatal/mesoaccumbens systems would lead to the release of abnormally

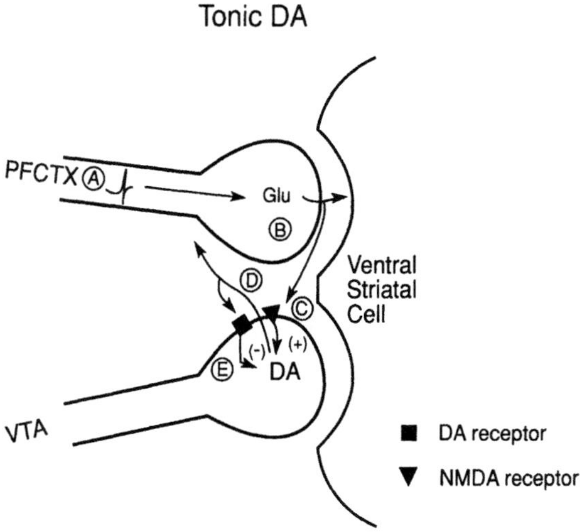

Fig. 6. The tonic component of DA release. In this model, extrasynaptic DA is proposed to depend on glutamate stimulation of *N*-methyl-D-aspartate (*NMDA*) receptors located on DA terminals within the striatal complex, causing an influx of calcium to trigger DA release. Thus, spike activity in corticostriatal fibers (*A*) evokes glutamate release in the striatal complex (*B*), where, in addition to stimulating glutamate receptors on postsynaptic cells, it escapes to the surrounding extrasynaptic space (*C*). As a result, the increase in glutamate levels in the vicinity of neighboring DA terminals is sufficient to provide tonic stimulation of NMDA receptors located on the DA terminals (*C*), resulting in spike-independent DA release into the extrasynaptic space (*D*). The prolonged time course of this tonic DA component may underlie the steady-state level of extracellular DA, which is known to exist in sufficient concentrations to activate presynaptic autoreceptors located on DA terminals (*E*). This would result in a presynaptic down-regulation of spike-dependent DA release (from GRACE 1992, with permission)

large amounts of DA and produce pathologically high degrees of postsynaptic receptor stimulation. Indeed, the hypofrontality has been associated primarily with negative symptoms (LIDDLE et al. 1992a,b; LIDDLE and MORRIS 1991; WOLKIN et al. 1992), which have been proposed to arise from decreased basal levels of DA (DAVIS et al. 1991; GRACE 1991a; WADDINGTON 1993; WYATT et al. 1988). Thus, a decrease in tonic DA receptor stimulation secondary to cortical pathology may underlie the negative symptoms, while at the same time the resultant increase in spike-dependent phasic DA release may contribute to the DA-dependent positive symptoms of schizophrenia (DAVIS et al. 1991; GRACE 1991a; WADDINGTON 1993; WYATT et al. 1988).

Although speculative, there are several therapeutic implications that may be drawn from this model. For example, the high level of plasticity in the DA system would predict that even pathologically large levels of spike-dependent DA release would not be effectively attenuated by a steady-state level of DA receptor blockade, since the system would respond by compensating for the blockade. On the other hand, APD-induced depolarization block would do more that simply attenuate the DA receptor stimulation: it would cause a qualitative change in the way the system responds to stimuli. Thus, if the DA cell is in depolarization block, it is already too depolarized to fire spikes. While in this state, stimuli that normally activate DA cell firing would be unable to cause any spike-dependent phasic DA release. However, since the normal state of DA cells is not depolarization block, this action of neuroleptics does not appear to be treating schizophrenia by restoring the system to normal. Instead, the induction of depolarization block may more accurately be portrayed as inducing an offsetting deficit to counter the pathological absence of cortical suppression of subcortical DA. This model could also account for the observation that blockade of glutamatergic actions by the NMDA antagonist phencyclidine appears to induce both positive and negative symptoms, whereas DA-releasing drugs like amphetamine and L-DOPA only produce positive symptoms (CUMMINGS 1992; SNYDER 1972; WOLKIN et al. 1994). Furthermore, the potential therapeutic benefit that this model would predict by restoring glutamate activity may account for the report that glycine administration may improve negative symptoms in schizophrenia (JAVITT et al. 1994), since glycine has been shown to potentiate NMDA receptor-related activity (JOHNSON and ASCHER 1987).

F. Antipsychotic Drug Treatment and Electrotonic Transmission Within the Basal Ganglia

I. Dopaminergic Control of Electrotonic Coupling in the Striatum

DA has been shown to exert modulatory actions within the striatal complex at several levels of complexity: (a) by direct postsynaptic actions on striatal neurons (CALABRESI et al. 1987), (b) by presynaptic stimulation of afferent terminals (O'DONNELL and GRACE 1994), or (c) through the modulation of cell:cell interactions at the regional network level by changing the degree of electrotonic coupling between neurons (CEPEDA et al. 1989; O'DONNELL and GRACE 1993, 1995a; ONN and GRACE 1994a). Electrotonic interactions in other systems have been shown to be mediated by gap junctions (SOTELO and KORN 1978), which enable both electrical signals and small molecules to cross between cells. The fluorescent dye Lucifer yellow has the ability to completely fill a cell after intracellular injection and is sufficiently compact to pass through gap junctions, causing cells that are coupled to the injected neuron to be labeled (STEWART 1981). Thus, provided sufficiently restrictive criteria are used (O'DONNELL and GRACE 1993), dye coupling can be a reliable indicator of electrotonic coupling in a number of brain regions (ANDREW et al. 1981; DUDEK and SNOW 1985; GRACE and BUNNEY 1983c; GUTNICK and PRINCE 1981; NÚÑEZ et al. 1990; O'DONNELL and GRACE 1993, 1995a; ONN and GRACE 1994a;

WALSH et al. 1989), although correlative electrophysiological evidence can provide an important level of substantiation of its presence (ANDREW et al. 1981; DUDEK and SNOW 1985; GRACE and BUNNEY 1983c; GUTNICK and PRINCE 1981; NÚÑEZ et al. 1990; O'DONNELL and GRACE 1993; ONN and GRACE 1994a).

As in other structures, the degree of coupling that is present in the striatal complex appears to be under dynamic regulation by DA. Following intracellular injection of Lucifer yellow into single neurons, pairs of cells labeled with Lucifer yellow were found to have somata separated by up to 100 μm but with a high degree of overlap between dendrites (O'DONNELL and GRACE 1993). The extent of dye coupling observed varies with the region of the striatal complex examined. Thus, in slices obtained from untreated animals and perfused with physiological saline, instances of dye coupling can be observed in 8% of the cells injected with dye in the dorsal striatum, 24% in the accumbens core, 14% in the caudal part of the accumbens shell, and 35% in the rostral aspect of the shell (O'DONNELL and GRACE 1995a). Anatomical studies have shown that the accumbens is composed of several distinct subregions; i.e., the core and the shell (HEIMER et al. 1991). The core region is believed to be related functionally to the striatum, whereas the shell region (and in particular the caudal aspects of this structure) have primarily limbic connections. In the accumbens core, D_1 receptor stimulation induces a decrease in the incidence of coupling between accumbens cells (Fig. 7). This effect is observed in brain slices containing the nucleus accumbens that had been perfused with the nonspecific DA agonist apomorphine, or the D_1 agonist SKF 38393; and it can be prevented by coadministration of the D_1 antagonists SCH 23390 or SCH 39166 (O'DONNELL and GRACE 1993). However, the effect of D_1 agonists on the decrease in coupling can also be prevented by administration of the D_2 antagonist sulpiride (O'DONNELL and GRACE 1993). Thus, activation of D_1 receptors induces a decrease in dye coupling in the core region of the accumbens, but an enabling activation mediated by tonic stimulation of D_2 receptors is also necessary. Since the accumbens slice apparently has sufficient basal levels of DA to activate the highly sensitive D_2 receptors (O'DONNELL and GRACE 1993, 1994), administration of a D_1 agonist alone is sufficient to induce this D_1/D_2-dependent decrease in coupling. Furthermore, in slices made from DA-depleted rats, administration of D_1 or D_2 agonists alone failed to induce changes in the incidence of dye coupling, whereas the co-administration of both D_1 and D_2 agonists to DA-depleted slices induced a decrease in the incidence of coupling in this region (O'DONNELL and GRACE 1993). There is also evidence that D_1 activation in the accumbens core can suppress fast prepotentials that are likely to arise from the activation of spike discharge and its electrotonic spread between coupled cells (Fig. 8; O'DONNELL and GRACE 1993).

In the shell region of the accumbens, the dopaminergic modulation of electrotonic coupling is not homogeneous. In the rostral aspect of the shell, activation of D_1 DA receptor results in a decrease in the levels of dye coupling

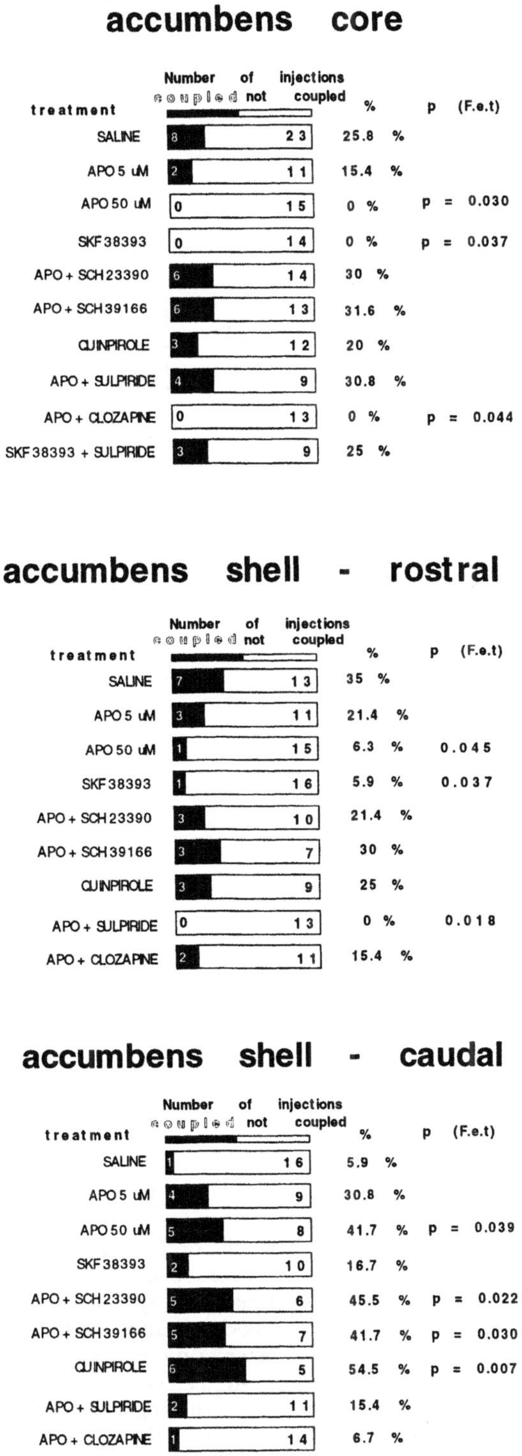

Fig. 7. Dopaminergic drugs modulate the incidence of dye coupling between neurons in the core, rostral shell, and caudal shell of the nucleus accumbens. The *horizontal bars* represent the results from the population of cells tested in each group, which is divided into those injections resulting in dye coupling (*dark bar, left*) and those which resulted in the labeling of single medium spiny neurons (*open bar, right*). The numbers in these bars are the cells showing either coupling (*left*) or absence of coupling (*right*) across the different treatments. The incidence of dye coupling is also shown for each group as percent of injections resulting in coupling, and the *p* value given is the result of a comparison with controls using Fisher's exact test (*F.e.t*)

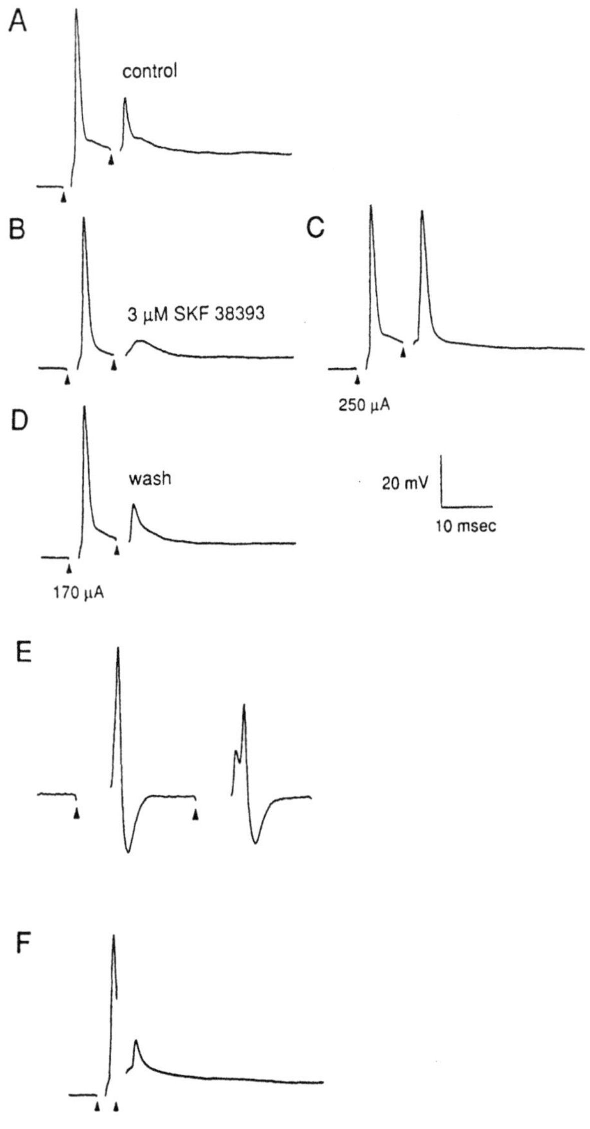

Fig. 8A–F. D_1 receptor stimulation blocks the occurrence of putative coupling potentials in neurons in the accumbens core recorded in vitro. **A** Stimulation of cortical afferents evokes spike discharge. When a second response is evoked during the inhibitory period following the first pulse, a small-amplitude, fast spikelet is observed. These fast prepotentials are believed to be the expression of spikes occurring in neurons that are coupled to the cell impaled and transmitted electrotonically to the recorded cell. **B** The addition of the D_1 agonist SKF 38393 to the superfusion fluid in doses that have been shown to reverse dye coupling also blocks the occurrence of the spikelets. **C** Increasing the stimulus intensity results in a full action potential in response to the second stimulus. **D** After washing out the drug, spikelets can again be observed. **E** Differentiation of the waveform recorded during stimulation in the presence of the agonist (shown in **C**) reveals an early component of the action potential evoked by the second stimulus. **F** Decreasing the interpulse interval to cause the second pulse to be delivered during the refractory period of the evoked action potential reveals the absence of collision of the spikelet with the initial spike, suggesting that they do not originate from the same cell (from O'DONNELL and GRACE 1993, with permission)

that is not dependent on D_2 coactivation (O'DONNELL and GRACE 1993). On the other hand, the DA regulation of dye coupling between neurons in the caudal aspect of the accumbens shell exhibits an entirely different pharmacology. In this subregion, administration of the D_2 agonist quinpirole or apomorphine at low doses (that are more likely to activate the higher-sensitivity D_2 receptors) induces an increase in the levels of coupling (Fig. 8). This effect is not mimicked by D_1 agonists, and it can be prevented by coadministration of D_2, but not D_1 antagonists.

In contrast to the actions of the D_2-specific classical APDs like HAL on coupling, the atypical APD clozapine, which is a weak antagonist at both D_1 and D_2 receptors (SEEMAN 1987), exerts unique actions with respect to the control of dye coupling in selected regions of the nucleus accumbens. Thus, in the core region CLZ administration fails to block the D_1 and D_2 agonist-dependent decrease in coupling, and therefore fails to mimic either a D_1 or a D_2 antagonist (O'DONNELL and GRACE 1993). However, this distinction may be significant in that both clinically effective APDs exert common actions in the caudal shell region, which is an area that several studies have implicated as the site of therapeutic action of APDs (DEUTCH et al. 1992; WHITE and WANG 1983).

In vitro intracellular injection of Lucifer yellow in the dorsal striatum also results in low levels of dye coupling that can be modulated by manipulations of the DA system. In animals with a 6-hydroxydopamine lesion, the incidence of coupling is higher (CEPEDA et al. 1989). This suggests an inhibitory action of DA on coupling between neurons in the dorsal striatum, and this action has been proposed to be mediated by D_1 receptor activation (LEVINE 1991; WALSH et al. 1991).

II. Effects of Subchronic Treatment with Haloperidol or Clozapine on Striatal Cell Dye Coupling Observed In Vitro

In vitro studies of striatal brain slices obtained from rats that had received long-term APD administration revealed that continuous administration of CLZ or HAL for 21 days induced sustained alterations in dye coupling. Thus, both CLZ and HAL pretreatment resulted in an increase in the incidence of dye coupling between neurons in the caudal accumbens shell, whereas only HAL caused an increase in coupling in the dorsal striatum (Fig. 9). In contrast, neither drug altered coupling in any region of the striatum when administered for only 1 day. Thus, as shown previously in DA-depleted animals (CEPEDA et al. 1989; ONN and GRACE 1993), the long-term compromise in DA cell firing induced by continuous haloperidol administration leads to increased coupling in the striatum.

None of the treatment paradigms tested produces significant changes in dye coupling in the core region of the accumbens or in the rostral part of the accumbens shell. In contrast, continuous treatment with either HAL or CLZ results in significantly higher levels of dye coupling in the caudal shell region

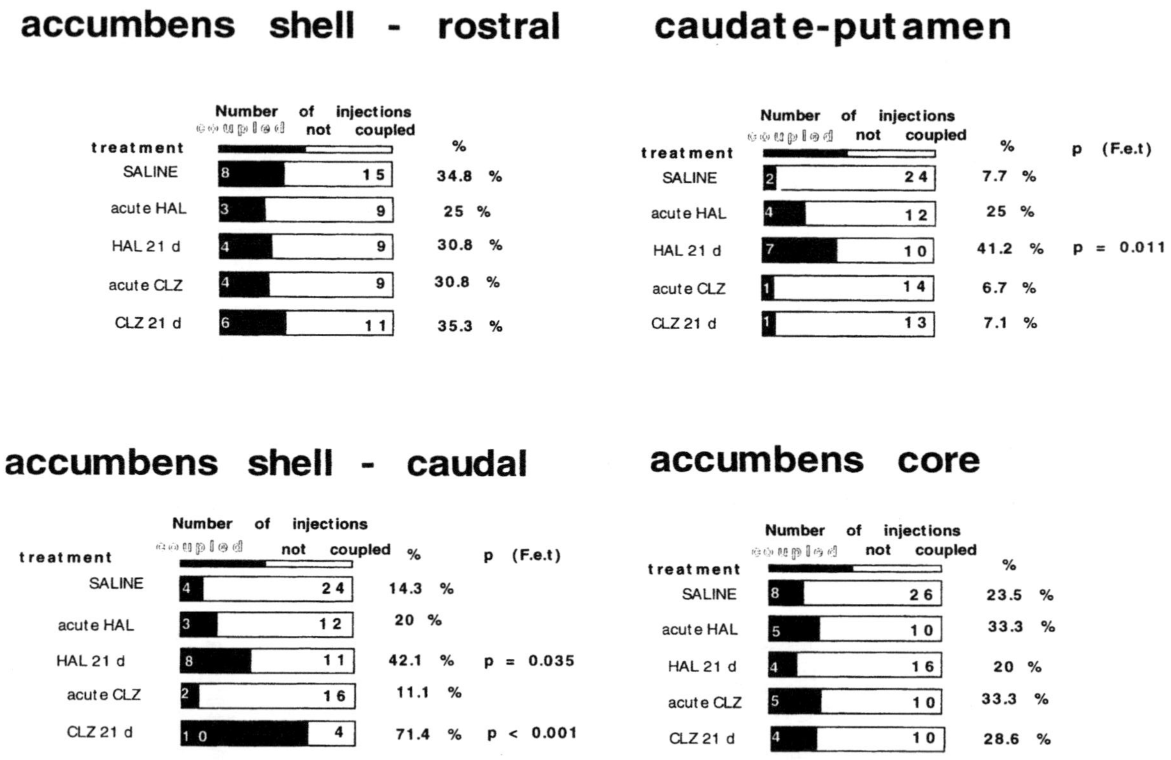

Fig. 9. Effects of subchronic treatments with HAL or clozapine (CLZ) on dye coupling between neurons in the striatal complex assessed in vitro. The bars and numbers represent the proportion and number of cells exhibiting coupling, as described in Fig. 7. Significant changes can be observed in the cuadal shell after 21 days of treatment with either HAL or CLZ, and in the caudate-putamen after long-term treatment with HAL

of the accumbens, but only if the drug is not present in the superfusion medium. In slices from naive rats, this region of the accumbens exhibits increased coupling upon D_2 receptor stimulation (O'DONNELL and GRACE 1993). One way to account for this data is that depolarization block of mesoaccumbens DA cell firing induced by long-term CLZ or HAL may have led to an increase in coupling secondary to the induction of D_2 receptor supersensitivity. The endogenous DA levels present in the slice (O'DONNELL and GRACE 1993, 1994) could thereby serve as the source of D_2 receptor stimulation following drug withdrawal.

These results indicate that the level of electrotonic coupling between accumbens neurons is a plastic phenomenon that can undergo alterations following long-term APD administration. The increase in coupling induced by HAL, but not CLZ, in the motor-related dorsal striatum may reflect the propensity of this classic neuroleptic to induce motor side effects in schizophrenics (GERLACH et al. 1974; KANE et al. 1988). In contrast, the observation that both drugs exert similar effects on electrotonic coupling in the caudal accumbens shell may reflect their common mode of therapeutic action in the alleviation of positive schizophrenic symptoms. The observed change in dye coupling with 3 weeks of APD treatment but not with acute drug administration is consistent with the delayed onset of DA cell depolarization block of DA neurons (WHITE and WANG 1983), suggesting that it occurs as a consequence of DA cell inactivation. Similarly, both HAL and CLZ have also been reported to exert common actions in the nucleus accumbens with respect to the induction of immediate early genes (DEUTCH et al. 1992; ROBERTSON and FIBIGER 1992), the expression of neurotensin mRNA (MERCHANT and DORSA 1993), and effects on basal DA levels (ICHIKAWA and MELTZER 1992) and DA turnover (CSERNANSKY et al. 1993), whereas only HAL affects these parameters in the striatum. Thus, the topography of changes in coupling induced by CLZ and HAL correlates with the time course and the proposed sites of therapeutic action of APDs (i.e., the caudal region of the nucleus accumbens shell) as well as for their motor side effects (i.e., the dorsal striatum) that are produced by continuous treatment (DEUTCH et al. 1992). These results suggest that interference with DA-mediated control of gap junction conductance within these regions may play a role in the clinical actions of antipsychotic drugs.

III. Effects of Subchronic Treatment with Haloperidol or Clozapine on Striatal Cell Dye Coupling Observed In Vivo

The effects of long-term administration of APDs on the incidence of dye coupling have also been examined in neurons of the striatal complex during intracellular recordings from intact, anesthetized rats in vivo (ONN and GRACE 1994b). As for the in vitro cases, repeated daily injections of CLZ or HAL for 1 month were found to result in a nearly sevenfold higher incidence of dye coupling between cells in the shell region of the nucleus accumbens. In contrast, repeated administration of HAL, but not CLZ, caused an increase in the incidence of dye coupling in the matrix compartment of the dorsal striatum after subchronic administration (ONN and GRACE 1994b). These results support the contention that CLZ- or HAL-mediated depolarization block of the VTA DA cells that project to the limbic aspect of the striatal complex may result in DA-dependent changes in the apparent permeability of gap junctions between dopaminoceptive neurons in this region. Similarly, the ability of HAL but not CLZ to induce depolarization block of SNc DA cells projecting to the motor-related matrix compartment of the striatal complex appears to underlie the alterations in gap junction permeability in these regions.

Although the typical seven-layered images characteristic of the presence of gap junctions have not been found to be prevalent in electron microscopic studies of these regions, studies have revealed close membrane apposition between neurons of the nucleus accumbens (PASKEVICH et al. 1991; VAN BOCKSTAELE et al. 1994a,b) and dorsal striatum (KITA 1993; SESACK et al. 1994). Furthermore, terminals stained with tyrosine hydroxylase antibodies (i.e., dopaminergic terminals) have been found to exist in close anatomical relationship to such membrane appositions (VAN BOCKSTAELE et al. 1994b), providing morphological evidence to support a role for DA in the control of electrical coupling. Furthermore, the protein connexin, which is a functional subunit of gap junctions, has been localized to these structures (MATSUMOTO et al. 1991). The difficulties in finding images characteristic of gap junctions in these regions may be related to the observation that dye coupling is rarely observed between somata (O'DONNELL and GRACE 1993; ONN and GRACE 1994a). Indeed, a recent electron microscopic study showed close membrane appositions between the dendritic processes arising from two distinct cells within the striatum (SESACK et al. 1994).

The functional relevance of coupling in the adult mammalian CNS is unclear at present. Nonetheless, its function is likely to be substantially different from its established role in developing systems or in invertebrate preparations. Thus, gap junctions in developing organisms are present between different classes of cells that are associated with a single substrate (PEINADO et al. 1993). In invertebrates, coupling is found to connect different cell types to enable the direct passage of currents in a synapse-like manner (DERMIETZEL and SPRAY 1993). In contrast, in adult mammalian nerve cells coupling has only been observed to occur between identical cell types. Thus, in the striatum, coupling has been invariably found between the GABAergic medium spiny neurons (O'DONNELL and GRACE 1993, 1995a; ONN and GRACE 1994a,b). Because these cells form inhibitory synapses with postsynaptic neurons, the presence of gap junctions in this structure has the unique ability to allow the transfer of excitation between clusters of these neurons. Thus, the transfer of information via gap junctions may be an effective way of synchronizing the activity of a population of neurons that could otherwise only exert inhibitory interactions. Indeed, by modulating cellular interactions occurring through gap junctions, DA may be positioned to exert unique yet powerful modulatory control over the types of complex neuronal interactions that occur within the striatum. Thus, unlike its actions on the activity of single cells or in neurotransmitter release from afferent terminals, the ability of DA to modulate gap junctional conductance may enable it to alter the information processing functions of this region. In turn, by impacting the relationship between DA and cellular interactions, APDs would be expected to produce complex modulatory actions on these structures that extend far beyond a simple blockade of an inhibitory neurotransmitter action.

G. Conclusions

A unique aspect of the DA system is the presence of a rather complex yet powerful homeostatic compensatory mechanism that controls DA release, which may be under the regulatory control of cortical afferent projections. A disruption in this cortical regulation of DA systems is proposed to lead to schizophrenia, with APDs exerting their therapeutic effects by inducing an offsetting disturbance in the DA system.

Acute administration of APDs exerts comparatively mild influences on the physiology of DA systems as well as on the symptoms present in schizophrenic patients, presumably due to the fact that the DA system can compensate for large reductions in DA transmission. Although introduction of these compensatory changes may be expected to lead to a tolerance in the actions of APDs, in reality there is an increase in therapeutic benefits of these drugs with repeated administration. This is distinctly different from the case with acutely active drugs, such as benzodiazepines, which are maximally effective upon acute administration and show a decrease in their potency with time. Thus, the acute APD actions do not appear to be directly related to their clinical actions. Instead, the therapeutic response to these drugs may depend on the pharmacologically induced compensatory changes they produce in the DA system. One such compensatory change is likely to be the depolarization block of DA cell firing, which is significant because, unlike the effect of acute APDs, it causes qualitative changes in the response of the system to what would otherwise be excitatory stimuli to DA cells. This result may actually counterbalance the cortically-induced pathological up-regulation of the DA system by introducing this deficit in DA cell firing.

However, to understand a disorder that involves complex cognitive pathologies such as those present in schizophrenia, one must be able to reach beyond the actions of DA on striatal cell activity to provide a link with the pathological changes that are likely to be present in cortically mediated cognitive processes. Such complex regulatory properties may actually be derived from the multiple actions of DA. Dopamine acts at several levels within the striatum: it (a) inhibits postsynaptic cell activity (HU and WANG 1988; WOODRUFF et al. 1976), (b) reduces the impact of cortical inputs by reducing the amplitude of corticoaccumbens or corticostriatal excitatory postsynaptic potentials (EPSPs) (O'DONNELL and GRACE 1994), and (c) changes the flow of information mediated via electrotonic transmission by affecting the permeability of gap junctions (O'DONNELL and GRACE 1993). Therefore, as a consequence of the imbalance in the DA system proposed to occur in the schizophrenic brain, changes in these actions of DA could ultimately contribute to the development of schizophrenic symptoms via their influence on accumbens-ventral pallidum-thalamic-cortical pathways (LAVIN and GRACE 1994).

The ability of APDs to modulate electrical coupling provides a basis for developing a model of psychopathological alterations of schizophrenia that relates the hyperdopaminergic state in animals to the symptoms observed in patients. Thus, a DA-dependent reduction in the levels of coupling between neurons in the core region of the accumbens may lead to a reduction of cortical integration, possibly causing a decrease in behavioral options, in a similar manner to the DA-related behavior of perseverance (EVENDEN and ROBBINS 1983), the inability to switch strategies (MATTHYSSE 1978), or the impaired attention (ROBBINS 1990) that are also a component of negative symptoms or thought disorders. Conversely, a pathologically high DA-dependent increase in coupling in the shell region of the accumbens could conceivably result in a spread of cortical afferent excitation to coupled neurons that may normally be unaffected by this input. Thus, an increase in intercellular coupling may be expressed as an overgeneralization in the type of affective responses that can be evoked by a given stimulus. Such a condition may thus contribute to the positive symptoms that may arise from a hyperactive DA system in the accumbens (GRAY et al. 1991; WYATT et al. 1988).

In addition to the modulation of accumbens cells by the prefrontal cortex, recent studies have provided evidence that the hippocampus may also play a major role in the integration of information within this region. Thus, hippocampal afferents to the nucleus accumbens have been shown to gate the cortical input to these cells by putting those accumbens cells that show a bistable membrane potential into their depolarized, active state (O'DONNELL and GRACE 1995b). Indeed, activation of prefrontal cortical afferents only triggers accumbens cell firing when it occurs in coincidence with the hippocampal afferent-mediated depolarized state (O'DONNELL and GRACE 1995b). Since a common finding in schizophrenics is the alteration in volume and even cell organization in the temporal lobe, in particular in the hippocampus (FALKAI and BOGERTS 1986; KOVELMAN and SCHEIBEL 1984), this loss of gating could itself induce a functional hypofrontality by preventing prefrontal cortical activation of the accumbens. Therefore, a disorganization of the context-sensitive hippocampal input may substantially disrupt the normal flow of cortical information through the nucleus accumbens and its ultimate influence of thalamocortical function.

References

Abercrombie ED, Hollerman JR, Grace AA (1989) In vivo biochemical correlates of acute depolarization inactivation in substantia nigra dopaminergic neurons. Soc Neurosci Abstr 15:1002

Ackerman JM, Johansen PA, Clark D, White FJ (1993) Electrophysiological effects of putative autoreceptor-selective dopamine agonists on A10 dopamine neurons. J Pharmacol Exp Ther 265:963–970

Andrew RD, MacVicar BA, Dudek FE, Hatton GI (1981) Dye transfer through gap junctions between neuroendocrine cells on rat hypothalamus. Science 211:1187–1189

Angrist B, van Kammen DP (1984) CNS stimulants as tools in the study of schizophrenia. Trends Neurosci 388–390

Angrist B, Shopin B, Gershon S (1971) Comparative psychotomimetic effects of stereoisomers of amphetamine. Nature 234:152–153

Angrist B, Santhananthan G, Wilk S, Gershon S (1974) Amphetamine psychosis: behavioral and biochemical aspects. J Psychiatr Res 11:13–23

Arnold SE, Hyman BT, Van Hoesen GW, Damasio AR (1991) Some cytoarchitectural abnormalities of the entorhinal cortex in schizophrenia. Arch Gen Psychiatry 48:625–632

Asencio H, Bustos G, Gysling K, Labarca R (1991) N-Methyl-D-aspartate receptors and release of newly-synthesized [^{3}H]dopamine in nucleus accumbens slices and its relationship with neocortical afferents. Prog Neuropsychopharmacol Biol Psychiatry 15:663–676

Barta PE, Pearlson GD, Powers RE, Richards SS, Tune LE (1990) Auditory hallucinations and smaller superior temporal gyral volume in schizophrenia. Am J Psychiatry 147:1457–1462

Berman KF, Zec RF, Weinberger DR (1986) Physiologic dysfunction of dorsolateral prefrontal cortex in schizophrenia. II. Role of neuroleptic treatment, attention, and mental effort. Arch Gen Psychiatry 43:126–135

Blaha CD, Lane RF (1987) Chronic treatment with classical and atypical antipsychotic drugs differentially decreases dopamine release in striatum and nucleus accumbens in vivo. Neurosci Lett 78:199–204

Bogerts B, Lieberman JA, Ashtari M, Bilder RM, Degreef G, Lerner G, Johns C, Masiar S (1993) Hippocampus-amygdala volumes and psychopathology in chronic schizophrenia. Biol Psychiatry 33:236–246

Bouyer JJ, Park DH, Joh TH, Pickel VM (1984) Chemical and structural analysis of the relation between cortical inputs and tyrosine hydroxylase-containing terminals in rat neostriatum. Brain Res 302:267–275

Bowery B, Rothwell LA, Seabrook GR (1994) Comparison between the pharmacology of dopamine receptors mediating the inhibition of cell firing in rat brain slices through the substantia nigra pars compacta and ventral tegmental area. Br J Pharmacol 112:873–880

Brodie MS, Dunwiddie TV (1987) Cholecystokinin potentiates dopamine inhibition of mesencephalic dopamine neurons in vitro. Brain Res 425:106–113

Buchsbaum MS, Ingvar DH, Kessler R, Waters RN, Cappelletti J, van Kammen DP, King C, Johnson JL, Manning RG, Flynn RW, Mann LS, Bunney WE, Sokoloff L (1982) Cerebral glucography with positron tomography. Use in normal subjects and in patients with schizophrenia. Arch Gen Psychiatry 39:251–259

Bunney BS, Aghajanian GK (1975) Antipsychotic drugs and central dopaminergic neurons: a model for predicting therapeutic efficacy and the incidence of extrapyramidal side effects. In: Sudilovsky A, Gershon S, Beer B (eds) Predictability in psychopharmacology: preclinical and clinical correlates. Raven, New York, pp 225–245

Bunney BS, Grace AA (1978) Acute and chronic haloperidol treatment: comparison of effects on nigral dopaminergic cell activity. Life Sci 23:1715–1728

Bunney BS, Walters JR, Roth RH, Aghajanian GK (1973) Dopaminergic neurons: effect of antipsychotic drugs and amphetamine on single cell activity. J Pharmacol Exp Ther 185:560–571

Cai N-S, Kiss B, Erdö SL (1991) Heterogeneity of *N*-methyl-D-aspartate receptors regulating the release of dopamine and acetylcholine from striatal slices. J Neurochem 57:2148–2151

Calabresi P, Mercuri N, Stanzione P, Stefani A, Bernardi G (1987) Intracellular studies on the dopamine-induced firing inhibition of neostriatal neurons in vitro: evidence for D_1 receptor involvement. Neuroscience 20:757–771

Carlsson A, Lindqvist M (1963) Effect of chlorpromazine or haloperidol on formation of 3-methoxytyramine and normetanephrine in mouse brain. Acta Pharmacol Toxicol 20:140–144

Carpenter WT, Buchanan RW (1994) Schizophrenia. N Engl J Med 330:681–690

Carrozza DP, Ferraro TN, Golden GT, Reyes PF, Hare TA (1991) Partial characterization of kainic acid-induced striatal dopamine release using in vivo microdialysis. Brain Res 543:69–76

Carrozza DP, Ferraro TN, Golden GT, Reyes PF, Hare TA (1992) In vivo modulation of excitatory amino acid receptors: microdialysis studies on *N*-methyl-D-aspartate-evoked striatal dopamine release and effects of antagonists. Brain Res 574:42–48

Carter CJ, L'Heurex R, Scatton B (1988) Differential control by *N*-methyl-D-aspartate and kainate of striatal dopamine release in vivo: a trans-striatal study. J Neurochem 51:462–468

Cepeda C, Walsh JP, Hull CD, Howard SG, Buchwald NA, Levine MS (1989) Dye-coupling in the neostriatum of the rat. I. Modulation by dopamine-depleting lesions. Synapse 4:229–237

Chen J, Paredes W, Gardner EL (1991) Chronic treatment with clozapine selectively decreases basal dopamine release in nucleus accumbens but not in caudate-putamen as measured by in vivo brain microdialysis: further evidence for depolarization block. Neurosci Lett 122:127–131

Chéramy A, Romo R, Godeheu G, Baruch P, Glowinski J (1986) In vivo presynaptic control of dopamine release in the cat caudate nucleus – II. Facilitatory or inhibitory influence of L-glutamate. Neuroscience 19:1081–1090

Chergui K, Charléty PJ, Akaoka H, Saunier CF, Brunet J-L, Buda M, Svensson T, Chouvet G (1993) Tonic activation of NMDA receptors causes spontaneous burst discharge of rat midbrain dopamine neurons in vivo. Eur J Neurosci 5:137–144

Chergui K, Akaoka H, Charléty PJ, Saunier CF, Buda M, Chouvet G (1994) Subthalamic nucleus modulates burst firing of nigral dopamine neurones via NMDA receptors. Neuroreport 5:1185–1188

Chiodo LA (1988) Dopamine-containing neurons in the mammalian central nervous system: Electrophysiology and pharmacology. Neurosci Biobehav Rev 12:49–91

Chiodo LA (1992) Dopamine autoreceptor signal transduction in the DA cell body: a 'current view". Neurochem Int 20 [Suppl]:81S–84S

Chiodo LA, Bunney BS (1983) Typical and atypical neuroleptics: differential effects of chronic administration on the activity of A9 and A10 midbrain dopaminergic neurons. J Neurosci 3:1607–1619

Chiodo LA, Bunney BS (1985) Possible mechanisms by which repeated clozapine administration differentially affects the activity of two subpopulations of midbrain dopamine neurons. J Neurosci 5:2539–2544

Chiodo LA, Kapatos G (1992) Membrane properties of identified mesencephalic dopamine neurons in primary dissociated cell culture. Synapse 11:294–309

Clow DW, Jhamandas K (1989) Characterization of L-glutamate action on the release of endogenous dopamine from the rat caudate-putamen. J Pharmacol Exp Ther 248:722–728

Creese I, Burt DR, Snyder SH (1976) Dopamine receptor binding predicts clinical and pharmacological potencies of antischizophrenic drugs. Science 192:596–598

Csernansky JG, Bellows EP, Barnes DE, Lombrozo L (1990) Sensitization to the dopamine turnover-elevating effects of haloperidol: the effect of regular-intermittent dosing. Psychopharmacology 101:519–524
Csernansky JG, Murphy GM, Faustman WO (1991) Limbic/mesolimbic connections and the pathogenesis of schizophrenia. Biol Psychiat 30:383–400
Csernansky JG, Wrona CT, Bardgett ME, Early TS, Newcomer JW (1993) Subcortical dopamine and serotonin turnover during acute and subchronic administration of typical and atypical neuroleptics. Psychopharmacology 110:145–151
Cummings JL (1992) Neuropsychiatric complications of drug treatment of Parkinson's disease. In: Huber SJ, Cummings JL (eds) Parkinson's disease. Neurobehavioral aspects. Oxford University Press, New York, pp 313–327
Davis KL, Kahn RS, Ko G, Davidson M (1991) Dopamine in schizophrenia: a review and reconceptualization. Am J Psychiatry 148:1474–1486
Dermietzel R, Spray DC (1993) Gap junctions in the brain: where, what type, how many and why? Trends Neurosci 16:186–192
Deutch AY, Roth RH (1988) Alterations in dopamine synthesis induced by chronic neuroleptic administration: a possible biochemical correlate of depolarization inactivation. Soc Neurosci Abstr 14:27
Deutch AY, Moghaddam B, Innis RB, Krystal JH, Aghajanian GK, Bunney BS, Charney DS (1991) Mechanisms of action of atypical antipsychotic drugs. Implications for novel therapeutic strategies for schizophrenia. Schizophr Res 4:121–156
Deutch AY, Lee MC, Iadarola MJ (1992) Regionally specific effects of atypical antipsychotic drugs on striatal fos expression: the nucleus accumbens shell as a locus of antipsychotic action. Mol Cell Neurosci 3:332–341
Doherty MD, Gratton A (1991) Behavioral evidence of depolarization block of mesencephalic dopamine neurons by acute haloperidol in partially 6-hydroxydopamine lesioned rats. Behav Neurosci 105:579–587
Dudek FE, Snow RW (1985) Electrical interactions and synchronization of cortical neurons: electrotonic coupling and field effects. In: Bennet MVL, Spray DC (eds) Gap junctions. Cold Spring Harbor Laboratories, Cold Spring Harbor, New York, pp 325–336
Engberg G, Kling-Petersen T, Nissbrandt H (1993) $GABA_B$-receptor activation alters the firing pattern of dopamine neurons in the rat substantia nigra. Synapse 15:229–238
Evenden JL, Robbins TW (1983) Increased response switching, perseveration, and perseverative switching following *d*-amphetamine in the rat. Psychopharmacology 80:67–73
Falkai P, Bogerts B (1986) Cell loss in the hippocampus of schizophrenics. Eur Arch Psychiatry Neurol Sci 236:154–161
Fey ET (1951) The performance of young schizophrenics and young normals on the Wisconsin Card Sorting Test. J Consult Psychol 15:311–319
Finlay JM, Jakubovic A, Fu S, Fibiger HC (1987) Tolerance to haloperidol-induced increases in dopamine and metabolites: fact or artifact? Eur J Pharmacol 137:117–121
Freeman AS, Chiodo LA (1987) Electrophysiological aspects of cholecystokinin/dopamine interactions in the central nervous system. Ann NY Acad Sci:205–236
French ED (1994) Phencyclidine and the midbrain dopamine system: electrophysiology and behavior. Neurotoxicol Teratol 16:355–362
Friston KJ (1992) The dorsolateral prefrontal cortex, schizophrenia and PET. J Neural Transm [Suppl] 37:79–93
Gariano RF, Groves PM (1988) Burst firing induced in midbrain dopamine neurons by stimulation of the medial prefrontal and anterior cingulate cortices. Brain Res 462:194–198
Gerlach J, Koppelhus P, Helweg E, Monrad A (1974) Clozapine and haloperidol in a single-blind cross-over trial: therapeutic and biochemical aspects in the treatment of schizophrenia. Acta Psychiat Scand 50:410–424

Gifford AN, Wang RY (1994) The effect of 5-HT_3 receptor antagonists on the morphine-induced excitation of A10 dopamine cells: electrophysiological studies. Brain Res 638:325–328

Giorguieff MF, Kemel ML, Glowinski J (1977) Presynaptic effect of L-glutamic acid on the release of dopamine in rat striatal slices. Neurosci Lett 6:73–77

Goldberg TE, Bigelow LB, Weinberger DR, Daniel DG, Kleinman JE (1991) Cognitive and behavioral effects of the coadministration of dextroamphetamine and haloperidol in schizophrenia. Am J Psychiatr 148:78–84

Goldstein JM, Litwin LC (1988) Spontaneous activity of A9 and A10 dopamine neurons after acute and chronic administration of the selective dopamine D_1 receptor antagonist SCH23390. Eur J Pharmacol 155:175–180

Goldstein JM, Litwin LC, Sutton EB, Malick JB (1989) Effects of ICI 169,369, a selective serotonin-2 antagonist, in electrophysiological tests predictive of antipsychotic activity. J Pharmacol Exp Ther 249:673–680

Gonon FG (1988) Nonlinear relationship between impulse flow and dopamine released by rat midbrain dopaminergic neurons as studied by in vivo electrochemistry. Neuroscience 24:19–28

Gonzalez Burgos GR, Biali FI, Cherksey BD, Sugimori M, Llinás RR, Uchitel O (1995) Different calcium channels mediate transmitter release evoked by transient or sustained depolarization at mammalian sympathetic ganglia. Neuroscience 64:117–123

Grace AA (1987) The regulation of dopamine neuron activity as determined by in vivo and in vitro intracellular recordings. In: Chiodo LA, Freeman AS (eds) Neurophysiology of dopaminergic systems – current status and clinical perspectives. Lakeshore, Chicago, pp 1–66

Grace AA (1988) In vivo and in vitro intracellular recordings from rat midbrain dopamine neurons. Ann NY Acad Sci 537:51–76

Grace AA (1990) Evidence for the functional compartmentalization of spike generating regions of rat midbrain dopamine neurons recorded in vitro. Brain Res 524:31–41

Grace AA (1991a) Phasic versus tonic dopamine release and the modulation of dopamine system responsivity: a hypothesis for the etiology of schizophrenia. Neuroscience 41:1–24

Grace AA (1991b) Regulation of spontaneous activity and oscillatory spike firing in rat midbrain dopamine neurons recorded in vitro. Synapse 7:221–234

Grace AA (1992) The depolarization block hypothesis of neuroleptic action: implications for the etiology and treatment of schizophrenia. J Neural Transm [Suppl] 36:91–131

Grace AA, Bunney BS (1980) Nigral dopamine neurons: intracellular recording and identification using L-DOPA injection combined with fluorescence histochemistry. Science 210:654–656

Grace AA, Bunney BS (1983a) Intracellular and extracellular electrophysiology of nigral dopaminergic neurons – 1. Identification and characterization. Neuroscience 10:301–315

Grace AA, Bunney BS (1983b) Intracellular and extracellular electrophysiology of nigral dopaminergic neurons – 2. Action potential generating mechanisms and morphological correlates. Neuroscience 10:317–331

Grace AA, Bunney BS (1983c) Intracellular and extracellular electrophysiology of nigral dopaminergic neurons – 3. Evidence for electrotonic coupling. Neuroscience 10:333–348

Grace AA, Bunney BS (1984a) The control of firing pattern in nigral dopamine neurons: burst firing. J Neurosci 4:2877–2890

Grace AA, Bunney BS (1984b) The control of firing pattern in nigral dopamine neurons: single spike firing. J Neurosci 4:2866–2876

Grace AA, Bunney BS (1986) Induction of depolarization block in midbrain dopamine neurons by repeated administration of haloperidol: analysis using in vivo intracellular recording. J Pharmacol Exp Ther 238:1092–1100

Grace AA, Onn S-P (1989) Morphology and electrophysiological properties of immunocytochemically identified rat dopamine neurons recorded in vitro. J Neurosci 9:3463–3481
Gray JA, Feldon J, Rawlins JNP, Hemsley DR, Smith AD (1991) The neuropsychology of schizophrenia. Behav Brain Sci 14:1–84
Groves PM, Wilson CJ, Young SJ, Rebec GV (1975) Self-inhibition by dopaminergic neurons. Science 190:522–529
Gur RE, Pearlson GD (1993) Neuroimaging in schizophrenia research. Schizophrenia Bull 19:337–353
Gutnick MJ, Prince DA (1981) Dye coupling and possible electrotonic coupling in the guinea pig neocortical slices. Science 211:67–70
Harden DG, Grace AA (1994) Electrophysiological examination of feedback pathways to A9 and A10 dopamine neurons originating in the core and shell regions of the nucleus accumbens. Soc Neurosci Abstr 20:566
Harris NC, Greenfield SA (1991) The electrophysiological properties of substantia nigra pars compacta neurones recorded from 6-hydroxydopamine lesioned guinea-pigs in vitro. J Neural Transm [P-D Sect] 3:89–98
Heimer L, Zahm DS, Churchill L, Kalivas PW, Wohltmann C (1991) Specificity in the projection patterns of accumbal core and shell in the rat. Neuroscience 41:89–125
Henry DJ, Wise RA, Rompré P-R, White FJ (1992) Acute depolarization block of A10 dopamine neurons: interactions of morphine with dopamine antagonists. Brain Res 596:231–237
Heritch AJ (1990) Evidence for reduced and dysregulated turnover of dopamine in schizophrenia. Schizophr Bull 16:605–615
Hollerman JR, Grace AA (1989) Acute haloperidol administration induces depolarization block of nigral dopamine neurons in rats after partial dopamine lesions. Neurosci Lett 96:82–88
Hollerman JR, Grace AA (1990) The effects of dopamine-depleting brain lesions on the electrophysiological activity of rat substantia nigra dopamine neurons. Brain Res 533:203–212
Hollerman JR, Abercrombie ED, Grace AA (1992) Electrophysiological, biochemical, and behavioral studies of acute haloperidol-induced depolarization block of nigral dopamine neurons. Neuroscience 47:589–601
Hommer DW, Palkovits M, Crawley JN, Paul SM, Skirboll LR (1985) Cholecystokinin-induced excitation in the substantia nigra: evidence for peripheral and central components. J Neurosci 5:1387–1392
Hounsgaard J, Nedergaard S, Greenfield SA (1992) Electrophysiological localization of distinct calcium potentials at selective somatodendritic sites in the substantia nigra. Neuroscience 50:513–518
Hu X-T, Wang RY (1988) Comparison of effects of D_1 and D_2 dopamine receptor agonists on neurons in the rat caudate putamen: an electrophysiological study. J Neurosci 8:4340–4348
Ichikawa J, Meltzer HY (1992) The effect of chronic atypical antipsychotic drugs and haloperidol on amphetamine-induced dopamine release in vivo. Brain Res 574:98–114
Imperato A, DiChiara G (1985) Dopamine release and metabolism in awake rats after systemic neuroleptics as studied by trans-striatal dialysis. J Neurosci 5:297–306
Imperato A, Honore T, Jensen LH (1990) Dopamine release in the nucleus caudatus and in the nucleus accumbens is under glutamatergic control through non-NMDA receptors: a study in freely-moving rats. Brain Res 530:223–228
Innis RB, Aghajanian GK (1987) Pertussis toxin blocks autoreceptor-mediated inhibition of dopaminergic neurons in rat substantia nigra. Brain Res 411:139–143
Iversen LL (1975) Uptake processes for biogenic amines. In: Iversen LL, Iversen SD, Snyder SH (eds) Handbook of psychopharmacology. Plenum, New York, pp 381–392

Iversen SD, Wilkinson S, Simpson B (1971) Enhanced amphetamine responses after frontal cortex lesions in the rat. Eur J Pharmacol 13:387
Iwatsubo K, Clouet DH (1977) Effects of morphine and haloperidol on the electrical activity of rat nigrostriatal neurons. J Pharmacol Exp Ther 202:429–436
Jaskiw GC, Karoum F, Freed WJ, Phillips I, Kleinman JE, Weinberger DR (1990) Effect of ibotenic acid lesions of the medial prefrontal cortex on amphetamine-induced locomotion and regional brain catecholamines concentrations in the rat. Brain Res 534:263–272
Javitt DC, Zukin SR (1991) Recent advances in the phencyclidine model of schizophrenia. Am J Psychiatry 148:1301–1308
Javitt DC, Zylberman I, Zukin SR, Heresco-Levy U, Lindenmayer J-P (1994) Amelioration of negative symptoms in schizophrenia by glycine. Am J Psychiatry 151:1234–1236
Jenkins RB, Groh RH (1970) Mental symptoms in parkinsonian patients treated with L-DOPA. Lancet ii:177–179
Jiang LH, Tsai M, Wang RY (1988) Chronic treatment with high doses of haloperidol fails to decrease the time course for the development of depolarization inactivation of midbrain dopamine neurons. Life Sci 43:75–81
Johnson JW, Ascher P (1987) Glycine potentiates the NMDA response in cultured mouse brain neurons. Nature 325:529–531
Johnson KM, Jeng Y-J (1991) Pharmacological evidence for *N*-methyl-D-aspartate receptors on nigrostriatal dopaminergic nerve terminals. Can J Physiol Pharmacol 69:1416–1421
Johnson SW, North RA (1992) Two types of neurone in the rat ventral tegmental area and their synaptic inputs. J Physiol (Lond) 450:455–468
Johnson SW, Seutin V, North RA (1992) Burst firing in dopamine neurons induced by *N*-methyl-D-aspartate: role of electrogenic pump. Science 258:665–667
Jones SM, Snell LD, Johnson KM (1987) Inhibition by phencyclidine of excitatory amino acid-stimulated release of neurotransmitter in the nucleus accumbens. Neuropharmacology 26:173–179
Kabzinski AM, Szewczak MR, Cornfeldt ML, Fielding S (1987) Differential effects of dopamine agonists and antagonists on the spontaneous electrical activity of A9 and A10 dopamine neurons. Soc Neurosci Abstr 13:908
Kalivas PW, Duffy P (1991) A comparison of axonal and somatodendritic dopamine release using in vivo dialysis. J Neurochem 56:961–967
Kamata K, Rebec GV (1984) Nigral dopaminergic neurons: differential sensitivity to apomorphine following long-term treatment with low and high doses of amphetamine. Brain Res 321:147–150
Kane J, Honigfeld G, Singer J, Meltzer H (1988) Clozapine for the treatment-resistant schizophrenic. Arch Gen Psychiatry 45:789–796
Karoum F, Chrapusta SJ, Egan MF (1994) 3-methoxytyramine is the major metabolite of released dopamine in the rat frontal cortex: reassessment of the effects of antipsychotics on the dynamics of dopamine release and metabolism in the frontal cortex, nucleus accumbens, and striatum by a simple two pool mode. J Neurochem 63:972–979
Kawasaki Y, Maeda Y, Urata K, Higashima M, Yamaguchi N, Suzuki M, Takashima T, Ide Y (1993) A quantitative magnetic resonance imaging study of patients with schizophrenia. Eur Arch Psychiatry Clin Neurosci 242:268–272
Kebabian JW, Calne R (1979) Multiple receptors for dopamine. Nature 277:93–96
Keefe KA, Sved AF, Zigmond MJ, Abercrombie ED (1993) Stress-induced dopamine release in the neostriatum: evaluation of the role of action potentials in nigrostriatal dopamine neurons or local initiation by endogenous excitatory amino acids. J Neurochem 61:1943–1052
Kelland M, Chiodo LA, Freeman AS (1990) Anesthetic influences on the basal activity and pharmacological responsiveness of nigrostriatal dopamine neurons. Synapse 6:207–209

Kita H (1993) GABAergic circuits of the striatum. In: Arbuthnott GW, Emson PC (eds) Chemical signaling in the basal ganglia. Elsevier, Amsterdam, pp 51–72
Klawans HL, Margolin DI (1975) Amphetamine-induced dopaminergic hypersensitivity in guinea pigs. Arch Gen Psychiatry 32:725–732
Kovelman JA, Scheibel AB (1984) A neurohistological correlate of schizophrenia. Biol Psychiatry 19:1601–1917
Krebs M-O, Trovero F, Desban M, Gauchy C, Glowinski J, Kemel M-L (1991a) Distinct presynaptic regulation of dopamine release through NMDA receptors in striosome- and matrix-enriched areas of the rat striatum. J Neurosci 11:1256–1262
Krebs M-O, Desce JM, Kemel ML, Gauchy C, Godeheu G, Chéramy A, Glowinski J (1991b) Glutamatergic control of dopamine release in the rat striatum: evidence for presynaptic N-methyl-D-aspartate receptors on dopaminergic nerve terminals. J Neurochem 56:81–85
Lacey MG, Mercuri NB, North RA (1987) Dopamine acts on D2 receptors to increase potassium conductance in neurons of the rat substantia nigra zona compacta. J Physiol (Lond) 392:397–416
Lahti AC, Laporte DJ, Mokciski B, Tamminga CA (1993) Effect of the NMDA antagonist ketamine in schizophrenic patients. Soc Neurosci Abstr 19:1351
Lane RF, Blaha CD (1987) Chronic haloperidol decreases dopamine release in striatum and nucleus accumbens in vivo: depolarization block as a possible mechanism of action. Brain Res Bull 18:135–138
Lavin A, Grace AA (1994) Modulation of dorsal thalamic cell activity by the ventral pallidum: its role in the regulation of thalamocortical activity by the basal ganglia. Synapse 18:104–127
Leviel V, Gobert A, Guibert B (1990) The glutamate-mediated release of dopamine in the rat striatum: further characterization the dual excitatory-inhibitory function. Neuroscience 39:305–312
Levine MS (1991) Dopamine modulates coupling in the striatum. In: Gap junction protein (connexin) and electrical synapses in the central nervous system. 3rd IBRO world congress of neuroscience, p W35
Lichtensteiger W, Felix D, Lienhart R, Hefti F (1976) A quantitative correlation between single unit activity and fluorescence intensity of dopamine neurones in zona compacta of substantia nigra, as demonstrated under the influence of nicotine and physostigmine. Brain Res 117:85–103
Liddle PF, Morris DL (1991) Schizophrenic syndromes and frontal lobe performance. Br J Psychiatry 158:340–345
Liddle PF, Friston KJ, Frith CD, Frachowiak RSJ (1992a) Cerebral blood flow and mental processes in schizophrenia. J R Soc Med 85:224–227
Liddle PF, Friston KJ, Frith CD, Hirsch SR, Jones T, Frachowiak RSJ (1992b) Patterns of cerebral blood flow in schizophrenia. Br J Psychiatry 160:179–186
Liu L, Shen R-Y, Kapatos G, Chiodo LA (1994) Dopamine neuron membrane physiology: characterization of the transient outward current (I_A) and demonstration of a common signal transduction pathway for I_A and I_K. Synapse 17:230–240
MacGibbon GA, Lawlor PA, Bravo R, Dragunow M (1994) Clozapine and haloperidol produce a differential pattern of immediate early gene expression in rat caudate-putamen, nucleus accumbens, lateral septum and islands of Calleja. Mol Brain Res 23:21–32
Martínez-Fong D, Rosales MG, Góngora-Alfaro JL, Hernández S, Aceves J (1992) NMDA receptor mediates dopamine release in the striatum of unanesthetized rats as measured by brain microdialysis. Brain Res 595:309–315
Matsumoto A, Arai Y, Urano A, Hyodo S (1991) Cellular localization of gap junction mRNA in the neonatal rat brain. Neurosci Lett 124:225–228
Matthysse S (1978) A theory of the relation between dopamine and attention. In: Wynne LC, Cromwell RL, Matthysse S (eds) The nature of schizophrenia. New approaches to research and treatment. Wiley, New York, pp 307–310

Meltzer LT, Christoffersen CL, Heffner TG, Freeman AS, Chiodo LA (1989) CI-943, a potential antipsychotic agent. III. Evaluation of effects on dopamine neuronal activity. J Pharmacol Exp Ther 251:123–130
Merchant KM, Dorsa DM (1993) Differential induction of neurotensin and c-fos gene expression by typical versus atypical antipsychotics. Proc Natl Acad Sci USA 90:3447–3451
Mercuri NB, Stratta F, Calabresi P, Bernardi G (1992) A voltage-clamp analysis of NMDA-induced responses on dopaminergic neurons of the rat substantia nigra zona compacta and ventral tegmental area. Brain Res 593:51–56
Mercuri NB, Stratta F, Calabresi P, Bernardi G (1993) Neurotensin induces an inward current in rat mesencephalic dopaminergic neurons. Neurosci Lett 153:192–196
Mereu G, Casu M, Gessa GL (1983) (–) Sulpiride activates the firing rate and tyrosine hydroxylase activity of dopaminergic neurons in unanesthetized rats. Brain Res 264:105–110
Mereu G, Fanni B, Gessa GL (1984) General anesthetics prevent dargic neuron stimulation by neuroleptics. In: Usdin E, Carlsson A, Dahlström F, Engel J (eds) Catecholamines, neuropharmacology and central nervous system, theoretical aspects. Liss, New York, p 353
Mereu G, Lilliu V, Vargiu P, Muntoni AL, Diana M, Gessa GL (1994) Failure of chronic haloperidol to induce depolarization inactivation of dopamine neurons in unanesthetized rats. Eur J Pharmacol 264:449–453
Miller JD, Sanghera MK, German DC (1981) Mesencephalic dopaminergic unit activity in the behaviorally conditioned rat. Life Sci 29:1255–1263
Moghaddam B, Bolinao ML (1994) Glutamatergic antagonists attenuate ability of dopamine uptake blockers to increase extracellular levels of dopamine: implications for tonic influence of glutamate on dopamine release. Synapse 18:337–342
Moghaddam B, Gruen R, Roth RH, Bunney BS, Adams RN (1990) Effect of L-glutamate on the release of striatal dopamine: in vivo dialysis and electrochemical studies. Brain Res 518:55–60
Mueller AL, Brodie MS (1989) Intracellular recording from putative dopamine-containing neurons in the ventral tegmental area of Tsai in a brain slice preparation. J Neurosci Methods 28:15–22
Murase S, Grenhoff J, Chouvet G, Gonon F, Svensson T (1993) Prefrontal cortex regulates burst firing and transmitter release in rat mesolimbic dopamine neurons studied in vivo. Neurosci Lett 157:53–56
Nguyen TV, Kosofsky BE, Birnbaum R, Cohen BM, Hyman SE (1992) Differential expression of c-fos and zif268 in rat striatum after haloperidol, clozapine, and amphetamine. Proc Natl Acad Sci U S A 89:4270–4274
Nieoullon A, Chéramy A, Glowinski J (1978) Release of dopamine evoked by electrical stimulation of the motor and visual areas of the cerebral cortex in both caudate nuclei and in the substantia nigra in the cat. Brain Res 145:69–83
Nisenbaum ES, Berger TW, Grace AA (1993) Depression of glutamatergic and GABAergic synaptic responses in striatal spiny neurons by stimulation of presynaptic $GABA_B$ receptors. Synapse 14:221–242
Nissbrandt H, Elverfors, A, Engberg G (1994) Pharmacologically induced cessation of burst activity in nigral dopamine neurons: significance for the terminal dopamine efflux. Synapse 17:217–224
Núñez A, Garcia-Austt E and Buño W (1990) In vivo electrophysiological analysis of Lucifer yellow-coupled hippocampal pyramids. Exp Neurol 108:76–82
O'Donnell P, Grace AA (1993) Dopaminergic modulation of dye coupling between neurons in the core and shell regions of the nucleus accumbens. J Neurosci 13:3456–3471
O'Donnell P, Grace AA (1994) Tonic D_2-mediated attenuation of cortical excitation in nucleus accumbens neurons recorded in vitro. Brain Res 634:105–112
O'Donnell P, Grace AA (1995a) Different effects of subchronic clozapine and haloperidol on dye coupling between neurons in the rat striatal complex. Neuroscience 66:763–767

O'Donnell P, Grace AA (1995b) Synaptic interactions among excitatory afferents to nucleus accumbens neurons: hippocampal gating of prefrontal cortical input. J Neurosci 15:3622–3639

Onn S-P, Grace AA (1993) Effects of dopamine depletion on dye- and tracer-coupling between spiny cells and between aspiny cells in striatum. Soc Neurocsi Abstr 19:997

Onn S-P, Grace AA (1994a) Dye coupling between rat striatal neurons recorded in vivo: compartmental organization and modulation by dopamine. J Neurophysiol 71:1917–1934

Onn S-P, Grace AA (1994b) Repeated treatment with typical and atypical neuroleptics enhances electrotonic neurotransmission in the ventral corticostriatal regions. Soc Neurosci Abstr 20:565

Overton P, Clark D (1992) Iontophoretically administered drugs acting at the N-methyl-D-aspartate receptor modulate burst firing in A9 dopamine neurons in the rat. Synapse 10:431–440

Paskevich PA, Evans KH, Domesick VB (1991) Morphological assessment of neuronal aggregates in the striatum of the rat. J Comp Neurol 305:361–369

Peinado A, Yuste R, Katz LC (1993) Extensive dye coupling between rat neocortical neurons during the period of circuit formation. Neuron 10:103–114

Pettegrew JW, Keshavan M, Panchalingbam K, Strychor S, Kaplan DB, Tretta MG, Allen M (1991) Alterations in brain high-energy phosphate and membrane phospholipid metabolism in first-episode, drug-naive schizophrenics. Arch Gen Psychiatry 48:563–568

Pickar D, Labarca R, Linnoila M, Roy A, Hommer D, Everett D, Paul SM (1984) Neuroleptic-induced decrease in plasma homovanillic acid and antipsychotic activity in schizophrenic patients. Science 225:954–957

Post RM, Fink E, Carpenter WT, Goodwin FK (1975) Cerebrospinal fluid amine metabolites in acute schizophrenia. Arch Gen Psychiatry 32:1063–1069

Pucak ML, Grace AA (1991) Partial dopamine depletions result in an enhanced sensitivity of residual dopamine neurons to apomorphine. Synapse 9:144–155

Pucak ML, Grace AA (1994) Evidence that systemically administered dopamine antagonists activate dopamine neuron firing primarily by blockade of somatodendritic autoreceptors. J Pharmacol Exp Ther 271:1181–1192

Raine A, Lencz T, Reynolds GP, Harrison G, Sheard C, Medley I, Reynolds LM, Cooper JE (1992) An evaluation of structural and functional prefrontal deficits in schizophrenia: MRI and neuropsychological measures. Psychiatry Res Neuroimaging 45:123–137

Riederer P, Wuketich S (1976) Time course of nigrostriatal degeneration in Parkinson's disease. J Neural Transm 38:277–301

Robbins TW (1990) The case for frontostriatal dysfunction in schizophrenia. Schizophr Bull 16:391–402

Roberts PJ, Anderson SD (1979) Stimulatory effect of L-glutamate and related amino acids on [^{3}H]dopamine release from rat striatum: an in vitro model for glutamate actions. J Neurochem 32:1539–1545

Robertson GS, Fibiger HC (1992) Neuroleptics increase c-fos expression in the forebrain: contrasting effects of haloperidol and clozapine. Neuroscience 46:315–328

Rompré P-P, Wise RA (1989) Behavioral evidence for midbrain dopamine neuron depolarization block. Brain Res 477:152–156

Sanghera MK, Trulson ME, German DC (1984) Electrophysiological properties of mouse dopamine neurons: in vivo and in vitro studies. Neuroscience 12:793–801

Santiago M, Westerink BHC (1991) Characterization and pharmacological responsiveness of dopamine release recorded by microdialysis in the substantia nigra of conscious rats. J Neurochem 57:738–747

Santiago M, Westernik BHC (1992) Simultaneous recording of the release of nigral and striatal dopamine in the awake rat. Neurochem Int 20 [Suppl]:107S–110S

Schwartz J-C, Sokoloff P, Giros B, Martres MP, Bouthenet ML (1992) The dopamine D_3 receptor as a target for antipsychotics. In: Meltzer HY (ed) Novel antipsychotic drugs. Raven, New York, pp 135–144

Sedvall G, Farde L, Persson A, Weisel FA (1986) Imaging of neurotransmitter receptors in the living human brain. Arch Gen Psychiatry 43:995–1006

Seeman P (1987) Dopamine receptors and the dopamine hypothesis of schizophrenia. Synapse 1:113–152

Sesack SR, Pickel VM (1990) In the rat medial nucleus accumbens, hippocampal and catecholaminergic terminals converge on spiny neurons and are in apposition to each other. Brain Res 527:266–279

Sesack SR, Pickel VM (1992) Prefrontal cortical efferents in the rat synapse on unlabeled neuronal targets of catecholamine terminals in the nucleus accumbens septi and on dopamine neurons in the ventral tegmental area. J Comp Neurol 320:145–160

Sesack Sr, Aoki C, Pickel VM (1994) Ultrastructural localization of D_2 receptor-like immunoreactivity in midbrain dopamine neurons and their striatal targets. J Neurosci 14:88–106

Seutin V, Verbank P, Massotte L, Dresse A (1990) Evidence for the presence of *N*-methyl-D-aspartate receptors in the ventral tegmental area of the rat: an electrophysiological in vitro study. Brain Res 514:147–150

Seutin V, Johnson SW, North RA (1993) Apamin increases NMDA-induced burst-firing of rat mesencephalic dopamine neurons. Brain Res 630:341–344

Shen R-Y, Altar CA, Chiodo LA (1994) Brain-derived neurotrophic factor increases the electrical activity of pars compacta dopamine neurons in vivo. Proc Natl Acad Sci USA 91:8920–8924

Shenton ME, Kikinis R, Jolesz FA, Pollak SD, LeMay M, Wible CG, Hokama H, Martin J, Matcalf D, Coleman M, McCarley RW (1992) Abnormalities of the left temporal lobe and thought disorder in schizophrenia. A quantitative magnetic resonance imaging study. N Engl J Med 327:604–612

Shepard PD, Bunney BS (1991) Repetitive firing properties of putative dopamine-containing neurons in vitro: regulation by an apamin-sensitive Ca^{2+}-activated K^+ conductance. Exp Brain Res 86:141–150

Sherman AD, Davidson AT, Baruah S, Hegwood TS, Waziri R (1991) Evidence of glutamatergic deficiency in schizophrenia. Neurosci Lett 121:77–80

Shi W-X, Bunney BS (1992) Roles of intracellular cAMP and protein kinase A in the actions of dopamine and neurotensin on midbrain dopamine neurons. J Neurosci 12:2433–2438

Shimizu N, Duan S, Hori T, Oomura Y (1990) Glutamate modulates dopamine release in the striatum as measured by brain microdialysis. Brain Res Bull 25:99–102

Skarsfeldt T (1988) Differential effects after repeated treatment with haloperidol, clozapine, thioridazine, and tefludazine on SNC and VTA dopamine neurons in rats. Life Sci 42:1037–1044

Skarsfeldt T (1993) Comparison of the effect of substituted benzamides on midbrain dopamine neurones after treatment of rats for 21 days. Eur J Pharmacol 240:269–275

Skarsfeldt T (1994) Comparison of short-term administration of sertindole, clozapine and haloperidol on the inactivation of midbrain dopamine neurons in the rat. Eur J Pharmacol 254:291–294

Smith ID, Grace AA (1992) Role of the subthalamic nucleus in the regulation of nigral dopamine neuron activity. Synapse 12:287–303

Snyder SH (1972) Catecholamines in the brain as mediators of amphetamine psychosis. Arch Gen Psychiat 27:169–179

Snyder SH (1973) Amphetamine psychosis: a model of schizophrenia mediated by catecholamines. Am J Psychiatry 130:61–67

Sorensen SM, Humphreys TM, Palfreyman MG (1989) Effect of acute and chronic MDL 73147EF, a 5-HT_3 receptor antagonist, on A9 and A10 dopamine neurons. Eur J Pharmacol 1163:115–118

Sotelo C, Korn H (1978) Morphological correlates of electrical and other interactions through low-resistance pathways between neurons of the vertebrate central nervous system. Int Rev Citol 55:67–107

Stewart WW (1981) Lucifer dyes. Highly fluorescent dyes for biological tracing. Nature 292:17–21

Suaud-Chagny MF, Chergui K, Chouvet G, Gonon F (1992) relationship between dopamine release in the rat nucleus accumbens and the discharge activity of dopaminergic neurons during local in vivo application of amino acids in the ventral tegmental area. Neuroscience 49:63–72

Suddath RL, Christison GW, Torrey EF, Casanova MF, Weinberger DR (1990) Anatomical abnormalities in the brains of monozygotic twins discordant for schizophrenia. N Engl J Med 322:789–794

Szewczak MR, Dunn RW, Corbett R, Geyer HM, Rush DK, Wilker JC, Strupczewski JT, Helsely GC, Cornfeldt ML (1990) The in vivo pharmacology of the novel antipsychotic HP 873. Soc Neurosci Abstr 16:249

Tamminga CA, Thaker GK, Buchanan R, Kirkpatrick B, Alphs LD, Chase TN, Carpenter WT (1992) Limbic system abnormalities identified in schizophrenia using positron emission tomography with fluorodeoxyglucose and neocortical alterations with deficit syndrome. Arch Gen Psychiatry 49:522–530

Tepper JM, Nakamura S, Young ST, Groves PM (1984) Autoreceptor-mediated changes in dopaminergic terminal excitability: effects of striatal drug infusions. Brain Res 309:317–333

Todorova A, Dimpfel W (1994) Multiunit activity from the A9 and A10 areas in rats following chronic treatment with different neuroleptic drugs. Eur J Neuropsychopharmacol 4:491–501

Tung CS, Grenhoff J, Svensson T (1991) Kynurenate blocks the acute effects of haloperidol on midbrain dopamine neurons recorded in vivo. J Neural Transm 84:53–64

van Bockstaele EJ, Sesack SR, Pickel VM (1994a) Dynorphin-immunoreactive terminals in the rat nucleus accumbens: cellular sites for modulation of target neurons and interactions with catecholamine afferents. J Comp Neurol 341:1–15

van Bockstaele EJ, Gracy KN, Pickel VM (1994b) Ultrastructure of dynorphine immunoreactive perikarya and terminals in the nucleus accumbens: relations to dopamine and substance P. Soc Neurosci Abstr 20:1733

van Kammen DP, Bok van Kammen W, Mann LS, Seppala T, Linnoila M (1986) Dopamine metabolism in the cerebrospinal fluid of drug-free schizophrenic patients with and without cortical atrophy. Arch Gen Psychiatry 43:978–983

Wachtel ST, White FJ (1992) The effect of continuous and repeated administration of D_1 dopamine receptor antagonist on midbrain dopamine neurons. Neurochem Int 20 [Suppl]:129S–133S

Waddington JL (1993) Schizophrenia: developmental neuroscience and pathobiology. Lancet 341:531–536

Walsh JP, Cepeda C, Hull CD, Fisher RS, Levine MS, Buchwald NA (1989) Dye-coupling in the neostriatum of the rat: II. Decreased coupling between neurons during development. Synapse 4:238–247

Walsh JP, Cepeda C, Buchwald NA, Levine MS (1991) Neurophysiological maturation of cat substantia nigra neurons: evidence from in vitro studies. Synapse 7:291–300

Wang JKT (1991) Presynaptic glutamate receptors modulate dopamine release from striatal synaptosomes. J Neurochem 57:819–822

Wang T, French ED (1993a) Effects of phencyclidine on spontaneous and excitatory amino acid-induced activity of ventral tegmental dopamine neurons: an extracellular in vitro study. Life Sci 53:49–56

Wang T, French ED (1993b) L-glutamate excitation of A10 dopamine neurons is preferentially mediated by activation of NMDA receptors: extra- and intracellular electrophysiological studies in brain slices. Brain Res 627:299–306

Weinberger DR (1987) Implications of normal brain development for the pathogenesis of schizophrenia. Arch Gen Psychiatry 44:660–669

Weinberger DR, Berman KF, Suddath R, Torrey EF (1992) Evidence of dysfunction of a prefrontal-limbic network in schizophrenia: a magnetic resonance imaging and regional cerebral blood flow study of discordant monozygotic twins. Am J Psychiatry 149:890–897

White FJ, Wang RY (1983) Differential effects of classical and atypical antipsychotic drugs on A9 and A10 dopamine neurons. Science 221:1054–1057

Wolkin A, Sanfilipo M, Wolf AP, Angrist B, Brodie JD, Rotrosen J (1992) Negative symptoms and hypofrontality in chronic schizophrenia. Arch Gen Psychiatry 49:959–965

Wolkin A, Sanfilipo M, Angrist B, Duncan E, Wieland S, Wolf AP, Brodie JD, Cooper TB, Laska E, Rotrosen JP (1994) Acute *d*-amphetamine challenge in schizophrenia: effects on cerebral glucose utilization and clinical symptomatology. Biol Psychiatry 36:317–325

Woodruff GN, McCarthy PS, Walker RJ (1976) Studies on the pharmacology of neurones in the nucleus accumbens of the rat. Brain Res 115:233–242

Wu H-Q, Schwarcz R, Shepard PD (1994) Excitatory amino acid-induced excitation of dopamine-containing neurons in the rat substantia nigra: modulation by kynurenic acid. Synapse 16:219–230

Wyatt RJ (1986) The dopamine hypothesis: variations on a theme (II). Psychopharmacol Bull 22:923–927

Wyatt RJ, Alexander RC, Egan MF, Kirch DG (1988) Schizophrenia, just the facts. What do we know, how well do we know it? Schizophr Res 1:3–18

Youngren KD, Daly DA, Moghaddam B (1993) Distinct actions of endogenous excitatory amino acids on the outflow of dopamine in the nucleus accumbens. J Pharmacol Exp Ther 264:289–293

Yung WH, Häusser MA, Jack JJB (1991) Electrophysiology of dopaminergic and non-dopaminergic neurones of the guinea-pig substantia nigra pars compacta in vitro. J Physiol (Lond) 436:643–667

Zetterström T, Sharp T, Ungerstedt U (1985) Effect of neuroleptic drugs on striatal dopamine release and metabolism in the awake rat studied by intracerebral dialysis. Eur J Pharmacol 106:27–37

Zhang H, Kiyatkin EA, Stein EA (1994) Behavioral and pharmacological modulation of ventral tegmental dendritic dopamine release. Brain Res 656:59–70

Zieglgansberger W, Puil EA (1983) Actions of glutamic acid on spinal neurones. Exp Brain Res 17:35–49

Zigmond MJ, Abercrombie ED, Berger TW, Grace AA, Stricker EM (1990) Compensations after lesions of central dopaminergic neurons: some clinical and basic implications. Trends Neurosci 13:290–296

CHAPTER 7

Tolerance and Sensitization to the Effects of Antipsychotic Drugs on Dopamine Transmission

RONALD E. SEE and PETER W. KALIVAS

A. Introduction

Beginning with the finding by CARLSSON and LINDQVIST (1963) that acute administration of chlorpromazine increases dopamine (DA) turnover, an extensive literature has developed concerning the effects of antipsychotic drug (APD) administration on multiple brain DA systems. Although APDs of different chemical classes can vary markedly in their molecular structure and receptor binding profiles (HYTTEL et al. 1985), they all act to some degree as DA receptor antagonists (SEEMAN et al. 1976; FARDE et al. 1988). A wide variety of neuronal mechanisms have been studied in relation to acute and prolonged APD effects on dopaminergic function, including changes in DA receptors, DA receptor-linked second messenger activity, and DA neuron electrophysiology (SEE and CHAPMAN 1994, for review). Among the many effects of APDs on DA function, the release and turnover of forebrain DA continues to serve as the primary reflection of dynamic APD-induced alterations in neural activity. One aim of the present chapter is to review findings on APD effects on DA release and metabolism.

Because of the frequent necessity to administer APDs chronically in the treatment of schizophrenia, many investigators have sought to identify the effects of long-term APD administration on DA transmission and related motor behaviors. In these studies, it has been generally observed that continuous or frequent drug administration produces some degree of tolerance, while more intermittent injections can result in augmentation of some of the acute effects of the APDs. This augmentation has been termed sensitization and is of particular interest since the gradual development of some extrapyramidal motor side effects may result from the development of sensitization. In addition, it has been proposed that the progressive exacerbation of schizophrenia in some individuals is a sensitization-like phenomenon, and that the symptoms manifested during sensitization are attenuated by intervention with APDs. Thus, the delayed onset of therapeutic benefit following the initiation of APD therapy may reflect a reversal of endogenous sensitization-like phenomena. The second aim of this chapter is to critically evaluate the literature demonstrating sensitization and tolerance to chronic APD administration and speculate on the role these processes may have in the therapeutic benefit and side effects of APD therapy.

B. Effects of Single Administration on Dopamine and Related Behaviors

Acute administration of APDs results in increased firing rates of DA neurons (BUNNEY et al. 1973), increased DA turnover rates (CARLSSON and LINDQVIST 1963), and increased tyrosine hydroxylase activity (KUCZENSKI 1980). The increase in DA turnover following acute treatment is evident across several brain regions, including the striatum, cortex, and limbic regions (SCATTON et al. 1975, 1976; MATSUMOTO et al. 1983). Recent microdialysis experiments have further demonstrated that acute APD administration significantly increases levels of extracellular striatal DA and the DA metabolites 3,4-dihydroxyphenylacetic acid (DOPAC) and homovanillic acid (HVA) (DICHIARA and IMPERATO 1985; ZETTERSTROM et al. 1985; SEE et al. 1991). This increase in DA overflow and turnover appears to be primarily regulated by DA D_2 receptor activation in that selective DA D_2 antagonists are more potent than DA D_1 antagonists when administered either systemically (SEE et al. 1991) or directly into DA terminal fields (WESTERINK and DE VRIES 1989).

The acute APD-induced increase in DA release has been attributed to blockade of presynaptic DA autoreceptors as well as postsynaptic receptors leading to feedback on midbrain DA neurons (CARLSSON and LINDQVIST 1963; KEHR et al. 1972; BUNNEY et al. 1973). WESTERINK and DEVRIES (1989) have presented evidence that the increased DA release is due almost exclusively to blockade of DA autoreceptors at the nerve terminals. They based this conclusion on the fact that rats lesioned with kainic acid (which produces degeneration of postsynaptic neurons) show a similar increase in DA and DA metabolite in response to sulpiride when compared to nonlesioned rats, thus suggesting that postsynaptic feedback is not required for mediating APD effects on DA release. In addition, changes in nerve impulse activity appeared not to be involved since both systemic administration or direct infusion produced similar increases in DA. Further studies assessing both nigral and striatal DA receptor activation have strengthened the proposition that activation of DA release by APDs is largely mediated by autoreceptors on nerve terminal (SANTIAGO and WESTERINK 1991). While the data from these microdialysis studies is compelling, it must be considered in light of evidence that DA release occurs in two distinct and independently regulated forms: (a) phasic DA release related directly to DA neuron firing and (b) tonic DA release regulated largely by cortical afferents (GRACE 1991). Acute APD administration is clearly associated with activation of DA neuron firing (BUNNEY and GRACE 1978), which is then reflected in an enhanced phasic DA release (GRACE 1992).

Attempts to relate the effects of acute APD administration on DA release or turnover to temporally related behavioral measures have generally been unsatisfactory. Although HONMA and FUKUSHIMA (1976) did report a correlation between APD-induced decreases in striatal tissue DA content and catalepsy, measurement of extracellular DA and DA metabolites indicates no direct relationship to catalepsy (ZETTERSTROM et al. 1985; MEIL and SEE 1994). Given that most of the clinical benefits as well as many of the motor side effects of APDs are not evidenced until prolonged exposure to APDs has occurred, most studies have focused on the relationship between DA activity and behavioral measures following repeated drug administration.

C. Effects of Repeated Administration and the Induction of Sensitization and Tolerance

An examination of the effects of repeated administration of APDs is essential for several reasons. Since the drugs generally do not exert their clinical effects until they have been administered over periods lasting from days to weeks, prolonged time frames allow for the study of the appropriate mechanisms by which APDs exert their antipsychotic effects. In addition, individuals treated with APDs are often maintained on continuous or intermittent treatment regimens that may span decades during the course of the individual's illness, particularly in the case of schizophrenia. Finally, repeated APD administration can lead to several types of movement disorders appearing at various times during the course of treatment, including dystonia and tardive dyskinesia (CASEY 1987). Thus, understanding the alterations in brain function produced by prolonged APD administration can provide insight into both the clinical efficacy of these compounds as well as the neurological side effects produced by long-term drug administration.

Tolerance and sensitization refer to a decrease or increase, respectively, in the response to a stimulus as a result of previous exposure to that stimulus. When the altered responsiveness results from previous exposure to a different stimulus this is referred to as cross-tolerance or cross-sensitization. Sensitization and tolerance can be demonstrated using both behavioral and biochemical measures. Repeated or continuous exposure to a variety of pharmacological treatments has been shown to produce sensitization and/or tolerance (STEWART and BADDIANI 1993). Although pharmacological tolerance is a more widely studied phenomenon, in recent years the development of sensitization has received increasing attention. This is due primarily to two factors: (1) There has been a rapid emergence of mechanistic information associated with sensitization processes such as long-term potentiation in hippocampal slices and sensitization to the locomotor-stimulant properties of amphetamine-like psychostimulants. (2) Based in part upon this emerging information, an important role for sensitization-like processes in the development and maintenance of a variety of neuropsychiatric disorders has been proposed (ANTELMAN et al. 1986; POST and WEISS 1988). Examples of sensitization-based psychiatric disorders may be amphetamine-induced psychosis, cocaine-induced panic attacks, and posttraumatic stress syndrome. In addition, other psychotic, affective, and anxiety disorders may be, in part, the result of sensitization processes in the nervous system.

The current literature showing both tolerance and sensitization to APDs is critically reviewed below. In addition, because of the well-developed literature surrounding sensitization and tolerance to amphetamine-like psychostimulants, we will reflect on studies with psychostimulants as a guide for interpreting some of the APD literature. Two general principles apply for the development of tolerance versus sensitization to psychostimulants, and as detailed below, also apply for APDs: (1) More intermittent repeated treatment regimens are required to produce behavioral and neurochemical sensitization, while more frequently repeated or continuous administration yields tolerance. (2) Dopamine projections can differentially manifest tolerance or sensitization depending upon the treatment regimens and biochemical measurements employed.

I. Tolerance

In contrast to the increase in DA release and turnover produced by acute injection of APDs, repeated APD treatment produces tolerance as evidenced by an attenuation of APD-induced increases in DOPAC and HVA content in striatal homogenates (ASPER et al. 1973; SAYERS et al. 1975; SCATTON et al. 1975; BOWERS and HOFFMAN 1986). In addition, decreases in striatal DA metabolite levels can persist following withdrawal from subchronic APD treatment (LERNER et al. 1977; FINLAY et al. 1987), although tolerance may dissipate by 1 week after drug withdrawal (GIANUTSOS et al. 1975). However, it has been noted that both tolerance and lower DA metabolite levels may not occur in basal ganglia nuclei of subchronic APD-treated primates (BACOPOULOS et al. 1982). Tolerance can exhibit a relatively early onset since many studies have demonstrated changes in DA turnover within the first week of repeated drug administration (ASPER et al. 1973; LERNER et al. 1977; SALLER and SALAMA 1985). Tolerance to the DA release-enhancing effects of APDs has even been reported to occur 4–12h after an initial dose (DICHIARA and IMPERATO 1985).

Several other approaches have verified that a reduced response of DA release and/or turnover occurs following prolonged exposure to APDs. In vitro studies following subchronic haloperidol treatment indicate significantly reduced spontaneously and electrically evoked DA release in striatal slices (UMEDA and SUMI 1990; YAMADA et al. 1993), although one report showed no effect of subchronic haloperidol on either amphetamine- or electrically stimulated striatal DA release (COMPTON and JOHNSON 1989). In vivo microdialysis and in vivo voltammetry have also been used to determine changes in basal extracellular DA levels and DA metabolites produced by repeated treatment.

In regards to basal DA metabolism, several studies have reported significant decreases in striatal extracellular DOPAC and HVA concentrations following subchronic typical APD administration (HERNANDEZ and HOEBEL 1989; ICHIKAWA and MELTZER 1990, 1991). There have also been some studies reporting decreases in basal striatal DA levels using microdialysis (ICHIKAWA and MELTZER 1990, 1991) and voltammetry (BLAHA and LANE 1987). In contrast, other findings have indicated increased extracellular DA following subchronic haloperidol (ZHANG et al. 1989) or no changes (HERNANDEZ and HOEBEL 1989; SEE and MURRAY 1992; YAMAMOTO and COOPERMAN 1994). Microdialysis experiments have also shown tolerance to acute D_2 antagonist administration on DA release after repeated APD treatment (HERNANDEZ and HOEBEL 1989; ICHIKAWA and MELTZER 1990, 1991; SEE and MURRAY 1992).

The reduced response to acute APD challenge following repeated APD exposure has been reported to show regional specificity in that it is more pronounced in the striatum than in limbic or cortical regions (BOWERS and ROZITIS 1974; SCATTON et al. 1975, 1976; BOWERS and HOFFMAN 1986; CSERNANSKY et al. 1990) and also develops more rapidly in the striatum than limbic regions (SCATTON 1977). Although repeated APD administration has been reported to produce tolerance in the cortex (WHEELER and ROTH 1980; RAO and WOOD 1990), most studies show no changes (GIANUTSOS et al. 1975; SCATTON et al. 1976; MATSUMOTO et al. 1983; LINDEFORS et al. 1986) or even enhanced metabolism (MEFFORD et al. 1988) in cortical regions. Subchronic haloperidol administration has generally been reported not to produce tolerance in DA cell body regions (NICOLAOU 1980; LINDEFORS et al. 1986; CSERNANSKY et al. 1990), although MELLER et al. (1980) reported an increase in DOPAC levels in the substantia nigra in response to subsequent haloperidol challenge.

A number of mechanisms have been described which may account for the tolerance that can develop with repeated APD treatment. FINLAY et al. (1987) have suggested that the apparent tolerance may simply be an artifact of reduced basal levels of DA metabolites seen after repeated exposure to APDs and that the absolute response to acute APD challenge is maintained in the presence of reduced basal levels of DA activity. However, this cannot fully explain the effects of repeated APDs, since tolerance has been demonstrated in the absence of changes in basal extracellular levels of DA, DOPAC, and HVA (SEE 1991; SEE and MURRAY 1992). Most explanations for APD-induced tolerance have centered on alterations in pre- and postsynaptic DA mechanisms developing after subchronic APD administration. An increase in presynaptic autoreceptor sensitivity appears to play a significant role for decreased DA turnover (SALLER and SALAMA 1985) as well as decreased DA synthesis (NOWYCKY and ROTH 1977). In vitro studies have provided evidence for APD-induced autoreceptor supersensitivity as demonstrated by enhanced inhibition of tyrosine hydroxylase activity (BOOTH et al. 1991) and enhanced reduction of DA release by apomorphine (YAMADA et al. 1993). Increased sensitivity of postsynaptic receptors leading to long loop negative feedback has also been implicated in altered DA turnover (BUNNEY et al. 1973, 1991). Depolarization blockade of DA neurons has also been explored as an explanation for the changes in terminal release of DA following prolonged treatment (BUNNEY and GRACE 1978; WHITE and WANG 1983; GRACE 1992). Several studies confirm the role of depolarization block, including studies utilizing in vivo voltammetry (BLAHA and LANE 1987) and microdialysis (CHEN et al. 1991). However, some findings question a direct correlation between changes in DA neuron firing and terminal release and metabolism (EGAN et al. 1991; ICHIKAWA and MELTZER 1991).

The effects of prolonged APD treatment on DA turnover can be strongly influenced by dosing regimen, assay methods, and type of APD. Manipulation of all of these variables has been found to result in situations where no tolerance occurs. The use of different dosing regimens may even produce reverse tolerance or sensitization (see below). Different indices of DA metabolism may also indicate differential effects of APD treatment. A recent study using accumulation of 3-methoxytyramine as a measure of DA metabolism reported that no tolerance developed in the striatum after subchronic haloperidol treatment (EGAN et al. 1991). Finally, atypical APDs such as clozapine may result in a lack of tolerance in specific brain regions (see below).

Chronic APD administration of several months' duration has been reported to produce effects on DA release and metabolism that are uniquely different from those seen after subchronic APD administration of several days' or weeks' duration. MESSIHA (1974) reported increased levels of DA and DA metabolites in cerebrospinal fluid and urine of APD-treated monkeys after 12, but not 4 months of chlorpromazine administration. These increases in indices of DA metabolism were also positively associated with the incidence of oral dyskinesia. In rats treated with chronic trifluoperazine, striatal DA and DA metabolite levels were not different from control at 6 months of treatment (CLOW et al. 1979), but DA and DOPAC levels were elevated after 1 year of administration (CLOW et al. 1980). In contrast to most reported changes in extracellular DA and DA metabolites found after subchronic APD treatment, long-term APD administration (6–8 months) has been shown to increase extracellular DOPAC and HVA concentrations in striatal but not limbic regions (SEE 1991, 1993; SEE et al. 1992b). In addition, even though basal levels were elevated, profound tolerance was seen following acute challenge with a DA D_2 receptor antagonist to a degree similar to the tolerance seen after subchronic haloperidol administration (SEE 1991; SEE and MURRAY 1992). These results suggest that administration of APDs for periods of time analogous to long-term pharmacotherapy in the clinic may result in a selective reversal of some of the effects seen after subchronic treatment.

Several behavioral measures have been found to show tolerance following repeated APD treatment, including catalepsy and blockade of apomorphine-induced behavior (ASPER et al. 1973; EZRIN-WATERS and SEEMAN 1977). However, few studies have actually attempted to discern the relationship between tolerance to DA turnover or release and behavioral measures. Changes over time in catalepsy, the most commonly used measure of APD-induced motor effects in animals, have been shown to be dissociated from the development of tolerance to APD-induced DA turnover (COYLE et al. 1985; DE GRAAF and KORF 1986). Catalepsy suffers from being a limited test since it is subject to many variables that can impact measurement outcome. Nonetheless, BARNES et al. (1990) used catalepsy to provide a clear demonstration that chronic haloperidol administration can lead to either sensitization or tolerance depending on dose, treatment schedule, and test apparatus. An example of these data is shown in modified form in Fig. 1. While measures of catalepsy are useful, other behavioral measures may also serve as indicators of APD effects related to tolerance. For example, one study suggested that rats exhibiting a higher degree of oral movements following chronic haloperidol treatment showed greater tolerance to the effects of acute haloperidol challenge on striatal HVA levels (KOLENIK et al. 1988).

II. Sensitization

While most studies have focused on the apparent tolerance that occurs for DA release and turnover with repeated APDs, several lines of evidence suggest that certain forms of sensitization can also occur with repeated APD administration. These data include various measures of motor activity as well as biochemical indices. In a manner analogous to psychostimulant treatment, regimen differences appear to determine (at least in part) whether sensitization will develop.

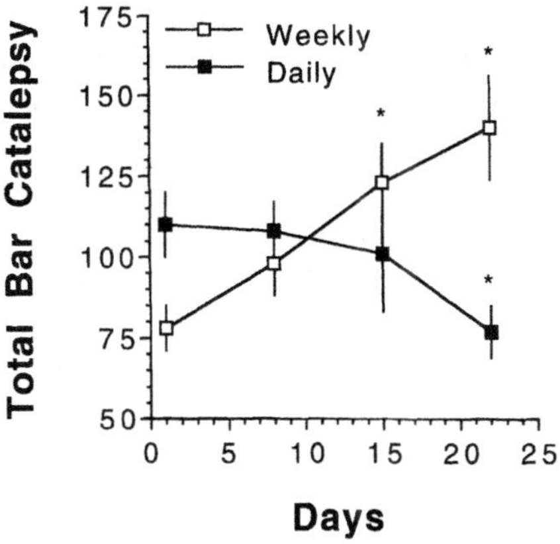

Fig. 1. Daily or weekly injections of haloperidol (1.5 mg/kg, sc) produce tolerance or sensitization, respectively. Animals were tested for catalepsy on days 1, 8, 15, and 22 using a horizontal bar. Data are shown as mean ± SEM. These data were adapted with permission from BARNES et al. (1990). *$p < 0.05$ compared to day 1

Differences in APD administration regimens has been shown to determine the altered responsivity of dopaminergic pathways. CAREY and DEVEAUGH-GEISS (1984), measuring spontaneous motor activity, found a trend towards behavioral sensitization in rats treated every other day with haloperidol, but behavioral tolerance in rats given twice daily injections. This pattern was mirrored by changes in HVA levels in the two groups, with the twice daily group showing tolerance (reduced levels) and the every other day group showing neurochemical sensitization (elevated levels). CSERNANSKY et al. (1990) found that subchronic haloperidol given daily produced the oft-reported tolerance, while weekly injections produced the opposite effect of sensitization. COSTALL et al. (1985) reported that continuous DA receptor antagonism prevented the development of increased sensitivity to DA agonist infusion, while intermittent dosing enhanced the effects on locomotor activity. Finally, KASHIHARA et al. (1986) administered continuous haloperidol (osmotic minipumps) or intermittent haloperidol (daily injections) for 14 days, followed by a 7-day withdrawal period. Following a challenge dose of haloperidol on the seventh day, they found a greater tolerance response to DA turnover and higher [^{3}H]spiperone binding in the continuous treatment group. These studies support the contention that intermittent APD treatment regimens may lead to sensitization to the DA turnover-elevating properties of APDs and dopaminergic related behaviors, while continuous regimens lead to tolerance on both measures.

Sensitization has been explored as a possible mechanism in the development of motor side effects produced by repeated APD exposure. As discussed above (MASUDA et al. 1982; CSERNANSKY et al. 1990; SEE et al. 1992a), repeated subchronic or chronic administration of APDs in rodents can lead to sensitization, rather than tolerance in measures of catalepsy. The sensitization to catalepsy-producing effects seems to be dependent on several factors, including schedule, dose, and conditioning (MASUDA et al. 1982; CSERNANSKY et al. 1990). In primates, prolonged, intermittent APD administration reliably produces symptoms that can best be described as dystonic in nature (RUPNIAK et al. 1986; CASEY 1992). This pattern in the primate model of "priming" acute motor symptoms becomes most severe after repeated drug dosing (WEISS and SANTELLI 1978; LIEBMAN and NEALE 1980). WEISS and SANTELLI (1978) found that weekly doses of haloperidol produced dyskinetic episodes of a greater magnitude than those seen in animals given equivalent cumulative doses over a week's time. This approach of intermittent drug administration in which monkeys are primed to show acute dyskinetic reactions may be viewed as a clear example of sensitization. However, the relationship of DA release and turnover to this behavioral sensitization remains unclear. Acute dystonic reactions have been speculated to occur due to greater DA release onto receptors made supersensitive by APD exposure (KOLBE et al. 1981). Further support for the role of enhanced DA release includes findings that L-DOPA increases dystonic movements (PARKES et al. 1976) and that DA depletion can reduce the incidence of dystonia (MELDRUM et al. 1977). However, a hyperfunctional

DA release explanation for dystonia remains oversimplified, since the etiology clearly involves additional mechanisms, particularly excess cholinergic activity (RUPNIAK et al. 1986).

In addition to dystonic reactions, some evidence suggests that tardive dyskinesia may be a form of kindling by repeated on-off APD treatment (POST 1980; GLENTHOJ et al. 1990). The practice of intermittent interruption of treatment ("drug holidays") was actually initiated based on the theoretical possibility of preventing the development of motor side effects (AYD 1967). However, there is some suggestion that the degree of severity of tardive dyskinesia is related to the number of prior discontinuities in drug treatment (JESTE et al. 1979) and that drug holidays may actually worsen tardive dyskinesia (JESTE and WYATT 1982). Animal studies have demonstrated that differences in dosing regimen during long-term APD administration can profoundly influence the nature of late-onset changes in motor behavior produced by long-term APD administration. SEE and ELLISON (1990) demonstrated that weekly high-dose injections of haloperidol and continuous daily oral administration produced different forms of oral dyskinesias, with intermittent treatment causing more pronounced rapid openings and closings of the jaws, while continuous treatment produced slower, low-frequency jaw movements. These results suggest that regimen differences produce quite different motor syndromes which may model various clinical forms of oral dyskinesia. GLENTHOJ and colleagues have also found differences in oral behaviors produced by different long-term regimens of haloperidol. Intermittent dosing produced a greater magnitude of increased oral movements and this increase was found to persist for long periods following drug withdrawal (GLENTHOJ and HEMMINGSEN 1989; GLENTHOJ et al. 1990). The authors suggest that this form of oral activity sensitization may serve as an animal model of tardive dyskinesia. Thus, the general results from animal models suggest that chronic intermittent exposure acts to sensitize DA-mediated motor systems in ways uniquely different from chronic continuous exposure.

A series of studies by ANTELMAN et al. (1986, 1992b) have focused on the phenomenon of time-dependent sensitization for a variety of psychoactive compounds, including APDs. They have reported that a single exposure to a neuroleptic can lead to enhancement of catalepsy upon subsequent neuroleptic administration at time points many weeks after the initial injection. They suggest that single drug exposure initiates a process of sensitization which may be due to the foreign stressor nature of the agent. In support of this hypothesis, ANTELMAN et al. (1992a) have presented data that single pretreatment with amphetamine or haloperidol both produce a blunting of amphetamine-induced cortisol elevation 2 weeks later, indicating that drugs with opposing pharmacological actions can produce the same lasting effect. While these results suggest that time-dependent changes play a critical role in determining the development of behavioral and neurochemical indices of sensitization and/or tolerance, it remains unclear as to how this form of single APD treatment induced sensitization relates to dopaminergic function. MEIL and

SEE (1994) recently reported that single pretreatment with fluphenazine produced sensitization for catalepsy upon subsequent injection. In contrast, this single preexposure produced significant tolerance in the capacity of fluphenazine to elevate DA. Further, the temporal pattern of changes in DA release and catalepsy were not correlated.

D. Atypical Antipsychotic Drugs

Atypical APDs are generally defined as compounds that exhibit antipsychotic properties, but which do not show a high degree of neurological side effects. Recent elaboration of the concept of atypical APDs includes several defining characteristics that may qualify a compound as atypical, including lack of extrapyramidal side effects, minimized risk for development of tardive dyskinesia, greater efficacy in treatment-resistant patients, and greater improvement in negative symptoms (COWARD 1993). The best characterized atypical APD, clozapine, is the one APD that is generally agreed upon as lacking the relatively high incidence of motor syndromes seen with all typical APDs (CASEY 1989). In addition to its very low incidence of motor side effects, clozapine also shows improved efficacy in the treatment of patients who have been nonresponsive to typical APD administration (see Chap. 8 by S.-O. ÖGREN, this volume). Although clozapine has been intensely studied, the mechanisms of action that might account for its atypical profile remain unclear. Some of the more feasible explanations offered include a possible selectivity for mesolimbic pathways (BARTHOLINI 1977; WHITE and WANG 1983; COWARD et al. 1989), a relatively higher affinity for serotonin receptors (MELTZER 1989), and higher relative binding affinity to selective DA receptor subtypes such as the DA D_3 and DA D_4 receptors (SEEMAN 1991; SOKOLOFF et al. 1992).

Several lines of evidence support a unique profile for clozapine in regards to DA release and turnover. Although most typical APDs have been reported to produce robust tolerance following prolonged exposure, tolerance has been either shown not to occur with subchronic clozapine treatment (SAYERS et al. 1975) or to preferentially affect limbic DA turnover (ANDEN and STOCK 1973; CSERNANSKY et al. 1993). It has been generally noted that clozapine fails to affect measures of basal DA release and metabolism in the striatum (BLAHA and LANE 1987; CHAI and MELTZER 1992; INVERNIZZI et al. 1990; ICHIKAWA and MELTZER 1991; CHEN et al. 1991), with only one report of an increase in basal DA (ICHIKAWA and MELTZER 1992). In line with a possible selective effect on mesolimbic DA pathways, some reports indicate that while clozapine has no effect on striatal DA, there is a selective reduction of DA levels in the nucleus accumbens (BLAHA and LANE 1987; CHEN et al. 1991). However, this effect has not been seen in several other studies (CHAI and MELTZER 1992; INVERNIZZI et al. 1990; ICHIKAWA and MELTZER 1991). Some support for mesocortical selectivity of clozapine comes from studies showing that it preferentially increases DA levels in the cortex when compared to the striatum following acute and chronic treatment (MOGHADDAM and BUNNEY 1990; PEHEK et al. 1993; YAMAMOTO and COOPERMAN 1994), although one report has not found this to be the case (IMPERATO and ANGELUCCI 1989).

In addition to clozapine, several other compounds exhibiting putative atypical profiles have been tested for their effects on DA release and metabolism. The highly DA D_2-specific substituted benzamides exhibit an apparent preferential action on limbic-mediated motor behaviors relative to nigrostriatal-mediated motor behaviors (ÖGREN et al. 1986). While acute administration of these drugs produces an increase in DA, DOPAC, and HVA similar to that seen with typical APDs (ZETTERSTROM et al. 1985; SEE et al. 1991), studies utilizing repeated administration of drugs such as raclopride and remoxipride have reported both similarities and differences when compared with typical APDs such as haloperidol. Subchronic (AHLENIUS et al. 1991) and chronic (FOWLER et al. 1987) raclopride administration produced tolerance to DA turnover, but a comparatively lesser degree of DA receptor-mediated supersensitivity (AHLENIUS et al. 1991). In addition, subchronic administration of remoxipride suggested that tolerance to its effects on DA turnover develop much slower than tolerance seen with haloperidol (MAGNUSSON et al. 1987). However, chronic administration of raclopride shows a very similar profile to haloperidol on several measures including an enhanced response to DA D_2 agonist-mediated decreases in extracellular DA and alterations in synaptic ultrastructure (SEE et al. 1992a).

Several newer compounds exhibiting higher affinity for serotonin receptors relative to DA receptors have been recently developed and show much more promise as atypical APDs. These drugs all appear to differ from typical APDs in regards to both their acute and repeated effects on DA release and metabolism. When compared to haloperidol, acute treatment with the serotonin-dopamine antagonist risperidone produces different dose–response curves and relatively more potent effects on limbic and cortical nuclei (LEYSEN et al. 1992). The compound ICI 204636 (Seroquel) shows a "clozapine-like" effect in that chronic treatment fails to produce tolerance in the striatum when DOPAC and HVA are measured (SALLER and SALAMA 1993). Finally, the novel APD amperozide has been shown to preferentially increase DA efflux in the cortex when compared to the striatum, an effect similar to that seen with clozapine (PEHEK et al. 1993; NOMIKOS et al. 1994). Thus, newer classes of atypical APDs may be distinguished in part from typical APDs based on their regionally selective effects on DA release and metabolism. Current and future development of novel APDs show promise for even greater selectivity of action, particularly highly selective DA D_3 and D_4 receptor selective compounds which are now becoming available.

E. Conclusions

Pharmacological sensitization of DA pathways induced by psychostimulants has served as a widely utilized model for psychosis (SEGAL and SCHUKITT 1983;

Robinson and Becker 1986; Kalivas et al. 1993). In addition to stimulant-induced sensitization, electrical sensitization of DA cell bodies (Glenthoj et al. 1993) and stress-induced sensitization (Antelman 1988; Kalivas and Duffy 1989) have also been explored as models of psychosis. Models based on DA cell electrophysiology (Grace 1991, 1992) suggesting hyperactivation of phasic DA release in schizophrenia also fit well into the concept of sensitization of dopaminergic function. Although the neural substrates of psychosis are immensely complex, DA sensitization models provide a valid working hypothesis, particularly for examining the substrates of positive symptoms and the impact of APD treatment on symptoms. Recent clinical evidence continues to confirm the importance of the primary impact of APDs on DA function in regards to both antipsychotic efficacy and motor side effects (Farde et al. 1992; Nordstrom et al. 1993).

If it is assumed that sensitization plays a role in the etiology of psychosis, how might prolonged APD exposure act upon sensitized DA pathways to attenuate psychotic symptoms? Based on the data reviewed above, we hypothesize that endogenous sensitization of specific DA pathways occurring during idiopathic psychosis may be reversed or attenuated by the capacity of APDs to induce tolerance in DA transmission. Indeed, clinical measurement of peripheral HVA has provided some evidence that tolerance development is related to clinical improvement. Specifically, reductions in posttreatment HVA levels (tolerance) have been correlated with greater clinical response, while patients showing no decreases in HVA (nontolerance) slow poorer clinical response (Bowers 1984; Davidson et al. 1991; Sharma et al. 1993). More direct experimental testing of the attenuation of sensitization comes from assessment of the effects of DA receptor antagonists on stimulant-induced sensitization. Sensitization produced by repeated psychostimulant administration is expressed through augmented release of mesolimbic DA (Kalivas et al. 1993). Behavioral indices of both sensitization (Kuczenski and Leith 1981) and tolerance (Barrett and White 1980) to repeated amphetamine indicate that coadministration of haloperidol will block their development. More recently, coadministration of DA D_1 and D_2 receptor antagonists has been found to block the development of methamphetamine-induced sensitization of both DA release and stereotypy (Hamamura et al. 1991). Systemic administration of the DA D_1 antagonist SCH23390 has also been found to prevent amphetamine-induced behavioral sensitization (Vezina and Stewart 1989), while direct infusion of SCH23390 into midbrain DA nuclei attenuates amphetamine-induced sensitization (Stewart and Vezina 1989). Based on these results of an attenuation of DA agonist-induced sensitization by APDs, it may be the case that the drugs also act to reverse psychotic symptoms arising from sensitized DA pathways. However, some discrepancies exist in the literature regarding the capacity of selective DA D_2 antagonists to prevent psychostimulant-induced sensitization (Kalivas and Stewart 1991). It will be useful to further examine several current animal models of schizophrenic symptoms to evaluate the role of APD-induced reversal of sensitization, particularly models of sensorimotor gating deficits such as those seen with prepulse inhibition (SWERDLOW et al. 1991) and latent inhibition (DUNN et al. 1993).

It appears clear that regional differences occur in the development of sensitization and tolerance to APDs and that these differences may reflect the multiple behavioral changes wrought by chronic APD exposure. APD-induced reversal of endogenous sensitization most likely occurs in areas implicated in the pathology of schizophrenia, particularly temporal-limbic regions such as the hippocampus, temporal cortices, amygdala, and prefrontal cortex (REYNOLDS 1989; KRIECKHAUS et al. 1992; LIPSKA et al. 1993). In this regard, it is of interest to note a report of increased amygdalar DA in nonmedicated schizophrenics (REYNOLDS 1983) and the finding that the highest degree of tolerance across several brain regions observed after subchronic haloperidol administration in rats occurred in the amygdala (MATSUMOTO et al. 1983). However, it would be premature to rule out a role for striatal regions in mediating such effects, based on the number of studies showing the greatest magnitude of APD-induced tolerance in striatal regions (BOWERS and ROZITIS 1974; SCATTON et al. 1975, 1976; BOWERS and HOFFMAN 1986; CSERNANSKY et al. 1990). It is also worth nothing that an integral role of the striatum in the pathophysiology of psychosis is a core feature of many current models of schizophrenia (CARLSSON 1988; GRAY et al. 1991). While reversal of sensitized limbic/striatal regions may be one possible outcome of APD treatment, mesocortical DA pathways may show further differentiation from these regions in regards to tolerance and sensitization. It has been hypothesized that hyperactive mesolimbic DA pathways may underlie many of the positive symptoms, while concurrent hypofunction of cortical DA pathways may relate to negative symptoms (WEINBERGER 1987). Evidence showing that repeated APD administration enhances DA metabolism (BANNON and ROTH 1983; MEFFORD et al. 1988) and extracellular DA levels (YAMAMOTO and COOPERMAN 1994) in the prefrontal cortex supports the role of differential actions of APDs in this region when compared to limbic or striatal areas.

While APDs may attenuate endogenous/exogenous processes of sensitization, they may simultaneously produce sensitization in other DA pathways. Based on the reviewed literature, this seems to be particularly true of motor pathways, including the nigrostriatal DA system. Thus, repeated exposure to typical APDs, particularly in an intermittent fashion, may heighten dopaminergic activity and be manifested in the form of hyperkinetic behaviors. While the emphasis of the present review is on APD effects on DA pathways, the role of other neurotransmitters is particularly important in considering the etiology of motor side effects that may develop as a form of sensitization to repeated treatment. Recent speculation includes a primary role for basal ganglia amino acid neurotransmitters such as γ-aminobutyric acid (GABA) and glutamate (GUNNE and ANDREN 1993). Further understanding of the process of APD-induced sensitization within these nuclei will better elucidate the propensity of drugs to induce motor side effects.

The variable effects of APDs on producing regionally selective forms of sensitization and tolerance may help account for the different profiles obtained with clozapine and other atypical drugs. Clozapine has now been well established as exhibiting preferential actions on mesolimbic/mesocortical pathways and this preferential activation may underlie its ability to alleviate positive symptoms via blockade in the mesolimbic activity and alleviate negative symptoms via sustained enhancement of mesocortical DA activity (MOGHADDAM 1994). The failure of clozapine to exhibit significant motor side effects may be attributable to its relative lack of long-term changes in striatal DA release and metabolism. Further exploration of clozapine and other atypical APD effects on multiple DA pathways will help in our understanding of their favorable clinical profiles and direct future drug development.

Results from differential drug administration schedules in animal studies may have implications for patterns of APD dosing in patients. As described above, intermittent treatment in the form of "drug holidays" has been suggested as a means of blocking the development of motor side effects (AYD 1967). However, the effects of intermittent dosing in animal models suggests that such practices may actually increase the incidence of motor side effects. Current clinical data on the issue of drug side effects indicates that intermittent medication practices have either no differential effect when compared to continuous medication regimens (HERZ et al. 1991) or that they do indeed increase the likelihood and/or severity of motor side effects, including tardive dyskinesia (JESTE et al. 1979; JESTE and WYATT 1982; GOLDMAN and LUCHINS 1984). If the greater degree of tolerance seen after continuous treatment regimens in animal models relates favorably to the mitigation of sensitized dopaminergic pathways, then one might also predict that more continuous treatment regimens would lead to better clinical outcome. This possibility is supported by several clinical studies showing that the use of continuous maintenance regimens results in greater long-term treatment efficacy when compared to intermittent dosing practices (for review, see JANICAK et al. 1993).

Finally, while the present review has focused on mechanisms of APD effects on DA release and metabolism, clearly other adaptive changes may be involved in the processes of tolerance and sensitization that can occur with repeated APD treatment. These alterations include many nondopaminergic systems that interact with midbrain DA pathways (see Chap. 3 by O. CIVELLI, this volume). Of particular interest is the growing focus on dopaminergic-serotonergic interactions in mediating the clinical efficacy of typical and atypical APDs (MELTZER 1989; KAHN et al. 1993). Continued study of dopaminergic interactions with other neurotransmitters in the context of tolerance and sensitization may further our understanding of both the beneficial and deleterious effects of chronic APD treatment.

Acknowledgements. The authors' work was supported by United States Public Health Service grants DE-09678 (RES), MH-40817 (PWK), DA-03506 (PWK), and the State of Washington.

References

Ahlenius S, Ericson EL, Hogberg K, Wijkstrom A (1991) Behavioural and biochemical effects of subchronic treatment with raclopride in the rat: tolerance and brain monoamine receptor sensitivity. Pharmacol Toxicol 68:302–309

Anden NE, Stock G (1973) Effects of clozapine on the turnover of dopamine in the corpus striatum and in the limbic system. J Pharm Pharmacol 25:346–348

Antelman SM (1988) Stressor-induced sensitization to subsequent stress: implications for the development and treatment of clinical disorders. In: Kalivas PW, Barnes CD (eds) Sensitization in the nervous system. Telford, Caldwell, pp 227–256

Antelman SM, Kocan D, Edwards DJ, Knopf S, Perel JM, Stiller R (1986) Behavioral effects of a single neuroleptic treatment grow with the passage of time. Brain Res 385:58–67

Antelman SM, Caggiula AR, Knopf S, Kocan DJ, Edwards DJ (1992a) Amphetamine or haloperidol 2 weeks earlier antagonize plasma corticosterone response to amphetamine; evidence for the stressful/foreign nature of drugs. Psychopharmacology (Berl) 107:331–336

Antelman SM, Kocan D, Knopf S, Edwards DJ, Caggiula AR (1992b) One brief exposure to a psychological stressor induces long-lasting, time-dependent sensitization of both the cataleptic and neurochemical responses to haloperidol. Life Sci 51:261–266

Asper H, Baggiolini M, Burki HR, Lauener H, Ruch W, Stille G (1973) Tolerance phenomena with neuroleptics: catalepsy, apomorphine stereotypies and striatal dopamine metabolism in the rat after single and repeated administration of loxapine and haloperidol. Eur J Pharmacol 22:287–294

Ayd FJ (1967) Drug holidays: intermittent pharmacotherapy for psychiatric patients. Med Sci 1:59–62

Bacopoulos NG, Bustos G, Redmond DE, Roth RH (1982) Chronic treatment with haloperidol or fluphenazine decanoate: regional effects on dopamine and serotonin metabolism in primate brain. J Pharmacol Exp Ther 221:22–28

Bannon MJ, Roth RH (1983) Pharmacology of mesocortical dopamine neurons. Pharmacol Rev 35:53–68

Barnes DE, Robinson B, Csernansky JG, Bellows EP (1990) Sensitization versus tolerance to haloperidol-induced catalepsy: multiple determinants. Pharmacol Biochem Behav 36:883–887

Barrett RJ, White DK (1980) Reward system depression following chronic amphetamine: reversal by haloperidol. Pharmacol Biochem Behav 13:555–559

Bartholini G (1977) Preferential effect of noncataleptogenic neuroleptics on mesolimbic dopaminergic function. Adv Biochem Psychopharmacol 16:607–611

Blaha CD, Lane RF (1987) Chronic treatment with classical and atypical antipsychotic drugs differentially decreases dopamine release in striatum and nucleus accumbens in vivo. Neurosci Lett 78:199–204

Booth RG, Baldessarini RJ, Campbell A (1991) Inhibition of dopamine synthesis in rat striatal minces: evidence of dopamine autoreceptor supersensitivity to S(+)- but not R(−)-N-n-propylnorapomorphine after pretreatment with fluphenazine. Biochem Pharmacol 41(12):2040–2043

Bowers MB (1984) Family history and CSF homovanillic acid pattern during neuroleptic treatment. Am J Psychiatry 141:296–298

Bowers MB, Hoffman FJ (1986) Homovanillic acid in caudate and pre-frontal cortex following acute and chronic neuroleptic administration. Psychopharmacology (Berl) 88:63–65

Bowers MB, Rozitis A (1974) Regional differences in homovanillic acid concentrations after acute and chronic administration of antipsychotic drugs. J Pharm Pharmacol 26:743–745
Bunney BS, Grace AA (1978) Acute and chronic haloperidol treatment: comparison of effects on nigral dopaminergic cell activity. Life Sci 23:1715–1728
Bunney BS, Walters JR, Roth RH, Aghajanian GK (1973) Dopaminergic neurons: effect of antipsychotic drugs and amphetamine on single cell activity. J Pharmacol Exp Ther 185:560–571
Bunney BS, Chiodo LA, Grace AA (1991) Midbrain dopamine system electrophysiological functioning: a review and new hypothesis. Synapse 9:79–94
Carey RJ, De Veaugh-Geiss J (1984) Treatment schedule as a determinant of the development of tolerance to haloperidol. Psychopharmacology (Berl) 82:164–167
Carlsson A (1988) The current status of the dopamine hypothesis of schizophrenia. Neuropsychopharmacology 1:179–186
Carlsson A, Lindqvist M (1963) Effect of chlorpromazine or haloperidol on the formation of 3-methoxytyramine and normetanephrine in mouse brain. Acta Pharmacol (Kbh) 20:140–144
Casey DE (1987) Tardive dyskinesia. In: Meltzer HY (ed) Psychopharmacology: the third generation of progress. Raven, New York, pp 1411–1420
Casey DE (1989) Clozapine: neuroleptic-induced EPS and tardive dyskinesia. Psychopharmacology (Berl) 99:S47–S53
Casey DE (1992) Dopamine D_1 (SCH23390) and D_2 (haloperidol) antagonists in drug-naive monkeys. Psychopharmacology (Berl) 107:18–22
Chai B, Meltzer HY (1992) The effect of chronic clozapine on basal dopamine release and apomorphine-induced DA release in the striatum and nucleus accumbens as measured by in vivo brain microdialysis. Neurosci Lett 136:47–50
Chen J, Paredes W, Gardner EL (1991) Chronic treatment with clozapine selectively decreases basal dopamine release in nucleus accumbens but not in caudate-putamen as measured by in vivo brain microdialysis: further evidence for depolarization block. Neurosci Lett 122:127–131
Clow A, Jenner P, Theodorou A, Marsden CD (1979) Striatal dopamine receptors become supersensitive while rats are given trifluoperazine for six months. Nature 278:59–61
Clow A, Jenner P, Theodorou A, Marsden CD (1980) Changes in cerebral dopamine metabolism and receptors during one-year neuroleptic administration and subsequent withdrawal: relevance to brain biochemistry in schizophrenia. Adv Biochem Psychopharmacol 24:53–55
Compton DR, Johnson KM (1989) Effects of acute and chronic clozapine and haloperidol on in vitro release of acetylcholine and dopamine from striatum and nucleus accumbens. J Pharmacol Exp Ther 248:521–530
Costall B, Domeney AM, Naylor RJ (1985) The continuity of dopamine receptor antagonism can dictate the long-term behavioural consequences of a mesolimbic infusion of dopamine. Neuropharmacology 24(3):193–197
Coward DM (1993) The pharmacology of clozapine-like, atypical antipsychotics. In: Barnes TM (ed) Antipsychotic drugs and their side effects. Academic, London, pp 27–44
Coward DM, Imperato A, Urwyler S, White TG (1989) Biochemical and behavioral properties of clozapine. Psychopharmacology (Berl) 99:S6–S12
Coyle S, Napier TC, Breese GR (1985) Ontogeny of tolerance to haloperidol: behavioral and biochemical measures. Dev Brain Res 23:27–38
Csernansky JG, Bellows EP, Barnes DE, Lombrozo L (1990) Sensitization versus tolerance to the dopamine turnover-elevating effects of haloperidol: the effect of regular/intermittent dosing. Psychopharmacology (Berl) 100:519–524
Csernansky JG, Wrona CT, Bardgett ME, Early TS, Newcomer JN (1993) Subcortical dopamine and serotonin turnover during acute and subchronic administration of typical and atypical neuroleptics. Psychopharmacology (Berl) 110:145–151

Davidson M, Kahn RS, Knott P, Kaminsky R, Cooper M, DuMont K, Apter S, Davis KL (1991) Effects of neuroleptic treatment on symptoms of schizophrenia and plasma homovanillic acid concentrations. Arch Gen Psychiatry 48:910–913
De Graaf CJ, Korf J (1986) Conditional tolerance to haloperidol-induced catalepsy is not caused by striatal dopamine receptor supersensitivity. Psychopharmacology (Berl) 90:54–57
DiChiara G, Imperato A (1985) Rapid tolerance to neuroleptic-induced stimulation of dopamine release in freely moving rats. J Pharmacol Exp Ther 235(2):487–494
Dunn LA, Atwater GE, Kilts CD (1993) Effects of antipsychotic drugs on latent inhibition: sensitivity and specificity of an animal behavioral model of clinical drug action. Psychopharmacology (Berl) 112:315–323
Egan MF, Karoum F, Wyatt RJ (1991) Effects of acute and chronic clozapine and haloperidol administration on 3-methoxytyramine accumulation in rat prefrontal cortex, nucleus accumbens and striatum. Eur J Pharmacol 199:191–199
Ezrin-Waters C, Seeman P (1977) Tolerance to haloperidol catalepsy. Eur J Pharmacol 41:321–327
Farde L, Wiesel FA, Halldin C, Sedvall G (1988) Central D2-dopamine receptor occupancy in schizophrenic patients treated with antipsychotic drugs. Arch Gen Psychiatry 45:71–76
Farde L, Nordstrom AL, Wiesel FA, Pauli S, Halldin C, Sedvall G (1992) Positron emission tomographic analysis of central D_1 and D_2 dopamine receptor occupancy in patients treated with classical neuroleptics and clozapine. Arch Gen Psychiatry 49:538–544
Finlay JM, Jakubovic A, Fu DS, Fibiger HC (1987) Tolerance to haloperidol-induced increases in dopamine metabolites: fact or artifact? Eur J Pharmacol 137:117–121
Fowler CJ, Magnusson O, Thorell G, Mohringe B, Huang RB (1987) Dopamine turnover and glutamate decarboxylase activity in the rat brain after acute and chronic treatment with raclopride, a dopamine D_2-selective antagonist. Neuropharmacology 26(4):339–345
Gianutsos G, Hynes MD, Lal H (1975) Enhancement of apomorphine-induced inhibition of striatal dopamine-turnover following chronic haloperidol. Biochem Pharmacol 24:581–582
Glenthoj B, Hemmingsen R (1989) Intermittent neuroleptic treatment induces long-lasting abnormal mouthing in the rat. Eur J Pharmacol 164:393–396
Glenthoj B, Hemmingsen R, Allerup P, Bolwig TG (1990) Intermittent versus continuous neuroleptic treatment in a rat model. Eur J Pharmacol 190:275–286
Glenthoj B, Mogensen J, Laursen H, Holm S, Hemmingsen R (1993) Electrical sensitization of the meso-limbic dopaminergic system in rats: a pathogenetic model for schizophrenia. Brain Res 619:39–54
Goldman MB, Luchins DJ (1984) Intermittent neuroleptic therapy and tardive dyskinesia: a literature review. Hosp Community Psychiatry 35:1215–1219
Grace AA (1991) Phasic versus tonic dopamine release and the modulation of dopamine system responsivity: a hypothesis for the etiology of schizophrenia. Neuroscience 41(1):1–24
Grace AA (1992) The depolarization block hypothesis of neuroleptic action: implications for the etiology and treatment of schizophrenia. J Neural Transm 36:91–131
Gray JA, Feldon J, Rawlins JNP, Hemsley DR, Smith AD (1991) The neuropsychology of schizophrenia. Behav Brain Sci 14:1–20
Gunne LM, Andren PE (1993) An animal model for coexisting tardive dyskinesia and tardive parkinsonism: a glutamate hypothesis for tardive dyskinesia. Clin Neuropharmacol 16:90–95
Hamamura T, Akiyama K, Akimoto K, Kashihara K, Okumara K, Ujike H, Otsuki S (1991) Co-administration of either a selective D_1 or D_2 dopamine antagonist with methamphetamine prevents methamphetamine-induced behavioral sensitization and neurochemical change, studied by in vivo intracerebral dialysis. Brain Res 546:40–46

Hernandez L, Hoebel BG (1989) Haloperidol given chronically decreases basal dopamine in the prefrontal cortex more than the striatum or nucleus accumbens as simultaneously measured by microdialysis. Brain Res Bull 22:763–769

Herz MI, Glazer WM, Mostert MA, Sheard MA, Szymanski HV, Hafez H, Mirza M, Vana J (1991) Intermittent vs maintenance medication in schizophrenia. Arch Gen Psychiatry 48:333–339

Honma T, Fukushima H (1976) Correlation between catalepsy and dopamine decrease in the rat striatum induced by neuroleptics. Neuropharmacology 15:601–607

Hyttel J, Larsen JJ, Christensen AV, Arnt J (1985) Receptor-binding profiles of neuroleptics. In: Casey DE, Chase TN, Christensen AV, Gerlach J (eds) Dyskinesia – research and treatment. Springer, Berlin Heidelberg New York, pp 9–18

Ichikawa J, Meltzer HY (1990) Apomorphine does not reverse reduced basal dopamine release in rat striatum and nucleus accumbens after chronic haloperidol. Brain Res 507:138–142

Ichikawa J, Meltzer HY (1991) Differential effects of repeated treatment with haloperidol and clozapine on dopamine release and metabolism in the striatum and the nucleus accumbens. J Pharmacol Exp Ther 256(1):348–357

Ichikawa J, Meltzer HY (1992) The effect of chronic atypical antipsychotic drugs and haloperidol on amphetamine-induced dopamine release in vivo. Brain Res 574:98–104

Imperato A, Angelucci L (1989) The effects of clozapine and fluperlapine on the in vivo release and metabolism of dopamine in the striatum and in the prefrontal cortex of freely moving rats. Psychopharmacol Bull 25(3):383–389

Imperato A, DiChiara G (1985) Dopamine release and metabolism in awake rats after systemic neuroleptics as studied by trans-striatal dialysis. J Neurosci 5:297–306

Invernizzi R, Morali F, Pozzi L, Samanin R (1990) Effects of acute and chronic clozapine on dopamine release and metabolism in the striatum and nucleus accumbens of conscious rats. Br J Pharmacol 100:774–778

Janicak PG, Davis JM, Preskorn SH, Ayd FJ (1993) Principles and practice of psychopharmacotherapy. Williams and Wilkins, Baltimore, pp 93–184

Jeste DV, Wyatt RJW (1982) Therapeutic strategies against tardive dyskinesia: two decades of experience. Arch Gen Psychiatry 39:803–816

Jeste DV, Potkin SG, Sinha S, Feder S, Wyatt RJ (1979) Tardive dyskinesia – reversible and persistent. Arch Gen Psychiatry 36:585–590

Kahn RS, Davidson M, Knott P, Stern RG, Apter S, Davis KL (1993) Effect of neuroleptic medication on cerebrospinal fluid monoamine metabolite concentrations in schizophrenia. Serotonin-dopamine interactions as a target for treatment. Arch Gen Psychiatry 50:599–605

Kalivas PW, Duffy P (1989) Similar effects of daily cocaine and stress on mesocorticolimbic dopamine neurotransmission in the rat. Biol Psychiatry 25:913–928

Kalivas PW, Stewart J (1991) Dopamine transmission in the initiation and suppression of drug- and stress-induced sensitization of motor activity. Brain Res Rev 16:223–244

Kalivas PW, Sorg BA, Hooks MS (1993) The pharmacology and neural circuitry of sensitization to psychostimulants. Behav Pharmacol 4:315–334

Kashihara K, Sato M, Fujiwara Y, Harada T, Ogawa T, Otsuki S (1986) Effects of intermittent and continuous haloperidol administration on the dopaminergic system in the rat brain. Biol Psychiatry 21:650–656

Kehr W, Carlsson A, Lindqvist M, Magnusson T, Atack CV (1972) Evidence for a receptor mediated feedback control of striatal tyrosine hydroxylase activity. J Pharm Pharmacol 24:744–747

Kolbe H, Clow A, Jenner P, Marsden CD (1981) Neuroleptic-induced acute dystonic reactions may be due to enhanced release on to supersensitive postsynaptic receptors. Neurology 31:434–439

Kolenik SA, Hoffman FJ, Bowers MB (1989) Regional homovanillic acid levels and oral movements in rats following chronic haloperidol treatment. Psychopharmacology (Berl) 98:430–431

Kriekhaus EE, Donahoe JW, Morgan MA (1992) Paranoid schizophrenia may be caused by dopamine hyperactivity of CA1 hippocampus. Biol Psychiatry 31:560–570

Kuczenski R (1980) Amphetamine-haloperidol interactions on striatal and mesolimbic tyrosine hydroxylase activity and dopamine metabolism. J Pharmacol Exp Ther 215:135–142

Kuczenski R, Leith NJ (1981) Chronic amphetamine: is dopamine a link in or a mediator of the development of tolerance and reverse tolerance? Pharmacol Biochem Behav 15:405–413

Lerner P, Nose PN, Gordon EK, Lovenberg W (1977) Haloperidol: effect of long-term treatment on rat striatal dopamine synthesis and turnover. Science 197:181–183

Leysen JE, Janssen PMF, Gommeren W, Wynants J, Pauwels PJ, Janssen PAJ (1992) In vitro and in vivo receptor binding and effects on monoamine turnover in rat brain regions of the novel antipsychotics risperidone and ocaperidone. Mol Pharmacol 41:494–508

Liebman J, Neale R (1980) Neuroleptic-induced acute dyskinesias in squirrel monkeys: correlation with propensity to cause extrapyramidal side effects. Psychopharmacology (Berl) 68:25–29

Lindefors N, Sharp T, Ungerstedt U (1986) Effects of subchronic haloperidol and sulpiride treatment on regional brain dopamine metabolism in the rat. Eur J Pharmacol 129:401–404

Lipska BK, Weinberger DR (1993) Cortical regulation of the mesolimbic dopamine system: implications for schizophrenia. In: Kalivas PW, Barnes CD (eds) Limbic motor circuits and neuropsychiatry. CRC, Boca Raton, pp 329–350

Magnusson O, Mohringe B, Thorell G, Lake-Bakaar DM (1987) Effects of the dopamine D_2 selective receptor antagonist remoxipride on dopamine turnover in the rat brain after acute and repeated administration. Pharmacol Toxicol 60:368–373

Masuda Y, Murai S, Itoh T (1982) Tolerance and reverse tolerance to haloperidol catalepsy induced by the difference of administration interval in mice. Jpn J Pharmacol 32:1186–1188

Matsumoto T, Uchimura H, Hirano M, Kim JS, Yokoo H, Shimomura M, Nakahara T, Inoue K, Oomagari K (1983) Differential effects of acute and chronic administration of haloperidol on homovanillic acid levels in discrete dopaminergic areas of rat brain. Eur J Pharmacol 89:27–33

Mefford IN, Roth KA, Agren H, Barchas JD (1988) Enhancement of dopamine metabolism in rat brain frontal cortex: a common effect of chronically administered antipsychotic drugs. Brain Res 475:380–384

Meil W, See RE (1994) Single pre-exposure to fluphenazine produces persisting behavioral sensitization accompanied by tolerance to fluphenazine-induced striatal dopamine overflow in rats. Pharmacol Biochem Behav 48(3):605–612

Meldrum BS, Anlezark GM, Marsden CD (1977) Acute dystonia as an idiosyncratic response to neuroleptics in baboons. Brain 100:313–326

Meller E, Friedhoff AJ, Friedman E (1980) Differential effects of acute and chronic haloperidol treatment on striatal and nigral 3,4-dihydroxyphenylacetic acid (DOPAC) levels. Life Sci 26:541–547

Meltzer HY (1989) Clinical studies on the mechanism of action of clozapine: the dopamine-serotonin hypothesis of schizophrenia. Psychopharmacology (Berl) 99:S18–S27

Messiha FS (1974) A study of biogenic amine metabolites in the cerebrospinal fluid and urine of monkeys with chlorpromazine-induced dyskinesia. J Neurol Sci 21:39–46

Moghaddam B (1994) Preferential activation of cortical dopamine neurotransmission by clozapine: functional significance. J Clin Psychiatry 55(9):27–29

Moghaddam B, Bunney BS (1990) Acute effects of typical and atypical antipsychotic drugs on the release of dopamine from prefrontal cortex, nucleus accumbens, and striatum of the rat: an in vivo microdialysis study. J Neurochem 54:1755–1760

Nicolaou NM (1980) Acute and chronic effects of neuroleptics and acute effects of apomorphine and amphetamine on dopamine turnover in corpus striatum and substantia nigra of the rat brain. Eur J Pharmacol 64:123–132

Nomikos GG, Iurlo M, Andersson JL, Kimura K, Svensson TH (1994) Systemic administration of amperozide, a new atypical antipsychotic drug, preferentially increases dopamine release in the rat medial prefrontal cortex. Psychopharmacology (Berl) 115:147–156

Nordstrom A, Farde L, Wiesel F, Forslund, Pauli S, Halldin C, Uppfeldt G (1993) Central D_2-dopamine receptor occupancy in relation to antipsychotic drug effects: a double-blind PET study of schizophrenic patients. Biol Psychiatry 33:227–235

Nowycky MC, Roth RH (1977) Presynaptic dopamine receptors. Development of supersensitivity following treatment with fluphenazine decanoate. Naunyn Schmiedebergs Arch Parmacol 300:247–254

Ogren SV, Hall H, Kohler C, Magnusson O, Sjostrand SE (1986) The selective dopamine D_2 receptor antagonist raclopride discriminates between dopamine-mediated motor functions. Psychopharmacology (Berl) 90:287–294

Parkes JD, Bedard P, Marsden CD (1976) Chorea and torsion in Parkinsonism. Lancet 1:155

Pehek EA, Meltzer HY, Yamamoto BK (1993) The atypical antipsychotic drug amperozide enhances rat cortical and striatal dopamine efflux. Eur J Pharmacol 240:107–109

Post RM (1980) Intermittent versus continuous stimulation: effect of time interval on the development of sensitization or tolerance. Life Sci 26:1275–1282

Post RM, Weiss SRB (1988) Sensitization and kindling: implications for the evolution of psychiatric symptomatology. In: Kalivas PW, Barnes CD (eds) Sensitization in the nervous system. Telford, Caldwell, pp 257–292

Rao TS, Wood PL (1990) Haloperidol decanoate administration induces differential tolerance to striatal and pyriform cortical dopamine metabolism. Neurosci Res Commun 7(2):83–88

Reynolds GP (1983) Increased concentrations and lateral asymmetry of amygdala dopamine in schizophrenia. Nature 305:527–529

Reynolds GP (1989) Beyond the dopamine hypothesis: the neurochemical pathology of schizophrenia. Br J Psychiatry 155:305–316

Robinson TE, Becker JB (1986) Enduring changes in brain and behavior produced by chronic amphetamine administration: a review and evaluation of animal models of amphetamine psychosis. Brain Res Rev 11:157–198

Rupniak NMJ, Jenner P, Marsden CD (1986) Acute dystonia induced by neuroleptic drugs. Psychopharmacology (Berl) 88:403–419

Saller CF, Salama AI (1985) Alterations in dopamine metabolism after chronic administration of haloperidol. Possible role of increased autoreceptor sensitivity. Neuropharmacology 24:123–129

Saller CF, Salama AI (1993) Seroquel: biochemical profile of a potential atypical antipsychotic. Psychopharmacology (Berl) 112:285–292

Santiago M, Westerink BHC (1991) The regulation of dopamine release from nigrostriatal neurons in conscious rats: the role of somatodendritic autoreceptors. Eur J Pharmacol 204:79–85

Sayers AC, Burki HR, Ruch W, Asper H (1975) Neuroleptic-induced hypersensitivity of striatal dopamine receptors in the rat as a model of tardive dyskinesias. Effects of clozapine, haloperidol, loxapine and chlorpromazine. Psychopharmacologia 41:97–104

Scatton B (1977) Differential regional development of tolerance to increase in dopamine turnover upon repeated neuroleptic administration. Eur J Pharmacol 46:363–369

Scatton B, Garrett C, Julou L (1975) Acute and subacute effects of neuroleptics on dopamine synthesis and release in the rat striatum. Naunyn Schmiedebergs Arch Pharmacol 289:419–434

Scatton B, Glowinski J, Julou L (1976) Dopamine metabolism in the mesolimbic and mesocortical dopaminergic systems after single or repeated administration of neuroleptics. Brain Res 109:184–189

See RE (1991) Striatal dopamine metabolism increases during long-term haloperidol administration in rats but shows tolerance to acute challenge with raclopride. Neurosci Lett 129:265–268

See RE (1993) Assessment of striatal extracellular dopamine and dopamine metabolites in haloperidol-treated rats exhibiting oral dyskinesia. Neuropsychopharmacology 9:101–109

See RE, Chapman MA (1994) The consequences of long-term antipsychotic drug administration on basal ganglia neuronal function in laboratory animals. Crit Rev Neurobiol 8(1/2):85–124

See RE, Ellison G (1990) Intermittent and continuous haloperidol regimens produce different types of oral dyskinesias in rats. Psychopharmacology (Berl) 100:404–412

See RE, Murray CE (1992) Changes in striatal dopamine release and metabolism during and after subchronic haloperidol administration in rats. Neurosci Lett 142:100–104

See RE, Sorg BA, Chapman MA, Kalivas PW (1991) In vivo assessment of release and metabolism of dopamine in the ventrolateral striatum of awake rats following administration of dopamine D_1 and D_2 receptor agonists and antagonists. Neuropharmacology 30:1269–1274

See RE, Chapman MA, Meshul CK (1992a) Comparison of chronic intermittent haloperidol and raclopride effects on striatal dopamine release and synaptic ultrastructure in rats. Synapse 12:147–154

See RE, Chapman MA, Murray CE, Aravagiri M (1992b) Regional differences in chronic neuroleptic effects on extracellular dopamine activity. Brain Res Bull 29:473–478

Seeman P (1992) Dopamine receptor sequences. Therapeutic levels of neuroleptics occupy D_2 receptors, clozapine occupies D4. Neuropsychopharmacology 7:261–284

Seeman P, Lee T, Chau-Wong M, Wong K (1976) Antipsychotic drug doses and neuroleptic/dopamine receptors. Nature 261:717–719

Segal DS, Schukitt MA (1983) Animal models of stimulant-induced psychosis. In: Creese I (ed) Stimulants: neurochemical, behavioral, and clinical perspectives. Raven, New York, pp 131–167

Sharma RP, Javaid JI, Janicak PG, Davis JM, Faull K (1993) Homovanillic acid in the cerebrospinal fluid: patterns of response after four weeks of neuroleptic treatment. Biol Psychiatry 34:128–134

Sokoloff P, Martres MP, Giros B, Bouthenet ML, Schwartz JC (1992) The third dopamine receptor (D_3) as a novel target for antipsychotics. Biochem Pharmacol 43(4):659–666

Stewart J, Badiani A (1993) Tolerance and sensitization to the behavioral effects of drugs. Behav Pharmacol 4:289–312

Stewart J, Vezina P (1989) Microinjections of SCH23390 into the ventral tegmental area and substantia nigra pars reticulata attenuate the development of sensitization to the locomotor activating effects of systemic amphetamine. Brain Res 495:401–406

Swerdlow NR, Keith VA, Braff DL, Geyer MA (1991) Effects of spiperone, raclopride, SCH23390 and clozapine on apomorphine inhibition of sensorimotor gating of the startle response in the rat. J Pharmacol Exp Ther 256:530–536

Umeda Y, Sumi T (1990) Decrease in the evoked release of endogenous dopamine and dihydroxyphenylacetic acid from rat striatal slices after withdrawal from repeated haloperidol. Eur J Pharmacol 191:149–155

Vezina P, Stewart J (1989) The effect of dopamine receptor blockade on the development of sensitization to the locomotor activating effects of amphetamine and dopamine. Brain Res 499:108–120

Weinberger DR (1987) Implications of normal brain development for the pathogenesis of schizophrenia. Arch Gen Psychiatry 44:660–669

Weiss B, Santelli S (1978) Dyskinesias evoked in monkeys by weekly administration of haloperidol. Science 200:799–801

Westerink BHC, deVries JB (1989) On the mechanism of neuroleptic induced increase in striatal dopamine release: brain dialysis provides direct evidence for mediation by autoreceptors localized on nerve terminals. Neurosci Lett 99:197–202

Wheeler SC, Roth RH (1980) Tolerance to fluphenazine and supersensitivity to apomorphine in central dopaminergic systems after chronic fluphenazine decanoate treatment. Naunyn Schmiedebergs Arch Pharmacol 312:151–159

White FJ, Wang RY (1983) Differential effects of classical and atypical antipsychotic drugs on A9 and A10 dopamine neurons. Science 221:1054–1057

Yamada S, Yokoo H, Nishi, S (1993) Chronic treatment with haloperidol modifies the sensitivity of autoreceptors that modulate dopamine release in rat striatum. Eur J Pharmacol 232:1–6

Yamamoto BK, Cooperman MA (1994) Differential effects of chronic antipsychotic drug treatment on extracellular glutamate and dopamine concentrations. J Neurosci 14(7):4159–4166

Zetterstrom T, Sharp T, Ungerstedt U (1985) Effect of neuroleptic drugs on striatal dopamine release and metabolism in the awake rat studied by intracerebral dialysis. Eur J Pharmacol 106:27–37

Zhang W, Tilson H, Stachowiak MK, Hong JS (1989) Repeated haloperidol administration changes basal release of striatal dopamine and subsequent response to haloperidol challenge. Brain Res 484:389–392

CHAPTER 8

The Behavioural Pharmacology of Typical and Atypical Antipsychotic Drugs

SVEN OVE ÖGREN

A. Introduction

The introduction of the first antipsychotic drugs during the 1950s initiated a search for the mechanism of their therapeutic action to gain insight into the etiology of schizophrenia and related psychoses. Studies in the 1960s suggested that chlorpromazine and haloperidol (the two "prototype" antipsychotic drugs) may act via blockade of postulated dopamine (DA) receptors (the "DA receptor blockade hypothesis") (CARLSSON and LINDQVIST 1963; VAN ROSSUM 1966). This hypothesis received additional support from behavioural, biochemical and receptor ligand studies in the 1960s and 1970s (SEEMAN 1987). Of major importance was the observation that clinically active antipsychotic agents, irrespective of chemical structure, behaved as DA receptor antagonists in vivo since they antagonized the behavioural effects induced by direct or indirect DA receptor stimulation (VAN ROSSUM 1966; JANSSEN et al. 1965). The observation that antipsychotic drugs block the acute paranoid symptoms of schizophrenia induced by *d*-amphetamine (RANDRUP and MUNKVAD 1967) established the clinical significance and the predictive validity of these findings. Thus, the ability of antipsychotic drugs to inhibit *d*-amphetamine-induced stereotypies in the rat was found to correlate with their antipsychotic efficacy (JANSSEN et al. 1965).

Subsequent studies identified and categorized the behavioural actions of a large number of antipsychotic drugs and established a set of behavioural effects which were believed to be predictive of their therapeutic effects. These behavioural properties characterized antipsychotic drugs as a group and distinguished them from other psychotropic agents (see NIEMEGEERS and JANSSEN 1979). These developments were reviewed by KREISKOTT (1980) in an earlier series of this publication. He defined the task of behavioural pharmacology as follows: "to find and to differentiate substance actions in animals and in man by observation or instrumental analysis of spontaneous individual and social behaviour, of the antagonistic and/or synergistic influence on stimulus and/or drug-induced behavioural changes as well as of effects on learning and memorizing capacity". The term antipsychotic was used synonymously with neuroleptic drugs and defined as "chemical compounds which are able to prevent, to suppress or to abolish processes of extreme excitation in man". In this context the wide range of behavioural effects produced by different types of antipsychotics in different models were extensively described as they related particularly to the acute phase of schizophrenia. This review led to the conclusion that on the basis of results from animal testing, it is possible to characterize the profile of action of various antipsychotic drugs and to predict their properties in the clinic (Kreiskott 1980).

B. Typical and Atypical Antipsychotic Drugs

I. Concepts and Nomenclature

Following the introduction of antipsychotic drugs it was soon realized that they all produced side effects varying from impairments of mental functions (e.g., cognitive deficits) to different types of motor side effects (from parkinsonism and dystonia to severe, irreversible forms of tardive dyskinesia). (For a review, see Chap. 13). In fact, the induction of extrapyramidal symptoms (EPS) or catalepsy in the animal that was observed after the administration of most antipsychotic drugs was interpreted as an inevitable consequence of their mechanism of action (Stille and Hippius 1971). The discovery of the unique clinical properties of clozapine during the early 1970s resulted in a re-evaluation of this hypothesis and the subsequent development of the concept of "atypical antipsychotic drugs" (Stille and Hippius 1971; Baldessarini and Frankenburg 1991). In addition to clozapine, thioridazine and sulpiride were later identified as putatively atypical by some authors (see Gerlach 1991).

Although antipsychotic drugs are currently categorized as either "typical" or "atypical", mainly depending on their clinical properties (see Meltzer 1992), the criteria for this categorization have never been clearly defined or validated. Both groups of compounds are effective in treating schizophrenia. However, in the clinical setting the concept of atypicality serves mostly as a label for compounds (like clozapine) that are effective antipsychotic drugs with a low propensity to cause EPS (including tardive dyskinesia) and with a limited effect on prolactin secretion (Baldessarini and Frankenburg 1991; Meltzer 1992; Casey 1989). Other definitions include efficacy in treatment-resistant patients or effectiveness against primary negative symptoms (Meltzer 1992; Kane et al. 1988). Since aspects on negative symptomatology and therapeutic resistance are rarely incorporated into animal studies, there are clear discrepancies between the definitions used in clinical and preclinical research. The definitions used in animal research are more limited, and an "atypical" drug has come to mean a compound which is effective in models for "psychosis" despite a weak tendency to induce catalepsy or EPS (see Lowe et al. 1988; Ögren et al. 1984; Ögren and Högberg 1988). In animal models, atypical drugs can therefore be operationally defined as compounds capable of discriminating between doses which block various pharmacological measures for "model psychosis" in the rat (e.g., DA agonist-induced hyperactivity) and doses which induce "extrapyramidal side effects" (e.g., catalepsy in the rat) (ÖGREN et al. 1984, 1986, 1990) (Fig. 1).

Neuropharmacological research and testing has therefore been directed not only towards analyzing behavioural effects predictive of therapeutic activity but particularly at those behavioural measures which may predict EPS potential (IVERSEN 1986; ÖGREN et al. 1984). However, the failure to incorporate the multidimensional aspects of atypicality remains a major limitation in pharmacological testing (see below). Moreover, since behavioural analysis depends on the theoretical construct by which atypical antipsychotic activity is defined, different types of drugs will be categorized as "atypical" depending on the criteria used (see MOORE and GERSHON 1989). At least three theoretical possibilities can be envisaged depending on the criteria used to define EPS. The first possibility is a compound that is effective in the "psychosis" model and shows significantly less EPS induction than the prototype "typical" drug. However, since the "antipsychotic effect" and EPS induction may not follow

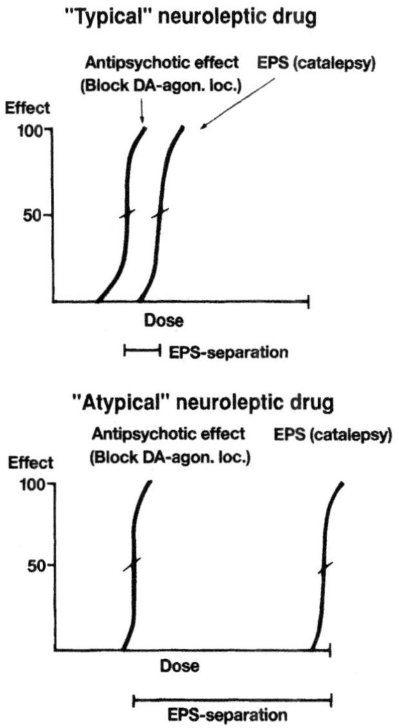

Fig. 1. A theoretical account for the difference between typical and atypical neuroleptic drugs. In atypical neuroleptics, unlike typical neuroleptics, there is a wide separation between the dose levels that block dopamine (DA) agonist-induced locomotion (the "psychosis model") and those that induce catalepsy [the "model for extrapyramidal syndromes (EPS)"]. *agon.*, agonist; *loc.*, locomotion

the same dose-dependency it is possible that the "atypical profile" may only be shown in a limited dose range. Above a certain dose range the compound may behave as the "typical" haloperidol. A second possibility is a compound which displays a very wide "separation" between the estimated dose range for "antipsychotic effect" and the dose range for "EPS induction". As pointed out by CASEY (1992), the critical question is: how much separation is required? No a priori answer can be given to this question and the issue can only be settled empirically. The third possibility is a compound with activity in models for "psychosis" but with no effect in the EPS model. If antipsychotic activity can be established, this compound is a truly atypical drug with respect to the propensity to induce EPS. Thus, as the concept is presently applied, we are not dealing with classes of typical or atypical drugs with defined properties. This creates a problem in the pharmacological analysis and in the validation of atypical drugs, particularly if there is a lack of correspondence between results from different animal studies. This is illustrated by results obtained by CASEY when he showed that putatively "atypical" drugs based on the catalepsy test in the rat were found to elicit dystonia in a monkey model (CASEY 1992). The lack of correspondence between the results from these two tests of EPS induction led CASEY to question the utility of the concept of atypicality (CASEY 1992).

II. Atypical Antipsychotic Drugs

The concept of atypicality has become a new vista for research on novel antipsychotic drugs. The analysis of the in vitro affinity of clozapine for dopaminergic and non-dopaminergic receptors has resulted in a number of hypotheses to explain mechanisms of atypicality (COWARD et al. 1989; SEEMAN 1992) (Table 1). However, no hypothesis has so far been able to explain satisfactorily the behavioural and therapeutic effects of clozapine (see Chap. 12). A number of potentially atypical antipsychotic drugs based on different pharmacological approaches have been developed during the past decade (COWARD 1991; LOWE et al. 1988) (see Chap. 12). Representatives of some of these compounds are categorized in Table 2 according to their receptor affinities. One approach has focussed on compounds with selective affinities for

Table 1. Current hypotheses to account for "atypical" antipsychotic activity

- Combined actions at muscarinic and DA D_2 receptors
- Combined actions at 5-HT_2 and DA D_2 receptors
- Optimal balance between DA D_1 and D_2 receptors
- Selective effects on structural subtypes of DA D_2-like receptors, e.g., D_3 or D_4 receptors
- Regional selectivity with preference for DA receptors in the mesolimbic/mesocortical DA systems

DA, dopamine.

either D_1 or D_2 receptors. One representative from this group, the D_2 antagonist remoxipride, is an effective antipsychotic drug with a weak tendency to induce all forms of EPS, including akathisia, compared to haloperidol or placebo (LEWANDER et al. 1990; KING et al. 1990). A recent double-blind study has shown that remoxipride is superior to haloperidol in "treatment-resistant" schizophrenia patients (CONLEY et al. 1993). Another group of potentially atypical agents with a multitransmitter interaction includes compounds such as risperidone (JANSSEN et al. 1988), sertindole (SÁNCHEZ et al. 1991) and amperozide (GUSTAFSSON and CHRISTENSSON 1990). These compounds have a high affinity for 5-HT_2 receptors, a moderate to low affinity for α_1-receptors and a relatively low affinity for D_2, D_1 or D_3 receptors (GERLACH 1991). The atypical profile is suggested to be due to combined 5-HT_2 and D_2 antagonistic properties (MELTZER et al. 1989). Risperidone has been shown to be an effective antipsychotic drug with a significantly lower incidence of EPS than haloperidol (BORISON et al. 1992; CHOUINARD et al. 1993). Attempts to mimic the pharmacological properties of clozapine have resulted in the development of clozapine-like agents such as olanzapine (MOORE et al. 1992). Among the

Table 2. Compounds with an "atypical" profile in various behavioural paradigms

1. Selective D_1 receptor antagonists
   SCH 23390
   SCH 39166
   NNC756, NNC687 (NO756)
2. Selective D_2/D_3 receptor antagonists
   Amisulpiride
   Raclopride
   Remoxipride
   Sulpiride
3. Partial D_2 receptor agonists
   (−)-3-PPP
   Terguride
   Roxindole
   SDZ 208-911
4. Combined 5-HT_2 and D_2 receptor antagonists
   Risperidone
   Sertindole
   Amperozide
5. Clozapine-like agents (D_4 + 5-HT_2 + D_1 + D_2 receptor antagonists)
   Clozapine
   Olanzapine
6. Nondopaminergic receptor agonists or antagonists
   Serotonin-active drugs
      5-HT_{1A} agonists (buspirone)
      5-HT_2 antagonists (ritanserin)
      5-HT_3 antagonists (ondansetron)
   Sigma-antagonists
      Rimcazole
      BMY 14802

non-dopaminergic approaches some of the compounds tested seem to be ineffective in "classical" models for antipsychotic effects (COWARDS 1991). Since the clinical experience with these approaches is very limited, the validity of the findings reported so far cannot be evaluated.

Despite intensive research efforts, the mechanisms underlying the behavioural properties of atypical antipsychotic drugs are still largely unknown. The purpose of this section is to review selective aspects of the behavioural neuropharmacology of typical and atypical antipsychotic drugs as they may relate to their therapeutic action. Furthermore, the complex neuroanatomical and neurochemical basis for the behavioural methods used will be examined. A particularly critical issue is whether compounds characterized as typical or atypical differ in any significant degree in behavioural models believed to predict antipsychotic effects and EPS. Another equally important question relates to the mechanisms underlying such differences. In view of the lack of data on many purported atypical drugs (see Table 2), the emphasis will be on compounds with documented clinical efficacy in schizophrenic patients.

III. Behavioural Testing Procedures for Typical and Atypical Antipsychotic Drugs

Since there exists no animal model for schizophrenia, the behavioural analysis of antipsychotic drugs has mainly relied on methods that are DA mechanism-oriented (ELLENBROEK and COOLS 1990). The ability of antipsychotic drugs to interfere with pharmacologically induced behaviours elicited by various DA receptor agonists has been extensively employed. The use of such models is based on the (tacit) assumption that they "simulate" certain features of schizophrenia and that efficacy in such models is predictive of antipsychotic activity. However, the actual measure for the "antipsychotic" effect has varied throughout the years. Earlier studies used antagonism of stereotypies induced by *d*-amphetamine or apomorphine (JANSSEN et al. 1965, 1988) or later methylphenidate (CHRISTENSEN et al. 1984) as the criterion for antipsychotic activity.

However, the selection of appropriate behavioural tests changed in the attempt to assess drugs with an "atypical" profile. The major change in conceptualization was brought about by the hypothesis that antipsychotic action correlates with effectiveness in blocking mesolimbic or mesocortical DA D_2 transmission (see CROW et al. 1977; MELTZER and STAHL 1976; ÖGREN et al. 1984), while DA (D_2) receptor blockade in the (dorsal) striatum is responsible for the induction of EPS. This conceptual change resulted in behavioural test paradigms aiming at a comparison of limbic and striatal activity. A combination of various experimental models is presently used to characterize atypical neuroleptic properties (see Table 3). A blockade of locomotion elicited by "DA agonists" such as *d*-amphetamine or apomorphine became the measure for therapeutic efficacy (ÖGREN et al. 1984) based on the assumption that stimulation of DA receptors within parts of the mesolimbic DA system, e.g., the nucleus accumbens, is responsible for the locomotor response (KELLY et al. 1975; JONES and ROBBINS 1992). Other predictive measures for antipsychotic effects include, for example, inhibition of apomorphine-induced climbing, inhibition of the conditioned avoidance response (CAR) or a differential action in the paw test (see below). The paw test is a new animal model which has been proposed to have predictive validity for both antipsychotic efficacy and induction of EPS (ELLENBROEK et al. 1987). However, measures for the potential of inducing EPS have also varied. The ability to induce catalepsy in relationship to potency in the "antipsychotic" test has been used to assess the potential for EPS (ÖGREN et al. 1984). Inhibition of oral stereotypies, e.g., gnawing, elicited by high doses of DA agonists was used by some authors as an indication of later motor side effects (LJUNGBERG and UNGERSTEDT 1978, 1985). Recently, the relative ability to differentially affect the response pattern in the paw test (see below) has been proposed to characterize atypical drugs (ELLENBROEK et al. 1987). An atypical compound should consequently have a greater ability to antagonize DA agonist-induced locomotor stimulation (and/or inhibition of apomorphine-induced climbing; inhibition of CAR) compared to stereotyped behaviour and it should have a low potential for producing catalepsy. Thus, a high ratio between the ED_{50} value for induction of catalepsy and the ED_{50} value in the "psychosis" model would indicate an atypical profile.

Although activity in these behavioural tests may be predictive of atypical antipsychotic activity, conclusions on neuroanatomical and neurochemical mechanisms on the basis of results obtained in behavioural tests must be interpreted with great caution. The subdivision of function, as reported above,

Table 3. Behavioural methods used to evaluate typical and atypical antipsychotic drugs in rodents

Antipsychotic efficacy
- Inhibition of DA agonist-induced hyperlocomotion (unselective $D_1/D_2/D_3$ DA "agonists": apomorphine, amphetamine; selective D_2/D_3 agonists: quinpirole)
- Inhibition of methylphenidate-induced hypermotility
- Inhibition of apomorphine-induced climbing
- Inhibition of conditioned avoidance
- Activity in the paw test

Potential for causing EPS including tardive dyskinesia
- Comparison of the blockade of DA agonist-induced hyperlocomotion (stimulation of mesolimbic DA receptors) and oral stereotypies (stimulation of striatal DA receptors)
- Inhibition of methylphenidate-induced gnawing
- Relationship between blockade of DA agonist-induced hyperlocomotion and induction of catalepsy
- Differential activity in the paw test
- Induction of behavioural supersensitivity to DA agonists following chronic treatment with neuroleptics ("tardive dyskinesia")

EPS, extrapyramidal syndromes.

is a crude approximation of the complexity involving striatal and limbic behavioural functions. Dopamine neurons and receptors within the caudate putamen and nucleus accumbens appear to play the major role in the behavioural response to systemically administered DA receptor agonists (see JOYCE 1983). However, neuroanatomical, neurochemical and behavioural studies have shown that these brain areas are functionally heterogeneous and organized into functional subsystems (ALEXANDER and CRUTCHER 1990; GERFEN 1992; GRAYBIEL 1990; SCHEEL-KRÜGER and WILLNER 1991). This functional heterogeneity has important implications for behavioural testing since neuronal circuits within subdivisions of each brain area may be involved in distinct aspects of behavioural function. Alternatively, distinct sites within either the striatum or the nucleus accumbens may be part of the neuronal circuits involved in the same behavioural function (see below). Studies of selective components of DA-stimulated behaviours, e.g., selective items of oral stereotypies (FRAY et al. 1980), may thus provide clues as to the mechanisms of action of antipsychotic drugs.

C. Basic Actions of Antipsychotic Drugs and Their Behavioural Effects

Studies during the past two decades have demonstrated that brain DA receptors play a key role in antipsychotic drug action (SEEMAN 1987). Originally, two DA receptors, D_1 and D_2, were though to account for all of the pharmacological action of DA (KEBABIAN and CALNE 1979). The close correlation between clinical efficacy and in vitro affinity for DA D_2 but not for D_1 receptors (SEEMAN 1987) led to the conclusion that the DA D_2 receptors in the brain are the major targets for antipsychotic action. On the basis of correlational studies, most of the behavioural effects of antipsychotic drugs as well as their effects on DA-mediated responses were exclusively ascribed to blockade of D_2 receptors (CREESE et al. 1983). The cataleptogenic effects associated with most typical "neuroleptic drugs" were also linked to D_2 receptor blockade (JENNER and MARSDEN 1983). However, with the identification of the first selective D_1 antagonist SCH23390 in 1983 (IORIO et al. 1983; HYTTEL 1983), it became clear that both D_1 and D_2 receptor blockade can produce similar effects on DA-mediated behaviours (see CHRISTENSEN et al. 1984; WADDINGTON 1988).

Today, the relationship between DA-receptor subtypes and behavioural function is intensively being studied due to developments in molecular biology (see Chap. 3). There are several subtypes of DA in the brain which underlie the action of DA in producing its diverse behavioural effects (see JOYCE 1983). The recent classification of dopamine receptors into dopamine D_1 (D_1 and D_5) and dopamine D_2-like receptors (D_2, D_3, D_4) (see Chap. 3) with overlapping distributions and mRNA expressions in different brain areas has created a major problem in the behavioural analysis of antipsychotic drugs. Therefore, in the present context, D_1 and D_2 receptors will refer to D_1- and D_2-like receptors. All five DA receptors are in varying degrees targets for currently used antipsychotic drugs. For instance, most antipsychotic drugs do not discriminate between D_2 receptor subtypes (D_2 and D_3). Since the behavioural function of DA subtypes is still largely unknown (see Table 4), their relative contribution to the overall behavioural effects of antipsychotic drugs is presently unclear. Furthermore, much of the knowledge of DA-mediated behavioural effects is based on the use of DA agonists which have affinities for several DA receptor subtypes (see Chap. 3). The results based on unselective DA agonists must therefore be interpreted with caution. This is particularly important since there are important interactions between different DA subtypes, e.g., D_1 and D_2 receptors, in the expression of many DA agonist-related behaviours (see BRAUN and CHASE 1986; WADDINGTON and DALY 1993). For instance, tonic activity via D_1 receptor stimulation appears to exert an "enabling" or "permissive" role in the mediation of D_2 receptor-mediated responses (CLARK and WHITE 1987; WADDINGTON and DALY 1993). Finally, most antipsychotic drugs also have potent effects on nondopaminergic receptors which may contribute in various degrees to their behavioural actions (see Table 4).

Table 4. Possible roles of dopaminergic and nondopaminergic receptors in the behavioural effects of antipsychotic drugs

Receptor	Behavioural effects
D_1	Antipsychotic? Arousal Locomotor activity Extrapyramidal
D_2	Antipsychotic Locomotor activity Extrapyramidal
D_3	Antipsychotic? Locomotor activity
D_4	Antipsychotic?
Muscarinic (M_1, M_2, M_3, M_4)	Antimuscarinic Fewer EPS Locomotor activity Peripheral side effects (dry mouth, tachycardia, blurred vision, constipation, micturition and sexual dysfunction) Central side effects (confusion, agitation)
α_1-Adrenergic	Antimanic effects Sedation Motor side effects – catalepsy?
Histamine – H_1	Sedation
Serotonin 5-HT_2	Sleep Fewer EPS? Negative symptoms?
Sigma	Not known

D. Effects on Motor Function

I. Spontaneous Locomotor Activity

Analysis of various aspects of locomotor activity or exploratory behaviour has been extensively used to assess the properties of antipsychotic drugs. However, as pointed out by ROBBINS, the results of such studies are critically dependent on the technique used to measure locomotion (ROBBINS 1977). In addition, several factors can affect base-line locomotor activity; they include strain, sex, age, handling, route of drug injection, diurnal factors and dietary factors (HARRO 1993). In many studies motor activity has been recorded only in an either/or manner, i.e., in photocell-equipped cages, in which the actual definition of the motor response or movement is highly arbitrary. Recording done in locomotor cages which permit measurements of several defined components of locomotion such as motility, locomotion and rearing is preferable to methods which only record one aspect of locomotion (LJUNGBERG and UNGERSTEDT 1978). However, even automatic recordings of selective items cannot adequately describe the complex pattern of behavioural effects induced by many drug treatments (see FRAY et al. 1980).

When placed in a novel environment, e.g., a locomotor cage, rodents initially display a high level of locomotor activity. This initial phase characterized by spontaneous exploratory activity lasts approximately 10–20 min in the rat and is followed by a phase with a low level of all aspects of motor activity, i.e., the habituation phase. The effect of drugs on spontaneous locomotion should preferably be examined during the exploratory phase. Studies on drug-elicited activity, on the other hand, are preferably performed in habituated rats, i.e., rats with negligible spontaneous activity. By comparing the effects on spontaneous and drug-elicited activity it is possible to analyze the contribution of nonspecific suppression on locomotion (see below).

Dopamine neurons play an important role in various aspects of locomotor activity (BENINGER 1983). Although effects on locomotor activity have been related to the main action of neuroleptic drugs, e.g., DA receptor blockade (BENTALL and HERBERG 1980), much evidence supports the view that antipsychotic drugs may have differential effects on the neuronal basis for locomotion. Thus, it is possible that these compounds could be discriminated on the basis of their effectiveness in modulating selective components of locomotor activity. Spontaneous locomotor activity or exploratory activity in the rat has mostly been related to mesolimbic DA transmission (FINK and SMITH 1980). Since several DA-containing areas of the brain such as the olfactory tubercle (COOLS 1986), prefrontal cortex (JONES and ROBBINS 1992) and the striatum (BENINGER 1983) are involved in the regulation of spontaneous locomotion, the regional effects of DA receptor blockade must be taken into account (see COOLS et al. 1994). Furthermore, the mechanisms involved in basal and DA-receptor elicited locomotor activity may differ depending on the relative degree of DA release or DA receptor stimulation. This probably explains why the DA receptor mechanisms involved in *d*-amphetamine-induced and spontaneous locomotion appear to differ (VAN DEN BOSS et al. 1988; JONES and ROBBINS 1992).

Most neuroleptic or antipsychotic drugs have been shown to produce a dose-dependent reduction of spontaneous locomotor activity in rodents (NIEMEGEERS and JANSSEN 1979; KAISER and SETLER 1981) and this effect is often regarded as analogous to "sedation" or "hypokinesia" and possibly related to cortical EEG synchronization (ONGINI and LONGO 1989). However, the quantitative changes in various aspects of motor performance depend on several factors, most importantly the dose and type of neuroleptic. Therefore, the effect of motor activity can range from stimulating activity at very low doses to reduction in locomotor activity, akinesia and induction of tonic immobility or catalepsy at higher doses (COSTALL et al. 1983; PUECH et al. 1981). Typical neuroleptic drugs such as chlorpromazine, *cis*-flupenthixol and haloperidol produce a marked and full suppression of all aspects of motor activity at doses close to those inducing catalepsy. Clozapine and thioridazine also caused a marked suppression of motor activity, which was clearly dose-dependent (ÖGREN and ÄNGEBY-MÖLLER 1996a; JACKSON et al. 1994a). On the other hand, the two benzamides remoxipride and sulpiride did not cause "sedation" or ataxia in the dose range relevant to their antipsychotic effects in behavioural studies (ÖGREN et al. 1984, 1990). Only very high doses of remoxipride (40 μmol/kg) or sulpiride (300 μmol/kg) were able to suppress the exploratory activity, whereas the low doses tested did not show any apparent effects; neither enhancement nor suppression was seen (ÖGREN and ÄNGEBY-MÖLLER 1996a). In contrast to these selective D_2 antagonists, the D_1 antagonist SCH23390 caused a potent reduction in rearing and locomotor activity at low doses (BOYCE et al. 1985; CHRISTENSEN et al. 1984; HOFFMAN and BENINGER 1985). There are marked differences among the compounds believed to act through blockade of the 5-HT_2 receptor subtypes. Risperidone caused a marked and dose-dependent inhibition of locomotion (MEGENS et al. 1992) while amperizode and sertindole caused a weak motor suppression (GUSTAFSSON and CHRISTENSSON 1990; JACKSON et al. 1994a). In nonhabituated rats amperozide induced a dose-dependent increase in locomotor activity unlike other neuroleptic drugs (WATERS et al. 1989). Partial DA agonists such as (–)3-PPP (HJORTH et al. 1983; ARNT et al. 1983) caused a 40%–50% reduction in locomotor activity even at high doses in contrast to the full inhibition seen with most neuroleptic drugs.

It is presently not clear whether the various effects observed after the administration of different compounds relate to region-specific actions on DA transmission. Since most antipsychotic drugs have an about equal affinity for both the D_2 and D_3 receptors, it is important to note that D_3 agonists suppress locomotion while D_3 receptor antagonists have been reported to enhance locomotor activity (WATERS et al. 1993). Since the D_3 receptors are enriched in mesolimbic areas of the brain (see Chap. 3) it is possible that postsynaptic actions at D_3 receptors may contribute (in addition to D_2 receptor blockade) to the locomotor response. Actions at DA D_2 autoreceptors also appear to play a significant role since local application of (–)-3-PPP into the nucleus accumbens markedly suppressed exploratory activity in the rat (SVENSSON and AHLENIUS 1983). On the other hand, locomotor suppression by SCH23390 has been related to blockade of striatal D_1 receptor mechanisms (BOYCE et al. 1985). However, since stimulation of D_1 receptors induces cortical EEG desynchronization and behavioural arousal (ONGINI and LONGO 1989), effects in other brain regions (cortical) probably contribute as well.

A major problem in the interpretation of the locomotor effects of clozapine, chlorpromazine, risperidone and thioridazine is that in addition to DA, they also affect α-adrenergic, histaminergic, serotonergic and muscarinic cholinergic receptors (see Chap. 4). Thus, mechanisms other than DA receptor blockade could at least in part underlie the motor suppression induced by some of these compounds. In view of its receptor-binding profile, the contribution of α-adrenergic and histaminergic receptors can probably partly explain the marked motor suppression caused by clozapine while α-receptor blockade probably contributes to the action of risperidone (MEGENS et al. 1992).

Taken together, the different effects of antipsychotic drugs on spontaneous locomotor activity appear to reflect different types of DA and non-DA receptor involvement. Therefore, locomotor activity cannot be used simply as a behavioural measure indicative of atypical activity. However, although lack of suppression of locomotion or "sedation" is not usually included as one of the criteria for atypical drugs, this property may be beneficial particularly in young schizophrenic patients. It is possible that the mechanisms underlying "sedative" properties, e.g., α-receptor blockade, may be partly responsible for impairments of mental functioning, e.g., cognitive functions seen after the administration of typical antipsychotic drugs (see Chap. 15).

II. Catalepsy and the Paw Test

After the introduction of neuroleptics in the early 1950s it was soon discovered that compounds such as chlorpromazine in higher doses caused a state of tonic immobility in rodents which was labelled catalepsy. The catalepsy test was traditionally used to screen neuroleptic drugs, and a positive response in this model was regarded as predictive for antipsychotic activity (MORPURGO and THEOBALD 1964). Later studies showed, however, that the "antipsychotic" action in animal models could be differentiated from the ability to induce EPS by the use of anticholinergic agents. Today, catalepsy is regarded as analogous to EPS in man, particularly parkinsonism (NIEMEEGERS and JANSSEN 1979), and the degree of catalepsy is often used as the measure for predicting the probable incidence of EPS (LOWE et al. 1988).

Catalepsy is characterized by akinesia and an inability of the rat to correct its position when placed in an awkward or unusual posture. Two tests of catalepsy are commonly used, the horizontal bar test and the vertical grid test, and there are several variations of these test models. In the most commonly used test model (the bar test) the forepaws of the rats (or mice) are placed on a horizontal bar or wooden block located 5–10 cm above the cage floor while the hind paws remain on the floor. In the grid test (often a vertical grid test) the rats are placed on a wire mesh grid placed at an angle of 70°–90° to the supporting bench. The animals are often placed on the grid with their heads up and the four legs abducted and extended. An unnatural posture is essential for the specificity of the catalepsy tests to rule out the influence of nonspecific akinesia (see MORELLI and DICHIARA 1985).

Although seemingly simple, catalepsy tests have been shown to be influenced by a number of variables, and the degree of catalepsy is highly test-dependent (SANBERG et al. 1988), which probably explains the large number of contradictory findings reported (see ELLENBROEK and COOLS 1990). A combination of tests such as the vertical grid test and the horizontal bar test is preferable to minimize the influence of test dependence. An important, as yet unanswered question is whether the different catalepsy tests measure similar or different aspects of motor functions. Repeated testing instead of single testing is required in order to properly assess cataleptic thresholds and temporal effects of different antipsychotic drugs (FERRÉ et al. 1990; ÖGREN et al. 1996) and to avoid the stress-induced inhibition of catalepsy caused by the experience of a new environment (FERRÉ et al. 1990). Furthermore, control groups must be included to evaluate the possible contribution of learned "pseudo-catalepsy" which varies in different catalepsy tests (FERRÉ et al. 1990; ÖGREN et al. 1996). Another critical issue involves the various definitions of the degree of catalepsy. Some investigators use a threshold criterion, i.e., catalepsy is considered to be present if the rat remains on the bar or grid for more than 15 s (see ARNT 1982; MORELLI and DICHIARA 1985; ÖGREN et al. 1996). The use of a 15-s period is based on the empirical finding that control animals never exceed this limit. Other investigators have used a 30-s or 60-s criterion (see SANBERG et al. 1988). Needless to say, the sensitivity of the test used will have an important impact on the dose–response curve for catalepsy.

Most neuroleptic drugs such as the butyrophenones, phenothiazines, diphenylbutyl-amines and thioxanthenes have been reported to produce dose-dependent catalepsy in the rat (NIEMEGEERS and JANSSEN 1979). However, the effectiveness (ED_{50} values) as well as the temporal effects in the induction of catalepsy were found to differ markedly between different compounds (ÖGREN et al. 1996). Compounds such as chlorpromazine, *cis*-flupenthixol and haloperidol produced catalepsy at low doses while thioridazine was found to cause catalepsy only at high doses. Moreover, clozapine failed to induce dose-dependent catalepsy while olanzapine was cataleptogenic only at very high doses (Table 5). The benzamides remoxipride (ÖGREN et al. 1984, 1990) and sulpiride (WAMBEBE 1987) also induced an "atypical" form of catalepsy at high doses only, while metoclopromide (another benzamide) caused catalepsy at relatively low doses. Risperidone caused catalepsy in both the bar and grid test at relatively low doses (Janssen et al. 1988; Megens et al. 1992) while sertindole and amperozide appears not to be cataleptogenic even at high doses (Sánchez et al. 1991; Gustafsson and Christensson 1990). Moreover, selective autoreceptor agonists such as (–)3-PPP or SDZ 208-911 are incapable of inducing catalepsy even when given at very high doses (Arnt et al. 1983; Hjort et al. 1983; Coward et al. 1990). Thus, the compounds characterized as atypical differ clearly in their cataleptogenic properties. However, a more important measure is the relative effectiveness in "psychosis" models in comparison with the ability to induce catalepsy (see below).

Although catalepsy was related mainly to a blocking action at postsynaptic striatal DA D_2 receptors (Jenner and Marsden 1983), recent investigations indicate that both D_1 and D_2 receptors are involved in catalepsy. Many investigators (Christensen et al. 1984; Morelli and DiChiara 1985; Meller et al. 1985; Ögren and Fuxe 1988), but not all (Iorio et al. 1983) have reported that SCH23390 produces a profound catalepsy in the rat after systemic administration. However, the mechanisms behind the D_1 and D_2 receptor involvement in catalepsy are not the same and appear to involve different efferent striatal pathways (Ögren and Fuxe 1988). Moreover, D_1 and D_2 receptor antagonists were found to be synergistic in their effects on catalepsy (Ögren and Fuxe 1988; Parashos et al. 1989; Wanibuchi and Usuda 1990), probably partly by interactions at the striatal levels since both D_1 and D_2 receptor antagonists are cataleptogenic after intrastriatal injection (Fletcher and Starr 1988; Ossowska et al. 1990). It is therefore difficult to explain the noncataleptogenic action of clozapine by the hypothesis that there may exist an optimal balance between D_1 and D_2 receptor blockade (Coward et al. 1989; Farde et al. 1992).

In the paw test, rats are placed on a platform with their fore- and hindlimbs lowered through four separate holes (Ellenbroek et al. 1991). Rats placed in this uncomfortable situation withdraw their limbs from the holes, and the time the animal needs to withdraw its forelimbs and hindlimbs is recorded. Rats treated with antipsychotic drugs such as haloperidol show an increase in both the hindlimb reaction time (HRT) and the forelimb reaction time (FRT) in about the same dose range (Ellenbroek et al. 1987). In contrast, clozapine, thioridazine and the D_2 antagonist remoxipride, as well as SCH23390 and SCH39166 have been found to enhance the HRT at clearly lower doses than FRT (Ellenbroek et al. 1987, 1991, 1994). Based on these findings the increase in FRT has been proposed to model EPS while an increase in HRT models antipsychotic potential (Ellenbroek et al. 1991). Although action at the D_1 receptor plays an important role in the effect of clozapine, this property alone appears not to be sufficient to explain the action of clozapine in the paw test (Ellenbroek et al. 1994). Thus, similar to the catalepsy test, both D_1 and D_2 receptors are involved in the paw test but the neuroanatomical and neurochemical basis for the two tests differs (see below). Furthermore, the validity of the paw test remains unclear in view of the possible lack of antipsychotic action of D_1 receptor antagonists such as SCH39166 (see Karlsson et al. 1995) and the observation that D_1 receptor antagonists may not have atypical antipsychotic properties (see "Conclusion").

Some evidence has suggested that neuroleptic drug-induced catalepsy is primarily mediated via blockade of DA receptors in the rostro-dorsal part of the neostriatum (ELLENBROEK et al. 1985). Moreover, ELLENBROEK et al. (1985) failed to observe any cataleptic effects following microinjections of haloperidol (0.5–2.5 μg) into the medial part of the nucleus accumbens of the rat. Other results also based on intracranial injections of haloperidol and D_2 and D_1 receptor antagonists indicate that the catalepsy induced by haloperidol is due to a DA blocking action in the ventro-rostral but not in the dorso-rostal part of the striatium (OSSOWSKA et al. 1990; HARTGROVES and KELLY 1984). In the ventro-rostal part of the striatum both D_1 and D_2 receptors appear to mediate the cataleptogenic action of neuroleptic drugs (OSSOWSKA et al. 1990). However, the nucleus accumbens also appears to play an important role in neuroleptic-induced catalepsy mediated primarily via D_1 receptors (OSSOWSKA et al. 1990; HARTGROVES and KELLY 1984). The dorsal striatum as well as the nucleus accumbens have been proposed to play a role in the effects of DA antagonists in the paw test (ELLENBROEK and COOLS 1990). Based on this background it is notable that the D_1 antagonist SCH23390 behaves as an atypical agent in the paw test but not in the catalepsy test (see Tables 5 and 6). Thus, selective actions on subtypes of DA D_1 and D_2 receptors within mesolimbic and mesostriatal neuronal circuits will determine the cataleptogenic action of antipsychotic drugs (ÖGREN et al. 1994).

The role of serotonin receptor subtypes also appears to differ in the catalepsy and the paw tests. The combination of D_2 and 5-HT_2 antagonism has been suggested to result in a weak propensity to induce catalepsy in rodents (MELTZER et al. 1989). Some studies have also shown that 5-HT_2 antagonists can reduce haloperidol-induced catalepsy in rats (BALSARA et al. 1979) while other studies have failed to confirm this finding using the 5-HT_{1c}/5-HT_2 antagonist ritanserin (see WADENBERG 1992). Ritanserin did not antagonise catalepsy caused by the blockade of D_1 and D_2 receptors with SCH23390 and raclopride, respectively. In addition, ritanserin did not attenuate haloperidol-induced dystonia in monkeys while clozapine has been shown to do so in this monkey model for neuroleptic-induced parkinsonism (CASEY 1991a). However, in the paw test ritanserin inhibited the increase in FRT induced by haloperidol while it did not influence the increase in HRT (ELLENBROEK et al. 1994). On the other hand, the action of clozapine in the paw test was not altered by ritanserin. Furthermore, activation of 5-HT_{1A} receptors by the selective 5-HT_{1A} agonist 8-OH-DPAT is able to reverse the cataleptic state induced by the DA antagonist haloperidol in rats (HICKS 1990; INVERNIZZI et al. 1988), and therefore would attenuate the extrapyramidal side effects of classical neuroleptics. 8-OH-DPAT also reduced the effects of haloperidol in the paw test, while it increased the effects of clozapine (ELLENBROEK et al. 1994). These findings indicate that the role of the 5-HT receptor subtypes in the action of atypical antipsychotic drugs is complex and awaits further evaluation.

E. Effects of Antipsychotic Drugs on Behavioural Effects Elicited by DA Receptor Agonists or Glutamate Antagonists

I. *d*-Amphetamine (Methylphenidate)-Induced Behavioural Effects

The relevance of the use of indirect DA agonists such as *d*-amphetamine or methylphenidate as a behavioural model for certain forms of schizophrenia, e.g., positive or paranoid forms, has been extensively discussed (see ANGRIST 1983; LIEBERMANN et al. 1987). *d*-Amphetamine can produce psychotic symptoms in normal individuals (RANDRUP and MUNKVAD 1967) and precipitate psychotic symptoms in schizophrenic patients which are blocked by antipsychotic drugs (see MELTZER and STAHL 1976). Although the pattern of behavioural effects of amphetamine and methylphenidate are similar, methylphenidate appears to have greater psychotogenic potency than amphetamine (LIEBERMANN et al. 1987).

Both the locomotor stimulation and the stereotypies induced by *d*-amphetamine appear to be mediated mainly by enhanced release of DA from presynaptic terminals (CARLSSON 1970; ERNST 1967). The locomotor-stimulating effects of *d*-amphetamine are dose-dependent in the rat. In the lower dose range (0.5–1.5 mg/kg) *d*-amphetamine increases most aspects of locomotor activity, e.g., forward activity, motility and rearing but also low-level stereotypies such as sniffing. Higher doses (2.5–10 mg/kg) of *d*-amphetamine elicit, in addition, compulsive motor behaviours such as head bobbing and repetitive stereotypies consisting of perseverative orofacial movements (licking, gnawing) often combined with a suppression of forward locomotor activity (ERNST 1967; RANDRUP and MUNKVAD 1967; KELLY et al. 1975). However, as pointed out by REBEC and BASHORE (1984) the behavioural effects elicited by amphetamine even at low doses (1 mg/kg) indicate both mesolimbic (locomotion) and mesostriatal (sniffing, head bobbing) involvement (see TSCHANZ and REBEC 1989). Since most investigators have used different routes of administration and doses of *d*-amphetamine ranging from 1 mg/kg s.c. or 1.25 mg/kg i.v. to 10 mg/kg i.v. (KELLY et al. 1975; CHRISTENSEN et al. 1984; JANSSEN et al. 1988; ARNT 1982) the pattern of behavioural effects has differed. Thus, the ability of antipsychotic compounds to affect the behavioural actions of *d*-amphetamine depend critically on the dose of this indirect DA agonist.

The mesolimbic and mesocortical DA system, which includes the nucleus accumbens and the olfactory tubercle (BJÖRKLUND and LINDVALL 1984), appears to be the neuronal substrate for the motor hyperactivity induced by small doses of *d*-amphetamine (CLARKE et al. 1988; COOLS 1986; KELLY et al. 1975; SHARP et al. 1987). The induction of stereotypies, on the other hand, seems to be mediated mainly via the nigrostriatal DA system (KELLY et al. 1975; SHARP et al. 1987; STATON and SOLOMON 1984). Moreover, a recent study has shown that microinjection of *d*-amphetamine into the ventro-lateral striatum elicits intense oral stereotypies (KELLEY et al. 1988). However, the exact role of the neuroanatomical subdivisions of the DA system in the behavioural actions of *d*-amphetamine is still unresolved. A recent study based on ibotenic acid lesions indicates that the dorsal striatum does not play a significant role in amphetamine-induced stereotypies while it is involved in locomotion (ANTONIOU and KAFETZOPOULUS 1992).

Drugs such as haloperidol, chlorpromazine and spiperone have been shown to inhibit a variety of *d*-amphetamine-induced behavioural effects including locomotor activity dose-dependently (BENTALL and HERBERG 1980; NIEMEGEERS and JANSSEN 1979). The results with clozapine and thioridazine are not consistent since some studies have reported a blockade of *d*-amphetamine-induced locomotor activity (IVERSEN and KOOB 1977; NIEMEGEERS and JANSSEN 1979; LJUNGBERG and UNGERSTEDT 1985), while others have failed to find any inhibition (BENTALL and HERBERG 1980; SCHAEFER and MICHAEL 1984). These inconsistent findings may relate to the techniques used for measurements (see REBEC and BASHORE 1984) or the dose of the amphetamine used. Recent results using a relatively low dose of *d*-amphetamine, e.g., the 1.5 mg/kg dose, which has been shown to result in a marked increase in locomotor activity, indicate, however, that both clozapine and thioridazine produce a dose-dependent inhibition of *d*-amphetamine-induced locomotor activity (ÖGREN et al. 1990; ÖGREN and ÄNGEBY-MÖLLER 1996a) (see Table 5). The available data show that all the compounds characterized as antipsychotic drugs or DA receptor antagonists block *d*-amphetamine-induced locomotor activity. This blockade was found with the selective DA D_2 antagonists remoxipride and sulpiride, the relatively selective D_2 antagonists haloperidol and the mixed D_1 and D_2 antagonists chlorpromazine, clozapine, *cis*-flupenthixol and thioridazine. However, sulpiride is a weak antagonist of *d*-amphetamine as noted previously (see ARNT 1982). Blockade of DA D_1 receptors by SCH23390, produced similar antiamphetamine effects as the D_2 receptor antagonists (CHRISTENSEN et al. 1984; JACKSON et al. 1994a). Stimulation by DA autoreceptors by (−)-3-PPP also reduced *d*-amphetamine-induced hyperactivity at doses somewhat higher than those inhibiting spontaneous locomotor activity (ARNT et al. 1983; HJORTH et al. 1983). The combined 5-HT_2 and D_2 antagonist risperidone also effectively blocked the action of *d*-amphetamine (JANSSEN et al. 1988; MEGENS et al. 1992) while sertindole showed a low potency (JACKSON et al. 1994a). Amperozide, on the other hand, has been reported to be ineffective in one study (JACKSON et al. 1994b) while another study showed that it effectively blocked the locomotor stimulation by a 1.5 mg/kg s.c. dose of *d*-amphetamine (WATERS et al. 1989). These data can be interpreted to mean that most aspects of the locomotor effects of *d*-amphetamine involve stimulation of both D_1 and D_2 receptors although the relative contributions of these receptor types probably differ (see ROSS et al. 1989). For instance, a detailed time sampling observation procedure combined with 2 h of automatic activity assessment revealed subtle differences between the ability of SCH23390 and raclopride to inhibit *d*-amphetamine-induced behavioural changes (PETRY et al. 1993).

As shown in Table 5 there are clear differences between various groups of the compounds examined with regard to induction of catalepsy and inhibition

Table 5. Effectiveness of various types of antipsychotic drugs in blocking *d*-amphetamine-induced locomotion and inducing catalepsy in the rat

Compounds	ED_{50}, μmol/kg			
	Blockade of amphetamine	Catalepsy	Ratio	Reference
Reference antipsychotic drugs				
Chlorpromazine	11	11	~1	a
	5.1	18.3	3.6	b
	1	>120	>120	c
cis-Flupenthixol	0.12	0.30	2.5	a
	0.33	0.63	2	b
Haloperidol	0.2	0.9	4.5	a
	0.11	0.48	4.5	b
	0.2	0.3	1.5	c
Pimozide	0.22	0.39	1.8	b
Thioridazine	11	36	4.5	a
	55	85	1.5	b
D_1-antagonists				
SCH 23390	0.05	0.12	2.5	a
D_2/D_3-antagonists				
Metoclopramide	3.4	13	4	a
Remoxipride	2.9	38	14	a
	1 (s.c.)	>100 (s.c.)	>100	a
Sulpiride	270	280	1	a
5-HT_2- and D_2-antagonists				
Risperidone	1.4	7.4	5	b
Sertindole	29	>91	>3	c
Amperozide	>120	>100	n.a.	c
	<11		n.a.	d
Clozapine-like agents				
Clozapine	2	>40	>20	a
	4.3	>40	>9	b
Olanzapine	19	145	8	e

n.a., not applicable; a, tested 1 h after i.p. administration and catalepsy in the bar test (ÖGREN et al. 1990, 1994; ÖGREN and ARCHER 1994; ÖGREN and ÄNGEBY-MÖLLER 1996a).
b, tested 1 h after s.c. administration (JANSSEN et al. 1965, 1988; and MEGENS et al. 1988).
c, tested 1 h after s.c. administration (JACKSON et al. 1994a).
d, tested 30 min after s.c. administration (WATERS et al. 1989).
e, tested 1 h after oral administration (MOORE et al. 1992 and unpublished) with cocaine as stimulant. The ED_{50}-values (mean or median values) were estimated from log-dose-response curves.

of *d*-amphetamine-induced hyperactivity. One group of compounds, represented by *cis*-flupenthixol, haloperidol and pimozide, does not discriminate much between these two measures. A recent study has reported an "atypical" profile with chlorpromazine and thioridazine (JACKSON et al. 1994b). However, the overall results with chlorpromazine indicate that this compound does not display any marked separation between blockade of *d*-amphetamine-induced motor activity and catalepsy similar to the D_1 antagonist SCH23390. In contrast, clozapine, olanzapine and remoxipride, but surprisingly not sulpiride, are "atypical" in the sense that they were clearly more effective in blocking the *d*-amphetamine-induced hyperstimulation than in producing catalepsy. The results of studies in antipsychotic-primed monkeys were consistent with the prediction from the rat study since clozapine and remoxipride, unlike haloperidol, exhibited a low effectiveness in inducing dyskinesia (CASEY 1991a). Risperidone, sertindole and amperozide also behaved as atypical drugs in the *d*-amphetamine model as they clearly separated between the measure for antipsychotic effect and catalepsy.

Drugs such as *cis*-flupenthixol, chlorpromazine, haloperidol, pimoxide and spiperone also readily antagonize or reverse the focussed, repetitive stereotypies induced by higher doses (5–10 mg/kg) of *d*-amphetamine (ARNT et al. 1981; ARNT 1982; BENTALL and HERBERG 1980; IVERSEN and KOOB 1977). In contrast, clozapine and thioridazine, even when given at high doses, were reported to be incapable of blocking amphetamine-induced stereotypies (AL-SHABIBI and DOGGETT 1980; BENTALL and HERBERG 1980; IVERSEN and KOOB 1977). Remoxipride, amperozide and sertindole, even when given at very high doses, failed to inhibit *d*-amphetamine-induced stereotypies (WATERS et al. 1989; ÖGREN and ÄNGEBY-MÖLLER 1996a; SÁNCHEZ et al. 1991). However, a study based on the use of a low and high dose of *d*-amphetamine (1 and 5 mg/kg) showed that clozapine and thioridazine blocked selective items of *d*-amphetamine-induced stereotypies while they both failed to block the *d*-amphetamine-induced forward locomotion (TSCHANZ and REBEC 1989). Moreover, clozapine and thioridazine did not influence *d*-amphetamine-induced behaviours equally, indicating differences in their mechanism of action. Haloperidol, in contrast, was about equally effective in blocking all components of the induced stereotypies including the locomotor activity. Taken together, although most of the available results support current views that atypical drugs such as clozapine are generally more efficient in blocking the locomotor activity produced by *d*-amphetamine than the focussed stereotyped behaviour (see IVERSEN and KOOB 1977; LJUNGBERG and UNGERSTEDT 1985), there is evidence that selective components of *d*-amphetamine-induced stereotypies are influenced at low doses of clozapine.

Under certain circumstances, clozapine, thioridazine and sulpiride all appear to be able to enhance amphetamine-induced stereotypies in the rat (ROBERTSON and MACDONALD 1984, 1985), probably by selectively blocking one of two competing responses. A similar enhancement has been noted with a number of discriminant benzamide derivatives in mice (VASSE et al. 1985). On the other hand, the benzamide metoclopromide has been reported to both block stereotypies (ROBERTSON and MACDONALD 1985) and enhance stereotypies and hyperactivity in the rat (HOWARD et al. 1987). The mechanistic basis for this potentiation is presently not known.

Comparing the effects on spontaneous and *d*-amphetamine-elicited motor activity makes it possible to investigate the effects of DA antagonists on a hyperfunctioning and a normally functioning mesolimbic system. The estimated ED_{50} values of several antipsychotic drugs for inhibition of spontaneous locomotor activity were compared with the corresponding ED_{50} values required to block *d*-amphetamine-induced motor activity. It is clear that most antipsychotic drugs do not show much difference (two- to three-fold separation) between the effect on spontaneous locomotion and effects on locomotion elicited by *d*-amphetamine (ÖGREN and ÄNGEBY-MÖLLER 1996a). Also, the D_1 antagonist SCH23390 did not discriminate much between motility inhibition, blockade of *d*-amphetamine and induction of catalepsy (CHRISTENSEN et al. 1984). Notably, clozapine and risperidone caused suppression of spontaneous locomotion in about the same dose range that also blocked *d*-amphetamine-induced hyperactivity (MEGENS et al. 1992; ÖGREN and ÄNGEBY-MÖLLER 1996a). Remoxipride, unlike the other tested compounds (including sulpiride), blocked amphetamine-induced locomotion at doses which differed widely from those (20-fold separation) affecting the spontaneous locomotion (ÖGREN et al. 1990).

These results are in agreement with recent notions that the DA-ergic mechanisms responsible for locomotor activity (motility, locomotion and rearing) induced by DA agonists like amphetamine may differ from the mechanism responsible for spontaneous motor activity or exploratory behaviour (VAN DEN BOSS 1988; JONES and ROBBINS 1992). Thus, it is possible that postsynaptic DA D_2 receptors in the mesolimbic system are less critically involved in motor activity of an exploratory nature than in amphetamine-induced motor activity (VAN DEN BOSS 1988). This could explain the relative lack of motor suppression induced by selective D_2 receptor antagonists such as remoxipride and sulpiride. Unlike haloperidol, remoxipride had a neglible ability to interfere with sleep–waking patterns and EEG activity in rats and rabbits, and it did not significantly reduce the arousal induced by sensory stimuli (ONGINI et al. 1992). These data further support the notion that DA D_2 receptors play a negligible role in the regulation of states of sedation and sleep.

The hyperactivity induced by methylphenidate in mice is similar to that caused by *d*-amphetamine, and it is thought to involve the mesolimbic DA pathway, while the stereotyped gnawing behaviour is considered to involve the nigrostriatal DA pathway (CHRISTENSEN et al. 1984). All antipsychotic drugs appear to inhibit the methylphenidate-induced hyperactivity but there are marked differences between inhibition of gnawing vs inhibition of hypermotility (NIELSEN et al. 1996). SCH23390 also blocked methylphenidate-induced gnawing in mice at doses related to blockade of amphetamine and apomorphine-induced stereotypies in the rat (CHRISTENSEN et al. 1984). Most classical drugs, with the exception of chlorpromazine (haloperidol, spiperone,

α-flupenthixol), do not show much discrimination between ED_{50} values for motility and gnawing. Interestingly, a similar profile is seen with clozapine, sulpiride and thioridazine while risperidone and sertindole showed a marked degree of separation (NIELSEN et al. 1996). Furthermore, the very low potency of clozapine given by the oral route (ED_{50} = 40 mg/kg) to block methylphenidate-induced hyperactivity indicates that its actions are non-specific. Another study showed that both clozapine and sertindole were incapable of blocking methylphenidate-induced gnawing in mice (SÁNCHEZ et al. 1991). Gnawing induced by a 40 mg/kg dose of methylphenidate in the rat was inhibited by most neuroleptic drugs in a dose-dependent manner which correlated highly with the ability to block apomorphine-induced agitation (KOEK and COLPAERT 1993). However, in contrast to most classical neuroleptics, clozapine, which inhibited methylphenidate-induced gnawing at high doses, failed to normalize the behaviour of methylphenidate-treated animals. Clozapine displayed a pattern of effects resembling that of low doses of typical neuroleptics (KOEK and COLPAERT 1993). Thus, the pattern of effects of different drugs with a purported atypical profile differs in the *d*-amphetamine and methylphenidate tests. This indicates that the neuroanatomical and neurochemical bases for these two behavioural models most likely differ.

II. Apomorphine-Induced Behavioural Effects

Unlike *d*-amphetamine, apomorphine affects behaviour via direct action on DA receptors (ANDÉN et al. 1967). The action of apomorphine is complex since it has an about equal affinity for the DA D_1, D_2 and D_3 receptors (SEEMAN 1987). After systemic administration at low doses (0.01–0.1 mg/kg) it causes a decrease in locomotor activity; at higher doses (1 mg/kg) it elicits an increase in locomotor activity and repetitive sniffing as well as various types of oral (licking, chewing-like movements, biting) and body stereotypies (CAMPBELL et al. 1986; KELLY et al. 1975; ÖGREN et al. 1984). At high doses of apomorphine (5 mg/kg) the behavioural syndrome is mainly characterized by oral stereotypies (gnawing and intense chewing/licking) with almost no locomotor activity (activity is limited to a small area in which the animal is in continous snout contact with the floor). It should be pointed out that the behavioural patterns induced by apomorphine and *d*-amphetamine differ (FRAY et al. 1980; ROBBINS and SAHAKIAN 1981; ANTONIOU and KAFETZOPOLOUS 1991). This indicates that, if stereotypies induced by apomorphine and *d*-amphetamine have predictive validity for positive symptoms in schizophrenia (see RANDRUP and MUNKVAD 1967), the two DA agonists probably act on overlapping but not identical neuronal circuits.

The reduction in locomotor behaviour induced by DA agonists such as apomorphine at low doses has been related to activation of D_2 receptors located presynaptically, e.g., autoreceptors, resulting in a decrease in synaptic levels of DA (STRÖMBERG 1976; DICHIARA et al. 1976). Other evidence suggests, on the other hand, that postsynaptic DA D_2 receptors mediate DA agonist-induced suppression of locomotor activity (Ståhle 1992; Scheel-Krüger 1986). However, the exact role of the DA subtypes for the behavioural effects of apomorphine as well their neuroanatomical basis are still unresolved (see Antoniou and Kafetzopolous 1992). The increases in the behavioural responses after apomorphine appear to be mediated by stimulation of postsynaptic DA receptors of both the DA D_1 and D_2 type (Walters et al. 1987) located in the nucleus accumbens and striatum, respectively (Kelly et al. 1975; Arnt 1987; Bordi et al. 1989). The full expression of stereotyped behaviours observed after systemic administration of apomorphine and amphetamine appears to require the activation of both D_1 and D_2 receptors in mesolimbic as well as mesostriatal pathways (see Arnt 1987; Cooper and Dourish 1990). The nucleus accumbens appears to be involved in the hyperlocomotion caused by apomorphine, while the oral and sniffing responses have been related to anterior ventral striatal sites (Bordi et al. 1989). On the other hand, the characteristic head-down sniffing response seen after systemic administration of apomorphine appears to be located in a discrete region of the ventro-lateral striatum and the dorso-lateral nucleus accumbens (Cameron and Crocker 1989).

Measurements of the behavioural effects elicited by apomorphine have varied considerably. In some studies the intensity of the stereotypies (sniffing, chewing, licking, biting) and the hyperactivity elicited by apomorphine were scored using various rating scales (see Campbell et al. 1986; Fray et al. 1980; Ögren et al. 1984). Other studies were designed to distinguish between some of the behavioural components (gnawing and locomotor activity) elicited by apomorphine by employing locomotor cages (Ljungberg and Ungerstedt 1978, 1985). Since the behavioural effects of apomorphine, like those of other DA stimulants, are context-dependent (Robbins 1977; Ögren et al. 1979), the conditions under which the experiments are performed will influence the results.

Early studies with apomorphine or *d*-amphetamine in the rat showed that most antipsychotic drugs did not differentially block the behavioural components induced by apomorphine, e.g., agitation and stereotypies (see Janssen et al. 1965, 1967). A study in mice failed to show any differences between the effects of a number of antipsychotic drugs including clozapine on apomorphine-induced hypermotility or stereotypies (Puech et al. 1981). Later studies using automatic recording of locomotor activity classified DA receptor antagonists as typical (e.g., chlorpromazine, haloperidol) or atypical neuroleptic drugs (e.g., clozapine, sulpiride and thioridazine) based on their ability to antagonise locomotion and one aspect of oral stereotypy, e.g., gnawing, induced by a high dose of apomorphine (5 mg/kg s.c.) (Ljungberg and Ungerstedt 1978, 1985). The drugs defined as atypical were more efficient in blocking the locomotor component than the gnawing response while haloperidol blocked both behavioural symptoms at about the same dose level. Using this model, remoxipride, unlike haloperidol, did not affect gnawing or locomotion if not given at extremely high doses (Ståhle et al. 1987). Similar results were obtained with several discriminant benzamide derivatives including (−)-sulpiride, while metoclopramide behaved similarly to haloperidol (SCHWARTZ et al. 1984). On the other hand, the D_1 antagonist SCH23390 effectively blocked stereotypies (gnawing, licking or head movements) elicited by apomorphine in the rat (CHRISTENSEN et al. 1984). After intrastriatal injections, both D_1 and D_2 receptor antagonists antagonised stereotypies caused by systemic apomorphine, supporting the critical role of the mesostriatal system for this response (ARNT 1985). Similar to classical antipsychotic drugs, some partial dopamine agonists such as SDZ 208-911 or terguride (but not (−)-3-PPP) inhibited apomorphine (2 mg/kg i.v.)-induced gnawing at low doses but were essentially devoid of cataleptogenic activity (COWARD 1991; COWARD et al. 1990). Thus, based on the assumption that gnawing predicts EPS (LJUNGBERG and UNGERSTEDT 1978, 1985), SDZ 208-911 would be defined as a typical neuroleptic drug. The differences between partial dopamine agonists such as SDZ 208-911 and (−)-3-PPP are probably due to varying degrees of partial agonistic activity at D_2 receptors (see CARLSSON 1983).

In subsequent studies using a behavioural scoring technique in the rat, all compounds characterised as antipsychotic drugs blocked the hyperactivity resulting from the administration of apomorphine (1 mg/kg s.c.) (Table 6). This procedure results in both oral stereotypies (chewing, biting, licking but no gnawing) and increases in locomotor activity. However, clozapine, raclopride, remoxipride, sulpiride and thioridazine (ÖGREN et al. 1984, 1986, 1990; ÖGREN and ÄNGEBY-MÖLLER 1996b) preferentially blocked the locomotor hyperactivity induced by apomorphine while higher doses were required to block the oral stereotypies (chewing, biting and licking). Chlorpromazine, haloperidol or spiperone did not differentiate between the induced stereotypies and hyperactivity. Also (−)-3-PPP antagonised apomorphine-induced hyperactivity (HJORTH et al. 1983) in a dose range which was about four times higher than that blocking spontaneous locomotor activity. The relatively weak ability of clozapine to antagonise the behavioural effects induced by apomorphine, particularly the oral stereotypies, has been noted previously (see PUECH et al. 1981; COWARD et al. 1989). This observation is intriguing since both D_1 and D_2 receptor antagonists such as SCH 23390 and raclopride appear to be equally effective in blocking the oral stereotypies induced by apomorphine (CHRISTENSEN et al. 1984; ÖGREN et al. 1986). Moreover, the low effectiveness of clozapine appears not to involve 5-HT_2 receptor functions since risperidone effectively blocked apomorphine-induced stereotypies and agitation (see MEGENS et al. 1992).

The effects of various compounds on apomorphine-induced hyperactivity and induction of catalepsy are compared in Table 6. Chlorpromazine and haloperidol showed little difference while thioridazine displayed a much greater separation between the doses blocking the motor-stimulating action of apomorphine and the doses inducing significant catalepsy. In contrast, with clozapine and remoxipride, unlike haloperidol, there was a large gap between the doses causing EPS, e.g., bar test catalepsy, in the rat and the doses which blocked apomorphine- or amphetamine-induced hyperactivity. Surprisingly, both risperidone and sulpiride showed a small separation between the DA agonist-induced hyperactivity and catalepsy. These results indicate that mechanistic differences between antipsychotic compounds can be revealed by using apomorphine and *d*-amphetamine as the DA agonists (compare Tables 5 and 6).

Unlike the blocking action seen with all antipsychotic drugs, some antipsychotic drugs, e.g., haloperidol, pimozide and sulpiride, have been shown to increase the locomotor response to apomorphine in mice (Puech et al. 1981). This enhancement is mostly seen at the lower dose range and disappears at the higher dose range of these compounds. The ability of some antipsychotic compounds to potentiate DA-agonist-induced hypermotility has been related to a disinhibitory action in the clinic (Lecrubier et al. 1980) possibly mediated via a preferential action at DA D_2 autoreceptors (Puech et al. 1981). However, the mechanism and clinical significance of this motor hyperstimulation are presently not known.

The increase in climbing behaviour elicited by apomorphine in mice has been used by some authors to investigate antipsychotic potential (Protais et al. 1976; Moore and Axton 1988). The neuroanatomical basis for the climbing behaviour induced by apomorphine has remained controversial since both the striatum and the nucleus accumbens have been implicated (Costall et al. 1980). Some investigators have speculated that a "selective" blocking effect of apomorphine-induced climbing will mean an atypical profile (Moore and Axton 1988). Since this response to apomorphine appears to be dependent on activation of both D_1 and D_2 receptors (Moore and Axton 1988), it is not

Table 6. Relationship between induction of bar test catalepsy in the rat and potency in blocking apomorphine-induced hyperactivity

Compounds	Catalepsy	DA agonist hyperactivity	Ratio
	(ED_{50}, μmol/kg)		
Chlorpromazine	11	6	1.8[a]
Clozapine	>40[c]	31	>2[a]
Haloperidol	0.9	0.3	3[a]
Remoxipride	38	0.9	42[a]
Risperidone	1.4	0.45; 0.7	2–9[b]
Sulpiride	280	66	4.2[a]
Thioridazine	55	9.8	6

Ratios are calculated from the ED_{50}-values.

[a] Data modified from Ögren et al. 1984, 1986, 1990 and 1996. Results (based on i.p. administration) refer to blockade of apomorphine-induced hyperactivity.

[b] Results with risperidone modified from Janssen et al. (1988); Megens et al. (1992). Results (based on s.c. administration) refer to a blockade of apomorphine-induced agitation.

[c] Highest dose tested without marked behavioural effects.

surprising that most antipsychotic drugs are effective in this model (ZIVKOVICS et al. 1983) at doses which are related to other measures of brain DA receptor blockade. Surprisingly, however, clozapine must be given at high i.p. or oral doses to be effective (MOORE et al. 1992; ZIVKOVICS et al. 1983). In contrast, olanzapine was effective at low doses and equipotent in antagonising climbing and in blocking conditioned aoidance response (MOORE et al. 1992). Also, the selective D_2/D_3 antagonist amisulpiride has been shown to block apomorphine-induced climbing in mice at doses much lower than those inducing catalepsy or inhibiting apomorphine-induced stereotypies (SCATTON et al. 1994). Several discriminate benzamides including (–)-sulpiride but not metoclopramide were also found to inhibit apomorphine-induced yawning at much lower doses than those blocking the induced oral stereotypies (SCHWARTZ et al. 1984). The D_1 receptor antagonists SCH23390 and SCH39166 have also been found to block apomorphine-induced climbing at doses clearly lower than those affecting apomorphine-induced sniffing (a presumed striatal response) (GERHARDT et al. 1985; CHIPKIN et al. 1988). These results indicate that some compounds with an atypical profile (but not clozapine) may exert a preferential effect on the climbing response in mice by mechanisms which remain to be elucidated.

III. Phencyclidine-Induced Behavioural Effects

Evidence accumulated during the past decade has led to the hypothesis that dysfunctions in the actions of the excitatory amino acid (EAA) glutamate may be part of the pathogenesis of schizophrenia (see KIM and KORNHUBER 1982). This hypothesis is partly based on anatomical, pharmacological and electrophysiological findings suggesting that glutamatergic and dopaminergic neurons may interact reciprocally in structures of the brain implicated in schizophrenia, e.g., in the dorsal and ventral striatum (see CARLSSON and CARLSSON 1990). Further evidence for glutamate/DA interactions is provided by the observation that the dissociative anaesthetic phencyclidine (PCP), which interferes with glutamate neurotransmission, produces a syndrome in humans which is similar to both the positive and negative symptoms of schizophrenia (JAVITT and ZUKIN 1991). PCP, which acts as a noncompetitive antagonist at the ion channel associated with the *N*-methyl-D-aspartate (NMDA) glutamate receptors induces motor stimulation and psychotomimetic effects, which have been suggested to be a result of NMDA receptor blockade or due to an indirect facilitation of dopaminergic (DA) transmission (JAVITT and ZUKIN 1991; FRENCH et al. 1991). This has led to the suggestion that the effect in this test model may be indicative of antipsychotic effects (see MCKINNEY 1989).

The locomotor stimulation caused by PCP in the rat has been found to be blocked by both typical and atypical antipsychotic drugs including clozapine (JACKSON et al. 1994a; ÖGREN and GOLDSTEIN 1994; WILLETS et al. 1990). Similar to haloperidol and clozapine, selective DA D_2 receptor antagonists such as remoxipride and raclopride were found to block the hyperactivity induced by low doses of PCP dose-dependently at about the same dose levels that inhibited *d*-amphetamine-induced hyperactivity (Jackson et al. 1994a; Ögren and Goldstein 1994). In contrast, selective D_1 receptor antagonists were clearly more potent in blocking the *d*-amphetamine than the PCP-induced hyperlocomotion (Jackson et al. 1994a). Unlike postsynaptic DA receptor blockade, stimulation of the DA D_2 autoreceptor by the D_2 receptor agonist quinpirole did not antagonize the motor stimulation caused by PCP (Ögren and Goldstein 1994). In addition, the motor stimulation produced by a high dose of PCP (3 mg/kg) was reduced by haloperidol only at cataleptogenic doses, whereas remoxipride fully blocked the effects of this PCP dose at a noncataleptogenic dose, indicating differences between the compounds with an atypical profile and haloperidol. These studies indicate that DA neurotransmission, mediated via DA D_1 and D_2 receptors in the brain, plays an essential role in the locomotor stimulatory response induced by PCP.

F. Action on Conditioned Avoidance

Procedures based on instrumental conditioning, e.g., active avoidance, have traditionally been among the most important methods used to study antipsychotic drug action in rodents (see Cook and Davidson 1978). The different rules (contingencies) for the presentation of shock (reinforcer) in relationship to the required response distinguishes classical and instrumental conditioning. Instrumental (aversive) conditioning refers to contingencies in which the conditioned stimulus (CS) (often a buzzer) is followed after a short interval (5–10 s) by the aversive, unconditioned stimulus (UCS) (a weak foot shock). In the active avoidance procedure the animal can either avoid the shock by responding (running to the safe side of the box) during the CS-UCS interval, or it can respond during the USC presentation, e.g., it escapes the shock or it can fail to respond during shock presentation (response failure). The response of the animal is thus "instrumental" in determining whether or not the unconditioned stimulus (UCS) or "reinforcing stimulus" will occur (Domjan and Burkhard 1982). It has generally been believed that the avoidance response provides reinforcement, the nature of which has been a central question in theories on avoidance learning (see Rescorla and Solomon 1967).

The required response differs in the active avoidance task. In some avoidance tasks the rat has to perform a discrete low-probability response, e.g., climbing a pole (see Cook and Davidson 1978) or, most commonly, running from one side of a two-compartment box to the other or pressing a lever (Sidman avoidance procedure), at the presentation of the conditioned stimulus (CS) in order to escape or avoid the noxious stimulus (shock). These procedures are different from passive avoidance tasks in which the animal learns to suppress a motor response to avoid the area (mostly a black box) which earlier has been associated with the foot shock. The features of the avoidance task are important, particularly when determining the influence of nonlearning factors (e.g., changes in motor activity and motivations).

Psychotropic drugs have been classified according to their effects on avoidance or escape response. Most studies on avoidance learning are conducted in the two-way active avoidance task, mostly referred to as the conditioned avoidance response (CAR) task. In the majority of these studies, testing is conducted with animals that have acquired the avoidance response, i.e., the rats are trained to a learning criterion (e.g., 9 out of 10 consecutive avoidance responses) and then treated with the antipsychotic drug in question. The number of avoidance responses in a test period out of ten trials is the measure most commonly used for performance, e.g., the retention of the CAR. A few studies have also studied drug effects on acquisition and extinction. Since prior training markedly attenuates the effects of neuroleptics on CAR (BENINGER et al. 1983), it is important to compare effects on both acquisition and retention.

Most antipsychotic drugs have been shown to produce a blockade of retention (and in some studies acquisition) of two-way conditioned (active) avoidance responses (CARs) in rodents (JANSSEN et al. 1965; ARNT 1982). Since potencies in the CAR test correlate with clinical efficacy (JANSSEN et al. 1965; ARNT 1982), the blockade of CARs is assumed to define an important characteristic of antipsychotic drugs and be of predictive validity for antipsychotic effect (see KAISER and SETLER 1981). However, as pointed out by ARNT (1982) it is not clear whether the impairment of CAR by neuroleptics is related to their antipsychotic efficacy or the ability to induce EPS, or both.

The available data demonstrate that all antipsychotic compounds belonging to different chemical classes produce a dose-dependent inhibition of performance in the CAR test (ARNT 1982). Table 7 shows the effect of representative antipsychotics on CAR and their effectiveness in inducing catalepsy. The results with clozapine on CAR appear to depend on the type of learning procedures since low doses of the compound were reported to increase the suppressed response in mice and squirrel monkeys by punishment (BARRETT 1982; SPEALMAN et al. 1983) while high doses blocked the response. Other studies have also shown that clozapine blocks CAR or performance in the rat only at high doses (COWARD et al. 1989; BLACKBURN and PHILLIPS 1989) which may produce a non-specific disruption of performance. It is clear that at high doses of clozapine (>50 μmol/kg) there are signs of serious motor disturbances combined with behavioural toxicity which interfere with the escape response (ÖGREN and ARCHER 1994). Other results indicate, however, that clozapine can block CAR in the rat at low doses which do not cause any apparent motor deficits (see ARNT 1982; ÖGREN and ARCHER 1994). Sulpiride was reported to be highly potent in affecting CAR performance following intracerebral administration (NISHIBE et al. 1982), but displayed a low potency following systemic administration (see ARNT 1982; ÖGREN and ARCHER 1994), probably attributable to a low penetration into relevant brain areas. The inability of (–)-3-PPP given at doses selective for the DA autoreceptor to affect both CAR acquisition and retention is intriguing (AHLENIUS et al. 1984), in view of the claim for predictive validity of the CAR test. On the other hand,

Table 7. Effectiveness of different types of antipsychotic drugs in blocking CAR and inducing catalepsy in the rat

Compounds	ED_{50}, μmol/kg (i.p., s.c. or p.o.)			
	CAR	Catalepsy	Ratio	Reference
Reference antipsychotic drugs				
Chlorpromazine	2.5	18.6	7.3	a
	16	11	~1	b
	>28	70	<2.5	c
	28.5	28	~1	d
cis-Flupenthixol	0.06	0.16	2.7	a
Haloperidol	0.11	0.35	3	a
	0.3	0.9	3	b
	1.3	2.9	2.2	c
	0.5	2.1	4	d
Pimozide	0.28	1.7	6	a
Thioridazine	32	88	2.8	a
D_1-antagonists				
SCH 39166	32	>956	>30	e
D_2/D_3-antagonists				
Remoxipride	5.6	38	7	b
	32	159	5	c
Sulpiride	152	527	3.5	a
5-HT_2- and D_2-antagonists				
Risperidone	2.2	15.3	7	c
	6.6	18.8	3	d
Sertindole	6.8	>227	>33	d
Clozapine-like agents				
Clozapine	8.9	428	>48	a
	4.3	>40	>9	b
	64	>490	>8	c
	34	>196	>5.8	d
Olanzapine	17.5	145	8	c

CAR, conditioned avoidance response.
a, CAR tested 2h after s.c. administration. Catalepsy was examined in the grid test (ARNT 1982).
b, CAR tested 1h after i.p. administration and catalepsy in the bar test (ÖGREN and ARCHER 1994).
c, CAR tested 1h after oral administration and catalepsy in the bar test (MOORE et al. 1992).
d. CAR and catalepsy (grid test) were tested 2h after i.p. administration (NIELSEN et al. 1996).
e, CAR tested 1h after p.o. administration and catalepsy in the bar test (CHIPKIN et al. 1988).
The ED_{50}-values were estimated from log-dose-response curves at peak effect of each drug.

the partial DA agonist SDZ 208-911 reduced CAR performance in the dose range inhibiting apomorphine-induced gnawing (Coward et al. 1990) indicating that degrees of intrinsic activity at DA receptors are critical for this class of compounds (see Carlsson 1983).

Taken together, although inhibition of CAR is a property shared by both typical and atypical antipsychotic drugs, there may exist qualitative differences in their relative effectiveness on acquisition and performance. In a one-way CAR test, clozapine and thioridazine disrupted both acquisition and performance equally while haloperidol and metoclopramide preferentially blocked acquisition (Blackburn and Phillips 1989). However, since both typical and atypical drugs appear to impair CAR acquisition more strongly than performance in the two-way task (Beninger et al. 1983; Ögren and Archer 1994), it is possible that response factors may underlie the differences observed in the one-way task.

Depending on the dose of a particular neuroleptic drug, nonspecific effects on behaviour (akinesia and excessive "sedation") or induction of extrapyramidal side effects (e.g., catalepsy) could interfere with the ability to escape or respond. The available results indicate, however, that antipsychotic drugs impair CAR at doses which are clearly lower than those interfering with escape responses (see Arnt 1982; Beninger 1983; Nielsen et al. 1996; Ögren and Archer 1994). Other classes of drugs such as the benzodiazepines (e.g., diazepam) appear to reduce CAR performance at doses causing sedation and muscle relaxation (Arnt 1982). However, the relationship between CAR and sedation is complex. Thus, (–)-3-PPP given at doses causing a marked sedative action or motor suppression failed to inhibit CAR (Ahlenius et al. 1984). On the other hand, relatively high doses of α-adrenergic antagonists such as prazosin, which induces "sedation", inhibit CAR without influencing escape response (Arnt 1982; Tabaoda et al. 1979). Therefore, it cannot be ruled out that α-receptor blockade can contribute to the action of clozapine (Arnt 1982).

These results indicate that subtle impairments in motor function, e.g., a deficit in the ability to initiate motor responses, can be important (see Beninger 1983). The impairment of CAR by neuroleptic drugs has often been related to an inability to perform or initiate the required motor response; this is termed the "motor-incapacity hypothesis". It is notable that the inhibition of CAR by most antipsychotic drugs with the exception of clozapine shows a reasonable correlation to the dose–response data for blockade of methylphenidate- or *d*-amphetamine-induced stereotypies (Arnt 1982) or apomorphine-induced stereotypies (Ögren and Archer 1994). However, antipsychotic drugs differ clearly when comparing the relative potency for blocking CAR and inducing catalepsy by calculating the ratio of catalepsy induction to inhibition of CAR (Table 7). Drugs such as chlorpromazine, haloperidol, *cis*-fluphentixol and pimozide displayed small ratios when dividing the ED_{50} for catalepsy by the ED_{50} for CAR, suggesting that the inhibition of CAR occurred at doses that were somewhat lower than those inducing catalepsy. In contrast, clozapine, olanzapine, remoxipride and sertindole displayed a clear-cut difference between the doses that blocked CAR and those that induced catalepsy while the results with risperidone were not conclusive (see Table 7). Surprisingly, neither sulpiride and thioridazine exhibited a high degree of selectivity. SCH23390 given s.c. has been reported to cause catalepsy at doses about equal to (MORELLI and DICHIARA 1985) or twice (CHRISTENSEN et al. 1984) those found to be active in the CAR test (IORIO et al. 1983). The novel D_1 antagonist SCH39166 showed, on the other hand, a 30-fold separation following oral administration between inhibition of CAR and induction of catalepsy (CHIPKIN et al. 1988). It remains to be determined whether this marked difference is due to kinetic factors. These findings, taken together, suggest that some compounds with a putatively high degree of mesolimbic selectivity apparently block CAR by mechanisms not directly related to those striatal mechanisms believed to be responsible for catalepsy.

Alternative explanations for the selective CAR impairment induced by antipsychotic drugs have also been suggested. However, the available evidence does not support the view that the CAR blockade is due to a motivational deficit related to lack of reinforcement (BENINGER 1983), termed "the anhedonia hypothesis". Moreover, the acquisition and retention deficits in active avoidance caused by neuroleptic drugs appear not to reflect any impairment of the associative learning procedure (BENINGER et al. 1983).

Both mesolimbic and mesostriatal DA systems play an important role in various aspects of learning and memory including conditioned avoidance learning (see BENINGER 1983; KOOB et al. 1984). However, it is not known whether typical and atypical drugs differ in the manner by which they affect the DA pathways involved in CAR performance. Studies using the DA neurotoxin 6-hydroxydopamine (6-OHDA) have demonstrated that both the nigroneostriatal DA neurons and the mesolimbic DA neurons (see BJÖRKLUND and LINDVALL 1984) play a significant role in the neuroleptic-induced disruption of CAR (KOOB et al. 1984). Based on intracerebral injections of (−)-sulpiride (a selective D_2 antagonist), the ventral (necleus accumbens) but not the dorsal part of the striatum was implicated in CAR (WADENBERG et al. 1990). However, the relationships between CAR performance and different brain structures are complex. This has been illustrated with studies showing that injections of antipsychotic drugs, including those injected into the amygdala, cause severe CAR deficits (PETTY et al. 1984). Thus, the most likely explanation at the present stage is that both striatal and mesolimbic DA neurons as well as other limbic structures, e.g., the amygdala, belong to the neuronal circuits involved in CAR. The observed differences between typical and atypical antipsychotic drugs may reflect differential modulations of subpopulation of D_1 and D_2 receptors localized within striatal and mesolimbic circuits.

This interpretation is consistent with the observation that DA D_1 antagonists such as SCH23390 and SCH39166 block CAR by a mechanism which appears to differ from that of the DA D_2 antagonists (IORIO et al. 1991). In keeping with this theory, the muscarinic antagonist scopolamine enhanced the CAR impairment induced by SCH23390 and SCH39166 while scopolamine attenuated the CAR blockade produced by D_2 receptor antagonists such as haloperidol and raclopride (IORIO et al. 1991). Scopolamine also reduced the CAR blockade caused by clozapine, indicating that the action of clozapine on CAR is probably mediated via blockade of DA D_2 receptors. This is contrary to the suggestion that clozapine exerts its antipsychotic effect via a preferential blockade of D_1 receptors in vivo (COWARD et al. 1989). These and additional findings indicate that D_1 and D_2 receptors differentially modulate cholinergic mechanisms important for CAR performance (see ARNT et al. 1981; IORIO et al. 1991).

G. Animal Models of Psychosis and Antipsychotic Drugs

During recent years much attention has been devoted to the development of animal models for schizophrenia (MCKINNEY 1989). The criteria for assessments of the validity of some of these animal models have been extensively discussed by ELLENBROEK and COOLS (1990). These models are based on the assumption that there may exist an analogous or homologous relationship between neuronal mechanisms underlying animal behaviour and the putative disturbance in the schizophrenic patient. Such models include the latent inhibition and blocking paradigm (see LUBOW et al. 1987) which is primarily used to analyse attentional processes. In addition, in some models it is hypothesised there may exist analogous behavioural functions in animals and in man. Thus, a model for "negative symptoms" is based on the ability of *d*-amphetamine to increase social isolation in monkeys (ELLENBROEK and COOLS 1990). However, in most of these models there is limited knowledge at present on the effects of both typical and atypical drugs. More information is available in another animal model aimed at studying sensorimotor integration, e.g., the startle response and prepulse inhibition. This model is believed to have construct validity for schizophrenia (see ELLENBROEK and COOLS 1990).

The glutamate antagonist phencyclidine seems to be one of the few drugs which affect the startle response in a manner similar to that found in schizophrenic patients. Thus, systemic administration of phencyclidine to the rat retards habituation and impairs prepulse inhibition in the acoustic startle paradigm (MANSBACH and GEYER 1989). The effects of PCP on prepulse inhibition appear to involve primarily D_2 receptors. Thus, haloperidol was found to attenuate the disruptive effects of PCP on prepulse inhibition based on an electrophysiological study of sensory gating (ADLER et al. 1986). In contrast, a recent study using the acoustic startle response in the rat failed to find any significant effect of haloperidol on the PCP-induced disruption of prepulse inhibition. Also, the selective D_2 antagonist raclopride as well as clozapine were ineffective, while remoxipride caused a dose-dependent block of the PCP-induced disruption (JOHANSSON et al. 1994). These findings indicate that both typical and atypical drugs may differ in their action in some animal "models" for schizophrenia, suggesting differential effects on the neuronal mechanisms underlying sensorimotor integration.

H. Conclusions

The accumulated results from the present studies illustrate the major problem that arises in the attempt to assess both antipsychotic efficacy as well as EPS liability by using different types of behavioural tests. The outcome will critically depend on the reliability of the test results as well as the validity of the test paradigms used as predictors of an atypical antipsychotic profile in the clinic. Furthermore, since the definition of atypical antipsychotic activity is mainly based on the ability to discriminate between different motor functions, multidimensional aspects of atypicality are not assessed. In addition, a direct measurement of mesolimbic and nigrostriatal selectivity is not possible with most existing behavioural models. The available data base also suffers from other limitations which make interpretations and comparisons difficult. Very few studies have examined different routes of administration and measurements of plasma/brain concentrations in combination with estimations of receptor occupancy in relevant brain areas to shed light on the receptor basis for the behavioural effects observed.

Despite these obvious limitations there are marked similarities between various types of antipsychotic drugs. It is clear that most antipsychotic compounds, irrespective of pharmacological profile, are active in tests believed to be valid predictors of antipsychotic activity. Of particular interest is the observation that all compounds characterized as antipsychotic drugs block DA agonist-induced locomotor activity. Moreover, the results also clearly indicate that typical and atypical drugs do not represent two distinct classes of drugs. However, based on the overall spectrum of activity in different behavioural models, compounds with an atypical profile can be separated from those with a typical profile. Also, within the group of compounds with an atypical profile there are considerable differences, indicating that they do not constitute a functionally homogeneous group. This is also consistent with results obtained from biochemical and receptor binding studies (see Deutch et al. 1991).

The results of studies using classical neuroleptic drugs consistently demonstrate a typical pattern of effect, i.e., a small separation between measures for antipsychotic effects and measures predictive for EPS. Compounds with an atypical profile, on the other hand, preferentially act on behavioural responses in rodents believed to be mediated via the ventral striatum (mesolimbic DA pathway) compared to those behaviours associated with the neostriatal pathway. However, this difference is not always seen, even with clozapine, suggesting that the striatal–limbic dichotomy (see Köhler et al. 1981) is not fully adequate to explain atypicality. An alternative hypothesis is that atypical compounds may act via actions on subtypes of DA D_2 receptors localized within restricted neuronal circuits belonging to both the mesolimbic and mesostriatal systems (Ögren et al. 1994). The further analysis of atypical antipsychotic drugs must focus on the concept of functional subdivisions within both the caudate putamen and the nucleus accumbens. A recent study has shown that clozapine and remoxipride, unlike haloperidol, exert regionally selective effects on c-fos expression within subdivisions of the striatum and nucleus accumbens (shell vs core) (DEUTCH et al. 1992). Thus, behavioural atypicality may be paralleled by functional heterogeneity at the neuronal circuit level.

The present results also demonstrate that compounds with an atypical profile can be differentiated in their relative propensity for inducing EPS. Some compounds, e.g., the 5-HT_2/D_2 antagonists amperozide and sertindole, as well as the clozapine-analogue olanzapine and the two substituted benzamides remoxipride and amisulpiride, have a profile comparable to clozapine in the sense that they show a large EPS separation. Furthermore, if ratios between activity in tests for psychosis and catalepsy are valid predictors, this would indicate that the D_1 antagonist SCH23390 has a high potential for EPS similar to the benzamide metoclopramide, and that the 5-HT_2/D_2 antagonist risperidone has an intermediate potential, while olanzapine has a low potential. Partial DA receptor agonists such as (–)-3-PPP are also atypical since they do not induce the appearance of EPS. Since representatives of this group of compounds appear to have a low ability to block CAR, the therapeutic efficacy of partial DA agonists in schizophrenia may be questioned. Thus, the available results do not support the hypothesis that atypical neuroleptics work at least partly via DA D_1 receptors while classical neuroleptics act mainly via the DA D_2 receptors (see ELLENBROEK et al. 1991). However, even the drugs which display a high EPS separation elicit catalepsy at high doses, indicating that they will not be entirely free of EPS. Moreover, although some compounds with atypical properties show functional similarities with clozapine, they differ markedly with respect to their receptor binding profile from clozapine. This indicates that compounds with an atypical profile act via a variety of mechanisms and that they do not constitute a single class of compounds.

References

Adler L, Rose G, Freedman R (1986) Neurophysiological studies of sensory gating in rats: effects of amphetamine, phencyclidine and haloperidol. Biol Psychiatry 21:787–798

Ahlenius S, Archer T, Tandberg B, Hillegaart V (1984) Effects of (–)3-PPP on acquisition and retention of a conditioned avoidance response in the rat. Psychopharmacology 84:441–445

Alexander GE, Crutcher MD (1990) Functional architecture of basal ganglia circuits: neural substrates of parallel processing. Trends Neurosci 13:266–271

Al-Shabibi UMH, Doggett NS (1980) Clozapine's anti-acetylcholine property modulates its antistereotypic action in the mesolimbic system. J Pharm Pharmacol 32:359–361

Andén N-E, Rubenson A, Fuxe K, Hökfelt T (1967) Evidence for dopamine receptor stimulation by apomorphine. J Pharm Pharmacol 19:627–629

Angrist B (1983) Psychoses induced by central nervous system stimulants and related drugs. In: Creese I (ed) Stimulants: neurochemical, behavioral, and clinical perspectives. Raven, New York

Antoniou K, Kafetzopoulos E (1991) A comparative study of the behavioral effects of *d*-amphetamine and apomorphine in the rat. Pharmacol Biochem Behav 39:61–70

Antoniou K, Kafetzopoulos E (1992) Behavioral effects of amphetamine and apomorphine after striatal lesions in the rat. Pharmacol Biochem Behav 43:705–722
Arnt J (1982) Pharmacological specificity of conditioned avoidance response inhibition in rats: inhibition by neuroleptics and correlation to dopamine receptor blockade. Acta Pharmacol Toxicol 51:321–329
Arnt J (1985) Antistereotypic effects of dopamine D_1 and D_2 antagonists after intrastriatal injection in rats. Pharmacological and regional specificity. Naunyn-Schmiedebergs Arch Pharmacol 330:97–104
Arnt J (1987) Behavioral studies of dopamine receptors: evidence for regional selectivity and receptor multiplicity. In: Creese I, Breese G (eds) Dopamine receptors. Liss, New York, pp 199–231
Arnt J, Christensen AV, Hyttel J (1981) Differential reversal by scopolamine of effects of neuroleptics in rats. Relevance for evaluation of therapeutic and extrapyramidal side-effect potential. Neuropharmacology 20:1331–1334
Arnt J, Christensen AV, Hyttel J, Larsen J-J, Svendsen O (1983) Effects of putative dopamine autoreceptor agonists in pharmacological models related to dopaminergic and neuroleptic activity. Eur J Pharmacol 86:185–198
Baldessarini RJ, Frankenburg FR (1991) Clozapine – a novel antipsychotic agent. N Engl J Med 324:746–754
Balsara JJ, Jadhav JH, Chandorkar AG (1979) Effect of drugs influencing central serotonergic mechanisms on haloperidol-induced catalepsy. Psychopharmacology 62:67–69
Barrett J (1982) Antipsychotic drug effect on the behavior of squirrel monkeys differentially controlled by noxious stimuli. Psychopharmacology 77:1–8
Beninger RJ (1983) The role of dopamine in locomotor activity and learning. Brain Res Rev 6:173–196
Beninger RJ, Phillips AG, Fibiger HC (1983) Prior training and intermittent retraining attenuate pimozide-induced avoidance deficits. Pharmacol Biochem Behav 18:619–624
Bentall ACC, Herberg LJ (1980) Blockade of amphetamine-induced locomotor activity and stereotypy in rats by spiroperidol but not by an atypical neuroleptic thioridazine. Neuropharmacology 19:699–703
Björklund A, Lindvall O (1984) Dopamine-containing systems in the CNS. In: Björklund A, Hökfelt T (eds) Classical transmitters in the CNS. Elsevier, Amsterdam, pp 55–122 (Handbook of chemical neuroanatomy, vol 2/I)
Blackburn JR, Phillips AG (1989) Blockade of acquisition of one-way conditioned avoidance responding by haloperidol and metoclopramide but not by thioridazine or clozapine: implications for screening new antipsychotic drugs. Psychopharmacology 98:453–459
Bordi F, Carr KD, Meller E (1989) Stereotypies elicited by injection of N-propylnorapomorphine into striatal subregions and nucleus accumbens. Brain Res 489:205–215
Borison RL, Pathiraya AP, Diamond BF, Meibach RC (1992) Risperidone clinical safety and efficacy in schizophrenia. Psychopharmacol Bull 281:213–218
Boyce S, Kelly E, Davis A, Fleminger S, Jenner P, Marsden CD (1985) SCH23390 may alter dopamine-mediated motor behaviour via striatal D-1 receptors. Biochem Pharmacol 34:1665–1669
Braun AR, Chase TN (1986) Obligatory D_1/D_2 receptor interaction in the generation of dopamine agonist related behaviours. Eur J Pharmacol 131:301–306
Cameron DL, Crocker AD (1989) Localization of striatal dopamine receptor function by central injection of an irreversible receptor antagonist. Neuroscience 32:769–778
Campbell A, Baldessarini RJ, Teicher MH, Neumeyer JL (1986) Behavioral effects of apomorphine isomers in the rat: selective locomotor-inhibitory effects of S(+)N-n-prophylnorapomorphine. Psychopharmacology 88:158–164

Carlsson A (1970) Amphetamine and brain catecholamines. In: Costa E, Garattini S (eds) Amphetamines and related compounds. Raven, New York, pp 289–300
Carlsson A (1983) Dopamine receptor agonists: intrinsic activity vs. state of receptor. J Neural Transm 57:309–315
Carlsson A, Lindqvist M (1963) Effect of chlorpromazine or haloperidol on formation of 3-methoxytyramine and normetanephrine in mouse brain. Acta Pharmacol Toxicol 20:140–144
Carlsson M, Carlsson A (1990) Interactions between glutamatergic and monoaminergic systems within the basal ganglia – implications for schizophrenia and Parkinson's disease. Trends Neurosci 13:272–276
Casey DE (1989) Clozapine: neuroleptic-induced EPS and tardive dyskinesia. Psychopharmacology 99:S47–S53
Casey DE (1991a) Antipsychotic drugs in schizophrenia: new compounds and differential outcomes. Psychopharmacol Bull 27:47–50
Casey ED (1991b) The effect of a serotonin S2 antagonist, ritanserine, and anticholinergic benztropine on haloperidol-induced dystonia in nonhuman primates. Am Coll Neuropsychopharmacol 30:127
Casey (1992) What makes a neuroleptic atypical? In: Meltzer HY (ed) Novel antipsychotic drugs. Raven, New York, pp 241–251
Chipkin RE, Iorio LC, Coffin VL, McQuade RD, Berger JG, Barnett A (1988) Pharmacological profile of SCH39166: a dopamine D_1 selective benzonaphthazepine with potential antipsychotic activity. J Pharmacol Exp Ther 247:1093–1102
Chouinard G, Jones B, Remington G et al. (1993) A Canadian multicentre placebo-controlled study of fixed doses of risperidone and haloperidol in the treatment of schizophrenic patients. J Clin Psychopharmacol 13:25–40
Christensen AV, Arnt J, Hyttel J, Larsen J-J, Svendsen O (1984) Pharmacological effects of a specific dopamine D_1 antagonist SCH23390 in comparison with neuroleptics. Life Sci 34:1529–1540
Clark PBS, Jakubovic A, Fibiger HC (1988) Anatomical analysis of the involvement of mesolimbocortical dopamine in the locomotor stimulant actions of *d*-amphetamine and apomorphine. Psychopharmacology (Berl) 96:511–520
Clarke D, White FJ (1987) Review: D_1 dopamine receptor – the search for a function: a critical evaluation of the D_1/D_2 dopamine receptor classification and its functional implications. Synapse 1:347–388
Conley R, Tamminga C, An Nguyen J, Hain R (1993) Remoxipride therapy in treatment resistant schizophrenia. Schizophr Res 9:235–236
Cook L, Davidson AB (1978) Behavioral pharmacology: animal models involving aversive control of behavior. In: Lipton A, Dimascio A, Killam K (eds) Psychopharmacology, a generation of progress. Raven, New York, pp 563–567
Cools AR (1986) Mesolimbic dopamine and its control of locomotor activity in rats: differences in pharmacology and light/dark periodicity between the olfactory tubercle and the nucleus accumbens. Psychopharmacology 88:451–459
Cools A, Prinssen E, Ellenbroek B, Heeren D (1994) Role of olfactory tubercle and nucleus accumbens in the effects of classical and atypical neuroleptics: search for regional specificity. In: Polama T, Seiden L, Archer T (eds) Strategies for studying CNS active compounds. Farrand, London, pp 33–53
Cooper SJ, Dourish CT (1990) An introduction to the concept of stereotypy and a historical perspective on the role of brain dopamine. In: Cooper SJ, Dourish CT (eds) Neurobiology of stereotyped behaviour. Clarendon, Oxford, pp 1–24
Costall B, Fortune DH, Naylor RJ, Nohria V (1980) The mesolimbic system, denervation and the climbing response in the mouse. Eur J Pharmacol 66:207–215
Costall B, Domeney AM, Naylor RJ (1983) Stimulation of rat spontaneous locomotion by low doses of haloperidol and (–)-sulpiride: importance of animal selection and measurement technique. Eur J Pharmacol 90:307–314
Coward DM (1991) Pharmacological approaches to the development of atypical antipsychotics. In: Tamminga CA, Schulz SC (eds) Schizophrenia research. Raven,

New York, pp 297–305 (Advances in neuropsychiatry and psychopharmacology, vol 1)
Coward DM, Imperato A, Urwyler S, White TG (1989) Biochemical and behavioural properties of clozapine. Psychopharmacology 99:S6–S12
Coward DM, Dixon AK, Urwyler S, White TG, Enz A, Karobath M, Shearman G (1990) Partial dopamine-agonistic and atypical neuroleptic properties of the amino-ergolines SDZ 208-911 and SDZ 208-912. J Pharmacol Exp Ther 252:279–285
Creese I, Sibley DR, Hamblin MW and Leff SE (1983) The classification of dopamine receptors: relationships to radioligand binding. Annu Rev Neuroscience 6:43–71
Crow TJ, Deakin JFW, Longden A (1977) The nucleus accumbens – possible site of antipsychotic action of neuroleptic drugs? Psychol Med 7:213–221
Deutch AY, Moghaddam B, Innis RB, Krystal JH, Aghajanian GK, Bunney BS, Charney DS (1991) Mechanisms of action of atypical antipsychotic drugs. Implications for novel therapeutic strategies for schizophrenia. Schizophr Res 4:121–156
Deutch AY, Lee MC, Iadarola MJ (1992) Regionally specific effects of the atypical antipsychotic drugs on striatal Fos expression: the nucleus accumbens shell as a locus of antipsychotic action. Mol Cell Neurosci 3:332–341
DiChiara G, Porceddu ML, Vargiu L, Argiolas A, Gessa GL (1976) Evidence for dopamine receptors mediating sedation in the mouse brain. Nature 264:564–565
Domjan M, Burkhard B (1982) The principles of learning and behavior. Brooks/Cole, Monterey
Ellenbroek BA, Cools AR (1990) Animal models with construct validity for schizophrenia. Behav Pharmacol 1:469–490
Ellenbroek B, Schwarz M, Sontag K-H, Jaspers R, Cools A (1985) Muscular rigidity and delineation of a dopamine-specific neostriatal subregion: tonic EMG activity in rats. Brain Res 345:132–140
Ellenbroek B, Peeters B, Honig W, Cools A (1987) The paw test: A behavioural paradigm for differentiating between classical and atypical neuroleptic drugs. Psychopharmacology 93:343–348
Ellenbroek BA, Artz MT, Cools AR (1991) The involvement of dopamine D_1 and D_2 receptors in the effects of the classical neuroleptic haloperidol and the atypical neuroleptic clozapine. Eur J Pharmacol 196:103–108
Ellenbroek BA, Prinssen EPM, Cools AR (1994) The role of serotonin receptor subtypes in the behavioural effects of neuroleptic drugs. A paw test study in rats. Eur J Neurosci 6:1–8
Ernst AM (1967) Mode of action of apomorphine and dexamphetamine in gnawing compulsion in rats. Psychopharmacologia 10:316–323
Farde L, Nordström A-L, Wiesel F-A, Pauli S, Halldin C, Sedvall G (1992) Positron emission tomographic analysis of central D_1 and D_2 dopamine receptor occupancy in patients treated with classical neuroleptics and clozapine. Arch Gen Psychiatry 49:538–544
Ferré S, Guix T, Prat G, Jane F, Casas M (1990) Is experimental catalepsy properly measured? Pharmacol Biochem Behav 35:735–757
Fink JS, Smith GP (1980) Mesolimbic and mesocortical dopaminergic neurons are necessary for normal exploratory behavior in rats. Neurosci Lett 17:61–65
Fletcher GH, Starr MS (1988) Intracerebral SCH 23390 and catalepsy in the rat. Eur J Pharmacol 149:175–178
Fray PJ, Sahakian BJ, Robbins TW, Koob GF, Iversen SD (1980) An observational method for quantifying the behavioural effects of dopamine agonists: contrasting effects of *d*-amphetamine and apomorphine. Psychopharmacology 69:253–259
French ED, Ferkany J, Abreu M, Levenson S (1991) Effects of competitive *N*-methyl-D-aspartate antagonists on midbrain dopamine neurons: an electrophysiological and behavioural comparison to phencyclidine. Neuropharmacology 30:1039–1046
Gerfen CR (1992) The neostriatal mosaic: multiple levels of compartmental organization in the basal ganglia. Annu Rev Neurosci 15:285–320

Gerhardt S, Gerber R, Liebman JM (1985) SCH23390 dissociated from conventional neuroleptics in apomorphine climbing and primate acute dyskinesia models. Life Sci 37:2355–2363

Gerlach (1991) New antipsychotics: classification, efficacy, and adverse effects. Schizophr Bull 17(2):289–309

Graybiel AM (1990) Neurotransmitters and neuromodulators in the basal ganglia. Trends Neurosci 13:244–254

Gustafsson B, Christensson E (1990) Amperozide – a new putatively antipsychotic drug with a limbic mode of action on dopamine-mediated behaviour. Pharmacol Toxicol 66 [Suppl 1]:12–17

Harro J (1993) Measurement of exploratory behavior in rodents. In: Conn PM (ed) Methods in neurosciences, vol 14. Academic, San Diego, pp 359–377

Hartgraves SL, Kelly PH (1984) Role of mesencephalic reticular formation in cholinergic-induced catalepsy and anticholinergic reversal of neuroleptic-induced catalepsy. Brain Res 307:47–54

Hicks PB (1990) The effect of serotonergic agents on haloperidol-induced catalepsy. Life Sci 47:1609–1615

Hjorth S, Carlsson A, Clark D, Svensson K, Wikström H, Sanchez D, Lindberg P, Hacksell U, Arvidsson L-E, Johansson A, Nilsson JLG (1983) Central dopamine receptor agonist and antagonist actions of the enantiomers of 3-PPP Psychopharmacology 81:89–99

Hoffman DC, Beninger RJ (1985) The D_1 dopamine receptor antagonist, SCH23390, reduces locomotor activity and rearing in rats. Pharmacol Biochem Behav 22:341–342

Howard JL, Pollard GT, Craft RM, Rohrbach KW (1987) Metoclopramide potentiates d-amphetamine-induced hypermotility and stereotypy in rat. Pharmacol Biochem Behav 27:165–169

Hyttel J (1983) SCH23390–the first selective dopamine D_1 antagonist. Eur J Pharmacol 91:153–154

Invernizzi RW, Cervo L, Saminin R (1988) 8-hydroxy-2-(di-n-propylamino) tetralin, a selective serotonin 1A receptor agonist, blocks haloperidol-induced catalepsy by an action on raphe nuclei medianus and dorsalis. Neuropharmacology 27:515–518

Iorio LC, Barnett A, Leitz FH, Houser VP, Korduba CA (1983) SCH23390, a potential benzazepine antipsychotic with unique interactions on dopamine systems. J Pharmacol Exp Ther 226:462–468

Iorio LC, Cohen M, Coffin VL (1991) Anticholinergic drugs potentiate dopamine D_1 but not D_2 antagonists on a conditioned avoidance task in rats. J Pharmacol Exp Ther 258(1):118–123

Iversen SD (1986) Animal models of schizophrenia. In: Bradley PB, Hirsch SR (eds) The psychopharmacology and treatment of schizophrenia. Oxford University Press, Oxford, pp 71–102

Iversen SD, Koob GF (1977) Behavioral implications of dopaminergic neurons in the mesolimbic system. In: Costa E, Gessa GL (eds) Nonstriatal dopaminergic neurons. Adv Biochem Psychopharmacol 16:209–214

Jackson DM, Johansson C, Lindgren L-M, Bengtsson A (1994a) Dopamine receptor antagonists block amphetamine and phencyclidine-induced motor stimulation in rats. Pharmacol Biochem Behav 48:465–471

Jackson DM, Ryan C, Evenden J, Mohell N (1994b) Preclinical findings with new antipsychotic agents: what makes them atypical? Acta Psychiatr Scand 89 [Suppl 380]:41–48

Janssen PAJ, Niemegeers CJE, Schellekens KHL (1965) Is it possible to predict the clinical effects of neuroleptic drugs (major tranquilizers) from animal data? Part I: "neuroleptic activity spectra" for rats. Arzneimittelforschung 15:104–117

Janssen P, Niemegeers C, Schellekens K, Lenaerts F (1967) Is it possible to predict the clinical effects of neuroleptic drugs (major tranquillizers) from animal data? Part IV. An improved experimental design for measuring the inhibitory effects of

neuroleptic drugs on amphetamine-induced "chewing" and "agitation" in rats. Arzneimittelforschung 17:841–854
Janssen PAJ, Niemegeers CJE, Awouters F, Schellekens KHL, Megens AAHP, Meert TF (1988) Pharmacology of risperidone (R 64766), a new antipsychotic with serotonin-S_2 and dopamine-D_2 antagonistic properties. J Pharmacol Exp Ther 244:685–693
Javitt DC, Zukin SR (1991) Recent advances in the phencyclidine model of schizophrenia. Am J Psychiatry 148:1301–1308
Jenner P, Marsden CD (1983) Neuroleptics. In: Grahame-Smith DG, Cowen PJ (eds) Preclinical psychopharmacology. Excerpta Medica, Amsterdam, pp 180–247 (Psychopharmacology, vol 1/1)
Johansson C, Jackson DM, Svensson L (1994) The atypical antipsychotic, remoxipride, blocks phencyclidine-induced disruption of prepulse inhibition in the rat. Psychopharmacology 16:437–442
Jones GH, Robbins TW (1992) Differential effects of mesocortical, mesolimbic and mesostriatal dopamine depletion on spontaneous, conditioned and drug-induced locomotor activity. Pharmacol Biochem Behav 43:887–895
Joyce JN (1983) Multiple dopamine receptors and behavior. Neurosci Biobehav Rev 7:227–256
Kaiser C, Setler P (1981) Antipsychotic agents. In: Wolff M (ed) Burger's medicinal chemistry. Wiley, New York, pp 859–980
Kane J, Honigfeld G, Singer J, Meltzer H (1988) Clozapine for the treatment-resistant schizophrenic. Arch Gen Psychiatry 45:789–796
Karlsson P, Smith L, Farde L, Härnryd C, Sedvall G, Wiesel F-A (1995) Lack of apparent antipsychotic effect of the D_1-dopamine receptor antagonist SCH39166 in acutely ill schizophrenic patients. Psychopharmacology 121:309–316
Kebabian JW, Calne DB (1979) Multiple receptor mechanisms for dopamine. Nature 277:93–96
Kelley AE, Lang CG, Gauthier AM (1988) Induction of oral stereotypy following amphetamine microinjection into a discrete subregion of the striatum. Psychopharmacology 95:556–559
Kelly PH, Seviour PW, Iversen SD (1975) Amphetamine and apomorphine responses in the rat following 6-OHDA lesions of the nucleus accumbens septi and corpus striatum. Brain Res 94:507–522
Kim JS, Kornhuber HH (1982) The glutamate theory in schizophrenia: clinical and experimental evidence. In: Namba M, Kaiya H (eds) Psychobiology of schizophrenia. Pergamon, Oxford, pp 221–234
King DJ, Devaney N, Cooper SJ, Blomqvist M, Mitchell MJ (1990) Pharmacokinetics and antipsychotic effect of remoxipride in chronic schizophrenic patients. J Psychopharmacol 4:83–89
Koek W, Colpaert FC (1993) Inhibition of methylphenidate-induced behaviors in rats: differences among neuroleptics. J Pharmacol Exp Ther 267:181–191
Köhler C, Haglund L, Ögren SO, Ängeby T (1981) Regional blockade by neuroleptic drugs of in vivo 3H-spiperone binding in the rat brain. Relation to blockade of apomorphine-induced hyperactivity and stereotypies. J Neural Transm 52:163–173
Koob GF, Hervé S, Herman JP, Le Moal M (1984) Neuroleptic-like disruption of the conditioned avoidance response requires destruction of both the mesolimbic and nigrostriatal dopamine systems. Brain Res 303:319–329
Kreiskott H (1980) Behavioral pharmacology of antipsychotics. In: Hoffmeister F, Stille G (eds) Antipsychotics and antidepressants. Springer, Berlin Heidelberg New York, pp 59–88 (Handbook of experimental pharmacology, vol 55/3)
Lecrubier Y, Puech AJ, Simon P, Widlocher D (1980) Schizophrenic: hyper-ou hypofunctionnement du système dopaminergique? Une hypothèse bipolaire. Psychol Méd 12:2431–2441

Lewander T, Westerbergh SE, Morrison D (1990) Clinical profile of remoxipride – a combined analysis of a comparative double-blind multicentre trial programme. Acta Psychiatr Scand 82 [Suppl 358]:92–98

Lieberman JA, Kane JM, Alvir J (1987) Provocative tests with psychostimulant drugs in schizophrenia. Psychopharmacology 91:415–433

Ljungberg T, Ungerstedt U (1978) Classification of neuroleptic drugs according to their ability to inhibit apomorphine-induced locomotion and gnawing: evidence for two different mechanisms of action. Psychopharmacology 56:239–247

Ljungberg T, Ungerstedt U (1985) A rapid and simple behavioural screening method for simultaneous assessment of limbic and striatal blocking effects of neuroleptic drugs. Pharmacol Biochem Behav 23:479–485

Lowe JA, Seeger TF, Vineck FJ (1988) Atypical antipsychotics – recent findings and new perspectives. Med Res Rev 8:475–497

Lubow R, Weiner I, Schlossberg A, Baruch I (1987) Latent inhibition and schizophrenia. Bull Psych Soc 25:464–467

Mansbach R, Geyer M (1989) Effects of phencyclidine and phencyclidine analogs on sensorimotor gating in the rat. Neuropsychopharmacology 2:299–308

McKinney WT (1989) Animal models of schizophrenic disorders. In: Schulz SC, Tamminga CA (eds) Schizophrenia: scientific Progress. Oxford University Press, New York, pp 141–154

Megens AAHP, Awouters FHL, Meert TF, Schellekens KHL, Niemegeers CJE, Janssen PAJ (1992) Pharmacological profile of the new potent neuroleptic ocaperidone (R 79 598). J Pharmacol Exp Ther 260: 146–159

Meller E, Kuga S, Friedhoff AJ, Goldstein M (1985) Selective D_2 dopamine receptor agonists prevent catalepsy induced by SCH23390, a selective D_1 antagonist. Life Sci 36:1857–1864

Meltzer HY (1992) The mechanism of action of clozapine in relation to its clinical advantages. In: Melzer HY (ed) Novel neuroleptic drugs. Raven, New York, pp 1–13

Meltzer HY, Stahl SM (1976) The dopamine hypothesis of schizophrenia. A review. Schizophr Bull 2:19–76

Meltzer HY, Matsubara S, Lee J-C (1989) Classification of typical and atypical antipsychotic drugs on the basis of dopamine D_1, D_2 and serotonin$_2$ pK_i values. J Pharmacol Exp Ther 251:238–246

Moore NA, Axton MS (1988) Production of climbing behaviour in mice requires both D1 and D2 receptor activation. Psychopharmacology 94:263–266

Moore NA, Tye NC, Axton MS, Risius FC (1992) The bahavioral pharmacology of olanzapine, a novel "atypical" antipsychotic agent. J Pharmacol Exp Ther 262:545–551

Moore NC, Gershon S (1989) Which atypical antipsychotics are identified by screening tests? Clin Neuropharmacol 12:167–184

Morelli M, DiChiara G (1985) Catalepsy induced by SCH23390 in rats. Eur J Pharmacol 117:179–185

Morpugo C, Theobald W (1964) Influence of antiparkinson drugs and amphetamine on some pharmacological effects of phenothiazine derivatives used as neuroleptics. Psychopharmacologia 6:178–191

Nielsen EB, Bondo Hansen J, Grønvald FC, Swedberg DBM, Scheideler M (1996) NNC 19-1228 and 22-0031, novel neuroleptics with a "mesolimbic-selective" behavioral profile. Psychopharmacology (submitted)

Niemegeers CJE, Janssen PAJ (1979) A systematic study of the pharmacological activities of dopamine antagonists. Life Sci 24:2201–2216

Nishibe Y, Matsuo Y, Yoshizaki T, Eigyo M, Shiomi T, Hirose K (1982) Differential effects of sulpiride and metoclopramide on brain homovanillic acid levels and shuttle box avoidance after systemic and intracerebral administration. Naunyn Schmiedebergs Arch Pharmacol 321:190–194

Ögren SO, Archer T (1994) Effects of typical and atypical antipsychotic drugs on two-way active avoidance. Relationship to DA receptor blocking profile. Psychopharmacology 114:383–391
Ögren SO, Ängeby-Möller K (1996a) Effects of remoxipride and different types of dopamine receptor antagonists on spontaneous and *d*-amphetamine-induced changes in locomotor activity in the rat. J Neural Transm (submitted)
Ögren SO, Ängeby-Möller K (1996b) Clozapine and remoxipride: comparison of their atypical antipsychotic profile. J Pharmacol Exp Ther (submitted)
Ögren SO, Högberg T (1988) Novel dopamine D_2 antagonists for the treatment of schizophrenia. ISI atlas of science. Pharmacology 2(2):141–147
Ögren SO, Fuxe K (1988) D_1 and D_2 receptor antagonists induce catalepsy via different efferent striatal pathways. Neurosci Lett 85:333–338
Ögren SO, Goldstein M (1994) Phencyclidine- and dizocilpine-induced hyperlocomotion are differentially mediated. Neuropsychopharmacology 11:167–177
Ögren SO, Köhler C, Fuxe K, Ängeby K (1979) Behavioural effects of ergot drugs. In: Fuxe K, Calne DB (eds) Dopamine ergot derivatives and motor function. Pergamon, Oxford, pp 187–205
Ögren SO, Hall H, Köhler C, Magnusson O, Lindbom LO, Ängeby K, Florvall L (1984) Remoxipride, a new potential antipsychotic compound with selective antidopaminergic actions in the rat brain. Eur J Pharmacol 102:459–474
Ögren SO, Hall H, Köhler C, Magnusson O, Sjöstrand SE (1986) The selective dopamine D_2 receptor antagonist raclopride discriminates between dopamine-mediated motor functions. Psychopharmacology 90:287–294
Ögren SO, Florvall L, Hall H, Magnusson O, Ängeby-Möller K (1990) Neuropharmacological and behavioural properties or remoxipride in the rat. Acta Psychiatr Scand 82:21–27
Ögren SO, Rosén L, Fuxe K (1994) The dopamine D_2 antagonist remoxipride acts in vivo on a subpopulation of functionally coupled DA D_2 receptors. Neuroscience 61(2):269–283
Ögren SO, Thyréen G, Lindeberg A, Amkéus E (1996a) Remoxipride and other dopamine receptor antagonists in two tests of catalepsy in the rat. Differential effects on dopamine receptor functions. Psychopharmacology (submitted)
Ongini E, Longo VG (1989) Dopamine receptor subtypes and arousal. Int Rev Neurobiol 31:239–255
Ongini E, Bo P, Dionisotti S, Trampus M, Savoldi F (1992) Effects of remoxipride, a dopamine D_2 antagonist antipsychotic, on sleep-waking patterns and EEG activity in rats and rabbits. Psychopharmacology 107:236–242
Ossowska K, Karcz M, Wardas J, Wolfarth S (1990) Striatal and nucleus accumbens D_1/D_2 dopamine receptors in neuroleptic catalepsy. Eur J Pharmacol 182:327–334
Parashos S, Marin C, Chase T (1989) Synergy between a selective D_1 antagonist and a selective D_2 antagonist in the induction of catalepsy. Neurosci Lett 105:169–173
Petry N, Furmidge L, Tong Z-Y, Martin C, Clark D (1993) Time sampling observation procedure for studying drug effects: interaction between *d*-amphetamine and selective dopamine receptor antagonists in the rat. Pharmacol Biochem Behav 44:167–180
Petty F, Mott J, Sherman AD (1984) Potential locus and mechanism of blockade of conditioned avoidance responding by neuroleptics. Neuropharmacology 23(1):73–78
Protais P, Costentin J, Schwartz JC (1976) Climbing behavior induced by apomorphine in mice – a simple test for the study of dopamine receptors in striatum. Psychopharmacology (Berlin) 50:1–6
Puech AJ, Rioux P, Poncelet M, Brochet D, Chermat R, Simon P (1981) Pharmacological properties of new antipsychotic agents: use of animal models. Neuropharmacology 20:1279–1284
Randrup A, Munkvad I (1967) Stereotyped activities produced by amphetamine in several animal species and man. Psychopharmacologia 11:300–310

Rebec GV, Bashore TR (1984) Critical issues in assessing the behavioral effects of amphetamine. Neurosci Biobehav Rev 8:153–159

Rescorla RA, Solomon RL (1967) Two-process learning theory: relationships between Pavlovian conditioning and instrumental learning. Psychol Rev 74:151–182

Robbins TW (1977) A critique of the methods available for the measurement of spontaneous motor activity. In: Iversen LL, Iversen SD, Snyder SD (eds) Handbook of psychopharmacology, vol 7. Plenum, New York, pp 37–81

Robbins TW, Sahakian BJ (1981) Behavioural and neurochemical determinants of drug-induced stereotypy. In: Clifford F (ed) Metabolic disorders of the nervous system. Pitman, pp 244–291

Robertson A, MacDonald C (1984) Atypical neuroleptics clozapine and thioridazine enhance amphetamine-induced stereotypy. Pharmacol Biochem Behav 21:97–101

Robertson A, MacDonald C (1985) Opposite effects of sulpiride and metoclopramide on amphetamine-induced stereotypy. Eur J Pharmacol 109:81–89

Ross SB, Jackson DM, Edwards SR (1989) The involvement of dopamine D_1 and D_2 receptors in the locomotor stimulation produced by (+)-amphetamine in naive and dopamine-depleted mice. Pharmacol Toxicol 64:72–77

Sanberg PR, Giordano M, Bunsey MD, Norman AB (1988) The catalepsy test: its ups and downs. Behav Neurosci 102:748–759

Sánchez C, Arnt J, Dragsted N, Hyttel J, Lembol HL, Meier E, Perregard, Skarsfeldt T (1991) Neurochemical and in vivo pharmacological profile of sertindole, a limbic-selective neuroleptic compound. Drug Dev Res 22:239–250

Scatton B, Perrault G, Sanger DJ, Schoemaker H, Carter C, Fage D, Gonon F, Chergui K, Cudennec A, Benavides J (1994) Pharmacological profile of amisulpride, an atypical neuroleptic which preferentially blocks presynaptic D_2/D_3 receptors. Neuropsychopharmacology 10:2423

Schaefer GJ, Michael RP (1984) Drug interactions on spontaneous locomotor activity in rats: neuroleptics and amphetamine-induced hyperactivity. Neuropharmacology 23:909–914

Scheel-Krüger J (1986) The syndrome of sedation and yawning behaviour in the rat is dependent on postsynaptic dopamine D_2 receptors. Psychopharmacology 89:S32

Scheel-Krüger J, Willner P (1991) The mesolimbic system: principles of operation. In: Willner P, Scheel-Krüger J (eds) The mesolimbic dopamine system: from motivation to action. Wiley, Chichester, pp 559–597

Schwartz J-C, Delandre M, Martres MP, Sokoloff P, Protais P, Vasse M, Costentin J, Laibe P, Mann A, Wermuth CG, Gulat C, Laffite A (1984) Biochemical and behavioral identification of discriminant benzamide derivatives: new tools to differentiate subclasses of dopamine receptors. In: Usdin E, Carlsson A, Dahström A, Engel J (eds) Catecholamines. Part B: neuropharmacology and central nervous system – theoretical aspects. Liss, New York, pp 59–72

Seeman P (1987) Dopamine receptors and the dopamine hypothesis of schizophrenia. Synapse 1:133–152

Seeman P (1992) Dopamine receptor sequences. Therapeutic levels of neuroleptics occupy D_2 receptors, clozapine occupies D4. Neuropsychopharmacology 7:261–284

Sharp T, Zetterström T, Ljungberg T, Ungerstedt U (1987) A direct comparison of amphetamine- induced behaviours and regional brain dopamine release in the rat using intracerebral dialysis. Brain Res 401:322–330

Spealman R, Kelleher R, Goldberg S-O, DeWeese J, Goldberg D (1983) Behavioral effects of clozapine: comparison with thioridazine, chlorpromazine, haloperidol and chlordiazepoxide in squirrel monkeys. J Pharmacol Exp Ther 224:127–134

Ståhle L (1992) Do autoreceptors mediate dopamine agonist-induced yawning and suppression of exploration? A critical review. Psychopharmacology 106:1–13

Ståhle L, Ljungberg T, Rodebjer A, Ögren SO, Ungerstedt U (1987) Differential effects of the dopamine antagonist remoxipride on apomorphine induced behavidor in the rat. Pharmacol Toxicol 60:227–232

Staton DM, Solomon PR (1984) Microinjections of *d*-amphetamine into the nucleus accumbens and caudate-putamen differentially affect stereotypy and locomotion in the rat. Physiol Psychrl 12:159–162

Stille G, Hippius H (1971) Kritische Stellungnahme zum Begriff der Neuroleptika (anhand von pharmakologischen und klinischen Befunden mit Clozapin). Pharmakopsychiatry 4:182–191

Strömbom U (1976) Catecholamine receptor agonists: effects on motor activity and tyrosine hydroxylation in mouse brain. Naunyn Schmiedebergs Arch Pharmacol 292:167–176

Svensson L, Ahlenius S (1983) Suppression of exploratory locomotor activity in the rat by the local application of 3-PPP enantiomers into the nucleus accumbens. Eur J Pharmacol 88:393–397

Tabaoda ME, Souto M, Hawkins H, Monti JM (1979) The actions of dopaminergic and noradrenergic antagonists on conditioned avoidance responses in intact and 6-hydroxydopamine-treated rats. Psychopharmacology 62:83–88

Tschanz JT, Rebec GV (1989) Atypical antipsychotic drugs block selective components of amphetamine-induced stereotypy. Pharmacol Biochem Behav 31:519–522

van den Boss R, Cools AR, Ögren SO (1988) Differential effects of the selective D_2-antagonist raclopride in the nucleus accumbens of the rat on spontaneous and *d*-amphetamine-induced activity. Psychopharmacology 95:447–451

van Rossum JM (1966) The significance of dopamine-receptor blockade for the mechanism of action of neuroleptic drugs. Arch Int Pharmacodyn Ther 160:492–494

Vasse M, Protais P, Costentin J, Schwartz J (1985) Unexpected potentiation by discriminant benzamide derivatives of stereotyped behaviours elicited by dopamine agonists in mice. Naunyn Schmiedebergs Arch Pharmacol 329:108–116

Wadenberg M (1992) Antagonism by 8-OH-DPAT, but not ritanserin, of catalepsy induced by SCH23390 in the rat. J Neural Transm 89:49–59

Wadenberg M-L, Ericson EL, Magnusson O, Ahlenius S (1990) Suppression of conditioned avoidance behavior by the local application of (–) sulpiride into the ventral, but not the dorsal striatum of the rat. Biol Psychiatry 28:297–307

Waddington JL (1988) Therapeutic potential of selective D_1 dopamine receptor agonists and antagonists psychiatry and neurology. Gen Pharmacol 19:55–60

Waddington JL, Daly SA (1993) Regulation of unconditioned motor behaviour by $D_1:D_2$ interactions. In: Waddington JL (ed) $D_1:D_2$ Dopamine receptor interactions. Academic, London, pp 51–78

Walters JR, Bergstrom DA, Carlson JH, Chase TN, Braun AR (1987) D_1 dopamine receptor activation required for postsynaptic expression of D_2 agonist effects. Science 236:719–722

Wambebe C (1987) Influence of (–)-sulpiride and YM-09151-2 on stereotyped behavior in chicks and catalepsy in rats. Jpn J Pharmacol 43:121–128

Wanibuchi F, Usuda S (1990) Synergistic effects between D_1 and D_2 dopamine antagonists on catalepsy in rats. Psychopharmacology 102:339–342

Waters N, Pettersson G, Carlsson A, Svensson K (1989) The putatively antipsychotic agent amperozide produces behavioural stimulation in the rat. A behavioural and biochemical characterization. Neunyn Schmiedebergs Arch Pharmacol 340:161–169

Waters N, Svensson K, Haadsma-Svensson SR, Smith MW, Carlsson A (1993) The dopamine D_3-receptor: a postsynaptic receptor inhibitory on rat locomotor activity. J Neural Transm [GenSect] 94:11–19

Willets J, Balster RL, Leander D (1990) The behavioral pharmacology of NMDA receptor antagonists. Trends Pharmacol Sci 11:423–428

Zivkovic B, Worms P, Scatton B, Dedek J, Oblin A, Lloyd KG, Bartholini G (1983) Functional similarities between benzamides and other neuroleptics. In: Costa E, Biggio G (eds) Advances in biochemical psychopharmacology. Raven, New York, pp 155–170

CHAPTER 9

Antipsychotic Drug Action After Lesions to the Hippocampus or Frontal Cortex

MARK E. BARDGETT and JOHN G. CSERNANSKY

A. Introduction

Numerous studies have attempted to define the biological mechanisms of antipsychotic drug action. After 40 years of research, the effectiveness of antipsychotic drugs, as well as their side effects, continues to be interpreted in the context of the neurotransmitter dopamine. While antipsychotic action on central dopaminergic pathways remains a focus of preclinical research, advances in functional neuroanatomy have widened the scope of investigation. For instance, dopaminergic pathways are anatomically and functionally integrated with neuronal processes emanating from the hippocampus and frontal cortex. These brain regions can exhibit gross neuropathology in some individuals with schizophrenia or Alzheimer's disease. Although antipsychotics are commonly prescribed for the symptoms of both disorders, little empirical attention has addressed the possibility that hippocampal or prefrontal cortical neuropathology may qualitatively alter normative antipsychotic drug action in the brain.

Animal studies have begun to examine the impact of site-selective neuroanatomical disturbances on antipsychotic drug action. Many of these studies have employed lesions to the hippocampus or frontal cortex because of their link to the neuropathology of schizophrenia or Alzheimer's disease. This chapter will review and integrate studies of hippocampal or frontal cortical lesions relevant to antipsychotic drug action. First, several reliable neurochemical and behavioral correlates of antipsychotic drug action will be identified. Then, these correlates will be used as a framework to examine how hippocampal or frontal cortical lesions affect antipsychotic drug action.

B. Mechanisms of Antipsychotic Drug Action

I. Dopaminergic Mechanisms

Despite the relevance of other neurotransmitters and brain regions to antipsychotic pharmacology, central dopamine pathways are still regarded as the primary sites of antipsychotic drug action. The effects of antipsychotic drugs on these pathways are not homogeneous; various drugs can have different effects on dopamine receptor subtypes and anatomically distinct dopamine pathways. These receptor- and site-specific actions are manifested by differential changes in dopamine turnover or release, firing rates of dopaminergic neurons, and dopamine-mediated behaviors (see Table 1 for a summary).

Antipsychotic efficacy has been linked to the response of the mesolimbic dopaminergic pathway to drug treatment. Neurons in the mesolimbic pathway originate in the ventral tegmental area (region A10) and terminate in the nucleus accumbens, olfactory tubercle, septum, amygdala, and prefrontal, cingulate, and piriform cortices (BJORKLUND and LINDVALL 1984; COOPER et al. 1986). The extrapyramidal side effects and tardive dyskinesia associated with antipsychotic drugs have been linked to the response of the nigrostriatal dopaminergic pathway to drug treatment. Neurons in this pathway emanate from the substantia nigra (region A9) and terminate predominantly in the caudate nucleus in a topographically organized manner (BJORKLUND and LINDVALL 1984).

Dopamine receptors in these pathways serve as the principal receptor sites for most antipsychotic drugs. In particular, most antipsychotics bind with high affinity to the D_2 receptor, and several aspects of D_2 receptor binding are associated with antipsychotic drug action. The affinity of most antipsychotics

Table 1. Dopaminergic mechanisms of antipsychotic drug action

Dopamine anatomy
Antipsychotic drug action in mesolimbic pathways is related to clinical efficacy
Antipsychotic drug action in nigrostriatal pathways is related to clinical side effects
Dopamine receptors
Affinity of most antipsychotics for D_2 receptor correlates with clinically effective dose
Antipsychotic efficacy is achieved with 65%–80% D_2 receptor occupancy in vivo
Antipsychotic-induced extrapyramidal symptoms achieved with more than 80% D_2 receptor occupancy
Upregulation of D_2 receptors occurs after long-term antipsychotic administration
Dopamine turnover
Acute antipsychotic administration increases dopamine turnover; partial tolerance to this effect is observed with long-term antipsychotic administration
Dopamine release
Acute antipsychotic administration increases dopamine release; tolerance to this effect is observed after long-term treatment
Midbrain dopamine neuronal firing
Acute antipsychotic treatment increases firing rate of A9 and A10 neurons
Chronic antipsychotic treatment inhibits spontaneous firing rate of A9 and A10; the effect of atypical antipsychotics is limited to A10
Dopamine-mediated behaviors
Dopamine agonist-induced hyperlocomotion and stereotypy are equaly sensitive to inhibition by single dose of typical antipsychotics; stereotypy induced by dopamine agonists is less sensitive to atypical antipsychotics
Antipsychotic dose required to reverse apomorphine-induced deficits in prepulse inhibition correlates with clinically effective dose

for the D_2 receptor can be correlated with their clinically effective dose (SEEMAN 1992), although the atypical antipsychotic clozapine is a notable exception. In addition, positron emission tomography (PET) imaging studies have indicated that antipsychotic efficacy is achieved when 65%–80% of striatal D_2 receptors are occupied (FARDE et al. 1988, 1992; NORDSTROM et al. 1993), with clozapine again proving the exception to the rule. The appearance of extrapyramidal side effects has been observed when more than 80% of striatal D_2 receptors are occupied by antipsychotic drugs (FARDE et al. 1992; NORDSTROM et al. 1993). Finally, antipsychotic drugs upregulate D_2 receptor density in the caudate nucleus, the nucleus accumbens, and the medial prefrontal cortex (O'DELL et al. 1990; JANOWSKY et al. 1992; MESHUL et al. 1992). A site-specific effect has been observed after long-term clozapine administration, with increased D_2 receptor binding observed only in the prefrontal cortex (O'DELL et al. 1990; JANOWSKY et al. 1992; MESHUL et al. 1992).

The binding of antipsychotics to presynaptic dopamine receptors is associated with elevated dopamine turnover and release. Single injections of haloperidol, clozapine, and putative antipsychotic drugs such as sulpiride, raclopride, remoxipride, and (–) 3-PPP increase dopamine turnover in the caudate putamen, nucleus accumbens, and prefrontal cortex (KOHLER et al. 1990; DEUTCH and CAMERON 1992; CHRAPUSTRA et al. 1993; CSERNANSKY et al. 1993). Striatal dopamine release is increased in vitro after acute haloperidol, clozapine, and risperidone treatment (COMPTOM and JOHNSON 1989; LEYSEN et al. 1988). With repeated administration of antipsychotic drugs, tolerance to the acute elevating effects on dopamine turnover can be reliably observed in the caudate nucleus and amygdala and to a lesser degree in the nucleus accumbens; however, tolerance is not observed in the prefrontal cortex (see REYNOLDS 1992 for a review). Microdialysis studies have indicated that striatal dopamine release is not altered after chronic haloperidol or clozapine administration (CHAI and MELTZER 1992; MOGHADDAM and BUNNEY 1993; YAMAMOTO and COOPERMAN 1994), although basal dopamine levels in the prefrontal cortex are elevated after long-term clozapine treatment (YAMAMOTO and COOPERMAN 1994).

The acute elevations in dopamine turnover and release induced by antipsychotics have been attributed to antipsychotic-induced increases in the spontaneous firing rate of midbrain dopamine neurons. Long-term antipsychotic administration, however, decreases the spontaneous firing rate of roughly 80% of A9 and A10 neurons (WHITE and WANG 1983; see GRACE 1991 for a review). This phenomenon of "depolarization blockade" has two important implications for antipsychotic drug action: (1) the delayed development of depolarization blockade is temporally contiguous with the delayed onset of antipsychotic efficacy, and (2) depolarization blockade is not observed in A9 neurons after long-term administration of atypical antipsychotics, such as clozapine (WHITE and WANG 1983). This latter finding suggests that depolarization blockade of A9 neurons may be involved in antipsychotic-induced extrapyramidal symptoms.

Finally, the dopaminergic actions of antipsychotics can be revealed in several behavioral paradigms. These include alterations in the expression of apomorphine-induced emesis, apomorphine- or amphetamine-induced locomotion and stereotypy, conditioned avoidance, and intracranial self-stimulation (see Hollister and Csernansky 1990 for a review). Quite simply, most antipsychotics inhibit activity in each of these paradigms. Furthermore, amphetamine-induced locomotion and stereotypy have been used to discriminate between typical and atypical antipsychotic drugs. Clinically effective doses of many atypical compounds have been found to inhibit locomotion without altering stereotypy, whereas clinically effective doses of typical drugs often inhibit both behaviors (Ljundberg and Ungerstedt 1978; Ogren et al. 1990). The differential sensitivity of locomotion and stereotypy to atypical drug treatment has been attributed to the respective effects of atypical drugs on mesolimbic and nigrostriatal dopamine systems.

Classical conditioning paradigms have also been used to identify potential antipsychotic drugs (Hollister and Csernansky 1990; Weiner 1990), and the prepulse inhibition paradigm has evolved as a particularly sensitive assay of antipsychotic efficacy (see Swerdlow et al., this volume). Prepulse inhibition is impaired in individuals with schizophrenia and is disrupted by dopamine agonists in rats. Deficits in prepulse inhibition induced by dopamine agonists can be reversed by antipsychotic drugs at doses which correlate with their clinical potency and D_2 receptor affinity (Swerdlow et al. 1994).

In summary, antipsychotic drug action in dopamine pathways has been characterized by numerous approaches (see Table 1 for a summary). The efficacy and side effect liability of most antipsychotics appear to depend on their occupancy of D_2 receptors, with the exception of the atypical antipsychotic clozapine. The binding of antipsychotics to D_2 receptors contributes to their acute elevating effects on dopamine turnover, dopamine release, and midbrain neuronal firing. These biochemical alterations induced by antipsychotics may underlie their activity in behavioral assays, which have served as useful screens for atypical compounds. While the dopamine hypothesis of antipsychotic drug action has provided many mechanistic insights, many investigators have nonetheless turned to other neurotransmitter systems such as glutamate, in order to develop a greater understanding of antipsychotic drug action.

II. Glutamatergic Mechanisms

Two facets of glutamate neurotransmission evince its involvement in antipsychotic drug action (see Table 2 for a summary). First, terminals emanating from glutamate neurons in the cortex, limbic system, and thalamus are found in close proximity to dopaminergic terminals in the accumbens and caudate. Second, drugs which block *N*-methyl-D-aspartate (NMDA)-type glutamate receptors produce psychosis in humans, and their behavioral effects in animals can be reversed by antipsychotics.

Table 2. Glutamatergic mechanisms of antipsychotic drug action

Glutamate metabolism	
	Acute atypical treatment decreases whole-tissue levels of glutamate in accumbens
Glutamate release	
	Acute clozapine treatment increases glutamate release in prefrontal cortex
	Chronic haloperidol treatment increases basal glutamate release
	Chronic clozapine treatment increases stimulated-glutamate release in accumbens
Glutamate-mediated behaviors	
	Antipsychotics attenuate hyperlocomotion, stereotypy, and learning deficits produced by NMDA receptor antagonists
	Antipsychotics attenuate hyperlocomotion produced by glutamate or non-NMDA receptor agonists.

NMDA, *N*-methyl-D-aspartate.

Glutamatergic neurons are the predominant neuronal type in the cortex and limbic system. These neurons project in a topographically ordered manner to the caudate nucleus and the nucleus accumbens (GRAYBIEL et al. 1994). Glutamatergic neurons from the prefrontal cortex and limbic system terminate primarily in the accumbens (SESACK and PICKEL 1990, 1992), while the caudate receives a broad glutamatergic input from the neocortex (GRAYBIEL et al. 1994). Glutamatergic terminals are also found in the ventral tegmental area and substantia nigra (see KALIVAS et al. 1993 for a review), and some of these terminals emanate from the prefrontal cortex (SESACK and PICKEL 1992).

The binding and affinity of antipsychotic drugs for glutamate receptors has not been well characterized. Nonetheless, both haloperidol and clozapine have been reported to displace the binding of the NMDA receptor antagonist and psychotomimetic drug MK-801 from striatal and cortical preparations (LIDSKY et al. 1993), with clozapine displacing MK-801 binding at nanomolar concentrations (ULAS and COTMAN 1993). Long-term haloperidol administration increases NMDA binding sites in the parietal cortex by 20%, without altering NMDA binding in the hippocampus or thalamus (ULAS and COTMAN 1993). Antipsychotic drug treatment does not change the level of non-NMDA glutamate receptor binding in the brain (ULAS and COTMAN 1993).

By binding to presynaptic dopamine receptors on glutamatergic terminals (YANG and MOGENSON 1984, 1986), antipsychotics may alter striatal glutamate metabolism and release. Whole-tissue glutamate levels, which probably reflect glutamate metabolism more so than neurotransmitter pools, are decreased in the accumbens following acute administration of clozapine, but not haloperidol (BARDGETT et al. 1993). In contrast, microdialysis studies, which presumably measure the release of neurotransmitter pools, have demonstrated that acute haloperidol or clozapine administration has no effect on basal extracellular glutamate in the striatum (DALY and MOGHADDAM 1993). However, acute clozapine treatment has been found to increase basal extracellular levels of aspartate and glutamate in the prefrontal cortex (DALY and MOGHADDAM 1993). Long-term haloperidol treatment increases basal extracellular levels of glutamate in the striatum (MOGHADDAM and BUNNEY 1993). YAMAMOTO and COOPERMAN (1994) have also shown that long-term haloperidol treatment increases basal levels of extracellular glutamate in the caudate nucleus and nucleus accumbens, without altering glutamate levels in the prefrontal cortex. In this latter study, long-term clozapine treatment had no effect on basal glutamate levels in any of the examined brain regions, but did enhance potassium-stimulated glutamate release exclusively in the accumbens.

The behavioral hyperactivity elicited by NMDA receptor antagonists has received considerable attention as an animal model of psychosis and as a tool for assaying new antipsychotic compounds (see CARLSSON and CARLSSON 1990 for a review). Systemic injections of haloperidol, clozapine, or the D_2 antagonist eticlopride can block the expression of MK-801-induced locomotion, stereotypy, and learning deficits (BEHRENS and GATTAZ 1992; WILLINS et al. 1993; HAUBER 1993). These results indicate that antipsychotics with a high affinity for D_2 receptors can block the expression of MK-801–induced behaviors. It is possible that the effects of haloperidol are mediated at extrastriatal sites, since direct striatal injections of haloperidol are ineffective in blocking the locomotor effects of MK-801 (CARLSSON and CARLSSON 1990).

Although the glutamatergic mechanisms of antipsychotic drug action (see Table 2 for a summary) may not seem as numerous as the dopaminergic mechanisms, this paucity probably reflects the still-evolving story of normative glutamate neurotransmission. As the regulation of glutamate metabolism and release, glutamate receptor subtype expression, and glutamate-associated behaviors continues to be revealed, it is likely that more glutamatergic mechanisms of antipsychotic drug action will be discovered. At present, the existing data simply suggest that antipsychotics increase glutamate release in the striatum and prefrontal cortex. This elevation in glutamate release may explain the ability of antipsychotics to reverse the behavioral alterations elicited by NMDA antagonists.

C. Neuropathology and Antipsychotic Drug Action

Given the possibility of neuroanatomical disturbances in neuropsychiatric disorders, it has become imperative to define antipsychotic drug action in the context of neuropathology. The remainder of this review will focus on preclinical studies relevant to this issue. The discussion will be limited to the effects of hippocampal and frontal cortical lesions, since many imaging and neuropathological studies of schizophrenia and Alzheimer's disease have focused on these two regions. In addition, numerous animal studies have examined the impact of these two areas on various aspects of basal ganglia function. This limited discussion is not meant to imply that other regionally specific neuroanatomical disturbances do not exist in schizophrenia, Alzheimer's disease, or other disorders or that such disturbances would not affect antipsychotic drug action.

Many of the studies reviewed here have examined the consequences of circumscribed brain lesions for basal ganglia function. Two caveats must be considered in interpreting these data. First, few studies have examined a direct causal relation between hippocampal or frontal lesions and antipsychotic drug action. However, a larger literature exists that has examined the effects of lesions on biological or behavioral parameters which are also affected by antipsychotics (e.g., amphetamine-induced locomotion). These studies are considered in this review, since they may provide important insights into antipsychotic drug action after hippocampal or frontal lesions.

The second caveat pertains to the relevance of animal lesion models to human neuropathology. Complete lesions of an entire brain region are commonly used to define the region's role in a specific neurobiological function or behavior. However, using these lesions to model human neuropathology is problematic, since few clinical disorders involve the complete absence of a single brain region. Such models ignore the possibility that the modest hippocampal or cortical cell loss associated with neuropsychiatric disorders such as schizophrenia could have qualitatively different biochemical and behavioral outcomes in comparison to complete hippocampal destruction. Perhaps the first conclusion one can draw from this review is the need for more relevant animal models of such neuropathology.

D. Effect of Hippocampal Neuropathology on Mechanisms of Antipsychotic Drug Action

I. Anatomy

The hippocampal projection to the nucleus accumbens provides an anatomical substrate for hippocampal modulation of antipsychotic pharmacology (see Fig. 1). Hippocampal terminals account for approximately 10% of all presynaptic terminals and 30% of all asymmetric (i.e., excitatory) terminals in the accumbens (SESACK and PICKEL 1990). These terminals emanate from glutamatergic pyramidal neurons in the subiculum, the CA1 cell field of the hippocampus, and the entorhinal cortex and are located along the rostral–caudal extent of medial shell area of the nucleus accumbens (KELLEY and DOMESICK 1982; PHILLIPSON and GRIFFITHS 1985; FULLER et al. 1987). This medial shell area of the accumbens also contains dopaminergic afferents from the interfascicular nucleus and nucleus paranigralis of the ventral tegmental area (HEIMER et al. 1991). The dopaminergic terminals in the accumbens are located on the dendrites or dendritic spines of medium spiny neurons. Hippocampal terminals can be found on the same dendrites as dopaminergic terminals, and hippocampal axons make axo-axonal contacts with dopaminergic axons (SESACK and PICKEL 1990). While hippocampal terminals form asymmetric (excitatory) synapses on the base of dendritic spine heads, dopaminergic terminals form symmetric (inhibitory)

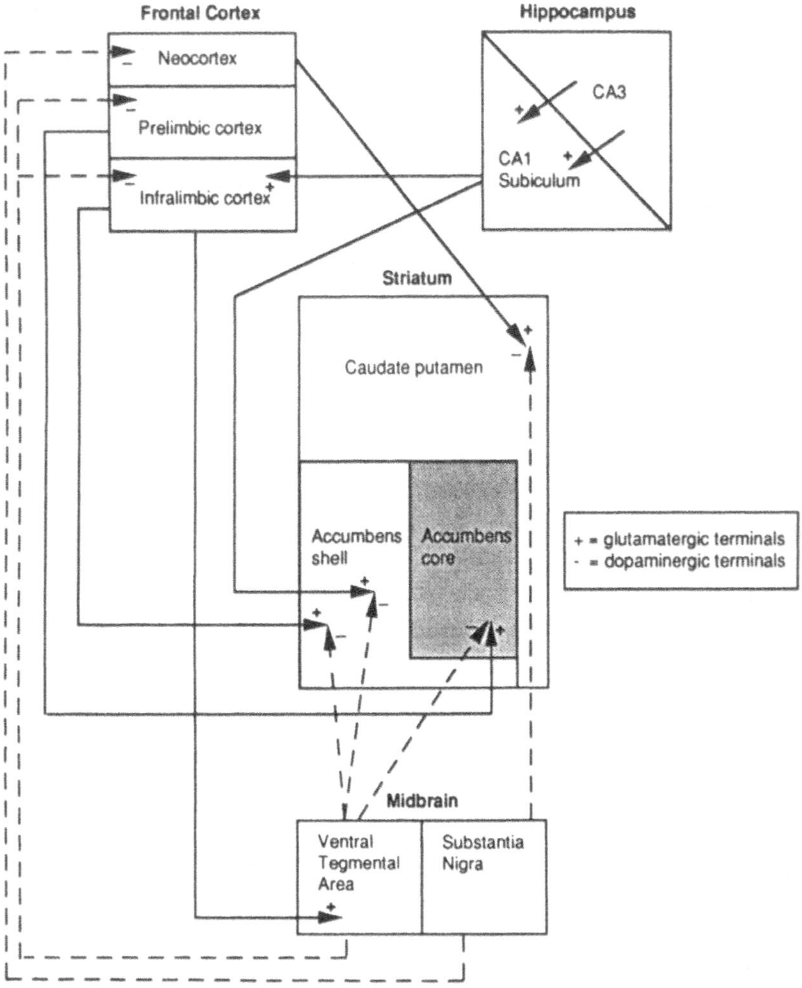

Fig. 1. Hippocampal and frontal cortical connections with striatal regions important in antipsychotic drug action. The *solid lines* represent glutamatergic projections from the hippocampus or cortex. The *dashed lines* represent dopaminergic projections from the midbrain. See text for detailed description of anatomy

contacts on dendritic shafts and the heads and necks of spines (SESACK and PICKEL 1990).

The expression of antipsychotic drug efficacy may be regulated by spiny neurons in the medial shell. DEUTCH et al. (1992; also see DEUTCH et al. this volume) have demonstrated that most typical antipsychotic drugs, such as haloperidol, elevate immediate-early gene expression throughout the nucleus accumbens and caudate nucleus. However, clozapine, an atypical drug which lacks extrapyramidal side effects, increases *fos* expression only in the medial shell. This finding indicates that the medial shell serves as a cardinal site for *fos* expression induced by clinically effective antipsychotics. More lateral and dorsal sectors of the striatum serve as sites for *fos* expression induced by antipsychotics possessing extrapyramidal liability. If the efficacy of antipsychotic drugs is mediated within the medial shell of the accumbens, then glutamatergic afferents from the hippocampus are ideally situated to modulate the expression of antipsychotic efficacy.

II. Consequences for Glutamatergic Mechanisms of Antipsychotic Drug Action

Neuroanatomical disturbances in the hippocampus or related temporal lobe structures could generate a primary aberration in glutamatergic tone in the nucleus accumbens and other afferent structures. Other biochemical and behavioral changes, such as those associated with dopaminergic function, would be predicated on initial deficiencies in glutamatergic function (see Table 3 for a summary). Therefore, it is important to begin a discussion of lesion-induced changes in antipsychotic drug action by first examining the status of glutamate neurotransmission. Unfortunately, little is known about glutamatergic function in afferent structures after hippocampal lesions. NITSCH et al. (1979) found that hippocampal extirpation reduced tissue glutamate content in the accumbens, but not in the caudate. A similar effect was observed after bilateral transsections of the fornix (ZAZCEK et al. 1979), the major fiber tract from the hippocampus to the accumbens. However, more moderate forms of hippocampal disruption (see Fig. 2), such as kainic acid-induced pyramidal cell loss, do not alter whole-tissue glutamate levels in the accumbens (BARDGETT et al. 1995b). Furthermore, while fornix transections may reduce tissue

Table 3. Effects of hippocampal lesions on mechanisms of antipsychotic drug action

Glutamate levels
Fornix transection decreases glutamate levels in accumbens
Dopamine receptors
Kainic acid lesion elevates [^{3}H] spiperone binding in the accumbens
Dopamine turnover and release
Hippocampal extirpation/ventral hippocampal lesion increase dopamine turnover in accumbens
Kainic acid/colchicine lesion enhances amphetamine-induced dopamine release in accumbens
Kainic acid lesion attenuates striatal DOPAC response to haloperidol
Dopamine-mediated behaviors
Various lesions increase basal locomotor activity, without corresponding increase in stereotypy
Various lesions augment locomotor response to amphetamine and D_1 agonists
Various lesions disrupt prepulse and latent inhibition conditioning
Neonatal ventral hippocampal lesions enhance apomorphine-induced stereotypy
Neonatal ventral hippocampal lesions reduce cataleptic response to haloperidol

DOPAC, 3,4-dihydroxyphenylacetic acid.

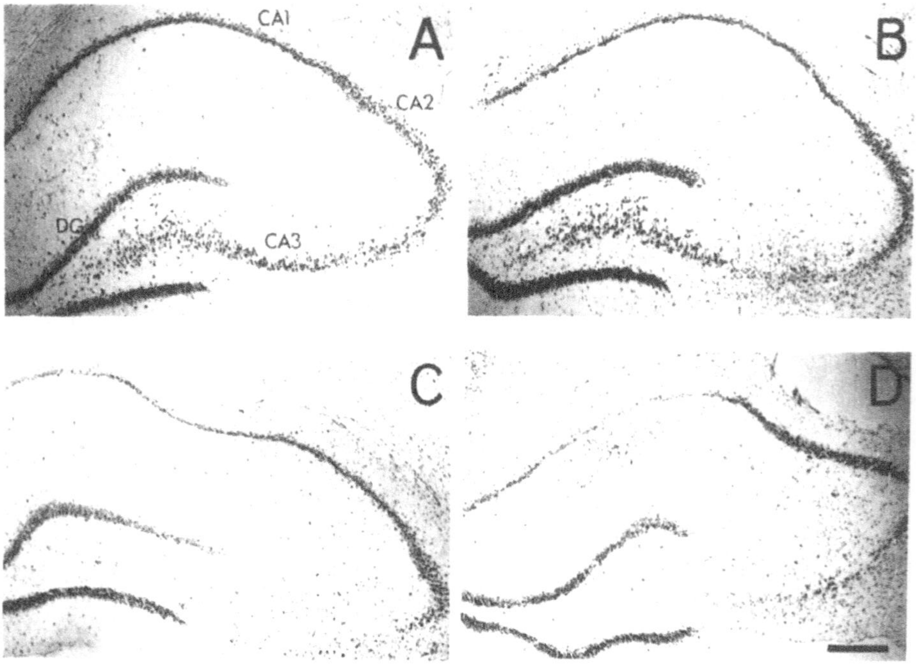

Fig. 2A–D. Photomicrograph of graded hippocampal cell loss in the rat dorsal hippocampus induced by intracerebroventricular administration of the excitotoxin kainic acid. **A** An unlesioned hippocampus. **B** The animal received 1.5 nmol kainic acid into each lateral ventricle and was lesioned 7 days later. Note the partial neuronal loss in the CA3 region of the hippocampus. **C, D** Each animal received 4.5 and 6.6 nmol kainic acid, respectively. With the increasing doses, the entire CA3 cell field was lesioned, and some neuronal loss is apparent in the CA1 region (**D**). Despite this neuronal loss in CA3 and CA1, the dentate gyrus (*DG*) and CA2 region of the hippocampus remain intact. These lesions may better approximate the neuropathology associated with psychiatric disorders than gross tissue destruction or cavitation. From BARDGETT et al. (1995b)

glutamate concentrations in the accumbens, they do not reduce accumbens glutamate uptake (JASKIW et al. 1991).

Obviously, further research is needed in this area. Microdialysis monitoring of tonic and phasic glutamate release could provide more meaningful data regarding lesion effects on glutamate neurotransmission in afferent structures. In addition, it will be important to define the effects of hippocampal lesions on other aspects of afferent glutamatergic function, such as glutamate receptor expression and glutamate-mediated behaviors. These studies could provide meaningful data regarding the impact of hippocampal lesions on normative parameters of glutamatergic neurotransmission and establish a foundation for examining glutamatergic mechanisms of antipsychotic drug action in lesioned animals.

III. Consequences for Dopaminergic Mechanisms of Antipsychotic Drug Action

1. Dopamine Receptors

If glutamatergic input to the accumbens is disrupted by hippocampal damage, then secondary changes may arise in dopaminergic processes critical to antipsychotic drug action. As described above, the binding of antipsychotic drugs to D_2 receptors remains the strongest biological correlate of antipsychotic drug action. Some forms of hippocampal neuropathology in rats have been associated with increased D_2 receptor expression in the nucleus accumbens. For instance, kainic acid-induced hippocampal cell loss is accompanied an increased density of D_2 receptors in the accumbens (Bardgett et al. 1995b). Other research has shown that hippocampal kindling, which is associated with subtle morphological changes in the hippocampus, also elevates D_2 receptor binding (Csernansky et al. 1988). These studies indicate that a modest degree of hippocampal neuropathology, which may be more relevant to the neuropathology associated with psychiatric disorders, can elevate D_2 receptor density in the accumbens. Future studies need to address the impact of hippocampal cell loss on the density of other dopamine and nondopaminergic receptors which exhibit affinity for antipsychotic drugs.

2. Dopamine Turnover and Release

Numerous studies have examined basal levels of dopamine turnover and release in the striatum following hippocampal lesions, while fewer studies have addressed the effects of antipsychotic drug treatment on these indices in lesioned rats. Elevations in accumbens dopamine content have been demonstrated at 7 days following complete hippocampal aspiration (Springer and Issacson 1982) and at 28 days following ibotenic acid lesions of the ventral hippocampus (Lipska et al. 1992). While these relatively large lesions produce a hyperdopaminergic state in the accumbens after hippocampal lesions, the more modest form of hippocampal neuropathology observed after intracerebroventricular kainic acid administration does not alter ambient levels of dopamine in the accumbens or the caudate at 7, 14, or 36 days after treatment (Bardgett et al. 1995a,b). Wilkinson et al. (1993), using microdialysis techniques, have also shown that ambient dopamine release in the accumbens is not affected by hippocampal lesions produced by kainic acid and colchicine, although these lesions do enhance amphetamine-induced dopamine efflux in the accumbens.

Since the hippocampal projection to the accumbens terminates primarily in the accumbens shell, it is possible that dopamine function in the shell could be selectively altered by hippocampal lesions. We have examined dopaminergic function in the accumbens shell and core and in the dorsolateral and mediodorsal stratum after acute and long-term haloperidol treatment in kainic acid-lesioned rats (Bardgett et al. 1995a). While basal levels of dopamine and the dopamine metabolites, dihydroxyphenylacetic acid (DOPAC) and homovanillic acid (HVA), in each brain region were not altered by kainic acid lesions, haloperidol-induced increases in DOPAC and HVA were blunted in lesioned rats. While control rats demonstrated robust increases in DOPAC and HVA in each brain region after acute (1-day) haloperidol (1.5 mg/kg, s.c.) treatment, lesioned rats exhibited a decrease (25%) in dopamine and a blunted DOPAC response after acute haloperidol, primarily within the accumbens shell. Dopamine was also reduced in the accumbens core (25%), and the DOPAC response was blunted in the dorsolateral striatum of the lesioned rats following acute haloperidol. Lesioned and control animals displayed similar dopaminergic responses to long-term (21-day) administration of haloperidol; however, lesioned animals had lower levels of DOPAC (26% less) and HVA (30% less) in the dorsolateral striatum than controls. These findings indicate that the striatal dopaminergic response to haloperidol, particularly within the accumbens shell, is blunted in rats after hippocampal cell loss. Further studies of the dopaminergic response to antipsychotics in lesioned animals are clearly merited, especially those involving different drugs and doses and those comparing lesion-induced changes on a variety of indices such as dopamine turnover, dopamine release, and the firing rate of midbrain dopamine neurons.

3. Dopamine-Mediated Behaviors

Behavioral studies have yielded consistent findings regarding the effects of hippocampal lesions on dopamine-mediated behaviors. Generally, animals with hippocampal lesions display enhanced sensitivity to dopamine agonists. Numerous studies have shown that amphetamine-induced locomotion, as well as basal locomotion, is heightened in lesioned rats (EMERICH and WALSH 1990; PORT et al. 1991; WHISHAW and MITTLEMAN 1991; LIPSKA et al. 1992; MITTLEMAN et al. 1993; WILKINSON et al. 1993; BARDGETT et al. 1995b). These behavioral effects may reflect changes in dopaminergic function within the accumbens, since: (a) lesion-induced hyperactivity is not accompanied by alterations in stereotypy (HANNIGAN et al. 1984; WHISHAW and MITTLEMAN 1991; BARDGETT et al. 1995b), which may reflect dopamine function in the caudate, and (b) lesion-induced hyperactivity is blocked by lesions of the dopaminergic input to the accumbens (EMERICH and WALSH 1990; WHISHAW and MITTLEMAN 1991).

The enhanced basal and amphetamine-induced locomotion observed in lesioned animals may depend on D_2 receptor-specific changes in the accumbens (BARDGETT et al. 1995b). MITTLEMAN et al. (1993) have demonstrated that quinpirole, a D_2 receptor agonist, exacerbates basal hyperactivity in lesioned animals, while the D_1 receptor agonist SKF 38393 has no effect. These findings suggest that antipsychotic drugs with a high affinity for D_2 receptors should reverse lesion-induced increases in locomotion. Indeed, haloperidol administration has been found to reverse the effect of hippocampal damage on locomotor behavior (DEVENPORT et al. 1981). STEELE and WILLIAMS (1991) have also demonstrated that sulpiride, a putative antipsychotic with a high affinity for D_2 receptors, prevents the hyperlocomotion induced by fimbria-fornix lesions. LIPSKA et al. (1993) have shown that the hyperlocomotion observed in rats after neonatal ventral hippocampal lesions can be blocked by 3 weeks of haloperidol treatment. While these studies demonstrate that antipsychotics can reverse lesion-induced hyperlocomotion, it will be important to determine whether they reverse the enhancement in dopamine agonist-induced locomotion. In addition, future studies will need to examine the dose–response curves of antipsychotics in behavioral assays in order to determine any lesion-induced shifts in the response curve.

Neonatal hippocampal lesions have also been found to alter the expression of agonist-induced stereotypical behaviors and the cataleptic effects of haloperidol. LIPSKA and WEINBERGER (1993) have demonstrated that adult rats in which ventral hippocampal lesions were induced as neonates exhibit a greater sensitivity to stereotypy induced by the dopamine agonist apomorphine, but less sensitivity to the cataleptic effects of haloperidol. These findings indicate that hippocampal lesions enhance sensitivity to dopamine agonists, yet reduce sensitivity to dopamine antagonists. This view would be consistent with the blunted dopaminergic response to haloperidol observed in animals after kainic acid lesions of the hippocampus. These findings also indicate that dopaminergic function in the caudate is altered after hippocampal lesions, since apomorphine-induced stereotypy was disrupted. As suggested by LIPSKA and WEINBERGER (1993), hippocampal lesions may disrupt the hippocampal projection to frontal cortical regions which innervate the caudate.

Hippocampal lesions and perturbations have been found to disrupt prepulse inhibition. Neonatal ventral hippocampal lesions impair prepulse inhibition and enhance the disruptive effects of apomorphine on prepulse inhibition (SWERDLOW et al. 1994). In adult rats, prepulse inhibition can be disrupted by the infusion of the cholinergic agonist carbachol into the hippocampus (SWERDLOW et al. 1994). This effect suggests that neurochemical disruption of hippocampal activity may have the same effect on prepulse inhibition as that observed after hippocampal lesions. In addition, conditioning in another paradigm sensitive to dopamine agonists, latent inhibition, is attenuated in rats with hippocampal lesions (see WEINER 1991 for a review). It will be important to determine if antipsychotics can reverse the deficits in prepulse and latent inhibition observed in lesioned animals.

IV. Summary

Hippocampal lesions have been found to alter activity in various behavioral assays of antipsychotic activity. Specifically, they augment locomotor responses to dopamine agonists and reduce behavioral sensitivity to dopamine antagonists (see Table 3 for a summary). It is now incumbent on researchers to elucidate the biochemical mechanisms accounting for these effects. While some studies have demonstrated increases in dopamine turnover, release, and receptor binding after hippocampal lesions (Table 3), these data need to be replicated and other facets of dopaminergic neurotransmission examined after such lesions. In addition, investigators must examine the glutamatergic consequences of hippocampal lesions in afferent structures in order to fully interpret lesion-induced alterations in dopaminergic function.

E. Effect of Frontal Cortical Neuropathology on Mechanisms of Antipsychotic Drug Action

Two types of lesions to the frontal cortex have been studied in detail over the past 20 years. First, the 6-hydroxydopamine (6-OHDA) lesion, which destroys dopaminergic terminals in the frontal cortex without damaging intrinsic cortical neurons, has been used to examine the role of dopamine terminals in the frontal cortex. The presence of these terminals within the frontal cortex suggests a direct role for the frontal cortex in antipsychotic drug action, independent of drug effects in the basal ganglia. Second, other studies have lesioned intrinsic neurons of the frontal cortex, using either ablation, electrolytic, or toxic methods, in order to determine the role of cortical neurons in specific biochemical and behavioral functions. The main focus of this section will be on these latter lesions, since deficits in frontal cortical morphology and metabolism have been implicated in several psychiatric disorders.

I. Anatomy

Afferents from the prelimbic and infralimbic portion of the medial prefrontal cortex in the rat form asymmetric synapses on the spines or small dendrites of medium spiny neurons in the accumbens (SESACK and PICKEL 1992). Neurons from the prelimbic, anterior cingulate, and dorsal agranular insular cortex project mainly to the accumbens core, while neurons from the infralimbic, ventral agranular insular, and piriform cortices project mainly to the accumbens shell (ZAHM and BROG 1992). Medial prefrontal cortical neurons also send excitatory projections to the ventral tegmental area (SESACK and PICKEL 1992). In both the nucleus accumbens and the ventral tegmental area, dopaminergic terminals form symmetric contacts with the dendritic shafts, spines, or heads of neurons which are also apposed by terminals from the prefrontal cortex (SESACK and PICKEL 1992). In addition to these specific projections from the medial prefrontal cortex to accumbens, glutamatergic projections from other regions of the frontal cortex innervate remaining areas of the basal ganglia in a topographically organized manner (see GRAYBIEL et al. 1994 for a review).

II. Consequences for Glutamatergic Mechanisms of Antipsychotic Drug Action

As hypothesized for hippocampal lesions, changes in antipsychotic drug action after frontal cortical lesions would presumably reflect a primary deficit in the cortical glutamatergic input to the basal ganglia (see Table 4 for a summary). JASKIW et al. (1991) have demonstrated that bilateral frontal cortical transection temporarily decreases glutamate uptake, a putative marker of glutamatergic terminals, in septal and accumbens homogenates. This potential for a lesion-induced hypoglutamatergic state could modify antipsychotic drug action in the accumbens. Clearly, more research is needed to elucidate the glutamatergic status (i.e., release, receptors, and behavior) of the accumbens and remaining basal ganglia following frontal cortical lesions.

III. Consequences for Dopaminergic Mechanisms of Antipsychotic Drug Action

1. Dopamine Receptors

Initial studies by PYCOCK et al. (1980) indicated that D_2 receptors were increased in the caudate and accumbens after 6-OHDA lesions of the frontal cortex. However, subsequent studies have not confirmed these findings (see CSERNANSKY et al. 1991; GRACE 1991 for a review). SCHWARTZ et al. (1978) reported that cortical ablation decreases [^{3}H]haloperidol binding in the striatum, although their lesion was not specific to the frontal cortex. Given the paucity of data, further empirical effort is needed to characterize the effects of frontal cortical lesions on dopamine receptor binding in afferent structures.

2. Dopamine Turnover and Release

Numerous studies have characterized the functional consequences of 6-OHDA lesions of the frontal cortex. These lesions produce: (a) increased

Table 4. Effects of frontal lesions on mechanisms of antipsychotic drug action

Glutamate levels
Transection produces transient decrease in glutamate uptake
Dopamine turnover and release
Aspiration in adults has no effect on striatal dopamine levels
Partial neonatal ablation increases dopamine and its metabolites in remaining prefrontal cortex
Dopamine-mediated behaviors
Lesions increase basal locomotor activity, without corresponding increase in stereotypy
Lesions augment locomotor response to amphetamine
Excitotoxic lesions enhance apomorphine-induced stereotypy
Transection reduces cataleptic response to haloperidol

levels of dopamine and its metabolites in the accumbens and the caudate nucleus (see DAVIS et al. 1991 for a review), (b) increased potassium-stimulated dopamine release in the striatum (ROBERTS et al. 1994), and (c) increased tyrosine-hydroxylase activity after haloperidol administration in the accumbens (ROSIN et al. 1992). The elevations in striatal dopamine function observed after 6-OHDA lesions may result from disinhibition of corticostriatal glutamatergic neurons (see CSERNANSKY et al. 1991; GRACE 1991 for reviews). Cortical dopamine terminals are believed to provide tonic inhibition of glutamatergic neurons in the frontal cortex. This inhibition would be lost following 6-OHDA lesions, ultimately resulting in increased glutamate release from corticostriatal neurons. Since glutamate may stimulate dopamine release in the striatum, the disinhibition of cortical neurons induced by a 6-OHDA lesion would serve to elevate striatal dopamine function.

In contrast to 6-OHDA lesions, lesions of intrinsic neurons of the frontal cortex do not alter basal dopamine levels in the nucleus accumbens or caudate putamen (WHISHAW et al. 1992), although neonatal lesions may increase dopamine levels in the frontal cortex itself (DE BRABANDER et al. 1993). It will be important to determine whether frontal lesions alter phasic parameters of dopamine turnover and release in the basal ganglia and disrupt the striatal dopaminergic response to antipsychotic drugs.

3. Dopamine-Mediated Behaviors

The loss of frontal cortical neurons induces a pattern of behavioral changes remarkably similar to that observed after hippocampal cell loss (see Table 4). Lesions of the frontal cortex potentiate the locomotor effects of amphetamine and the stereotypical effects of apomorphine (reviewed by GRACE 1991). Basal locomotor activity is increased following frontal lesions, but stereotypy is not (WHISHAW et al. 1992). The enhancing effect on locomotor activity can be reversed by removal of the dopaminergic input to the nucleus accumbens (WHISHAW et al. 1992), suggesting that the dopaminergic status of the accumbens is essential in the expression of this effect.

JASKIW et al. (1993) have recently studied the behavioral response of animals with frontal lesions to antipsychotics. Analogous to the effects of ventral hippocampal lesions (LIPSKA and WEINBERGER 1993), frontal lesions reduce the cataleptic effects of haloperidol. In addition, JASKIW et al. (1993) examined the effect of frontal lesions on cataleptic responses to clozapine, since some investigators have speculated that clozapine's low potential for producing catalepsy depends on intact frontal function. However, lesions of the frontal cortex did not produce catalepsy after administration of relatively high doses of clozapine.

While frontal deficits may not alter clozapine's noncataleptogenic properties, a clinical study by FRIEDMAN et al. (1991) indicated that prefrontal cortical atrophy may impair the treatment response to clozapine. The authors treated patients with clozapine for 6 weeks and then divided individuals into three groups based on their clinical status: good responders, moderate responders, and nonresponders. Using computed tomography (CT) scans to quantify prefrontal sulcal size, they demonstrated that nonresponse to clozapine was associated with a larger prefrontal sulcus. This study suggests that a prefrontal cortical deficit may impair clinical responsiveness to clozapine treatment. Preclinical studies are warranted to examine the possible biochemical explanations of this finding, in order to develop new treatment strategies which could circumvent prefrontal deficits.

IV. Summary

Frontal cortical lesions produce many of the same behavioral effects observed after hippocampal lesions, for example enhanced responses to dopamine agonists and blunted responses to dopamine antagonists (see Table 4 for a summary). However, unlike hippocampal lesions, frontal lesions do not appear to alter dopamine concentrations in the accumbens. In addition, few studies have addressed the effects of these lesions on dopamine release and receptor binding. The impact of frontal lesions on these basic aspects of dopamine neurotransmission, as well as on glutamatergic neurotransmission and antipsychotic pharmacology, merits further investigation.

F. Conclusions

We have reviewed the effects of hippocampal and frontal cortical lesions on antipsychotic drug action by considering these effects in the context of the normative mechanisms of antipsychotic drug action. Unfortunately, few studies have directly addressed the impact of hippocampal and frontal cortical lesions on antipsychotic pharmacology. If an exception exists to this rule, it is in the area of behavioral pharmacology. These studies have provided extensive examinations of lesion-induced changes in dopamine-mediated behaviors and generally indicate that both hippocampal and frontal lesions enhance behavioral sensitivity to dopamine agonists, but decrease sensitivity to dopamine antagonists. These findings clearly suggest that a neuroanatomical disturbance in the hippocampus or frontal cortex would blunt responsiveness to antipsychotics, since most antipsychotics are potent D_2 receptor antagonists. Obviously, more biochemical studies are needed to identify the physiological mechanisms which promote lesion-induced changes in behavioral sensitivity to antipsychotics, and other dopamine antagonists and agonists. In this regard, biochemical studies of hippocampal lesions presently indicate that basal markers of dopamine function in the accumbens are elevated after lesions, but are decreased in response to antipsychotics.

Although the field of antipsychotic pharmacology has been primarily concerned with dopaminergic neurotransmission, investigations of lesion-induced changes in antipsychotic pharmacology must consider alterations in glutamatergic neurotransmission. The hippocampus and frontal cortex are primarily comprised of glutamatergic neurons which project directly to sites of antipsychotic action, such as the accumbens or ventral tegmental area. Alterations in the dopaminergic (or serotonergic, cholinergic, peptidergic, etc.) response to antipsychotics induced by limbic or cortical lesions should presumably reflect a primary deficit in glutamatergic input. However, it cannot be simply assumed that hippocampal or cortical lesions reduce glutamatergic tone in afferent regions. Many studies have demonstrated that some forms of neuronal cell loss, such as kainic acid-induced cell loss, produce hyperexcitability in remaining hippocampal neurons (see Turner and Wheal 1991 for review). Thus, dynamic changes in glutamate release from lesioned brain regions, or glutamate receptor expression in afferent structures, should be considered when formulating mechanisms of antipsychotic drug action in animal models of neuropathology.

Future preclinical research on the impact of neuropathology on antipsychotic drug action will be increasingly driven by the emerging clinical literature on neuropathology and antipsychotic drug response (i.e., Friedman et al. 1991). If antipsychotic responsiveness is altered in patients with specific neuroanatomical abnormalities on CT or magnetic resonance imaging (MRI) scans, animal models could be used to provide insights into the neurobiological mechanisms accounting for these alterations. These insights may suggest novel therapeutic approaches which circumvent treatment nonresponsiveness produced by specific neuroanatomical disturbances.

A greater understanding of limbic-cortical neuropathology and antipsychotic drug action will depend not only on advances in clinical research, but on the ongoing identification of preclinical markers of antipsychotic drug action. Many putative biological and behavioral markers have been proposed; several may be recognized as false positives, while a few will emerge as reliable and robust correlates of antipsychotic drug action. These correlates will offer preclinical investigators new tools for defining the impact of neuropathology on antipsychotic drug response. More importantly, these markers may facilitate the development of safer, more effective antipsychotic drugs.

Acknowledgments. The preparation of this work and our own research was supported by grants from the Scottish Rite Schizophrenia Research Program, the Theodore and Vada Stanley Foundation, the National Institute of Mental Health (grant nos. MH14677 and MH01109), and the National Institute of Alcohol Abuse and Alcoholism (grant no. AA07144).

References

Bardgett ME, Wrona CT, Newcomer JW, Csernansky JG (1993) Subcortical excitatory amino acid levels after acute and subchronic administration of typical and atypical neuroleptics. Eur J Pharmacol 230:245–300

Bardgett ME, Jackson JL, Csernansky JG (1995a) Schizophrenia-like hippocampal neuropathy in rats: Changes in accumbens core and shell dopamine after haloperidol treatment. Schizophr Bull (in press)

Bardgett ME, Jackson JL, Taylor GT, Csernansky JG (1995b) Kainic acid decreases hippocampal neuronal number and increases dopamine receptor binding in the nucleus accumbens: an animal model of schizophrenia. Behav Brain Res (in press)

Behrens S, Gattaz WF (1992) MK-801 induced stereotypies in rats are decreased by haloperidol and increased by diazepam. J Neural Transm 90:219–224

Bjorklund A, Lindvall O (1984) Dopamine-containing systems in the CNS. In: Bjorklund A, Hokfelt T (eds) Handbook of chemical neuroanatomy, 2nd edn. Elsevier, Amsterdam, p 55

Braun AR, Jaskiw GE, Vladar K, Sexton RH, Kolachana BS, Weinberger DR (1993) Effects of ibotenic acid lesion of the medial prefrontal cortex on dopamine agonist-related behaviors in the rat. Pharmacol Biochem Behav 46:51–60

Carlsson M, Carlsson A (1990) Interactions between glutamatergic and monoaminergic systems within the basal ganglia: implications for schizophrenia and Parkinson's disease. Trends Neurosci 13:272–276

Chai B, Meltzer HY (1992) The effect of chronic clozapine on basal dopamine release and apomorphine-induced DA release in the striatum and nucleus accumbens as measured by in vivo microdialysis. Neurosci Lett 136:47–50

Chrapusta SJ, Karoum F, Egan MF, Wyatt RJ (1993) Haloperidol and clozapine increase intraneuronal dopamine metabolism, but not g-butyrolactone-resistant dopamine release. Eur J Pharmacol 233:135–142

Compton DR, Johnson KM (1989) Effects of acute and chronic clozapine and haloperidol on in vitro release of acetylcholine and dopamine from striatum and nucleus accumbens. J Pharmacol Exp Ther 248:521–530

Cooper JR, Bloom FE, Roth RH (1986) The biochemical basis of neuropharmacology, 5th edn. Oxford University Press, New York

Csernansky JG, Kerr S, Pruthi R, Prosser ES (1988) Mesolimbic dopamine receptor increases two weeks following hippocampal kindling. Brain Res 449:357–360

Csernansky JG, Murphy GM, Faustman WO (1991) Limbic/mesolimbic connections and the pathogenesis of schizophrenia. Biol Psychiatry 30:383–400

Csernansky JG, Wrona CT, Bardgett ME, Early TS, Newcomer JW (1993) Subcortical dopamine and serotonin turnover during acute and chronic administration of typical and atypical neuroleptics. Psychopharmacology (Berl) 110:145–151

Daly DA, Moghaddam B (1993) Actions of clozapine and haloperidol on extracellular levels of excitatory amino acids in the prefrontal cortex and striatum of conscious rats. Neurosci Lett 152:61–64

Davis KL, Kahn RS, Ko G, Davidson M (1991) Dopamine in schizophrenia: a review and reconceptualization. Am J Psychiatry 148:1474–1486

de Brabander JM, van Eden CG, de Bruin JPC, Feenstra MGP (1993) Monoamine concentrations in the rat prefrontal cortex and other mesolimbocortical structures in response to partial neonatal lesions of the medial prefrontal cortex. Brain Res 601:20–27

Deutch AY, Cameron DS (1992) Pharmacological characterization of dopamine systems in the nucleus accumbens core and shell. Neurosci 46:49–56

Deutch AY, Lee MC, Iadarola MJ (1992) Regionally specific effects of atypical antipsychotic drugs on striatal fos expression: the nucleus accumbens shell as a locus of antipsychotic action. Mol Cell Neurosci 3:332–341

Devenport LD, Devenport JA, Holloway FA (1981) Reward-induced stereotypy: modulation by the hippocampus. Science 212:1288–1289

Emerich DF, Walsh TJ (1990) Hyperactivity following intradentate injection of colchicine: a role for dopamine systems in the nucleus accumbens. Pharmacol Biochem Behav 37:149–154

Farde L, Nordstrom AL, Wiesel FA, Pauli S, Halldin C, Sedvall G (1992) Positron emission tomographic analysis of central D_1 and D_2 dopamine receptor occupancy in patients treated with classical neuroleptics and clozapine: relation to extrapyramidal side effects. Arch Gen Psychiatry 49:538–544

Farde L, Wiesel FA, Halldin C, Sedvall G (1988) Central D_2-dopamine receptor occupancy in schizophernic patients treated with antipsychotic drugs. Arch Gen Psychiatry 45:71–76
Friedman L, Knutson L, Shurell M, Meltzer HY (1991) Prefrontal sulcal prominence is inversely related to response to clozapine in schizophrenia. Biol Psychiatry 29:865–877
Fuller TA, Russchen FT, Price JL (1987) Sources of presumptive glutamatergic/aspartergic afferents to the rat ventral striatopallidal region. J Comp Neurol 258:317–338
Grace AA (1991) Phasic versus tonic dopamine release and the modulation of dopamine system responsivity: a hypothesis for the etiology of schizophrenia. Neuroscience 41:1–24
Graybiel AM, Aosaki T, Flaherty AW, Kimura M (1994) The basal ganglia and adaptive motor control. Science 265:1826–1831
Hannigan JH Jr, Springer JE, Issacson RL (1984) Differentiation of basal ganglia dopaminergic involvement in behavior after hippocampectomy. Brain Res 291:83–91
Hauber W (1993) Clozapine improves dizocilpine-induced delayed alternation impairment in rats. J Neural Transm 94:223–233
Heimer L, Zahm DS, Churchill L, Kalivas PW, Wohltmann C (1991) Specificity in the projection patterns of accumbal core and shell in the rat. Neuroscience 41:89–125
Hollister LE, Csernansky JG (1990) Clinical pharmacology of psychotherapeutic drugs, 3rd edn. Churchill Livingstone, New York
Janowsky A, Neve KA, Kinzie JM, Taylor B, de Paulis T, Belknap JK (1992) Extrastriatal dopamine D2 receptor: distribution, pharmacological characterization and region-specific regulation by clozapine. J Pharmacol Exp Ther 261:1282–1290
Jaskiw GE, Tizabi Y, Lipska BK, Kolachana BS, Wyatt RJ, Gilad GM (1991) Evidence for a frontocortical-septal glutamatergic pathway and compensatory changes in septal glutamate uptake after cortical and fornix lesions. Brain Res 550:7–10
Jaskiw GE, Hussain G, Meltzer HY (1993) Frontal cortex lesion modify the cataleptogenic properties of haloperidol but not of clozapine. Biol Psychiatry 34:188–190
Kalivas PW (1993) Neurotransmitter regulation of dopamine neurons in the ventral tegmental area. Brain Res Rev 18:75–113
Kane JM, Marder SR (1993) Psychopharmacologic treatment of schizophrenia. Schizophr Bull 19:287–302
Kelley AE, Domesick VB (1982) The distribution of the projection from the hippocampal formation to the nucleus accumbens in the rat: An anterograde- and retrograde-horseradish peroxidase study. Neuroscience 7:2321–2335
Kohler C, Hall H, Magnusson O, Lewander T, Gustafsson K (1990) Biochemical pharmacology of the atypical neuroleptic remoxipride. Acta Psychiatr Scand 82:27–36
Lancaster B, Wheal HV (1984) Chronic failure of inhibition of CA1 area of the hippocampus following kainic acid lesion of CA3/CA4 area. Brain Res 295:317–324
Leysen JE, Gommeren W, Eens A, de Chaffoy de Courcelles D, Stoof JC, Janssen PAJ (1988) Biochemical profile of risperidone, a new antipsychotic. J Pharmacol Exp Ther 247:661–670
Lidsky TI, Yablonsky-Alter E, Zuck L, Banerjee SP (1993) Anti-glutamatergic effects of clozapine. Neurosci Lett 163:155–158
Lipska BK, Weinberger DR (1993) Delayed effects of neonatal hippocampal damage on haloperidol-induced catalepsy and apomorphine-induced stereotypic behaviors in the rat. Dev Brain Res 75:213–222
Lipska BK, Jaskiw GE, Chrapusta S, Karoum F, Weinberger DR (1992) Ibotenic lesion of the ventral hippocampus differentially affects dopamine and its metabolites in the nucleus accumbens and prefrontal cortex in the rat. Brain Res 585:1–6

Lipska BK, Jaskiw GE, Weinberger DR (1993) Postpubertal emergence of hyperresponsiveness to stress and to amphetamine after neonatal excitotoxic hippocampal damage: A potential animal model of schizophrenia. Neuropsychopharmacology 9:67–75
Ljundberg T, Ungerstedt U (1978) Classification of neuroleptic drugs according to their ability to antagonize apomorphine-induced locomotion and gnawing: Evidence for two mechanisms of action. Psychopharmacology (Berl) 56:239–247
Meshul CK, Janowsky A, Casey DE, Stallbaumer RK, Taylor B (1992) Effect of haloperidol and clozapine on the density of "perforated" synapses in caudate, nucleus accumbens, and medial prefrontal cortex. Psychopharmacology (Berl) 106:45–52
Mittleman G, LeDuc PA, Whishaw IQ (1993) The role of D1 and D2 receptors in the heightened locomotion induced by direct and indirect dopamine agonists in rats with hippocampal damage: an animal analogue of schizophrenia. Behav Brain Res 55:253–267
Moghaddam B, Bunney BS (1993) Depolarization inactivation of dopamine neurons: terminal release characteristics. Synapse 14:195–200
Nitsch C, Kim JK, Shimade C, Okada Y (1979) Effect of hippocampus extirpation in the rat on glutamate levels in target structures of hippocampal efferents. Neurosci Lett 11:295–299
Nordstrom AL, Farde L, Wiesel FA, Forslund K, Pauli S, Halldin C, Uppfedlt G (1993) Central D2-dopamine receptor occupancy in relation to antipsychotic drug effects: a double-blind PET study of schizophrenic patients. Biol Psychiatry 33:227–235
O'Dell SJ, LaHoste GJ, Widmark CB, Shapiro RM, Potkin SJ, Marshall JF (1990) Chronic treatment with clozapine or haloperidol differentially regulates dopamine and serotonin receptors in rat brain. Synapse 6:146–153
Ogren SO, Florvall L, Hall H, Magnusson O, Angeby-Moller K (1990) Neuropharmacological and behavioral properties of remoxipride in the rat. Acta Psychiatr Scand 82:21–26
Phillipson OT, Griffiths AC (1985) The topographic order of inputs to the nucleus accumbens in the rat. Neuroscience 2:275–296
Port RL, Sample JA, Seybold KS (1991) Partial hippocampal pyramidal cell loss alters behavior in rats: implications for an animal model of schizophrenia. Brain Res Bull 26:993–996
Pycock CJ, Kerwin RW, Carter CJ (1980) Effect of lesion of cortical dopamine terminals on subcortical dopamine in rats. Nature 286:74–77
Reynolds GP (1992) Developments in the drug treatment of schizophrenia. Trends Pharmacol 13:116–121
Roberts AC, De Salvia, MA, Wilkinson LS, Collins P, Muir JL, Everitt BJ, Robbins TW (1994) 6-hydroxydopamine lesions of the prefrontal cortex in monkeys enhance performance on an analog of the Wisconsin Card Sort Test: possible interactions with subcortical dopamine. J Neurosci 14:2531–2544
Rosin DL, Clark WA, Goldstein M, Roth RH, Deutch AY (1992) Effects of 6-hydroxydopamine lesions of the prefrontal cortex on tyrosine hydroxylase activity in mesolimbic and nigrostriatal dopamine systems. Neuroscience 48:831–839
Schwartz R, Creese I, Coyle JT, Snyder SH (1978) Dopamine receptors localized on cerebral cortical afferents to rat corpus striatum. Nature 271:766–768
Seeman P (1992) Dopamine receptor sequences: therapeutic levels of neuroleptics occupy D_2 receptors, clozapine occupies D_4. Neuropsychopharmacol 7:261–284
Sesack SR, Pickel VM (1990) In the rat nucleus accumbens, hippocampal and catecholaminergic terminals converge on spiny neurons and are in apposition to each other. Brain Res 527:266–279
Sesack SR, Pickel VM (1992) Prefrontal cortical efferents in the rat synapse on unlabeled neuronal targets of catecholamine terminals in the nucleus accumbens septi and on dopamine neurons in the ventral tegmental area. J Comp Neurol 320:145–160

Springer JE, Issacson RL (1982) Catecholamine alterations in basal ganglia after hippocampal lesions. Brain Res 252:185–188

Steele TL, Williams M (1991) Effects of amphetamine and sulpiride on exploratory behaviors in rats with fornix knife cuts. Soc Neurosci Abstr 344:11

Swanson LW (1981) A direct projection from Ammon's horn to prefrontal cortex in the rat. Brain Res 217:150–154

Swerdlow NR, Braff DL, Taaid N, Geyer MA (1994) Assessing the validity of an animal model of deficient sensorimotor gating in schizophrenic patients. Arch Gen Psychiatry 51:139–154

Turner DA, Wheal HV (1991) Excitatory synaptic potentials in kainic acid-denervated rat CA1 pyramidal neurons. J Neurosci 11:2786–2794

Ulas J, Cotman CW (1993) Excitatory amino acid receptors in schizophrenia. Schizophr Bull 19:105–117

Weiner I (1991) Neural substrates of latent inhibition: the switching model. Psychol Bull 108:442–461

Whishaw IQ, Mittleman G (1991) Hippocampal modulation of the nucleus accumbens: behavioral evidence from amphetamine-induced activity profiles. Behav Neural Biol 55:289–306

Whishaw IQ, Fiorino D, Mittleman G, Castaneda E (1992) Do forebrain structures compete for behavioral expression? Evidence from amphetamine-induced behavior, microdialysis, and caudate-accumbens lesions in the medial frontal cortex damaged rats. Brain Res 576:1–11

White FJ, Wang RY (1983) Differential effects of classical and atypical antipsychotic drugs on A9 and A10 dopamine neurons. Science 221:1054–1057

Wilkinson LS, Mittleman G, Torres E, Humby T, Hall FS, Robbins TW (1993) Enhancement of amphetamine-induced locomotor activity and dopamine release in the nucleus accumbens following excitotoxic lesions of the hippocampus. Behav Brain Res 55:143–150

Willins DL, Narayanan S, Wallace LJ, Uretsky NJ (1993) The role of dopamine and AMPA/kainate receptors in the nucleus accumbens in the hypermotility response to MK801. Pharmacol Biochem Behav 46:881–887

Yamamoto BK, Cooperman MA (1994) Differential effects of chronic antipsychotic drug treatment on extracellular glutamate and dopamine concentrations. J Neurosci 14:4159–4166

Yang CR, Mogenson GJ (1984) Electrophysiological responses of neurones in the nucleus accumbens to hippocampal stimulation and the attenuation of the excitatory responses by the mesolimbic dopaminergic system. Brain Res 324:69–84

Yang CR, Mogenson GJ (1986) Dopamine enhances terminal excitability of hippocampal-accumbens neurons via D_2 receptor: role of dopamine in presynaptic inhibition. J Neurosci 6:2470–2478

Zahm DS, Brog JS (1992) On the significance of subterritories in the "accumbens" part of the rat ventral striatum. Neuroscience 50:751–767

Zazcek R, Hedreen JC, Coyle JT (1979) Evidence for a hippocampal-septal glutamatergic pathway in the rat. Exp Neurol 65:145–156

CHAPTER 10

An Animal Model of Sensorimotor Gating Deficits in Schizophrenia Predicts Antipsychotic Drug Action

N.R. Swerdlow, D.L. Braff, V.P. Bakshi, and M.A. Geyer

A. Introduction

Psychiatric researchers have searched for animal models of schizophrenia in order to better understand the neurobiological basis of this group of disorders. Since it is unlikely that a "model" of the entire schizophrenia syndrome can be created, much energy has been devoted to identifying models of specific dysfunctions that characterize schizophrenia patients. Schizophrenia patients are deficient in the normal inhibition of the startle reflex that occurs when the startling stimulus is preceded by a weak prestimulus (Braff et al. 1978, 1992; Braff and Geyer 1990) (Fig. 1A). This loss of normal "prepulse inhibition" (PPI) is thought to be a measure of the deficient sensorimotor gating (Braff and Geyer 1990) that underlies sensory flooding and cognitive fragmentation in these patients (McGhie and Chapman 1961). The neural basis of deficient sensorimotor gating (Swerdlow et al. 1992a) is strikingly consistent with findings from neuroimaging (Wong et al. 1986; Early et al. 1987; Wu et al. 1990) and neuropathological studies (Alishuler et al. 1987, 1990; Pakkenberg 1990; Conrad et al. 1991; Seeman et al. 1984; Bogerts et al. 1985) in schizophrenia patients that have identified abnormalities within limbic cortico-striato-pallido-thalamic (CSPT) circuitry. By elucidating the neural substrates of impaired PPI, it might be possible to gain critical information about the contribution of specific brain abnormalities to sensorimotor gating deficits in schizophrenia patients.

Thus, a valid animal model of deficient PPI in schizophrenia patients might be critically important for understanding the probable neural substrates of impaired sensorimotor gating in schizophrenia. PPI deficits are produced in rats treated with dopamine (DA) agonists such as apomorphine (APO) (Swerdlow et al. 1986a, 1991c), amphetamine (Mansbach et al. 1988; Swerdlow et al. 1990c), or the D2 agonist quinpirole, but not in rats treated with D1 agonists (Peng et al. 1990; Wan and Swerdlow 1994; Wan et al. 1994a). PPI is also reduced or eliminated by systemic treatment with glutamate antagonists such as phencyclidine (PCP) or dizocilpine (MK-801) (Mansbach and Geyer 1989). The loss of PPI in rats treated with DA agonists or glutamate antagonists may satisfy some or all levels of validity as animal models of sensorimotor gating deficits in schizophrenia patients. First, the loss of PPI in DA agonist- or PCP-treated rats may be animal models with face validity for the sensorimotor gating deficits exhibited by schizophrenia patients, since PPI is measured in rats and humans using nearly identical stimulus parameters that yield very similar response characteristics. Second, the loss of PPI in DA agonist- or PCP-treated rats may be animal models with predictive validity, since the APO-disruption of PPI in rats is reversed by both typical and atypical antipsychotic agents (SWERDLOW et al. 1994a,d). This APO-disruption of PPI is not reversed by drugs with clinical antidepressant or anxiolytic properties (RIGDON and VIIK 1991). The PCP-disruption of PPI in rats appears to be sensitive to reversal by atypical antipsychotics, but not the typical antipsychotic haloperidol (KEITH et al. 1991; BAKSHI et al. 1994). Third, the loss of PPI in rats models a psychophysiological construct for the pathophysiology of schizophrenia, that is, impaired sensorimotor gating (MCGHIE and CHAPMAN 1961). In addition, PPI appears to be modulated by neural activity in limbic and mesolimbic CSPT circuitry implicated in the pathophysiology of schizophrenia and in the therapeutic mechanisms of antipsychotic agents: PPI can be predictably disrupted or restored in rats by pharmacologic and surgical manipulations of the hippocampus (CAINE et al. 1991; SWERDLOW et al. 1993b), the nucleus accumbens (NAC) (SWERDLOW et al. 1986a, 1990b, 1992b; KODSI and SWERDLOW 1994; WAN and SWERDLOW 1994; WAN et al. 1994a), and the subpallidum (SWERDLOW et al. 1990a; KODSI and SWERDLOW 1994). Furthermore, PPI is disrupted in adult rats that sustained neonatal hippocampal damage or that were raised in social isolation, consistent with existing neurodevelopmental theories for the pathogenesis of schizophrenia. These findings suggest that deficient PPI in these rats may be an animal model with construct validity in psychophysiological, neurochemical, and neurodevelopmental domains for sensorimotor gating deficits in schizophrenia patients. Importantly, DA agonist- or PCP-induced PPI deficits in rats are clearly not animal models of schizophrenia per se, but these deficits may provide models of sensorimotor gating deficits in schizophrenia patients that have face, predictive, and construct validity.

This chapter will review findings that support the validity of PPI deficits in rats as an animal model of sensorimotor gating deficits in schizophrenia patients. First, in support of the face validity of these models, we will describe measurements of PPI in humans and rats using near-identical stimulus parameters. These parameters produce strikingly parallel patterns of startle and of prepulse latency- and amplitude-modulation across species. Using this paradigm, PPI is disrupted in an orderly and dose-dependent fashion by DA agonists and glutamate antagonists. Second, in support of the predictive value of this model, we will describe the ability of a wide range of antipsychotic agents to restore PPI in APO-treated rats. The potency of antipsychotic agents – including clozapine – to prevent the PPI-disruptive effects of APO correlate significantly with their clinical potency. In contrast, we will describe the apparent sensitivity of glutamate antagonist-disrupted PPI to "atypical," but not "typical" antipsychotic agents. Finally, the construct validity of this animal model is studied by measuring PPI in rats after pharmacological, surgical, or developmental manipulations that are consistent with existing theories regarding the neurochemical, neuroanatomical, and neurodevelopmental etiologies of schizophrenia.

I. Face Validity: Are the Eliciting Stimuli and Response Patterns of Prepulse Inhibition (PPI) Similar Across Species?

Previous studies have demonstrated that the reduction in startle amplitude caused by an acoustic prepulse increases with prepulse intensity (ISON et al. 1973; GRAHAM 1975; SWERDLOW et al. 1993a). In a recent study (SWERDLOW et al. 1994a), PPI was measured in rats and humans using identical startle stimuli and using prepulse intensities over a similar range of 2–17dB above background. These results are seen in Fig. 1B.

In separate work, we have reported on the reduction in PPI in rats treated systemically with either DAergic agonists such as APO (SWERDLOW et al. 1994a), or in rats treated with glutamate antagonists such as PCP (MANSBACH and GEYER 1989). Examples of PPI in rats treated with APO or PCP, and in human control and schizophrenia patients (BRAFF et al. 1992), are seen in Fig. 2.

These findings have several implications. First, there are clear parallels between the human startle response and the rat startle response to similar eliciting stimuli: both species exhibit a graded increase in PPI with increasing prepulse intensities; both species appear to exhibit a "threshold" range for startle gating at prepulse intensities approximately 4 dB above background; and both species exhibit latency facilitation at the weakest prepulse intensities, even though these weak prepulses do not significantly reduce startle amplitude. Second, PPI in rats is decreased in an orderly and dose-dependent manner by very low doses of APO or PCP, and in humans with schizophrenia. Startle gating in rats appears to be quite sensitive to DAergic tone, since a decrease in PPI is evident even at doses of APO (e.g., 0.1mg/kg) that are considered "threshold" doses for postsynaptic DA receptor activation, and which do not stimulate other obvious forms of behavioral activation (SWERDLOW et al. 1986a). The doses of PCP that reduce PPI are in a range that is known to produce behavioral activation in other behavioral paradigms (MANSBACH and GEYER 1989).

II. Predictive Validity: Can This Model Be Used as a Sensitive and Specific Screen of Antipsychotic Potency?

We have previously reported that the APO-induced loss of PPI in rats is antagonized by antipsychotic agents, including clozapine (SWERDLOW et al. 1991c; SWERDLOW and GEYER 1992, 1993a) and the putative atypical antipsychotic ICI 204,636 (Seroquel) (SWERDLOW et al. 1994d). Others have

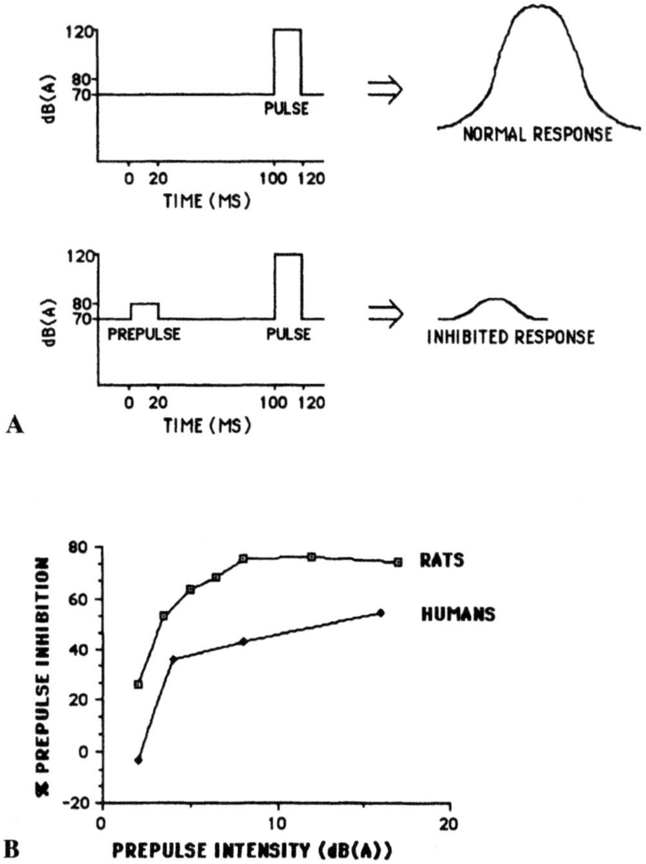

Fig. 1. **A** Schematic description of the prepulse inhibition (PPI) paradigm used in both humans and rats. Typically, there is a 65–70 dB(A) background noise, over which combinations of startle pulses (20–40 ms 118 dB(A) noise bursts) or prepulses (2–16 dB(A) above background 20-ms noise bursts) are superimposed. **B** PPI in normal humans and rats, as previously reported, demonstrating the close parallels between responses elicited using identical stimuli, despite quite different measurement techniques in humans (EMG of orbicularis oculi) and rats (whole-body startle measured using a piezoelectric device) (Swerdlow et al. 1994a)

reported that the novel antipsychotic risperidone also restores PPI in APO-treated rats (Rigdon and Viik 1991). It is thus possible that the ability to restore startle gating in APO-treated rats might serve as a useful screening measure for antipsychotic agents. Several other reports suggest that the APO-disruption of PPI might have predictive validity, i.e., be a sensitive and specific measure of antipsychotic action. While clinically potent antipsychotics antagonize the effects of APO on PPI, psychoactive agents without antipsychotic properties such as buspirone, diazepam, and imipramine do not restore PPI in APO-treated rats (Rigdon and Viik 1991), and neither does the opiate antagonist naloxone (Swerdlow et al. 1991a) or the $5HT_2$- and beta-adrenergic antagonist *d,l*-propranolol (Swerdlow et al. 1994a). These observations sug-

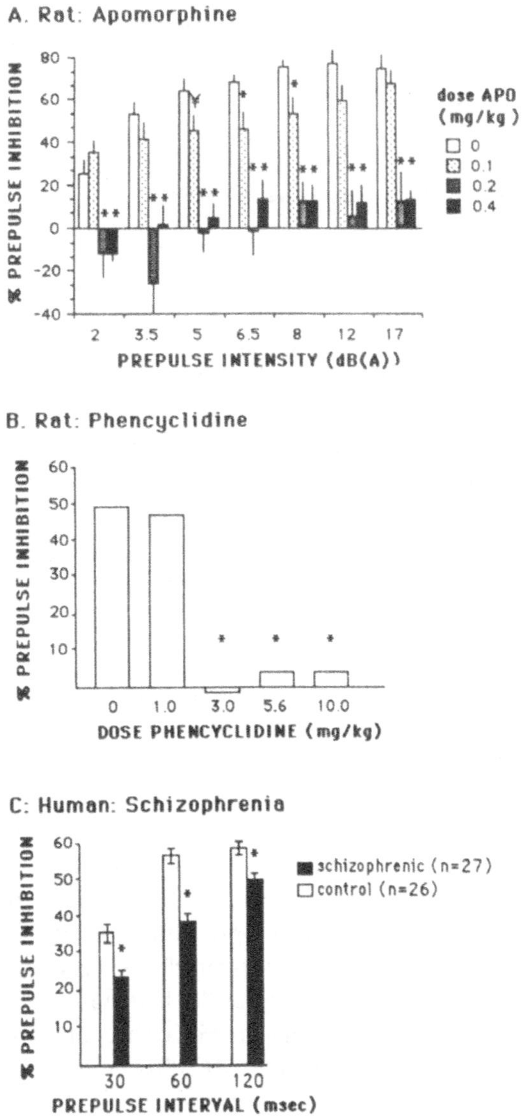

Fig. 2. PPI reduced in rats by apomorphine (APO) (**A**) or phencyclidine (PCP) (**B**), or in humans with schizophrenia (**C**), from previously published studies (SWERDLOW et al. 1994a; MANSBACH and GEYER 1989; BRAFF et al. 1992). These data, as well as those shown in Fig. 1B support the face validity of this PPI model

gest that the ability of a drug to reverse the APO-disruption of PPI might predict antipsychotic properties, and we recently reported that this model accurately predicts antipsychotic potency, a prerequisite of predictive validity (ELLENBROEK and COOLS 1990). Data summarized from dose–response studies of several drugs, supporting the sensitivity and specificity of this PPI model for predicting antipsychotic potency, are shown in Fig. 3.

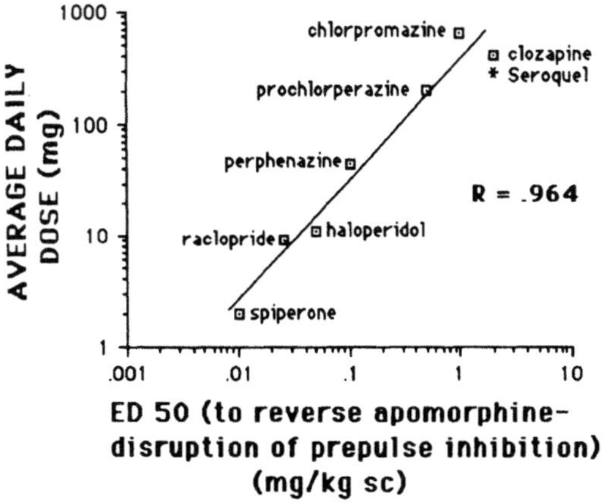

Fig. 3. ED_{50} to restore PPI in apomorphine-treated rats, as previously reported (SWERDLOW et al. 1994a), plotted against average daily dose in milligrams used to treat schizophrenia patients. These data support the predictive validity of the PPI model. In recent studies, the putative atypical antipsychotic ICI204,636 (Seroquel) was noted to exhibit clozapine-like potency in this paradigm; clinical trials with this compound are currently in progress. (*, predicted average daily dose approximately equal to clozapine based on potency in the PPI model)

In contrast to the effects of antipsychotic agents on APO-disrupted PPI, neither haloperidol (KEITH et al. 1991) nor raclopride (BAKSHI et al. 1994) restore PPI in rats treated with PCP. The PCP-induced organic brain syndrome has been described by some to share some features of schizophrenia (JAVITT and ZUKIN 1991), although clearly – in contrast to DA agonist-induced psychosis – the anesthesia, acute confusion, fluctuating consciousness, extreme aggressiveness, violence, and autonomic activation associated with PCP intoxication bear little resemblance to most presentations of schizophrenia. Still, some features of PCP-psychosis – particularly aspects related to impaired sensory or sensorimotor gating, as well as "deficit state" symptoms – may be quite relevant to related phenomena in schizophrenia. For this reason it is intriguing that both the PCP- and the MK-801-induced disruptions of PPI appear to be antagonized by the atypical antipsychotic clozapine (BAKSHI et al. 1994), as seen for PCP in Fig. 4. Other compounds that lack antipsychotic potency, such as ritanserin and ketanserin, fail to restore PPI in PCP-treated rats (BAKSHI et al. 1994).

These findings suggest that the APO- and PCP-induced disruptions of PPI may be sensitive and specific tools for screening drugs for their antipsychotic properties, and even perhaps for their relative antipsychotic potency. The "APO-PPI" model appears to be sensitive to antipsychotic compounds in general, since the ability of a wide range of antipsychotics to restore PPI in APO-treated rats is highly predictive of their clinical potency. This correlation was noted with antipsychotics from several different chemical classes, including the dibenzodiazepine clozapine (SWERDLOW et al. 1991c, 1994d; SWERDLOW and GEYER 1993a). The fact that numerous drugs that lack antipsychotic po-

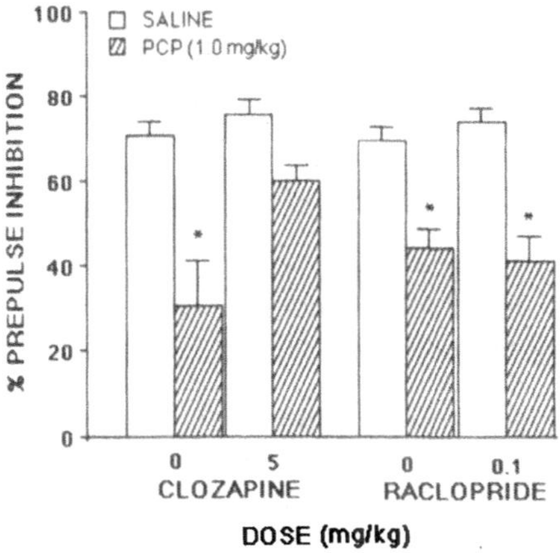

Fig. 4. Partial restoration of PPI in PCP-treated rats by clozapine, but not raclopride (Bakshi et al. 1994). A similar pattern was noted in dizocilpine (MK-801)-treated rats. Haloperidol also failed to restore PPI in PCP-treated rats (Keith et al. 1991). The sensitivity of this model to non-DAergic psychotogens may prove particularly valuable in elucidating the mechanisms of action of atypical antipsychotic agents

tency, do *not* restore PPI in APO-treated rats supports the "specificity" of this model; such "negative" compounds include *d,l*-propranolol (Swerdlow et al. 1994a), diazepam, buspirone, imipramine (Rigdon and Viik 1991), and naloxone (Swerdlow et al. 1991a). These findings all support the notion that the loss of PPI in APO-treated rats may have predictive validity as an animal model of deficient startle gating in schizophrenia patients.

The observation that clozaphine (Bakshi et al. 1994), but not haloperidol (Keith et al. 1991) or raclopride (Bakshi et al. 1994), restores PPI in PCP-treated rats may also have important implications. These findings suggest that the "PCP-PPI" model might be able to differentiate compounds with "typical" and "atypical" antipsychotic properties. Such a model would have obvious utility in the development and screening of novel compounds with putative antipsychotic properties and "atypical" profiles.

III. Construct Validity: Is This Model Consistent with Existing Constructs of Schizophrenia?

A model has construct validity if it demonstrates both empirical and theoretical consistency with a proposed physiological construct: either related to the etiology of a disorder or to the therapeutic mechanism of drug action (Ellenbroek and Cools 1990). Since impaired sensorimotor gating is proposed to be a critical psychopathological construct in schizophrenia patients (McGhie and Chapman 1961), and since impaired sensorimotor gating is exhibited by schizophrenia patients (Braff et al. 1978, 1992; Braff and Geyer 1990) and by rats after manipulations of limbic corticostriatopallidopontine circuitry (Swerdlow et al. 1992a), this animal model of impaired sensorimotor gating, by definition, has construct validity (ELLENBROEK and COOLS 1990). We have extended this level of validity by studying the loss of sensorimotor gating after neurochemical, neuroanatomical, and developmental manipulations that are consistent with constructs for the pathophysiology of schizophrenia, and for the mechanism of action of antipsychotic agents.

1. Construct Validity in a Neurochemical Domain: Mesolimbic Hyperdopaminergia

Mesolimbic hyperdopaminergia has long been implicated in the pathophysiology of psychosis and schizophrenia, either as a primary causative insult (STEVENS 1973; WONG 1986) or as a secondary consequence of prefrontal or limbic-cortical dysfunction (PYCOCK et al. 1990; WEINBERGER et al. 1987a,b; DAVIS et al. 1991). Other evidence suggests that the blockade of mesolimbic DA receptors may be responsible for at least some of the beneficial properties of antipsychotic agents (WHITE and WANG 1984; RUPNIAK et al. 1985; TAMMINGA et al. 1988). Mesolimbic hyperdopaminergia disrupts PPI in rats: we have reported (SWERDLOW et al. 1990b) and replicated in three reports (SWERDLOW et al. 1990a, 1992b, 1994a) the finding that PPI in rats is decreased significantly by direct infusion of DA into the NAC, which is the terminal region of mesolimbic DA fibers. We recently used mesolimbic hyperdopaminergia as a neurochemical manipulation that disrupts sensorimotor gating in rats and which is consistent with a major construct for the pathophysiology of schizophrenia – the "dopamine hypothesis" of schizophrenia. We further explored whether the reduction in PPI after intra-NAC DA or quinpirole infusion is reversed by antipsychotics, since such an effect would be consistent with a major construct for the therapeutic mechanism of antipsychotic action (WHITE and WANG 1984; RUPNIAK et al. 1985; TAMMINGA et al. 1988). The aim of this work was *not* to demonstrate the mesolimbic hyperdopaminergia is a singular cause of deficient sensorimotor gating in schizophrenia, or to imply that blockade of mesolimbic DA is the major therapeutic mechanism of antipsychotic action. In fact, we have reported that sensorimotor gating is modulated by neural circuitry linking the hippocampus, ventral striatum, subpallidum, and pontine tegmentum (SWERDLOW et al. 1992a), and dysfunction at various levels of this circuitry probably accounts for the loss of sensorimotor gating in schizophrenia patients (see below). Rather, this work simply assessed whether a model with construct validity in a psychophysiological domain also satisfied criteria for construct validity in a neurochemical domain.

As is seen in Fig. 5, PPI is disrupted by intra-NAC infusion of either DA or the D_2 receptor agonist quinpirole. Also seen in this figure is our finding that both the DA and quinpirole effects on PPI are reversed by pretreatment with the D_2 antagonist haloperidol. In a separate study, we found that the effects of intra-NAC DA infusion on PPI are reversed by the atypical antipsychotic clozapine (SWERDLOW and GEYER 1992).

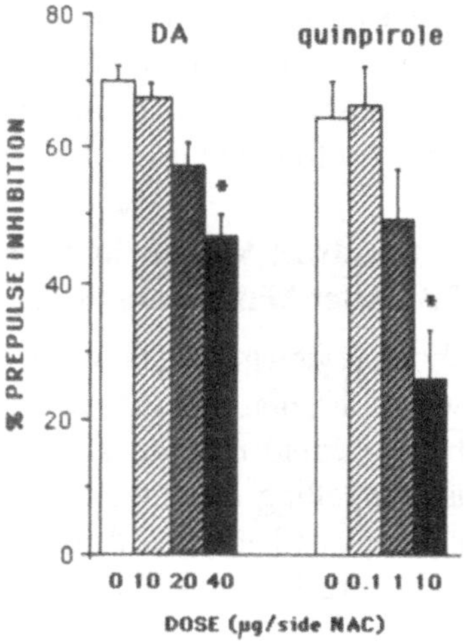

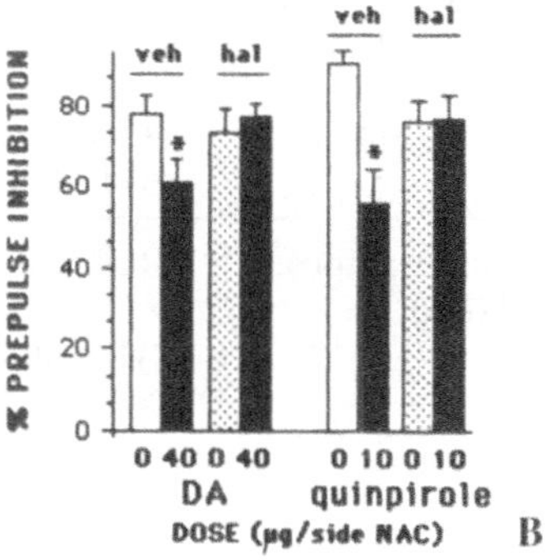

Fig. 5. **A** PPI is reduced after infusion of dopamine (*DA*) (SWERDLOW et al. 1992b) or quinpirole (WAN and SWERDLOW 1994) into the nucleus accumbens (*NAC*). **B** Both of these effects are reversed by treatment with haloperidol (*hal*) (SWERDLOW et al. 1994a; WAN and SWERDLOW 1994). Clearly, D_2 receptors in the ventral striatum serve a critical role in the modulation of PPI in rats, and this substrate is consistent with a major neurochemical construct for the pathophysiology of schizophrenia: the "DA hypothesis' (STEVENS 1973). *veh*, vehicle

These results are consistent with our previous reports (SWERDLOW et al. 1986a, 1990a,b, 1992b) that PPI is disrupted by specific manipulations that produce mesolimbic hyperdopaminergia. Here, we have demonstrated that mesolimbic hyperdopaminergia results in a reduction of PPI that is reversible by typical or atypical antipsychotic agents. These data are also consistent with the idea that the decrease in PPI after intra-NAC DA or quinpirole infusion reflects DAergic activation of D_2 receptors, since this effect is blocked by pretreatment with haloperidol. Thus, while we cannot be certain of the cause(s) of deficient sensorimotor gating in schizophrenia patients, the present findings suggest that the substrates of the DA-disruption of PPI in rats are consistent with one major – if not predominant – hypothesis for the pathophysiology of schizophrenia: overactivation of mesolimbic D_2 receptors (STEVENS 1973; WONG et al. 1986). Furthermore, haloperidol appears to "normalize" startle gating in dopamine-activated rats via blockade of mesolimbic D_2 receptors, a mechanism that is consistent with a major hypothesis for the therapeutic activity of these agents (WHITE and WANG 1984; RUPNIAK et al. 1985; TAMMINGA et al. 1988). This observation does not identify mesolimbic hyperdopaminergia as the single neural substrate that modulates sensorimotor gating. Instead, these data add to a series of findings suggesting that PPI is modulated by activity in interconnected limbic corticostriatopallidopontine circuitry (SWERDLOW et al. 1992a).

2. Construct Validity in Neuroanatomical Domain: PPI After Manipulations of Limbic Corticostriatopallidopontine Circuitry

When reviewing the literature regarding the neuroanatomical abnormalities in schizophrenia, one is struck by the multitude of brain regions that appear to be abnormal in at least some patients with this disorder (Table 1). There is compelling evidence to suggest that these regional abnormalities are not random and unrelated, but rather that the symptoms of schizophrenia reflect disturbances within anatomically and functionally interconnected cortico-

Table 1. Brain regions implicated n the pathophysiology of schizophrenia and in the neural modulation of PPI

Brain region	Abnormality in schizophrenia	Relationship to the modulation of PPI
1. Hippocampus (HPC)	Reduced volume [1]; cytoarchitectural abnormalities [2]	Reduced PPI after carbachol infusion into dentate gyrus, reversed by atropine coinfusion [3]; reduced PPI after NMDA infusion into ventral HPC, reversed by AP5 coinfusion [4]; ventral HPC lesions in adult rats produce a supersensitive apomorphine (APO) reduction of PPI [5]; ventral HPC lesions in neonatal rats reduce PPI post puberty, and increase the sensitivity to the APO reduction of PPI [6].
2. Prefrontal cortex (PFC)	Hypometabolism [7]; related neuropsychological deficits [8]	Lesions of the medial PFC in adult rats produce a supersensitive APO reduction of PPI [5].
3. Ventral striatum (VS)	40% reduction in cell number [9]; elevated D2 receptor density [10]; increased AMPA receptor number [11]	VS lesions reduce PPI [12], as does intra-VS infusion of direct or indirect DA agonists [13]; the latter effect is reversed by typical or atypical antipsychotics [14]; PPI is also reduced by AMPA infusion into the nucleus accumbens (NAC), and this effect is reversed by D2 blockade or 6-OHDA denervation [15]; NAC AMPA receptor blockade reverses the PPI-disruptive effects of amphetamine, but not quinpirole [15]. PPI is eliminated in patients with Huntington's disease who experience striatal spiny I cell degeneration [16].

Table 1. *Continued*

Brain region	Abnormality in schizophrenia	Relationship to the modulation of PPI
4. Ventral pallidum (VP)	Reduced cell number in internal pallidum [17]; abnormal metabolism in left globus pallidus [18]; enlarged lenticular nucleus [19]	VP infusion of muscimol restores PPI after NAC DA infusion [13] or NAC lesions [12]; VP infusion of picrotoxin [13], but not saclofen [20] reduces PPI; lenticular enlargement significantly correlates with PPI deficits in schizophrenia patients [21].
5. DM thalamus	Reduced cell number [9]	DMT is innervated by the VP [22], which modulates PPI (see above).
6. Pedunculopontine nucleus (PPTg)	Cytoarchitectural abnormalities [23]	PPTg lesions markedly reduce PPI [24].

PPI, prepulse inhibition; DM, dorsomedial; AMPA, α-amino-3-hydroxy-5-methyl-4-isoxazoleprionic acid; NMDA, *N*-methyl-D-aspartate; DA, dopamine; 6-OHDA, 6-hydroxydopamine.

References:

1. Suddath et al. (1990)
2. Altshuler et al. (1987)
3. Caine et al. (1991)
4. Swerdlow et al. (1993b)
5. Swerdlow et al. (1994b)
6. Lipska et al. (1994)
7. Weinberger et al. (1987a)
8. Weinberger et al. (1987b)
9. Pakkenberg (1990)
10. Wong et al. (1986)
11. Noga et al. (1994)
12. Kodsi and Swerdlow (1994)
13. Wan and Swerdlow (1994)
14. Swerdlow and Geyer (1992)
15. Wan et al. (1994b)
16. Swerdlow et al. (1995)
17. Bogerts et al. (1985)
18. Early et al. (1987)
19. Jernigan et al. (1991)
20. Kodsi et al. (1994)
21. Braff et al. (1993)
22. Groenewegen et al. (1993)
23. Karson et al. (1991)
24. Swerdlow and Geyer (1993b)

striato-pallido-thalamic (CSPT) circuitry. Furthermore, we believe that one of many functions of this interconnected CSPT circuitry, and its pallidal output via the pontine tegmentum, involves the modulation of sensorimotor gating. This belief comes from over 15 years of studies documenting the reduction of PPI in rats after manipulations of limbic CSPT circuitry, and in nonschizophrenic humans with known CSPT pathology. A noninclusive list of brain regions with documented abnormalities in schizophrenia patients, together with the results of PPI measures in related laboratory or clinical studies, is seen in Table 1.

a) The Hippocampus

We have reported that the neural modulation of PPI in rats is effected by circuitry linking the hippocampus (HPC), the NAC, the subpallidum, and the pontine reticular formation. A role for the HPC in controlling PPI is suggested by our findings that PPI is blocked by infusion of the cholinergic agonist carbachol into several HPC regions (including the dentate gyrus, area CA1, and the ventral subiculum (CAINE et al. 1991, 1992), but not by carbachol infusion into the parietal cortex (CAINE et al. 1991). PPI is also disrupted by infusion of *N*-methyl-D-aspartate (NMDA) into the ventral subiculum, and this effect is reversed by co-infusion of the NMDA receptor antagonist AP5 (SWERDLOW et al. 1993b,c). Intra-HPC carbachol or NMDA may impair PPI by stimulating glutamate release in the NAC via HPC-NAC glutamate fibers; such a mechanism is consistent with reports that intra-HPC carbachol infusions cause behavioral activation that is blocked by intra-NAC infusion of glutamate antagonists (MOGENSEN and NIELSEN 1984), and with our preliminary observations that the reduction in PPI after ventral subiculum NMDA infusion is antagonized by glutamate receptor blockade within the NAC (SWERDLOW et al. 1993b). Finally, a hippocampal-NAC link in the modulation of PPI is supported by our finding that lesions of the ventral hippocampus or medial prefrontal cortex in adult rats result in a "supersensitive" APO-disruption of PPI, presumably reflecting changes in DA receptor sensitivity within the NAC (SWERDLOW et al. 1994b) (Fig. 6).

The hippocampal substrates of deficient PPI identified by this animal model may add a functional context for interpreting reports of hippocampal dysfunction in schizophrenia based on neuropsychological (GRAY et al. 1991), neuropathological (ALTSHULER et al. 1987, 1990), and neuroimaging studies (SUDDATH et al. 1990). Furthermore, the hippocampal substrate links PPI with electrophysiological measures such as P50 gating, which is believed to reflect hippocampal function and is deficient in schizophrenia patients and asymptomatic first-degree relatives (SIEGEL et al. 1984), and which is significantly correlated with PPI within individual subjects (SCHWARZKOPF et al. 1992).

b) The Ventral Striatum

Evidence reviewed above suggests that PPI is modulated in part (but not exclusively) by brain DA activity. In rats, PPI is reduced by drugs that directly or indirectly activate D_2 receptors, but not D_1 receptors (PENG et al. 1990; WAN and SWERDLOW 1994; WAN et al. 1994a). Conversely, these effects are reversed by blockade of D_2 receptors, but not D_1 receptors (SWERDLOW et al. 1991c). There is evidence that the effects of DA agonists on PPI are mediated in part by increased DA activity in the NAC. First, low doses of APO that do not decrease PPI in control rats potently disrupt PPI in rats that are regionally altered to have "supersensitive" DA receptors in the NAC (SWERDLOW et al. 1986a). Second, the loss of PPI induced by the indirect DA agonist amphetamine (AMPH) is reversed by depletion of DA in the NAC (SWERDLOW et al. 1990b). Third, PPI is disrupted in rats by DA infusion into the NAC or anteromedial striatum (but not the orbital cortex, amygdala, or posterior striatum) (SWERDLOW et al. 1990a,b, 1992b, 1994a). Fourth, as described

A

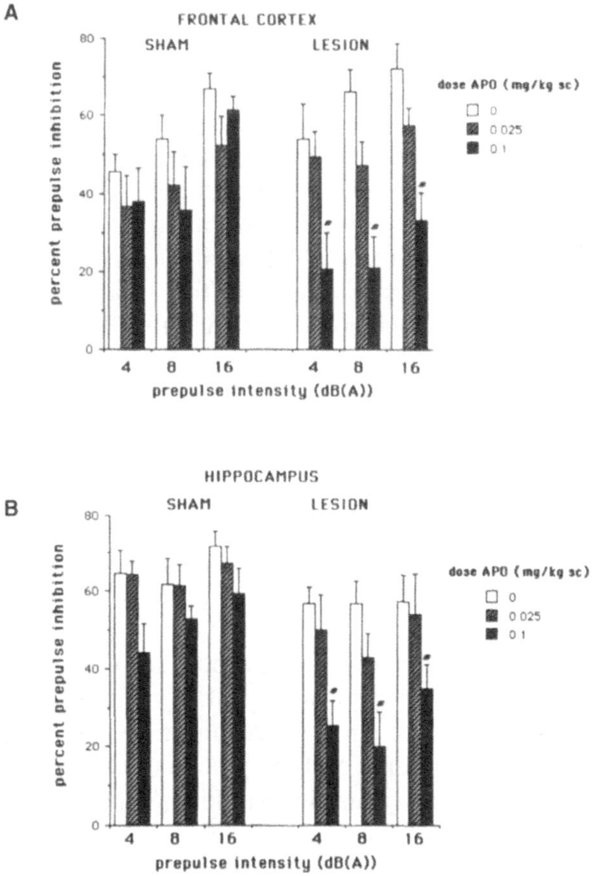

Fig. 6. Increased sensitivity to the apomorphine-disruption of PPI in rats after ibotenic acid lesions of the medial prefrontal cortex (**A**) or ventral hippocampus (**B**) (SWERDLOW et al. 1994b). Dysfunction in both of these regions has been implicated in the pathophysiology of schizophrenia. This observation is consistent with an important anatomical construct related to this illness, that limbic cortical damage results in subcortical DA hypersensitivity. WEINBERGER (1987) and others have hypothesized that such a condition, combined with a stress-induced release of subcortical DA, might be a critical precipitant of cognitive dysfunction in schizophrenia

above, the effects of intra-NAC DA or quinpirole infusion on PPI are reversed by systemic treatment with the D_2 antagonist haloperidol, and thus probably reflect activity at NAC D2 receptors.

c) *The Ventral Pallidum and Pontine Reticular Formation*

We have proposed that decreased PPI after NAC DA activation reflects decreased activity in γ-aminobutyric acid (GABA) ergic fibers projecting from the NAC to the subpallidal regions that include the ventral pallidum and the substantia innominata (SWERDLOW et al. 1990a). The PPI-disruptive effects of NAC DA infusion (SWERDLOW et al. 1990a) or quinolinic acid lesions of the

NAC (KODSI and SWERDLOW 1994) in rats are reversed by infusion of the GABA agonist muscimol into the subpallidum, and are reproduced by subpallidal infusion of the GABA antagonist picrotoxin (SWERDLOW et al. 1990a; SWERDLOW and GEYER 1993b). This NAC-subpallidal GABAergic projection is a substrate for other behavioral effects of NAC DA activation (JONES and MOGENSEN 1980; SWERDLOW and KOOB 1985; SWERDLOW et al. 1986b), and may translate the effects of activity in the amygdala and HPC to lower motor circuitry (MOGENSON 1987). While it is not clear how decreased subpallidal GABA activity is translated to the "primary startle circuit" to modulate PPI, one possible route is via subpallidal efferents to the pedunculopontine tegmental nucleus (PPTg) (SWANSON et al. 1984). Others (LEITNER et al. 1981) have suggested that the pontine tegmentum is a critical substrate of startle gating. Similarly, we have reported that electrolytic or quinolinic acid lesions of the PPTg significantly reduce PPI in rats (SWERDLOW and GEYER 1993a; SWERDLOW et al. 1993c). Preliminary reports suggest that there may be PPTg abnormalities in some schizophrenia patients (KARSON et al. 1991), indicating that dysfunction even in this most "distal" end of the sensorimotor "gating circuitry" might contribute to reduced PPI in schizophrenia patients.

The findings reviewed above give support for the role of HPC-NAC-subpallidal-PPTg circuitry in modulating startle gating in rats. The utility of PPI in elaborating the functional interconnections of limbic-cortical and mesolimbic-subpallidal circuitry, and the potential relevance of this circuitry to the pathophysiology of psychiatric disorders (SWERDLOW and KOOB 1987; GRAY et al. 1991), underscores the importance of assessing the validity of this animal model of sensorimotor gating deficits in schizophrenia. Specifically, if the disruption of PPI in rats by manipulations of this circuitry is a valid model of deficient sensorimotor gating in schizophrenia patients, then gating deficits in schizophrenia patients might reflect pathology in this same neural circuitry. The PPI model would thus be a critical tool for understanding the development and pathophysiology of deficient sensorimotor gating in humans, for identifying general strategies for normalizing gating, and for screening specific compounds for gating-restorative properties.

3. Construct Validity in a Neurodevelopmental Domain: PPI After Isolation Rearing or Neonatal Brain Lesions

Sensorimotor gating deficits in schizophrenia patients – unlike PPI deficits in APO-treated rats – are not an acute drug response, but instead reflect longitudinal and complex interactions of genetic, developmental, social, and environmental forces. For this reason, it is important that startle gating in rats appears to be sensitive to these same forces. Genetic factors may be critical determinants of sensorimotor gating in rats, since strain-related differences in the dopaminergic modulation of PPI have been reported (RIGDON 1990). Other studies have begun to systematically characterize differences in startle response properties in approximately fifty different rat strains (GLOWA and HANSEN 1994). If susceptibility to the gating-disruptive effects of DA agonists is genetically controlled in rats, this model might offer critical insight into genetic factors mediating the susceptibility to, and development of schizophrenia in humans (WEINBERGER 1987). Studies also indicate that development and stress may significantly alter startle gating in rats. For example, recent studies (GEYER et al. 1993) indicate that isolation-reared rats that exhibit elevated NAC DA activity also demonstrate a neuroleptic-reversible deficiency in PPI compared to group-reared controls (Fig. 7).

Developmental issues may be particularly relevant to understanding PPI deficits in humans: since PPI appears to develop in children between ages 5 and 8 (ORNITZ et al. 1991), it is conceivable that abnormal developmental processes may contribute to PPI deficits in adult schizophrenia patients. In a preliminary finding of direct relevance to neurodevelopmental theories of schizophrenia (WEINBERGER 1987), we have noted impaired PPI and enhanced sensitivity to the PPI-disruptive effects of APO in postpubescent rats that had received neurotoxin lesions of the hippocampus as neonates (LIPSKA et al. 1994). Other findings suggest that social influences may also critically impact the DAergic modulation of PPI: the APO-disruption of PPI is dramatically reduced in rats that are tested in the presence of other rats, compared to rats that are tested along, an effect which may be attributable to changes in sen-

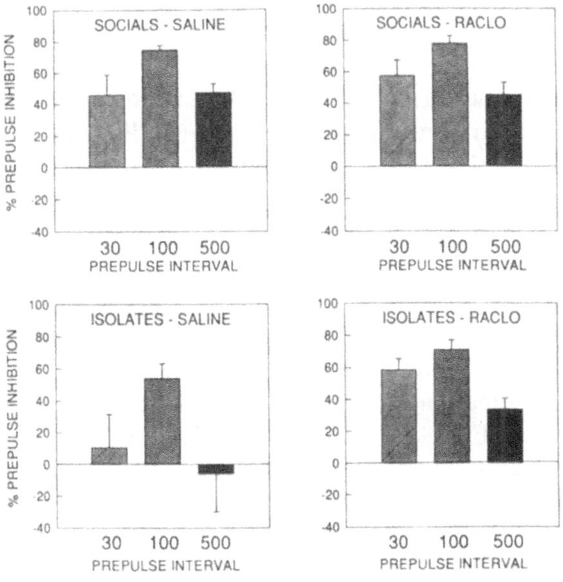

Fig. 7. Reduced PPI in isolation-reared rats compared to rats raised in groups. PPI is restored in isolation-reared rats by the D_2 antagonist raclopride (*RACLO*) (GEYER et al. 1993). This model, together with a model of reduced PPI and increased sensitivity to apomorphine-disruption of PPI in neonatal hippocampal lesioned rats (LIPSKA et al. 1994), demonstrate the construct validity of this model with critical neurodevelopmental hypotheses of schizophrenia

sorimotor gating caused by high-frequency vocal communications between rats (MICZEK et al. 1992; DAVIS, personal communication). Thus, this model can potentially be used to study the contribution of genetic, developmental, and social forces to the pathophysiology of sensorimotor gating deficits in schizophrenia.

In summary, impaired PPI in rats appears to be a valid model for a psychophysiological construct – impaired sensorimotor gating in schizophrenia – and it also appears to be a valid model for neurochemical, neuroanatomical, and neurodevelopmental constructs for the pathophysiology of schizophrenia and for the mechanism of antipsychotic action. Studies supporting construct validity for this model in neurochemical, neuroanatomical, and neurodevelopmental domains have added to our understanding of several issues in the neurosciences, ranging from the connectivity and behavioral functions of integrated CSPT circuitries to the neurobiological consequences of isolation rearing.

B. Discussion

We have argued that the loss of PPI of acoustic startle in rats is a model with face, predictive, and construct validity for the loss of sensorimotor gating in schizophrenia patients. This model may be critically important for understanding the neurobiological substrates of schizophrenia, and it may assist in designing novel antipsychotic agents and pharmacological treatment strategies. The PPI model differs from existing animal models since (1) it uses identical methods across species to measure a deficit that is integral to the schizophrenia syndrome (impaired sensorimotor gating), (2) PPI predicts antipsychotic potency for both typical and atypical antipsychotics, and (3) PPI is controlled by a defined neural circuit that has repeatedly been implicated in the pathophysiology of schizophrenia and which can be manipulated and studied in animals.

The search for robust and valid animal models of psychopathological states and abnormal cognitive function is a critically important area of research. Among the features that make the startle reflex a valuable behavioral measure of CNS function (DAVIS 1980), two features are most critical to the present discussion. First, this reflex is modulated by forebrain circuitry, and hence can be used to study the organization and pathophysiology of brain structures that may be relevant to human psychopathology. In a series of investigations, we have studied the modulation of startle by weak prepulses in rats as a means of understanding the neural substrates of sensorimotor gating (cf. BRAFF and GEYER 1990; SWERDLOW et al. 1992a). By delineating the CNS control of sensorimotor gating, it might be possible to identify a neural basis for deficient sensorimotor gating in certain disorders, including schizophrenia. Second, startle can be elicited across species using similar or identical stimulus parameters that elicit similar responses. The cross-species comparability of startle is striking: for example, despite the great disparity in body size, the onset latency of acoustic startle in rats (time from auditory nerve stimulation to first EMG response) is 8ms (DAVIS et al. 1982), while in humans it is approximately 14ms (MEIER-EWERT et al. 1974). This cross-species comparability largely reflects the fact that the "primary" startle response is controlled by simple neural circuitry – at or below the pontine reticular formation (DAVIS et al. 1982) – that is similar in rats and humans. But it is the inhibition of startle, rather than the reflex itself, that offers special promise in understanding the neurobiology of schizophrenia.

I. Model Limitations

The findings reviewed above support the validity of the PPI animal model in three domains: face, predictive, and construct validity. Nonetheless, there are clear limitations and caveats to our ability to validate each level of this model. First, startle in humans was measured using EMG recordings from orbicularis oculi, while rat startle was measured using whole-body displacement. While substantial data supports the comparability of startle measures from a wide range of muscle groups (DAVIS 1984), the face validity of this model might be better assessed using identical startle measures in rats and humans. Second, the studies reported here compared the ability of drugs to restore PPI in rats to their ability to decrease schizophrenia-linked symptoms in humans. While one "core" element of the pathology of schizophrenia is an inability to "inhibit" or "gate" sensory stimuli (MCGHIE and CHAPMAN 1961; SHAGASS 1976) and schizophrenia patients exhibit deficient PPI (BRAFF et al. 1978), the predictive validity of the PPI model would be better assessed by comparing the ability of drugs to restore PPI in DA-stimulated rats to their ability to restore PPI in schizophrenia patients. Importantly, recent studies have reported that clozapine and haloperidol significantly increase PPI in schizophrenia patients (WU et al. 1992).

Finally, there are several limitations to assessing the construct validity of this model. These limitations stem from our ignorance regarding the pathophysiology of PPI deficits in schizophrenia or the therapeutic mechanisms of antipsychotic agents, from the likelihood that several variations of this disorder reflect different forms of neuropathology that may differ in the extent and substrates of deficient gating, and from the construct limitations inherent to modeling cognitive symptoms in infrahumans. For example, the inability of schizophrenia patients to "gate" stimuli may contribute to cognitive flooding and fragmentation (MCGHIE and CHAPMAN 1961), the consequences of which are probably fundamental to psychotic thought processes. Obviously, parallel flooding and fragmentation in rats must have a very different impact on a rat's "consciousness", as well as its integrative social functioning. Furthermore, PPI is deficient in patients with "schizotypal personality disorder" (CADENHEAD et al. 1993) and in "psychosis-prone" individuals (SIMMONS 1990). Thus, deficient PPI may reflect pathology that exists beyond the "nuclear" schizophrenias, and which is expressed in dimensions of personality, temperament, and "ego" function (Perry et al. 1992). Particularly in these domains, it will be difficult to demonstrate the construct validity of an animal model.

II. Model Strengths

Despite the above caveats, this model has clear strengths, particularly in the domain of predictive validity. Several assets of the PPI model simply reflect the utility of using the startle reflex as a dependent measure: it is an easily and objectively quantified graded response that is under tight stimulus control, is comparable across species, has a nonzero baseline, and is controlled by a simple defined neural circuit. All of these characteristics distinguish the PPI model from other preclinical models of antipsychotic potency, such as the ability of drugs to prevent APO-induced canine emesis (Creese et al. 1976; Freedman and Giarman 1956) and rodent stereotypy (Creese et al. 1976). Furthermore, the PPI model demonstrates sensitivity to both typical antipsychotics and atypical antipsychotics. In contrast, clozapine fails to reverse amphetamine- and APO-induced stereotypy in rats, or APO-induced emesis in dogs (Creese et al. 1976).

In addition to its sensitivity, the specificity of the PPI model is supported by findings that it predicts no antipsychotic potency for buspirone, diazepam, imipramine, naloxone, or propranolol. We are presently studying the role of noradrenergic and serotonergic systems in the modulation of startle gating, and we will extend these studies to examine interactions of these systems with the DAergic control of PPI. Indeed, while our data suggests that D_1 dopamine receptors do not modulate PPI in intact rats, findings from others suggest that these receptors may be important in the control of PPI in dopamine-denervated rats (Schwartzkopf et al. 1991), and perhaps in intact animals (Hoffman and Donovan 1994). Our preclinical studies indicate that PPI is regulated by a complex set of forebrain circuitries, with interactions between multiple neurotransmitters and receptors, including prominent cholinergic and GABAergic elements. These interactions may be important in a model of schizophrenic pathology since we have moved, in general, from "one transmitter – one locus" theories of the pathophysiology of schizophrenia towards understanding multiple neurotransmitters in complex neural circuit distributions. Certainly, this level of complexity suggests that drugs with many different pharmacological properties might be able, in certain conditions, to restore "gating circuit" functions.

The strength of this model and its exciting long-term benefits in the predictive domain is enhanced by our ability to define the neural control of PPI "upstream" and "downstream" from the DA receptor. The risks of tardive dyskinesia and the prominence of "typical" antipsychotic-resistant symptoms in schizophrenics have stimulated efforts to develop antipsychotic compounds that do not act primarily through DA receptor blockade. Since the antipsychotic potency of these drugs will not be detected by traditional screening models of DA receptor activation (Creese et al. 1976), it will be important to develop screening models that measure processes that are central to the pathophysiology of the schizophrenias, but that are not linked exclusively to DA receptor activation. In this context, the PPI model seems quite useful, since PPI is not controlled exclusively by DA receptor activation, but instead is mediated via integrated corticolimbic and basal ganglia circuitry, of which the DA receptor is only one – albeit a central – element. Thus, it is conceivable that variations of the PPI model might be used to predict antipsychotic potency of compounds that act at non-DAergic receptors at several levels of limbic, mesolimbic, and subpallidal circuitry.

C. Conclusions

Sensorimotor gating of the startle reflex in rats can be reliably disrupted by several different experimental manipulations, including systemic administration of DA agonists or glutamate antagonists, isolation rearing, or surgical alterations of neural activity within limbic cortico-striato-pallido-pontine circuitry. Sensorimotor gating of startle is also reduced in schizophrenia spectrum patients, and by systematically studying deficient sensorimotor gating in specific animal models, we hope to gain insight into the neurobiological substrates of gating, deficits in schizophrenia, and into strategies for antipsychotic drug development. These animal models of impaired gating offer both strengths and weaknesses, but collectively they may be particularly valuable in their ability to identify compounds with clinical antipsychotic potency, and in their utility for addressing mechanistic questions at the level of the underlying neural circuitry.

Acknowledgements. This work was supported by National Institute of Mental Health grant nos. R29 MH48381 (NRS) and MH-42228 (DLB, MAG, NRS), and Research Scientist Development Award grant no MH-00188 (MAG). The authors gratefully acknowledge the assistance of Ms. Pamela Auerbach in manuscript preparation. All data in this paper have been previously published in the referenced material.

References

Altshuler LL, Conrad A, Kovelman JA, Scheibel A (1987) Hippocampal pyramidal cell orientation in schizophrenia. Arch Gen Psychiatry 44:1094–1098
Altshuler LL, Casanova MF, Goldberg TE, Kleinman JE (1990) The hippocampus and parahippocampus in schizophrenic, suicide, and control brains. Arch Gen Psychiatry 47:1029
Bakshi VP, Swerdlow NR, Geyer MA (1994) Clozapine antgaonizes phencyclidine-induced deficits in sensorimotor gating of the startle response. J Pharmacol Exp Ther 271:787–794
Black JL, Richelson E, Richardson JW (1985) Antipsychotic agents: a clinical update. Mayo Clin Proc 60:777–789
Bogerts B, Meertz E, Schonfeldt-Bausch R (1985) ganglia and limbic system pathology in schizophrenia: a morphometric study of brain volume and shrinkage. Arch Gen Psychiatry 42:784–791

Braff DL, Geyer MA (1990) Sensorimotor gating and schizophrenia: Human and animal model studies. Arch Gen Psychiatry 47:181–188
Braff D, Stone C, Callaway E, Geyer M, Glick I, Bali L (1978) Prestimulus effects on human startle reflex in normals and schizophrenics. Psychophysiology 15:339–343
Braff DL, Heaton R, Kuck J, Cullum M, Moranville J, Grant I, Zisook S (1991) The generalized pattern of neuropsychological deficits in outpatients with chronic schizophrenia with heterogeneous Wisconsin Card Sorting Test results. Arch Gen Psychiatry 48:889–898
Braff DL, Geyer MA, Swerdlow NR, Jernigan T (1993) Gating deficits, neural circuits, and schizophrenia. Biol Psychiatry 33:88A
Cadenhead KS, Geyer MA, Braff DL (1993) Impaired startle prepulse inhibition and habituation in schizotypal patients. Am J Psychiatry 150:1862–1867
Caine SB, Geyer MA, Swerdlow NR (1991) Decreased startle gating in the rat after carbachol infusion into CA1, dentate gyrus and subiculum. Soc Neurosci Abstr 17:1453
Caine SB, Geyer MA, Swerdlow NR (1992) Carbachol infusion into the dentate gyrus disrupts sensorimotor gating of startle in rats. Psychopharmacology (Berl) 105:347–354
Casey DE (1991) Raclopride versus haloperidol: comparative efficacy in a double-blind multicenter investigation. Biol Psychiatry 29:110A
Conrad AJ, Abebe T, Austin R, Forsythe S, Scheibel AB (1991) Hippocampal pyramidal cell disarray in schizophrenia as a bilateral phenomenon. Arch Gen Psychiatry 48:413–417
Creese I, Burt DR, Snyder SH (1976) Dopamine receptor binding predicts clinical and pharmacological potencies of antischizophrenic drugs. Science 192:481–483
Davis KL, Kahn RS, Ko G, Davidson M (1991) Dopamine in schizophrenia: a review and reconceptualization. Am J Psychiatry 148:1474–1486
Davis M (1980) Neurochemical modulation of sensory-motor reactivity: acoustic and tactile startle reflexes. Neurosci Biobehav Rev 4:241–263
Davis M (1984) The mammalian startle response. In: Eaton RC (ed) Neural mechanisms of startle behavior. Plenum, New York, pp 287–342
Davis M, Aghajanian GK (1976) Effects of apomorphine and haloperidol on the acoustic startle response in rats. Psychopharmacology (Berl) 47:217–223
Davis M, Gendelman D, Tischler M, Gendelman P (1982) A primary acoustic startle circuit: lesion and stimulation studies. J Neurosci 2:791–805
Early TS, Relman EM, Raichle ME, Spitzmagel EL (1987) Left globus pallidus abnormality in newly medicated patients with schizophrenia. Proc Natl Acad Sci USA 84:561–563
Ellenbroek BA, Cools AR (1990) Animal models with construct validity for schizophrenia. Behav Pharmacol 1:469–490
Freedman DX, Giarman NJ (1956) Apomorphine test for tranquilizing drugs: effect of dibenamine. Science 124:264
Geyer MA, Tapson GS (1988) Habituation of tactile startle is altered by drugs acting on serotonin-2 receptors. Neuropsychopharmacology 1:135–147
Geyer MA, Wilkinson LS, Humby T, Robbins TW (1993) Isolation rearing of rats produces a deficit in prepulse inhibition of acoustic startle similar to that in schizophrenia. Biol Psychiatry 34:361–372
Glowa JR, Hansen CT (1994) Differences in response to an acoustic startle stimulus among forty-six rat strains. Behav Genet 24:79–84
Graham F (1975) The more or less startling effects of weak prestimulation. Psychophysiology 12:238–248
Gray JA, Feldon J, Rawlins JNP, Hemsley DR, Smith AD (1991) The neuropsychology of schizophrenia. Behav Brain Sci 14:1–84
Groenewegen HJ, Berendse HW, Haber SN (1993) Organization of the output of the ventral striatopallidal system in the rat: ventral pallidal efferents. Neuroscience 57:113–142

Hall H, Farde L, Sedvall G (1988a) Human dopamine receptor subtypes – in vitro binding analysis using 3H-SCH 23390 and 3H-raclopride. J Neural Transm 73:7–21

Hall H, Kohler C, Gawell L, Farde L, Sedvall G (1988b) Raclopride, a new selective ligand for the dopamine D_2 receptors. Prog Neuropsychopharmacol Biol Psychiatry 12:559–568

Heimer L, Zahm DS, Churchill L, Kalivas PW, Wohltmann C (1991) Specificity in the projection patterns of accumbal core and shell in the rat. Neuroscience 41:89–125

Hoffman DC, Donovan H (1994) D_1 and D_2 dopamine receptor antagonists reverse prepulse inhibition deficits in an animal model of schizophrenia. Psychopharmacology (Berl) 115:447–453

Ison JR, McAdam DW, Hammond GR (1973) Latency and amplitude changes in the acoustic startle reflex produced by variations in auditory prestimulation. Physiol Behav 10:1035–1039

Javitt DC, Zukin SR (1991) Recent advances in the phencyclidine model of schizophrenia. Am J Psychiatry 148:1301–1308

Jernigan TL, Zisook S, Heaton RK, Moranville JT, Hesselink JR, Braff DL (1991) Magnetic resonance imaging abnormalities in lenticular nuclei and cerebral cortex in schizophrenia. Arch Gen Psychiatry 48:881–890

Jones DL, Mogenson GJ (1980) Nucleus accumbens to globus pallidus GABA projection subserving ambulatory activity. Am J Physiol 238:R63–R69

Karson CN, Garcia-Rill E, Biedermann J, Mrak RE, Husain MM, Skinner RD (1991) The brainstem reticular formation in schizophrenia. Psychiatry Res 40:31–48

Keith VA, Mansbach RS, Geyer MA (1991) Failure of haloperidol to block the effects of phencyclidine and dizocilpine on prepulse inhibition of startle. Biol Psychiatry 30:557–566

Kodsi MH, Swerdlow NR (1994) Quinolinic acid lesions of the ventral striatum reduce sensorimotor gating of acoustic startle in rats. Brain Res 643:59–65

Kodsi MH, Wan FJ, Swerdlow NR (1994) Prepulse inhibition of startle: anatomy and neurochemistry of striato-pallidal modulation. Soc Neurosci Abstr 20:781

Leitner DS, Powers AS, Stitt CL, Hoffman HS (1981) Midbrain reticular formation involvement in the inhibition of acoustic startle. Physiol Behav 26:259–268

Lipska BK, Swerdlow NR, Jaskiw GE, Geyer MA, Braff DL, Weinberger DR (1994) Neonatal hippocampal damage disrupts sensorimotor gating in postpubertal rats. Biol Psychiatry 35:631–632

Madreas BK, Fahey MA, Canfield DR, Spealman RD (1988) D_1 and D_2 dopamine receptors in caudate-putamen of nonhuman primates (Macaca fascicularis). J Neurochem 51:934–943

Mansbach RS, Geyer MA, Braff DL (1988) Dopaminergic stimulation disrupts sensorimotor gating in the rat. Psychopharmacology (Berl) 94:507–514

Mansbach RS, Geyer MA (1989) Effects of phencyclidine and phencyclidine biologs on sensorimotor gating in the rat. Neuropsychopharmacology 2:299–308

McGhie A, Chapman J (1961) Disorders of attention and perception in early schizophrenia. Br J Med Psychol 34:102–116

Meier-Ewert K, Gleitsmann K, Reiter F (1974) Acoustic jaw reflex in man: its relationship to other brain stem and microreflexes. Electroencephalogr Clin Neurophysiol 36:629–638

Miczek KA, Vivian JA, Tornatzky W, Farrell WJ, Sapperstein SB (1992) Withdrawal from diazepam in rats: ultrasonic vocalizations and acoustic startle reflex. J Psychopharmacol 187:A47

Mogenson GJ (1987) Limbic-motor integration. Prog Psychobiol Physiol Psychol 12:117–170

Mogenson GJ, Nielsen M (1984) A study of the contribution of hippocampal-accumbenssubpallidal projections to locomotor activity. Behav Neural Biol 42:52–60

Noga JT, Hyde TM, Herman M, Bigelow LL, Weinberger DR, Kleinman JE (1994) Glutamate receptor autoradiography of postmortem basal ganglia in schizophrenia. Biol Psychiatry 35:735
Ornitz EM, Guthrie D, Sadeghpour M, Sugiyama T (1991) Maturation of prestimulation-induced startle modulation in girls. Psychophysiology 28:11–20
Pakkenberg B (1990) Pronounced reduction of total neuron number in mediodorsal thalamic nucleus and nucleus accumbens in schizophrenics. Arch Gen Psychiatry 47:1023–1028
Pellegrino LJ, Pellegrino AS, Cushman AJ (1979) A stereotaxic atlas of the rat brain, 2nd edn. Plenum, New York
Peng RY, Mansbach RS, Braff DL, Geyer MA (1990) A D2 dopamine receptor agonist disrupts sensorimotor gating in rats: implications for dopaminergic abnormalities in schizophrenia. Neuropsychopharmacology 3:211–218
Perry W, Cadenhead K, Braff DL (1992) Thought disorder in the group of schizophrenias. Biol Psychiatry 31:117A
Pycock PJ, Kerwin RW, Carter CJ (1990) Effect of lesions of cortical dopamine terminals on subcortical dopamine receptors in rats. Nature 286:74–77
Richelson E (1985) Pharmacology of neuroleptics in use in the United States. J Clin Psychiatry 46:9–14
Rigdon GC (1990) Differential effects of apomorphine on prepulse inhibition of acoustic startle reflex in two rat strains. Psychopharmacology (Berl) 102:419–421
Rigdon GC, Viik K (1991) Prepulse inhibition as a screening test for potential antipsychotics. Drug Dev Res 23:91–99
Rupniak NMJ, Hall MD, Kelly E., Fleminger S, Kilpatrick G, Jenner P, Marsden CD (1985) Mesolimbic dopamine function is not altered during continuous chronic treatment with typical or atypical neuroleptic drugs. J Neural Transm 62:249–266
Schwartzkopf SB, Bruno JP, Mitra T (1991) Dopamine D_1/D_2 receptor effects on sensory gating in controls and neonatally dopamine depleted animals. Abstr Am Coll Neuropsychopharmacol 208
Schwartzkopf SB, Lamberti JS, Crilly SF, Martin R, Hirt J, Holley L (1992) Combined startle and P50 measures of sensory gating: support for a shared neurophysiology. Biol Psychiatry 31:147A
Seeman P, Lee T, Chau-Wong M, Wong K (1976) Antipsychotic drug doses and neuroleptic/dopamine receptors. Nature 261:717–719
Seeman P, Ulpian C, Bergeron C, Riederer P, Jellinger K, Babriel E, Reynolds G, Tourtellotte W (1984) Bimodal distribution of dopamine receptor densities in brain of schizophrenics. Science 225:728–731
Shagass C (1976) An electrophysiological view of schizophrenia. Biol Psychiatry 11:3–29
Siegel C, Waldo M, Mizner G, Adler LE, Freedman R (1984) Deficits in sensory gating in schizophrenic patients and their relatives: evidence obtained with auditory evoked responses. Arch Gen Psychiatry 41:607–612
Simmons RF (1990) Schizotypy and startle prepulse inhibition. Psychophysiology 27 [Suppl]:S6
Sprouse JS, Aghajanian GK (1986) (–)-Propranolol blocks the inhibition of serotonergic dorsal raphe cell firing by 5HT1A agonists. Eur J Pharmacol 128:295–298
Stevens JR (1973) An anatomy of schizophrenia? Arch Gen Psychiatry 29:177–189
Suddath RL, Christison MD, Torrey EF, Cassanova M, Weinberger DR (1990) Anatomic abnormalities in the brains of monozygotic twins discordant for schizophrenia. N Engl J Med 322:789–794
Swanson LW, Mogenson GJ, Gerfen CR, Robinson P (1984) Evidence for a projection from the lateral preoptic area and substantia innominata to the "mesencephalic locomotor region" in the rat. Brain Res 295:161–178
Swerdlow NR, Geyer MA (1992) Clozapine and haloperidol in a novel model for assessing antipsychotic potency. Soc Neurosci Abstr 18:375

Swerdlow NR, Geyer MA (1993a) Clozapine and haloperidol in an animal model of sensorimotor gating deficits in schizophrenia. Pharmacol Biochem Behav 44:741–744
Swerdlow NR, Geyer MA (1993b) Prepulse inhibition of acoustic startle in rats after lesions of the pedunculopontine nucleus. Behav Neurosci 107:104–117
Swerdlow NR, Koob GF (1985) Separate neural substrates of amphetamine-, caffeine, and CRF-stimulated locomotor activation. Pharmacol Biochem Behav 23:303–307
Swerdlow NR, Koob GF (1987) Dopamine, schizophrenia, mania and depression: toward a unified hypothesis of cortico-striato-pallido-thalamic function. Behav Brain Sci 10:197–245
Swerdlow NR, Geyer M, Braff D, Koob GF (1986a) central dopamine hyperactivity in rats mimics abnormal acoustic startle in schizophrenics. Biol Psychiatry 21:23–33
Swerdlow NR, Vaccarino FJ, Amalric M, Koob GF (1986b) Neural substrates for the motor-activating properties of psychostimulants: A review of recent findings. Pharmacol Biochem Behav 25:233–248
Swerdlow NR, Braff DL, Geyer MA (1990a) GABAergic projection from nucleus accumbens to ventral pallidum mediates dopamine-induced sensorimotor gating deficits of acoustic startle in rats. Brain Res 532:146–150
Swerdlow NR, Braff DL, Masten VL, Geyer MA (1990b) Schizophrenic-like sensorimotor gating abnormalities in rats following dopamine infusion into the nucleus accumbens. Psychopharmacology (Berl) 101:414–420
Swerdlow NR, Mansbach RS, Geyer MA, Pulvirenti L, Koob GF, Braff DL (1990c) Amphetamine disruption of prepulse inhibition of acoustic startle is reversed by depletion of mesolimbic dopamine. Psychopharmacology (Berl) 100:413–416
Swerdlow NR, Caine B, Geyer MA (1991a) Opiate-dopamine interactions in the neural substrates of acoustic startle gating in the rat. Prog Neuropsychopharmacol Biol Psychiatry 15:415–426
Swerdlow NR, Caine SB, Geyer MA, Swenson MR, Downs NS, Braff DL (1991b) sensorimotor gating deficits in schizophrenia and Huntington's disease and in rats after manipulation of hippocampus-accumbens circuitry (Abstr). International Congress on Schizophrenia Research
Swerdlow NR, Keith VA, Braff DL, Geyer MA (1991c) The effects of spiperone, raclopride, SCH 23390 and clozapine on apomorphine-inhibition of sensorimotor gating of the startle response in the rat. J Pharmacol Exp Ther 256:530–536
Swerdlow NR, Caine SB, Braff DL, Geyer MA (1992a) Neural substrates of sensorimotor gating of the startle reflex: A review of recent findings and their implications. J Psychopharmacol 6:176–190
Swerdlow NR, Caine BC, Geyer MA (1992b) Regionally selective effects of intracerebral dopamine infusion on sensorimotor gating of the startle reflex in rats. Psychopharmacology (Berl) 108:189–195
Swerdlow NR, Benbow CH, Zisook S, Geyer MA, Braff DL (1993a) A preliminary assessment of sensorimotor gating in patients with obsessive compulsive disorder (OCD). Biol Psychiatry 33:298–301
Swerdlow NR, Wan FJ, Caine SB (1993b) Modulation of prepulse inhibition of acoustic startle through NMDA receptors in the ventral subiculum: neurochemical and neuroanatomical substrates. Soc Neurosci Abstr 19:144
Swerdlow NR, Wan F, Kodsi M, Hartston H, Caine S (1993c) Limbic cortico-striato-pallido-pontine (CSPP) substrates of startle gating. Biol Psychiatry 33:62A
Swerdlow NR, Braff DL, Taaid N, Geyer MA (1994a) Assessing the validity of an animal model of sensorimotor gating deficits in schizophrenic patients. Arch Gen Psychiatry 51:139–154
Swerdlow NR, Lipska B, Weinberger D, Braff DL, Jaskiw G, Geyer MA (1994b) Effects of frontal cortex and hippocampal lesions on apomorphine-disrupted startle gating. Biol Psychiatry 35:707

Swerdlow NR, Paulsen J, Braff DL, Butters N, Geyer MA, Swenson MR (1995) Impaired prepulse inhibition of acoustic and tactile startle in patients with Huntington's disease. J Neurol Neurosurg Psychiatry 58:192–200
Swerdlow NR, Zisook D, Taaid N (1994d) Seroquel (ICI 204,636) restores prepulse inhibition of acoustic startle in apomorphine-treated rats: similarities to clozapine. Psychopharmacology (Berl) 4:675–657
Tamminga CA, Burrows GH, Chase TN, Alphs LD, Thaker GK (1988) Dopamine neuronal tracts in schizophrenia: their pharmacology and in vivo glucose metabolism. In: Kalivas PW, Nemeroff CB (eds) The mesocorticolimbic dopamine system. New York Academy of Sciences, New York, pp 443–450
Wan FJ, Swerdlow NR (1994) Intra-accumbens infusion of quinpirole impairs sensorimotor gating of acoustic startle in rats. Psychopharmacology (Berl) 113:103–109
Wan FJ, Geyer MA, Swerdlow NR (1994a) Accumbens D2 substrates of sensorimotor gating: assessing anatomical localization. Pharmacol Biochem Behav 48:155–163
Wan FJ, Kodsi MH, Swerdlow NR (1994b) Nucleus accumbens dopamine-glutamate interactions in the modulation of prepulse inhibition in rats. Soc Neurosci Abstr 20:824
Weinberger DR (1987) Implications of normal brain development for the pathogenesis of schizophrenia. Arch Gen Psychiatry 44:660–669
Weinberger DR, Berman KF, Zec RF (1987a) Physiologic dysfunction of dorsolateral prefrontal cortex in schizophrenia. I. Regional cerebral blood flow evidence. Arch Gen Psychiatry 44:660–669
Weinberger DR, Berman KF, Illowsky BP (1987b) Physiologic dysfunction of dorsolateral prefrontal cortex in schizophrenia. III. A new cohort and evidence of a monoaminergic mechanism. Arch Gen Psychiatry 44:609–615
White FJ, Wang RY (1984) A10 dopamine neurons: role of autoreceptors in determining firing rate and sensitivity to dopamine agonists. Life Sci 34:1161–1170
Wong D, Wagner H, Tune L, Dannals R, Pearlson G, Links J, Tamminga C, Broussolle E, Ravert H, Wilson A, Toung J, Malat J, Williams J, O'Tuama L, Snyder S, Kuhar M, Gjedde A (1986) Positron emission tomography reveals elevated D_2 dopamine receptors in drug-naive schizophrenics. Science 234:1558–1563
Wu JC, Siegel BV, Haier RJ, Buchsbaum MS (1990) Testing the Swerdlow/Koob model of schizophrenia pathophysiology using positron emission tomography. Behav Brain Sci 13:169–170
Wu JC, Potkin SG, Ploszaj DI, Lau V, Telford J, Richmond G (1992) Clozapine improves sensory gating more than haldol. APA Abstr 156

CHAPTER 11

Patterns of Clinical Efficacy for Antipsychotic Drugs

JULIA GOLIER and MICHAEL DAVIDSON

A. Introduction

Antipsychotic agents are beneficial in the treatment of many medical and psychiatric disorders such as agitation, delirium, and affective psychoses; however, the bulk of the research on their efficacy has been conducted on schizophrenic patients. As psychosis is only one of many clinical manifestations of schizophrenia, the efficacy of a medication as an antipsychotic agent per se and as a treatment for schizophrenia are by no means the same. Besides delusions and hallucinations, other signs and symptoms of schizophrenia include flattened or inappropriate affect, disordered thought processes, and impaired social and occupational functioning. Although psychosis has been the primary target symptom in neuroleptic treatment studies, a maximally effective medication would ideally target the full range of signs and symptoms associated with schizophrenia. This chapter will review the scales used to measure the signs and symptoms of schizophrenia and the efficacy of currently available neuroleptics in treating them.

B. Rating Scales Used to Measure Neuroleptic Effect

Rating instruments that reliably measure the signs and symptoms of a disorder and their changes over time are crucial for clinical trials. Over the past two decades significant advances have been made in the development of standardized scales used to diagnose schizophrenia and measure symptom change. Initially such scales focused primarily on psychotic or positive symptoms of schizophrenia; more recently, they have incorporated the negative symptoms. The term "positive symptom" was first coined by Hughlings JACKSON to suggest symptoms that were potentially the result of a loss of inhibition. In schizophrenia, positive symptoms generally refer to delusions and hallucinations. "Negative symptoms" are those characterized by a loss of function or the absence of expected behavior, potentially as a result of brain damage. In schizophrenia a loss of social drive or flattened affect are among the symptoms classified as negative.

I. Brief Psychiatric Rating Scale

The Brief Psychiatric Rating Scale (BPRS) is frequently used to measure antipsychotic efficacy. It was developed in the early 1960s to assess treatment efficacy in an array of psychiatric symptoms including depression, anxiety, guilt, and hallucinations (OVERALL and GORHAM 1962). It was designed primarily for psychiatric inpatients, not for schizophrenic patients in particular. The sum of the 18 items which reflect signs, symptoms, and behaviors of psychiatric illnesses measures severity of illness. A 20% reduction in the total BPRS score is frequently used as one criterion for defining a schizophrenic patient as a treatment responder.

Factor analysis of the 18 BPRS items in schizophrenics results in four clusters that may reflect specific symptoms with a common pathophysiology. The following four empirically derived factors, with their contributing BPRS items in parentheses, are frequently used: (1) anxiety-depression (anxiety, guilt feelings, depressive mood) (2) thought disturbance (conceptual disorganization, hallucinatory behavior, unusual thought content) (3) withdrawal-retardation (emotional withdrawal, motor retardation, blunted affect) and (4) hostile-suspiciousness (hostility, suspiciousness, and uncooperativeness). Although the BPRS was designed before the renewed interest in classifying symptoms as positive and negative, the thought-disturbance and withdrawal-retardation factors are often used as indicative of these respective symptom clusters. Scales that measure a wider array of schizophrenic symptoms may gradually supplant the BPRS.

II. Clinical Global Impression

The Clinical Global Impression (CGI) scale is a commonly used measure of global efficacy in psychopharmacologic research. It is used for a variety of psychiatric populations to measure either severity of illness or degree of improvement. The global impression of severity scale asks the rater, "Considering your total clinical experience with this particular population, how mentally ill is the patient at this time?"; the answer is rated on a seven-point scale from "normal, not at all ill" to "among the most extremely ill."

For the global impression of change, the clinician is asked how much the patient has changed since the beginning of treatment; the answer ranges from "very much improved" to "very much worse." While the global impression is highly subjective and dependent on the breadth of a rater's experience, it incorporates the myriad factors upon which clinical judgments are based. When used together with structured measures, the global impression provides a clinically relevant backdrop against which changes in specific symptoms can be appreciated. Used in this way the global impression can help answer the question of whether a statistically significant change is also a clinically significant one. To avoid bias, a clinician who does not have knowledge of the patients' BPRS scores should rate the CGI.

III. Global Assessment Scale

The Global Assessment Scale (GAS) is another global measure used to assess overall severity of psychiatric disturbance (ENDICOTT et al. 1976). It is a 100-point scale, based on both symptoms and functioning, which encompasses the continuum from severe psychiatric illness to psychological well-being. The patients' lowest level of functioning within a specified time period is rated. A ten-point increase on the GAS is often considered indicative of significant clinical improvement. One advantage of the GAS is that clinicians are familiar with it as it is the basis for the global assessment of function on Axis V of DSM. Schizophrenics fall into a fairly restricted range on this scale. The use of a single score to measure both symptoms and functioning limits its sensitivity in measuring neuroleptic response.

IV. Negative Symptoms and Rating Scales

1. Methodologic Problems in the Measurement of Negative Symptoms

Attempts to define and measure negative symptoms have been fraught with numerous methodologic obstacles. Negative symptoms are more difficult to quantify than positive ones. The presence of a hallucination is unequivocally abnormal. The severity of a hallucination can be measured in terms of its intensity, frequency, and duration, or according to the degree to which it causes distress or influences behavior. In contrast, the nature and intensity of motivation, social interest, and emotional relatedness vary considerably over time and within the normal population. Without quantifiable social norms, deviations are difficult to measure (SOMMERS 1985). Another obstacle is that negative symptoms are not unique to schizophrenia. Psychomotor retardation and reduced social interest are also diagnostic features of depression and dementia. The side effects of medication can mimic negative symptoms. For example, the masked facies of neuroleptic-induced parkinsonism may be difficult to discriminate from blunted affect. Additionally, negative symptoms are usually inferred from observations which have multiple possible determinants. For example, social withdrawal may be due to lack of social drive, which is a negative symptom. Alternatively, it may be an adaptive means of reducing overstimulation or a response to paranoid delusions or auditory hallucinations.

2. Rating Scales Used to Measure Negative Symptoms and the Deficit Syndrome

Several scales have been developed to measure negative symptoms, each with their own means of addressing the methodologic difficulties encountered in eliciting and defining them. Among these scales are: the Positive and Negative Syndrome Scale (PANSS), the Scale for the Assessment of Negative Symptoms (SANS), the Schedule for the Deficit Syndrome (SDS) and the Quality of

Life Scale (QLS). Table 1 lists the individual negative signs and symptoms measured on each scale.

a) Positive and Negative Syndrome Scale

The Positive and Negative Syndrome Scale (PANSS) is a 30-item scale based on 18 items from the BPRS and 12 items from the Psychopathology Rating Scale (Kay et al. 1989). Individual items are grouped into three domains: positive, negative, and general symptoms, and rated on a seven-point severity

Table 1. Scales used to measure negative symptoms and/or the deficit syndrome in schizophrenia

Scale	Individual items
BPRS	Withdrawal-retardation factor – Emotional withdrawal – Motor retardation – Blunted affect
PANSS	Blunted affect Emotional withdrawal Poor rapport Social withdrawal Difficulty in abstraction Lack of spontaneity Stereotyped thinking
SANS	Apathy Attentional impairment Affective blunting Asociality Anhedonia
SDS	Restricted affect Diminished emotional range Diminished sense of purpose Diminished social drive
QLS	Interpersonal relationships – Household, friends, acquintances – Social activity, social network – Social initiative, withdrawal – Sociosexual functioning Instrumental role – Occupational role, work functioning – Work level, work satisfaction Intrapsychic foundations – Sense of purpose, motivation, curiosity – Anhedonia, aimless inactivity, empathy – Emotional interaction Common objects and activities

BPRS, Brief Psychiatric Rating Scale; PANSS, Positive and Negative Syndrome Scale; SANS, Scale for the Assessment of Negative Symptoms; SDS, Schedule for the Deficit Syndrome; QLS, Quality of Life Scale.

scale. Table 1 lists the symptoms considered as definitively belonging to the negative syndrome. Delusions, hallucinations, grandiosity, excitement, hostility, and conceptual disorganization belong to the positive symptom subscale. Symptoms which cannot be defined as clearly positive or negative (e.g., feelings of guilt, poor attention, and anxiety) are included on the global psychopathology subscale. This subscale serves as a measure of severity against which individual symptoms are evaluated. The PANSS is currently in common usage, and is reliable. Guidelines for conducting the interview and operationalized criteria for rating symptoms and their severity enhance the reliability of the PANSS compared with earlier scales (Kay et al. 1989).

b) Scale for the Assessment of Negative Symptoms

There are five negative symptom complexes measured by the Scale for the Assessment of Negative Symptoms (SANS) as shown in Table 1 (Andreasen 1982). The SANS has been shown to have good interrater reliability and the symptom complexes to have good internal consistency. Contributing to the high interrater reliability is the fact that the rating of each symptom is broken down into observable behavioral components. For example, individual behaviors contributing to the measure of affective flattening include unchanging facial expression, decreased spontaneous movements, paucity of expressive gestures, poor eye contact, affective nonresponsivity, inappropriate affect, and lack of vocal inflection. Unlike other scales, the SANS includes attentional impairment as a negative symptom. This symptom complex correlated less well than the others with the composite SANS score, and may be related to positive symptoms and severity of illness.

c) Schedule for the Deficit Syndrome

The Schedule for the Deficit Syndrome (SDS) (Kirkpatrick et al. 1989) was designed to aid classification of patients with the deficit syndrome. Patients with the deficit syndrome are those with chronic negative symptoms, symptoms thought to be a core feature of schizophrenia. In contrast to these primary negative symptoms, secondary negative symptoms are understood as symptoms that are a consequence of factors extrinsic to schizophrenia (e.g., medication side effects) or of other schizophrenic symptoms (e.g., psychosis) (Carpenter et al. 1985). To meet criteria for the deficit syndrome, patients must have two of the symptoms listed in Table 1 for at least 1 year. The presence of these symptoms cannot be limited to psychotic exacerbation but must also be present during periods of relative clinical stability. Additionally, depression, anxiety, medication side effects, and environmental deprivation must be ruled out as probable causes of the negative symptoms. The origins of negative symptoms are not always clearly discernible. However, classifying negative symptoms as either primary or secondary can help discriminate potentially treatable symptoms from those for which current treatment strategies may be of limited use.

d) Quality of Life Scale

The Quality of Life Scale (QLS) was designed to measure the deficit syndrome of schizophrenia (HEINRICHS et al. 1984). Compared with other scales of negative symptoms, the QLS emphasizes the functional disability related to this syndrome. It was intended to standardize the kinds of clinical judgments used to infer why certain behaviors are either manifest or absent. Though limited primarily to outpatients, the QLS is likely to be used more widely in the future as the deficit syndrome and social and occupational impairments are increasingly being recognized as targets for treatment.

V. Rating Scales and Phenomenology

The development of reliable scales to measure positive and negative signs and symptoms of schizophrenia has also facilitated research on the phenomenology of this disorder. There has been considerable debate about which psychopathological processes underlie schizophrenia, which symptoms contribute to these processes, and how they are related. CROW (1980) hypothesized that there were two syndromes in schizophrenia. Type I schizophrenia is characterized by psychotic symptoms, good response to treatment, good outcome, and minimal or no cognitive dysfunction. Type II is characterized by a predominance of negative symptoms, poor response to neuroleptics, poor outcome, and varying degrees of cognitive impairment. The two syndromes were also thought to have separate etiologies: type I was thought to reflect neurochemical abnormalities such as dopamine excess, and type II was thought to reflect structural brain abnormalities and cell loss. The testing of this hypothesis has generated much research. Others have thought that positive and negative symptoms represent opposite poles of a continuum (ANDREASEN and OLSEN 1982) or semiindependent processes (STRAUSS et al. 1974). The existence of semiindependent processes would be consonant with the frequent coexistence of these symptoms both longitudinally and cross-sectionally. While the precise relationship of these symptom clusters remains unclear, the validity of basic positive-negative distinction has gained support from multiple areas of research. It has been consistently shown that patients with significant negative or deficit symptomatology have a history of inferior premorbid social or sexual functioning, lower educational level, compromised work capacity and social adjustment, and a poorer prognosis (MCGLASHAN and FENTON et al. 1992; MEUSER et al. 1991).

While the positive and negative distinction has been supported, several factor-analytic studies suggest the presence of three rather than two domains of psychopathology in schizophrenia. The three domains are (1) psychomotor poverty (including blunted affect, poverty of speech), (2) disorganization (formal thought disorder, inappropriate affect and poverty of content of speech), and (3) reality distortion (delusions and hallucinations) (LIDDLE 1987; MALLA et al. 1993; KLIMIDIS et al. 1993). This solution is not inconsistent with the positive-negative distinction. Instead it emphasizes the possibility that thought disturbance, which had been variably considered as either positive or negative, may be a separate domain of psychopathology. Social dysfunction itself may be another another unique dimension of schizophrenia (KEEFE et al. 1992), and persecutory delusions and suspiciousness may be a separate symptom subset within the psychotic symptoms (MINAS et al. 1992). Although positive symptoms have traditionally been the primary target symptoms, negative symptoms, thought disorder, and cognitive deficits are increasingly being considered as targets for pharmacologic treatment.

C. Efficacy of Neuroleptics in Schizophrenia

I. Efficacy in the Treatment of Schizophrenic Exacerbation

1. Introduction of Chlorpromazine

Chlorpromazine, an aliphatic phenothiazine, was the first available antipsychotic medication. Its introduction in the 1950s was a turning point in the history of psychiatric treatment, and in the treatment of schizophrenia in particular. Chlorpromazine was first used in combination with barbiturates for prolonged sleep treatment or *hibernation artificielle*. Chlorpromazine was found to differ from barbiturates in that it decreased emotional reactivity without inducing significant somnolence. These properties of chlorpromazine were further explored at a time when somatic treatments in psychiatry were primarily limited to insulin shock, frontal leukotomy, and electroconvulsive treatment.

A landmark paper described the effects of chlorpromazine on general psychiatric inpatients in careful detail (KINROSS-WRIGHT 1955). Chlorpromazine preferentially benefitted the severely ill, psychotic, and delirious patients. Among the therapeutic effects described were increased contact with reality among severely withdrawn patients, and improved sleep and appetite. Delusions and hallucinations disappeared completely in some schizophrenics. Psychomotor agitation and excitement associated with various conditions responded to 200–400 mg/day of chlorpromazine, but the average dose for psychotic patients was 800 mg/day. Depressives and neurotic patients responded less well, and the latter group especially had difficulty tolerating the side effects. The notable side effects of chlorpromazine included lethargy and somnolence from which patients could be easily aroused, difficulty focusing on close objects, dry mouth and nose, and hypotension with compensatory tachycardia. At high doses a Parkinson's-like syndrome with coarse tremor, muscular rigidity, and immobile expression developed, which disappeared upon dose reduction. Almost as dramatic as the unprecedented response of psychotic patients to chlorpromazine was the variability of response itself. Among the 29 schizophrenics treated in this series, 8 went into remission, 13 were much improved, 4 were slightly improved, and 4 were unchanged.

2. Treatment of Psychosis

The primary goal of acute treatment is the alleviation of psychotic symptoms. Multiple studies in the 1960s and 1970s demonstrated the superior efficacy of chlorpromazine compared to placebo in treating psychosis, predominantly among schizophrenic patients, as well as its superiority to sedatives such as barbiturates and benzodiazepines (BALDESSARINI 1985). Since the discovery of chlorpromazine, numerous other phenothiazines and neuroleptics of other classes, such as the thioxanthenes and butyrophenones, have been developed. All have in common the propensity to cause extrapyramidal symptoms (EPS) and do not differ in their antipsychotic efficacy. The following neuroleptics were consistently found more efficacious than placebo and as efficacious as chlorpromazine or other standard reference neuroleptics: fluphenazine, mesoridazine, perphenazine, thioridazine, triflupromazine, trifluoperazine, chlorprothixene, haloperidol, and molindone (BALDESSARINI 1985). An example of such a comparison study is the NIMH-sponsored cooperative study of acute psychosis among schizophrenics. Overall, 75% of patients treated with phenothiazine improved, compared to only 25% on placebo. Among those receiving phenothiazines, no difference was found in the percentage improved between thioridazine, fluphenazine, or chlorpromazine (NIMH 1964). The finding in multiple studies that the percentage of patients improved did not differ by type or class of standard neuroleptic suggests that standard neuroleptics are equally efficacious. Whether or not standard neuroleptics are equally efficacious for a given individual has not been well studied. Clinical standards favor switching the type or class of neuroleptic for patients who respond poorly to a given neuroleptic. (See Chap. 12 by H.Y. MELTZER for a discussion of management of the treatment-refractory patient.)

3. Treatment of Negative Symptoms

While there is abundant evidence that neuroleptics alleviate psychotic exacerbation in the majority of schizophrenics patients, whether they also reduce negative symptoms has not been definitively answered. The question had been put forth in part to test CROW's hypothesis (CROW 1980) that schizophrenics with negative symptoms were poorly responsive to neuroleptics. One study frequently cited in support of CROW's hypothesis is a randomized double-blind comparison of α-flupenthixol (a thiaxanthene compound with dopamine-blocking properties), β-flupenthixol (an isomer without significant dopamine-blocking activity), and placebo in the acute treatment of schizophrenia (JOHNSTONE et al. 1978). (As needed, doses of chlorpromazine were also used to treat behavioral disturbances in all three treatment groups.) α-flupenthixol was superior to β-flupenthixol and placebo in reducing delusions, hallucinations, and thought disorder, but not flattened affect, muteness, depression, anxiety, or retardation. The small sample size ($n = 15$ in each treatment group) and the significantly more frequent use of chlorpromazine in the placebo and β-fluphenthixol group may have obscured differences in drug effects. Similarly, in an open-label study of 21 schizophrenics, positive symptoms (BPRS thought disturbance, activation, and hostile-suspiciousness factors) but not negative symptoms (BPRS withdrawal-retardation factor) responded to neuroleptic treatment (ANGRIST et al. 1980).

Others have concluded that negative symptoms do respond to standard neuroleptics during acute treatment. A review of five large-scale studies conducted in the 1960s, primarily on acute patients, provides evidence for this assertion (GOLDBERG 1985). Although these studies were conducted prior to the development of reliable scales of negative symptoms, conclusions based on them are enhanced by the use of large sample sizes and comparison placebo groups. For instance, in an NIMH-sponsored study, drug–placebo differences were found in negative and deficit-like symptoms (indifference to environment, slowed speech and movements, hebephrenic symptoms, social participation, poor self-care, confusion) in addition to positive symptoms (GOLDBERG 1985; GOLDBERG et al. 1965). During the study, the placebo group was more likely to develop these broadly defined negative or deficit symptoms, which highlights the importance of having such a comparison group (GOLDBERG 1985). Of the five large-scale studies reviewed, all found a reduction in thought disorder – a symptom which has been variably classified as negative – with drug treatment, and several found reductions in blunted affect and withdrawal-retardation.

Thus, while these clinical trials suggest that neuroleptics may reduce some negative symptoms and prevent their development in the context of an exacerbation, it is not clear whether this is due to the treatment of psychosis or to the treatment of an underlying pathophysiology common to both symptom clusters (KANE and MAYERHOFF 1989). Methodologic means of addressing this question include studying predominantly negative-symptom schizophrenics and accounting for the change in psychotic symptoms when measuring the change in negative symptoms. An example of such a design was a study of symptomatic, but not floridly psychotic, ambulatory schizophrenics who were selected based on having at least moderate negative symptoms (SERBAN et al. 1992). Patients improved in all five domains of the SANS as well as in positive symptoms during treatment with thiothixene. After controlling for the change in positive symptoms, the change in each of the five negative symptom domains remained statistically significant. Within the limits of an open-label design, these findings suggest that when symptoms coexist, reduction in negative symptoms may not wholly depend on reduction in psychotic symptoms.

4. Neuroleptic Dose and Efficacy

Given that multiple classes of standard neuroleptics have been found equally efficacious in treating psychosis, attempts have been made to maximize their efficacy by finding the optimal dose. In the 1970s, very large oral or parenteral doses of neuroleptics were used in an attempt to achieve this goal rapidly, a practice called "rapid neuroleptization." The use of very high doses of neuroleptics became possible with the development of higher-potency drugs than chlorpromazine such as haloperidol and fluphenazine. Higher-potency drugs are less likely to cause hypotension, sedation, and other dose-limiting side effects. Rapid neuroleptization and other very high dose strategies have subsequently been discredited as causing unnecessary side effects without improving efficacy.

In a review of 19 controlled studies of dose–effect relationships in acute treatment, high doses of neuroleptics [mean chlorpromazine equivalents (CPZE) 2035 ± 782 mg] provided rates or degrees of response similar to moderate doses (CPZE 505 ± 114 mg) (BALDESSARINI et al. 1988). Further analyses suggested a dose–effect relationship: 250 mg CPZE/day or less was suboptimal for some patients, 500–700 mg CPZE/day was optimal, and doses higher than 700 mg/day were excessive. That doses greater than CPZE 500–700 mg/day do not increase the likelihood of response has been largely confirmed in subsequent studies. For example, in a double-blind comparison of 10, 30, and 80 mg/day of haloperidol (2 mg/day of haloperidol is roughly equivalent to 100 mg/day of chlorpromazine) in newly admitted schizophrenics, two-thirds of the patients went into remission within 6 weeks. No difference was found among the three dosage groups in terms of psychotic symptoms or clinical impression. All patients were prophylactically treated with benztropine 2 mg three times a day. While there was a trend towards increased EPS over the course of study, the three dosage groups did not differ with respect to EPS (RIFKIN et al. 1991). Negative symptoms were not measured. Similarly, in a double-blind comparison of an average of 20 mg versus 58 mg/day of haloperidol, there were no differences between the groups with respect to reduction in positive and negative symptoms or in the emergence of EPS (MODESTIN et al. 1983). In contrast, there are conflicting results about the dose–effect relationship at the lower end of the standard neuroleptic range. In a 4-week single-blind study of three doses of haloperidol (5, 10, and 20 mg/day) in newly admitted schizophrenics, 20 mg was superior to 5 mg in reducing BPRS psychotics symptoms throughout the 4 weeks, and was marginally superior to the 10-mg dose after 2 weeks (VAN PUTTEN et al. 1990). However, the neuroleptic dose greatly affected the BPRS withdrawal-retardation factor. The 5- and 10-mg groups showed slight improvement with treatment compared with their baseline. The 20-mg group scored significantly worse than the lower-dose groups on the withdrawal-retardation factor, and deteriorated with respect to their baseline. They also had greater akinesia and akathisia and were more likely to sign out against medical advice. The discrepancy between these studies with respect to dose-dependent EPS and exacerbation of negative symptoms is striking. The most likely explanation is that the studies that did not find dose-dependent differences used anticholinergics more aggressively and were double-blind. Although dose–effect studies are unlikely to find a single optimal dose for all patients, the above study suggests that roughly 10 mg/day of haloperidol may be the optimal minimal dose for most patients.

While the aforementioned studies provide useful information about group response rates, treating all patients in the standard range will expose a proportion of them to more than is needed, with the possible risk of additional side effects and discomfort. Ideally, the clinician would be able to find the lowest effective dose for a given individual. Such a method was proposed by HAASE in 1961. He suggested that the lowest dose at which muscular rigidity first becomes apparent, the "neuroleptic threshold," is the dose at which maximum therapeutic benefit is attained, and that doses above the threshold do not lead to greater benefit (cited in McEVOY et al. 1991). This hypothesis was tested in a large sample of schizophrenic and schizoaffective patients (McEVOY et al. 1991). The mean neuroleptic threshold dose of haloperidol was found to be 3.7 ± 2.3 mg/day. During 2 weeks of open-label treatment with the threshold dose of haloperidol, 52% of patients responded by a priori response criteria. Patients were then randomized to either continue at the threshold dose or at a dose two to ten times greater than the threshold dose (11.6 ± 4.7 mg/day). Among the initial non-responders, the percentage that responded to the higher dose (44%) did not differ from the percentage that responded to continued treatment at the threshold dose (39%). The initial responders remained responders, whether they continued the threshold dose or a higher dose. Of the patients only exposed to the neuroleptic threshold dose, 72% responded within the 5-week period. The only benefit seen at higher doses was a greater reduction is scores of hostility, but not a greater reduction in psychotic symptoms. However those on the higher dose experienced more EPS, including bradykinesia and akathisia, anxiety, and dysphoria, despite the use of anticholinergics. These results suggest that relatively low individualized doses of haloperidol (<6 mg/day) provide the maximum therapeutic benefit attainable during acute treatment. Higher dosages did not confer any advantage to patients who responded to the threshold dose. The success of low-dose treatment in this group may be related to the fact that 30% of the sample was neuroleptic-naive and patients with known treatment resistance were excluded, rather than being due to treatment with the "neuroleptic threshold." Given the importance of developing methods for individualized treatment, studies comparing individualized dosage with fixed dosage are needed.

II. Efficacy of Maintenance Neuroleptic Treatment

1. Relapse Prevention

Following remission of acute symptoms, the next goal of treatment is to prevent relapse. Continued medication is a critical component of maintenance treatment, which should be supplemented with psychosocial interventions such as psychoeducation, social skills training, rehabilitation, supportive housing, and family treatment.

Maintenance medicaion significantly reduces but by no means eliminates the risk of relapse among schizophrenics. This is demonstrated in a review of

24 placebo-controlled double-blind comparisons of maintenance neuroleptic treatment versus placebo. In this pooled sample ($n = 3195$), 65% of patients on placebo relapsed, as compared to only 30% on maintenance therapy (DAVIS 1975). Multiple double-blind studies with follow-up periods of a year or greater have subsequently confirmed higher rates of relapse in patients maintained off neuroleptics. Inasmuch as the data show that maintenance neuroleptic treatment protects against relapse, a substantial percentage of patients off neuroleptics did not relapse. Given the serious side effects of neuroleptics, including tardive dyskinesia, extrapyramidal reactions, and neuroleptic malignant syndrome, the question of whether to begin maintenance treatment is a difficult one. Even among remitted, acute first-episode schizophrenics, in whom a good short-term prognosis might be expected, 41% on placebo deteriorated clinically within a year compared to 0% maintained on fluphenazine (KANE et al. 1982). Ideally, the clinician would have the ability to predict which patients are unlikely to relapse off medication to avoid exposing them to neuroleptics unnecessarily. Unfortunately, predictors of relapse are not well established. In the aforementioned first-episode study, poorer social adjustment predicted relapse. Other predictors of relapse found by several authors include: younger age, early age of onset, and higher baseline neuroleptic dose (PRIEN et al. 1968; RASSIDAKIS et al. 1970; RUSKIN and NYMAN 1991). In the absence of clear guidelines, the risk–benefit ratio of maintenance neuroleptic treatment is determined by the clinician on a case-by-case basis, taking into account possible predictors, and the individual's history of relapse and its consequences. Although further research in this area is enormously important, it will probably be delayed until guidelines for the ethical conduct of neuroleptic-withdrawal studies are established.

2. Neuroleptic Dose and Efficacy

As with acute treatment, much work has been done to find the optimal dose of neuroleptic for maintenance treatment and relapse prevention. Very high dose neuroleptic treatment has not been found more effective than standard dose in preventing relapse; rather, it results in more side effects. The search for the optimal dose has thus more recently been based on attempts to find the minimum effective dose. One rationale for dosage reduction is that lower doses may be needed to prevent relapse than are required to treat exacerbation. Dosage reduction may also decrease reversible side effects, and enhance compliance and subjective sense of well-being (SCHOOLER 1991). Supporting this idea is a survey which found that 46% of schizophrenics took less medication than prescribed. Noncompliance was related to a dysphoric response to neuroleptics and EPS, most notably akathisia (VAN PUTTEN 1974). It is also hoped, but not established, that using lower doses of maintenance neuroleptic can reduce the incidence of tardive dyskinesia. The incidence of tardive dyskinesia, a stigmatizing and potentially irreversible involuntary movement disorder, increases with age and years of exposure to neuroleptic.

Several methods for studying dose–response are used. Patients undergoing dosage reduction can be compared with controls maintained on a stable dose of medication. Alternatively, patients can be randomly assigned to different dose levels for comparison of different dose ranges (SCHOOLER 1991). One of the earliest of such studies compared dosage reduction among hospitalized inpatients to continued treatment at a stable dose and to neuroleptic discontinuation. Following stabilization, 5% of those continued on the same dose relapsed within 4 months, compared to 15% of those maintained on $^3/_7$ of the original dose and 45% of those switched to placebo (CAFFEY et al. 1964). The inclusion of a placebo group highlights the fact that even though those on a reduced dose had an elevated relapse rate, treatment with reduced dose was superior to neurolepic withcrawal (SCHOOLER 1991).

Several well-controlled studies of low-dose maintenance treatment have been conducted using fixed dose ranges of neuroleptic. In a 2-year double-blind study of low versus standard dose of fluphenazine decanoate (5 mg vs 25 mg every 2 weeks), the low-dose group had a significantly higher rate of psychotic exacerbation (64% versus 31%), but not of relapse or hospital readmission (MARDER et al. 1984). (Relapses were defined as psychotic exacerbations not controlled by an increase in dose of up to twice the maintenance dose.) This study did not measure negative symptoms; however, the low-dose group had lower subjective scores of obsessive-compulsiveness, interpersonal sensitivity, depression, and phobic anxiety. These subjective scores were highly correlated with akathisia and retardation. Another study did not find a difference in either relapse rates or minor psychotic exacerbations between standard and low-dose fluphenazine decanoate treatment (25 ± 25.7 mg vs 3.8 ± 2.1 mg every 2 weeks, respectively) (HOGARTY et al. 1988). In a double-blind comparison of very low dose (1.25–5 mg every 2 weeks) versus standard dose (12.5–50 mg every 2 weeks) of fluphenazine decanoate, 56% of the low dose group and only 7% of the standard dose group relapsed during the 1-year follow-up. Although the low-dose group relapsed more often, during periods of clinical stability it had fewer negative symptoms as measured by the BPRS than the standard-dose group. Since the groups did not differ on scores of EPS, this would be an unlikely explanation for the difference in negative symptoms. Even though very low doses of maintenance neuroleptic may increase the likelihood of exacerbation, low-dose decanoate treatment is superior to placebo in reducing relapse among schizophrenics (EKLUND and FORSMAN 1991). The above studies in well-stabilized outpatient schizophrenics are not entirely consistent but suggest that very low doses are accompanied by an increased risk or relapse. Doses as low as 5 mg of fluphenazine decanoate every 2 weeks appear to be as efficacious as standard doses in reducing relapse if exacerbations are treated promptly. However, this dose may be associated with an increase in exacerbations. Compared with studies of acute treatment, in maintenance studies, dose-dependent increases in EPS and subjective discomfort are more apparent. Particular attention should be paid to minimizing and treating these side effects during maintenance treatment.

III. Neuroleptics with Special Profiles of Efficacy

It has long been hoped that neuroleptics with unique profiles would be developed: those without short- or long-term neurologic side effects, those that treat primary enduring negative symptoms, and those that treat patients who do not respond to standard neuroleptics. Several antipsychotics have shown promise in this regard.

Pimozide and other diphenylbutylpiperidines as well as benzamides have shown promise in being differentially effective in reducing negative symptoms. For example, in a sample of patients selected in part because of high levels of emotional withdrawal, pimozide was superior to chlorpromazine in reducing this symptom (KOLVAKIS et al. 1974). Unfortunately, on the whole, the advantages of diphenylbutylpiperidines have been subtle and may be explained by differences in dose and EPS (KANE and MAYERHOFF 1989). While the possible advantage of amisulpiride and sulpiride have generally not been significant (GERLACH et al. 1985), low-dose amisulpiride was recently shown to significantly reduce negative symptoms in a neuroleptic-naive sample (PAILLÈRE-MARTINOT et al. 1995). This study was unique in that patients were chosen on the basis of having primary negative symptoms. Additionally, positive symptoms were not reduced compared with placebo and the treated group had more EPS. That these two factors did not confound interpretation raises the possiblity that amisulpiride had a specific therapeutic effect on negative symptoms and deserves further study.

Clozapine, the prototypical "atypical" antipsychotic, has received much attention as a treatment of negative symptoms. It is unique in that it is superior to standard neuroleptics in the treatment of otherwise refractory patients and does not cause EPS or tardive dyskinesia. (See Chap. 12 by H.Y. MELTZER for a discussion of the treatment of refractory patients.) Given its side-effect profile, which includes agranulocytosis and seizures, its use is limited primarily to treatment-refractory schizophrenics who can comply with weekly hematologic monitoring.

In a pilot study of newly admitted schizophrenics, clozapine did not improve negative symptoms but did improve thought disturbance, hostile-suspiciousness, and psychosis; however, this study was limited by an open-label design, a small sample size, and low doses of clozapine (CHOUINARD and ANNABLE 1976). In contrast, in a double-blind comparison of chlorpromazine and clozapine (up to 900 mg/day) in acute schizophrenic patients, clozapine was superior to chlorpromazine in all factors of the BPRS and on the CGI (CLAGHORN et al. 1987). In a treatment-refracory population, clozapine (up to 900 mg/day) was superior to chlorpromazine on all positive symptom items, emotional withdrawal, blunted affect, psychomotor retardation, and disorientation (KANE et al. 1988). In another study of treatment-refractory schizophrenics, a crossover placebo-controlled double-blind trial of fluphenazine and clozapine, 38% of patients showed a superior response to clozapine compared to fluphenazine; 48% showed an equivalent response (PICKAR et al. 1992). Both treatments were superior to placebo in treating positive and negative symptoms. Treatment at an optimal dose of clozapine (mean 543 ± 207 mg/day) but not a moderate dose (mean 374 ± 111 mg/day) was also superior to fluphenazine in reducing BPRS negative symptoms. Neither dose range of clozapine was superior in reducing the total SANS score. This discrepancy may be due to the higher levels of EPS associated with fluphenazine, which may be more likely to influence the BPRS negative symptom score than the SANS score. Overall the data suggest that clozapine reduces negative symptoms even in otherwise refractory patients, and the effect may be dose-dependent. Whether it is superior to standard neuroleptics in treating negative symptoms in treatment-responsive populations or in patients with primary negative symptoms requires further study.

Whether clozapine specifically improves negative symptoms during long-term treatment is unclear. In a year-long open-label study of clozapine in schizophrenic outpatients with a history of residual positive or negative symptomatology, patients had significant improvement in their level of functioning at 6- and 12-month follow-up compared to their preclozapine baseline, as well as in illness severity and positive symptoms. However, improvement measured on the SANS and the QLS was not significant (BREIER et al. 1993). In another follow-up study of treatment-refractory patients at 6 months, clozapine improved thought disturbance and BPRS positive symptoms but not negative symptoms. However, improvement on all quality-of-life factors was noted (MELTZER et al. 1990). Thus, while both open-label studies found improvement in functioning with long-term clozapine treatment, albeit on different scales, it is not clear to what degree, if any, this is mediated by a reduction in negative symptoms.

Risperidone is a novel antipsychotic agent which was recently approved by the U.S. Food and Drug Administration (FDA). Like clozapine it is a D2 and 5HT2 antagonist; in contrast to clozapine, however, it does not have serious adverse effects. The property which most distinguishes risperidone from standard neuroleptics is its reduced propensity for causing EPS at therapeutic doses. Whether it is superior to standard neuroleptics in reducing negative symptoms is unclear. Two multicenter studies have compared several fixed doses of risperidone to placebo and to a single dose of haloperidol (20 mg/day). In the first study (CHOUINARD et al. 1993), of all the treatment groups only the risperidone 6-mg/day group had a significant reduction in negative symptoms compared to the placebo group. The 6-mg/day group did not differ significantly from the haloperidol group with respect to negative symptoms. None of the doses of risperidone was superior to haloperidol in the treatment of positive symptoms. In the second multicenter trial, the 6- and 16-mg/day group, but not the 2- or 10-mg/day risperidone group or the haloperidol group, showed significant reduction in negative symptoms (MARDER and MEIBACH 1994). These studies provide partial support for the hypothesis that risperidone treats negative symptoms which coexist with positive symptoms. These studies cannot answer whether risperidone is superior to haloperidol in reducing negative symptoms, given that only a single, moderately high dose of haloperidol was used, and that this dose was associated with significantly more EPS in both studies.

Although encouraging research has shown a reduction in some negative symptoms with neuroleptic treatment, many patients continue to suffer from negative and deficit symptoms despite adequate treatment of exacerbation. Patients with the deficit syndrome require further study, preferably during the maintenance treatment. To address the methodologic pitfalls, study populations should be chosen on the basis of persistent negative symptoms, excluding those with considerable positive symptoms or evidence of depression. Those treated with neuroleptics should be on minimally effective doses of neuroleptics and adequately treated with anticholinergics to avoid the confounding effects of EPS and akathisia. One possible design is the randomized discontinuation trial (RDT), in which all patients are treated first with a standard reference compound. Depending on the goal, patients could then be randomized to continue standard treatment or placebo, or to receive a study drug thought to have enhanced efficacy in diminishing negative symptoms. Compared to randomized clinical trials, the efficiency of an RDT can be enhanced by excluding noncompliant patients, treatment nonresponders, and those with adverse reactions during the run-in phase (Kopec et al. 1993). With improved efficiency a smaller sample size may be needed than that used in classic trials of the same power carried out in the general population (Kopec et al. 1993). As applied to schizophrenia research, patients who remained psychotic despite neuroleptic treatment or whose negative symptoms resolved with acute treatment would not proceed with the study. This design could enhance the ability to detect differential effects of neuroleptics in treating negative symptoms. The RDT may also be used to test treatments that may be effective during maintenance but less so in acute treatment.

The introduction of chlorpromazine and related compounds made possible the wave of deinstitutionalization of the chronically mentally ill. While this development was revolutionary, today, discharge from the hospital is the least that the majority of patients, their families, and clinicians expect and deserve from therapeutic strategies in schizophrenia. Over the last 40 years clear improvements in standardized and reliable assessments of schizophrenia have improved our understanding of the phenomenology of this complex disease. This development has helped refine our measurement of the efficacy of neuroleptics, and of the risks, benefits, and limitations of treatment. Much work has been done to find the optimal and the minimally effective doses of neuroleptics during acute and maintenance treatment, with the goal of making treatment safer and more comfortable, and of reducing iatrogenesis. Equally important, this work has brought into sharp focus the limitations of existing treatments and the goals for future drug development. There is a compelling need for treatments which reduce social and occupational disability, cognitive impairment, and the negative symptoms associated with the deficit syndrome. Additionally, there is a need for new compounds for patients who respond partially or not at all to existing medications. The recent development of antipsychotics with unique profiles of efficacy provides hope that these goals will be attained.

References

Andreasen NC (1982) Negative symptoms in schizophrenia. Definition and reliability. Arch Gen Psychiatry 39:784–788

Andreasen NC, Olsen S (1982) Negative vs. positive schizophrenia: definition and validation. Arch Gen Psychiatry 39:789–794

Angrist B, Retrusion J, Gershon S (1980) Differential effects of amphetamine and neuroleptics on negative vs. positive symptoms in schizophrenia. Psychopharmacology (Berl) 72:17–19

Baldessarini RJ (1985) Antipsychotic agents. In: Chemotherapy in Psychiatry, 2nd edn. Harvard University Press, Cambridge, MA

Baldessarini RJ, Cohen BM, Teicher MH (1988) Significance of neuroleptic dose and plasma level in the pharmacological treatment of psychoses. Arch Gen Psychiatry 45:79–91

Breier A, Buchanan RW, Irish D, Carpenter WT Jr (1993) Clozapine treatment of outpatients with schizophrenia: outcome and long-term response patterns. Hosp Community Psychiatry 44:1145–1149

Caffey EM, Diamond LS, Frank TV, Grasberger JC, Herman L, Klett CJ, Rothstein G (1964) Discontinuation or reduction of chemotherapy in chronic schizophrenics. J Chronic Dis 17:347–358

Carpenter WT Jr, Heinrichs DW, Alphs LD (1985) Treatment of negative symptoms. Schizophr Bull 11(3):440–452

Carpenter WT Jr, Heinrichs DW, Wagman AMT (1988) Deficit and nondeficit forms of schizophrenia: the concept. Am J Psychiatry 145:578–583

Chouinard G, Annable L (1976) Clozapine in the treatment of newly admitted schizophrenic patients: a pilot study. J Clin Pharmacol May–June:289–297

Chouinard G, Jones B, Remington G, Bloom D, Addington D, MacEwan Gw, Labelle A, Beauclair L, Arnott W (1993) A Canadian multicenter placebo-controlled study of fixed doses of risperidone and haloperidol in the treatment of chronic schizophrenic patients. J Clin Psychopharmacol 13(1):25–40

Claghorn J, Honigfeld G, Abuzzanhab, Wang R, Steinbook R, Tuason V, Klerman G (1987) The risk and benefits of clozapine versus chlorpromazine. J Clin Psychopharmacol 7:377–384

Crow TJ (1980) Molecular pathology of schizophrenia: more than one disease process? Br Med J 280:66–86

Davis JM (1975) Overview: maintenance therapy in psychiatry. I. Schizophrenia. Am J Psychiatry 132(12):1237–1245

Eklund K, Forsman A (1991) Minimal effective dose and relapse-double-blind trial: haloperidol decanoate vs. placebo. Clin Neuropharmacol 14 [Suppl 2]:S7–S15

Endicott J, Spitzer RL, Fleiss JL, Cohen J (1976) The Global Assessment Scale, a procedure for measuring overall severity of psychiatric disturbances. Arch Gen Psychiatry 33:766–771

Gerlach J, Behnke K, Heltberg J, Munk-Andersen E, Nielsen H (1985) Sulpiride and haloperidol in schizophrenia: a double-blind crossover study of therapeutic effect, side effects and plasma concentrations. Br J Psychiatry 147:283–288

Goldberg SC (1985) Negative and deficit symptoms in schizophrenia do respond to neuroleptics. Schizophr Bull 11(3):453–456

Goldberg SC, Klerman GL, Cole JO (1965) Changes in schizophrenic psychopathology and ward behavior as a function of phenothiazine treatment. Br J Psychiatry 111:120–133

Guy W (1976) Clinical global impression scale. In: Guy W (ed) ECDEU assessment manual. Department of Health, Education, and Welfare, Washington, pp 583–585 (Publication (ADM) 76-338)

Heinrichs DW, Hanlon TE, Carpenter WT (1984) The Quality of Life Scale: an instrument for rating the schizophrenic deficit syndrome. Schizophr Bull 10(3):388–398

Hogarty GE, McEvoy JP, Munetz M, DiBarry AL, Bartone P, Cather R, Cooley SJ, Ulrich RD, Carter M, Madonia MJ (1988) Dose of fluphenazine, familial expressed emotion, and outcome in schizophrenia: results of a two-year controlled study. Arch Gen Psychiatry 45:797–805

Johnstone EC, Crow TJ, Frith CD, Carney MWP, Price JS (1978) Mechanism of the antipsychotic effect in the treatment of acute schizophrenia. Lancet 848–851

Kane JM, Mayerhoff D (1989) Do negative symptoms respond to pharmacological treatment? Br J Psychiatry 155 [Suppl 7]:115–118

Kane JM, Rifkin A, Quitkin F, Nayak D, Ramos-Lorenzi J (1982) Fluphenazine vs placebo in patients with remitted, acute first-episode schizophrenia. Arch Gen Psychiatry 39:70–73

Kane JM, Honigfeld G, Singer J, Meltzer H, the Clozaril Collaborative Study Group (1988) Clozapine for the treatment-resistant schizophrenic. A double-blind comparison with chlorpromazine. Arch Gen Psychiatry 45:789–796

Kay SR, Opler LA, Lindenmayer JP (1989) The positive and negative syndrome scale (PANSS): rationale and standardization. Br J Psychiatry 155 [Suppl 7]:59–65

Keefe RSE, Harvey PD, Lenzenweger MF, Davidson M, Apter SH, Schmeidler J, Mohs RC, Davis KL (1992) Empirical assessment of the factorial structure of clinical symptoms in schizophrenia: negative symptoms. Psychiatry Res 44:153–165

Kinross-Wright V (1955) Chlorpromazine treatment of mental disorders. Am J Psychiatry 3:12 [Reprinted Am J Psychiatry (1994) 151 [Suppl 6]:268–272]

Kirkpatrick B, Buchanan RW, McKinney PD, Alphs LD, Carpenter WT Jr (1989) The schedule for the deficit syndrome: an instrument for research in schizophrenia. Psychiatry Res 30:119–123

Klimidis S, Stuart GW, Minas IH, Copolov DL, Singh BS (1993) Positive and negative symptoms in the psychoses, reanalysis of published SAPS and SANS global ratings. Schizophr Res 9:11–18

Kolivakis T, Azim H, Kingstone E (1974) A double-blind comparison of pimozide and chlorpromazine in the maintenance care of chronic schizophrenic outpatients. Curr Ther Res 16:998–1004

Kopec JA, Abrahamowicz M, Esdaile JM (1993) Randomized discontinuation trials: utility and efficiency. J Clin Epidemiol 46(9):959–971

Liddle PF (1987) The symptoms of chronic schizophrenia: a reexamination of the positive-negative dichotomy. Br J Psychiatry 151:145–151

Malla AK, Norman RMG, Williamson P, Cortese L, Diaz F (1993) Three syndrome concept of schiaophrenia: a factor analytic study. Schizophr Res 10:143–150

Marder SR, Meibach RC (1994) Risperidone in the treatment of schizophrenia. Am J Psychiatry 151:825–835

Marder SR, Van Putten T, Mintz J, Levell M, McKenzie J, May PRA (1987) Low- and conventional-dose maintenance therapy with fluphenazine decanoate. Two-year outcome. Arch Gen Psychiatry 44:518–521

McEvoy J, Hogarty GE, Steingard S (1991) Optimal dose of neuroleptic in acute schizophrenia. A controlled study of the neuroleptic threshold and higher haloperidol dose. Arch Gen Psychiatry 48:739–745

McGlashan TH, Fenton WS (1992) The positive-negative distinction in schizophrenia. Review of natural history validators. Arch Gen Psychiarty 49:63–72

Meltzer HY, Burnett S, Bastani B, Ramirez LF (1990) Effects of six months of clozapine treatment on the quality of life of chronic schizophrenic patients. Hosp Community Psychiatry 4:892–897

Meuser KT, Douglas MS, Bellack AS, Morrison RL (1991) Assessment of enduring deficit and negative symptom subtypes in schizophrenia. Schizophr Bull 17:565–582

Minas IH, Stuart GW, Klimidis S, Jackson HJ, Singh BS, Copolov DL (1992) Positive and negative symptoms in the psychoses: multidimensional scaling of SAPS and SANS items. Schizophr Res 8:143–156

Modestin J, Toffler G, Pia M, Grent E (1983) Haloperidol in acute schizophrenic inpatients. A double-blind comparison of two dosage regimens. Pharmacopsychiatry 16:121–126

Möller HJ (1993) Neuroleptic treatment of negative symptoms in schizophrenic patients. Efficacy problems and methodological difficulties. Eur Neuropsychopharmacol 3:1–11

National Institute of Mental Health Psychopharmacology Service, Center Collaborative Study Group (1964) Phenothiazine treatment in acute schizophrenia. Arch Gen Psychiatry 10:246–260

Overall JE, Gorham DR (1962) The brief psychiatric rating scale. Psychol Rep 10:799–812

Paillère-Martinot ML, Lecrubier Y, Marinot JL, Aubin F (1995) Improvement of some schizophrenic deficit symptoms with low doses of amisulpiride. Am J Psychiatry 152:130–133

Pickar D, Owen RR, Litman RE, Konicki PE, Gutierrez R, Rapaport MH (1992) Clinical and biologic response to clozapine in patients with schizophrenia, crossover comparison with fluphenazine. Arch Gen Psychiatry 49:345–353

Pinder RM, Brogden RN, Sawyer PR, Speight TM, Spencer R, Avery GS (1976) Pimozide: a review of its pharmacological properties and therapeutic uses in psychiatry. Australian Drug Information Services. Drugs 12:1–40

Prien RF, Cole JO, Belkin NF (1968) Relapse in chronic schizophrenics following abrupt withdrawal of tranquilizing medication. Br J Psychiatry 115:679–686

Rassidakis C, Kondakis X, Papanastassiou A, Michalakeas A (1970) Withdrawal of antipsychotic drugs from chronic psychiatric patients. Bull Menninger Clin 34:216–222

Rifkin A, Doddi S, Karajgi B, Borenstein M, Wachspress M (1991) Dosage of haloperidol for schizophrenia. Arch Gen Psychiatry 48:166–170

Ruskin PE, Nyman G (1991) Discontinuation of neuroleptic medication in older, outpatient schizophrenics: a placebo-controlled, double-blind trial. J Nerv Ment Dis 179:212–214

Schooler NR (1991) Maintenance medication for schizophrenia: strategies for dose reduction. Schizophr Bull 17:311–324

Serban G, Siegel S, Gaffney M (1992) Response of negative symptoms of schizophrenia to neuroleptic treatment. J Clin Psychiatry 53:229–234

Sommers AA (1985) "Negative symptoms": conceptual and methodological problems. Schizophr Bull 11(3):364–379

Strauss JS, Carpenter WT Jr, Bartko JJ (1974) The diagnosis and understanding of schizophrenia. III. Speculations on the processes that underlie schizophrenic symptoms and signs. Schizophr Bull 11:61–69

Van Putten T (1974) Why do schizophrenic patients refuse to take drugs? Arch Gen Psychiatry 31:67–72

Van Putten T, Marder SR, Mintz J (1990) A controlled dose comparison of haloperidol in newly admitted schizophrenic patients. Arch Gen Psychiatry 47:754–758

CHAPTER 12

Efficacy of Novel Antipsychotic Drugs in Treatment-Refractory Schizophrenia

HERBERT Y. MELTZER and RAKESH RANJAN

A. Introduction

The limitations in efficacy and side effects of typical neuroleptic drugs as treatments for schizophrenia are well-known (DAVIS et al. 1980; MELTZER 1992). Although major positive symptoms and disorganization remit during neuroleptic drug treatment in as many as 60%–70% of patients with schizophrenia, negative symptoms, e.g., withdrawal, flat affect, and lack of initiative, usually respond only partially or not at all to these agents (CROW 1980; MELTZER and ZUREICK 1989; KAY and SINGH 1989) and contribute significantly to caretaker burden. Despite the fact that most schizophrenic patients only have mild negative symptoms to begin with, the lack of effect of typical neuroleptic drugs on patients with severe negative symptoms is a serious problem. The social function and quality of life of patients with schizophrenia treated with typical neuroleptics is generally poor. Low employment and marital rates are evidence for the limitations of current therapy. The high rate of noncompliance with the neuroleptic drugs (FLEISHHACKER et al. 1994b) and the high suicide rate (9%–13% lifetime, 0.4%–0.8% per year; DRAKE et al. 1985) also provide evidence of the limitations of typical neuroleptic drugs in the treatment of schizophrenia. Noncompliance with typical neuroleptics is the result of many factors, including untreated psychopathology, inadequate patient preparation, and familial and environmental influences (FLEISHHACKER et al. 1994b). Extrapyramidal symptoms (EPS), especially akathisia, that are not adequately relieved by antiparkinsonian medication have been suggested to be the most important reason for noncompliance (VAN PUTTEN 1974). Suicide, too, is the result of many factors, and is commonly precipitated by depression and hopelessness (DRAKE et al. 1985). The suicide attempt rate appears to be similar in patients regardless of whether they are neuroleptic-responsive or neuroleptic-resistant (MELTZER and OKAYLI 1995).

Schizophrenic patients who have persistent, moderate-to-severe, positive, disorganization, or negative symptoms, or some combination thereof, despite trials of neuroleptic drugs at doses and for durations which are usually effective, are generally referred to as being treatment-resistant (TR), treatment-refractory, or neuroleptic-resistant. Estimates of the proportion of TR schizophrenic patients vary, depending on the criteria for treatment resistance and the way in which the sample was obtained, e.g., a valid community sample, a sample consisting of clients who attend community mental health centers, or of recently or chronically hospitalized patients, etc. Using criteria for clozapine eligibility on the basis of neuroleptic resistance (failure to have an adequate response to at least two trials of standard neuroleptics), approximately 36% of a random sample of 293 patients from a public mental health system in California met criteria for neuroleptic resistance (JUAREZ-REYES et al. 1994). Dimensions of outcome that are relevant to defining treatment resistance include: (1) symptomatology (positive, negative, disorganization, affective, irritability, anxiety, etc.), (2) cognitive function, (3) social and work function, (4) quality of life, (5) suicidality, (6) compliance, and (7) rehospitalization (MELTZER 1992). Treatment resistance may vary from very slight to essentially total in one or more of these dimensions (BRENNER et al. 1990). Two trials of a neuroleptic drug for at least 6 weeks at adequate dosages, e.g., 10–20 mg/day haloperidol or its equivalent, are usually required to assess neuroleptic responsivity, but there is evidence that one trial may be sufficient to determine whether a patient is treatment-refractory (LIEBERMAN et al. 1993). Although there is no single accepted standard of neuroleptic resistance, the most widely accepted criterion is persistence of moderate-to-severe positive symptoms (KANE et al. 1988) and about 15%–30% of schizophrenia patients meet this criterion. Severe negative symptoms, without significant positive symptoms, may occur in 5% of schizophrenia patients and is thus a minor type of TR schizophrenia.

Since treatment resistance is, by definition, failure to respond adequately to typical neuroleptic drugs, the need for somatic treatments which are more effective to treat this group of patients is clear. As will be discussed, clozapine was the first agent shown to be effective in some but not all TR patients (KANE et al. 1988). Clozapine is considered an atypical antipsychotic because it produces fewer EPS, not because of its greater antipsychotic efficacy. Melperone, a butyrophenone, is also effective in some TR cases (MELTZER et al. 1990a) and causes fewer EPS than typical neuroleptics (BJERKENSTEDT 1989). The same may be true for risperidone (CLAUS et al. 1992) but it has not been definitively established that low EPS and enhanced efficacy to treat positive symptoms in treatment-resistant schizophrenia are linked (MELTZER 1995).

B. The Concept of Atypicality

Typical antipsychotic drugs are called neuroleptics because they produce catalepsy in rodents (DENIKER 1984) and EPS in humans, at doses which produce the animal equivalents of antipsychotic action (e.g., blockade of the conditioned avoidance response) or antipsychotic effects in humans, respectively. By contrast, antipsychotic drugs which either do not produce catalepsy in rodents or produce them only weakly at clinically effective doses have been designated as atypical (MELTZER 1994). However, because clozapine is generally regarded as the prototypic atypical antipsychotic drug, and because it appears to differ in a number of other clinically important ways from other antipsychotic drugs designated as "atypical," the usefulness of this concept of atypicality has been questioned (GERLACH and CASEY 1994). For example, clozapine does not increase serum prolactin levels in humans (MELTZER et al. 1979) while other atypical antipsychotic drugs may (e.g., risperidone) or may not (e.g., melperone) (MELTZER et al. 1994). Furthermore, clozapine has superior efficacy for some TR schizophrenic patients while the evidence for other atypical antipsychotic drugs in this regard is less robust or nonexistent. The same is true with regard to the potential to cause tardive dyskinesia, which is very weak for clozapine and melperone but not known for other drugs (MELTZER et al. 1994). Nevertheless, until the mechanism(s) of the pharmacologic actions of various atypical antipsychotic drugs are elucidated in detail, the previously mentioned definition of atypical antipsychotic drugs serves a useful purpose. At this time, it is important that atypical antipsychotic drugs not be defined as "clozapine-like" since the pharmacology of clozapine is so diverse that no other drug is likely to mimic it exactly, and because the characteristics which account for its various advantages are not known for certain (MELTZER 1995; MELTZER et al. 1995b). With expansion of our current knowledge about atypical antipsychotic drugs, it is quite possible that atypicality will be defined differently in the future. For the purpose of the following discussion, the definition of an atypical antipsychotic drug will be restricted to that described above: the limited capacity to cause EPS at antipsychotic doses.

C. Atypical Antipsychotic Drugs

Currently, clozapine, risperidone, and thioridazine are the only approved atypical antipsychotic drugs in the USA. It should be clear that there is no evidence that thioridazine has any advantages over typical neuroleptics other than for producing fewer EPS. Thioridazine can cause a severe exacerbation of Parkinson's disease whereas clozapine does not (SCHOLZ and DICHGANS 1985) and may cause tardive dyskinesia when given long-term. Risperidone at low doses is tolerated in patients with PARKINSON's disease (MECO et al. 1994). However, other atypical antipsychotic drugs are in various stages of clinical development and also appear to produce fewer EPS than typical neuroleptics and to be closer in efficacy to clozapine. These include melperone, olanzapine, sertindole, amperozide, seroquel, and ziprasidone. Remoxipride was in development but has been permanently withdrawn because it caused aplastic anemia at an unacceptable rate. A more complete list of putative atypical antipsychotic drugs based on their pharmacologic profile is provided in Table 1. Clinical studies will be reviewed after discussing clozapine. The diversity in putative mechanism of action is noteworthy.

Table 1. Putative atypical antipsychotic drugs

$5\text{-}TH_{2A}/D_2$ antagonists		DA autoreceptor agonists/partial D_2 agonists	
Clozapine	Seroquel	BHT-920	3-PPP
Risperidone	Sertindole	CGS 15873A	SND 919
HP 873	SM-9018	DD 118717E	U-66444B
Melperone	Ziprasidone	Pramipexole	Roxindole
Olanzapine	Zotepine		
	Fluperlapine[a]		
	Tiosperone[a]		
$5\text{-}HT_{2A}$ antagonists		Partial DA agonists	
Amperozide	ICI 169369[a]	MER-327	
MDL 100,907[b]		SDZ-208-911	
Ritanserin[a]		SD-208-912	
Sigma antagonists		Glutamate receptor agonists	
DUP 734	Cinuperone	Glycine	
BMY 13980	NPC 16377	Milacemide	
BMY 14802	Umespirone	Umespirone	
Rimcazole	Tiosperone		
Remoxipride			
Selective D_2/D_3 antagonists		CCK antagonists	
Remoxipride[a]	Amisulpride	LY 262691[a]	
Sulpiride[a]	Raclopride	Caerulin[a]	

[a] No longer in development in U.S.
[b] Not yet in clinical trials.

I. Clozapine

Clozapine, a dibenzodiazepine chemically related to loxapine, was first synthesized in 1959 and clinically tested in Europe in the 1960s (SCHMUTZ and EICHENBERGER 1982). A series of early, relatively small-scale clinical trials established its efficacy as an antipsychotic drug without EPS (for review, see BALDESSARINI and FRANKENBERG 1991). At least six controlled studies demonstrated that clozapine had superior efficacy compared to typical antipsychotic drugs for the treatment of neuroleptic-responsive schizophrenia (e.g., EKBLÖM and HAAGSTRÖM 1974; GERLACH et al. 1974; FISHER-CORNELLSEN et al. 1974). However, in 1975, 16 cases of granulocytopenia(-agranulocytosis), 8 of which were fatal, in six hospitals in southwestern Finland were attributed to clozapine (AMSLER et al. 1977). The overall incidence was 1:100. No contributing factors were identified. All cases occurred within the first 4 months of clozapine treatment. Prior to this, there had been no appreciation that the risk of agranulocytosis due to clozapine was significantly greater than that due to other neuroleptics. The latter rate is not precisely known but is approximately one per 2000 cases (KRUPP and BARNES 1989).

Clozapine was subsequently withdrawn from general use in Europe and further clinical trials in the USA. However, because many of the schizophrenic patients who were withdrawn from clozapine had a marked relapse that did not respond well to reinstitution of typical neuroleptics, use of clozapine was permitted on a restricted basis in several western European countries (e.g., Sweden, Denmark, Germany, Switzerland). Clinical experience with clozapine over the next decade suggested that clozapine was more effective than typical neuroleptics in some patients (JUUL-POVLSEN et al. 1985; LINDSTRÖM 1988; MELTZER and LUCHINS 1984) and that it not only caused fewer EPS but was also more effective in blocking tardive dyskinesia while it did not cause tardive dyskinesia (CASEY 1989).

Several retrospective studies which reported long-term efficacy and tolerability of clozapine in TR schizophrenia have been published. TR schizophrenic inpatients ($n = 216$) treated with either clozapine alone ($n = 85$) or clozapine plus typical neuroleptics ($n = 131$) were found to be significantly improved compared to when they were treated with typical neuroleptics alone (JUUL-POVLSEN et al. 1985). KUHA and MIETTINAN (1986) found that 25 of 108 (23.1%) chronic schizophrenic inpatients resistant to traditional neuroleptics were able to be discharged and another 33% were significantly improved after 15 years of clozapine treatment. LINDSTRÖM (1988) reported a follow-up of 96 inpatients with schizophrenia or schizoaffective disorder. Patients were either resistant (85%) or intolerant (15%) to typical neuroleptics. The mean duration of clozapine treatment was 3.7 years. Marked or moderate improvement was reported in 43% and 38% of the patients, respectively.

The first prospective controlled study of clozapine in TR schizophrenia was that of KANE et al. (1988). This was a double-blind, 6-week trial comparing clozapine with chlorpromazine in nearly 300 hospitalized TR patients. Treatment resistance, in this study, was defined as the failure to respond to at least three separate trials of neuroleptics, from at least two different chemical classes, over a 5-year period, at dosages equivalent to or greater than 1000mg/day of chlorpromazine for a period of 6 weeks. To be included in the study, patients also had to have a score ≥45 on the Brief Psychiatric Rating Scale (BPRS; OVERALL and GORHAM 1962) with the 18 items rated 1–7, and a minimum rating of 4 (moderate) on the Clinical Global Impression (CGI) scale. During the initial open-label phase of the study, patients were treated with haloperidol (up to 60mg/day or more) for 6 weeks. Responders were excluded from the double-blind phase of the study. Nonresponders to haloperidol were randomly assigned to a 6-week, double-blind trial with either clozapine (up to 900mg/day) or chlorpromazine (up to 1800mg/day) and benztropine (up to 6mg/day). Improvement in this study was defined as reduction of >20% from baseline in the BPRS total score and either a posttreatment CGI score of 3 or less or a posttreatment BPRS total score of 35 or lower. Of all patients who completed at least 1 week of double-blind treatment, only 4% of those treated with chlorpromazine and benztropine were responders, compared to 30% of clozapine-treated patients. As a result of this study, clozapine was approved for use by the U.S. Food and Drug Administration (FDA) in neuroleptic-responsive and resistant schizophrenic patients in 1989. By the end of 1994, it had been administered to approximately 100000 patients in the

USA with about 60000 remaining in treatment for at least 6 months. Numerous other countries also approved clozapine as a result of this study.

KANE et al. (1988) reported the results of only 6 weeks of treatment. The results of that study were confirmed in a crossover, placebo-controlled, double-blind comparison of long-term fluphenazine and clozapine treatment in 21 schizophrenic patients who were either neuroleptic-resistant or -intolerant (PICKAR et al. 1992). Clozapine, in comparison with both fluphenazine and placebo, was significantly superior in reducing psychopathology. In a 10-week, double-blind, parallel-group comparison of clozapine and haloperidol, BREIER et al. (1994) also reported that clozapine was superior to haloperidol in reducing symptoms among schizophrenic outpatients who were partial responders to typical neuroleptics.

Subsequent experience has indicated that the response to clozapine may be slow in onset in a substantial proportion of patients. Thus, MELTZER et al. (1989) reported an open study of 51 TR schizophrenics in which 31 (60.8%) of patients showed at least a 20% decrease in BPRS total score. The mean duration of treatment at the time of the report was 10.3 ± 8.1 months. Half of the responders did not respond until 3 months of treatment or beyond. In a follow-up report (MELTZER 1992) on 85 TR schizophrenics which included 51 patients from the preliminary study, 81.5% of patients who remained on clozapine at the end of 12 months showed a decrease of 20% or more in the BPRS total. LIEBERMAN et al. (1994) reported that the probability of response to clozapine in 66 TR schizophrenic or schizoaffective patients was about 0.35 through 6 weeks of treatment. It declined to 0.22 between 12 and 26 weeks and 0.19 between 26 and 39 weeks. There were additional responders between weeks 26 and 39. Thirty-three (50%) of the patients in this study were responders by 52 weeks. It should be noted that there may be some outcome measures which show early response, e.g., negative symptoms, mood, grooming, before there is an overall response as noted by the BPRS total score or Global Assessment Score. Patients and clinicians need to be sensitive to a variety of outcome measures. Before stopping a clozapine trial because of apparent poor response, a variety of issues needs to be considered (see below).

1. Effect on Psychopathology

a) Effect on Positive Symptoms

Clozapine has been found to be significantly superior to chlorpromazine in reducing the total score on the BPRS Positive Symptom Subscale as well as the scores on individual items that constitute this subscale: conceptual disorganization, hallucinatory behavior, suspiciousness, and unusual thought content in TR (KANE et al. 1988) and typical schizophrenia (CLAGHORN et al. 1987). The superior effect of clozapine on positive symptoms was evident in TR schizophrenia as early as the first week of treatment (KANE et al. 1988). PICKAR et al. (1992) noted that clozapine was significantly more effective than both placebo and fluphenazine in reducing BPRS positive symptoms and scores on the Bunney-Hamberg psychosis scale among a group of TR schizophrenics. Finally, Breier et al. (1994) found clozapine to be significantly more efficacious than haloperidol for treating BPRS positive symptoms in TR schizophrenia. The effect of clozapine on positive symptoms may be delayed, but once in place is usually enduring (Meltzer 1992).

b) Effect on Negative Symptoms

Negative schizophrenic symptoms include blunted affect, anhedonia, alogia, attentional impairment, and asociality (Andreasen and Olsen 1982). These symptoms are known to be more resistant than positive symptoms to typical antipsychotic drugs (Meltzer and Zureick 1989; Kay and Singh 1989; Crow 1990). The study by Kane et al. (1988) found clozapine to be clearly more effective than chlorpromazine in reducing the BPRS Negative Symptom Subscale, i.e., emotional withdrawal, blunted affect, psychomotor-retardation, and disorientation. As was the case for positive symptoms, the effect of clozapine on negative symptoms was evident after the second week of treatment.

Clozapine has been reported to significantly decrease the BPRS Withdrawal/Retardation Subscale scores (Meltzer 1989). This differs from the Negative Symptom Subscale in that it does not include disorientation, which is more appropriately part of the disorganization cluster. Furthermore, improvement in the Withdrawal/Retardation Subscale score was independent of the improvement in the Paranoid Disturbance Subscale scores, an indication that the improvement in negative symptoms was independent of the improvement in positive symptoms (Meltzer, in preparation). Clozapine was also effective in ameliorating the negative symptoms in patients who do not have notable positive symptoms (Meltzer and Zureick 1989). In a larger, follow-up report by Meltzer (1992), significant improvement over a 12-month period was noted on the Schedule for Affective Disorders and Schizophrenia (SADS)-C (Endicott and Spitzer 1978) Negative Symptom Subscale ratings in 54 patients, and on the Scale for the Assessment of Negative Symptoms (SANS) ratings (Andreasen 1983) in 28 patients. Again, positive and negative symptom measures on the BPRS and SADS-C were not significantly correlated at any time point during a 12-month period. The study by Pickar et al. (1992) found significantly more improvement on the BPRS Negative Symptom Subscale but not the SANS in clozapine-treated compared to fluphenazine-treated patients. Breier et al. (1994), in their comparison of clozapine and haloperidol treatment in partially neuroleptic-responsive outpatients, did find significantly greater improvement with clozapine treatment on the SANS total score. However, they found no differences between the effects of clozapine and haloperidol on negative symptoms among patients designated as having deficit schizophrenia, based on the Schedule for the Deficit Syndrome (SDS) (Kirkpatrick et al. 1989). This raises the issue of whether clozapine's effect on negative symptoms may be due to changes in associated features such as depression and EPS. KANE et al. (1994) reported significant correlations between BPRS Anergia and Akinesia scores at baseline and through 39 weeks. They concluded it was unclear but not of great practical importance whether the beneficial effect of clozapine on negative symptoms was secondary to changes in EPS or positive symptoms.

c) Effect on Disorganization Symptoms

Clozapine has been reported to produce a significant decrease in ratings on SADS-C disorganization factor, which includes loose association, poverty of thought content, incoherence, and inappropriate affect (MELTZER 1992). The effect of clozapine on the disorganization cluster was superior to its effect on both the Positive and Negative Subscales of the SADS-C.

d) Effect on Cognitive Function

Cognitive dysfunction is a key characteristic of schizophrenia (KENNY and MELTZER 1991). It is already present at the onset of illness in most patients (BOLDEN et al. 1992; HOFF et al. 1992). Various types of cognitive deficits are present, including deficits in attention, memory recall, conceptual sorting, and executive function (SEIDMAN 1983; KENNY and MELTZER 1991). Typical antipsychotic drugs are generally ineffective in ameliorating the cognitive deficits in schizophrenic patients (MEDALIA et al. 1988; SPOHN and STRAUSS 1989; CASSENS et al. 1990; KING 1990). These cognitive deficits are important contributors to the social and occupational disabilities associated with schizophrenia.

There is evidence that clozapine is effective in improving some of these deficits in TR schizophrenics. In a study of 25 TR schizophrenic patients (KENNY and MELTZER 1991), clozapine produced a significant improvement in semantic memory at 6 weeks and 6 months, and in secondary memory at 6 months of treatment. The improvement in cognitive function was independent of the change in psychopathology. In another study by HAGGER et al. (1993), cognitive function and psychopathology were assessed in 36 TR schizophrenic patients before initiation of clozapine, and at 6 weeks and 6 months of clozapine treatment. Compared to 26 normal controls, schizophrenic patients were noted to have deficits in measures of memory, attention, and executive function. With clozapine treatment, significant improvement occurred in retrieval from reference memory at 6 weeks and 6 months in some measures of executive function (revised Wechsler Intelligence Scale for Children, WISC-R, Maze Test), attention (revised Wechsler Adult Intelligence Scale, WAIS-R, Digit Symbol Subtest), and recall memory (Verbal List Learning Test) at 6 months. CLASSEN and LAUX (1988) reported that 7 days of treatment with clozapine tended to produce better performance on the Stroop color and color word tests than haloperidol or fluphenthixol. GOLDBERG et al. (1993) reported that 3 to 24 months of clozapine treatment had no beneficial effect on attention, memory, and higher-level problem solving in 15 psychotic, primarily schizophrenic patients. Some of these patients also received a variety of other psychotropic drugs. Recently, the results of MELTZER et al. (1995a) were confirmed by BUCHANAN et al. (1994). In that study, two groups of 19 patients were treated with clozapine or haloperidol for 10 weeks. Clozapine produced significantly greater improvement than haloperidol in two tests: Category Fluency and WAIS-R Block Design. Thirty-three patients were treated with clozapine for a year. Improvement in both of these tests as well as the Mooney Faces Closure Test were found during this time period.

e) Effect on Mood Symptoms

Clozapine has been reported to be effective in decreasing both manic and depressive symptoms in TR schizoaffective patients (OWEN et al. 1989; NABER et al. 1989, 1992; NABER and HIPPIUS 1990; MCELROY et al. 1991). Overall, clozapine was found to be more effective in schizoaffective patients than in schizophrenic patients. Thus, NABER and coworkers (1989) found that 45% of 229 patients with schizophrenia and 65% of 55 patients with schizoaffective disorder had marked improvement in symptoms with clozapine treatment. OWEN and colleagues (1989) also reported significantly greater decreases in BPRS total score in 112 schizoaffective patients, compared to 37 schizophrenic patients. In a retrospective study of 39 patients with schizophrenia, 25 patients with schizoaffective disorder and 14 patients with bipolar disorder and psychotic feelings, the affective-disorder patients showed significantly higher response rates to clozapine than did the schizophrenic patients (MCELROY et al. 1991).

In a recent prospective study, clozapine markedly reduced suicidality, especially low- and high-probability suicide attempts, in TR schizophrenia and schizoaffective disorder (MELTZER and OKAYLI 1995). This decrease in suicidality was associated with improvement in depression and hopelessness, as measured by the Hamilton Depression Rating Scale (HAMILTON 1960). These findings could be the basis for re-evaluation of the risk–benefit assessment of clozapine. Thus, the overall morbidity and mortality for TR schizophrenia could be lower with clozapine than with typical neuroleptic drugs because of decreased suicidality. Suicide has been reported to occur in 9%–13% of schizophrenic patients, while risk of agranulocytosis from clozapine is less than 1% and the mortality is approximately 0.0% (ALVIR et al. 1993).

f) Effect on Quality of Life

The concept of quality of life as a distinct and significant dimension of schizophrenic pathology was proposed by LEHMAN (1983). Schizophrenia causes a progressive decline in quality of life. Amelioration of psychotic symptoms may be insufficient to significantly improve quality of life. The effect of clozapine on the Quality of Life Scale (QLS) (HEINRICHS et al. 1984, 1990) has been assessed in two studies. This scale has four factors: (1) intrapsychic foundations, (2) interpersonal relations, (3) instrumental role, and (4) common objects and activities. MELTZER and colleagues (1989) were the first to report the effect of clozapine on the quality of life in 33 TR schizophrenics. Clozapine treatment (mean duration 10.3 ± 8.1 months) in 33 TR patients significantly improved in each of the 21 items of the OLS. In subsequent studies, MELTZER (1992; MELTZER et al. 1990b) assessed the effect of 6 and 12 months of clozapine treatment on the quality of life of 38 TR schizophrenic patients. Significant improvement was noted on the OLS total score and on all four factors of the scale, with the greatest improvement occurring in interpersonal role function and intrapsychic function. BUCHANAN et al. (1994) reported a significant correlation between quality of life and some cognitive measures. Similar results were observed by MELTZER et al. (1995b).

2. Duration of Clozapine Trials

In the study by KANE et al. (1988), the double-blind comparison of clozapine with chlorpromazine showed that 30% of the clozapine-treated patients met response criteria at the end of a 6-week study period. Subsequently, MELTZER and colleagues (1989), in a prospective, open trial of clozapine for a mean duration of 10.3 ± 8.1 months, found that 60.8% of patients responded. About half of the responders did so between 3 and 6 months of treatment. The criterion for improvement or drug response was the same in both studies: 20% decrease in BPRS total score. Based on these findings, MELTZER and coworkers have suggested that an adequate duration of clozapine trial for a given patient should be at least 6–12 months. Dosage should be raised to 900 mg/day, if tolerated, in poor responders. It is notable that the risk of agranulocytosis is highest between 4 and 18 weeks, so the risk–benefit ratio improves after that time. The reasons for delayed response to clozapine in some patients are unknown.

3. Clozapine Dosage and Administration

The recommended starting dose of clozapine is 12.5 mg/day to test for possible hypotension. The dose of clozapine may be increased by 25 mg every other day until it reaches 100 mg. This can be done on an outpatient basis if the patient is assessed to be capable of adhering to the prescribed schedule or if there are family members or case managers who can assist the patient. It can then be increased by 50 mg every other day until a dose of 300–450 mg/day is reached. Twice-daily dosage is recommended because the half-life of clozapine is 16 h (CHOC et al. 1987; JANN et al. 1993). The dose need not exceed 450–600 mg/day in most adults ≤60 years in the initial phase of treatment. However, if response at 600 mg/day is unsatisfactory, the dosage should be further increased up to a maximum of 900 mg/day. The dosage of clozapine required in the elderly is usually 200–300 mg/day, but may be as low as 5–100 mg/day. There are no data available as to whether lower doses of clozapine are needed for maintenance treatment.

To date, no fixed dose studies have been done to determine the optimal dosage in TR schizophrenia. In the USA, doses of 400–600 mg/day are most common (mean dose 444 mg), while European psychiatrists mostly use doses in the range of 200–500 mg/day or lower (mean dose 284 mg/day) (NABER et al. 1989; FLEISHHACKER et al. 1994a). The reasons for this discrepancy are unknown at this time, but could be related to more common use of concomitant neuroleptic drugs in Europe.

4. Relationship Between Clinical Efficacy and Plasma Concentrations of Clozapine

The half-life of clozapine after twice-daily dosage at the steady state is about 16 h (range 6–33). Thus, a steady state is achieved about 1 week after constant-dose twice-daily administration (CHOC et al. 1987). Current evidence suggests that the major metabolites of clozapine, norclozapine and clozapine-*N*-oxide, are pharmacologically inactive; however, this issue needs further elucidation.

There is some evidence that plasma levels of clozapine may be a useful guide to identify the optimal dose. It appears that dosages that achieve plasma levels of greater than 350–370 ng/ml are maximally effective in TR schizophrenia (PERRY et al. 1991; HASEGAWA et al. 1993). However, there is no evidence of a therapeutic window, and some patients do respond at lower plasma levels. Clozapine plasma levels are most commonly measured by high-pressure liquid chromatography (LOVDAHL et al. 1991).

Plasma levels of clozapine should be obtained in patients who have poor response. Also, it may be useful to check clozapine levels once the dose exceeds 600 mg/day since the incidence of seizures increases significantly with doses of more than 600 mg/day and since high plasma concentrations of clozapine are associated with seizures (SIMPSON and COOPER 1978). Finally, it may be possible, in selected cases, to exceed the recommended limit of 900 mg/day if there is a poor response, the plasma level is less than 370 ng/ml, and there are no serious or troublesome side effects.

Factors that have been found to affect plasma levels of clozapine include gender, smoking, and certain drugs. Smoking was found to decrease clozapine concentrations in men in one study (HARING et al. 1990), whereas another study found no difference in clozapine concentrations among male smokers and nonsmokers (HASEGAWA et al. 1993). Phenytoin has been reported to lower clozapine concentrations.

5. Predictors of Response to Clozapine

Various demographic, clinical, neuroanatomical, neurophysiological, and neurochemical variables have been claimed to be the predictors of response to clozapine in TR schizophrenic patients. HONIGFELD and PATIN (1989) reported that a paranoid subtype according to DSM-III criteria, a greater number of previous hospitalizations, and low grandiosity predicted good response. However, these three variables accounted for only 13.3% of outcome variance. Furthermore, a discriminant function analysis using these three variables produced unsatisfactory predictive power. PICKAR et al. (1992) found that later age of onset and greater decreases in parkinsonian side effects in patients on clozapine predicted a good response to clozapine. LAMBERTI et al. (1992) found that weight gain (which is a side effect of clozapine) by patients who were under clozapine treatment was inversely correlated with the decrease in BPRS scores. We have found that female gender, shorter duration of illness, and presence of tardive dyskinesia prior to clozapine therapy may predict better response to clozapine (unpublished data).

As described in the preceding section, PERRY et al. (1991) and HASEGAWA et al. (1993) have found that plasma clozapine concentrations above 350–370 ng/ml may be a useful predictor of favorable response. Raising plasma levels to ≥350 ng/ml increases the response rate (MILLER et al. 1994). Similar relations between plasma levels and response have been found by others (LIEBERMAN et al. 1994; POTKIN et al. 1994).

Increased prefrontal cortical sulcal widening (PFSW) has been found to predict a poor response to clozapine at 6 weeks of treatment (FRIEDMAN et al. 1991). In the study by TENKE et al. (1993), which involved 50 schizophrenic patients including 21 treated with clozapine, a significant negative correlation was found between decrease in total BPRS score during a mean 30-day treatment period, and average width of the three largest cortical sulci. However, BILDER et al. (1994) found no relationship between MRI measures such as total brain or sulcal volume and response to clozapine.

SMALL et al. (1987) reported that clozapine responders have higher EEG amplitudes in frontal and temporal areas than nonresponders.

PICKAR et al. (1992) found that plasma homovanillinic acid (HVA) levels were significantly lower during clozapine treatment in superior responders to clozapine. GREEN et al. (1993) reported that higher pretreatment plasma HVA levels correlated with the clinical response to clozapine at 3 months. On the other hand, DAVIDSON et al. (1993) found no effect of clozapine on plasma HVA and no relationship between change in plasma HVA levels and response to clozapine. PICKAR et al. (1992) also noted that superior responders to clozapine had lower cerebrospinal fluid (CSF) HVA/5-HIAA ratios while receiving typical neuroleptics, placebo, or clozapine. Similarly, SZYMANSKI et al. (1993) reported that lower pretreatment CSF HVA/5-HIAA ratios predicted response to clozapine. This finding is consistent with the hypothesis that serotonin – dopamine imbalance in the brain is a substrate for clozapine's action (MELTZER 1989).

6. Nonresponders to Clozapine Therapy and Augmentation Strategies

Partial response to clozapine may be approached in a variety of ways. Addition of a low dose of high-potency neuroleptic drug has been found to be helpful in ameliorating positive symptoms in some of these patients. Concomitant electroconvulsive therapy may also be useful in some of these patients (Sajatovic and Meltzer 1993). Recently, addition of an H_2-blocker, ranitidine, has been found to be effective in treating persistent negative symptoms. Antidepressants and mood stabilizers have also been used as adjuncts in treating depressive or manic symptoms among schizoaffective patients. Selective serotonin reuptake inhibitors like fluoxetine, sertraline, or paroxetine are usually better tolerated than tricyclic antidepressants (Cassady and Thaker 1992). It may be prudent not to initiate clozapine and lithium simultaneously, as there have been a few reports of increased neurotoxicity (Pope et al. 1991; Blake et al. 1992), but once a stable dose of clozapine has been attained, lithium can safely be added in most cases. Valproic acid may also be used. Carbamazepine should be avoided because of increased risk of agranulocytosis. For persistent anxiety symptoms, benzodiazepines or buspirone may be added. A few earlier reports of cardiorespiratory collapse with concomitant use of benzodiazepines had raised a scare; however, subsequent clinical experience has shown that this combination is safe in more medically healthy patients. Again, benzodiazepines should preferably be added only after a stable dose of clozapine has been achieved.

Systematic studies assessing the usefulness and safety of various psychotropics as an adjunct to clozapine therapy in patients with persistent symptoms are clearly lacking and needed at this time.

7. Dropouts and Withdrawal from Clozapine

The reasons for dropout from clozapine therapy include lack of efficacy, noncompliance, and intolerable side effects other than granulocytopenia (agranulocytosis). There is some evidence that abrupt withdrawal of clozapine may be associated with severe exacerbation of symptoms (Ekblöm, et al. 1984; Parsa et al. 1993; Alphs and Lee 1991). Hence, in cases of noncompliance, clozapine should be reinstituted as soon as possible. In cases of agranulocytosis, prolonged treatment with standard neuroleptics may be needed once the total white blood cell count (WBC) drops to 2000–3000/mm^3.

8. Cost-Effectiveness

The cost-effectiveness of clozapine in TR schizophrenia has been demonstrated by two studies. Revicki et al. (1990) conducted a retrospective study with 86 TR schizophrenic patients, comparing estimated costs before and during clozapine treatment with those for a comparable group of patients who received conventional treatments. During the first year, total costs in the clozapine group were higher than in the neuroleptic group; however, by the second year of treatment with clozapine, a savings of more than U.S. $33000 per patient compared to the neuroleptic group was noted. In a prospective study, Meltzer and colleagues (1993) compared the total costs of treatment for 2 years before and after clozapine treatment in 96 TR schizophrenic patients. The cost of treatment was significantly decreased (by about U.S. $23 000/year) for the 37 patients who continued clozapine treatment for at least 2 years and for whom cost data were available. The reduction in cost was primarily due to a dramatic decrease in the frequency of rehospitalization. The total cost of treatment of all subjects, including dropouts, decreased significantly.

It is hoped that these findings will help dispel the misconception that the high cost of clozapine compared with typical neuroleptic drugs escalates the cost of treatment of schizophrenia. This change in perception could potentially maximize the use of clozapine in suitable patients, especially by the public mental health systems which treat the majority of schizophrenic patients.

II. Risperidone

Risperidone is a benzisoxazole derivative and is chemically unrelated to any other currently available antipsychotic drug. It was approved for clinical use in the USA in early 1994. Several open trials have shown that it is an effective antipsychotic drug (ROOSE et al. 1988; CASTELAO et al. 1989; MESOTTEN et al. 1989; GELDERS et al. 1990; MECO et al. 1989; BERSANI et al. 1990). These studies also showed its low propensity to cause EPS.

The first double-blind study of risperidone was performed by CLAUS et al. (1992). In this study 44 chronic schizophrenic inpatients participated in a multicenter, 12-week, parallel-group, double-blind trial. After a run-in period of 2 weeks and a single-blind placebo washout of 1 week, patients were randomly assigned to treatment with either risperidone or haloperidol. The mean daily dose was 12 mg for risperidone and 10 mg for haloperidol. Many of these patients had not responded to conventional neuroleptic treatment prior to their entry into the study. Furthermore, the risperidone group had more psychopathology at baseline than the haloperidol group. Thirty-three percent of the patients in the risperidone group, as opposed to 24% in the haloperidol group, had clinically significant improvement of the total Positive and Negative Syndrome Scale (PANSS) score. Furthermore, improvement in the PANSS total score was about three times greater in the risperidone group. Also, the use of antiparkinsonian drugs was 10 times less prevalent in risperidone-treated patients. Although this study was not designed to test the efficacy of risperidone specifically in TR schizophrenic patients, some apparently TR patients were included, thus raising the possibility that risperidone may be useful for some TR patients. This possibility has considerable appeal to the clinician because of the favorable side-effect profile of risperidone, compared to clozapine, with regard to agranulocytosis. However, to date, no trial of risperidone in TR schizophrenic patients has been completed or published. In three other double-blind studies, risperidone has been reported to be superior to haloperidol in treating symptoms of chronic schizophrenia (CHOUINARD et al. 1993; MARDER and MEIBACH 1994; SVESTKA et al. 1990). In these studies, at least a subgroup of subjects was probably TR. The studies of

Chouinard et al. (1993) and Marder and Meibach (1994) were the Canadian and U.S. American arms, respectively, of a multicenter U.S.-Canadian collaborative study. Both studies examined the efficacy of risperidone in schizophrenic patients using a fixed-dose, parallel-group, placebo-controlled, double-blind design. In the study of Chouinard et al. (1993), 135 inpatients were randomly assigned to 8 weeks of treatment with four different dosages of risperidone (2, 6, 10, and 16 mg/day), one dose of haloperidol (20 mg/day), and placebo. Double-blind drug assignments were preceded by 3–7 days of single-blind placebo treatment. Favorable outcome was defined as a 20% or greater reduction in total PANSS score. Compared to the placebo group, significantly more patients improved in the 2-mg, 6-mg, and 16-mg risperidone groups and in the haloperidol group. Only the 6-mg risperidone group had a significantly higher number of improved patients than the haloperidol group. Finally, at doses of 6 mg, 10 mg, and 16 mg, risperidone was found to have marked antidyskinetic effects as compared to placebo. In the study of Marder and Meibach (1994) there were 388 schizophrenic patients with slightly different results: significantly more patients receiving 6 mg, 10 mg, and 16 mg of risperidone had a good response than the patients receiving a placebo. Also, significantly more patients treated with 6 mg and 16 mg of risperidone had improved than patients receiving haloperidol. The incidence of EPS was significantly higher in patients receiving 16 mg of risperidone or 20 mg of haloperidol than in those given a placebo. The incidence of EPS in patients receiving 6 mg of risperidone was no higher than that in patients receiving a placebo.

1. Effect on Psychopathology

Before discussing the effects of risperidone on various aspects of schizophrenic pathology, it should be reiterated that even though most double-blind studies of risperidone included some patients who were refractory to conventional treatments, efficacy of risperidone in TR schizophrenia has yet to be examined in systematic studies.

In the study of Claus et al. (1992), improvement on the PANSS Subscales of Positive Symptoms and General Psychopathology was about three times greater in the risperidone group compared to the haloperidol group, both at week 6 and at endpoint. In the study of Chouinard et al. (1993), 2 mg, 6 mg, and 16 mg of risperidone were better than placebo on the General Psychopathology Subscale of the PANSS. Only the 6-mg dose of risperidone was superior to haloperidol on this subscale. On the Positive Symptom Subscale of the PANSS, 6 mg, 10 mg, and 16 mg of risperidone were shown to be superior to placebo, with the greatest degree of efficacy observed with 6 mg. However, no dose of risperidone was superior to haloperidol on this subscale. In the study of Marder and Meibach (1994), a similar pattern of efficacy for risperidone was observed on the Positive Symptom Subscale of the PANSS.

Improvement in negative symptoms was more pronounced with risperidone than haloperidol (CLAUS et al. 1992). In the study of CHOUINARD et al. (1993), only the 6-mg dose of risperidone was better than the placebo and had a tendency to produce greater improvement as compared with haloperidol on the PANSS Negative Symptom Subscale. In contrast, the study of MARDER and MEIBACH (1994), showed significant improvement as compared to placebo on the PANSS Negative Symptom Subscale in patients receiving 6mg and 16mg of risperidone. No significant improvement in negative symptoms was noted in patients treated with haloperidol.

The effect of risperidone on disorganization symptoms, cognitive function, mood symptoms, and quality of life has not been investigated as of yet.

2. Risperidone Dosage and Administration

The usual starting dose of risperidone is 1mg twice daily, increasing to 2mg twice daily and 3mg twice daily over the next 2 days. The mean optimal dose is 6mg/day, as was shown by the studies of CHOUINARD et al. (1993) and MARDER and MEIBACH (1994). Higher doses may be needed to control severe positive symptoms. Risperidone, like clozapine, should usually be given without additional neuroleptics to avoid EPS.

To date, no useful data are available regarding risperidone on the relationship between plasma levels and efficacy, predictors of response, or cost-effectiveness.

III. Melperone

Melperone, a butyrophenone, has been in clinical use in Europe for over two decades. It is more effective than placebo and equivalent to thiothixene in treating schizophrenia in an acute exacerbation (BJERKENSTEDT 1989; HÄRNYRD et al. 1989). It is similar to clozapine in a number of ways, e.g., low EPS, diminished liability to cause tardive dyskinesia, and no plasma prolactin increases (MELTZER et al. 1989).

Like clozapine, melperone has been found to have efficacy in some neuroleptic-resistant patients (MELTZER et al. 1990a). It appears to be comparable to risperidone in efficacy in this group (MELTZER HY, unpublished data) but is less effective than clozapine. Most of the studies have shown that the optimal dose of melperone is 300mg/day or higher.

IV. Remoxipride

Remoxipride, which is a substituted benzamide, is the first drug of this class to be developed for use in North America. This class of compounds, which includes sulpiride and amisulpride, has been considered to have advantages with regard to treating negative symptoms and EPS, although clinical trials have not unequivocally confirmed these claims.

Remoxipride has been extensively studied in large double-blind, multicenter trials in comparison with haloperidol. LEWANDER et al. (1990) provided a combined analysis of nine trials involving 667 remoxipride- and 437 haloperidol-treated patients. After 4–6 weeks of treatment, 55%–60% of patients in both groups were rated as much or very much improved. Haloperidol and remoxipride produced similar effects on positive and negative symptoms with approximately 60% and 30% improvement in each class of symptoms, using an intention-to-treat analysis. Several small-scale studies have produced less favorable results. CHOUINARD (1990) reported that chlorpromazine but not remoxipride was superior to placebo in a 4-week double-blind study of newly admitted, acutely exacerbated schizophrenic patients with about 20 patients in each group. Remoxipride was, however, more effective than placebo in patients with a history of good response to neuroleptic drugs and worse than placebo in patients with a history of poor response. At this time no systematic studies have been performed or published to demonstrate its efficacy in TR schizophrenia.

A dose of 300 mg/day (range 150–500 mg/day) in two divided doses seems to be optimal for short-term treatment in most cases (LEWANDER et al. 1990; LAPIERRE et al. 1990). For elderly patients, patients with hepatic or renal disease, and maintenance treatment, lower doses should be used.

In a composite summary of side effect data from nine multicenter trials, remoxipride was found to have clear advantages over haloperidol with regard to EPS (LEWANDER et al. 1990). This was especially true for interference with gait, elbow rigidity, fixation of position, head dropping, tremor, and salivation. However, 20% of remoxipride-treated patients required anticholinergic drugs, indicating it does have significant EPS liability for certain individuals. Whether remoxipride would cause tardive dyskinesia with long-term use is not known at this time. Drowsiness, tiredness, and difficulty concentrating are reported by 10%–30% of remoxipride-treated patients. Following the introduction of remoxipride in Europe, a significant number of cases of aplastic anemia were reported. The rate may be as high as 1 in 10000. For this reason, the further use of remoxipride has been stopped until the incidence, cause, and possible prevention of this side effect can be further studied.

V. Amperozide

Amperozide, a benzamide compound, has undergone clinical trials in both Europe and the USA. BJÖRK et al. (1992) performed a 4-week open trial of amperozide in 45 male inpatients. Thirty-six (69%) patients had at least 25% improvement of BPRS scores. The mean scores of the BPRS Positive Symptoms Subscale, BPRS Anergia Subscale, and BPRS General Symptoms Subscale showed improvements of 57%, 50%, and 68%, respectively. The patient sample included 12 patients who had a history of poor response to typical neuroleptics. Of these patients, nine (75%) showed significant improvement. These nine responders had a mean improvement of about 60% in the BPRS total score, positive symptoms, negative symptoms, and general symptoms.

VI. Others

There are numerous other antipsychotic drugs which have atypical characteristics and are in various stages of clinical testing. Some of these drugs have been shown to have efficacy in schizophrenic patients while for others the evidence is inconclusive. Also, many drugs with demonstrated efficacy in schizophrenia have yet to be tested in TR populations. Some of these drugs have been shown to have superior efficacy compared to the typical neuroleptics.

Ritanserin, a 5-$HT_{2A/2C}$ antagonist, has been shown to reduce EPS and negative symptoms of schizophrenia (BERSANI et al. 1990). In a placebo-controlled trial it decreased negative symptoms when added to typical neuroleptic treatment (DUINKERKE et al. 1993). Finally, WEISEL et al. (1994) found ritanserin to be effective in ameliorating positive and negative symptoms in a group of acutely exacerbated schizophrenic patients.

Mianserin, another 5-$HT_{2A/2C}$ antagonist, has also been found to effect improvement in both positive and negative symptoms of schizophrenia, when added to typical neuroleptics (ROGUE and ROGUE 1992).

In a double-blind, placebo-controlled trial, olanzapine (mean dose 15mg/day), a heterocyclic 5-HT_2/D_2 antagonist, has been found to be superior to haloperidol (mean dose 15mg/day) in decreasing BPRS Negative Symptom scores (BEASLEY et al. 1995). Effects on positive symptoms and total score were comparable. It had fewer EPS than haloperidol.

Zotepine, which, like clozapine, has high affinity for 5-HT_{2A}, 5-HT_{2C}, 5-HT_6, and 5-HT_7 receptors, was shown in a double-blind study to be superior to haloperidol in patients with predominantly negative symptoms.

D. Conclusions

About 30% of schizophrenic patients fail to respond adequately to typical neuroleptic drugs. Clozapine's demonstrated efficacy in these patients, coupled with its lack of liability to cause EPS, has spurred research directed at developing antipsychotic drugs with similar clinical and/or neuropharmacologic profile. Such drugs have been called "atypical" because of their low propensity to cause EPS.

Current evidence indicates that clozapine is effective in up to 60% of treatment-refractory (TR) patients. Clozapine ameliorates positive, negative, disorganization, and mood symptoms of schizophrenia. There is also some evidence for improvement in some cognitive function under clozapine treatment. In addition to the improvement in psychopathology, clozapine has also been shown to improve the quality of life for many schizophrenia patients. It has been suggested that the adequate duration of clozapine trials should be 6–

12 months. Despite the relatively high current cost of clozapine and the cost associated with required weekly WBC monitoring, clozapine has been found to be a cost-effective treatment option for TR patients.

None of the other atypical antipsychotic drugs have yet been tested in TR populations under double-blind, placebo-controlled conditions. In some of the double-blind studies involving these drugs, the patient sample included some apparently TR schizophrenic patients. Thus, some limited and indirect evidence of efficacy in TR schizophrenia exists for risperidone, melperone, remoxipride, and amperozide. Clearly, these observations need to be confirmed in systematic studies using a well-defined TR population.

To date, no studies comparing clozapine to other atypical antipsychotic drugs have been published. At this time, clozapine remains the cornerstone in the pharmacotherapy of TR schizophrenia.

References

Alphs LD, Lee HS (1991) Comparison of withdrawal of typical and atypical antipsychotic drugs: a case study. J Clin Psychiatry 52:346–348

Alvir JMJ, Lieberman JA, Safferman AZ, Schwimmer JL, Schaaf JA (1993) Clozapine-induced agranulocytosis incidence and risk factors in the United States. N Engl J Med 329:162–167

Amsler HA, Teerenhovi L, Bartha E, Harjula K, Vuopio P (1977) Agranulocytosis in patients treated with clozapine: a study of the Finnish epidemic. Acta Psychiatr Scand 56:241–248

Andreasen NC (1983) The Scale for the Assessment of Negative Symptoms (SANS). University of Iowa, Iowa City

Andreasen NC, Olsen S (1982) Negative vs. positive schizophrenia. Definition and validation. Arch Gen Psychiatry 39:789–794

Baldessarini R, Frankenberg F (1991) Clozapine: A novel antipsychotic agent. N Engl J Med 324:746–754

Beasley CM, Tollefson G, Tran P, Satterlee W, Sanger T, Hamilton S, The Olanzapine HGAD Stduy Group (1995) Olanzapine versus placebo and haloperidol: acute phase results of the North American double-blind olanzapine trial. Neuropsychopharmacology In press, 1996.

Bersani G, Bressa GM, Meco G, Pozzi F (1990) Mixed D_2 and S_2 antagonism in a preliminary study with risperidone (R 64 766). Hum Psychopharmacol 5:225–231

Bilder RM, Wu H Chakos MH, Bogerts B, Pollack S, Aronowitz J, Ashtari M, Degreef G, Kane JM, Lieberman JA (1994) Cerebral morphometry and clozapine treatment in schizophrenia. J Clin Psychiatry [Suppl B]:53–56

Bjerkenstedt L (1989) Melperone in the treatment of schizophrenia. Acta Psychiatr Scand [Suppl 352]:35–39

Björk A, Bergman I, Gustavsson G (1992) Amperozide in the treatment of schizophrenic patients: a preliminary report. In: Meltzer HY (ed) Novel antipsychotic drugs. Raven, New York, pp 59–66

Blake LM, Marks RC, Luchins DJ (1992) Reversible neuroleptic symptoms with clozapine. J Clin Psychopharmacol 12:297–299

Bolden C, Cusack B, Richelson E (1992) Antagonism by antimuscarinic and neuroleptic compounds at the five cloned human muscarinic cholinergic receptors expressed in Chinese hamster ovary cells. J Pharmacol Exp Ther 260:576–580

Breier A, Buchanan RW, Kirkpatrick B, Davis OR, Irish D, Summerfelt A, Carpenter WT Jr (1994) Effects of clozapine on positive and negative symptoms in outpatients with schizophrenia. Am J Psychiatry 151:20–26

Brenner HD, Dencker SJ, Goldstein MJ, Hubbard JW, Keegan DL, Kruger G, Kulhanek F, Liberman RP, Malm U, Midha KK (1990) Defining treatment refractoriness in schizophrenia. Schizophr Bull 17:551–562

Buchanan RW, Holstein C, Breier A (1994) The comparative efficacy and long-term effect of clozapine treatment of neuropsychological test performance. Biol Psychiatry 36:717–725

Casey DE (1989) Clozapine: neuroleptic-induced EPS and tardive dyskinesia. Psychopharmacology 99 [Suppl]:S47–S53

Cassady SL, Thacker GK (1992) Addition of fluoxetine to clozapine. Am J Psychiatry 149:1274

Cassens G, Inglis AK, Appelbaum PS, Gutheil TG (1990) Neuroleptics: effects on neuropsychological function in chronic schizophrenic patients. Schizophr Bull 16:477–499

Castelao JF, Ferrerira L, Gelders YG, Heylen SLE (1989) The efficacy of the D_2 and 5-HT_2 antagonist risperidone (R 64 766) in the treatment of chronic psychoses. An open dose finding study. Schizophr Res 2:411–415

Choc MG, Lehr RG, Hsuan F, Honigfeld G, Smith HT, Borison R, Volavka J (1987) Multiple-dose pharmacokinetics of clozapine in patients. Pharmaceut Res 4:402–405

Chouinard G (1990) A placebo-controlled clinical trial of remoxipride and chlorpromazine in newly admitted schizophrenic patients with acute exacerbation. Acta Psychiatr Scand Suppl 358 82:111–119

Chouinard G, Jones B, Remington G, Bloom D, Addington D, MacEwan GW, Labelle A, Beauclair L, Arnott W (1993) A Canadian multicenter placebo-controlled study of fixed doses of risperidone and haloperidol in the treatment of chronic schizophrenic patients. J Clin Psychopharmacol 13:25–40

Claghorn J, Honigfeld G, Abuzzahab FS, Wang R, Steinbook R, Tuason V, Klerman G (1987) The risks and benefits of clozapine versus chlorpromazine. J Clin Psychopharmacol 7:377–384

Classen W, Laux G (1988) Sensorimotor and cognitive performance of schizophrenic inpatients treated with haloperidol, fluphenthixol or clozpaine. Pharmacopsychiatry 21:295–197

Claus A, Bollen J, De Cuyper H, Eneman M, Malfroid M, Peuskens J, Heylen S (1992) Risperidone versus haloperidol in the treatment of chronic schizophrenic inpatients: a multicentre double-blind comparative study. Acta Psychiatr Scand 85:295–305

Crow TJ (1990) Temporal lobe asymmetries as the key to the etiology of schizophrenia. Schizophr Bull 16:433–443

Crow TJ (1980) Molecular pathology of schizophrenia: more than one disease process? BMJ 280:66–68

Davidson M, Kahn RS, Stern RG, Hirschowitz J, Apter S, Knott P, Davis KL (1993) Treatment with clozapine and its effect on plasma homovanillic acid and norepinephrine concentrations in schizophrenia. Psychiatr Res 46:151–164

Davis JM, Schaffer CB, Killian GA, Kinard C, Chan C (1980) Important issues in the drug treatment of schizophrenia. Schizophr Bull 6:70–87

Deniker P (1984) Introduction of neuroleptic chemotherapy into psychiatry. In: Ayd FG, Blackwell B (eds) Discoveries in biological psychiatry, Ayd Medical, Baltimore, pp 155–164

Drake GE, Gates C, Whitaker A, Cotton PG (1985) Suicide among schizophrenics: a review. Compr Psychiatry 26:90–100

Duinkerke SJ, Botter PA, Jansen AAI, Van Donegen PAM, Van Haaften AJ, Boom AJ, Van Laarhoven JHM, Busard HLSM (1993) Ritanserin, a selective 5-$HT_{2/1C}$ antagonist, and negative symptoms in schizophrenia: a placebo-controlled double-blind trial. Br J Psychiatry 163:451–455

Ekblöm B, Haggström JE (1974) Clozapine (Leponex) compared with chlorpromazine: A double-blind evaluation of pharmacological and clinical properties. Curr Ther Res 16:945–957

Ekblom B, Eriksson K, Lindström LH (1984) Supersensitivity psychosis in schizophrenic patients after sudden clozapine withdrawal. Psychopharmacology 83:293–294

Endicott J, Spitzer RL (1978) A diagnostic interview: the Schedule for Affective Disorders and Schizophrenia. Arch Gen Psychiatry 35:837–844

Fischer-Cornelssen KA, Ferner UJ, Steiner H (1974) Multifokale Psychopharmakaprüfung (Multihospital trial). Arzneimittelforschung 24:1706–1724

Fleischhacker WW, Hummer M, Kurz M, Kurzthaler I, Lieberman JA, Pollack S, Safferman AZ, Kane JM (1994a) Clozapine dose in the United States and Europe: implications for therapeutic and diverse effects. J Clin Psychiatry 55 [Suppl B]:78–81

Fleishhacker WW, Meise U, Günther V, Kurz M (1994b) Compliance with antipsychotic drug treatment: influence of side effects. Acta Psychiatr Scand 89 [Suppl 382]:11–15

Friedman L, Knutson L, Shurell M, Meltzer HY (1991) Prefrontal sulcal prominence is inversely related to response to clozapine in schizophrenic patients. Biol Psychiatry 29:865–877

Gelders YG, Heylen SL, Vanden BG, Reyntjens AJ, Janssen PA (1990) Pilot clinical investigation of risperidone in the treatment of psychotic patients. Pharmacopsychiatry 23:206–211

Gerlach J, Casey DE (1994) Drug treatment of schizophrenia: myths and realities. Curr Opin in Psychiatry 7:65–70

Gerlach J, Koppelhus P, Helweg E, Manrad A (1974) Clozapine and haloperidol in a single-blind cross-over trial. Therapeutic and biochemical aspects in the treatment of schizophrenia. Acta Psychiatr Scand 50:410–424

Goldberg TE, Greenberg RD, Griffin SJ, Gold JM, Kleinman JE, Pickar D, Schulz SC, Weinberger DR (1993) The effect of clozapine on cognition and psychiatric symptoms in patients with schizophrenia. Br J Psychiatry 162:43–48

Green AI, Alam MY, Sobieraj JT, Pappalardo KM, Waternaux C, Salzman C, Schatzberg AF, Schildraut JJ (1993) Clozapine response and plasma catecholamines and their metabolites. Psychiatry Res 46:139–150

Hagger C, Buckley P, Kenny JT, Friedman L, Ubogy D, Meltzer HY (1993) Improvement in cognitive functions and psychiatric symptoms in treatment-refractory schizophrenic patients receiving clozapine. Biol Psychiatry 34:702–712

Hamilton M (1960) A rating scale for depression. J Neurol Neurosurg Psychiatr 23:56–62

Haring C, Fleishhacker W, Schett P, Humpel C, Barnas C, Saria A (1990) Influence of patient-related variables on clozapine plasma levels. Am J Psychiatry 147:1471–1475

Härnyrd C, Bjerkenstedt L, Gullberg B (1989) A clinical comparison of melperone and placebo in schizophrenic women on a milieu therapeutic ward. Acta Psychiatr Scand 352:40–47

Hasegawa M, Gutierrez-Esteinou R, Way L, Meltzer HY (1993) Relationship between clinical efficacy and clozapine plasma concentrations in schizophrenia: effect of smoking. J Clin Psychopharmacol 13:383–390

Heinrichs DW, Hanlon TE, Carpenter WT Jr (1984) The Quality of Life Scale: an instrument for rating the schizophrenic deficit syndrome. Schizophr Bull 10:388–398

Heinrichs K, Klieser E, Lehmann E, Kinzler E (1990) Experimental comparison of the efficacy and compatibility of clozapine and ripseridone in acute schizophrenia. Satellite Symposium "Risperidone", 17th Cong CINP, Kyoto, Japan, p 22

Hoff AL, Riordan H, O'Donnell DW, Morris L, DeLisi LE (1992) Neuropsychological functioning of first-episode schizophreniform patients. Am J Psychiatry 149:898–903

Honigfeld G, Patin J (1989) Predictors of response to clozapine therapy. Psychopharmacology 99 [Suppl]:S64–S67

Jann MW, Grimstey SR, Gray EC, Change WH (1993) Pharmacokinetics and pharmacodynamics of clozapine. Clin Pharmacokinetics 24:161–176
Juarez-Reyes MG, Shumway M, Battle C, Bacchetti P, Hansen MS, Hargreaves WA (1994) Restricting clozapine use: the impact of stringent eligibility criteria. Abstract presented ACNP Meeting, Honolulu, 1993 and NCDEU Meeting, Ft. Myers, 1994
Juul-Povlsen U, Noring U, Fog R, Gerlach J (1985) Tolerability of therapeutic effect of clozapine. Acta Psychiatr Scand 71:176–185
Kane J, Honigfeld G, Singer J, Meltzer HY, Clozaril Collaborative Study Group (1988) Clozapine for the treatment-resistant schizophrenic: a double-blind comparison with chlorpromazine. Arch Gen Psychiatry 45:789–796
Kane JM, Safferman AZ, Pollack S, Johns C, Szymanski S, Kronig M, Lieberman JA (1994) Clozapine, negative symptoms, and extrapyramidal side effects. J Clin Psychiatry 55 [Suppl B]:74–77
Kay SR, Singh MM (1989) The positive-negative distinction in drug-free schizophrenic patients. Arch Gen Psychiatry 46:711–718
Kenny J, Meltzer HY (1991) Attention and higher cortical functions in schizophrenia. J Neuropsychiatry Clin Neurosci 3:269–275
King DJ (1990) The effect of neuroleptics on cognitive and psychomotor function. Br J Psychiatry 157:799–811
Kirkpatrick B, Buchanan RW, McKenney PD, Alphs LD, Carpenter WT Jr (1989) The Schedule for the Deficit Syndrome: an instrument for research in schizophrenia. Psychiatry Res 30:119–123
Krupp P, Barnes P (1989) Leponex-associated granulocytopenia: a review of the situation. Psychopharmacology (Berl) 99 [Suppl]:S118–S121
Kuha S, Miettinen E (1986) Long-term effect of clozapine in schizophrenia: a retrospective study of 108 chronic schizophrenics treated with clozapine for up to 7 years. Nord Psychiatr Tidskr 40:225–230
Lamberti JS, Bellnier T, Schwarzkopf SB (1992) Weight gain among schizophrenic patients treated with clozapine. Am J Psychiatry 149:689–690
Lapierre YD, Nair NPV, Chouinard G, et al. (1990) A controlled dose-ranging study of remoxipride versus haloperidol in schizophrenia. Acta Psychiatr Scand 82 [Suppl 358]:72–76
Lehman AF (1983) The well-being of chronic mental patients: assessing their quality of life. Arch Gen Psychiatry 40:369–373
Lewander T, Westerbergh SE, Morrison D (1990) Clinical profile of remoxipride – a combined analysis of a comparative double-blind, multicentre trial programme. Acta Psychiatr Scand 82 [Suppl 358]:92–98
Lieberman J, Jody D, Geisler S, Alvir J, Loebel A, Szumanski S, Woerner M, Borenstein M (1993) Time course and biological correlates of treatment response in first episode schizophrenia. Arch Gen Psychiatry 50:369–376
Lieberman JA, Kane JM, Safferman AZ, Pollack S, Howard A, Szymanski S, Masiar SJ, Kronig MH, Corper T, Novacenko H (1994) Predictors of response to clozapine. J Clin Psychiatry 55 [Suppl B]:126–128
Lewander T, Westerbergh SE, Morrison D (1990) Clinical profile of remoxipride – a combined analysis of a comparative double-blind, multicentre trial programme. Acta Psychiatr Scand 82 [Suppl 358]:92–98
Lindström LH (1988) The effect of long-term treatment with clozapine in schizophrenia: A retrospective study in 96 patients treated with clozapine for up to 13 years. Acta Psychiatr Scand 77:524–529
Lovdahl MJ, Perry PJ, Miller DD (1991) The assay of clozapine and N-desmethylclozapine in human plasma by high-performance liquid chromatography. Ther Drug Monit 13:69–72
Marder SR, Meibach RC (1994) Risperidone in the treatment of schizophrenia. Am J Psychiatry 151:825–835
McElroy SL, Dessain EC, Pope HG, Cole JO, Keck PE, Frankenberg FR, Aizley HG,

O'Brien S (1991) Clozapine in the treatment of psychotic mood disorders, schizoaffective disorder, and schizophrenia. J Clin Psychiatry 52:411–414

Meco G, Alessandria A, Bonifati V, Giustini P (1994) Risperidone for hallucinations in levodopa-treated Parkinson's disease patients. Lancet 343:1370–1371

Meco G, Bedini L, Bonfati V, Sonini U (1989) Risperidone in the treatment of chronic schizophrenia with tardive dyskinesia: a single blind crossover study versus placebo. Curr Ther Res 46:876–883

Medalia A, Gold J, Merriam A (1988) The effects of neuroleptics on neuropsychological test results of schizophrenics. Arch Clin Neuropsychol 3:249–271

Meltzer HY (1989) Clinical studies on the mechanism of action of clozapine: the dopamine-serotonin hypothesis of schizophrenia. Psychopharmacology 99:S18–S27

Meltzer HY (1992) Dimensions of outcome with clozapine. Br J Psychiatry 160 [Suppl 17]:46–53

Meltzer HY (1994) An overview of the mechanism of action of clozapine. J Clin Psychiatry 55 [Suppl B]:47–52

Meltzer HY (1995) The concept of atypical antipsychotics. In: den Boer JA, Westenberg HGM, van Praag HM (eds) Advances in the neurobiology of schizophrenia. Wiley, Chichester

Meltzer HY, Luchins DJ (1984) Effect of clozapine in severe tardive dyskinesia: A case report. J Clin Psychopharmacol 4:286–287

Meltzer HY, Zureick JL (1989) Negative symptoms in schizophrenia. A target for new drug development. In: Dahl SG, Gram LF (eds) Clinical pharmacology in psychiatry, vol 1. Springer, Berlin Heidelberg New York, pp 68–77 (Psychopharmacology series 7)

Meltzer HY, Okayli G (1995) The reduction of suicidality during clozapine treatment in neuroleptic-resistant schizophrenia: impact on risk-benefit assessment. Am J Psychiatry 152:183–190

Meltzer HY, Goode DJ, Schyve PM, Young M, Fang VS (1979) Effect of clozapine on human serum prolactin levels. Am J Psychiatry 136:1550–1555

Meltzer HY, Kolakowska T, Fang VS, Fogg L, Robertson A, Lewine R, Strahilevitz M, Busch D (1985) Section 1: Biological psychiatry. Growth hormone and prolactin response to apomorphine in schizophrenia and the major affective disorders: relation to duration of illness and depressive symptoms. In: Freedman DX, Lourie RS, Meltzer HY, Nemiah JC, Talbott JA, Weiner H (eds) The year book of psychiatry and applied mental health. Mosby Yearbook, Chicago, pp 16–17

Meltzer HY, Koenig JI, Nash JF, Gudelsky GA (1989) Melperone and clozapine: neuroendocrine effects of atypical neuroleptic drugs. In: G Sedvall (ed) Acta Psychiatr Scand 80(S 352):24–26

Meltzer HY, Alphs LD, Bastani B, Ramirez L (1990a) Effect of melperone in treatment-resistant schizophrenia. In: Stefanis CN, Soldatos CR, Rabavilas AD (eds) Psychiatry today, accomplishments and promises, VIII World Cong of Psychiatry Abstracts, Excerpta Medica Intl Cong Srs 899, Amsterdam, p 502

Meltzer HY, Burnett S, Bastani B, Ramirez LF (1990b) Effect of six months of clozapine treatment on the quality of life of chronic schizophrenic patients. Hosp Community Psychiatry 41:892–897

Meltzer HY, Cola P, Way L, Thompson PA, Bastani B, Davies MA, Snitz B (1993) Cost effectiveness of clozapine in neuroleptic-resistant schizophrenia. Am J Psychiatry 150:1630–1638

Meltzer HY, Maes M, Elkis H (1994) The biological basis of refractory depression. In: Nolen W, Zohar J, Roose S, Amsterdam J (eds) Refractory depression current strategies and future directions. Wiley, Chichester pp 177–198

Meltzer HY, Lee MA, Ranjan R (1995a) Recent advances in the pharmacotherapy of schizophrenia. Acta Psychiatr Scand 90 [Suppl 384]:95–101

Meltzer HY, Yamamoto BK, Lowy MT, Stockmeier CA (1995b) The mechanism of action of atypical antipsychotic drugs: an update. In: Watson SJ, Akil H (eds)

Biology of schizophrenia and affective disease – ARNMD Series. Raven, New York, 73
Mesotten F, Suy E, Pictquin M, Burton P, Heylen S, Gelders Y (1989) Therapeutic effect and safety of increasing doses of risperidone (R 64 766) in psychotic patients. Psychopharmacology 99:445–449
Miller DD, Fleming F, Holman TL, Perry PJ (1994) Plasma clozapine concentrations as a predictor of clinical response: a follow-up study. J Clin Psychiatry 55 [Suppl B]:117–121
Naber D, Leppig M, Grohman R, Hippius H (1989) Efficacy and adverse effects of clozapine in the treatment of schizophrenia and tardive dyskinesia – a retrospective study of 387 patients. Psychopharmacology 99:S73–S76
Naber D, Hippius H (1990) The European experience with use of clozapine. Hosp Community Psychiatry 41:886–890
Naber D, Holzbach R, Perro C, Hippius H (1992) Clinical management of clozapine patients in relation to efficacy and side-effects. Br J Psychiatry 160:54–59
Overall JE, Gorham D (1962) The brief psychiatric rating scale. Psychol Rep 10:149–165
Owen RR, Beake BJ, Marby D, Dessain EC, Cole JO (1989) Response to clozapine in chronic psychotic patients. Psychopharmacol Bull 25:253–256
Parsa MA, Al-Lanhram Y, Ramirez LF, Meltzer HY (1993) Prolonged psychotic relapse after abrupt clozapine withdrawal. J Clin Psychopharmacol 13:154–155
Perry PJ, Miller DD, Arndt SV, Cadoret RJ (1991) Clozapine and norclozapine plasma concentrations and clinical response of treatment-refractory schizophrenic patients. Am J Psychiatry 148:231–235
Pickar D, Owen RR, Litman RE, Konicki PE, Gutierrez L, Rapaport MH (1992) Clinical and biological response to clozapine in patients with schizophrenia. Arch Gen Psychiatry 49:345–353
Pope HG Jr, McElroy SL, Keck PE, Hudson JI (1991) Valproate in the treatment of acute mania. Arch Gen Psychiatry 44:113–118
Potkin SG, Bera R, Gulasekaram B, Costa J, Hayes S, Ju Y et al. (1994) Plasma clozapine concentrations predict clinical response in treatment-resistant schizophrenia. J Clin Psychiatry 55 [Suppl B]:133–136
Revicki DA, Luce BR, Wechsler JM, Brown RE, Adler MA (1990) Cost-effectiveness of – clozapine for treatment-resistant schizophrenic patients. Hosp Community Psychiatry 41:850–855
Rogue A, Rogue P (1992) Mianserin in the management of schizophrenia. Schizophrenia 1992 Abs Book, Intl Conf, Vancouver, British Columbia, p 135
Roose K, Gelders Y, Heylen S (1988) Risperidone (R 64766) in psychotic patients: a first clinical therapeutic exploration. Acta Psychiatr Belg 88:233–241
Sajatovic M, Meltzer HY (1993) The effect of short-term electroconvulsive treatment plus neuroleptic in treatment-resistant schizophrenia and schizoaffective disorder. Convuls Ther 9:167–175
Schmutz J, Eichenberger E (1982) Clozapine. In: Bindra JS, Lednicer D (eds) Chronicles in drug discovery, vol 1. Wiley, New York, pp 39–58
Scholz E, Dichgans J (1985) Treatment of drug-induced exogenous psychosis in parkinsonism with clozapine and fluperlapine. Eur Arch Psychiatry Neurol Sci 235:60–64
Seidman LJ (1983) Schizophrenic and brain dysfunction: an integration of recent neurodiagnostic findings. Psychol Bull 94:195–238
Simpson GM, Cooper TA (1978) Clozapine plasma levels and convulsions. Am J Psychiatry 135:99–100
Small J, Milstein V, Marhenke JD, Hall DD, Kellams JJ (1987) Treatment outcome with clozapine in tardive dyskinesia, neuroleptic sensitivity and treatment-resistant psychosis. J Clin Psychiatry 48:263–267
Spohn HE, Strauss ME (1989) Relation of neuroleptic and anticholinergic medication to cognitive function in schizophrenia. J Abnorm Psychol 98:367–380
Svestka J, Ceskova E, Rysanek R, Obrovsko V (1990) Double-blind clinical compari-

son of risperidone and haloperidol in acute schizophrenic and schizoaffective psychoses. Activitas Nervosa Suppl 32:237–238

Szymanski S, Lieberman J, Pollack S, Mienne R, Safferman A, Kane J, Kronig M, Cooper T (1993) The dopamine-serotonin relationship in clozapine response. Psychopharmacology 112:S85–S89

Tenke CE, Schroeder CE, Arezzo JC, Vaughan HG Jr (1993) Interpretation of high-resolution current source density profiles: a stimulation of sublaminar contributions to the visual evoked potential. Exp Brain Res 94:183–192

Van Putten T (1974) Why do schizophrenic patients refuse to take their drugs? Arch Gen Psychiatry 31:67–72

Wiesel F-A, Nordström A-L, Farde L, Eriksson B (1994) An open clinical and biochemical study of ritanserin in acute patients with schizophrenia. Psychopharmacology 114:31–38

CHAPTER 13

Antipsychotic-Induced Side Effects Related to Receptor Affinity

LINDA PEACOCK and JES GERLACH

A. Introduction

Treatment with antipsychotics is often complicated by side effects, whether slight and harmless or severe and potentially lethal. Severe side effects are a major cause of poor compliance, which in turn has implications for relapse, hospitalization, and morbidity. They cause suffering for patients and relatives, limit possibilities for rehabilitation and leisure activities, and deter social integration. At times, they counteract therapeutic efficacy and may even lead to suicide. Finally, some side effects may become irreversible, thereby entailing a risk of persistent damage to the central nervous system.

With this background in mind, it is of great import to become familiar with these side effects, to understand their pathophysiology, and to limit them as much as possible. The primary goal of research is to develop new antipsychotics with as few and slight side effects as possible. In this decade, many significant innovations are underway, with fewer side effects being a main theme.

B. Predictable and Unpredictable Side Effects

The side effects of antipsychotic drugs can be attributed to a great extent to their affinities for the various central and peripheral nervous system receptors – dopamine, acetylcholine, norepinephrine, and histamine receptors, in particular. Thus, extrapyramidal side effects (EPS) are correlated with anti-dopaminergic actions, autonomic side effects with anticholinergic actions, cardiovascular side effects with anticholinergic and antinoradrenergic actions, and sedation with antinoradrenergic and antihistaminergic effects. It is therefore essential to realize the pharmacological profiles of antipsychotic drugs to select the drug most appropriate to an individual patient.

The most common side effects are therefore predictable based on the biochemical characteristics of the antipsychotic in question. Some side effects, however, are unpredictable and not immediately related to receptor affinities. This is true of allergic reactions, the neuroleptic malignant syndrome, and blood dyscrasias. Other, less predictable side effects are long-term side effects, slowly induced over weeks, months, or years. Examples of these side effects are tardive dyskinesia and tardive dystonia.

The predictable side effects are dose-related and can usually be limited or totally prevented by utilizing as low a dose as possible. The unpredictable side effects, on the other hand, are not dose-related and cannot be prevented but may be minimized by early diagnosis and discontinuation. The less predictable, tardive side effects assume a middle position (see Sect. G).

As the majority of side effects are thus determined by pharmacological profiles, antipsychotics both old and new will be classified in the following section according to their receptor affinities. Thereafter, a description will be provided of individual side effects as related to the blockade of different receptors.

C. Classification of Antipsychotic Drugs

In recent years there has been a growing interest in the development of so-called atypical antipsychotics, although a clear-cut or logical definition of the term "atypicality" remains to be established. In stringent terms, a typical antipsychotic is a compound producing catalepsy in rodents and acute extrapyramidal symptoms (EPS) in humans (i.e., a neuroleptic), while an atypical antipsychotic is without these properties. In this strict sense, there is no (or only one) atypical antipsychotic on the market. Even clozapine can produce atypical catalepsy in rodents and mild acute EPS in humans. However, broadening the definition of atypicality to include compounds displaying a low propensity to induce catalepsy and EPS, there have long been a number of atypical antipsychotics in existence, with clozapine as the present prototype, and many others underway.

It can be questioned, however, whether this atypicality concept, with sole emphasis on acute EPS, is relevant or fruitful. The group of "atypical antipsychotics" encompasses compounds displaying a wide variety of pharmacological effects, and ranging from pure dopamine D_2 antagonists such as sulpiride to multiple receptor antagonists such as clozapine. These compounds are thus a heterogeneous group of drugs with various therapeutic effects and side effects, and only one common feature, namely a low level of acute EPS. Such a concept is misleading to clinicians (who believe that all these drugs can be used for the same indications as clozapine) as well as biochemists (who believe they can find common properties for "atypicals" versus "typicals"). Also included in this group of "atypical antipsychotics" are some antipsychotics from the 1960s such as chlorprothixene and levomepromazine which have a strong anticholinergic (and antiserotonergic $5HT_2$) effect and therefore induce only mild EPS. It is further-more important to note that a weaker acute EPS potential does not necessarily reflect less of a propensity to induce chronic EPS such as tardive dyskinesia (see following), and that side effects other than acute and chronic EPS (e.g., sedation, orthostatic hypotension, weight gain) may be more distressing to the patient.

With the above considerations in mind, the general distinction "typical/atypical" antipsychotics will not be utilized in this chapter. Instead, a more precise classification based on the compounds' pharmacological nuances has been chosen, although older and newer antipsychotics will be treated separately. This classification has not precluded a subcategorization of antipsychotics with typical (high) versus atypical (low) EPS profiles.

Table 1 shows a selection of the most important antipsychotics, the old drugs above and the newer below. The old antipsychotics constitute a spectrum beginning with the dopamine D_2 selective drug pimozide and ending with the multireceptor drugs chlorpromazine and thioridazine. This receptor spectrum corresponds to a clinical dose and side effect spectrum, as described later (Sect. E, Table 3).

The newer antipsychotics are depicted at the bottom of Table 1. These are all drugs which are clinically available or expected to become so within a year or two. They represent a more heterogeneous group than the old drugs, and a systematization according to the aforementioned receptor affinity-dose relationship cannot be maintained. For example, the multireceptor drug olanzapine (which corresponds to clozapine with respect to receptor binding)

Table 1. Receptor-binding profiles* of older and newer antipsychotics

	D_1	D_2	$5HT_2$	Alpha-1	Chol	Hist
Old antipsychotics						
Pimozide	–	+++	–	–	–	–
Flupentixol	+ (+)	+++	+	+	–	–
Haloperidol	–	+++	+	+	–	–
Fluphenazine	–	+++	+	+	–	–
Zuclopenthixol	+	+++	+	++	–	–
Perphenazine	–	+++	++	++	–	++
Melperone	–	+++	++	+++	–	–
Chlorprothixene	+	+++	+++	+++	++	–
Thioridazine	+ (+)	+++	++	+++	++	+
Chlorpromazine	–	+++	++	+++	++	++
Levomepromazine	–	+++	++	+++	++	++
New antipsychotics						
Sulpiride	–	+++	–	–	–	–
Risperidone	–	+++	+++	+++	–	–
Ziprasidone	+	+++	+++	++	–	–
Sertindole	–	+++	+++	++	–	–
Clozapine	++	++	+++	+++	+++	++
Olanzapine	+(+)	++	+++	++	+++	++
Seroquel	(+)	+	+	+++	–	++
Zotepine	–	+++	+++	++	++	–

*Antipsychotics also bind to D_3 and D_4 receptors and to different types of 5HT receptors, but because the clinical significance is uncertain, these affinities are not included in the table.
Chol, acetylcholine; Hist, histamine.

has to be given in low doses as is the case with the dopamine-selective drug pimozide. It should also be observed that some of the new drugs have rather traditional antidopaminergic D_2 receptor actions (sulpiride, risperidone, and ziprasidone) while others, such as clozapine and seroquel, cause a relatively low dopamine D_2 receptor blockade.

By means of these receptor profiles it may be possible to predict fairly precisely the individual compound's side effects when the relation between specific side effects to the blockade of receptor types is appreciated.

D. Side Effects in Relation to Receptor Affinities

In this section, side effects will be related to receptor affinities. The most common acute and chronic clinical side effects (defined as side effects affecting 10% or more of patients) and less common but serious side effects will be outlined, with brief clinical descriptions.

I. Side Effects Related to Dopamine D_2 Receptor Binding

Dopamine D_2 receptor antagonism is still the most significant and unifying mechanism of action of all older and newer antipsychotics. Therefore, all antipsychotics share the side effects associated with this mechanism, though to a lesser degree in the case of certain older drugs with inherent anticholinergic properties and possibly in those with antiserotonergic ($5HT_2$) effects, as well as in some of the newer drugs. These D_2 receptor-related side effects, which include psychological, neurological, and endocrine side effects, will therefore be treated in detail in the following.

1. Psychological Side Effects

Shortly after the discovery of the antipsychotic chlorpromazine, it was recognized that neuroleptics could cause a state of psychological indifference (for reviews of the early history of neuroleptics see DENIKER 1983 and HAASE 1985). Furthermore, it was recognized at an early stage of research that neuroleptics could induce depression (BOWERS and ASTRACHAN 1967; HELMCHEN and HIPPIUS 1967; DE ALARCON and CARNEY 1969). As early as 1974, VAN PUTTEN concluded that 45% of schizophrenic patients were noncompliant due to psychological side effects. Nevertheless, it is only recently that the psychological side effects of antipsychotics have received due attention, especially in conjunction with the recognition of the so-called neuroleptic-induced deficit syndrome (NIDS) (for reviews see AWAD 1993; LADER 1993; CASEY 1994; LEWANDER 1994). This syndrome of psychological inhibition produces discomfort for both the patients and their relatives who note the change, and can in many ways be considered a parallel to the well-known motor inhibition. Just as one differentiates between subjective (mental) and objective (motor) akathisia, one can speak of subjective (mental) and objective (motor) parkinsonism.

This "mental parkinsonism" includes emotional indifference, blunted affect, and anhedonia in the emotional realm; apathy, reduced initiative, and lack of drive in the social realm; and sluggish thought processes and poor concentration in the cognitive realm (see Table 2). Expressions such as "feeling like a zombie" and "being in a mental straightjacket" have been used in a somewhat critical manner to characterize this "mental parkinsonism," which , together with subjective akathisia, is probably the primary cause of patients' (and relatives') negative attitude towards antipsychotics. The exploration of this syndrome, which is still in its prime, is especially difficult because it may be hard to differentiate between side effects and primary negative symptoms. As of yet, there are no systematic comparative clinical studies as to prevalence.

Depression can be promoted by antidopaminergic treatment and is characterized by lowered mood, a sense of being overwhelmed, and despair. Again, it may be difficult to distinguish this side effect from the primary condition.

Psychological akathisia is an uncomfortable feeling of restlessness and uneasiness. The patients feel jittery, as though they are about to crawl out of their skins. Psychological akathisia is often associated with dysphoria and anxiety, reports indicating a strong relation with a poor treatment response (VAN PUTTEN and MAY 1978; NEWCOMER et al. 1994). Motor akathisia is described in the following section.

Table 2. Neuroleptic-induced psychological side effects related to receptor binding

Psychological side effects related to dopamine receptor blockade	
Emotional indifference Blunted affect Anhedonia, no pleasure	"Emotional parkinsonism"
Reduced initiative, apathy Lack of energy Reduced social drive Reduced sexual drive	"Social parkinsonism"
Slowing of thought processes Feeling of being empty Concentration difficulties Less inspiration and fewer ideas	"Cognitive parkinsonism"
Inhibition of sensations, e.g., hearing Depression	
Subjective akathisia	
Psychological side effects related to noradrenaline and histamine Sedation	
Psychological side effects related to acetylcholine Confusion	

2. Acute Extrapyramidal Side Effects (EPS)

The classic tetrad of acute EPS is comprised of parkinsonism, dystonia, akathisia, and acute dyskinesia (for reviews see CASEY 1991; BARNES and EDWARDS 1993; GERLACH and PEACOCK 1995).

The three essential elements of neuroleptic-induced parkinsonism are hypo- or bradykinesia, rigidity, and tremor. Hypokinesia, which is the most common of all acute EPS, varies in signs from reduced facial expression (with or without hypersalivation), through slowed movements (e.g., reduced arm swing and stride).

Acute dystonia is characterized by briefly sustained or fixed abnormal postures and tonic-clonic muscular contractions. It may involve isolated muscle groups (e.g., oculogyric crisis, i.e., upward rotation of the eyes; forced opening of the mouth with protrusion/distortion of the tongue; and spasms of the neck/shoulder or plantar muscles) or, in the worst case, result in opisthotonus, a dramatic and distressful experience for the patient. Dystonia may also be present in a discrete form, expressed as splaying of one or more fingers in a fixed position, which sometimes goes unrecognized by both the patients and their physicians.

Acute dyskinesia consists of choreoathetoid movements of the limbs and/or orofacial musculature. In contrast to the tardive condition, the extremities are more often involved and the patient experiences distress (due to associated akathisia). Furthermore, in contrast to the tardive syndrome, acute dyskinesia is often immediately alleviated by reduction of the neuroleptic dose and/or by anticholinergic treatment; for this reason the term "paradoxical tardive dyskinesia (TD)" has been coined (GERLACH et al. 1974; CASEY and DENNEY 1977). In most studies, acute and tardive forms of dyskinesia are not distinguished.

Acute akathisia is comprised of a psychological component, as mentioned earlier, and a motor component, varying from stereotypical restless movements of the upper/lower extremities to involuntary stepping in place and/or pacing. These components may be seen either in isolation or in combination.

While patients usually notice discomfort in conjunction with dystonia, acute dyskinesia, or akathisia, most are not aware of the presence of hypokinesia (LARSEN and GERLACH, unpublished). This requires special attention on the part of the physician to hypokinesia as it can both hinder activities and lead to tardive effects.

3. Tardive EPS

All forms of acute EPS may be seen in tardive forms, although with a slightly modified expression.

Of the tardive EPS syndromes, tardive dyskinesia has received the most attention. Tardive dyskinesia differs from acute dyskinesia in its tendency to cause less subjective distress and to be localized more often to the orofacial area. As the term implies, the tardive syndromes generally occur in a later phase during or after treatment (months to years). Furthermore, as opposed to acute EPS, neuroleptic dose reduction/discontinuation and/or addition of an anticholinergic does not provide immediate relief, but rather, in the first instance, tends to worsen the symptoms by reducing parkinsonian hypokinesia and rigidity.

Tardive dystonia is an extremely serious, often irreversible side effect which can affect discrete muscle groups (e.g., splaying of the fingers) or the entire body, including the oral region, neck, back, and extremities. The patients may assume a leaning posture ("Pisa syndrome") and be severely disabled.

Tardive akathisia resembles acute akathisia and consists of a motor and/or psychological component. In contrast to the acute syndrome, tardive akathisia, like other tardive syndromes, often first appears on antipsychotic dose reduction or withdrawal.

Despite intensive efforts, the detailed pathophysiology of tardive dyskinesia and other tardive EPS is still poorly understood. Blockade of dopamine D_2 receptors is the primary prerequisite and most robust culprit (McClelland et al. 1991), and adaptive, secondary mechanisms are still hypothetical. One viable hypothesis as to the pathophysiological mechanism underlying tardive dyskinesia is a modified version of the original dopamine supersensitivity theory, the dopamine D_1/D_2 imbalance hypothesis (Rosengarten et al. 1983; Gerlach et al. 1991; Peacock and Gerlach, in Press). Briefly, this hypothesis states that dyskinesia is due to a preponderance of dopamine D_1 relative to dopamine D_2 activity. At treatment start, blockade of dopamine D_2 receptors allows a relative overstimulation of dopamine D_1 receptors, resulting in acute dyskinesia. Tardive dyskinesia is proposed to be due to the development of a dopamine D_1 supersensitivity and/or dysfunction of dopamine D_2 receptor efficacy. The latter proposed development of a dysfunction of dopamine D_2 efficacy might also underlie the other tardive syndromes.

4. Hormonal Side Effects

Antipsychotics exert a number of endocrine effects via influence on the hypothalamo-pituitary-mammary gland and hypothalamo-hypophysial-gonadal axes (for review see Edwards and Barnes 1993). Thus, they suppress corticotropin, growth hormone, thyrotropin, follicle-stimulating hormone, luteinizing hormone, and testosterone while increasing the release of melanocyte-stimulating hormone, antidiuretic hormone, and prolactin. The latter tendency to promote hyperprolactinemia is the clinically most important effect, sometimes leading to gynecomastia, galactorrhea, and amenorrhea in younger women and contributing to a loss of sexual drive.

II. Side Effects Related to Dopamine D_1 Receptor Blockade

Animal studies suggest that D_1 antagonists may have antipsychotic effects. In rodents, D_1 antagonists are cataleptogenic and antagonize locomotion and stereotypies induced by DA agonists (WADDINGTON 1993). Studies in primates suggest that D_1 antagonists have a more benign EPS profile than D_2 antagonists. Thus, in drug-naive Cebus monkeys, D_1 antagonists given in gradually increasing doses do not produce dystonia, while D_2 antagonists produce dystonia at low doses (GERLACH and HANSEN 1993). When given to monkeys which have previously received D_2 antagonists, D_1 antagonists initially produce dystonia corresponding to D_2 antagonists, but during continual treatment, tolerance rapidly develops, and the D_1 antagonist can be given in increasing doses without causing dystonia (LUBLIN et al. 1993). When a D_1 antagonist is added to a D_2 antagonist it is able to convey some tolerance to the D_2 antagonist so this may also be given in higher doses without causing dystonia (COFFIN et al. 1991; own unpublished data). D_1 antagonists also lead to less oral dyskinesia than D_2 antagonists, both during treatment (acute dyskinesia) and following withdrawal (tardive dyskinesia) (for review, see GERLACH et al. 1995).

A few open clinical trials have been performed with D_1 antagonists in schizophrenia (KARLSSON et al. 1995; DEN BOER et al., in press). The results have been disappointing as only a slight or no antipsychotic effect has been found, although improvement in negative symptoms has been described (DEN BOER et al., in press). No sedation or dampening of psychomotor unrest were seen. Side effects of the D_1 antagonist SCH 39116 included restlessness, anxiety, and emesis (KARLSSON et al. 1995) as well as dizziness, hypokinesia, nausea, anxiety, insomnia, and somnolence (DEN BOER et al., in press). No increase in prolactin level was found and there were no adverse effects in regards to any safety parameters.

It appears that the antipsychotic efficacy of D_1 antagonists given alone is limited, but if added to a traditional D_2 receptor treatment, they may improve the therapeutic effect, both with respect to the therapeutic effect and EPS.

III. Side Effects Related to Binding to Receptors Other Than Dopamine Receptors

Blockade of serotonin receptors ($5HT_2$) does not seem to entail important side effects (for review see EDWARDS and BARNES 1993). A weakening of appetite regulation may be seen, leading to weight gain. However, weight gain may also be due to other direct (noradrenergic blockade) and indirect side effects of antipsychotics (e.g., inactivity).

Blockade of $5HT_2$ receptors may counteract akathisia and parkinsonism (FLEISCHHACKER et al. 1990; BERSANI et al. 1986), but the documentation is insufficient. Studies with combined D_2-$5HT_2$ blockers such as risperidone have been advanced to support the assumption of an EPS-antagonistic effect of $5HT_2$ antagonists (MARDER and MEIBACH 1994; CHOUINARD et al. 1993). However, there are two essential flaws in this reasoning: for one, these studies are imbalanced in regards to drug doses (relatively high doses of the control drug while doses of the test drug were variable and relatively low) and secondly, they provide only indirect evidence and can therefore not be used as proof of an anti-EPS effect of $5HT_2$ antagonism. For further discussion, see GERLACH 1991 and GERLACH and CASEY 1994.

Blockade of acetylcholine (muscarine) receptors leads to autonomic side effects such as dry mouth, accommodation difficulties, decreased intestinal motility (constipation), urinary retention (and overflow incontinence), delayed ejaculation, orthostatic hypotension, and tachycardia. These side effects are especially severe in the elderly and in other vegetatively labile patients, and especially when several anticholinergic drugs are combined (low-potency antipsychotic, antiparkinsonian drug, and tricyclic antidepressant), they can result in an anticholinergic psychosyndrome with confusion and sometimes hallucinations.

Muscarine receptor blockade has an antiparkinsonian effect, which may also entail an unmasking of potential hyperkinesia.

Blockade of noradrenergic receptors leads especially to sedation and slight orthostatic hypotension, while blockade of histamine receptors promotes sedation and, as a potentially therapeutic component, has an added antiemetic effect.

IV. Side Effects with No Certain Relation to Receptor Binding

This category includes unexpected reactions such as allergic eczema, malignant neuroleptic syndrome, and granulocytopenia/agranulocytosis. As the latter two side effects are the most specific and serious, only they will be covered in the following.

1. Neuroleptic Malignant Syndrome

The neuroleptic malignant syndrome consists of two or more of the following symptoms: muscular rigidity, hyperpyrexia, autonomic dysregulation (lability), and delirium; laboratory tests may reveal elevated creatine kinase and leukocyte counts (for reviews see DAVIS 1990; BARNES and EDWARDS 1993). The incidence is unknown, the maximum estimate being 2%. Unrecognized, the condition has an estimated mortality of up to 30%. The syndrome most often occurs within the first week of initiation of antipsychotic treatment or dose increase, but may occur at any time during treatment. Rapid increases in dose are an important factor. Upon rechallenge there is about a 50% chance of a new episode, though the risk may be limited by beginning with a low dose and slow titration. All antipsychotics pose a risk, and it would be natural to assume a dopamine D_2 antagonistic mechanism, but the pathophysiology is unknown. Clozapine may entail a lesser risk, but clozapine is usually slowly and carefully titrated, so any true difference from other antipsychotics is speculative.

2. Agranulocytosis

Agranulocytosis is another unexpected and poorly understood complication of antipsychotic treatment. Among the older antipsychotics, phenothiazines (e.g., chlorpromazine, thioridazine, and fluphenazine) have long been recognized as entailing the greatest risk (1/2000) (LITVAK and KAELBLING 1971). As clozapine entails a considerable risk (1%–2%), there have been considerable efforts to clarify the pathophysiology of clozapine-induced agranulocytosis (KRUPP and BARNES 1992; VEYS et al. 1992; PISCIOTTA et al. 1992). To date, it appears that in some patients an allergic mechanism is involved, while in others there is an individual susceptibility to a direct toxicity. So far, efforts to predict individuals at risk (e.g., according to HLA subtypes) have been fruitless.

E. Older Antipsychotics with Typical and Atypical EPS Profiles

With reference to the preceding, in this and the following section, a short and largely schematic presentation of which side effects may occur during treatment with both older and newer antipsychotics will be provided. It must be emphasized that both among the older and newer compounds, representatives of antipsychotics with typical and atypical EPS profiles can be found.

The classical antipsychotics from the 1950s–'60s can be subdivided into low-, medium-, and high-potency (or high-, medium-, and low-dose) antipsychotics, a classification inspired by LAMBERT and REVOL (1960) (Table 3). This scheme follows the receptor binding profiles, with high-potency/low-dose antipsychotics preferentially binding to dopamine D_2 receptors whereas at the other extreme, low-potency/high-dose antipsychotics also bind to other receptor systems (e.g., adrenergic, serotonergic, and cholinergic receptors) (Table 1) (GERLACH 1991).

This system offers the advantage of giving the clinician a simple guideline as to which side effects may be expected. Thus, it can be seen from Table 3 that, as one moves from high- to low-potency neuroleptics, the risk of side effects other than acute EPS generally increases, the risk of acute EPS showing the opposite tendency. It is among the low-potency neuroleptics that one finds old drugs exhibiting atypical acute EPS profiles, e.g., thioridazine and levomepromazine.

It must be mentioned that these side effect profiles are most useful in the acute treatment phase whereas, during chronic maintenance treatment, differ-

Table 3. Clinical classification, daily dose levels for antipsychotic effects, and side effect profiles of old antipsychotic drugs

Drug	Dose mg/day	Side effects		
		Sedation	Autonomic side effects	EPS
Pimozide	2–20	(+)	+	++
Flupentixol	2–20	+	+	++
Haloperidol	2–20	+	+	+++
Fluphenazine	2–20	+	+	++(+)
Zuclopenthixol	4–50	++	++	++
Perphenazine	8–64	++	+ (+)	++
Melperone	100–600	++ (+)	+	(+)
Chlorpromazine	100–600	++ (+)	++	+
Thioridazine	100–600	++ (+)	++ (+)	(+)
Chlorprothixene	100–600	+++	++ (+)	(+)
Levomepromazine	100–600	+++	+++	(+)

EPS, extrapyramidal symptoms.
Pluses represent a semiquantitative estimation of the degree of sedation, autonomic side effects, and EPS.

ences recede because of tolerance to acute side effects. Other factors leading to a lessening of side effects in general include changing to other neuroleptics which are better tolerated by the patient and/or addition of side effect medication/concomitant psychotropics (e.g., anticholinergics, benzodiazepines, beta-blockers, antidepressants, and antiepileptics) (LINGJÆRDE et al. 1987).

In judging the prevalence of side effects of antipsychotics, it is important to consider the patient sample, study methods, and doses used. Thus, the reported prevalence of acute EPS induced by typical antipsychotics varies from 2% to 90% (for review see CASEY 1991). Most of this variance must be sought in methodological factors such as whether the study was based on archive material or direct ratings and, in the latter case, whether there was a mere registration of patients' self-reported "adverse events" or whether detailed rating scales were used. In schizophrenic patients it is especially important to perform detailed ratings, as these patients, because of their basic symptoms (e.g., thought disturbance and disconnectedness from their own physical sensations) are often poor spontaneous reporters. A newly concluded study by LARSEN and GERLACH (unpublished) found that while 30% of schizophrenics on depot maintenanance therapy said that they did not have any side effects, 90% of these patients did indeed have side effects, especially hypo- and hyperkinesia, on specific questioning and objective examination.

On the basis of the above considerations, prevalence figures for individual side effects caused by antipsychotics are necessarily only rough estimates.

Table 4 represents a best estimate using data from several studies utilizing the Udvalg for Kliniske Undersøgelser (UKU) side effects rating scale (studies using this scale provide the broadest direct itemization of individual side effects) (LINGJÆRDE et al. 1987; RENDTORFF et al. 1990; PEACOCK et al., in press). Only the most common side effects are included in the table. Because of the aforementioned fading of differences between individual side effects during chronic treatment, all older antipsychotics have been lumped together. Only EPS will be briefly mentioned:

The highest reported incidence of acute dystonia is 39% (KEEPERS et al. 1983). Young men appear to be especially prone to this acute EPS, the exact figure being dependent on treatment traditions as well as dosage and tempo of dose increases. The figures in Table 4 represent the prevalence of tardive dystonia among patients in long-term therapy, all cases being mild. The prevalence of severe, irreversible forms is not known but is probably on the order of 0.1%.

The reported incidence of acute akathisia ranges from 25%–41% (MARSDEN and JENNER 1980; BRAUDE et al. 1983; KEEPERS et al. 1983) while the prevalence of the chronic form is on the order of 25%–33% (Table 4, and BARNES and BRAUDE 1985).

Table 4. Prevalence of side effects during long-term treatment with classical antipsychotics

Side effect	No. of patients (Percent)
Psychological	
Asthenia	21–36
Depression	1–9
Emotional indifference	2–17
Psychological akathisia	13–40
Neurological	
Dystonia	8–15
Hypokinesia	35–55
Rigidity	17–22
Tremor	8–21
Dyskinesia	17–57
Motor akathisia	25–29
Autonomic	
Accommodation difficulties	9–14
Hyposalivation	20–27
Constipation	4–8
Orthostatic hypotension	12–22
Other	
Weight gain	9–26
Galactorrhea	3–6
Amenorrhea	2–15
Decreased sexual drive	20–28

Data from LINGJÆRDE et al. 1987; RENDTORFF et al. 1990; PEACOCK et al. (in press).

The best estimate of the incidence of acute parkinsonism is on the order of 35%–40% (KEEPERS et al. 1983; BARNES and EDWARDS 1993), while the prevalence of chronic parkinsonism is 35%–55% (including mild forms only represented by slight hypokinesia).

The estimated incidence per treatment year of tardive dyskinesia is 4%–5%, the estimated prevalence after 10 and 25 years of treatment being 50% and 68%, respectively (KANE et al. 1984; GERLACH and CASEY 1988; GLAZER et al. 1993).

F. Newer Antipsychotics with Typical and Atypical EPS Profiles

This group includes the deviant and relatively new antipsychotics which are usually designated as atypical. They consist of a heterogeneous group of drugs with varying receptor affinities (Table 1) and varying side effect profiles (Table 5).

For the new compounds in development such as sertindole and seroquel, the available information as to clinical effects is limited, often based on open studies, and often in the form of abstracts and posters. Therefore, some of the information in this section should be treated with reservation.

I. Dopamine-Selective Antipsychotics

1. Sulpiride

Sulpiride is the oldest and most well-known of a series of substituted benzamides with selective binding to dopamine D_2 receptors (Table 1), and

Table 5. Clinical classification, daily dose levels, and side effect profiles of new antipsychotics

Drug	Dose mg/day	Side effects		
		Sedation	Autonomic side effects	EPS
Sulpiride	600–1800	+	+	++
Risperidone	4–12	++	++	++
Ziprasidone	80–160	++	++	++
Sertindole	8–24	+	++	+
Clozapine	200–600	+++	+++	(+)
Olanzapine	10–30	+	+ (+)	+
Seroquel	300–900	++	++	+
Zotepine	100–600	+++	+++	+

Pluses represent semiquantitative estimations of the degree of side effects of these antipshchotics based on the available, relatively limited literature.

possibly a subgroup of such receptors (PEUCH et al. 1976; KOHLER et al. 1979; CASTRO and STRANGE 1993). In sensitized Cebus monkeys, sulpiride has a relatively high threshold dose for induction of dystonia (10mg/kg, corresponding to 2000–8000mg/day in humans, which is above the 600–1800mg/day necessary to obtain antipsychotic effects) (Table 6) (CASEY 1995).

In double-blind studies, sulpiride has been shown to be comparable to haloperidol and chlorpromazine in antipsychotic efficacy, although according to European experience, it has proven more useful in the treatment of milder psychoses dominated by negative symptoms rather than more agitated conditions where positive symptoms predominate (probably due to its rather specific receptor-binding profile) (for sulpiride reviews, see GERLACH 1993; WAGSTAFF et al. 1994).

Sulpiride largely causes only dopamine D_2 receptor side effects, i.e., EPS and hormonal disturbances. It induces slightly less acute EPS (especially parkinsonism) than older high-potency antipsychotics such as haloperidol (EDWARDS et al. 1980; GERLACH et al. 1985) (Table 5). It may also have a lower tardive dyskinesia liability, although this has not been adequately investigated. Due to its low penetration of the blood-brain barrier, sulpiride must be administered in the relatively high doses mentioned above. This leads to high occupancy of peripheral pituitary dopamine D_2 receptors, to hyperprolactinemia, and thereby to a relatively high prevalence (around 25%) of galactorrhea and/or amenorrhea in younger women (HÄRNRYD et al. 1984). Because of its lack of antagonism of other categories of receptors than dopamine receptors, sulpiride is virtually free of nonspecific sedative and autonomic side effects.

Table 6. Threshold dose for dystonia in Cebus monkeys and the estimated dystonia dose in schizophrenic patients compared to the clinically effective doses

Drug	Dystonia-inducing threshold dose in monkeys (mg/kg)	Estimated dystonia-inducing dose in humans (mg/day)[a]	Clinical dose mg/day
Haloperidol	0.025	5–20	5–20
Sulpiride	10	2000–8000	600–1800
Risperidone	0.025	5–20	4–16
Sertindole	0.5	100–400	16–30
Haloperidol	0.015	4.5–18	5–20
Clozapine	>5[b]	>1500	100–800
Olanzapine	0.08	24–96	10–30
Seroquel	4	1200–4800	300–900

[a] Conversion factor from monkey to humans is 200–800 for the first four drugs (CASEY 1995) and 300–1200 for the last 4 (PEACOCK and GERLACH, unpublished) (different due to different sensitivity of the monkey groups), the difference in the absolute values of the dystonia-inducing thresholds of haloperidol in the two groups of monkeys also being explained by the different sensitivities in the two groups.
[b] No clear dystonia is induced by clozapine, but an anticholinergic ataxic syndrome occurs which limits further dose increases.

2. Other Substituted Benzamides

A series of other substituted benzamides with a relatively preferential antagonism of limbic dopamine D_2 receptors has been developed (for review, see GERLACH 1993). While the most promising compound in this group, remoxipride, showed antipsychotic efficacy and less acute EPS liability than haloperidol (SEDVALL 1990), subsequent reports of aplastic anemia have led to its withdrawal from the market. Raclopride, which showed a promising profile in animal behavorial studies (ÖGREN et al. 1986), has been found to have a rather traditional clinical therapeutic effect in schizophrenia, including elevation of clinical indices of increased brain dopamine turnover (CSERNANSKY et al. 1994). This drug has therefore been dropped.

Amisulpride appears to possess antipsychotic efficacy and slightly fewer acute EPS, much resembling sulpiride. Because of a disinhibitory effect in rodents, the drug has especially been given to patients with negative symptoms, where relatively small doses seem to have a beneficial effect (BOYER et al. 1995) or very few side effects of any kind. In higher doses amisulpiride may induce some EPS, but still less than haloperidol (DELCKER et al. 1990; for reviews see BORENSTEIN et al. 1989; VANELLE et al. 1994).

3. D_1 and Combined D_1/D_2 Dopamine Receptor Antagonists

As mentioned in Sect. D.II, there appear to be definite side effect advantages of D_1 antagonists (fewer EPS and tolerance to EPS caused by a concomitant D_2 antagonistic treatment), whereas the therapeutic antipsychotic effect may be limited. D_1 antagonists will therefore probably not be developed as pure antipsychotics though their development as conjunctive antipsychotics might prove of value.

Combining D_1 blockade with D_2 blockade may be a more fruitful approach to new antipsychotic drugs with low EPS potential. With classic neuroleptics such as flupenthixol and zuclopenthixol, which combine a strong D_2 blockade (approximately 80% occupancy) and a relatively lower D_1 blockade (occupancy of 15%–35%, FARDE et al. 1992), some minor clinical advantages may be seen. For flupenthixol, there is some evidence suggesting a mild mood-elevating or less of a mood-suppressing effect, possibly with fewer EPS, but a clinically significant EPS advantage with these drugs has not been documented (GERLACH et al. 1995). New antipsychotics such as olanzapine attempt to exploit the same potential advantage of a combined D_1–D_2 antagonism (see following). With clozapine, which has an equal and low blockade of D_1 and D_2 receptors (see following), more marked EPS advantages are seen, but due to many other receptor affinities, it is difficult to draw definite conclusions concerning the mechanism of action of clozapine.

II. Newer Antipsychotics with Dopamine-Serotonin-Norepinephrine Antagonism

1. Risperidone

Risperidone is a compound offering a greater serotonin 5HT2A blockade relative to dopamine D_2 receptor family blockade and a considerable noradrenergic alpha-1 blockade (Table 1) (Leysen et al. 1994; Moore et al. 1993). A rodent study indicates that risperidone at lower doses causes a partial D_2 receptor blockade which, together with the $5HT_2$ receptor antagonism, may lead to a relatively low EPS potential at lower doses (Leysen et al. 1994). On the other hand, in a positron emission tomography (PET) study, risperidone has shown a D_2 receptor blockade comparable to that of traditional neuroleptics (Nyberg et al. 1993), and in sensitized Cebus monkeys, risperidone has the same dystonia threshold dose as haloperidol (0.025 mg/kg, Table 6) (Casey 1995).

With this background, risperidone may or may not prove advantageous as to EPS, besides producing autonomic side effects. The clinical studies to date are positive, though not quite clear as to the degree of EPS.

The original double-blind, placebo-controlled studies compared several fixed doses of risperidone, ranging from 2–16 mg/day to a rather high dose of haloperidol (20 mg/day) (Chouinard et al. 1993; Marder and Meibach 1994). Risperidone 6 mg/day was found robustly advantageous over haloperidol 20 mg/day both in relation to the alleviation of positive and of negative symptoms and fewer EPS. With higher doses, no clear EPS advantage of risperidone over haloperidol was demonstrated.

Two recent studies comparing variable doses of risperidone (mean end dose risperidone 9 mg/day in each study) versus variable doses of either haloperidol (mean end dose of haloperidol 9 mg/day) (Ceskova and Svestka 1993) or perphenazine (mean end dose perphenazine 28 mg/day) (Høyberg et al. 1993) were unable to confirm any significant superiority of risperidone, whether in regards to clinical effects or side effects. The latter study did, however, indicate a possible advantage of risperidone in patients with predominately negative symptoms (greater number improved in the risperidone group, though no significant difference was seen in endpoint total, subtotal Positive and Negative Syndrome Scale, PANSS, or total Brief Psychiatric Rating Scale, BPRS, scores). In a small 4-week, double-blind study, risperidone 4 and 8 mg daily was, at endpoint, found comparable to clozapine 400 mg daily, both in regards to effects and EPS liability (Heinrich et al. 1994). However, with regards to antipsychotic effects, there were 5 dropouts due to therapeutic inefficacy in each of the two risperidone groups (total of 39 patients) versus 1 dropout (out of 20 patients) in the clozapine group. Furthermore, with regards to EPS, antiparkinson drugs were used in the risperidone groups though no details as to the extent of use were given.

HØYBERG et al. (1993) found a number of antiadrenergic side effects [sedation (40%), orthostatic hypotension (22%), and cardiac palpitations (18%)], as would be expected from risperidone's receptor-binding profile. The corresponding figures in the perphenazine group were sedation (24%), orthostatic hypotension (12%), and cardiac palpitations (10%), in accordance with this drug's weaker antiadrenergic effects. Furthermore, reflecting risperidone's strong antiserotonergic effect, 52% of the risperidone group had weight gain versus 24% in the perphenazine group.

The main conclusion of the results to date is that risperidone appears to be an effective antipsychotic. Low doses appear to induce relatively few EPS, while higher doses (which are necessary in severe cases of schizophrenia) cause the same degree of EPS as classical neuroleptics.

2. Ziprasidone

Ziprasidone is very similar to risperidone in its receptor-binding characteristics (a high ratio of $5HT_2$ to dopamine binding, and noradrenergic (alpha-1) affinity (ZORN et al. 1993) (Table 1). The compound also has some affinity towards $5HT_{1A/C/D}$ and dopamine D_1 receptors. In healthy volunteeers, PET studies have shown single doses of 40 mg ziprasidone to result in a maximal D_2 receptor occupancy of 76% at 4h post dose and a robust prolactin increase (BENCH et al. 1995). Binding potential correlated both with plasma levels of ziprasidone and with plasma prolactin concentrations. These findings predict traditional neuroleptic effects with side effects related to D_2 receptor blockade.

The results of the clinical studies to date have been conflicting. Thus, one double-blind trial of ziprasidone up to 80 mg b.i.d. versus haloperidol 15 mg daily showed equipotent antipsychotic effects, while a subsequent study of ziprasidone up to 40 mg b.i.d. versus placebo could not confirm the efficacy found in the former study.

The major side effects of ziprasidone were sedation (9%–40%), headache (6%–25%), agitation (7%–16%), and orthostatic hypotension/dizziness (7%–19%). Exact details as to relative EPS liability were not discernible, although ziprasidone did produce all forms of acute EPS. In general, the results have as yet not demonstrated an advantage of ziprasidone over haloperidol, neither as to side effects nor with respect to effects. Therefore, further investigations are indicated.

3. Sertindole

Sertindole is also a combined dopamine D_2, serotonin $5HT_2$, and norepinephrine alpha-1 antagonist, but in this case the dopamine antagonism is highly atypical. Although sertindole has a high affinity for D_2 receptors, it is not cataleptogenic and it only weakly inhibits dopamine-agonist induced rotation in 6-OH-dopamine lesioned rats. Furthermore, in an electrophysiological model, following long-term administration, it inhibits the spontaneous activity of dopamine neurons in the ventral tegmental area, while a 100-fold higher dose is needed to inhibit activity in the substantia nigra compacta neurons (like clozapine). Haloperidol induces an equal inhibition in both areas (SKARSFELDT 1992; HYTTEL et al. 1992). In agreement with these rodent observations, a study in sensitized Cebus monkeys has shown that sertindole has a relatively low propensity to induce dystonia [threshold dystonia dose for sertindole 0.5 mg/kg, for haloperidol 0.025 mg/kg (Table 6) (CASEY 1995)].

Sertindole has been studied in a double-blind, placebo-controlled, randomized, 6-week trial in three doses: 8, 12, and 20 mg daily (MARTIN et al. 1994; SCHULZ et al. 1994). A total of 205 schizophrenic patients were included. The highest doses of 20 mg/day significantly reduced the psychotic symptoms compared to placebo, while the lower doses had some, though not a significant effect. Virtually no EPS was recorded, and no antiparkinsonian medicine was given. In another double-blind, placebo-controlled study, sertindole was given in relatively low doses (4 and 12 mg/day) and compared to haloperidol in a relatively high dose (16 mg/day) (MARTIN et al. 1994). Sertindole induced no EPS, while 21% of the haloperidol-treated patients developed EPS and 28% required antiparkinsonian medicine. The main side effects of sertindole in these two studies were headache (26%), nasal congestion (26%), abnormal ejaculation (17%), and insomnia (17%).

Sertindole does, however, have the potential to produce traditional D_2 dopamine antagonistic side effects. This was seen in phase-1, double-blind, placebo-controlled studies in normal volunteers, especially when given in dose levels of 20 mg/day. Tremor, cogwheel rigidity, hypokinesia, akathisia, and dystonia developed in some patients, together with a significant prolactin increase. Of antiepinephrine effects, sertindole tends to produce orthostatic hypotension and sedation.

In conclusion, while the special mechanism of action of sertindole is attractive, and the animal data are very promising, the clinical data are still preliminary. Sertindole appears to be able to induce typical EPS in healthy volunteers, but in schizophrenic patients, when given in gradually increasing doses, the EPS appear to be minimal. As expected, alpha-1 side effects occur and may detract from the clinical impact of this drug.

III. Multiple Receptor Antagonists

Antipsychotics binding to more receptors than dopamine, serotonin, and norepinephrine alpha-1 receptors are included in this group, these being clozapine, olanzapine, seroquel, and zotepine. The first mentioned is the prototype of a low-EPS antipsychotic, while the last more closely resembles a classic neuroleptic. Olanzapine and seroquel appear to take a middle position.

1. Clozapine

With its therapeutic superiority and low level of EPS (FITTON and HEEL 1990), and its low dopamine D_2 and D_1 receptor occupancies (40%–60% in both cases) at clinically effective doses (FARDE et al. 1992), clozapine is the quintessential EPS atypical antipsychotic, the "lithmus test" for new antipsychotics. However, the occupancy of almost all other known receptors results in a large number of side effects, which may pose a serious problem to the patient in spite of therapeutic and EPS advantages.

Clozapine has a definite advantage over traditional antipsychotics not only in regards to acute, but also to chronic EPS. The reported prevalence of acute and chronic EPS found during clozapine treatment varies from 0% to 20% as compared to 0%–40% for chlorpromazine and 40%–85% for haloperidol (for review, see CASEY 1989). Clozapine's beneficial EPS profile has been found to extend to all subtypes of EPS, with the exception of sporadic reports of no superiority in relation to acute akathisia (CLAGHORN et al. 1987; COHEN et al. 1991). Our recently concluded study (PEACOCK et al., in press) comparing 100 schizophrenic patients in maintenance therapy with clozapine versus 100 in treatment with low-to-medium-dose neuroleptics (median length of treatment 5 years for both groups), showed an advantage of clozapine in regards to all EPS, in particular akathisia (objective akathisia 7% versus 29%; subjective akathisia 14% versus 40% in the clozapine and control groups, respectively; $p < 0.001$) (Fig. 1).

In contrast to other clinically effective antipsychotics, there is now considerable documentation for clozapine's minimal risk of producing TD, some

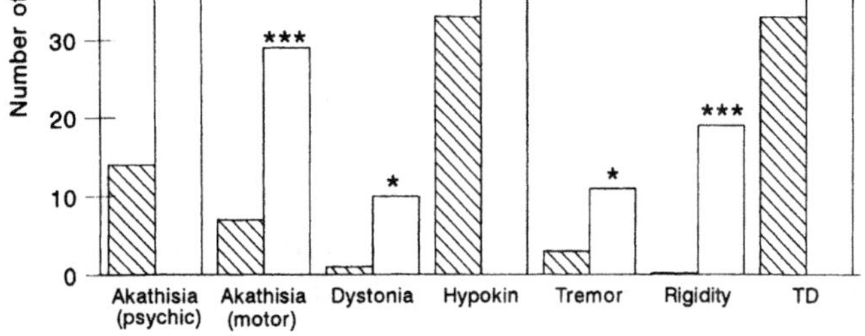

Fig. 1. Extrapyramidal symptoms seen during long-term antipsychotic monotherapy with clozapine (*hatched bars*; $n = 100$) and control neuroleptics (flupenthixol, perphenazine, and zuclopenthixol; *open bars*; $n = 100$). *Hypokin*, hypokinesia; *TD*, tardive dyskinesia. $*p < 0.05$; $**p < 0.01$; $***p < 0.001$ for differences between clozapine and control drug

investigators even proposing a curative effect (CASEY 1989; LIEBERMAN et al. 1991). Thus, relatively few cases of dystonia/dyskinesia have been reported in the literature. PEACOCK et al. (in press) confirmed fewer new TD cases (cases not described in the charts prior to treatment) during clozapine than during control treatment (14% versus 41%, $p < 0.001$). At one year's follow-up there were no new TD cases in the clozapine group and four new cases among controls. It must be noted that the clozapine-treated patients had received traditional neuroleptics for a median of 9 years, for which reason clozapine's contribution to TD remains uncertain, as all cases of TD are not necessarily noted in the charts. The study also found a tendency towards a greater disappearance of dyskinesia among clozapine-treated patients than among controls (disappearance in 18 of 33 patients versus 3 of 19, respectively; $p = 0.07$).

A final, important benefit of clozapine versus traditional antipsychotics from the patient's perpective is its weaker tendency to induce "mental parkinsonism," a syndrome marked by emotional indifference (4% clozapine-treated versus 17% control patients, $p < 0.05$), depression (1% clozapine-treated versus 9% control patients, $p = 0.07$), and sexual side effects (for the most part, sexual apathy) (24% clozapine-treated versus 45% control patients, $p < 0.01$).

Despite these clear advantages, there are a number of draw-backs to clozapine treatment. Though the risk of granulocytopenia(-agranulocytosis) has received the greatest attention, the major side effects posing a problem in the patients' daily living are hypersalivation and weight gain. Hypersalivation is usually nocturnal, and at its worst, the pillow may be soaked, some patients complaining of fear they will choke. Hypersalivation has been found in more than 70% of patients, while 18% of those on control neuroleptics have this complaint (PEACOCK et al., in press). A study examining 36 chronic schizophrenic patients switching from other neuroleptics to clozapine found that 33% gained at least 10 pounds and 42% 20 pounds or more during 6 months of treatment (LAMBERTI et al. 1992). PEACOCK et al. (in press) found weight gain as a side effect in 41% of clozapine-treated patients and 26% of controls ($p < 0.05$).

Other disturbing side effects are orthostatic hypotension (28% of clozapine-treated patients versus 14% of controls) and constipation (30% of clozapine-treated patients versus 4% of controls) (PEACOCK et al., in press).

While the majority of clozapine's side effects are autonomic in character and can be explained by its muscarine and noradrenergic receptor antagonism (as is the case with other high-dose neuroleptics), it still remains unexplained why clozapine, with its anticholinergic properties, causes paradoxical hypersalivation, Clozapine may act as a partial cholinergic agonist.

As to prevention/treatment of clozapine side effects, the autonomic side effects, including weight gain and hypersalivation, are dose-dependent and may be limited by dose reduction when possible. If, due to the psychiatric condition, clozapine dose reduction is not immediately feasible, it may sometimes be achieved by supplementing with a lesser dose of a traditional neuroleptic. Hypersalivation can, to a certain extent, be alleviated by the addition of an anticholinergic drug such as benztropine mesylate. Especially in motivated patients, weight gain can often be prevented and controlled by regular weighing and the provision of dietary counseling.

2. Olanzapine

Olanzapine is a new antipsychotic with a receptor-binding profile rather similar to that of clozapine, except that it has a slightly higher affinity than clozapine to all sites except alpha-1 receptors (MOORE et al. 1993). Thus olanzapine at high doses can produce catalepsy in rodents, but the ratio of the dose causing catalepsy (ED_{50} 23 mg/kg) versus the dose showing presumed antipsychotic efficacy (conditioned avoidance response) (ED_{50} 5.6 mg/kg) is favorable as compared to haloperidol, although less favorable than that of clozapine (MOORE et al. 1994). This corresponds to findings in sensitized Cebus monkeys, where the dystonia threshold dose for olanzapine is 0.08 mg/kg compared to 0.01 mg/kg for haloperidol (Table 6).

Preliminary double-blind, placebo-controlled dose-finding studies comparing olanzapine up to 17.5 mg/day with haloperidol 10–20 mg/day have found an antipsychotic efficacy equal to haloperidol at doses from 7.5 to 17.5 mg/day and a superior effect in regards to negative symptoms at doses from 12.5 to 17.5 mg/day (TRAN et al. 1994). The incidence of acute EPS did not significantly differ from that in placebo at any of the tested doses, whereas haloperidol produced significant akathisia, tremor, and dystonia. Significant side effects of olanzapine versus haloperidol included the expected anticholinergic reactions such as dizziness, constipation, and dry mouth, while mental CNS effects, i.e., somnolence, agitation, asthenia, and nervousness corresponded to those of haloperidol. No significant prolactin increase occurred, but olanzapine was associated with a significant dose-related increase in hepatic transaminases serum glutamic pyruric transaminase (SGPT) and alanine aminotransferase (ALT), that was sometimes transient.

In conclusion, olanzapine appears to be a potent antipsychotic with an advantageous EPS profile somewhere between that of clozapine and traditional neuroleptics. It is probably one of the most promising of the new antipsychotics.

3. Seroquel

Seroquel is a strong alpha-1 receptor blocker with a moderate antihistaminergic effect, a weak dopamine D_2 and serotonin $5HT_2$ antagonism, and a very slight antagonism of dopamine D_1 receptors (Table 1) (MOORE et al. 1993; SALLER and SALAMA 1993). In rodents, seroquel produces weak catalepsy after intraperitoneal administration of 20–40 mg/kg, while higher doses cause significant catalepsy, comparable to haloperidol 4 mg/kg (MIGLER et al. 1993). In sensitized Cebus monkeys, dyskinetic reactions were found in 0%–15% of animals treated with 2.5–40 mg/kg perorally versus 25%–100% of those receiving haloperidol 0.12–0.5 mg/kg (MIGLER et al. 1993). In our own sensitized Cebus monkeys, seroquel 6 mg/kg (subcutaneous injection) caused dystonia in 7 of 7 animals, 3 mg/kg causing dystonia in 1 of 7 (PEACOCK and GERLACH, manuscript in preparation). These data suggest that the risk of EPS during treatment with therapeutic doses of seroquel in humans is low (Table 5).

FABRE et al. (1990) reported results in 8 schizophrenics treated with seroquel up to 250 mg/day versus 4 treated with placebo. The group found a good antipsychotic effect with no EPS. Side effects included insomnia, sedation, and transient intermittent tachycardia. SZEGEDI et al. (1993), however, in an open trial treating 12 schizophrenic patients with doses up to 750 mg/day, found no EPS, but only 33% of patients improved.

HIRSCH (1994) reported the results of a 6-week double-blind, placebo-controlled study in 596 chronic or subchronic schizophrenic patients with acute exacerbations (109 in the placebo group, 201 in the chlorpromazine group, and 286 in a low-/high-dose seroquel group). The study found seroquel (mean dose 407 mg/day) to be an effective antipsychotic, comparable to chlorpromazine (mean dose 384 mg/day), and equally beneficial in the treatment of positive and negative symptoms. No difference in the level of EPS was found between seroquel and placebo and chlorpromazine, but details were not provided. For seroqual and chlorpromazine, the most common side effects were somnolence (14% vs 16%), insomnia (10% vs 16%), postural hypotension (5% vs 18%), agitation (7% vs 12%), and dry mouth (8% vs 6%), respectively. There was no prolactin increase in the seroquel patients.

Finally, WETZEL et al. (in press) treated 12 patients with schizophrenia or schizophreniform psychoses with predominantly positive symptomatology with seroquel at doses of 600–750 mg/day. The study was non-blind, with a treatment duration of 12–28 days. No EPS or prolactin increase was seen. Patients reported occasionally on mild sedation, dry mouth, or orthostatic hypotension during the first days of treatment. However, there were considerable interindividual differences in the antipsychotic response; some patients showed almost full remission, while about half of the patients showed only mild or no improvement. As patients showing a good antipsychotic response reported slight initial side effects such as mild sedation, the authors concluded that higher doses might be necessary, and recommended monitoring of plasma levels of the parent drug and active metabolites.

Seroquel represents a new type of antipsychotic drug (primarily norepinephrine antagonist) with a beneficial EPS profile. Although the clincial results are difficult to interpret and the optimal dose level has not been determined, seroquel may well have an antipsychotic effect comparable to existing antipsychotics. The results, however, require considerable elaboration before definite conclusions can be made.

4. Zotepine

Zotepine's receptor-binding profile resembles that of traditional low-potency neuroleptics such as chlorprothixene (UCHIDA et al. 1979; SHIMOMURA et al. 1982) (Table 1). Therefore, the question is whether this drug promises a different therapeutic effect/side effect profile.

The earliest double-blind studies in schizophrenics with productive symptoms found no advantage of zotepine in doses up to 450 mg/day compared to perazine up to 675 mg/day, while zotepine up to 309 mg/day compared to haloperidol up to 14.5 mg/day was found to have a significant EPS advantage (DIETERLE et al. 1991; FLEISCHHACKER et al. 1991; KLIESER et al. 1991; WETZEL et al. 1991). NISHIZONO (1994) reported the results of a multicenter double-blind comparison of 169 schizophrenic patients treated with zotepine (maximum dose 450 mg/day), chlorpromazine (maximum dose 450 mg/day), and haloperidol (maximum dose 21 mg/day). No differences were found as to antipsychotic effect or side effects among the three groups.

Though zotepine appears to induce relatively few acute EPS, it does not appear to possess any significant advantage over known low-potency antipsychotics.

G. Management of Side Effects

I. Prevention

Predictable side effects can, to a certain extent, be prevented by using the lowest therapeutic dose. Special attention is required in the treatment of elderly patients and patients who have previously demonstrated a particular sensitivity to side effects. It is characteristic of side effects that they depend as much on individual predisposition as on the individual drug. Some patients are prone to develop EPS, while others are more prone to develop cardiovascular side effects. Therefore, the choice of therapy depends on knowledge of both the individual patient's and the antipsychotic's side effect profile.

Benzodiazepines may be used as concomitant therapy to avoid extreme antipsychotic doses, expecially in the presence of anxiety and sleep disturbances. The same precept is true in regards to lithium or carbamazepine, in the case of affective components or aggression. If the patient's history denotes the appearance of EPS at a relatively low dose of an antipsychotic, it may be indicated to order clozapine directly (see Sect. F.III.1.) or another low-dopamine antispychotic (Table 1). The sooner such a treatment is begun the better, both in regards to the therapeutic and the EPS benefit (MELTZER 1992; PEACOCK et al., in press).

II. Early Detection of Side Effects

It is of utmost importance to identify side effects as early as possible. Many side effects, especially psychological side effects, are subjectively uncomfortable and lead to noncompliance and a poor doctor-patient relationsip. Furthermore, many side effects reduce the patient's level of social functioning and compromise the therapeutic benefits (see "Introduction").

It is therefore essential for both therapists and other medical personnel at all levels to be well informed as to these side effects in order to report them to the treating physician. It is especially important to take the patient's complaints over psychological side effects seriously and to register motor side effects such as hypokinesia and hyperkinesia, as patients often do not spontaneously complain about these side effects and they may lead to tardive syndromes. To ensure adequate record-keeping of side effects, a simple standardized examination procedure or rating scale is recommended (GERLACH and PEACOCK 1995). A simple handwriting test can also be used to detect early bradykinesia (HAASE 1985).

Early diagnosis is especially important in relation to unexpected side effects (blood dyscrasia, neuroleptic malignant syndrome). These can develop rapidly over a few days and may be fatal. With early diagnosis, discontinuation and symptomatic treatment, the prognosis is good.

Also, in regards to tardive side effects, early diagnosis is essential. Tardive dystonia, dyskinesia, and akathisia are often irreversible, and only early diagnosis may prevent the development of disturbances which may result in lifelong disability.

III. Treatment of Side Effects

As soon as the diagnosis of a side effect such as parkinsonism is made, the primary treatment is a reduction of the given dose. If this is not feasible because of persisting psychotic symptoms, a change to another antipsychotic or the addition of side effect medication may be attempted.

In the case of acute or tardive EPS, the most rational switch would be to clozapine, which is still the one antipsychotic offering significantly less EPS than other antipsychotics (see Sect. III. 1.). If such a change is not practicable, e.g., because of resistance to bloood tests, other possibilities should be considered with reference to the antipsychotics' receptor and side effect profiles (Tables 1 and 3).

Anticholinergics such as biperiden and orphenadrin effectively counteract acute dystonia, reduce parkinsonism, and lessen akathisia somewhat, including the dysphoria and anxiety associated with akathisia. Prophylactic use of anticholinergics, however, cannot be recommended, expect in cases where an EPS reaction is probable based on experience from earlier treatment series. Anticholinergics possess their own side effects, including a potential for abuse, and can also serve to mask what may become a chronic problem.

Benzodiazepines or beta-blockers (e.g., propranolol 30–90 mg/day) can also be used to treat akathisia.

H. Conclusions

Despite almost 40 years' experience, we have still not found the ideal antipsychotic, i.e., a compound providing antipsychotic efficacy with neither acute nor chronic EPS nor other disturbing side effects. Among the oldest neuroleptics, there are drugs with a low EPS profile, e.g., chlorprothixene and levomepromazine, but their advantages as to acute EPS liability are limited by other side effects (e.g., sedation, weight gain, and orthostatic hypotension), and they have never been proven to offer any advantage as to tardive EPS liability. While clozapine provides clear advantages, in terms of acute and chronic EPS liability and lesser induction of apathy and depression, its usefulness is also limited by other side effects (e.g., weight gain and hypersalivation, besides the risk of agranulocytosis). Sulpiride, though offering a lesser acute EPS potential with no autonomic side effects, is not a very potent antipsychotic, and for this reason, its greatest force is in the treatment of milder psychoses.

Clozapine is still the most potent antipsychotic and least potent EPS inductor. New drugs such as olanzapine, seroquel, and sertindole represent the further development of clozapine's positive qualities. Definitive clinical studies with these drugs have not yet been completed, but there is a good chance that, in the near future, potent antipsychotics with fewer side effects of all types will become available. The "new age" of antipsychotic treatment will require either the involvement of new transmitter systems, or an unexpected, serendipitous discovery such as that which originally led to the recognition of chlorpromazine's antipsychotic effect in 1952!

References

Awad AG (1993) Subjective response to neuroleptics in schizophrenia. Schizophr Bull 19:609–618

Barnas C, Stuppack CH, Miller C, Haring C, Sperner-Unterweger B, Fleischhacker WW (1992) Zotepine in the treatment of schizophrenic patients with prevailingly negative symptoms: a double-blind trial vs. haloperidol. Int Clin Psychopharmacol 7:23–27

Barnes TRE, Braude WM (1985) Akathisia variants and tardive dyskinesia. Arch Gen Psychiatry 42:874–878

Barnes TRE, Edwards JG (1993) The side-effects of antipsychotic drugs. I. CNS and neuromuscular effects. In: Barnes TRE (ed) Antipsychotic drugs and their side-effects. Academic, London, pp 213–247

Bench CJ, Lammertsma AA, Grasby PM, Dolan RJ, Warrington SJ, Boyce M, Gunn KP, Brannick LY, Franckowiak RSJ (1995) Time course of occupancy of striatal dopamine D2 receptors by the neuroleptic Ziprasidone determined by positron emission tomography. Psychopharmacology (in press)

Bersani G, Grispini A, Marini S, Pasani A, Valducci M, Ciani N (1986) Neuroleptic-induced extrapyramidal side effects: clinical perspectives with ritanserin, a new selective 5HT2 receptor blocking agent. Curr Ther Res 40:492–499

Bøgesø KP, Andersen K, Arnt J, Frederiksen K, Hyttel J, Perregaard J, Skarsfeldt T (1995) Design of atypical antipsychotic drugs. In: Fog R, Gerlach J, Hemmingsen R (eds) Schizophrenia, an integrated view. Munksgaard, Copenhagen pp 361–374

Borenstein P, Boyer P, Braconnier P (1989) Amisulpiride. Expansion Scientifique Francaise, Paris

Bowers MB Jr, Astrachan BM (1967) Depression in acute schizophrenic psychosis. Am J Psychiatry 123:976–979

Boyer P, Lecrubier Y, Peuch AJ, Dewailly J, Auben F (1995) Treatment of negative symptoms of schizophrenia with amisulpride. Br J Psychiatry 166:68–72

Braude WM, Barnes TRE, Gore SM (1983) Clinical characteristics of akathisia. A systematic investigation of acute psychiatric inpatient admissions. Br J Psychiatry 143:858–861

Casey DE (1989) Clozapine: neuroleptic induced EPS and tardive dyskinesia. Psychopharmacology 99 [Suppl]:47–53

Casey DE (1991) Neuroleptic-induced extrapyramidal syndromes and tardive dyskinesia. Schizophr Res 4:109–120

Casey DE (1994) Motor and mental aspects of acute extrapyramidal syndromes. Acta Psychiatr Scand 89 [Suppl 380]:14–20

Casey DE (1995) The nonhuman primate model: focus on dopamine D2 and serotonin mechanisms. In: Fog R, Gerlach J, Hemmingsen R (eds) Schizophrenia, an integrated view. Munksgaard, Copenhagen, pp 287–297

Casey DE, Denney D (1977) Pharmacological characterization of tardive dyskinesia. Psychopharmacology 54:1–8

Castro SW, Strange PG (1993) Differences in the ligand binding properties of the short and long versions of the D2 dopamine receptor. J Neurochem 60:372–375

Ceskova E, Svestka J (1993) Double-blind comparison of risperidone and haloperidol in schizophrenic and schizoaffective psychoses. Pharmacopsychiatry 26:121–124

Chouinard G, Jones B, Remington G et al. (1993) A Canadian multicentre placebo-controlled study of fixed doses of risperidone and haloperidol in the treatment of chronic schizophrenic patients. J Clin Psychopharmacol 13:25–40

Claghorn J, Honingfeld G, Abuzzahab FS, Wang R, Steinbook R, Tuason V, Klerman G (1987) The risks and benefits of clozapine versus chlorpromazine. J Clin Psychoparmacol 7:377–384

Coffin VL, Barnett A, McHugh DT (1991) Reversal of extrapyramidal side effects with SCH 39166, a dopamine antagonist. Abstract ACNP 30th annual meeting, Puerto Rico

Cohen BM, Keck PE, Satlin A, Cole JO (1991) Prevalence and severity of akathisia in patients on clozapine. Biol Psychiatry 29:1215–1219

Csernansky JG, Newcomer JW, Jackson K, Lombrozo L, Faull KF, Zipursky R, Pfefferbaum A, Faustman WO (1994) Effects of raclopride treatment on plasma and CSF HVA: relationships with clinical improvement in male schizophrenics. Psychopharmacology 116:291–296

Davis JM (1990) Drug treatment of schizophrenia. Curr Opin Psychiatry 3:29–34

De Alarcon R, Carney MWP (1969) Severe depressive mood changes following slow-release intramuscular fluphenazine injection. Br Med J 3:564–567

Delcker A, Schoon ML, Oczkowski B, Gaertner HJ (1990) Amisulpride versus haloperidol in treatment of schizophrenic patients – results of a double-blind study. Pharmacopsychiatry 23:125–130

Den Boer JA, van Megen HJGM, Fleischhacker WW, Louwerens JW, Slaap BR, Westenberg HGM, Burrows GD, Srivastava ON (1995) Differential effects of D1 dopamine antagonist SCH 39166 on positive and negative symptoms of schizophrenia. Psychopharmacology (in press)

Deniker P (1983) Discovery of the clinical use of neuroleptics. In: Parnam MJ, Bruinvels J (eds) Discoveries in pharmacology, vol I: Psycho- and neuro-pharmacology. Elsevier, Amsterdam, pp 164–180
Dieterle DM, Muller-Spahn F, Ackenheil M (1991) Wirksamkeit und Verträglichkeit von Zotepin im Doppelblindvergleich mit Perazin bei schizophrenen Patienten. Fortschr Neurol Psychiatr 59 [Suppl 1]: 18–22
Edwards JG, Barnes TRE (1993) The side-effects of antipsychotic drugs. II. Effects on other physiological systems. In: Barnes TRE (ed) Antipsychotic drugs and their side-effects. Academic, London, pp 249–275
Edwards JG, Alexander JR, Alexander MS, Gordon A, Zutchi T (1980) Controlled trial of sulpiride in chronic schizophrenic patients. Br J Psychiatry 137:522–529
Fabre L, Slotnick V, Jones V, Murray G, Malick J (1990) ICI 204,636, a novel atypical antipsychotic: early indication for safety and efficacy in man. Abstracts of the 17th congress of CINP, vol II. Kyoto, Japan, p 223
Farde L, Nordstrom A-L, Wiesel F-A, Pauli S, Halldin C, Sedvall G (1992) Positron emission tomographic analysis of central D1 and D2 dopamine receptor occupancy in patients treated with classical neuroleptics and clozapine: relation to extrapyramidal side effects. Arch Gen Psychiatry 49:538–544
Fitton A, Heel RC (1990) Clozapine: a review of its pharmacological properties, and therapeutic use in schizophrenia. Drugs 40:722–747
Fleischhacker WW, Miller CH, Ehrmann H (1990) Ritanserin in the treatment of neuroleptic-induced akathisia. Schizophr Res 3:47
Fleischhacker WW, Barnas C, Stuppack CH, Sperner-Unterweger B, Miller C, Hinterhuber H (1991) Zotepin vs haloperidol bei paranoider Schizophrenie: ein Doppeltblindstudie. Fortschr Neurol Psychiatr 59 [Suppl 1]:10–13
Gerlach J (1991) New antipsychotics: classification, efficacy, and adverse effects. Schizophr Bull 17:289–309
Gerlach J (1993) Pharmacology and clinical properties of selective dopamine antagonists with focus on substituted benzamides. In: Barnes TRE (ed) Antipsychotic drugs and their side-effects. Academic, London, pp 45–63
Gerlach J, Casey DE (1988) Tardive dyskinesia. Acta Psychiatr Scand 77:369–378
Gerlach J, Casey DE (1994) Drug treatment of schizophrenia: myths and realities. Curr Opin Psychiatry 7:65–70
Gerlach J, Hansen L (1993) Effect of chronic treatment with NNC 756, a new D_1 receptor antagonist, or raclopride, a D_2 receptor antagonist, in drug-naive Cebus monkeys: dystonia, dyskinesia and D_1/D_2 supersensitivity. J Psychopharmacol 7:355–364
Gerlach J, Peacock L (1995) Intolerance to neuroleptic drugs: the art of avoiding EPS. Eur Psychiatry 10 [Suppl 1]:27–31
Gerlach J, Reisby N, Randrup A (1974) Dopaminergic hypersensitivity and cholinergic hypofunction in the pathophysiology of tardive dyskinesia. Psychopharmacology 34:21–45
Gerlach J, Behnke K, Heltberg J, Munk-Andersen E, Nielsen H (1985) Sulpiride and haloperidol in schizophrenia: a double-blind cross-over study of therapeutic effect, side effects and plasma concentrations. Br J Psychiatry 147:283–288
Gerlach J, Hansen L, Peacock L (1991) D_1 dopamine hypothesis in tardive dyskinesia. In: Racagni G, Brunello N, Fukada T (eds) Biological psychiatry. Elsevier, Amsterdam: S609–611
Gerlach J, Lublin H, Peacock L (1995) Dopamine D_1 receptor antagonists in rodents, in primates and in humans. In: Fog R, Gerlach J, Hemmingsen R (eds) Schizophrenia, an integrated view. Munksgaard, Copenhagen, pp 300–309
Glazer WM, Morgenstern H, Doucette JT (1993) Predicting long-term risk of tardive dyskinesia in outpatients maintained on neuroleptic medications. J Clin Psychiatry 54:133–139
Haase H-J (1985) Clinical observations on the actions of neuroleptics. In: Haase H-J, Janssen PAJ (eds) The action of neuroleptic drugs. Elsevier, Amsterdam, pp 1–284

Härnryd C, Bjerkenstedt L, Bjork K, Gullberg B, Oxenstierna G, Sedvall G, Wiesel FA, Wik C, Åberg-Wistedt A (1984) Clinical evaluation of sulpiride in schizophrenic patients. A double-blind comparison with chlorpromazine. Acta Psychiatr Scand 69 [Supppl 311]:7–30

Heinrich K, Kleiser E, Lehmann E, Kinzler E, Hruschka H (1994) Risperidone versus clozapine in the treatment of schizophrenic patients with acute symptoms: a double blind, randomized trial. Prog Neuropsychopharmacol Biol Psychiatry 18:129–137

Helmchen H, Hippius H (1967) Depressive Syndrome in Verlauf neuroleptischer Therapie. Nervenarzt 38:455–458

Hirsch SR (1994) Seroquel: an example of an atypical antipsychotic drug. Neuropsychopharmacology 3S (10/1):371S

Høyberg OJ, Fensbo C, Remvig J, Lingjærde O, Sloth-Nielsen M, Salvesen I (1993) Risperidone versus perphenazine in the treatment of chronic schizophrenic patients with acute exacerbations. Acta Psychiatr Scand 88:395–402

Hyttel J, Arnt J, Costall B, Domeney A, Dragsted N, Lembøl HL, Meier E, Naylor RJ, Nowak G, Sanchez C, Skarsfeldt T (1992) Pharmacological profile of the atypical neuroleptic sertindole. Clin Neuropharmacol 15 [Suppl 1]:267–268A

Kane JM, Woerner M, Weinhold P, Wegner J, Kinon B, Borenstein M (1984) Incidence of tardive dyskinesia: five-year data from a prospective study. Psychoparmacol Bull 20:387–389

Karlsson P, Smith L, Farde L, Härnryd C, Wiesel FA, Sedwall GC (1995) Lack of apparent antipsychotic effect of the dopamine-D1-receptor antagonist SCH39166 in schizophrenia. Psychopharmacology (in press)

Keepers GA, Clappison VJ, Casey DE (1983) Initial anticholinergic prophylaxis for neuroleptic-induced extrapyramidal syndromes. Arch Gen Psychiatry 40:1113–1117

Klieser E, Lehmann E, Tegeler J (1991) Doppeltblindvergleich von 3 × 75 mg Zotepin und 3 × 4 mg Haloperidol bei akut schizophrenen Patienten. Fortschr Neurol Psychiatr 59 [Suppl 1]:14–17

Kohler C, Ögren SO, Haglund L, Angeby T (1979) Regional displacement by sulpiride of [3H] spiperone binding in-vivo: biochemical evidence for a preferential action on limbic and nigral dopamine receptors. Neurosci Lett 13:51–56

Krupp P, Barnes P (1992) Clozapine-associated agranulocytosis: risk and aetiology. Br J Psychiatry 160 [Suppl 17]:38–40

Lader MH (1993) Neuroleptic-induced deficit syndrome (NIDS). J Clin Psychiatry 54:493–500

Lambert PA, Revol L (1960) Classification psychopharmacologique et clinique des différents neuroleptiques: indications thérapeutiques générales dans les psychoses. Presse Med 41:1509–1511

Lamberti JS, Bellnier T, Schwarzkoph SB (1992) Weight gain among schizophrenic patients treated with clozapine. Am J Psychiatry 149:689–690

Lewander T (1994) Neuroleptics and the neuroleptic-induced deficit syndrome. Acta Psychiatr Scand 89 [Suppl 380]:8–13

Leysen JE, Janssen PM, Megens AAHP, Schotte A (1994) Risperidone: A novel antipsychotic with balanced serotonin-dopamine antagonism, receptor occupancy profile, and pharmacologic activity. J Clin Psychiatry 55 [Suppl]:5–12

Lieberman JA, Saltz BL, Johns CA, Pollack S, Borenstein M, Kane J (1991) The effects of clozapine on tardive dyskinesia. Br J Psychiatry 158:503–510

Lingjærde O, Ahlfors UG, Bech P, Dencker SJ, Elgen S (1987) The UKU side effect scale: a new comprehensive rating scale for psychotropic drugs and a cross-sectional study of side effects in neuroleptic-treated patients. Acta Psychiatr Scand 76 [Suppl 334]:1–100

Litvak R, Kaelbling R (1971) Agranulocytosis, leucopenia, and psychotropic drugs. Arch Gen Psychiatry 24:265–267

Lublin H, Gerlach J, Peacock L (1993) Chronic treatment with the D1 receptor antagonist, SCH 23390, in Cebus monkeys withdrawn from previous haloperidol treat-

ment: extrapyramidal syndromes and dopaminergic supersensitivity. Psychopharmacology 112:389–397
Marder SR, Meibach RC (1994) Risperidone in the treatment of schizophrenia. Am J Psychiatry 151:825–835
Marsden CD, Jenner P (1980) The pathophysiology of extrapyramidal side-effects of neuroleptic drugs. Psychol Med 10:55–72
Martin PT, Grebb JA, Schmitz P, Sebree T, Kashkin K (1994) Efficacy and safety of sertindole in two double-blind, placebo-controlled trials of schizophrenic patients. Poster presented at Biennial Winter Workshop on Schizophrenia Research, Les Diablets, Switzerland 1994
McClelland HA, Metcalfe AV, Kerr A, Dutta D, Watson P (1991) Facial dyskinesia: a 16-year follow-up study. Br J Psychiatry 158:691–696
Meltzer HY (1992) Treatment of the neuroleptic-nonresponsive schizophrenic patient. Schizophr Bull 18:515–542
Migler BM, Warawa EJ, Malick JB (1993) Seroquel: behavioural effects in conventional and novel tests for atypical antipsychotic drugs. Psychopharmacology 112:299–307
Moore NA, Calligaro DO, Wong DT, Bymaster F, Tye NC (1993) The pharmacology of olanzapine and other new antipsychotic agents. Curr Opin Invest Drugs 2:281–293
Moore NA, Tupper DE, Hotten TM (1994) Olanzapine. Drugs Future 19:114–117
Newcomer JW, Miller LS, Faustman WO, Wetzel MW, Vogler GP, Csernansky JG (1994) Correlations between akathisia and residual psychopathology: a by product of neuroleptic-induced dysforia. Br J Psychiatry 164:834–838
Nishizono M (1994) A comparative trial of zotepine, chlorpromazine and haloperidol in schizophrenic patients. Neuropsychopharmacology 10 (3S/2):S30
Nyberg S, Farde L, Eriksson L, Halldin C, Eriksson B (1993) 5HT2 and D_2 dopamine receptor occupancy in the living human brain. A PET study with risperidone. Psychopharmacology 110:265–272
Ögren SO, Hall H, Kohler C, Magnussen O, Sjöstrand S-E (1986) The selective dopamine D2 receptor antagonist raclopride discriminates between dopamine-mediated motor functions. Psychopharmacology 90:287–294
Peacock L, Gerlach J (1995) A reanalysis of the dopamine theory of tardive dyskinesia: the dopamine D_1/D_2 imbalance hypothesis. In: Yassa R, Jeste D (eds) Neuroleptic-induced movement disorders: a comprehensive survey. Cambridge University Press, Cambridge (in press)
Peacock L, Solgaard T, Lublin H, Gerlach J (in press) Clozapine versus typical antipsychotics: a retro- and prospective study of extrapyramidal side effects. Psychopharmacology
Peuch AJ, Simon P, Boissier J-R (1976) Antagonism by sulpiride of three apomorphine-induced effects in rodents. Eur J Pharmacol 36:439–441
Pisciotta AV, Konings SA, Ciesmier LL, Cronkite CE, Lieberman JA (1992) On the possible mechanisms and predictability of clozapine-induced agranulocytosis. Drug Safety 7 [Suppl 1]:33–44
Rendtorff C, Hansen L, Jensen B, Jörgensen O, Munck D, Nissen G, Olstrup L, Olesen G, Peacock L, Sahl I, Gerlach J (1990) Side effects of neuroleptics: a study of 159 patients with UKU's side effects scale (in Danish). Nord Psykiatr Tiddskr 3:303–307
Rosengarten H, Schweitzer JW, Friedhoff AJ (1983) Induction of oral dyskinesia in naive rats by D1 stimulation. Life Sci 33:2479–2482
Saller CF, Salama AI (1993) Seroquel: biochemical profile of a potential atypical antipsychotic. Psychopharmacology 112:285–292
Schulz C, Staser J, Schmitz P, Mark R, Sebree T (1994) An open-label assessment of the long-term safety of sertindole in the treatment of schizophrenic patients. Poster presented at the Marcus Wallenberg Symposium on schizophrenia, Gothenborg, 9–12 Sept 1994

Sedvall G (1990) Development of a new antipsychotic, remoxipride. Acta Psychiatrica Scand 82 [Suppl 358]:1–188
Shimomura K, Satch H, Hirai P et al. (1982) The central antiserotonin activity of zotepine, a new neuroleptic in rats. Jpn J Pharmacol 32:405–412
Skarsfeldt T (1992) Electrophysiological profile of the new atypical neuroleptic, sertindole, on midbrain dopamine neurones in rats: acute and repeated treatment. Synapse 10:25–33
Szegedi A, Wiesner J, Hillert A, Hammes E, Wetzel H, Benkert O (1993) ICI 204636, a putative "atypical" antipsychotic, in the treatment of schizophrenia with positive symptomatology: an open clinical trial. Pharmacopsychiatry 26:197
Tran PV, Beasley CM, Toffelsen GD, Snager T, Satterlee WE (1994) Clinical efficacy and safety of olanzapine, a new atypical antipsychotic agent. Poster presented at the annual meeting of the American Pyschiatric Association, Philadelphia, 21–26 May 1994
Uchida S, Honda F, Otsuka M et al. (1979) Pharmacological study of [2-chloro-11-(2-dimethylaminoethoxy) dibenzo [b, f] thiepine] (zotepine). a new neuroleptic drug. Arzneimittelforschung/Drug Res 29:1588–1594
Vanelle JM, Olie JP, Levy-Soussan P (1994) New antipsychotics in schizophrenia: the French experience. Acta Psychiatr Scand 89 [Suppl 380]:59–63
Van Putten T (1974) Why do schizophrenics refuse to take their drugs? Arch Gen Psychiatry 31:67–72
Van Putten T, May PRA (1978) Subjective response as a predictor of outcome in pharmacotherapy. Arch Gen Psychiatry 35:477–480
Veys PA, Wilkes S, Shah S, Noyelle R, Hoffbrand AV (1992) Clinical experience of clozapine-induced neutropenia in the UK. Drug Safety 7 [Suppl 1]:26–32
Waddington JL (1993) Pre- and postsynaptic D_1 to D_5 dopamine receptor mechanisms in relation to antipsychotic activity. In: Barnes TRE (ed) Antipsychotic drugs and their side-effects. Academic, London, pp 65–85
Wagstaff AJ, Fitton A, Benfield P (1994) Sulpiride: a review of its pharmacodynamic and pharmacokinetic properties, and therapeutic efficacy in schizophrenia. CNS Drugs 2:313–333
Wetzel H, von Bardeleben U, Holsboer F, Benkert O (1991) Zotepin versus Perazin bei Patienten mit paranoider Schizophrenie: eine doppeltblind-kontrollierte Wirksamkeitsprüfung. Fortschr Neurol Psychiatr 59 [Suppl 1]:23–29
Wetzel H, Szegedi A, Hain C, Wiesner J, Schlegel S, Benkert O (in press) Seroquel, a putative "atypical" antipsychotic, in schizophrenia with postitive symptomatology: results of an open clinical trial and changes of neuroendocrinological and EEG parameters. Psychopharmacology
Zorn SH, Morrone JM, Seeger TF, Jackson C, Lebel L, Howard H, Heym J (1993) The antipsychotic drug CP-88,059 is an antagonist at both 5HT2 and 5HT1C receptors. Soc Neurosci Abstr 19:599

CHAPTER 14

Biological Predictors of Antipsychotic Treatment Response

AMY R. KOREEN, BRIAN SHEITMAN, and JEFFREY LIEBERMAN

A. Introduction

Schizophrenia is a heterogeneous disorder with variable symptomatology, course, treatment response, and outcome. Widely variable responses to antipsychotic medication in turn influence patients' course of illness and outcome. Despite substantial research efforts attempting to separate patients into meaningful phenomenologic or biological subtypes that correspond to illness course and outcome, no such subtypes have been firmly established. So far, we have no reliable biological markers or predictors of treatment response. Clearly, given the recent pressure to decrease inpatient lengths of stay, any predictors of outcome are not only of clinical but also of economic importance. Clinically, such predictors might afford clinicians with the knowledge of which antipsychotics to use (i.e., typical vs atypical), which dose may be most beneficial, and which patient may be the one who may be withdrawn successfully from medication.

Among the most consistent and robust predictors of treatment response and outcome have been historical and phenomenologic clinical measures including premorbid social adjustment and level of function, mode of illness onset, age of illness onset, duration of illness, duration of symptoms prior to first treatment, negative symptoms, number of prior episodes, and hospitalization (LIEBERMAN and SOBEL 1993). Recent research, however, has also begun to focus on biological measures which could provide an objective method of predicting treatment response and outcome. These biological measures can be divided into biochemical, neuroimaging, and electrophysiological measures.

B. Biochemical Predictors

In the search for the etiology of schizophrenia, a major line of investigation has been clinical neurochemistry. Initial strategies consisted of measuring concentrations of neuroactive substances (neurotransmitters, metabolites, precursors, metabolic enzymes) in various biological fluids. Subsequent strategies were devised that focused on indirect measures of neurotransmitter activity, e.g., neuroendocrine strategies such as measures of pituitary hormones which were, to a significant degree, under central nervous system (CNS) regulation (BROWN et al. 1988; GARVER 1988). These strategies continue to be utilized in clinical studies of predictors of antipsychotic treatment response and have examined the relationship between treatment response and various neurotransmitters [dopamine (DA), norepinephrine (NE), serotonin (5-HT)] and their metabolites [plasma homovanillic acid (pHVA), 3-methoxy-4-hydroxyphenylglycol(MHPG), 5-hydroxyindoleacetic acid (5-HIAA)], phenethylamine (PEA), neurohormones [prolactin (PRL), cortisol, growth hormone (GH), thyroid stimulating hormone (TSH)], peptides [cholecystokinin, somatostatin, neurotensin) and enzymes (dopamine β-hydroxylase (DBH), monoamine oxidase (MAO)].

I. Homovanillic Acid (HVA)

Years of extensive investigation have given rise to numerous hypotheses of the pathophysiology of schizophrenia with the most enduring being the DA hypothesis. The DA hypothesis of schizophrenia proposes that schizophrenia is a manifestation of an overactivity of DA neural systems (MATTHYSSE 1973); it was based on early observations that DA agonists can induce paranoid schizophreniform psychosis (RANDRUP and MUNKVAD 1965; SNYDER et al. 1974) and that neuroleptics inhibit dopaminergic activity (SEEMAN et al. 1974; CARLSSON, 1978). Despite its obvious limitations and predominantly indirect support, it remains a fruitful source of research, and many studies have examined the relationship between measures of CNS DA and antipsychotic treatment response. Specific emphasis has been on homovanillic acid (HVA), the DA metabolite (see Table 1).

Preclinical (BACOPOULUS et al. 1979; KENDLER et al. 1982; KENDLER and DAVIS 1984; KORPIN et al. 1988) and subsequent clinical studies (PICKAR et al. 1986, 1988; DAVIDSON et al. 1988; DAVILA et al. 1989; KAMINSKY et al. 1990) have suggested that measurement of plasma HVA (pHVA) may be a useful tool for studying the pharmacologic action of antipsychotic drugs and treatment response since it reflects CNS DA activity, increases with acute antipsychotic treatment (HARRIS et al. 1984; PICKAR et al. 1986) and discontinuation (GLAZER et al. 1989; DAVIDSON et al. 1991a), decreases with chronic treatment (BOWERS et al. 1984; PICKAR et al. 1984, 1985, 1986), and is correlated with psychopathology (PICKAR et al. 1984, 1986; DAVIS et al. 1985).

The initial increase and subsequent decrease of pHVA with antipsychotic drug treatment have been associated with clinical response (BOWERS et al. 1984; CHANG et al. 1988; SHARMA et al. 1989; ALFREDSSON and WIESEL 1990; DAVIDSON et al. 1991a). Several studies have also found that pHVA can differentiate responders and nonresponders and has prognostic significance. CHANG et al. (1988) have found a higher baseline level of pHVA and a more acute rise with treatment in responders compared to nonresponders. DAVILA et al. (1988) found that the pattern of pHVA in response to antipsychotic drugs predicted short-term treatment response, i.e., the patients who had the greatest increase and subsequent decrease in pHVA had the best outcome. VAN PUTTEN et al. (1989), CHANG et al. (1990), DAVIDSON et al. (1991a), and

Table 1. Plasma homovanillic acid (pHVA)

Author/year	Sample	Drug-free ≥ (wks)	Methods	Rating scale(s)	Assay	Day of predictive assessment	Outcome day	Results
Pickar et al. 1984	8 (3♂, 5♀) DSM III schizophrenic pts	2	Plasma HVA drawn at BL, then weekly during 5 wks of fluphenazine tx	"Behavioral ratings"	HPLC	BL wkly	35	Changes in pHVA levels correlated with changes in psychosis.
Pickar et al. 1986	11 DSM III schizophrenic pts	5					35	Changes in mean weekly levels of pHVA were predictive of response.
Bowers et al. 1987	47 (25♂, 12♀) DMS III nonorganic psychotic pts	1	pHVA drawn at BL; pts then tx'd with neuroleptic for 10 days	4-point scale	GCMS	BL	10	Good responders had higher pHVA at BL.
Chang et al. 1988	24 (12♂, 12♀) DSM III schizophrenic pts	4	1–3 pHVA samples drawn at BL, then on days 3, 7, & each wk of haloperidol tx for 42 days	BPRS Responders = ≥50% ↓ by day 42	HPLC	BL wkly	42	1. BL pHVA ↑ in responders. 2. Responders had initial ↑ then a ↓ to below BL by day 28. 3. Poor responders ↑ over 1st 2 wks & then ↓ but remained above pre-tx values.

Table 1. *Continued*

Author/year	Sample	Drug-free ≥ (wks)	Methods	Rating scale(s)	Assay	Day of predictive assessment	Outcome day	Results
Davila et al. 1988	14 (11♂, 3♀) DSM III schizophrenic pts	5	pHVA drawn pre-tx & then on days 4 & 28 of tx with haloperidol	BPRS	HPLC	BL, days 4 & 28	28	1. BL pHVA not correlated with improvement. 2. ↑ in pHVA by day 4 & ↓ in pHVA from day 4 to 28 correlated with improvement (pts with greatest ↓ & then ↓ correlated with best outcome).
Bowers et al. 1989	37 DSM III nonorganic psychotic pts	1	pHVA drawn pre-tx & then on days 7, 8, & 9 of tx (these averaged) with either haloperidol or perphenazine	4-point scale	GCMS	BL	10	1. Early response correlated with BL pHVA. 2. Change in pHVA correlated with early response.
Sharma et al. 1989	17 (9♂, 8♀) RDC nonorganic psychotic pts	mean of 19 days	pHVA drawn at BL & for 12 pts at end of 5 wks of uncontrolled neuroleptic tx	BPRS, GAS	HPLC	BL	35	↓ in pHVA correlated (trend) with a ↓ in sxs.
Van Putten et al. 1989	22♂ DSM III schizophrenic pts	4	pHVA drawn BL & wk 1 of tx with fluphenazine	BPRS, CGI Responders = ≤2 on CGI	HPLC	BL wk 1	42	1. Responders had ↑ BL pHVA (trend). 2. BL pHVA not correlated with % improvement. 3. pHVA at wk 3 predictive of outcome at wks 4–6.

ALFREDSSON and WIESEL 1990	24 (14♂, 10♀) DSM III R schizophrenic pts	4 (p.o) 24 (IM)	pHVA drawn BL & then wkly while on sulpiride	CPRS, NOSIE Responders = ≥50% ↓ by wk 6	GCMS	BL wkly	42	1. BL pHVA not different in responders & nonresponders. 2. Responders showed a ↓ in pHVA with tx. 3. Nonresponders showed an initial ↑ & then ↓ in pHVA with tx.
JAVAID et al. 1990	25 DSM III & RDC schizophrenic pts	3	pHVA drawn at BL and then wkly during trifluoperazine tx	BPRS, GAS	HPLC	BL	28	1. BL pHVA did not predict subsequent improvement. 2. Changes in pHVA did not correlate with changes in sxs.
CHANG et al. 1990	33 (21♂, 12♀) DSM III schizophrenic pts	4 (p.o.) 12 (IM)	1–3 sample pHVA drawn BL & then wkly while on 20 mg/day haloperidol for 6 wks	BPRS Responders = ≥50% ↓ by wk 6 on BPRS	HPLC	BL	42	1. BL pHVA ↑ in responders. 2. Responders showed a gradual ↓ in pHVA with tx. 3. Nonresponders showed a transient ↑ in pHVA with tx.

Table 1. *Continued*

Author/year	Sample	Drug-free ≥ (wks)	Methods	Rating scale(s)	Assay	Day of predictive assessment	Outcome day	Results
Petrie et al. 1990	16♂ RO6 schizophrenic or schizoaffective pts	2	pHVA drawn at wks 2, 3, & 4 of tx with 20mg/day haloperidol	BPRS	HPLC	wks 2, 3, 4	28	1. pHVA differentiated between rapid & nonrapid responders at wk 4, with rapid responders having ↓ levels. 2. pHVA or change in pHVA not correlated with psychopathology.
Bowers 1991	85 (45♂, 40♀) DMS III nonorganic psychotic pts	1	pHVA drawn at BL; pts then tx'd with haloperidol or perphenazine	BPRS	GCMS	BL	10	Pts with high BL pHVA tended to have better prognostic profile than pts with low BL pHVA.
Mazure et al. 1991	37 (12♂, 25♀) DSM III R nonorganic psychotic pts	1	pHVA drawn at BL and then on days 7, 8 & 9 of tx with perphenazine (0.5mg/kg per day)	BPRS Response = BPRS ≤14 and 33% ↓	GCMS	BL	10	1. ↑ BL pHVA associated with good response. 2. Change in pHVA correlated with a good response. 3. ↓ in pHVA & BPRS from BL to end of tx correlated.

Davidson et al. 1991	20♂ RDC & DSM III R schizophrenic pts	2	pHVA drawn at BL, then 2x/wk while on 20 mg/day haloperidol × 5 wks	BPRS, CGI Responders = ≥1 pt ↓ in CGI by wk 5	HPLC	BL	35	1. pHVA ↓ in responders but not in nonresponders. 2. BL ↑ initially only in responders. 3. Change in pHVA correlated with change in CGI.
Volavka et al. 1992	28 (22♂, 6♀) schizophrenic or schizoaffective pts	mean = 1	pHVA drawn at BL & wkly for 5 wks of haloperidol tx	BPRS, CGI Responders = ≥1 pt ↓ in CGIs, wk5	GCMS	BL	35	1. BL pHVA no different in responders & nonresponders. 2. pHVA level at 5 wks greater in responders than nonresponders.
Davila 1992	RDC schizophrenic pts	12	pHVA drawn at BL and then on days, 4, 7, & 28 of haloperidol (0.25 mg/kg per day) tx	BPRS, SAPS, SANS		BL day 4		A low prolactin/HVA ratio at BL and day 4 predicted good clinical response.
Labarca et al. 1993	15 (12♂, 3♀) DSM III R schizophrenic pts	drug-naive	serial pHVA drawn pre-tx and then at wks 1, 3, & 5 of tx with 5 mg/day of haloperidol	BPRS, SANS	HPLC	BL	35	1. BL pHVA levels correlated with SANS total score but not BPRS total or positive sxs. 2. BL pHVA predicted negative sx response but not positive sx response in that higher BL pHVA correlated with improvement in negative sxs.

Table 1. *Continued*

Author/year	Sample	Drug-free ≥ (wks)	Methods	Rating scale(s)	Assay	Day of predictive assessment	Outcome day	Results
								3. No correlations seen between changes in pHVA & changes in clinical sxs.
Chang et al. 1993	95 (59♂, 36♀) DSM III schizophrenic pts	4 (p.o.) 12 (IM)	pHVA drawn at BL, then pt tx'd with haloperidol or flupenthixol for 6 wks	BPRS Responders = ≥50% ↓	HPLC	BL	42	1. BL pHVA significantly correlated with tx response with responders having greater levels than nonresponders. 2. 72% of pts with pHVA ≥12 ng/ml responded, as opposed to only 35% with pHVA <12 ng/ml.
Koreen et al. 1994	41 (24♂, 17♀) RDC first-episode schizophrenic or schizoaffective pts	2	Serial pHVA drawn at BL, then wkly while on 20 mg/day fluphenazine for 4 wks	SADS, CGI, SANS	GCMS	BL wkly	28	1. BL pHVA associated with time to reach remission. 2. BL & wk 1 pHVA greater in responders than nonresponders.

pts, patients; BL, baseline; HPLC, high-performance liquid chromatography; tx, treatment; BPRS, Brief Psychiatric Rating Scale (Overall and Gorhan 1962); GCMS, gas chromography mass spectroscopy; ♂, male; ♀, female; GAS, Global Assessment Scale (Endicott et al. 1976); DSM III (R), Diagnostic and Statistical Manual of Mental Disorders (American Psychiatric Association 1980 [1987]); RDC, Research Diagnostic Criteria (Spitzer et al. 1977); CGI, Clinical Global Impression scale; CPRS, Comprehensive Psychiatric Rating Scale (Jacobsson et al. 1978); SANS, Scale for the Assessment of Negative Symptoms (Andreasen 1984); SADS, Schedule for Affective Disorders and Schizophrenia (Spitzer et al. 1978); NOSIE, Nurse's Observation Scale for Inpatient Evaluation (Honigfeld and Klett 1965); p.o., by mouth; I.M., intramuscular; sx, symptom.

KOREEN et al. (1994), reported that antipsychotic drug responders had greater baseline pHVA levels than nonresponders, but ALFREDSSON and WIESEL (1990) did not replicate this finding. DAVIDSON et al. (1991a) also found that pHVA decreased in response to neuroleptic treatment but only in the responders. KOREEN et al. (1994) found that higher baseline pHVA was associated with shorter time to remission. It thus appears that in most studies, baseline pHVA levels or the pattern of pHVA in response to treatment is associated with treatment response or outcome.

Both positive (DAVIS et al. 1985; PICKAR et al. 1986, 1988; MAAS et al. 1988; DAVIDSON et al. 1991a, 1991b) and negative (KIRCH et al. 1988; SHARMA et al. 1989; VAN PUTTEN et al. 1989; ALFREDSSON and WIESEL 1990; PETRIE et al. 1990; KOREEN et al. 1994) correlations between baseline pHVA or the change in pHVA with treatment and severity of psychopathology have been found. DAVIS et al. (1991), in their review of some of these studies, suggested that those studies that did not find correlations with psychopathology used only single samples for the HVA value, whereas those that found positive correlations used multiple samples. The presumed advantage of multiple samples is that it diminishes the variability of this physiologic measure. This has not been fully corroborated because at least one of the more recent studies that did not find correlations between pHVA levels and psychopathology used serial and multiple samples (KOREEN et al. 1994). However, KOREEN et al.'s (1994) findings in treatment-naive, first-episode patients could be caused by differences in the patient sample from earlier studies. Chronic neuroleptic treatment may change DA neuronal activity such that the correlation between psychopathology and pHVA is enhanced. In order to detect this relationship in chronic patients, multiple samples may be needed.

Clearly pHVA is influenced by multiple variables that may affect the strength of the relationship between pHVA and psychopathology, treatment response, and outcome; these variables include gender, diet, activity, prior treatment exposure, duration of drug-free period, patient population (acute vs chronic), renal clearance (KAMINSKI et al. 1990; AMIN et al. 1991), and diurnal variation (CONTRETAS et al. 1988). It is important to remember that the peripheral nervous system is the primary contributor of pHVA; only 17% of pHVA originates in the brain (LAMBERT et al. 1991). Nevertheless, HVA still represents the most reliable and least invasive method available to evaluate central DA activity. In fact, considering the many factors that influence pHVA that are unrelated to the CNS, it is surprising that the results are this consistent.

Since obtaining CSF HVA is a more invasive procedure than obtaining pHVA, there have been fewer studies looking at the relationship between CSF HVA and treatment response (Table 2). CSF HVA has also been found to increase acutely with antipsychotic drug treatment (BJERKENSTEDT et al. 1977), and in one study (SHARMA et al. 1989) but not another (KAHN et al. 1993) there was a trend for the decrease of CSF HVA with chronic (5-week) treatment to be correlated with a decrease in symptoms. Most of the recent studies of CSF

Table 2. Cerebrospinal fluid homovanillinic acid (CSF HVA)

Author/year	Sample	Drug-free ≥ (wks)	Methods	Rating scale(s)	Assay	Day of predictive assessment	Outcome day	Results
Sharma et al. 1989	17 (9♂, 8♀) RDC nonorganic psychotic pts	mean of 19 days	CSF HVA drawn at BL & fo 12 pts at end of 5 wks of uncontrolled neuroleptic tx	BPRS, GAS	HPLC	BL	35	↓ in CSF HVA correlated (trend) with a ↓ in sxs.
Hsaio et al. 1992	50 DSM III & RDC schizophrenic pts	2	LP done at BL & then in 41 pts after 3 wks of tx	BPRS	HPLC	BL	21	1. HVA/5HIAA or HVA/MHPG did not predict tx response. 2. Negative correlation between HVA/5HIAA & HVA/MHPG & sxs at 3 wks.
Pickar et al. 1992	21 (13♂, 8♀) DSM III R schizophrenic or schizoaffective pts	4 prior to clozapine tx	LP for CSF HVA & 5-HIAA done after 4 wks of fluphenazine, 4 wks of placebo period, & then while on clozapine	BPRS, BH Response = BPRS ↓ by >20% and total BPRS <36 or BH ≤6	HPLC	BL	100	1. Responders to clozapine tx had lower CSF HVA/5HIAA levels during fluphenazine tx, placebo period, and clozapine tx. 2. CSF HVA, was not significantly different between responders & nonresponders.

Kahn et al. 1993	12♂ DSM III R & RDC schizophrenic & schizoaffective pts	2	• LP done at baseline • Pt then tx'd with 20 mg/day haloperidol for 5 wks & repeat LP done	• BPRS • Responders = >20% ↓ in BPRS • Partial responders = 0%–20% ↓ in BPRS • Nonresponders = no ↓	HPLC	BL	35	1. Change in HVA/5HIAA correlated with change in BPRS but changes in HVA or 5HIAA did not. 2. Change in HVA/5HIAA was different between responders, partial responders, & nonresponders with ratio greatest in responders and lowest in nonresponders.

pts, patients; BL, baseline; ♂, male; ♀, female; DSM III R, Diagnostic and Statistical Manual of Mental Disorders (American Psychiatric Association 1980 [1987]); RDC, Research Diagnostic Criteria (Spitzer et al. 1977); BPRS, Brief Psychiatric Rating Scale (Overall and Gorhan 1962); GAS, Global Assessment Scale (Endicott et al. 1976); HPLC, high-performance liquid chromatography; CSF, cerebrospinal fluid; HVA, homovanillic acid; 5-HIAA, 5-hydroxyindoleacetic acid; LP, lumbar puncture; MHPG, 3-methoxy-4 hydroxy-phenylglycol; sxs, symptoms.

HVA have not looked at CSF HVA alone but instead have looked at the ratio of HVA to various other neurotransmitter metabolites. Hsiao et al. (1993) found that baseline HVA/5-HIAA or HVA/MHPG did not predict treatment response to typical neuroleptics. However, they did find that standard antipsychotic drug treatment induced changes in the correlations between CSF HVA, 5-HIAA, and MHPG that approximated those found in a control group of healthy volunteers, suggesting a "normalized" effect. Kahn et al. (1993) did find that baseline HVA/5-HIAA differentiated haloperidol responders, partial responders, and nonresponders, with the ratio the greatest in responders and lowest in nonresponders. Pickar et al. (1992) and Szymanski et al. (1994), however, found that clozapine responders had lower CSF HVA/5-HIAA levels (Table 3). Given the paucity of studies looking at CSF HVA or the ratio of CSF HVA to other metabolites as a prediction of treatment response and the varying methodologies of those studies, any conclusions that can be drawn are, at best, tentative.

II. Dopamine β-Hydroxylase (DBH)

Because of the interest in DA in the pathophysiology of schizophrenia, several studies have examined dopamine β-hydroxylase (DBH) since it is the enzyme that converts DA to NE. DBH activity has been studied in the serum and CSF. There are two studies which have looked at DBH and antipsychotic response: Sternberg et al. (1982) found lower baseline CSF DBH and Bartkó et al. (1990) found lower serum DBH in responders to antipsychotic treatment than in nonresponders (Table 4). Given the consistent results, albeit in different body fluids, further studies of this enzyme are warranted.

III. Serotonin (5-HT)

Next to DA, 5-HT and NE have historically been the neurotransmitters most often implicated in schizophrenia. 5-HT receptors are distributed in brain regions that mediate behavioral functions, including the frontal cortex. Moreover, standard antipsychotic drugs bind to 5-HT receptors, and the more recently developed atypical antipsychotic drugs have greater potency at the 5-HT receptors than D_2 receptors. This greater affinity for the 5-HT receptor and the combined effects on the DA and 5-HT systems have been proposed as the basis for their atypical effects (Meltzer 1988, 1991).

In recent years, a limited number of studies have examined the 5-HT system in patients with schizophrenia and its relationships to treatment response (Table 5); almost all of them have analyzed the 5-HT metabolite 5-HIAA in the CSF, with only one study (Alfredsson and Wiesel 1990) examining the relationship with plasma 5HIAA. Interestingly, despite the recent studies of blood serotonin levels, platelet 5HT, and imipramine binding to assess the 5-HT system in psychiatric illness, none has assessed these variables and how they may be related to antipsychotic treatment response. Also,

Table 3. Clozapine response and pHVA

Author/year	Sample	Drug-free ≥ (wks)	Methods	Rating scale(s)	Assay	Day of predictive assessment	Outcome day	Results
PICKAR et al. 1992	21 (13♂, 8♀) DSM III R schizophrenic or schizoaffective pts	4	pHVA collected during fluphenazine tx, during 4 wk washout period, & then during moderate dose (mean 393 mg/d) and optimal dose (mean 543 mg/d) of clozapine tx	BPRS, BH Response = BPRS ↓ by >20% and total BPRS <36 or BH ≤6	HPLC	BL	100	Clozapine responders had lower pHVA levels during clozapine tx.
DAVIDSON et al. 1993	28 (24♂, 4♀) RDC & DSM III R schizophrenic pts	1	pHVA drawn at BL & then 2x/wk for 5 wks of clozapine tx	BPRS, CGI	HPLC	BL	35	pHVA ↓ in clozapine responders (not significant).
GREEN et al. 1993	8 (3♂, 5♀) DSM III R schizophrenic pts	2	pHVA drawn at BL, 2x during 1st wk, & then wkly during 12 wks of clozapine tx	BPRS Response = ≥20% ↓ in BPRS at wks 11 & 12	HPLC	BL	84	1. BL pHVA significantly correlated with tx response. 2. % ↓ in pHVA from wks 2–8 was correlated with tx response at 12 wks 3. In responders (n = 6). pHVA ↓ quickly, whereas in nonresponders (n = 2) it rose from wk 1–5 before ↓.

pts, patients; BL, baseline; HPLC, high-performance liquid chromatography; tx, treatment; BPRS, Brief Psychiatric Rating Scale (OVERALL and GORHAN 1962); CGI, Clinical Global Impression Scale; ♂, male; ♀, female; DSM III R, Diagnostic and Statistical Manual of Mental Disorders [AMERICAN PSYCHIATRIC ASSOCIATION 1980 (1987)]; RDC, Research Diagnostic Criteria (SPITZER et al. 1977); pHVA, plasma homovanillic acid; BH, Bunney-Hamburg scale (BUNNEY and HAMBURG 1963).

Table 4. Dopamine β-hydroxylase (DBH)

Author/year	Sample	Drug-free ≥ (wks)	Methods	Rating scale(s)	Assay	Day of predictive assessment	Outcome day	Results
Sternberg et al. 1982	25 RDC schizophrenic pts	2	LP for CSF DBH done at BL and then 3x over 4-month period of open tx	BH		Mean of the 4 levels drawn	28	Pts who responded to neuroleptic tx had significantly lower mean CSF DBH than those who remained psychotic.
Bartkó et al. 1990	98 (56♂, 42♀) RDC schizophrenic pts		Serum DBH drawn at BL; pts then tx'd openly with neuroleptics for 21 days	4-point scale	Method of Nagatsu & Udenfriend	BL	28	Reduced serum DBH activity at BL predicted good response to neuroleptics.

pts, patients; BL, baseline; HPLC, high-performance liquid chromatography; tx, treatment; ♂, male; ♀, female; RDC, Research Diagnostic Criteria (Spitzer et al. 1977); BH, Bunney-Hamburg Scale (Bunney and Hamburg 1963); CSF, cerebrospinal fluid; DBH, dopamine β-hydroxylase.

it is unclear whether 5-HIAA is a valid peripheral marker for CNS serotonergic activity, as pHVA is believed to be for DA, because it is thought to have extensive peripheral contribution from the gastrointestinal tract. Therefore, it is not surprising that ALFREDSSON and WIESEL (1990), KAHN et al. (1993), and HSAIO et al. (1993) find that baseline 5-HIAA did not predict treatment response. The ratio of HVA/5-HIAA, however, may be more informative since PICKAR et al. (1992) and KAHN et al. (1993) did find that it differentiated responders and nonresponders to fluphenazine and haloperidol, respectively: clozapine responders had a lower ratio whereas haloperidol responders had a higher ratio. In addition, HSAIO et al. (1993) found that the standard antipsychotic drug treatment improved the correlations between CSF HVA and HIAA as well as MHPG to levels that approximated those found in healthy volunteers. Given clozapine's and other atypical antipsychotic drugs' serotonergic activity, further studies of 5-HT and response are particularly warranted.

IV. Norepinephrine (NE)

Beginning with the hypothesis of STEIN and WISE (1971) and more recently with the demonstrated interactions between DA and NE systems in the CNS, several investigators have postulated that NE plays an important role in the pathophysiology of schizophrenia (HORNYKIEWICZ 1982; VAN KAMMEN 1991). Since 1981 when STERNBERG et al. demonstrated that a decreased CSF NE with treatment correlated with a decrease in psychosis, a number of studies have looked at the relationship between the NE system and response (Table 6). Some studies found a decrease in the NE metabolite, MHPG, in plasma with treatment that was correlated with an improvement (ALFREDSSON and WIESEL 1990; CHANG et al. 1990; MAZURE et al. 1991; and GREEN et al. 1993), whereas baseline MHPG in the plasma (BOWERS et al. 1987; ALFREDSSON and WIESEL 1990; CHANG et al. 1990; MAZURE et al. 1991) or the CSF (HSAIO ET al. 1993) did not appear to have prognostic significance in various studies. Five studies examined NE response to clozapine (PICKAR et al. 1992; LIEBERMAN et al. 1992; DAVIDSON et al. 1992; and GREEN et al. 1993; BREIER et al. 1994). All found that NE increased with clozapine treatment. However, only GREEN et al. (1993) found that baseline plasma NE was lower in responders than nonresponders while BREIER et al. (1994) found that the increase in plasma NE was associated with antipsychotic response. GREEN et al. (1993) also found that plasma MHPG decreased only in responders but not in nonresponders during clozapine treatment. Although these data are far from conclusive, they are consistent with the hypothesized involvement of NE in the pathophysiology of schizophrenia and suggest that noradrenergic systems are involved in mediating the effects of atypical antipsychotic drugs. Although baseline MHPG does not appear to be a marker for treatment response, a decrease of MHPG with treatment does consistently appear to differentiate responders from nonresponders.

Table 5. Serotonin/5-HIAA

Author/year	Sample	Drug-free ≥ (wks)	Methods	Rating scale(s)	Assay	Day of predictive assessment	Outcome day	Results
ALFREDSSON and WIESEL 1990	24 (14♂, 10♀) DSM III R schizophrenic pts	2	Plasma 5HIAA drawn pre-tx and weekly during sulpiride tx	CPRS, NOSIE Responders = ≥50% ↓	GCMS	BL	42	1. 5-HIAA pre-tx not correlated with psychopathology. 2. Negative relationship between 5-HIAA and depressive and negative sxs. 3. 5-HIAA did not differentiate responders from nonresponders.
PICKAR et al. 1992	21 (13♂, 8♀) DSM III R schizophrenic or schizoaffective pts	4 prior to clozapine tx	LP for CSF, HVA, & 5HIAA done at end of 4 wks fluphenazine tx, and of 4-wk placebo period, & then while on clozapine	BPRS, BH Response = BPRS ↓ by 20% and total BPRS <26 or BH ≤6	HPLC	BL	100	1. Clozapine responders had lower CSF HVA/5HIAA levels during fluphenazine, placebo, and clozapine periods. 2. No difference in CSF 5HIAA in clozapine responders and nonresponders.

Kahn et al. 1993	12♂ DSM III R & RDC schizophrenic & schizoaffective pts	2	LP for CSF HVA & 5-HIAA done at BL & then after 5 wks of 20 mg/day of haloperidol tx	BPRS Responders = >20% ↓ Partial responders = 0% to 20% ↓ Nonresponders no ↓	HPLC	BL	35	1. Change in HVA/5-HIAA correlated with change in BPRS but 5-HIAA level alone not correlated. 2. Change in HVA/5-HIAA different in responders, partial responders, and non-responders with ratio greatest in responders and lowest in nonresponders.
Hsiao et al. 1993	50 DSM III & RDC schizophrenic pts	2	LP for CSF HVA & 5-HIAA done at BL & then after 5 wks of 20 mg/day of haloperidol tx	BPRS	HPLC	BL	21	1. BL HVA/5HIAA or 5HIAA levels did not predict response. 2. Negative relationship between HVA/5HIAA and sxs at 3 wks.

pts, patients; BL, baseline; HPLC, high performance liquid chromatography; tx, treatment; BPRS, Brief Psychiatric Rating Scale (Overall and Gorhan 1962); GCMS, gas chromography mass spectroscopy; ♂, male; ♀, female; DSM III (R), Diagnostic and Statistical Manual of Mental Disorders (American Psychiatric Association 1980 [1987]); RDC, Research Diagnostic Criteria (Spitzer et al. 1977); CPRS, Comprehensive Psychiatric Rating Scale (Jacobsson et al. 1978); NOSIE, Nurse's Observation Scale for Inpatient Evaluation (Honigfeld and Klett 1965); LP, lumbar puncture; CSF, cerebrospinal fluid; HVA, homovanillic acid; 5-HIAA, 5-hydroxyindoleactic acid; sxs, symptoms; BH, Bunney-Hamburg scale (Bunney and Hamburg 1963).

Table 6. Neuroepinephrine (NE)/MHPG

Author/year	Sample	Drug-free ≥ (wks)	Methods	Rating scale(s)	Assay	Day of predictive assessment	Outcome day	Results
Sternberg et al. 1981	33 (19♂, 14♀) RDC schizophrenic or schizoaffective pts	3	LP done at BL for CSF NE; pts then tx'd with pimozide & repeat LP done. in 11 pts	BH	Radioenzymatic technique	BL		↓ CSF NE with tx correlated with the ↓ in global psychosis.
Bowers et al. 1987	47 (25♂, 22♀) DSM III nonorganic schizophrenic pts	1	Plasma MHPG drawn at BL, pt then tx'd with neuroleptic for 10 days	4-point scale	GCMS	BL	10	BL MHPG higher in good responders (trend).
Alfredsson and Wiesel 1990	24 (14♂, 10♀) DSM III R schizophrenic pts	2	Plasma MHPG drawn pre-tx & weekly during sulpiride tx	CPRS, NOSIE Responders = ≥50% ↓	GCMS	BL	42	1. A ↓ in MHPG was correlated with an improvement. 2. MHPG levels did not differ between responders & nonresponders.
Chang et al. 1990	33 (21♂, 12♀) DSM III R schizophrenic pts	4 (p.o.) 12 (IM)	Plasma MHPG drawn at BL & then weekly for 6 wks of 20 mg/day of haloperidol tx	BPRS Good responders = ≥50% ↓	HPLC	BL	42	1. No difference in BL MHPG between good and poor responders. 2. Good responders had a ↓ in MHPG with tx, whereas poor responders had no change with tx.

Mazure et al. 1991	37 (12♂, 25♀) DSM III R nonorganic psychotic pts	1	Plasma MHPG drawn at BL & then on days 7, 8, & 9 of tx with perphenazine (0.5mg/kg per day)	BPRS Response = BPRS ≤ 14 and 33% ↓	GCMS	BL	10	1. BL MHPG did not predict response. 2. Responders had a ↓ in MHPG with tx whereas nonresponders had no change with tx.
Pickar et al. 1992	21 (13♂, 8♀) DSM III R schizophrenic or schizoaffective pts	4wks prior to clozapine tx	Plasma NE drawn and LP done for CSF NE at end of 4 wk fluphenazine tx, at end of 4 wks of placebo and during clozapine tx	BPRS, BH Response = BPRS ↓ by > 20% and total BPRS <26 or BH ≤6	HPLC	BL	100	Plasma NE or CSF NE was not different in clozapine responders or nonresponders.
Davidson et al. 1993	28 (24♂, 4♀) RDC & DSM III R schizophrenic pts	1	Plasma NE drawn at BL and 2x wk for 5 wks of clozapine tx	BPRS, CGI Responders = ≥1 pt ↓ in CGI	HPLC	BL	35	BL or weekly pNE did not differentiate responders from nonresponders.
Green et al. 1993	8 (3♂, 5♀) DSM III R schizophrenic pts	2	Plasma NE & MHPG drawn at BL, 2x wk over 1st week and then weekly for 12 wks of clozapine tx	BPRS Responders = ≥20% ↓	HPLC	BL	84	1. Plasma MHPG ↓ during initial wks in responders but not nonresponders. 2. Plasma NE ↑ in both responders & nonresponders. 3. BL plasma NE was lower in responders (not significant), whereas there was no difference in BL MHPG.

Table 6. *Continued*

Author/year	Sample	Drug-free ≥ (wks)	Methods	Rating scale(s)	Assay	Day of predictive assessment	Outcome day	Results
Hsiao et al. 1993	50 DSM III & RDC schizophrenic pts	2	LP done at BL and then after 5 wks of 20 mg of haloperidol tx	BPRS	HPLC	BL	21	1. BL HVA/MHPG or MHPG levels did not predict response. 2. There was a negative relationship between HVA/MHPG and sxs at 3 wks.
Breier et al. 1994	11 (7♂, 4♀) DSM III R chronic schizophrenic pts	ø	plasma NE drawn after 5 wks of fluphenazine tx (BL) and then after 5 wks of clozapine tx	BPRS	HPLC	BL	35	1. Clozapine markedly increased plasma NE. 2. Increase in plasma NE was positively related to improvement in positive sxs and global symptomatology.

pts, patients; BL, baseline; HPLC, high-performance liquid chromatography; tx, treatment; BPRS, Brief Psychiatric Rating Scale (Overall and Gorhan 1962); ♂, male; ♀, female; DSM III (R), Diagnostic and Statistical Manual of Mental Disorders (American Psychiatric Association 1980 [1987]); RDC, Research Diagnostic Criteria (Spitzer et al. 1977); CGI, Clinical Global Impression Scale; BH, Bunney Hamburg scale (Bunney and Hamburg 1963); CSF, cerebrospinal fluid; NE, norepinephrine; MHPG, 3-methoxy-4-hydroxyphenylglycol.

V. Peptides

Since neuropeptides may serve as neurotransmitters, cotransmitters, and neuromodulators, attempts have been made to explore the significance of these compounds in the pathophysiology of schizophrenia. The number and types of peptides studied have varied, but only four studies (Table 7) have made any attempt to look at peptides as potential biological predictors of treatment response. Overall, the studies have been limited and in need of replication. Among its functions, somatostatin stimulates DA release from the striatum, whereas neurotensin interacts with DA in physiological, anatomical, and behavioral studies and mimics the action of antipsychotics drugs when given centrally (Kasckow and Nemeroff 1991). Studies have not found either of these two peptides to be correlated with treatment response (Doran et al. 1989; Nemeroff et al. 1989; Garver et al. 1991). Garver et al. (1991) did, however, find a subgroup of psychotic female patients with low neurotensin levels that had a delayed haloperidol response. On the other hand, cholecystokinin (CCK), which interacts with DA at both the presynaptic and postsynaptic levels in the nucleus accumbens (Wang 1989), and whose mRNA has been shown to coexist in the central mesencephalon DA cells of a schizophrenic patient (Schalling et al. 1989), was found to have prognostic significance. Garver et al. (1990) found that schizophrenic patients who responded early to haloperidol (within a week) had greater pretreatment CSF CCK than slow or nonresponders. Clearly further research is warranted in this area.

C. Behavioral Response to Psychostimulants

Since loading doses of psychostimulants, which are DA agonists, have been observed to induce paranoid schizophreniform psychosis in normal subjects (Snyder et al. 1974), such psychostimulants have been used as a pharmacologic probe of schizophrenia and as a predictor of treatment response (Lieberman et al. 1987). A group of studies examined DA agonist or psychostimulant behavioral response as a predictor of outcome in patients withdrawn from maintenance antipsychotic drug treatment (Angrist et al. 1980; Van Kammen et al. 1982; Lieberman et al. 1987; Davidson et al. 1991a). All four studies had consistent results in that patients who experienced transient activation in psychotic symptoms had increased risk of relapse. These results suggested that instability in DA neurons required continued administration of antipsychotic drugs to maintain clinical/behavioral stability. A logical inference of this work in the context of the DA hypothesis of schizophrenia is that patients with DA neurons reactive to psychostimulant administration would be responsive to antipsychotic drug treatment during periods of acute exacerbations. In four studies (Angrist et al. 1980; Van Kammen et al. 1982; Little et al. 1989; Lieberman et al. 1993) the behavioral response of acutely psychotic patients to *d*-amphetamine and methylphenidate as examined in relation to the subsequent response to antipsychotic drug treatment (Table

Table 7. Peptides

Author/year peptide	Sample	Drug-free ≥ (wks)	Methods	Rating scale(s)	Assay	Day of predictive assessment	Outcome day	Results
DORAN et al. 1989 Somatostatin	46 (25♂, 21♀) DSM III schizophrenic pts	2	LP done pre-tx and then in 14 pts after 2 wks of placebo & 3 wks of neuroleptic tx	BH	RIA	BL	21	1. CSF somatostatin-like immunoreactivity (SLI) ↓ with tx. 2. No relationship between SLI and response.
NEMEROFF et al. 1989 Neurotensin	68 DSM III schizophrenic pts	3	LP done at BL and for 42 pts after 3 wks of neuroleptic tx	CPRS, BPRS	RIA	BL	21	No correlations between changes in psychopathology and neurotensin.
GARVER et al. 1990 Cholecystokinin (CCK)	11 (6♂, 5♀) DSM III schizophrenic pts	2	LP done at BL, pts then tx'd with lithium; if no response, changed to haloperidol 10–15 mg/day	SADS, NHSI Early responders = 35% ↓ in psychosis by wk 1 of haloperidol tx	RIA	BL	7	Early responders had greater CCK than slow or nonresponders.
GARVER et al. 1991 Neurotensin	20 (10♂, 10♀) DSM III nonorganic psychotic pts	2	LP done at BL, then pts tx'd with lithium for 2 wks; if no response, tx'd with haloperidol	SADS, NHSI Early responders = 35% ↓ in psychosis by wk 1 of haloperidol tx	RIA	BL	7	1. Decreased neurotensin found in subtype of psychotic ♀ pts who had delayed haloperidol response and increased psychotic sxs. 2. Neurotensin ↑ with tx.

pts, patients; BL, baseline; tx, treatment; BPRS, Brief Psychiatric Rating Scale (OVERALL and GORHAN 1962); ♂, male; ♀, female; DSM III (R), Diagnostic and Statistical Manual of Mental Disorders (AMERICAN PSYCHIATRIC ASSOCIATION 1980 [1987]); RDC, Research Diagnostic Criteria (SPITZER et al. 1977); CPRS, Comprehensive Psychiatric Rating Scale (JACOBSSON et al. 1978); SADS, Schedule for Affective Disorders and Schizophrenia (SPITZER et al. 1978); NHSI, New Haven Schizophrenia Index (ASTRACHEN et al. 1972); RIA, radioimmunoassay; LP, lumbar puncture; BH, Bunney-Hamburg scale (BUNNEY and HAMBURG 1963).

8). Overall, the results (Table 8) have not been consistent. The two early studies by ANGRIST in 1980 (1980) and VAN KAMMEN in 1982 (1982) directly contradicted each other. ANGRIST et al. (1980) found that those patients with a stimulant-induced increase in psychosis had a better treatment response, whereas VAN KAMMEN et al. (1982) found that those who improved with antipsychotic treatment had a stimulant-induced decrease in psychosis. The two more recent studies (LITTLE et al. 1989; LIEBERMAN et al. 1993a) have also not been conclusive, but overall they found no significant correlation between behavioral response to methylphenidate and treatment response. Given the variable methodologies (dose, psychostimulant used, sample differences, time drug-free, rating period) which may affect the results, more standardized protocols may yield more meaningful and consistent results. Furthermore, specific analysis of subgroups of patients or certain symptoms may prove to be more informative.

D. Neuroendocrine Studies

Another line of investigation has been neuroendocrine studies. Since the pituitary is anatomically linked and in proximity to the hypothalamus and CNS, and pituitary hormone secretion is predominantly regulated by neurochemical mechanisms, schizophrenia investigators have been able to measure indirectly the neurochemical activity of specific brain structures involved in a given hormone's regulation. The specific anatomic level, either pituitary, hypothalamic, or suprahypothalamic, can be determined through the use of pharmacologic probes, e.g., releasing factors, neurotransmitter agonists, and antagonists. Although recently neuroendocrine investigation is waning due to its failure to produce clear-cut pathophysiologic findings, it continues to be a potentially fruitful strategy.

Since DA is the major neurotransmitter involved in the regulation of prolactin (PRL) and growth hormone (GH), these hormones have been studied most extensively in schizophrenia. In the case of PRL, DA acts as a tonic inhibitor of lactotrophic cell secretion (REICHLIN 1981), whereas with GH, DA acts as a phasic stimulator at the arcuate nucleus of the hypothalamus (RIVIER et al. 1982; WASS 1983).

GH response to DA agonists has been studied as a predictor of treatment response in four studies (Table 9). The two earliest studies (GARVER et al. 1984; MELTZER et al. 1981) found that an increased GH response was predictive of a rapid response to antipsychotics. LIEBERMAN et al. (1992) found baseline GH but not GH response to apomorphine had prognostic significance; patients with higher baseline GH levels took longer to go into remission and had poorer outcomes. HIRSCHOWITZ et al. (1991), in a slightly different study, found that GH response was not predictive of treatment response but was useful in determining the minimum antipsychotic dose needed to attenuate psychotic symptoms. In the only study to examine GH in relation to clozapine treatment

Table 8. Behavioral response to psychostimulants

Author/ year	Sample	Drug-free ≥ (wks)	Methods	Rating Scale(s)	Day of predictive assessment	Outcome day	Results
ANGRIST et al. 1980	21 (20♂, 1♀) RDC schizophrenic pts	6 days	0.5 mg/kg amphetamine p.o. given and then pts tx'd openly with neuroleptics	BPRS, CGI	BL	35	1. ↑ psychopathology with amphetamine correlated with improvement after tx. 2. In the absence of significantly clinical change with amphetamine, response to neuroleptics rare.
VAN KAMMEN et al. 1982	22 (15♂, 7♀) RDC schizophrenic or schizoaffective pts	3	Double-blind, placebo-controlled infusion of 20 mg *d*-amphetamine, followed by double-blind pimozide trial for 4 wks	BPRS Amphetamine Interview Rating Scale	BL	28 35	1. Pts whose psychosis ↓ with *d*-amphetamine had better tx response at 4 wks. 2. Behavioral response at BL did not predict pimozide response at 5 wks.

Little et al. 1988	14 (11♂, 3♀) DSM III schizophrenic pts	ø	Infusion of 40 mg methylphenidate, followed by a 50% ↑ in their antipsychotic medication for 3 wks	BPRS	BL	21	Several pts with stimulant-induced improvements in hallucinations improved mildly after 3 wks of ↑'d meds but overall group did not improve.
Lieberman et al. 1993	51 RDC schizophrenic or schizoaffective pts	2	Infusion of methylphenidate (0.5 mg/kg) given at BL; pts then tx'd by standardized tx protocol for up to 3 yrs	SADS, CGI	BL	up to 3 yrs	Methylphenidate response was not significantly associated with level of remission.

pts, patients; BL, baseline; tx, treatment; BPRS, Brief Psychiatric Rating Scale (Overall and Gorhan 1962); ♂, male; ♀, female; DSM III (R) = Diagnostic and Statistical Manual of Mental Disorders (American Psychiatric Association 1980 [1987]); RDC, Research Diagnostic Criteria (Spitzer et al. 1977); CGI, Clinical Global Impression Scale; SADS, Schedule for Affective Disorders and Schizophrenia (Spitzer et al. 1978).

Table 9. Neuroendocrine studies

Author/year hormone	Sample	Drug-free ≥ (wks)	Methods	Rating Scale(s)	Assay	Day of predictive assessment	Outcome day	Results
Meltzer et al. 1981 GH PRL	18 (7♂, 11♀) RDC schizophrenic or schizoaffective pts	1	Apomorphine (0.75 mg) infusion done in AM; plasma GH and PRL drawn at –20, –10, 0, 30, 60, 90, 120, 150, 180, and 240 min	4 point scale		BL	Discharge	1. PRL response to apomorphine did not predict response. 2. Significant negative correlation found between GH response to apomorphine and discharge ratings of psychosis & depression.
Garver et al. 1984 GH	27 RDC & DSM III mood-incongruent psychotic pts	10 days	Apomorphine (0.75 mg) given s.q. pts, then tx'd with antipsychotics	NHSI		BL	49	Rapid responders (response a mean of 5.5 days) had ↑ GH response to apomorphine than delayed/nonresponders.
Hirschowitz et al. 1991 GH	16♂ DSM III R schizophrenic pts	ø (all pts on ≥ 20 mg/day of haloperidol)	Bromocryptine (50 mg/kg) p.o. given at BL and then q 2 wks until there was an escape from blockade of GH response; after BL challenge pts had haloperidol ↓ by 50% every 2 wks	SAPS, BPRS	RIA	BL q 2 wks	–	In some pts the "just" blockade dose was the minimum dose needed to attentuate psychotic sxs.
Lieberman et al. 1992 GH	52 (28♂, 15♀) DSM III R & RDC schizophrenic pts	2	Apomorphine (0.0075 mg/kg) given s.q. GH sampled at –30, –15, 0, 15, 30, 60, & 90 mins	SADS, CGI	RIA	BL		Pts with higher BL GH took longer to go into remission and had poorer outcomes.

Lieberman et al. 1990 PRL	56♂ RDC schizophrenic or schizoaffective pts	ø on stable meds	2 mg haloperidol given IM; pts obtained pre & post haloperidol; pts maintained on standard or low dose fluphenazine decanoate × 1 yr	SADS, CGI	RIA	BL	1 yr	1. ↓ PRL response associated with ↑ relapse, especially in low-dose group. 2. 75% of pts with PRL ≤50 ng/ml relapsed as compared to 33% with PRL >50 ng/ml.
Awad et al. 1990 PRL	148 (108♂, 40♀) DSM III R schizophrenic pts	3 days	PRL drawn at BL & then after 6 wks of tx with remoxipride or haloperidol	BPRS, CGI Response = > 50% ↓ in BPRS	RIA	BL	42	1. ♂ responders to either tx had significantly higher BL PRL than nonresponders. 2. ♀ had no difference in BL PRL between responders and nonresponders.
Van Putten et al. 1991 PRL	73♂ DSM III schizophrenic pts	2	PRL drawn at BL and then weekly for 4 wks of haloperidol tx (5, 10, or 20 mg/day)	BPRS	RIA	BL wkly	28	Correlation between ↑ PRL during tx and better response (correlation largest in 5 mg/day group).
Markianos et al. 1991 PRL	12 ♂ schixophrenic pts	several months	PRL drawn pre & post 5 mg haloperidol I.M. given at BL & then on day 30 of tx with haloperidol and chlorpromizine; PRL also drawn at day 15	BPRS	RIA	BL, day 30	30	Improvement with tx not correlated with BL PRL response but did correlate with ↑ PRL with tx.
Lahdelman et al. 1991 PRL	20 (10♂, 10♀) DSM III schizophrenic or schizoaffective pts	3	PRL drawn at BL and then biweekly during tx with remoxipride or haloperidol for 6 wks	CGI, BPRS	RIA	BL, bi-wkly	42	A difference in PRL-elevating pattern was observed between remoxipride and haloperidol despite the fact there was no difference in tx response between the medications.

Table 9. *Continued*

Author/year hormone	Sample	Drug-free ≥ (wks)	Methods	Rating Scale(s)	Assay	Day of predictive assessment	Outcome day	Results
Pickar et al. 1992 PRL	21 (13♂, 8♀) DSM III R schizophrenic or schizoaffective pts	4 prior to clozapine tx	Plasma PRL drawn at end of 4 wks of fluphenazine tx & 4 weeks of placebo & then during clozapine tx	BPRS, BH Response = BPRS ↓ by >20% and total BPRS <36 or BN ≤6	RIA	BL	100	No significant difference found between clozapine responders and nonresponders and PRL levels.
Coryell and Zimmerman 1989 Cortisol	88 (40♂, 48♀) DSM III non-organic psychotic pts		DST (1 mg) performed within 1 week of admission; cortisol drawn at 8,[A] 4,[P] & 11[P] next day; pt then tx'd openly	SADS, GAS	RIA	BL	1 yr	Higher postdexamethasone cortisol levels (≥6 μg/dl) predicted recovery from psychosis at 1 yr independent of diagnosis.
Tandon et al. 1991 Cortisol	44 (29♂, 15♀) DSM III R & RDC schizophrenic pts	2 (p.o.) 24 (I.M.)	Dexamethasone suppression test (DST, 1 mg) performed at BL & then after 4 wks of open neuroleptic tx	BPRS, SANS, Hamilton Strauss & Carpenter Scale	Murphy's competitive protein-binding method	BL wk 4	4 wks	1. Persistent nonsuppression associated with poor outcome. 2. BL post dexamethasone cortisol levels unrelated to outcome at 4 wks & 1 yr.

Langer et al. 1986 TSH	29 RDC schizophrenic or schizoaffective pts	ø	BL thyroid releasing hormone (TRH) test done et al & then at regular intervals while on haloperidol for 9 wks	BPRS	RIA	BL	63	A blunted TSH response (<5 μU/ml) at BL associated with recovery at 9 wks.
Beasley et al. 1988 TSH	19 DSM III non-organic psychotic pts	10 days	BL TRH test done and then pts tx'd with lithium for 2 wks; lithium nonresponders tx'd with haloperidol	NHSI Response = 55% ↓	RIA	BL	Day 7 of neuroleptic tx	A ↓ TSH response was a significant predictor of a positive rapid response to haloperidol (≤7 days).
Baumgartner et al. 1988 TSH	15 (8♂, 7♀) RDC schizophrenic pts	–	400 mg protirelin given at BL & then 4 wks after tx; TSH drawn 30 min postinfusion	CGI	–	BL 4 wks	18 mos	TRH response did not predict therapeutic response or outcome.
Siris et al. 1991 TSH	25 (15♂, 10♀) RDC schizophrenic or schizoaffective pts with post-psychotic depression	ø	BL TRH test done and then pts entered into a double-blind, placebo-controlled trial of imipramine	CGI, SADS	RIA	BL	28	TSH response did not predict response and was not correlated with any measure of psychopathology.

pts, patients; BL, baseline; tx, treatment; BPRS, Brief Psychiatric Rating Scale (Overall and Gorhan 1962); ♂, male; ♀, female; DSM III (R), Diagnostic and Statistical Manual of Mental Disorders (American Psychiatric Association 1980 [1987]); RDC, Research Diagnostic Criteria (Spitzer et al. 1977); SADS, Schedule for Affective Disorders and Schizophrenia (Spitzer et al. 1978); NHSI, New Haven Schizophrenic Indices (Astrachen et al. 1972); RIA, radioimmunoassay; TRH, thyroid-releasing hormone; CGI, Clinical Global Impression Scale; SANS, Scale for the Assessment of Negative Symptoms (Andreasen 1984); SAPS, Scale for the Assessment of Positive Symptoms (Andreasen 1984); PRL, prolactin; GH, growth hormone; TSH, thyroid-stimulating hormone.

response, SZYMANSKI et al. (1994) found that greater GH stimulation by apomorphine pretreatment was associated with therapeutic response to clozapine.

Basal and stimulated PRL and its relationship to treatment response has been studied in both acute and maintenance treatment studies (Table 9). PRL response to apomorphine did not predict acute treatment response in the study by MELTZER et al. (1981). VAN PUTTEN et al. (1991), however, found that increased PRL with acute treatment was correlated with a better response, and AWAD et al. (1991) found that it was predictive of acute response in males but not females. In the maintenance studies where patients had been on a stable dose of medication, baseline PRL was not correlated with 1-month to 1-year outcomes (MARKIONOS et al. 1991; LIEBERMAN et al. 1990; PICKAR et al. 1992). However, a blunted PRL response to acute parenteral haloperidol administration was correlated with increased relapse in a study by LIEBERMAN et al. (1990) and increased PRL with treatment was correlated with improvement in a study by MARKIANOS et al. (1991). Two studies that looked at atypical neuroleptics (LAHDELMAN et al. 1991; PICKAR et al. 1992) found no correlations with baseline PRL and outcome, while SZYMANSKI et al. (in press) found that a greater decrease in PRL to apomorphine pretreatment was associated with a therapeutic response to clozapine. It thus appears that the inconsistent results are due to methodological differences such as medication status, acute vs chronic treatment, type of antipsychotic medication, baseline PRL vs PRL response to acute antipsychotic challenge vs PRL response to treatment as well as heterogeneous patient groups. Any consistent results await replication of studies utilizing standardized methods.

Two recent studies examined cortisol and treatment response in the context of a dexamethasone suppression test. CORYELL and ZIMMERMAN (1989) found that higher (≥6 mg/dl) pretreatment cortisol levels were associated with recovery from psychosis at 1 year independent of episode chronicity or diagnosis. TANDON et al. (1991), on the other hand, did not find baseline post dexamethasone levels related to outcome but did find that persistent nonsuppression at week 4 was associated with poor outcome TANDON et al. (1989), in a letter in which he reviewed previous studies including CORYELL and ZIMMERMAN's study (1989), suggested that the difference was the result of a variation in the time when the DST was done. Overall he stated that the results can be reconciled by saying baseline DST nonsuppression is associated with a better outcome from the index psychotic episode whereas persistent DST nonsuppression is associated with a poor outcome.

Finally, three studies evaluated the hypothalamic-pituitary thyroid axis and its relationship to treatment response by using the thyroid-releasing hormone (TRH) test. LANGER et al. (1986) and BEASLEY et al. (1988) found that a decreased thyroid-stimulating hormone (TSH) response at baseline was a predictor of a good response to haloperidol, whereas BAUMGARTNER et al. (1988) and SIRIS et al. (1991) did not find this to be true. Since the drug-free period was variable (in fact, in at least one study, it was nonexistent) and patient populations also varied (schizophrenic, schizoaffective, nonorganic psychotic patients) a closer look may be warranted.

E. Neuroimaging

I. Structural Brain Imaging

The finding of structural brain abnormalities in schizophrenia, particularly enlarged lateral ventricles, in a significant percentage of patients has been consistently replicated. This was first demonstrated with pneumoencephalograms, later with computed tomography (CT) scans, and more recently with magnetic resonance imaging (MRI). HUBER et al. (1975) initially reported that larger ventricles were associated with poor long-term outcome, while WEINBERGER et al. (1980) was the first to report that enlarged ventricles were associated with a poor treatment response. However, the association of brain morphology and treatment outcome has not been consistently replicated.

FRIEDMAN et al. (1992) performed a metaanalysis of 33 studies which pertained to the relationship of structural brain imaging to neuroleptic treatment response. The most widely studied parameter was ventricular brain ratio (VBR). The results of the studies were very inconsistent. Approximately half did show a relationship between enlarged VBR and poor treatment response while the other half did not. Studies which examined other brain structures, i.e., frontal and global atrophy, third ventricular size, and sulcal width reported no consistent findings in their relationship to treatment response. They concluded that, overall, the hypothesis of structural brain abnormalities as a predictor of antipsychotic response was not supported. However, considering the marked methodological differences and heterogeneity of the patient samples, it was not surprising that within the metaanalysis there were individual variables that were significant predictors of a lack of treatment response in schizophrenic patients with structural brain abnormalities. These variables were: younger age of onset of illness, longer illness duration, and the greater severity of the structural brain abnormality (>2 SD beyond control mean).

Since the above review there have been a number of additional studies which have continued to investigate this relationship (Table 10). VITA et al. (1991) assessed VBR and cortical atrophy by CT scan and examined outcome at 2 years. The only significant association for an increased VBR was poor outcome on an employment scale, while cortical atrophy was associated with numerous deficits in psychosocial functioning and an increased length of hospital stay. DELISI et al. (1992) and LIEBERMAN et al. (1993a) both studied first-episode schizophrenic patients with MRI scans and found that ventricular enlargement at baseline was associated with a poorer long-term outcome, while abnormalities of the temporal lobe were not related to treatment outcome. In contrast, WILMS et al. (1992), at 3 months, with CT scans done at

Table 10. Structural brain imaging

Author/ year	Sample	Drug-free ≥ (wks)	Methods	Rating Scale(s)	Imaging technique and areas assessed	Day of predictive assessment	Outcome day	Results
Friedman et al. 1991	34 (27♂, 7♀) RDC schizophrenic pts	–	Open clozapine tx	BPRS, responders ≥40 ↓; moderate responders ≥15–40% ↓; nonresponders <15% ↓	CT slices 8 mm thick: (1) prefrontal sulcal prominence (frontal portion of the slice through the foramen of Munro), (2) ventricles, (3) whole brain area	BL	6 wks	A linear trend occurred with ↑ prefrontal sulcal prominence indicating lack of response to clozapine. This was not true for ventricular brain ratio.
Vita et al. 1991	18 (11♂, 7♀) DSM III R chronic schizophrenic pts	–	–	Strauss-Carpenter scale	CT: (1) ventricle sizes, (2) cortical atrophy (determined by width of cortical sulci, and silvian and interhemispheric fissures)	BL	2 yrs	1. Ventricular enlargement was associated with poorer outcome on an employment scale. There were no other significant associations for VBR. 2. Cortical atrophy was associated with a poor outcome regarding useful employment, hospital stay, social contacts, total Strauss-Carpenter scale, and intimacy of personal contacts.

DeLisi et al. 1992	29 DSM III R first-episode psychotic pts (9 schizophreniform, 16 schizophrenic, 9 schizoaffective)	–	Prospective study	GAS, Strauss-Carpenter outcome subscale, BPRS	MRI: 26 coronal sections, 5 mm thick with a 2-mm separation: (1) temporal lobes, (2) lateral ventricles, (3) caudate and lentiform nuclei	BL	2 yrs	1. Ventricular enlargement at BL assessment was associated with ↑ number of hospitalizations and ↑ time in the hospital. 2. Temporal lobe volume did not correlate with any measures of outcome.
Wilms et al. 1992	42 (29♂, 13♀) DSM III R schizophrenic pts	7 days or less if acute psychotic sxs erupted	Double-blind parallel group design tx with either haloperidol or risperidone	PANSS	CT: 8–14 contiguous scans, 8 mm thick; (1) ventricular sizes, (2) whole brain	BL	12 wks	There was a trend towards greater improvement in total PANSS score in those pts with larger ventricles at BL (not related to type of antipsychotic tx).
Lieberman et al. 1993b	66 RDC schizophrenic pts (♂, ♀)	2 wks	Open standardized tx protocol	SADS C + PD, SANS, CGI; Response = ≤ 3 on SADS positive psychotic sxs; CGI ≤3; CGI improved score 1 or 2	MRI: 63 contiguous 3.1-mm slices in the coronal plane: (1) lateral ventricles, (2) third ventricle, (3) frontal/parietal cortex, (4) mesotemporal lobe structures	BL	Survival analysis	1. Pts who had definite or questionable brain pathomorphologic features took longer to recover than pts with normal brain morphology. This was specific to the lateral and third ventricles. 2. Abnormalities of the cortical and mesotemporal lobes did not predict time to remission.

pts, patients; BL, baseline; tx, treatment; BPRS, Brief Psychiatric Rating Scale (Overall and Gorman 1962); CGI, Clinical Global Impression Scale; ♂, male; ♀, female; DSM III R, Diagnostic and Statistical Manual of Mental Disorders [American Psychiatric Association 1980 (1987)]; RDC, Research Diagnostic Criteria (Spitzer et al. 1977); SADS C + PD, Schedule for Affective Disorders and Schizophrenia Change + Psychotic Disorganization (Spitzer et al. 1978); SANS, Scale for the Assessment of Negative Symptoms (Andreasen 1984); GAS, Global Assessment Scale (Endicott et al. 1976); PANSS, Positive and Negative Syndrome Scale for Schizophrenia; VBR, ventricular brain ratio; CT, computer tomography; MRI, magnetic resonance imaging.

baseline, found that larger ventricles were associated with a trend toward greater improvement. The authors explained this disparity from other studies by the observation that very few of their patients had relatively high VBRs (>2 SD beyond the control mean).

FRIEDMAN et al. (1991) examined the relationship of prefrontal sulcal prominence assessed by CT scan to the treatment response to clozapine. At 6 weeks there was a trend toward greater sulcal prominence being associated with a lack of response to clozapine. This was not true for VBRs. However, ARONOWITZ et al. (1994), in a study of clozapine response in chronic schizophrenia, found no association between interhemispheric tissure volume and treatment response. Additional studies of clozapine and other atypical antipsychotics are needed to assess whether similar or different treatment response predictors will emerge.

Taken in their entirety, studies of structural brain abnormalities being associated with poor treatment response have produced inconsistent findings; however, there is considerable evidence that some relationship does indeed exist, particularly for the lateral ventricles. With continually improving study methodology, coupled with more advanced technology, this relationship can be expected to become clearer in the near future. We will then be able to determine how much more predictive value this modality will be able to add to our clinical assessments.

II. Functional Brain Imaging

The use of functional brain imaging, in particular positron emission tomography (PET) scans, as a modality to predict antipsychotic treatment response has focused on two particular methodologies (Table 11): (1) measurement of brain dopamine receptor occupancy, particularly D_2 dopamine receptors; and (2) energy metabolism in specific brain regions.

The extent of D_2-dopamine receptor occupancy during antipsychotic treatment was definitively investigated by FARDE et al. (1988) and FARDE et al. (1992). They found that neuroleptics, irrespective of class, occupied D_2-dopamine receptors at rates of 65%–89%; however, the atypical antipsychotic clozapine had lower rates of D_2 occupancy: 38%–65%. Whether the extent of D_2-dopamine receptor occupancy by antipsychotic medication would predict treatment response was investigated by WOLKIN et al. (1989), MARTINOT et al. (1990), COPPENS et al. (1991), and NORDSTROM et al. (1993) with PET scans, and by PILOWSKY et al. (1993) with single photon emission computed tomography (SPECT). With the exception of NORDSTROM et al., each study concluded that there was no difference between antipsychotic treatment responders and nonresponders in the extent of D_2-dopamine receptor occupancy.

The studies of brain energy metabolism have investigated baseline predictors of antipsychotic treatment response, and how energy metabolism changes in accord with antipsychotic treatment. BUCHSBAUM et al. 1992a found that a low relative metabolic rate in basal ganglia (caudate nucleus and putamen) at baseline predicted a favorable clinical response to haloperidol, and that haloperidol "normalized," i.e., increased, metabolism in this region. BUCHSBAUM et al. (1992b), in a different study, found that patients with relatively lower metabolic rates in basal ganglia had a better outcome when treated with clozapine; however, relatively higher metabolic rates predicted treatment response to thiothixene. CLEGHORN et al. (1991) and CLEGHORN et al. (1992) found that neuroleptic treatment response was associated with an increase in striatal metabolism and a reduced frontal/parietal metabolic ratio.

The utility of functional brain imaging as a predictor of antipsychotic treatment response is as yet premature, though it appears that the percentage of D_2 receptor occupancy will not be a valid predictor. Other paradigms require further investigation.

F. Electrophysiologic Predictors

I. Galvanic Skin Conductance and Heart Rate

Differences in autonomic nervous system (ANS) functioning, measured primarily by galvanic skin conductance (GSC) and heart rate (HR), between a large subgroup of schizophrenic patients and healthy control subjects have repeatedly been demonstrated. These differences are believed to represent an altered level of arousal in schizophrenia. In studies that examined the altered ANS functioning of schizophrenic patients in relation to treatment outcome the results have been inconsistent (Table 12).

Five of the eight studies which examined treatment outcome with typical neuroleptics [(FRITH et al. (1979); ZAHN et al. (1981); STRAUBE et al. (1987); STRAUBE et al. (1989); and DAWSON et al. (1992)] found that a higher level of autonomic arousal at baseline was a predictor of poor treatment response. However, SCHNEIDER (1982), ALM et al. (1989), and LINDSTRÖM et al. (1992) found the opposite, with high levels of arousal at baseline indicative of a good treatment response. Of particular interest are the patient samples in these latter three studies. The sample investigated in SCHNEIDER's study consisted of elderly (55- to 67-year-old) chronic male schizophrenics, while ALM et al. and LINDSTROM et al. had a majority of neuroleptic-naive first-episode schizophrenics. In fact, ALM et al. (1989) reported no difference in outcome measures if neuroleptic-naive first-episode patients were excluded.

ZAHN and PICKAR (1993) measured GSC and HR in a standardized treatment protocol which included clozapine. They found that a good response to clozapine was predicted by a low tonic ANS response while on fluphenazine, and a low HR response to task instructions while on placebo. To our knowledge there are no other studies of clozapine where ANS measures were used as a predictor of antipsychotic treatment response.

The use of galvanic skin conductance and heart rate to predict antipsychotic treatment response is potentially promising; however the lack of

Table 11. Functional brain imaging

Author/ year	Sample	Drug-free ≥ (wks)	Methods	Rating Scale(s)	Modality and Ligand	Day of predictive assessment	Outcome day	Results
Farde et al. 1988	14 DSM III schizophrenic pts	–	Subjects had been taking 1 of 11 antipsychotics exclusively during the month preceding the PET scan; 15 drug-naive DSM III schizophrenic pts were used as a reference group for the determination of D_2 receptor occupancy	N/A	PET ^{11}C Raclopride	–	–	Each of the anti-psychotic drugs induced a 65%–85% occupancy of D_2 dopamine receptor in the putamen when binding was compared with the mean value in drug-naive schizophrenic pts.
Wolkin et al. 1989	10♂ DSM III/RDC schizophrenic pts	9 days	Open Haldol tx up to a maximum of 100 mg/d to obtain a minimum blood level of 10 ng/ml; responders and neonresponders based on past response to neuroleptics	BPRS	PET ^{18}F *N*-methyspiroperidol		4 wks	Responders and nonresponders had the same level of ligand binding, indicating that failure to respond clinically was not a function of neuroleptic uptake or binding in the CNS.
Martinot et al. 1990	15 (9♂, 6♀) DSM III schizophrenic pts	6 mos	Sample divided into 2 groups based on predominance of positive or negative sxs; negative sx group tx'd with amisulpride, positive sx group tx'd with a variety of neuroleptics; scans done both at BL & at 3 wks	SAPS, SANS Response = ↓ 25% for either scale	PET ^{76}Br Bromolisuride	BL	3 wks	There were no differences between responders & nonresponders on either BL measures of D_2 receptor density, or measures of D_2 receptor occupancy at 3 wks.

Cleghorn et al. 1991	10 (7♂, 3♀) DSM III schizophrenic pts	Neurolepic-native	Pts tx'd with either oral ($n = 5$) or depot neuroleptics ($n = 5$) for 2–3 yrs post BL assessment	–	PET ^{18}F Fluorodeoxyglucose	BL	2–3 yrs	↓ in psychotic sxs was associated with a reduced relative frontal metabolism & increased relative parietal metabolism.
Coppens et al. 1991	6 (2♂, 4♀) DSM III R schizophrenic pts	–	Open tx with moderate to high doses of neuroleptics	PSE, BPRS	PET ^{11}C *N*-methylspiperone		6 wks	A lack of therapeutic response could not be attributed to incomplete blockade of D_2 receptors.
Buchsbaum et al. 1992a	12 (11♂, 1♀) schizophrenic pts (classification system?)	14 days	7 pts treated with clozapine, 5 with thiothixene; PET scans obtained at BL & after 28 and 49 days of tx; pts underwent the Continuous Performance Test during ligand uptake	BPRS	PET ^{18}F fluorodeoxyglucose	BL	–	Pts treated with clozapine had greater improvement if they had lower metabolic rates at BL in caudate putamen & posterior putamen, while the reverse was true for pts tx'd with navane. The largest difference at BL was in the rt interior caudate.
Buchsbaum et al. 1992b	25 (21♂, 4♀) schizophrenic pts (classification system?)	5 wks	Pts entered into a 10 wk double blind crossover trial of haloperidol & placebo; pts underwent the Continuous Performance Test during ligand uptake; scans obtained at 5 & 10 wks of tx	BPRS Response = ↓ BPRS Score Nonresponse = ↑ BPRS Score	PET ^{18}F Fluorodeoxyglucose	BL	weeks 5 & 10	1. While not receiving medication, a low metabolic rate in the caudate nucleus & putamen in schizophrenic pts was found to predict a favorable clinical response to haloperidol. 2. Among responders, haloperidol tx had a "normalizing" effect on metabolic activity in the striatum 3. Nonresponders were more likely to

Table 11. *Continued*

Author/ year	Sample	Drug-free ≥ (wks)	Methods	Rating Scale(s)	Modality and Ligand	Day of predictive assessment	Outcome day	Results
								show a worsening of hypofrontality while they were receiving medication and an absence of change in the striatum.
Cleghorn et al. 1992	12 (8♂, 4♀) DSM III schizophrenic pts actively hallucinating & 10 (8♂, 2♀) with a hx of hallucinations; 19 of 22 were neuroleptic-naive	3 mos	Subjects underwent a PET scan at BL; 9 pts were rescanned after 1 yr & assessed after 2 yrs of neuroleptic tx	SADS-C SANS SAPS	PET Fluorodeoxyglucose 5 m<	BL, 1 yr	1–2 yrs	Neuroleptic tx resulted in a significant increase in striatal metabolism and a reduced frontal-parietal ratio, which was significantly correlated with a ↓ in hallucination scores.
Farde et al. 1992	28 (18♂, 10♀) DSM III schizophreniform or schizophrenic pts	–	– Subjects received at least 4 wks of antipsychotic monotherapy & had responded to tx; 28 subjects had PET scans to determine % of D_2 dopamine receptor occupancy; 11 subjects had D_1 dopamine receptor occupancy determined. – PET scans done 6 h after AM dose of orally administered medication, 1 wk after depot formulation ($n = 5$);	CGI, RSTD, DIA	PET D_2-^{11}C-Raclopride D_1-^{11}C-SCH 23390	–	–	1. D_2 dopamine receptor occupancy was 70%–87% in pts treatee with conventional neuroleptics, and 38%–63% in clozapine-treated pts. 2. Pts with acute EPS syndromes had higher D_2 dopamine receptor occupancy than those without. 3. Classical neuroleptics did not cause any evident D_1 occupancy while clozapine (4 pts), thiothixene,

			2 scans done for pts receiving clozapine ($n = 5$) at 3 & 6 h post medication administration. – For the calculation of D_2-dopamine receptor occupancy, 18 neuroleptic-naive (10♂, 8♀) schizophrenic pts were examined. – To calculate D_1 dopamine receptor occupancy, 10 (7♂, 3♀) healthy subjects were examined					& flupenthixol (1 pt) indicated a 36% –52% occupancy.
Nordstrom et al. 1993	13 (6♂, 7♀) DSM III R schizophrenic pts	4 days	Open tx with raclopride for 4 wks at 2, 6, or 12 mg	BPRS, CGI	PET ^{11}C Raclopride		wk 3 or 4	A statistically significant relationship was found between antipsychotic drug effect and the degree of D_2 receptor occupancy ($p < 0.05$).
Pilowsky et al. 1993	18 (12♂, 6♀) DSM III R	–	Open neuroleptic tx	BPRS – Responders ≥45 GAS	SPECT ^{123}I IBZM (D_2 specific)	–	6–8 wks of neuro-leptics	No difference found in striatal D_2 receptor availability between antipsychotic responders and nonresponders.

pts, patients; BL, baseline; tx, treatment; BPRS, Brief Psychiatric Rating Scale (Overall and Gorhan 1962); CGI, Clinical Global Impression Scale; ♂, male; ♀, female; DSM III R, Diagnostic and Statistical Manual of Mental Disorders [American Psychiatric Association 1980 (1987)]; RDC, Research Diagnostic Criteria (Spitzer et al. 1977); PSE, Present State Exam (Wing et al. 1974); CNS, Central nervous system; SAPS, Scale for the Assessment of Positive Symptoms; SANS, Scale for the Assessment of Negative Symptoms; PET, positron emission tomography; SPECT, single photon emission computed tomography; RSTD, Rating Scale for Tardive Dyskinesia (Simpson and Angus 1970); DIA, Scale for Drug-Induced Akathisia (Barnes et al. 1989); hx, history.

Table 12. Galvanic skin conduction and heart rate

Author/year test	Sample	Drug-free ≥ (wks)	Methods	Rating Scale(s)	Day of predictive assessment	Outcome day	Results
Frith et al. 1979 Galvanic skin conduction	41 PSE nuclear Schizophrenic pts	–	3 tx groups α-flupenthixol (active antipsychotic) β-flupenthixol (not active) placebo	Krawiecka rating scale	BL	21–28	Nonhabituation of skin conductance to a series of tones at BL predicted poor outcome.
Zahn et al. 1981 Galvanic skin conduction, heart rate, skin temperature	35 DSM II schizophrenic pts	3–4 wks	Emphasized psychosocial tx; 44% of improved & 53% of not improved pts received open antipsychotic tx	38-point scales utilized: A-psychopathology B-global illness C-schizophrenic sxs	BL	d/c from hospital	High resting arousal, slow habituation, and attenuated ANS reactivity at BL were found in pts who remained clinically ill, but not in pts whose recovery was more complete, especially in ♂s.
Schneider 1982 Galvanic skin conduction	26 (24♂, 2♀) elderly RDC chronic schizophrenic pts	4 days	Open tx with either thiothixene or thioridazine	BPRS, CGI Responders had ↓ BPRS scores while nonresponders had ↑ BPRS scores	BL	at least 60 days	Neuroleptic nonresponders at BL displayed faster habituation, ↓ amplitude skin conduction response, ↓ spontaneous fluctuations, and ↓ skin conduction levels than responders.
Straube et al. 1987 Galvanic skin conduction, heart rate	19 (9♂, 10♀) ICD schizophrenic pts	–	28 days of open tx with perazine	BPRS, AMDP	BL	28 days	The low-improvement group had higher cardiovascular response to orienting stimuli, a higher heart rate during habituation, and a lower level of skin conduction under demand.

Alm et al. 1989 Galvanic skin conduction	35 ♂ DSM III schizophrenic pts, 1 schizoaffective ♂ pt, 1 schizophrenic ♂ pt	–	Open tx; outcome assessed by pt interview & confirmed by registry check & family interview	CPRS Straus-Carpenter Outcome scale	BL	1 & 2 yrs post electrophysiology testing	Pts with a non-skin conduction orienting response (SCOR) at BL had a significantly worse outcome. However, this effect was entirely due to neuroleptic-naive first-episode pts. For previously tx'd pts, poor outcomes were equally common irrespective of SCOR.
Straube et al. 1989 Galvanic skin conduction, heart rate	45 schizophrenic pts, Carpenter Flexible checklist	–	Standard haloperidol for 4 wks, then open tx	GAS	BL	4 wks (GAS); length of hosp; relapse 9 mos after admission, & 2 yrs post d/c	Only significant outcome criterion was length of hospitalization with ↑ ANS activity found at BL in schizophrenic pts that were hospitalized longer. 1. ↑ # of SCR until habituation & ↑ # of SF 2. ↑ height of the SC amplitude & degree of heart rate deceleration to the 1st orienting tone.
Lindstrom et al. 1992 Galvanic skin conduction	21 DSM III ♂ schizophrenic pts	14 days	Open tx with neuroleptics; tx response rated on a 5-point scale; responders either had significant improvement or were sx-free	CPRS	BL	21–180 days	Nonresponders had ↓ SCR and ↓ skin fluctuations at BL than responders. This was much more pronounced for neuroleptic-naive pts.
Dawson et al. 1992 Galvanic skin conduction	69 (56♂, 13♀) RDC schizophrenic pts	0	Tx'd openly with neuroleptics, then switched in the hospital to fluphenazine decanoate 12.5 mg q 2 wks	BPRS	BL	3 month post d/c	The most electrodermally active inpatients tended to be the most symptomatic outpatients approximately 3 months following hospital d/c (trials to habituation & frequency of nonspecific skin conductance response).

Table 12. *Continued*

Author /year test	Sample	Drug-free ≥ (wks)	Methods	Rating Scale(s)	Day of predictive assessment	Outcome day	Results
Zahn et al. 1993 Galvanic skin conduction, heart rate	20 (?) DSM III R schizophrenic pts	–	Standardized tx in the following order: fluphenazine, placebo, clozapine; antipsychotics given for 6 wks	BPRS, BH	3 wks	6 wks	1. On placebo, a low-HR response to task instructions was predictive of a good response to clozapine.
						11 wks	2. ↓ response to task instruction when on fluphenazine was predictive of a good response to clozapine (SCR).

pts, patients; BL, baseline; tx, treatment; BPRS, Brief Psychiatric Rating scale (Overall and Gorhan 1962); CGI, Clinical Global Impression Scale; ♂, male; ♀, female; DSM III R, Diagnostic and Statistical Manual of Mental Disorders [American Psychiatric Association 1980 (1987)]; RDC, Research Diagnostic Criteria (Spitzer et al. 1977); VBR, ventricular brain ratio; CPRS, Comprehensive Psychopathological Rating Scale; BH, Bunney-Hamburg scale (Bunney and Hamburg 1963); CNS, central nervous system; AMDP, Scale for Assessment of Psychopathologic Symptoms set up by Arbeitsgemeinschaft für Diagnostik und Methodologie in der Psychiatrie (Scharfetter 1972); ANS, autonomic nervous system; PSE, Present State Exam (Wing et al. 1974); ICD, International Classification of Diseases; HR, heart rate; SCR, skin conduction response; SF, spontaneous fluctuations; SC, skin conduction; d/c, discharge; GAS, Global Assessment Scale.

standardization in research methodology makes it impossible to draw any definitive conclusions. Further investigations with homogeneous patient samples evaluated with uniform electrophysiologic and outcome measures at specific time intervals, with standardized treatment protocols, may help to disentangle some of the conflicting data.

II. Electroencephalography (EEG)

An overview of studies on the use of electroencephalography (EEG) in the prediction of response to treatment with antipsychotic drugs is provided in Table 13. Computerized EEGs were first reported to be a useful predictor of antipsychotic treatment response by ITIL et al. (1975). This was followed by Itil et al. (1981), SMALL et al. (1987), ULRICH et al. (1988), and CZOBOR and VOLAVKA (1991). In the 4 of 5 studies that used typical neuroleptics, an increased baseline alpha was predictive of a poor treatment response. SMALL et al. (1987), in contrast, found that treatment responders had higher amplitudes in the alpha band. However, this study used clozapine, an atypical neuroleptic.

CZOBOR and VOLAVKA (1992 and 1993), in an alternative design, examined whether EEG changes would be predictive of an antipsychotic treatment response. One study used haloperidol, the other risperidone. Both studies reported positive findings, though the results differed between studies. Whether their results are indicative of the various psychopharmacological effects of typical and atypical neuroleptics requires further replication. GALDERISI et al. (1994), in a different type of study design, examined if a test dose of neuroleptic medication could discriminate treatment responders from nonresponders. They found that an increase in slow alpha activity at 6h post test dose was a strong predictor of antipsychotic treatment response.

The interpretation of computerized EEG results as predictors of antipsychotic treatment response requires due caution since a very large number of variables are generated with each recording. Though some of the above studies had specific a priori hypotheses, others did not. Therefore, the use of EEG as a modality to predict antipsychotic treatment response is at best premature at this time.

III. Event-Related Potential (ERP)

A decrease in the amplitude of the P3 component of brain event-related potential (ERP) in schizophrenic patients has been well replicated. The data available for ERP as a predictor of treatment response is limited; however, FORD et al. (1994), BLACKWOOD et al. (1987), and DUNCAN et al. (1987) have investigated the change in P3 ERP as a potential predictor of antipsychotic medication response. FORD et al. (1994) and BLACKWOOD et al. (1987) did not find any change in P3 as it related to treatment response, while DUNCAN et al. (1987) did find a change in visual but not auditory P3 amplitude. DUNCAN et

Table 13. Electroencephalography (EEG)

Author/year	Sample	Drug-free ≥ (wks)	Methods	Rating scale(s)	Day of predictive assessment	Outcome day	Results
Itil et al. 1975	62 (38♂, 24♀) DSM I schizophrenic pts	6–8 wks	Double-blind, placebo-controlled crossover study with 3 antipsychotics: thiothixene, fluphenazine, & haloperidol	BPRS, CGI	BL	3 mos	Schizophrenic pts with more high-frequency fast activity at BL, & a lesser degree of α & slow waves, had a better therapeutic response to neuroleptics.
Itil et al. 1981	13 ♂, therapy-resistant, DSM II schizophrenic pts	14 days (pts had been on depot neuroleptics	4 test doses of medication given in a random order in a double-blind crossover design (20 mg thiothixene, 10 mg fluphenazine, 5 mg haloperidol, and placebo)	BPRS, CGI	BL & 3 h	–	Computerized EEG recordings of tx resistant pts were characterized by a large amount of α activity & ↓ fast activity at BL.
Small et al. 1987	13 chronic schizophrenic pts with failed tx trials of chlorpromazine and haloperidol	7 days	Open tx with clozapine	BPRS Favorable response = ↓20% CGI	–	–	Responders had higher amplitudes in the α band in a number of brain regions.
Ulrich et al. 1988	34 (20♂, 14♀) RDC schizophrenic pts	3 days	Open tx with perazine	BPRS Response = ↓66%	BL	2 h, 24 h, 28 days	• Independent of outcome day, responders showed a tendency toward more low voltage desynchronized non-α epochs than nonresponders. • Responders had ↑ decreases in both absolute power and coefficients of variation in some of the α leads over the first 48 h.

Czobor and Volavka 1991	34 DSM III/RDC schizophrenic or schizoaffective pts	3 days	Double-blind crossover study of haloperidol for 6 wks at different plasma blood levels	BPRS	BL	3 & 6 wks	↑ BL α was associated with poorer response to tx. This was specific to anterior & temporal brain regions.
Czobor and Volavka 1992	37 DSM III RDC acutely exacerbated schizophrenic pts		Haloperidol in double-blind crossover design (blood level = 2–35 ng/ml	BPRS Responders ≥50% ↓	BL & 6 wks	6 wks	1. There was a significant relationship between HAL plasma levels & EEG θ activity for the tx.-responders, whereas no relationship was detected for the nonresponders. 2. There were EEG changes (in the δ and α bands) that depended on clinical response but did not show any relationship in either responders or nonresponders to HAL plasma levels.
Czobor and Volavka 1993	9♂, DSM III R schizophrenic pts	3 days	Placebo-controlled, double-blind study of risperidone for 9 weeks (2, 6, or 16 mg of risperidone)	PANSS, CGI	BL & 8 wks	8 wks	Overall clinical improvement on risperidone related to 2 EEG measures. Absolute power changes in β frequency band and changes in power asymmetry in θ and δ bands. More prominent in anterior brain regions. This study found that the clinical response to risperidone was related to EEG Δs.
Galderisi et al. 1994	29 (17♂, 12♀) DSM III R schizophrenic pts	15 days	A baseline EEG was obtained prior to a single dose of either haloperidol 3 mg or clopenthixol 5 mg. EEGs were then obtained at 3, 6, and 8 h after this test dose. Twenty pts were next treated with haloperidol, the other 9 with clopenthixol.	SANS, SAPS, CPRS, Responders = 50% reduction of the total score of SADS and SANS	BL	28 days	An increase in slow alpha activity 6 h after the administration of a test dose of a neuroleptic medication was able to discriminate between responders and nonresponders at the end of 4 wks of treatment.

pts, patients; BL, baseline; tx, treatment; BPRS, Brief Psychiatric Rating scale (Overall and Gorhan 1962); CGI, Clinical Global Impression Scale; ♂, male; ♀, female; DSM II DSM II R, Diagnostic and statistical Manual of Mental Disorders [American Psychiatric Association 1980 (1987)]; RDC, Research Diagnostic Criteria (Spitzer et al. 1977); PANSS, Positive and Negative Syndrome Scale for Schizophrenia.

al.'s (1987) sample consisted of only 7 patients, while the other two studies consisted of a combined total of 39 patients.

Overall, the existing data, which is quite limited, does not support a role for ERP as a predictor of antipsychotic response, with P3 appearing to be a trait and not a state marker of illness (Table 14).

IV. Saccadic Eye Movements

MACKERT and FLECHTNER (1989) examined saccadic reaction times in schizophrenic patients who received open neuroleptic treatment. They found that prolonged reaction times related to poor treatment outcome (Table 14).

G. Discussion

In the last two decades there have been extensive research efforts to understand the pathophysiology of schizophrenia, and we have witnessed a rapid growth in our knowledge of the neurobiology of this devastating disease. Recent technological advances have allowed investigators to study various aspects of brain structure and function as potential predictors of antipsychotic treatment response. In addition, there are now atypical antipsychotic drugs available which exert their influences at sites in the brain that are different from those influenced by the conventional neuroleptics. Despite these advances and the numerous studies and avenues of investigation, no valid biological predictors of antipsychotic treatment responsiveness have been firmly established. Though predictors such as baseline pHVA, structural brain imaging, and galvanic skin conductance all seem to have some relationship with treatment response, the nature and magnitude of this correlation is obscured due to marked differences in methodologic designs across studies, and weaknesses in the design of individual studies. The sophisticated technologies now available are extremely enticing as potential objective biological predictors, but without thoughtful study designs guiding this technology, it is unlikely that we will be able to discern their usefulness as predictors of treatment response.

To underscore the importance of high-quality research methodology, it is informative to review some of the many constructs involved in the syndrome of schizophrenia. Irrespective of whether the heterogeneity in symptoms, course, and outcome of schizophrenic patients reflects a variety of etiologies, or possibly a greater genetic liability, similar to what was found in Huntington's disease, it is doubtful that the various symptoms are all mediated by the same pathophysiologic mechanisms. Schizophrenia can affect every aspect of human functioning, although the quality of symptoms often varies markedly both between individuals and within a single individual over the course of the illness. It is therefore unlikely that a biological predictor of both short-term treatment response and long-term outcome which assesses different symptom realms (i.e., perceptual disturbances, cognitive impairment, social limitations, or motor dysfunction) will produce reliable results.

Author/ year	Sample	Drug-free ≥ (wks)	Methods	Rating scale(s)	Tests	Day of predictive assessment	Outcome day	Results
Event-Related Potentials								
Ford et al. 1994	21 DSM III R ♂ schizophrenic pts	10 days	To enter treatment phase, pts must have a BPRS score >36 with mod. severity on 2 psychotic items after placebo phase. *Protocol* • Antipsychotic meds withdrawal • Placebo 1 wk • Double-blind administration of raclopride (2, 6, or 12 mg/d) or haloperidol 15 mg/d	BPRS	Auditory & visual oddball paradigm	Post withdrawal from antipsychotic meds & post placebo	4 wks of meds tx	P3 amplitude and latency was not significantly different for the auditory or visual paradigm during placebo or antipsychotic medication irrespective of treatment.
Blackwood et al. 1987	18 RDC schizophrenic pts	3 months	14 pts had ERPs performed at BL and 1 & 4 wks after tx. 13 pts had ERPs at BL & again between 6 & 24 months (9 pts common to both). Pts treated with a variety of neuroleptics.	BPRS	2-tone discrimination task	BL	4 wks 6 months	P3 amplitude & latency did not significantly differ after 4 wks or long-term tx.
Duncan et al. 1987	7 (6♂, 1♀) RDC + DSM III schizophrenic pts	4 wks	Pts were tested twice: once at BL & after being "stabilized" on meds (haloperidol or thithixene) for at least 6.5 wks.	BPRS	Auditory and visual choice reaction time tasks	BL	6.5 wks	Clinical improvement was associated with an increase in visual but not auditory P3 amplitude.
Eye Movements								
Mackert et al. 1989	47 (21♂, 26♀) ICD 9 schizophrenic pts	3 days	Open neuroleptic tx	BPRS, Prognostic Scale	Saccadic Rxn Times	BL	–	Prolonged reaction times related to poor tx outcome particularly in pts with a long course of illness.

pts, patients; BL, baseline; tx, treatment; BPRS, Brief Psychiatric Rating Scale (Overall and Gorman, 1962); CGI, Clinical Global Impression Scale; ♂, male; ♀, female; RDC, Research Diagnostic Criteria (Spitzer et al. 1977); GAS, Global Assessment Scale; ICD, International Classification of Diseases; Prognostic Scale (Strauss and Carpenter 1977).

The design of long-term outcome studies need also take into account that the underlying pathophysiology of the disease may change over time. It is clear that medication may affect neurochemical measures as well as outcome assessments; however, less is known about the natural course of the illness. Schizophrenia is generally accepted as a neurodevelopmental disorder, but it may also have a neurodegenerative component. There is clinical evidence that the length of time an individual is psychotic and each successive relapse both result in an increased time to remission (Loebel et al. 1992). Whether this is indicative of a neurodegenerative process or possibly of a diminished CNS plasticity is not known, but it does highlight the need to assess patients at comparable stages of their illness.

Biological markers may be used to predict treatment response in two ways. First, they may distinguish discrete subtypes of patients who have different forms or levels of severity of illness that are associated with different illness courses and outcomes. These might be called trait markers and are exemplified by measures of eye movement dysfunction and lateral ventricular enlargement. The second type would reflect or be associated with a neurobiologic substrate that mediates treatment response, i.e., pHVA, CSF, NE, or behavioral response to psychostimulant administration.

H. Conclusions

In conclusion, extensive review of the literature and of relevant studies reveals no clearly reproducible or potentially useful biological predictors of short-term treatment response or long-term outcome. Biochemical, neuroimaging, and electrophysiological predictors have been examined, and although a few measures have shown some consistent findings, the nature and magnitude even of these remain obscure.

In an era of dwindling research and clinical funding, the attention to both greater specificity in sample selection (i.e., first-episode, second-episode, chronic) and high-quality methodology is absolutely essential if we are to ascertain biological predictors of treatment that can be used to reduce the morbidity of this debilitating illness. Continuing on the same path that has led to a mass of conflicting data seems ill advised, no matter how sophisticated the available technology becomes.

Acknowledgments. This study was supported by NIMH grant MH-41646, NIMH Research Scientist Development Award MH-00537 (J. Lieberman), and NIMH grant MH-41960 from the Mental Health Clinical Research Center for the Study of Schizophrenia.

References

Alfredsson G, Wiesel F (1990) Relationships between clinical effects and monoamine metabolites and amino acids in sulpiride-treated schizophrenic patients. Psychopharmacology 101:324–331

Alm T, Wieselgren IM, Öst LG, Lindström LH (1989) Electrodermal nonresponding, premorbid adjustment, and symptomatology as predictors of long term social functioning in schizophrenics. J Abnorm Psychol 98(4):426–435
American Psychiatric Association (1980) DSM-III: Diagnostic and statistical manual of mental disorders, 3rd edn. Washington DC: The Association
American Psychiatric Association. (1987) DSM-III-R: Diagnostic and statistical manual of mental disorders, 3rd edn., revised The Association, Washington DC:
Amin F, Davison M, Davis KL (1991) Homovanillic acid measurement in clinical research: A review of methodology. Schizophr Bull 18:123–148
Andreasen NC (1984), The Scale for the Assessment of Negative Symptoms (SANS). University of Iowa, Iowa City
Angrist B, Rotrosen J, Gershon S (1980) Responses to apomorphine, amphetamine, and neuroleptics in schizophrenic subjects. Psychopharmacology 67:31–88
Aronowitz JS, Wu H, Pollack S, Bilder R, Kane J, Safferman A, Lieberman JA (1994) Clozapine response in chronic schizophrenia and brain morphology on MRI: a preliminary report. Abstract presented at the American Psychiatric Association Annual Meeting, May, 1994
Astrachan BM, Harrow M, Adler D, Brauer L, Schwartz A, Schwartz C, Tucker G (1972) A checklist for the diagnosis of schizophrenia. Br J Psychiatry 121:529–539
Awad AG, Lapierre YD, Jostell KG, the Canadian Remoxipride Study Group (1991) Selective dopamine D_2 antagonist and prolactin response in acute schizophrenia – results from remoxipride studies. Prog Neuropsychopharmacol Biol Psychiatry 14:769–777
Bacopoulus NG, Hattox SE, Roth RH (1979) 3,4 Dihydroxyphenylacetic acid and homovanillic acid in rat plasma: possible indicators of central dopaminergic activity. Eur J Pharmacol 56:225–236
Baker NJ, Kirch DG, Waldo M, Bell J, Adler LE, Hattox S, Murphy R, Freedman R (1991) Plasma homovanillic acid and prognosis in schizophrenia. Biol Psychiatry 29:192–196
Bartkó G, Frecska E, Horváth S, Zádor G, Arató M (1990) Predicting neuroleptic response from a combination of multilevel variables in acute schizophrenic patients. Acta Psychiatr Scand 82:408–412
Barnes TRE (1989) A rating scale for drug-induced akathisia. Br J Psychiatry 154:672–676
Baumgartner A, Gräf KJ, Kürten I, Meinhold H (1988) The hypothalamic-pituitary-thyroid axis in psychiatric patients and healthy subjects: Parts 1–4 Psychiatry Res 24:271–332
Beasley CM Jr, Magnusson M, Garver DL (1988) TSH response to TRH and haloperidol response latency in psychoses. Biol Psychiatry 24:423–431
Bjerkenstedt L, Gullberg B, Hornryd C, Sedvell G (1977) Monoamine metabolite levels in cerebrospinal fluid of psychotic women with nelperone or thiothixene. Arch Psychiatr Nerverkr 224:107–118
Blackwood DHR, Whalley LJ, Christie JE, Blackburn, IM, St Clair DM, McInnes A (1987) Changes in auditory P3 event-related potential in schizophrenia and depression. Br J Psychiatry 150:154–160
Bowers MB Jr, (1974) Central dopamine turnover in schizophrenic syndromes. Arch Gen Psychiatry 31:50–54
Bowers MB Jr, Swigar ME, Jatlow PI, Hoffman FJ, Goicoechea N (1987) Early neuroleptic response: clinical profiles and plasma catecholamine metabolites. J Clin Psychopharmacol 7(2):83–86
Bowers MB Jr (1991) Characteristics of psychotic inpatients with high or low HVA levels at admission. Am J Psychiatry 148(2):240–243
Bowers MB, Swigar ME, Jatlow PI, Hoffman FJ (1984) Plasma catecholamine metabolites and early response to haloperidol. J Clin Psychiatry 45:248–251
Bowers MB, Swigar ME, Jatlow PI, Hoffman FJ (1989) Plasma catecholamine metabolites and treatment response at neuroleptic steady state. Biol Psychiatry 25:734–738

Breier A, Buchanan RW, Weltrip RW, Listwak S, Holmes C, Goldstein DS (1994) The effect of clozapine on plasma norepinephrine: relationship to clinical efficacy. Neuropsychopharmacology 10:1–7

Brown G, Cleghorn J, Kaplan R, Szechtman H, Brown P, Szechtman B, Mitton J (1988) Longitudinal growth hormone studies in schizophrenia. Psychiatry Res 24:123–136

Buchsbaum MS, Potkin SG, Siegel BU, Lohr J, Katz M, Gottschalk LA, Gulasekasam B, Marshall JF, Lottenberg S, Teng CY, Abel L, Plon L, Bunney WE (1992a) Striatal metabolic rate and clinical response to neuroleptics in schizophrenia. Arch Gen Psychiatry 49:966–974

Buchsbaum MS, Potkin SG, Marshall JF, Lottenberg S, Teng C, Heh CW, Tafalla R, Reynolds C, Abel L, Plon L, Bunney WE (1992b) Effects of clozapine and thiothixene on glucose metabolic rate in schizophrenia. Neuropsychopharmacology 6(3):155–163

Bunney WE Jr, Hamburg DS (1963) Methods for reliable longitudinal observation of behavior. Arch Gen Psychiatry 9:280–294

Carlsson A (1978) Antipsychotic drugs, neurotransmitters, and schizophrenia. Am J Psychiatry 135:164–173

Chang W, Chen T, Lin S, Lung F, Lin W, Hu W, Yeh E (1987) Plasma catecholamine metabolites in schizophrenics Evidence for the two-subtype concept. Biol Psychiatry 44:189–190a

Chang W, Chen T, Lee C, Hung J, Hu W, Yeh E (1988) Plasma homovanillic acid levels and subtyping of schizophrenia. Psychiatry Res 23:239–244

Chang WH, Chen TY, Lin SK, Lung FW, Lin WL, Hu WH, Yeh EK (1990) Plasma catecholamine metabolites in schizophrenics: evidence for the two-subtype concept. Biol Psychiatry 27:510–518

Chang WH, Hwu HG, Chen TY, Lin SK, Lung FW, Chen H, Lin WL, Hu WH, Lin HN, Chien CP (1993) Plasma homovanillic acid and treatment response in a large group of schizophrenic patients. Schizophr Res 10:259–265

Cleghorn JM, Franco S, Szechtman B, Kaplan RD, Szechtman H, Brown GM, Nahmias C, Garnett ES (1992) Toward a brain map of auditory hallucinations. Am J Psychiatry 149(8):1062–1069

Cleghorn JM, Szechtman H, Garnett ES, Brown GM, Nahmias C, Szechtman B, Kaplan R, Franco, S (1991) Neuroleptic effects on regional brain metabolism in first-episode schizophrenics. Schizophr Res 5:208–209

Contretas A, Maas J, Seleshi E, Bowden C (1988) Urine and plasma levels of dopamine metabilites in response to apomorphine and neuroleptics in schizophrenics. Biol Psychiatry 24:815–818

Coppens HJ, Slooff CF, Paans AMJ, Wiegman T, Vaalburg W, Korf J (1991) High central D_2-dopamine receptor occupancy as assessed with positron emission tomography in medicated but therapy-resistant schizophrenic patients. Biol Psychiatry 29:629–634

Coryell WH, Zimmerman M (1989) HPA axis hyperactivity and recovery from functional psychoses. Am J Psychiatry 146:473–477

Creese I, Burt DR, Snyder SH (1975) Dopamine receptor binding predicts clinical and pharmacological potencies of antischizophrenic drugs. Science, 192:482–483

Czobor P, Volavka J (1991) Pretreatment EEG predicts short-term response to haloperidol treatment. Biol Psychiatry, 30:927–942

Czobor P, Volavka J (1992) Level of haloperidol in plasma is related to electroencephalographic findings in patients who improve. Psychiatry Res 42:129–144

Czobor P, Volavka J (1993) Quantitative electroencephalogram examination of effects of risperidone in schizophrenic patients. J Clin Psychopharmacol 13(5):332–342

Davidson M, Losonczy M, Mohs R, Lester J, Powchik P, Fried L, Davis B, Mykytyn M, Davis K (1988) Effects of debrisoquin and haloperidol on plasma homovanillic acid concentrations in schizophrenic patients. Neuropsychopharmacology 1:17–23

Davidson M, Kahn RS, Powchik P, Warne P, Losonczy MF, Kaminsky R, Aptu S, Jaff S, Davis KL (1991a) Changes in plasma homovanillic acid concentrations in schizophrenic patients following neuroleptic discontinuation. Arch Gen Psychiatry 48:73b–76b

Davidson M, Kahn R, Knott P, Kaminsky R, Cooper M, DuMont K, Aptu S, Davis K (1991b) Effects of neuroleptic treatment on symptoms of schizophrenia and plasma homovanillic acid concentrations. Arch Gen Psychiatry 48:910a–913a

Davidson M, Kahn RS, Stern RG, Hirschowitz J, Apter S, Knott P, Davis KL (1993) Treatment with clozapine and its effect on plasma homovanillic acid and norepinephrine concentrations in schizophrenia. Psychiatry Res 46:151–163

Davila R, Zumarraga M, Andia I, Manero E, Retuerto F, Barcena B, Guimon, J (1989) Dopaminergic balance and subtypes of schizophrenia. Br J Psychiatry 154[Suppl]:57–60

Davila R, Manero E, Zumarraga M, Andia I, Schweitzer J, Friedhoff A (1988) Plasma homovanillic acid as a predictor of response to neuroleptics. Arch Gen Psychiatry 5:564–567

Davila R (1992) Plasma HVA and other predictors of neuroleptic response. Clin Neuropharmacol 15(1):519a–520a

Davis KL, Davison M, Mohs RC, Kendler KS, Davis BM, Johns C, DeNigris Y, Horvath, TB (1985) Plasma homovanillic acid concentration and the severity of schizophrenic illness. Science 29:1601–1602

Davis KL, Kann RS, Ko G, Davidson M (1991) Dopamine in schizophrenia: a review and reconceptualization. Am J Psychiatry 148:1474–1486

Dawson ME, Nuechterlein KH, Schell AM (1992) Electrodermal anomalies in recent onset schizophrenia: relationships to symptoms and prognosis. Schizophr Bull 18(2):295–311

Delisi LE, Stritzke P, Riordon H, Holan U, Boccia A, Kushner M, McClelland J, Van Eyl O, Anand A (1992) The timing of brain morphological changes in schizophrenia and their relationship to clinical outcome. Biol Psychiatry 31:241–254

Doran A, Rubinow DR, Wolkowitz O, Roy A, Breier A, Pickar D (1989) Fluphenazine treatment reduces CSF somatostatin in patients with schizophrenia: correlation with CSF HVA. Biol Psychiatry 25:431–439

Duncan CC, Morihisa JM, Fawcett RW, Kirch DG (1987) P300 in schizophrenia: state or trait marker? Psychopharmacol Bull 23:497–501

Endicott J, Spitzer RL, Fleiss JL, Cohen J (1976) The global assessment scale-A procedure for measuring overall severity of psychiatric disturbance. Arch Gen Psychiatry 33:766–771

Farde L, Wiesel FA, Halldin C, Sedvall G (1988) Central D_2-dopamine receptor occupancy in schizophrenic patients treated with antipsychotic drugs. Arch Gen Psychiatry 45:71–76

Farde L, Nordstrom AL, Wiesel FA, Pauli S, Halldin C, Sedvall G (1992) Positron emission tomographic analysis of central D_1 and D_2 dopamine receptor occupancy in patients treated with classical neuroleptics and clozapine. Arch Gen Psychiatry 49:538–544

Ford JM, White PM, Csernansky JG, Faustman WO, Roth WT Pfefferbaum A (1994) ERPs in schizophrenia: effects of antipsychotic medication. Biol Psychiatry 36:153–170

Friedman L, Knutson L, Shurell M, Meltzer HY (1991) Prefrontal sulcal prominence is inversely, related to response to clozapine in schizophrenia. Biol Psychiatry 29:865–877

Friedman L, Lys C, Schulz SC (1992) The relationship of structural brain imaging parameters to antipsychotic treatment response: a review. J Psychiatry Neurosci 17(2):42–54

Frith CD, Stevens M, Johnstone EC, Crow TJ (1979) Skin conductance responsivity during acute episodes of schizophrenia as a predictor of symptomatic improvement. Psychol Med 9:101–106

Galderisi S, Maj M, Mucci A, Bucci P, Kemali D (1994) QEEG alpha1 changes after a single dose of high potency neuroleptics as a predictor of short-term response to treatment in schizophrenic patients. Biol Psychiatry 35:367–374
Garver DL (1988) Neuroendocrine findings in the schizophrenias. Neurol Clin 6:103–109
Garver DL, Zernlan F, Hirshowitz J, Hitzeman R, Mauroidis M (1984) Dopamine and non-dopamine psychosis. Psychopharmacology 84:138–140
Garver DL, Beinfeld MC, Yao JK (1990) Cholecystokinin, dopamine, and schizophrenia. Psychopharmacol Bull 26:377–380
Garver DL, Bisette G, Yao JK, Nemeroff CB (1991) Relation of CSF neurotensin concentrations to symptoms and drug response of psychotic patients. Am J Psychiatry 148:484–488
Glazer WM, Bowers MB, Charney DS, Heninger GR (1989) The effects of neuroleptic discontinuation on psychopathology, involuntary movements, and biochemical measures in patients with persistent tardive dyskinesia. Biol Psychiatry 26:224–233
Green AI, Alam MY, Sobieraj JT, Pappalardo KM, Waternaux C, Salzman C, Schatzberg AF, Schildkraut JJ (1993) Clozapine response and plasma catecholamines and their metabolites. Psychiatry Res 46:139–149
Gusella JF, McDonald ME (1994) Huntington's disease and repeating trinucleotides. N Engl J Med 330:1450–1451
Harris PQ, Brown, SJ, Friedman J, Bacopoulus NG (1984) Plasma drug and homovanillic acid levels in psychotic patients receiving neuroleptics. Biol Psychiatry 19:849–860
Hirschowitz J, Hitzemann R, Burr G, Schwartz A (1991) A new approach to dose reduction in chronic schizophrenia. Neuropsychopharmacology 5:103–113
Honigfeld G, Klett CJ (1965) The Nurse's Observation Scale for inpatient evaluation. J Clin Psychol 21:65–71
Hornykiewicz O (1982) Brain catecholamines in schizophrenia: A good case for noradrenaline. Nature 299:484–486
Hsiao JK, Colison J, Bartko JJ, Doran AR, Konicki PR, Potter WZ, Pickar D (1993) Monoamine neurotransmitter interactions in drug-free and neuroleptic-treated schizophrenics. Arch Gen Psychiatry 50:606–614
Huber G, Gross G, Schuttler R (1975) A long term followup study of schizophrenia: psychiatric course of illness and prognosis. Acta Psychiatr Scand 52(1):49–57
Itil TM, Marasa J, Saletu B, Davis S, Mucciardi AN (1975) Computerized EEG – predictor of outcome in schizophrenia. J Nerv Ment Dis 160(3):188–203
Itil TM, Shapiro D, Schneider SJ, Francis IB (1981) Computerized EEG as a predictor of drug response in treatment resistant schizophrenics. J Nerv Ment Dis 269(10):629–637
Jacobssson L, von Knorring L, Mattisson B, Perris C, Edenius B, Kettner B, Magnusson KE, Villemoes P (1978) The Comprehensive Rating Scale (CPRS) in patients with schizophrenic syndromes. Inter-rater reliability and in relation to Martens' S-Scale. Acta Psychiatr Scand 271:34–44
Javaid JI, Sharma RP, Janicak PG, Davis JM (1990) Plasma HVA in psychiatric patients: longitudinal studies Psychopharmacol Bull 26(3):361–365
Kahn RS, Amin F, Powchik P, Knott P, Goldstein M, Apter S, Kerman B, Jaff S, Davison M (1990) Increments in plasma homovanillic acid concentrations after neuroleptic discontinuation are associated with worsening of schizophrenic symptoms. Prog Neuropsychopharmacology Biol Psychiatry 14:8790–884
Kahn RS, Davidson M, Knott P, Stern RG, Apter S, Davis KL (1993) Effect of neuroleptic medication on cerebrospinal fluid monoamine metabolite concentrations in schizophrenia. Arch Gen Psychiatry 50:599–605
Kaminsky R, Powchik P, Warne P, Goldstein M, McQueeney R, Davidson M (1990) Measurement of homovanillic acid concentrations in schizophrenic patients. Prog Neuropsychopharmacol Biol Psychiatry 14:271–287
Kaschow J, Nemeroff C (1991) The neurobiology of neurotensin: focus on neurotension-dopamine interactors. Regul Pept 36:153–164

Kendler KS, Heninger GR, Roth RH (1982) Influence of dopamine agonists on plasma and brain levels of homovanillic acid. Life Sci 30:2063–2069
Kendler KS, Davis KL (1984) Acute and chronic effects of neuroleptic drugs on plasma and brain homovanillic acid in the rat. Psychiatry Res 13:51–58
Kirsch DG, Jaskiw G, Linnoila M, Weinberger DR, Wyatt RJ (1988) Plasma amino metabolites before and after withdrawal from neuroleptic treatment in chronic schizophrenic inpatients. Psychiatry Res 25:233–242
Koreen AR, Lieberman J, Alvir J, Mayerhoff D, Loebel A, Chakos M, Amin F, Cooper T (1994) Plasma homovanillic acid in first-episode schizophrenia: psychopathology and treatment response. Arch Gen Psychiatry 51:132–138
Korpin IJ, Bankiewicz KS, Harvey-White J (1988) Assessment of brain dopamine metabolism from plasma HVA and MHPG during debrisoquin treatment: validation in monkeys treated with MPTP. Neuropsychopharmacology 2:119–125
Labarca R, Silva H, Jerez S, Ruiz A, Forray MI, Gysling K, Andres ME, Bustos G, Castillo Y, Hono J (1993) Differential effects of haloperidol on negative symptoms in drug-naive schizophrenic patients: effects on plasma homovanillic acid. Schizophr Res 9:29–34
Lahdelma RL, Appelberg B, Kuoppasalmi K, Katila H, Rimon R (1991) Plasma concentrations of remoxipride and haloperidol in relation to prolactin and short-term therapeutic outcome in schizophrenic patients. Eur Neuropsychopharmacol 1:535–540
Lambert GW, Eisenhofer G, Cox HS, Horne M, Kaliff V, Kelly M, Jennings GL, Esker MD (1991) Direct determination of homovanillic acid release from the human brain: an indicator of central dopaminergic activity. Life Sci 39:1061–1072
Langer G, Koinig G, Hatzinger R, Schönbeck G, Resch F, Aschauer H, Keshavan MS, Sieghart W (1986) Response of thyrotropin to thyrotropin-releasing hormone as predictor of treatment outcome. Arch Gen Psychiatry 43:861–868
Lieberman JA, Koreen AR (1993) Neurochemistry and neuroendocrinology of schizophrenia: a Selective Review. Schizophr Bull 19(2):371–429
Lieberman JA, Kane JM, Sarantakos S, Gadaleta D, Woerner M, Alvir J, Ramos-Lorenzi J (1987) Prediction of relapse in schizophrenia. Arch Gen Pschiatry 44:597–603
Lieberman JA, Sobel SN (1993) Predictors of treatment response and course of schizophrenia. Cur Opin Psychiatry 6:63–69
Lieberman JA, Kane JM, Woerner M, Alvir JMJ, Borenstein M, Novacenko H (1990) Prediction of relapse in schizophrenia. Clin Neuropharmacol 13:434–435
Lieberman JA, Alvir JMJ, Woerner M, Degreef G, Bilder RM, Ashtari M, Bogerts B, Mayerhoff DL, Geisler SH, Loebel A, Levy D, Hinrichsen G, Szymanski SR, Chakos M, Koreen A, Borenstein M, Kane JM (1992) Prospective study of psychobiology in first-episode schizophrenia at Hillside Hospital. Schizophr Bull 18:351–371
Lieberman JA, Jody D, Geisler S, Alvir J, Loebel A, Szymanski S, Woerner M, Borenstein M (1993a) Time course and biologic correlates of treatment response in first-episode schizophrenia. Arch Gen Psychiatry 50:369–375
Lieberman JA, Jody D, Alvir JMJ, Ashtari M, Levy D Bogerts B, Degreef G, Mayerhoff D, Cooper T (1993b) Brain morphology, dopamine and eye tracking abnormalities in first episode schizophrenia. Prevalence and clinical correlates. Arch Gen Psychiatry 50:357–368
Lieberman JA, Brown AS, Gorman JM (1994) In: Oldham R (ed) American psychiatric press review of psychiatry, vol 13. American Psychiatric Press. Washington DC, pp 133–170
Lindstrom EM, Öhlund LS, Lindstrom LH, Öhman A (1992) Symptomatology and electrodermal activity as predictors of neuroleptic response in young male schizophrenic inpatients. Psychiatry Res 42:145–158
Little KY, Gay TL, Vore M (1989) Predictors of response to high dose antipsychotics in chronic schizophrenics. Psychiatry Res 39:1–9

Loebel A, Lieberman J, Alvir J, Mayerhoff D, Geisler S, Szymanski S (1992) Duration of psychosis and outcome in first episode schizophrenia. Am J Psychiatry 149: 1183–1188
Maas J, Contretas S, Seleshi E, Bowden C (1988) Dopamine metabolism and disposition in schizophrenic patients. Arch Gen Psychiatry 45:553–559
Mackert A, Flechtner M (1989) Saccadic reaction times in acute and remitted schizophrenics. Eur Arch Psychiatry Neuro Sci 239:33–38
Markianos M, Sakellariou G, Bistolaki E (1991) Prolactin responses to haloperidol in drug-free and treated schizophrenic patients. J Neural Transm 83:37–42
Martinot JL, Paillere-Martinot ML, Loc'h C, Peron-Magnan P, Mazoyer B, Lecrubier Y, Hardy P, Beaufils B, Allilaire JF, Maziere B, Slama MF, Syrota A (1990) Central D_2 receptor blockade and antipsychotic effects of neuroleptics. Preliminary study with positron emission tomography. Psychiatry Psychobiol 5:231–240
Matthysse S (1973) Antipsychotic drug actions: a clue to a neuropathology of schizophrenia? Fed Proc 32:200–205
Mazure CM, Nelson JC, Jatlow PI, Bowers MB (1991) Plasma free homovanillic acid (HVA) as a predictor of clinical response in acute psychosis. Biol Psychiatry 30:475–482
Meltzer HY (1988) New insights into schizophrenia through atypical antipsychotic drugs. Neuropsychopharmacology 1:193–196
Meltzer HY (1991) The mechanism of action of novel antipsychotic drugs. Schizophr Bull 1:263–287
Meltzer HY, Busch D, Fang VS (1981) Hormones, dopamine receptors and schizophrenia. Psychoneuroendocrinology 6:17–36
Nemeroff CB, Bisette G, Widerlov E, Beckman H, Gerner R, Manberg PJ, Lindstrom L, Prange AJ, Gattaz WF (1989) Neurotensin-like immunoreactivity in cerebrospinal fluid of patients with schizophrenia, depression, anorexia nervosa-bulimia, and premenstrual syndrome. J Neuropsychiatry 1:16–20
Nordstrom AL, Farde L, Weisel FA, Forslund K, Pauli S, Halldin C, Uppfeldt G (1993) Central D_2-dopamine receptor occupancy in relation to antipsychotic drug effects: a double blind PET study of schizophrenic patients. Biol Psychiatry 33:227–235
Overall JE, Gorhan DR (1962) The brief psychiatric rating scale. Psychol Rep 10:799–912
Peters J, Van Kammen D, Gelernter J, Yao J, Shaw D (1990) Neuropeptide Y-like immunoreactivity in schizophrenia relationships with clinical measures. Schizophr Res 3:287–294
Petrie EC, Faustman WO, Moses JA, Lombrozo L, Csernansky JG (!990) Correlates of rapid neuroleptic response in male patients with schizophrenia. Psychiatry Res 33:171–177
Pickar D (1992) Plasma homovanillic acid and other predictors of clozapine response. Clin Neuropharmacol 15(1):516a–517a
Pickar D, Labarca R, Linnoila M, Roy A, Hommer D, Everett D, Paul S (1984) Neuroleptic-induced decrease in plasma homovanillic acid and antipsychotic activity in schizophrenic patients. Science 25:954–957
Pickar D, Doran AR, Wolkowitz OM, Labarca R, Breier A, Paul SM (1985) Studies of plasma homovanillic acid in schizophrenia: effects of neuroleptic treatment and evidence for a diurnal rhythm. New Tr Exp Clin Psychiatry 1:187–200
Pickar D, Labarca R, Doran AR, Wolkowitz OM, Roy A, Breier A, Linnoila M, Paul S (1986) Longitudinal measurement of plasma homovanillic acid levels in schizophrenic patients. Arch Gen Psychiatry 43:669–676
Pickar D, Brier A, Kelsoe J (1988) Plasma homovanillic acid as an index of central dopaminergic activity: studies in schizophrenia patients. Ann NY Acad Sci 537:339–346
Pickar D, Owen RR, Litman RE, Konicki E, Gutierrez R, Rapaport MH (1992) Clinical and biologic response to clozapine in patients with schizophrenia. Arch Gen Psychiatry 49:345–353
Pilowsky LS, Costa DC, Ell PJ, Murray RM, Verhoeff NPLG, Kerwin RW (1993) Antipsychotic medication, D_2 dopamine receptor blockade and clinical response:

a ^{123}I IBZM SPET (single photon emission tomography) study. Psychol Med 23:791–797
Randrup A, Munkvad I (1965) Special antagonism of amphetamine-induced abnormal behavior. Psychopharmacology 7:416–422
Reichlin S (1981) Neuroendocrinology. In: Williams RH (ed) Textbook of endocrinology, 6th edn. Williams and Wilkins, Baltimore, pp 589–645
Rivier J, Spiess J, Thorner M (1982) Characterization of a growth hormone releasing factor from a pancreatic islet tumor. Nature 300:1321–1328
Schalling M, Friberg K, Bird E, Goldstein M, Schiffman S, Mailleux P, Vanderhaeghen JJ (1989) Presence of cholecystokinin mRNA in dopamine cells in the ventral mesencephalon of a human with schizophrenia. Acta Physiol Scand 137:467–468
Scharfetter C (ed) (1972) Das AMP-System. Manual Zur Dokumentation Psychiatrisclear Befunde, 2nd edn. Springer, Berlin Heidelberg New York
Schneider SJ (1982) Electrodermal activity and therapeutic response to neuroleptic treatment in chronic schizophrenic inpatients. Psychol Med 12:607–613
Sedvall G (1992) The current status of PET scanning with respect to schizophrenia. Neuropsychopharmacol 7(1):41–54
Seeman P, Wong M, Lee T (1974) Dopamine receptor-block and nigral fiber impulse blockade by major tranquilizers. Fed Proc 33:246
Sharma R, Javaid JI, Janicak R, Faull K, Comaty J, Davis JM (1989) Plasma and CSF HVA before and after pharmacological treatment. Psychiatry Res 28:97–104
Simpson GM, Angus JWS (1970) A rating scale for extrapyramidal side effects. Acta Psychiatr Scand [Suppl 212]:11–19
Siris SG, Frechen K, Swahan A, Cutler J, Owen K, Alvir JMJ, McCorry T (1991) Thyroid-releasing-hormone test in schizophrenic patients with post-psychotic depression. Prog Neuropsychopharmacol Biol Psychiatry 15:369–378
Small JG, Milstein V, Small IF, Miller MJ, Kellams JJ, Corsaro CJ (1987) Computerized EEG profile of haloperidol, chlorpromazine, clozapine and placebo in treatment resistant schizophrenia. Clin Electroencephalogr 18(3):124–135
Snyder S, Banergee S, Yamamura H (1974) Drugs, neuro-transmitters and schizophrenia. Science 184:1243–1253
Spitzer RL, Endicott J, Robins E (1977) Research diagnostic critieria (RDC) for a selected group of functional disorders, 3rd edn. Biometrics Research Unit, New York State Psychiatric Institute
Spitzer RL, Endicott J, Robins E (1978) The schedule for affective disorders and schizophrenia (SADS). NY, NY Biometrics Research Unit, New York State Psychiatric Institute
Stein L, Wise CD (1971) Possible etiology of schizophrnia: progressive damage to the noradrenergic reward system by 5-hydroxydopamine. Science 171:1032–1036
Stern RG, Kahn RS, Davidson M (1993) Predictors of response to neuroleptic treatment in schizophrenia. Schizophrenia 16(2):313–338
Sternberg DE, Van Kammen DP, Lake CR, Ballenger JC, Marder SR, Bunney WE Jr (1981) Am J Psychiatry 138(8):1045–1051
Sternberg DE, Van Kammen DP, Lerner P, Bunney WE (1982) Schizophrenia dopamine β-hydroxylase activity and treatment response. Science 216:1423–1425
Straube ER, Schied HW, Rein W, Breyer-Pfaff U (1987) Autonomic nervous system differences as predictors of short-term outcome in schizophrenics. Pharmacopsychiatry 20:105–110
Straube ER, Wagner W, Foerster K, Heimann H (1989) Findings significant with respect to short and medium term outcome in schizophrenia – a preliminary report. Prog Neuropsychopharmacol Biol Psychiatry 13:185–197
Strauss JS, Carpenter WR Jr (1977) The prediction of outcome in schizophrenia. III. Five year outcome and its predictors. Arch Gen Psychiatry 34:159–163
Szymanski S, Masiar S, Mayerhoff D, Loebel A, Geisler S, Gonzalez S, Pollack S, Kane J, Lieberman J (1994) Clozapine response in treatment-refractory first episode schizophrenia. Biol Psychiatry 35(4):278–280
Tandon R, Mazzara C, Dequardo JR (1989) The DST and outcome in schizophrenia. Am J Psychiatry 146(12):1648–1649

Tandon R, Mazzara C, DeQuardo J, Craig KA, Meador-Woodruff JH, Goldman R, Greden JF (1991) Dexamethasone suppression test in schizophrenia: relationship to symptomatology, ventricular enlargement, and outcome. Biol Psychiatry 29:-953–964

Ulrich G, Gaebel A, Pietzcker A, Müller-Oerlinghausen B, Stieglitz RD (1988) Prediction of neuroleptic on-drug response in schizophrenic inpatients by EEG. Eur Arch Psychiatry Neurol Sci 237:144–155

Van Kammen D (1991) The biochemical basis of relapse and drug response in schizophrenia: review and hypothesis. Pschol Med 21:881–895

Van Kammen DP, Schooler N (1990) Are biochemical markers for treatment-resistant schizophrenia state dependent or traits? Clin Neuropharmacol 13(1):16–28

Van Kammen DP, Docherty JP, Marder SR, Schulz SC, Dalton L, Bunney WE Jr (1982) Antipsychotic effects of pimozide in schizophrenia. Treatment response prediction with acute dextroamphetamine response. Arch Gen Psychiatry 39(3):261–266

Van Kammen DP, Agren H, Yao JK, O'Connor DT, Gurklis J, Peters JL (1994) Noradrenergic activity and prediction of psychotic relapse following haloperidol withdrawal in schizophrenia. Am J Psychiatry 151:379–384

Van Praag HM, Korf J (1975) Neuroleptics, catecholamines, and psychoses: a study of their interrelations. Am J Psychiatry 132:6, 593–597

Van Putten T, Marder SR, Mintz J Serum prolactin as a correlate of clinical response to haloperidol. J Clin Psychopharmacol 11:357–361

Van Putten T, Marder SR, Aravagiri M, Chabert N, Mintz J (1989) Plasma homovanillic acid as a predictor of response to fluphenazine treatment. Psychopharmacol Bull 1:89–91

Vita A, Dieci M, Giobbio GM, Azzone P, Garbarini M, Sacchetti E, Cesana BM, Cazzullo CL (1991) CT scan abnormalities and outcome of chronic schizophrenia. Am J Psychiatry 148(11):1577–1579

Volovka J, Douyon R, Convit A, Czobor P, Cooper TB (1992) Neuroleptic treatment, symptoms of schizophrenia, and plasma homovanillic acid concentrations revisited. Arch Gen Psychiatry 49:999–1000

Wang R (1989) Cholecystokinin, dopamine and schizophrenia: recent progress and current problems. Acad Sci 362–379

Wass JAH (1983) Growth Hormone neuroregulation and the clinical relevance of somatostatin. Clin Endocrin Metab 12:695–724

Weinberger DR, Bigelow LB, Kleinman JE, Klein ST, Rosenblatt JE, Wyatt RJ (1980) Cerebral ventricular enlargement in chronic schizophrenia. An association with poor response to treatment. Arch Gen Psychiatry 37:11–13

Wilms G, Van Ongeval C, Baert AL, Claus A, Bollen J, De Cuyper H, Eneman M, Malfroid M, Peuskens J, Heylen S (1992) Ventricular enlargement, clinical correlates and treatment outcome in chronic schizophrenic inpatients. Acta Psychiatr Scand 85:306–312

Wing JK, Cooper JE, Sartorius N (1974) The measurement of psychiatric symptomatology. Cambridge University Press, Cambridge

Wolkin A, Borouche F, Wolf AP, Rotrosen J, Fowler JS, Shire CY, Cooper TB, Brodic JD (1989) Dopamine blockade and clinical response: evidence for two biological subgroups of schizophrenia. Am J Psychiatry 146(7):905–908

Zahn TP, Carpenter WT, McGlashan TH (1981) Autonomic nervous system activity in acute schizophrenia. II Relationships to short-term prognosis and clinical state. Arch Gen Psychiatry 38:260–266

Zahn TP, Pickar D (1993) Autonomic effects of clozapine in schizophrenia – comparison with placebo and fluphenazine. Biol Psychiatry 34:3–12

CHAPTER 15

Effects of Antipsychotic Drugs on Neuropsychological Measures

WILLIAM O. FAUSTMAN and ANNE L. HOFF

A. Introduction

Numerous studies (see LEVIN et al. 1989b, for a review) have demonstrated that schizophrenia is often associated with deficits on a range of cognitive abilities including, but not limited to, attentional, memory, and executive cognitive functions. These deficits may have a profound impact on a schizophrenic individual – for some individuals they may, in fact, be more disabling than hallucinations or delusions. The cognitive impairments related to schizophrenia are present at the onset of the disorder (HOFF et al. 1992a; SAYKIN et al. 1994) and may vary in severity independently of psychotic features of the illness such as hallucinations and delusions (FAUSTMAN et al. 1988). Cognitive deficits in schizophrenia relate to the quality of life (e.g., interpersonal relationships, vocational functioning) in individuals with the disorder (WAGMAN et al. 1987). The exact reasons for cognitive dysfunction in schizophrenia are likely to be complex and may include both core structural pathology and neurochemical dysregulation.

Antipsychotic medications are effective in reducing the symptoms that are used to define the diagnosis of schizophrenia (e.g., hallucinations). If cognitive impairments are a related feature of schizophrenia, medications that treat the core clinical symptoms may also be able to reduce cognitive deficits in schizophrenia. If antipsychotic medications fail to alter cognitive impairments in schizophrenia, one could argue that cognitive impairments are a separate and defining trait of the disorder. One could further be concerned that schizophrenic patients who have chronic baseline cognitive deficits would be susceptible to cognition-impairing properties that might be present in treatment medications. In sum, the study of the effects of antipsychotic medications on neuropsychological performance in schizophrenia has important theoretical and clinical implications.

This chapter provides a review of the effects of antipsychotic medications on cognitive functioning in schizophrenia. Common methodological problems found in this literature are also examined. Recent studies evaluating the effects of clozapine will be closely examined since such atypical antipsychotic medications have both superior efficacy and diminished side effects (e.g., bradykinesia, other extrapyramidal symptoms) that could confound neuropsychological test performance. A brief review of the nature of cognitive deficits in schizophrenia will also be provided. In addition, a review of the effects of anticholinergic drugs on cognition is offered since antipsychotics frequently possess anticholinergic properties or are used in conjunction with anticholinergic agents.

B. Biological Underpinnings of Cognitive Deficit in Schizophrenia

The past 20 years have noted a major shift in our understanding of the nature of schizophrenia. This change has significantly influenced the interpretation of observed cognitive impairments in the disorder. Schizophrenia was previously viewed as being "functional" in nature, and a frequent clinical problem was to distinguish schizophrenia from disorders of an "organic" nature. As noted by HEATON and CROWLEY (1981), the 1970s saw an emergence of neurochemical explanations such as the dopamine theory. In the early 1980s, extensive evidence for structural brain changes was found and fundamentally changed the interpretation of cognitive impairments in schizophrenia. Table 1 provides an outline of selected research findings that support a biological basis for cognitive impairments in schizophrenia.

Our current understanding of the etiology of schizophrenia remains limited, though a variety of types of evidence suggests it is a neurodevelopmental disorder with both genetic and environmental factors contributing to its expression. Strong evidence points towards a significant role for genetic mechanisms (see TSUANG et al. 1991, for a recent review), though the patterns of inheritance remain poorly understood. Penetrance of the disorder is fairly limited even in families with a high rate of the illness. Recent work (WEINBERGER et al. 1992) examining monozygotic twins who were discordant for the disorder provides evidence for brain insult and pathology as factors in

Table 1. Examples of findings supportive of a biological basis for cognitive deficits in schizophrenia

Research area	Selected findings
Neuroimaging (MRI, CT)	Increased cerebral ventricular fluid volume Increased sulcal fluid volume Cerebral gray matter deficits
Postmortem neuropathology	Hippocampal pyramidal cell abnormalities Volume reductions and cytoarchitectonic malformations of the entorhinal cortex
Etiological investigations	Obstetric complications at birth Risk of second-trimester viral exposure
Studies of monozygotic twins discordant for schizophrenia	Elevated cognitive impairments and neuroimaging pathology in the schizophrenic twin

the expression of the disorder. Affected twins were found to have elevated rates of structural pathology and cognitive deficits (GOLDBERG et al. 1990; WEINBERGER et al. 1992).

Genetic factors may either act solely or interact with early brain insults to alter the expression of the disorder. Prenatal viral exposure has been shown to be a risk factor for the development of schizophrenia in some (MEDNICK et al. 1988; TAKEI et al. 1994), but not all investigations (CROW and DONE 1992). Insults to the developing fetus during the second trimester are of interest since this is a period of brain cortical development and organization. Recent neuropathology studies have noted evidence for both structural and cellular abnormalities in subcortical regions such as the hippocampus and entorhinal cortex (BOGERTS et al. 1985; CONRAD et al. 1991). These findings are of interest in light of widely replicated observations of attentional and memory impairments in schizophrenia, functions that are thought be linked to hippocampal integrity (ROBERTS 1990).

Numerous neuroimaging studies using computed tomography (CT) and magnetic resonance imaging (MRI) have demonstrated morphological differences between schizophrenic samples and nonpsychiatric control samples (BIGLER 1987). Some studies have demonstrated evidence for lateral ventricular enlargement at the onset of the disorder (WEINBERGER et al. 1982; SCHULTZ et al. 1983), although others find few differences between first-episode patients and controls on these measures (DELISI et al. 1992). Further work suggests that the magnitude of differences between schizophrenic individuals and controls is fairly constant among individuals of different ages, suggesting that differences that are present at the onset of the disease do not progress at an accelerated rate with aging as is typical of various forms of progressive dementia (PFEFFERBAUM et al. 1988). More recent MRI work has suggested evidence for diffuse gray matter deficits (ZIPURSKY et al. 1992).

Numerous studies (reviewed by SEIDMAN 1983; ZEC and WEINBERGER 1986) have noted relationships between structural measures (e.g., lateral ventricle volume) and performance on neuropsychological measures, though it is interesting to note that these relationships can be subtle and elusive since some works have been unable to find such a relationship (OBIOLS et al. 1987). It is of interest to note that very few studies have provided correlation estimates between structural measures and cognitive performance in nonpsychiatric control samples. Accordingly, the uniqueness of relationships between neuroimaging-derived structural measures and cognitive functioning in schizophrenia is unclear.

C. Cognitive Impairment in Schizophrenia

Though neuroimaging and neuropathological studies have yielded evidence for neuropathology in schizophrenia, these studies offer no information on the behavioral expression of the illness. Traditional neuropsychological approaches can offer valuable information on the dysfunction of brain systems in individuals with known cortical lesions or degenerative processes. One hope for the use of neuropsychological testing in schizophrenic samples is the elucidation of the brain pathology that may underlie the disorder. However, there is strong evidence for impairments across a broad range of cognitive functions in schizophrenia. As noted previously, the neuropathology of schizophrenia may be related to early brain insults and abnormalities in neurodevelopment. Thus, there may be limited utility in cognitive testing approaches based on models of fixed cortical lesions in adults.

Nevertheless, there have been numerous studies addressing many facets of cognitive dysfunction in schizophrenia since KRAEPLIN (1913) first characterized schizophrenia as a "disorder of attention." Broadly speaking, there have been two basic approaches to the study of these impairments, studies which have arisen from the experimental psychology literature and those which have developed from the field of clinical neuropsychology. In general, the experimental psychology literature has been chiefly concerned with reducing cognitive processes into measurable components. The information processing model of BROADBENT (1959) was critical in the stimulation of studies of attentional processes in schizophrenia. Early reaction time studies by SHAKOW (1962) and development of the Continuous Performance Test (KORNETSKY and MIRSKY 1966) demonstrated that schizophrenic patients performed more poorly on tasks of elementary information processing which are likely to involve the sensory register or short-term store (ATKINSON and SHIFFRIN 1968). Theories which have been advanced to explain these deficits have suggested that schizophrenics have a reduction in processing capacity, abnormal allocation of attention, or poor regulation of arousal (KAHNEMAN 1973). For a review of the attentional literature, please see NUECHTERLEIN and DAWSON (1984).

Unlike the experimental psychology approach, clinical neuropsychology has developed tests that are less reductionistic and more approximate to those cognitive tasks executed in everyday life. The original focus of clinical neuropsychological studies was on the differentiation of cognitive performance in schizophrenic and other psychotic disorders from that in brain-damaged patients (reviewed by HEATON et al. 1978). As it became established that chronic schizophrenic cognitive performances were often indistinguishable from those of the brain-injured, this focus shifted to the establishment of neuropsychological performance patterns as criteria for distinguishing between diagnostic categories (i.e., schizophrenia versus affective disorders.) Schizophrenic patients were thought to be more impaired on purported measures of left-hemisphere function, whereas affectively disordered patients appeared more impaired on right-hemisphere measure (TAYLOR and ABRAMS 1984). Even though schizophrenics were noted to have a loss of abstract attitude, which is frequently seen in patients with frontal lobe injury (GOLDSTEIN 1959; STUSS and BENSON 1984), the focus on prefrontal cortical dysfunction in schizophrenia became more pronounced after WEINBERGER et al. (1986) published their findings on hypofrontality in cerebral metabolism during the performance of the Wisconsin Card Sorting Test (WCST). The WCST measures the ability to abstract a problem-solving strategy in correctly matching cards on the basis of color, shape, and number. They interpreted their findings as consistent with impairment of the dorsolateral prefrontal cortex as a core cognitive deficit in the schizophrenic syndrome.

Studies of memory function, both from the experimental and clinical neuropsychological literature, have provided evidence for impairment in schizophrenic patients. Some studies have found that schizophrenic patients are impaired in recall but not recognition (KOH 1978); however, chronically ill patients may be impaired in both (CALEV 1984). Defective organizational strategies and/or attentional impairment at the time of encoding may account for reduced memory on both verbal and spatial tasks (CALEV et al. 1987; GOLD et al. 1992a). Whereas schizophrenic individuals are thought to perform normally on tasks which are automatic (e.g., procedural learning, semantic priming, implicit learning), GOLD et al. (1992b) found deficits in both effortful and automatic processes. As is true of the cognitive literature in general, findings may vary as a function of the clinical features of the cohort studied.

Individuals with schizophrenia also demonstrate reduction in speech syntactic complexity (MORICE and INGRAM 1983) and slowing in motor speed (ROSOFSKY et al. 1982), although long-term administration of neuroleptics is also thought to cause motoric slowing (CASSENS et al. 1990). The literature is inconsistent with regard to visuospatial impairment, with some investigators finding no evidence of impairment on tasks sensitive to parietal or posterior cortical functions (KOLB and WHISHAW 1983), while others find visuoconstructional deficits in subgroups of patients (CRAFT et al. 1988). From review of the literature, it appears that schizophrenic patients are impaired on a wide variety of cognitive functions, but the important question becomes whether or not there are areas of preferential deficit signifying greater dysfunction in one region of the brain than in the other.

Using a summary scale approach, SAYKIN et al. (1991) found evidence of selective learning and memory deficits in the context of generalized impairment in unmedicated inpatient and outpatient schizophrenics. Using a similar approach, HOFF et al. (1992a, 1992b) did not find evidence for selective impairment of cognitive functions in samples of first-episode patients and patients undergoing long-term medication. BRAFF et al. (1991) likewise concluded that impairments were diffuse and far-reaching in their sample of outpatient schizophrenics. Unfortunately, the answer to this question in obscured by the methodological difficulties in matching tasks on the basis of difficulty and reliability, which is necessary in order to make accurate conclusions (CHAPMAN and CHAPMAN 1973) regarding preferential impairment. In addition, results from studies vary as a function of the characteristics of the patient sample and the particular tests administered.

Other issues in this area of research are whether cognitive deficits are static or progressive and whether they are state- or trait-related. Most longitudinal studies indicate that schizophrenic patients improve on neurocognitive measures over time, but that there may be a subgroup of patients who deteriorate (reviewed by HEATON and DREXLER 1987). In controlled studies of first-episode patients (HOFF et al. 1992a; DELISI et al. 1995), patients have improved from baseline in most areas of cognition after 2 years of illness; however, there is some mild evidence of a possible decline in verbal memory after 4 years of illness. The literature is mixed with regard to the effects of severity of illness on cognition, with some studies concluding no relationship (FAUSTMAN et al. 1988), while others report that negative symptoms are associated with more impairment (ANDREASEN and OLSEN 1982; BRAFF et al. 1991).

D. Anticholinergics and Cognition

A variety of both standard and atypical antipsychotic compounds possess anticholinergic activity. In general, lower-potency antipsychotics, such as thioridazine or chlorpromazine, have relatively greater anticholinergic effects. Medications with higher potency (e.g., haloperidol, fluphenazine) have less anticholinergic activity. However, high-potency compounds are often administered with varying doses of anticholinergics (e.g., benztropine mesylate) to minimize extrapyramidal motor side effects. Accordingly, a large percentage of patients receiving antipsychotic treatment are exposed to significant central anticholinergic activity.

The effects of anticholinergics on memory functions have been widely studied in animal models, normal human controls, and psychiatric/neurologic populations. Animal studies have produced findings that parallel the human studies implicating memory impairments with anticholinergics. For example, SALA et al. (1991) found that increasing doses of anticholinergics produced deficits in radial arm maze performance in rats consistent with impairments in working memory.

Several recent works have noted that anticholinergics impair memory functions in healthy individuals. One study (NAKRA et al. 1992) noted that a single 2-mg dose of trihexyphenidyl could impair some measures of memory function in healthy elderly individuals. FLICKER et al. (1990) found that scopolamine could produce impairments on a range of memory functions and psychomotor speed in healthy young individuals. Older individuals may show relatively greater impairments with anticholinergics than younger persons (MOLCHAN et al. 1992; ZEMISHLANY and THORNE 1991).

The effects of anticholinergics on schizophrenic patients have received attention in multiple studies over the past 15 years. An important work (TUNE et al. 1982) examined correlates of working or recent memory in 24 stable schizophrenic outpatients. Patients were taking a range of antipsychotic compounds at the time of testing and the majority was also taking anticholinergic drugs such as benztropine. The severity of clinical symptoms and estimated IQ levels were unrelated to memory performance. However, a significant inverse correlation was found between serum levels of anticholinergics and memory test performance. Another study (HITRI et al. 1987) prospectively administered benztropine, trihexyphenidyl, or amantadine to groups of neuroleptic-treated schizophrenic patients. Memory tests (e.g., list learning) were administered at various phases of drug administration. The poorest recall of words was observed when patients showed elevated serum anticholinergic activity during treatment with benztropine or trihexyphenidyl. The authors noted that memory performance in schizophrenia may be impaired from the outset and that anticholinergics may exacerbate deficits. Further work (PERLICK et al. 1986) found that anticholinergic levels in 17 schizophrenic patients were inversely correlated with verbal recall performance (e.g., items recalled in a learning list) but not with recognition memory.

One study (FAYEN et al. 1988) performed a double-blind, crossover comparison of the effects of amantadine and trihexyphenidyl on memory functions as assessed by the Rey-Auditory Verbal Learning Test. Though the sample size of schizophrenic patients was somewhat small ($n = 9$), the results suggested that patients showed poorer performance while receiving trihexyphenidyl compared to amantadine.

STRAUSS et al. (1990) examined the effects of varying serum anticholinergic levels on both verbal recall and reaction time in schizophrenia. Higher anticholinergic levels were associated with poorer performance in verbal memory functions while reaction time was somewhat improved at higher anticholinergic levels. Other work (SWEENEY et al. 1991) noted inverse relationships between anticholinergic dose levels and several cognitive domains (e.g., verbal learning, motor speed) in stable schizophrenic patients. EITAN et al. (1992) compared the effects on memory measures of four typical neuroleptics (chlopromazine, thioridazine, trifluoperazine, and haloperidol) that were administered to schizophrenic patients. The agents (chlorpromazine, thioridazine) with the greatest anticholinergic action produced impairments in short-term verbal memory.

Recently introduced antidepressant drugs (e.g., fluoxetine) have significantly less anticholinergic action than traditional tricyclic compounds. These new medications may be of significance in minimizing cognitive impairments related to medications. For example, RICHARDSON et al. (1994) compared verbal learning performance in depressed patients blindly treated with amitriptyline or fluoxetine. Though both groups showed similar clinical improvement during treatment, the amitriptyline patients showed higher serum anticholinergic activity and poorer performance on the verbal learning test.

Additional evidence for memory deficits induced by anticholinergic agents has been offered in the treatment literature on Parkinson's disease (PD). Though patients with PD may have illness-related cognitive impairments, anticholinergic agents induce even further impairments in these patients (VAN SPAENDONCK et al. 1993). One recent work (VAN HERWAARDEN et al. 1993) noted that patients with PD show better performance on verbal learning measures following the discontinuation of anticholinergic medications. A comparison between dopaminergic and anticholinergic agents in the treatment of PD suggested that anticholinergics produce impairments in the immediate registration of information while dopaminergic agents produce improvements in measures of working memory and cognitive sequencing (COOPER et al. 1992). These findings are consistent with prior work (SYNDULKO et al. 1981) noting that patients with PD who were receiving benztropine mesylate had memory decrements even though they were receiving relatively low-dose therapy. Other work (VAN SPAENDONCK et al. 1993) demonstrated that patients with PD who were receiving anticholinergic therapy had impaired set shifting on measures such as the Wisconsin Card Sort.

In sum, all of the available evidence suggests that central anticholinergic activity can produce impairments in memory functions. Schizophrenic individuals may have a particular sensitivity to these detrimental effects since memory impairments may be a characteristic of the illness itself. The deficits produced by anticholinergics serve as a confounder in the interpretation of the literature on the effects of antipsychotics on cognition. A review of the antipsychotic drug literature reveals that a large percentage of studies, especially those from early years, employed low-potency phenothiazines such as chlorpromazine as the test agent. Novel new antipsychotic agents that lack anticholinergic action may allow for refinements in research on the effects of antipsychotics on cognition in schizophrenia.

E. Dopamine and Cognition

Over the past 25 years, the dopamine system has remained a major focus in explanations of the neurochemical basis of schizophrenia. Growing evidence from plasma homovanillic acid (HVA) studies provide supportive data for a role of dopamine in schizophrenia. For example, several works (see LIEBERMAN and KOREEN 1993 for a review) have noted that the degree of clinical response to standard medications is related to acute and chronic changes in plasma HVA. However, it is also clear that schizophrenia is far more complicated than a simple dopaminergic dysfunction disorder. Though dopamine antagonists can diminish core psychotic symptoms in some schizophrenic patients, the drugs are not "curative" and a vast majority of patients continue to have a range of negative/deficit symptoms (e.g., blunted affect, amotivation, social withdrawal), as well as residual positive symptoms. As will be noted in later portions of this chapter, neuroleptics that produce significant postsynaptic dopaminergic antagonism typically do not produce profound effects on the cognitive impairments frequently observed in schizophrenia.

Recent theory (CSERNANSKY et al. 1991) has suggested that schizophrenia may be based in dopaminergic hyperactivity in subcortical areas. Some animal studies (e.g., PYCOCK et al. 1980) suggest that deenervation of dopaminergic projections to the frontal lobes in rats can produce an increase in subcortical dopaminergic activity. These observations and other data have led to the idea that dopamine may be differentially dysregulated in different brain regions, with an overactivity in dopaminergic systems in limbic subcortical regions and an underactivity of dopaminergic activity in frontal cortical projections. Observations (CALIGIURI et al. 1993) showing that neuroleptic naive schizophrenic patients also have parkinsonian features lend further support for the idea that there may be dopaminergic hypofunction in some brain regions and patients. Though the abuse of dopamine agonists can produce a syndrome that resembles schizophrenia in nonschizophrenic individuals and long-term use may worsen schizophrenic symptoms, there is some evidence that the acute use of these agents can actually produce mild symptomatic and cognitive improvements (e.g., brighten affect) in some schizophrenic patients (ANGRIST and VAN KAMMEN 1984; GOLDBERG et al. 1991).

I. Related Evidence for the Role of Dopamine in Cognition

Evidence from research areas unrelated to schizophrenia provides some support for the role of the dopaminergic system in cognitive functions. Animal studies implicate dopaminergic systems in a range of functions that relate to cognition and motivation. For example, dopaminergic manipulations have been shown to interact with cholinergic systems in altering radial-arm maze performance in rats (MCGURK et al. 1992). The establishment of classical conditioning has been related to dopaminergic release (YOUNG et al. 1993). A large literature (e.g., FAUSTMAN and FOWLER 1982) has noted that dopamine antagonists can alter behavior in rats in a manner that suggests that reinforcers have diminished reinforcing properties.

Another source of models for the role of dopamine in cognitive functioning is based on studies of PD. Though the exact etiology of PD remains unclear, the major biochemical pathology of the disease includes depletion of dopamine in the neostriatum (FAHN 1992). Cognitive impairment has been widely noted in PD, though it is possible that many aspects of dementia in PD are unrelated to dopaminergic dysfunction. A slowing of cognition (i.e., "bradyphrenia") has been observed in numerous works (FAHN 1992). A range of cognitive functions (e.g., verbal memory, facial recognition, set shifting tasks) may be impaired even at early stages of PD (LEVIN et al. 1989a). Some work (SULLIVAN et al. 1989) has found that PD patients showed a deficit in set formation on the Picture Arrangement subtest of the Wechsler Adult Intelligence Scale (Revised). More recently, STERN et al. (1993b) interpreted impairments on a range of measures in PD as evidence that dementia in PD involves changes in neurotransmitter systems other than the dopaminergic system. In sum, PD represents a model of an illness with frequent dementia and known dopaminergic degeneration, though it is possible that these two observations may not be directly related to each other. Further work may clarify the extent to which dopaminergic systems underlie the cognitive deficits in PD.

II. Dopamine Agonists, Prefrontal Function, and Cognition in Schizophrenia

Numerous studies over the past 10 years have examined prefrontal cortical dysfunction in schizophrenia (GOLDBERG and WEINBERGER 1988). These studies have often relied on performance on the Wisconsin Card Sorting Test (WCST) as an index of prefrontal cortical function. Cerebral blood flow studies during WCST performance have suggested an activation of the prefrontal cortex in normal controls. However, schizophrenic patients attempting the WCST may have a relative inability to activate the prefrontal cortex (WEINBERGER et al. 1986). Some recent methodologies have tested the degree to which dopaminergic hypofunction could explain at least some of the cognitive impairments noted in schizophrenia. These works have attempted to increase dopamine function in schizophrenia with dopamine agonists. In one work (GOLDBERG et al. 1991), haloperidol-treated patients were administed acute doses of dextroamphetamine (0.25 mg/kg) or placebo in a double-blind, crossover study. While the results demonstrated that amphetamine administration could produce a "positive but not robust effect" on WCST performance, amphetamine did appear to improve simple motor speed.

Several works have examined the effects of methylphenidate on cognitive processes. One work (KLORMAN et al. 1984) in healthy adults has suggested that methylphenidate can enhance cognitive performance under challenging conditions. Methylphenidate has been shown to induce increases in psychosis ratings in actively psychotic schizophrenic patients (JANOWSKY et al. 1977). BILDER et al. (1992a) administered methylphenidate to 13 schizophrenic or schizoaffective disorder patients. Methylphenidate was given both while the patients were medication-free and while they were receiving neuroleptics. Oral word production was used as the dependent measure. A complex pattern of results suggested that dopaminergic manipulations had some effect on verbal fluency. The highest levels of word production took place when there was "intermediate catecholaminergic tone" (e.g., on neuroleptic with methylphenidate), while high or low (e.g., on neuroleptic with placebo) "tone" was associated with lower levels of word production (BILDER et al. 1992a).

F. Effects of Antipsychotic Drugs on Cognition in Healthy Nonpsychiatric Samples

Several studies have evaluated the effects of neuroleptics on cognitive functioning in nonpsychiatric volunteers. These studies have theoretical advantages and disadvantages. Since nonpsychiatric patients do not possess cognitive impairments that are often found in schizophrenia, these studies allow for some determination of the cognitive effects of antipsychotics in individuals free from baseline impairments. One would suspect that such studies could better document impairments induced by neuroleptics (e.g., anticholinergic memory impairments), but these methods cannot detect possible disorder-specific improvements that would be found in schizophrenic individuals. For example, one could hypothesize that improvements in cognitive processes in schizophrenia could be due to pharmacological regulation of dopaminergic systems not impaired in normal samples.

One recent review (King 1990) examined the literature on the effects of antipsychotics on cognitive and motor function in normal controls. The author (King 1990) concludes that generalizations about the effects of antipsychotics are difficult to make given the inconsistencies (e.g., test selection, medication dosage) of the studies in the literature. A notable problem in studies of healthy controls is that the tolerability (e.g., susceptibility to motor side effects) of antipsychotics in individuals naive to these medications may be far less than in schizophrenic individuals (King 1990). This observation may have led to some findings (e.g., Kornetsky et al. 1957) that normal controls show results following chlorpromazine administration that are suggestive of motor impairments. Such impairments may well be dose-related. King and Henry (1992) compared the effect of single doses of haloperidol (1 mg), benzhexol (5 mg), diazepam (10 mg), and caffeine (400 mg) on measures of psychomotor function in healthy controls. Relative decreases in psychomotor performance were noted with diazepam and the anticholinergic agent benzhexol. The small dose of haloperidol was generally not associated with impairments in functioning, and some data (e.g., simple visual reaction time) seemed to improve with haloperidol administration.

Other work (Magliozzi et al. 1989) examined the psychomotor effects of higher doses of haloperidol (4 or 10 mg) administered to healthy adults. Cognitive measures included a symbol-digit substitution test (SDST), a good screening measure of attention and psychomotor information processing. Both haloperidol doses produced time-related decreases in SDST performance, the higher dose producing greater impairment. Performance decrements also paralleled elevations in prolactin levels.

DiMascio et al. (1963) administered several doses of chlorpromazine, promethazine, perphenazine, and trifluoperazine to healthy males. The two aliphatic phenothiazines (chlorpromazine and promethazine) impaired motor tasks such as tapping speed. In addition, these agents significantly impaired performance on serial addition and symbol copy tasks. Interestingly, there was some evidence that the piperazine compounds caused relatively few effects and were even related to improvements in some conditions.

Danion et al. (1992) examined the effects of small single doses of chlopromazine (12.5 and 25 mg) and lorazepam (2.5 mg) on a range of memory functions in healthy volunteers. Cognitive measures included tests of skill learning, explicit memory, and repetition priming. Chlorpromazine did not alter performance on the memory measures, but it did diminish skill learning performance. An opposite pattern of results was noted in the lorazepam-treated patients. These results were interpreted as suggesting that the different medications induced a double dissociation between priming and skill learning that may be related to dopaminergic functioning (DANION et al. 1992). However, some caution should be used in interpreting the findings given the nonspecificity of chlorpromazine, the acute dosage design, and the very small dose far below that typically given in schizophrenia.

One recent work (MATTILA and MATTILA 1990) examined the acute effects of remoxipride, ethanol, and diazepam on sensory (e.g., reaction time, critical flicker fusion), cognitive (e.g., digit symbol substitution, paired associate learning), and neuromotor functions (e.g., simple and complex tracking, tapping rate) in healthy volunteers. The use of remoxipride makes for an interesting test agent since it has high specificity for the D_2 receptor (e.g., lacks anticholinergic action) and lacks strong extrapyramidal motor side effects. Remoxipride had no effects on sensory functions and limited effects on other measures. It tended to impair a digit substitution task but did not alter memory, learning, or a letter cancellation task. In addition, remoxipride did not affect the tracking task or finger tapping. The authors noted that the acute effects of both ethanol and diazepam were generally more pronounced than that of remoxipride (MATILLA and MATILLA 1990).

In sum, a range of effects of antipsychotics has been noted in healthy controls. The general findings of these studies may parallel the schizophrenia literature, though normals may well be more susceptible to motor side effects. Antipsychotics with anticholinergic action impair learning and memory functions. One recent work (MATILLA and MATILLA 1990) found limited cognitive effects using a relatively specific dopamine antagonist lacking motor side effects. As noted by KING (1990), diverse methodologies make it difficult to establish generalizations in this literature.

G. Effects of Typical Antipsychotics in Schizophrenic Patient Samples

In the past 15 years, several articles (HEATON and CROWLEY 1981; MEDALIA et al. 1988; SPOHN and STRAUSS 1989; CASSENS et al. 1990; KING 1990; BILDER et al. 1992b) have reviewed the treatment literature or have discussed major issues in the cognitive effects of antipsychotic medications in schizophrenia. These major works have come to a fairly consistent consensus regarding the acute and chronic effects of antipsychotics in schizophrenia. In general, recent reviews suggest that neuroleptics produce both positive and negative effects on cognitive and motor performance. Negative effects tend to be related to acute dosage, pseudoparkinsonian motor side effects, and anticholinergic side effects common to low-potency antipsychotics (BILDER et al. 1992b). Acute administration of standard antipsychotics, which commonly produces sedating and pseudoparkinsonian side effects (HOLLISTER and CSERNANSKY 1990), has been related to deficits in motor functions and attentional measures (CASSENS et al. 1990; MEDALIA et al. 1988). However, chronic dosing of antipsychotics is typically associated with few, if any changes relative to baseline performance on a range of cognitive measures in schizophrenia. As has been recently noted (CASSENS et al. 1990; SPOHN and STRAUSS 1989), there may even be improvement in attentional processes following chronic administration of antipsychotics.

A wide range of research designs and methodologies has been used in the analysis of the effects of antispychotics on cognition. SPOHN and STRAUSS (1989) found that over 400 studies were relevant to the question of antipsychotics and cognition. However, as noted in several recent reviews (BILDER et al. 1992b; SPOHN and STRAUSS 1989), a relatively small number of works have employed methodologies that can accurately and powerfully assess the neuropsychological effects of neuroleptics. Methodological difficulties can arise from both the medication manipulations (e.g., lack of baseline medication-free measures, frequent study of low-potency medications with high anticholinergic properties, dosage and duration of treatment problems), and problems with assessment instruments and procedures (e.g., lack of a comprehensive set of tests, practice effects with repeated testing).

I. Methodological Limitations in Examining the Effects of Neuroleptics in Schizophrenia

A listing of common problems in the literature is provided in Table 2 and may be detailed as follows:

Table 2. Common problems in studies examining the effects of antipsychotics on cognition

1. Problems with study design and medication selection
   a. Nonrandom assignment to medicated and unmedicated patient groups
   b. Inability to attribute causation in studies examining correlations between antipsychotic dose and/or blood levels and cognitive performance
   c. Frequent evaluation of the effects of acute rather than chronic dosage
   d. Frequent study of antipsychotics with high anticholinergic properties
   e. Lack of studies contrasting multiple different antipsychotics
   f. Lack of research quality diagnostic procedures
2. Psychometric problems
   a. Frequent lack of a broad battery of tests
   b. Use of tests with unknown psychometric properties in psychiatric samples
   c. Failure to evaluate redundancy in tests, as measures are often intercorrelated
3. Problems in reporting of results
   a. Limited information on patient characteristics
   b. Limited information on details of the study design and procedures

1. Lack of Random Assigment to Groups in Studies Contrasting Groups of Medicated and Unmedicated Patients

Numerous studies have compared separate samples of patients who were either medication-free or tested while receiving antipsychotic medications. However, in many cases these were separate groups of patients who had been withdrawn from or maintained on medications in a nonrandom manner. It is common that subject characteristics of the patient make the individual a good or bad candidate for a brief period of antipsychotic medication withdrawal. For example, significantly symptomatic patients and those with a history of violence are much less likely to be even asked, let alone succeed in going medication-free for a period of time during an inpatient hospitalization. Those patients who give consent for a washout and successfully complete a medication-free period are more likely less symptomatic (i.e., able to give consent), more compliant, and less likely to have a history of agitation or acting out while medication-free.

Numerous authors (e.g., CROW 1980) have speculated that factors such as medication responsiveness may be associated with brain neuropathology and cognitive deficits. A recent review (STERN et al. 1993a) concluded that increased cognitive impairments at medication-free baseline testing tends to be associated with poorer antipsychotic medication response. In sum, by examining nonrandomly assigned separate groups of medicated and unmedicated patients, it is possible that cognitive testing results are confounded by subject characteristics inherent in the ability to give consent and to complete a medication-free period. Given the nature of schizophrenia, nonrandomized group designs may be of limited value.

2. Correlations Between Cognitive Impairments and Neuroleptic Dose or Blood Levels Does Not Equal Causation

Several studies (e.g., SWEENEY et al. 1991) have examined a single group of schizophrenic patients and performed correlative analyses between cognitive measures and neuroleptic dose or blood levels. In some cases, correlations have been noted between cognitive deficits and neuroleptic blood levels. However, it should be noted that such analyses share similar problems as noted above in regard to nonrandom assignment to groups. It is in the general nature of psychiatric treatment that patients who are the most refractory to treatment (i.e., have the most symptoms while on maintenance treatment) are likely to be given the highest doses of medications. As noted above, poor medication response may be related to other disorder features such as brain pathology, neurological soft signs, and most importantly, neuropsychological test impairments. Thus, correlative studies between neuroleptic levels and cognitive testing are probably confounded by subject characteristics and are therefore of limited value. Such studies would be of greater value if patients were randomly assigned to fixed doses of antipsychotic medications.

3. Lack of Attention to Sampling Characteristics of the Patients Studied

Patients with schizophrenia show a great deal of variability in cognitive deficits, ranging from little or no deficit to a level of impairment comparable to that found in neurological samples with documented cerebral insults. Differing degrees of neuropsychological deficits are probably related to the sampling technique employed in recruiting patients. For example, stable outpatients in a private hospital setting may be less likely to show impairments than chronically hospitalized, severely ill patients who are treatment-refractory. Given large differences in baseline levels of cognitive impairments, relationships between medication administration and cognitive state may differ based on the type of sample utilized. For example, patients with baseline impairments may be susceptible to cognition-impairing effects of anticholinergic action, and neuroleptic responsive patients may be more likely to show medication-related attentional improvements. Little or no work has addressed this issue, because schizophrenic patients are generally regarded the same though they may be recruited from greatly differing populations.

4. Lack of Studies Comparing Multiple Antipsychotics with Different Neuropharmacologic Profiles

Antipsychotic medications have receptor-binding profiles that range from selective (e.g., substituted benzamide compounds such as remoxipride) to complex (e.g., clozapine). Based on varying neuropharmacological profiles, one can develop hypotheses for why antipsychotics impair or improve cognitive functioning. Anticholinergics clearly impair memory functions (Medalia et al. 1988), but antipsychotics range from having no anticholinergic action to having substantial activity. Some authors (e.g., Meltzer et al. 1989) have suggested that antagonism of the serotonin-2 receptor may be important in producing the atypical (e.g., superior negative symptom treatment, diminished extrapyramidal effects) qualities of some compounds. One could hypothesize that serotonin-2 affinity could also be related to improvements in functions such as psychomotor speed and attention. If a specific dopaminergic dysfunction is important in schizophrenia, studies examining the effects of various dopaminergic manipulations (e.g., administration of autoreceptor agonists or specific postsynaptic antagonists such as substituted benzamides) would be of value. In sum, the ideal study would provide a double-bind, randomized evaluation of pharmacologically different typical and atypical antipsychotics. Such a study would be able to offer conclusions on the unique properties of antipsychotics that relate to both improvements and decrements in cognitive performance. This study has not yet been performed to our knowledge.

5. Use of Relatively Acute Dosage Strategies in Patients Receiving Long-Term Treatment

Multiple studies examining the effects of antipsychotics in schizophrenia have tested patients fairly early (e.g., 7–10 days after initiation of treatment). While such a design allows for an optimal assessment of some early side effects (e.g., pseudoparkinsonism, sedation) that may impair cognition, patients at such an early stage of treatment may be far from showing complete treatment response. Clinical improvements in schizophrenic patients may continue for 4 or more weeks (Lieberman and Koreen 1993). Accordingly, studies that seek potential positive as well deleterious effects of antipsychotics on cognition would be best designed to perform assessments during both the acute phase and following longer periods of optimal treatment (e.g., 6 weeks).

6. Lack of Information Regarding the Details of the Study

Omissions of information often include important features such as dose levels and duration of treatment (Medalia et al. 1988). Diagnostic procedures are often not detailed, and many studies were conducted prior to the development of modern diagnostic systems and structured clinical interviews.

7. Inconsistent or Inappropriate Neuropsychological Test Selection

Most studies employ a limited range of tests (e.g., a small selection of attentional measures) that lack the ability to assess a wide range of executive, memory, and attentional processes. Studies that offer a selection of measures often employ a "home-spun battery" of generally unknown reliability and validity that is unlikely to be used in its form by other investigators. In addition, as noted by Bilder et al. (1992b), the tests employed in many studies are often factorially complicated. Tests with such complex qualities may be affected by a range of specific cognitive functions, thus making it difficult to accurately interpret changes resulting from medication administration.

H. The Effects of Standard Antipsychotics in Schizophrenia

Several recent works (Medalia et al. 1988; Spohn and Strauss 1989; Cassens et al. 1990; King 1990) have provided comprehensive reviews of the literature on the effects of neuroleptics on neuropsychological measures. This section reviews major prior studies and highlights the relevant new studies completed in the past 5 years. Studies may be divided on the basis of the major focus of the psychometric instruments employed as dependent measures. In general, these studies can be divided into projects focusing on: (1) attention/information-processing measures, (2) memory functions, (3) standardized batteries, and (4) a combination of selected cognitive measures other than standardized batteries.

I. Attention/Information-Processing Measures

Diverse testing procedures have been used to examine the effects of antipsychotics on attention and information-processing measures. These studies range from traditional reaction time tasks to computer-generated measures of distractibility and early information processing. Some studies have examined single measures related to attentional processes while others have combined numerous forms of data collection.

Some early work (WYNNE and KORNETSKY 1960) suggested that chlorpromazine (400 mg/day during long-term treatment) did not greatly alter reaction time during acute or long-term treatment. Other work (HEILIZER 1959) noted that chlorpromazine may reduce the variability in the reaction time of schizophrenic individuals. PEARL (1962) failed to find phenothiazine effects on complex and simple reaction time tasks. EATON et al. (1979) examined the reaction time and accuracy of information processing in schizophrenic patients who were presented visual stimuli in the left or right visual field, or both at once. Patients were assessed under both an unmedicated baseline and following neuroleptic treatment; they responded significantly faster following medication treatment regardless of the hemisphere stimulated (EATON et al. 1979).

SPOHN et al. (1977) examined the effects of antipsychotic administration in a large sample of patients following an extended 6-week medication washout. Attention-related measures included reaction time and the continuous performance test. Results suggested a "normalization" (i.e., changes in the direction of improvement) on selected measures of reaction time and information processing. Further work by this same group (SPOHN et al. 1985) performed a cross-sectional correlational study examining cognition and medication relationships in 84 schizophrenic patients. The antipsychotic dose level was found to be associated with longer reaction times and a diminished "span of apprehension" (SPOHN et al. 1985). The severity of tardive dyskinesia was also associated with both increased reaction time and eye-tracking dysfunction. This study suffers from the methodological limitations of dose–behavior correlation studies noted in the previous section.

Studies examining the effects of antipsychotics on early stages of information processing have employed a technique termed "backward masking". BRAFF and SACCUZZO (1982) utilized the backward masking technique to examine the speed of information processing in groups of medicated and unmedicated schizophrenic patients and depressed comparison subjects. The results demonstrated that both groups were slower in processing information than the depressed comparison group. However, the medicated schizophrenic patients showed more rapid information processing when compared to the unmedicated schizophrenic patients.

The span of apprehension provides other data on early information processing abilities. The procedure evaluates the ability of patients to perceive rapidly presented information of varying degrees of complexity. SPOHN et al. (1977) found certain span of apprehension measures to be superior in medication-treated schizophrenic patients receiving double-blind treatment with either placebo or chlorpromazine (adjusted for optimal treatment at a dose above 200 mg/day). Other work using a span of apprehension task (MARDER et al. 1984) evaluated 13 acutely symptomatic schizophrenic patients at an unmedicated baseline and repeated the assessment following 2 weeks of haloperidol (10 mg/day) treatment. Improvements were noted in the span of apprehension task following medication, and such improvements were not noted in an independent group of stable patients who were tested twice while receiving fluphenazine decanoate.

The Continuous Performance Test (CPT) represents a measure of sustained attention. The CPT typically involves attending to a series of rapidly presented stimuli with the subject being instructed to perform a task (e.g., press a key) when a preselected target item appears (EARLE-BOYER et al. 1991). Extensive prior work has demonstrated CPT impairments in schizophrenia (MEDALIA et al. 1988). Recent work (EARLE-BOYER et al. 1991) compared 17 unmedicated to 17 haloperidol-treated inpatient schizophrenic patients on the CPT and motor proficiency measures. A control group of 19 volunteers was noted to make fewer CPT errors than the schizophrenic patients. The medication-free patients were noted to make more errors than the haloperidol-treated patients on nonlexical (i.e., nonsense-syllable) stimuli. The authors noted that the measure of motor proficiency was correlated with CPT measures in the medicated but not the unmedicated patient group. The results were interpreted as suggesting that "improved performance of medicated patients on the nonlexical condition probably reflects an enhanced ability to perform multiple attention-demanding operations concurrently, rather than a perceptual advantage" (EARLE-BOYER et al. 1991, p. 54).

An additional strategy for examining attentional parameters is to require patients to perform a task while being exposed to intentional distractors. For example, OLTMANNS et al. (1978) administered a verbal learning test while exposing schizophrenic patients to auditory distractors. Patients were tested during a medication washout and while receiving antipsychotic medication. The results suggested that schizophrenic patients were more easily distracted than normal controls or nonschizophrenic psychiatric patients. In addition, patients receiving antipsychotic medications were less susceptible to the influence of distracting stimuli. According to the authors, "the drugs appear to have a particular effect in assisting schizophrenics to screen out distracting stimuli. In their absence a genuine deficit in selective attention becomes apparent" (OLTMANNS et al. 1978, p. 86).

STRAUSS et al. (1985) evaluated distractibility (e.g., verbal interference during a digit span task) and reaction time (e.g., lifting a finger from a key following an auditory stimulus) in 25 neuroleptic-treated schizophrenic subjects. There was a statistically significant inverse correlation between serum neuroleptic levels and distractibility. There were no relationships between the distractibility measure and psychopathology measures, serum anticholinergic level, or reaction time scores.

Another recent work on distractibility (HARVEY and PEDLEY 1989) examined the effects of medication on both auditory and visual recall tasks. The results found that both medicated (haloperidol 20 mg/day) and unmedicated patients were distractible during the auditory task, but only the unmedicated patients were also found to be susceptible to distraction on the visually presented task.

In sum, the literature on the effects of antipsychotics on attentional and information processing measures provides little evidence for a deleterious effect of chronic treatment and some support for improvements resulting from treatment.

II. Effects of Typical Antipsychotics on Memory Functions

Memory impairment has been noted to be a common finding in schizophrenia and it has been recently proposed that memory is disproportionately impaired in schizophrenics compared to other cognitive functions (TAMLYN et al. 1992). Interestingly, as noted by CASSENS et al. (1990), only a few works have emphasized memory functions in assessing the effects of antispychotics on cognition. DASTON (1959) tested schizophrenic patients at a neuroleptic-free baseline and again following double-blind treatment with chlorpromazine (400 mg/day), promazine (a phenothiazine agent; 400 mg/day), or phenobarbital (3 g/day). Testing consisted of verbal memory measures derived from the Wechsler Memory Scale (WMS). Only the chlorpromazine group showed improvement in performance on paired associate learning. The authors concluded that phenothiazine treatment facilitated attention and may therefore facilitate verbal memory measures in schizophrenia.

PEARL (1962) included WMS subtests and the Benton Visual Retention test in a study examining phenothiazine effects. Medications did not significantly improve performance on either test. SPIEGEL and KIETH-SPIEGEL (1967) suggest that some measures (e.g., word association tests, digit span) that may be somewhat related to verbal learning and memory were improved following chronic antispsychotic treatment. CALEV et al. (1987) noted no relationship between neuroleptic dose and remote or recent memory in a sample of chronic schizophrenic patients who had a history of lengthy hospitalization.

Two recent works have examined the effects of antipsychotics on memory measures. EITAN et al. (1992) studied the acute effects of four antipsychotics (chlopromazine, thioridazine, trifluoperazine, and haloperidol) that vary greatly in anticholinergic properties. The results suggested that the agents with the most anticholinergic activity (chlorpromazine, thioridazine) impaired short-term verbal memory (e.g., recall of a 20-sentence story) 6 h after administration. There was some evidence that the higher-potency agents (e.g., haloperidol) improved the verbal memory measure. Other recent work (TAMLYN et al. 1992) noted no correlation between memory functions and chlorpromazine dose equivalents in patients given a variety of antipsychotic medications.

In sum, the anticholinergic activity of some neuroleptics or the administration of antiparkinsonian anticholinergics may produce deleterious effects on memory functions. For example, PERLICK et al. (1986) found that in a sample of neuroleptic-treated schizophrenics there was a correlation between serum anticholinergic levels and verbal list learning even though only 24% of patients were taking antiparkinsonian drugs. Interestingly, antipsychotic serum levels were not significantly related to memory functions. Other recent work (SWEENEY et al. 1991) has noted a relationship between anticholinergic dose and a verbal learning task. In sum, this interesting work suggests that memory performance may be more closely related to anticholinergic activity than to neuroleptic levels in medication-treated patients.

III. Antipsychotic Effects on Motor Tasks

Motor changes induced through dopaminergic effects in the extrapyramidal motor system are among that most notable of side effects observed with traditional antipsychotics. Accordingly, it is not surprising that some motor tasks or tasks which require a major motoric component may be impaired during antipsychotic treatment. These impairments may be particularly associated with acute dosage regimens. For example, KORNETSKY et al. (1959) found that acute administration of chlorpromazine (200 mg) produced impaired performance on measures of tapping speed, pursuit rotor performance, and the digit symbol test. However, following 12 days of continuous chlorpromazine treatment there were no detectable effects of chlorpromazine on any of these same measures. PUGH (1968) also failed to note chlorpormazine effects on tapping speed following long-term treatment. Other early work (PEARL 1962) administered a range of cognitive (e.g., Wechsler subtests, Wechsler Memory Scale measures) and motor measures to phenothiazine-treated patients. The results suggested that the medications selectively impaired Purdue Pegboard Assembly performance, a measure of fine motor control. More recent work (GOODE et al. 1981) failed to find antipsychotic medication effects on the Crawford Small Parts Dexterity Test, a measure of fine motor control. In sum, some evidence suggests that acute dosage of antipsychotics produces motor performance decrements related to the extrapyramidal side effects of these agents.

IV. Hemispheric Asymmetries in Cognitive Performance Patterns

Numerous papers have suggested that left-hemisphere hyperarousal may be important in the cognitive and clinical features of schizophrenia. This concept emerged from studies of lateral eye movements (LEM) in which schizophrenic patients tend to show a rightward LEM, therefore suggesting left-hemisphere hyperarousal (TOMER and FLOR-HENRY 1989). A variety of attention/information-processing tasks has been employed to examine the hypothesis of asymmetrical dysfunction.

Several studies have also examined whether antipsychotic medications alter the pattern of cognitive functioning in a manner consistent with these hemispheric asymmetries. TOMER and FLOR-HENRY (1989) examined asymmetries in attention using the Mesulam Cancellation Test. Schizophrenic patients were assessed at an unmedicated baseline and again following treatment with a variety of antipsychotic agents. Though overall test performance did not change as a result of treatment, the investigators suggested that the pattern of results demonstrated that asymmetry was related to medication status. Specifically, medication-free patients showed inattention to the right side, which changed to left-sided inattention when they were administered neuroleptics (TOMER and FLOR-HENRY 1989). The results were further interpreted as evidence that "neuroleptics may normalize left-hemisphere performance, at the expense of deteriorated right-hemisphere performance" (TOMER and FLOR-HENRY 1989, p. 852). TOMER (1989) employed a tactile discrimination task in patients who were tested both unmedicated and while receiving antipsychotics. Performance in the unmedicated state suggested relative dysfunction of the left hemisphere (e.g., more discrimination errors made with the right hand than the left hand). However, tactile discrimination by the left hand tended to decrease while patients were receiving medication, whereas performance by the right hand tended to improve. Once again, the results were interpreted as suggesting that neuroleptics produce different effects on the two hemispheres.

V. Standardized Neuropsychological Batteries and the Wechsler Scales

Studies employing standardized measures have some attractive methodological features. These works allow for a standardized assessment of cognitive functions, thus allowing for a greater ability to generalize across studies. Moreover, these works have the advantage of using tests with known reliability and validity in large psychiatric patient samples. The major drawback of these tests is that they contain measures that tap a range of cognitive functions. Ideally, research studies would include a selection of standardized measures along with other tests that are better able to assess specific cognitive functions.

The Halstead-Reitan Battery has been used to assess medication-related effects. For example, FREDERICKS and FICKEL (1978) compared Halstead-Reitan performance in schizophrenic patients administered either chlorpromazine, perphenazine, or placebo. On nearly all tests of the Halstead-Reitan Battery, the medicated and placebo patients showed a similar level of performance. Other work examining the differences between samples of medicated and unmedicated schizophrenic patients has employed the Luria-Nebraska Neuropsychological Battery (LNNB). The LNNB has certain advantages in that it is much shorter than the Halstead-Reitan. A recent review (MOSES and MARUISH 1988) suggests that the test has acceptable psychometric properties in psychiatric samples. MOSES (1984) found no differences between two groups (n = approx. 40 per group) of schizophrenic patients administered the LNNB. A more recent analysis (MILLER et al. 1993) examined a partially overlapping but expanded (total $n = 136$) sample of medicated ($n = 80$) and unmedicated ($n = 56$) schizophrenic patients. The two groups also showed a generally similar pattern of performance across the clinical scales of the LNNB.

A variety of studies have employed various versions of the Wechsler Adult Intelligence Scale (WAIS) to evaluate the effects of antipsychotics on cognitive functions. One double-blind design compared the effects of chlorpromazine (900 mg/day) or placebo on Wechsler Bellevue performance in 44 patients diagnosed as having catatonic schizophrenia. Thirty days after the start of medication administration there was a significant increase in Wechsler scores in the chlorpromazine-treated patients while there was a slight decrease in the scores of the placebo-treated patients. Further early research on the effects of chlorpromazine treatment was offered by SHATIN et al. (1956). Patients treated with 800 mg/day of chlorpromazine for 11 days showed a level of performance on a range of cognitive measures that was similar to baseline levels.

Another work (GOLD and HURT 1990) examined the severity of thought disorder and WAIS performance in 19 schizophrenic patients who were assessed at an unmedicated baseline and again after 26 days of haloperidol treatment. Thought disorder ratings declined significantly during treatment while WAIS performance increased at a level that was consistent with practice effects. The severity of thought disorder was inversely correlated with IQ in the unmedicated state, but no relationship was noted following haloperidol treatment. The authors conclude that thought disorder may be linked to dopaminergic functioning while cognitive deficits in schizophrenia may be more closely linked to structural brain abnormalities often noted in the disorder. Other work (PEARL 1962) administered a selection of WAIS subtests (Comprehension, Similarities, Digit Span, Digit Symbol, and Picture Completion) to patients at an unmedicated baseline and following 12 weeks of phenothiazine treatment. The results showed no treatment-related changes in performance other than a significant improvement in the Similarities subtest. Two other early works (JUDSON and MACCASLAND 1960; NICKOLS 1958) also noted no significant changes in Wechsler-Bellevue performance as a result of chlorpromazine treatment.

VI. Studies Employing a Combination of Selected Cognitive Measures

Recent research (e.g., Braff et al. 1991) suggests that schizophrenic patients show deficits on neuropsychological measures that tap a broad range of cognitive functions. Accordingly, investigations that evaluate medication effects on a limited number of measures have limited ability to detect treatment effects on the diverse cognitive deficits observed in schizophrenia. A minority of studies have used this broad battery approach in assessing antipsychotic effects.

Spohn et al. (1977) administered measures including reaction time, the Continuous Performance Test, the digit symbol substitution test, and a range of other physiological (e.g., skin conductance) and cognitive measures. The effects of antipsychotic medication administration was evaluated in a large sample of patients who had participated in an extended, 6-week medication washout. Results suggested general improvements on a range of measures related to attention and information processing. However, there were no demonstrated improvements on abstraction measures such as proverbs testing.

Wahba et al. (1981) administed selected tests (digit span, proverbs, vocabulary skills) to 44 acutely ill schizophrenic patients who were subsequently given double-blind treatment with haloperidol at various dose levels (10–100 mg/day). Clinical ratings revealed marked symptom improvement over 10 days of treatment and the authors noted statistically significant improvements on all cognitive testing measures.

One excellent study (Killian et al. 1984) administered a range of tests (WAIS subtests, Stroop Color-Word Interference Test, measures of perception) to carefully diagnosed schizophrenic patients. Patients were then evaluated following a 3-week medication-free period, subsequently treated with standard antipsychotics for a month, and then retested. The results noted that the medications did not have any effects on cognitive testing, and the presence or absence of symptom improvement was not related to performance on the cognitive measures (Killian et al. 1984).

Sweeney et al. (1991) also attempted to assess the relationship between medication treatments and neuropsychological performance. As noted previously, the anticholinergic dose was related to measures that included verbal learning. This project also noted that antipsychotic medication dosage in stable schizophrenic inpatients was correlated with perseveration errors on the Wisconsin Card Sort, attentional measures, and psychomotor speed (Sweeney et al. 1991). Once again, conclusions drawn from such cross-sectional data may be limited as such works may reveal more about schizophrenia and its treatment (i.e., that medication nonresponders receive high-dose therapy and also tend to have the most cognitive impairment and structural pathology) than about the specific effects of antipsychotics in schizophrenia. Correlation in such studies may not be related to causation.

A recent report (Smet et al. 1994) compared three different groups of schizophrenic patients that varied greatly in current medication treatment and prior treatment histories. A battery of tests was administered to separate groups of patients that were either medication-naive (n = 10), stable and receiving treatment with a standard neuroleptic (n = 19), or previously treated patients who participated in a medication washout ($n = 21$). Though some subtle between-group differences were noted, the three different groups generally showed a comparable level of performance on measures of executive functioning, verbal memory, and speech/language functioning (SMET et al. 1994). This represents one of the few studies contrasting neuroleptic-naive patients with those who were receiving treatment and undergoing a washout but who had a prior history of neuroleptic treatment. These results suggest that cognitive deficits are present at the outset of psychosis and are generally unaffected by medication treatment.

An alternative strategy for the assessment of medication effects in schizophrenia is to test patients at a stable medicated baseline and then retest the patients following a significant reduction (e.g., 80%–90%) in antipsychotic medication dosage. One recent study (SEIDMAN et al. 1993) performed this experiment in a small sample of schizophrenic and schizoaffective disorder inpatients. Performance on an extensive cognitive battery was generally unchanged following significant dose reduction.

I. Effects of Atypical and Recently Developed Antipsychotics on Cognition

The effects of substituted benzamides on cognition are of particular interest. As noted above (see healthy volunteer studies), such studies can assess the specific action and therapeutic effects of dopamine blockade in schizophrenia while minimizing parkinsonian side effects. One study (STRAUSS and KLEISER 1990) conducted in Germany compared baseline to treatment values for small groups of remoxipride- ($n = 10$) and haloperidol- ($n = 8$) treated schizophrenic patients. Patients were administered measures of concentration, memory functioning, and intellectual ability. Though the interpretation of the results is limited by the use of parametric statistics in such a small sample, the results suggest that remoxipride may have produced performance increments in all the cognitive measures The authors note that further work is warranted in larger samples of patients.

The evaluation of clozapine effects on cognition is of interest since this compound has a unique profile that could be hypothesized to produce both facilitated (e.g., superior treatment efficacy, lack of motor side effects) and impaired (e.g., relatively strong anticholinergic action, sedation with acute treatment) cognitive measures. There has been only limited work in which groups of patients were compared who were either not on neuroleptics or receiving their typical treatment and clozapine. GOLDBERG et al. (1993) evaluated patients in a typical neuroleptic phase and a clozapine phase using a group of neuropsychological tests, including attention, memory, spatial ability, intellectual, and executive function. Fifteen patients were retested an average of 15 months after clozapine treatment (range 3–24 months). In spite of their clinical improvement (Brief Psychiatric Rating Scale symptom ratings declined by 40% on average), there were no significant differences between cognitive performances on traditional versus clozapine treatment. On one measure of visual-spatial memory, their performance significantly worsened, which the authors attributed to the strong anticholinergic profile of clozapine. One of the limitations of this study is that patients were on adjunctive medications either before or after clozapine, and some patients were adjunctively medicated on both occasions, primarily with lithium, although the authors point out that there were approximately equal numbers of patients receiving them in both phases of the study.

HAGGER et al. (1993) evaluated 36 chronic schizophrenic patients during a baseline washout period, after 6 weeks, and after 6 months of clozapine treatment. Patients made improvements at 6 weeks and 6 months on measures of verbal fluency (word production); at 6 months, they improved on the Digit Symbol test (a measure of perceptual-motor speed), the Category Instance Generation Test, verbal list learning–immediate recall, and the WISC-R Maze. Since the effect of clozapine is compared to a no-treatment condition, one cannot conclude that clozapine has better efficacy in improving cognitive functions than do other neuroleptics, which also improve performances on measures of attention and concentration.

In a follow-up study to the HAGGER et al. (1993) work, LEE et al. (1994) randomly assigned nontreatment-resistant schizophrenic patients to typical neuroleptic (n = 23) and clozapine treatment (n = 24) and evaluated the patients with neuropsychological measures at 6 weeks, 6 months, and 12 months. The clozapine-treated group was superior to the typical neuroleptic group on the Digit Symbol and Controlled Oral Word Association Test at all time points, suggesting to the authors that clozapine had a preferential effect on dopamine regulation in the frontal cortex.

HOFF et al. (1993) compared 17 refractory schizophrenic patients receiving "traditional" neuroleptics and after 12 weeks of clozapine monotherapy on a variety of neuropsychological measure. Statistically significant improvements were noted on measures of attention/concentration, motor speed (Trails A, Symbol-Digit Modalities Test–Written, Finger Tapping for nondominant hand) and confrontation naming (Boston Naming Test). These improvements were related to symptom improvement; however, it was relatively poor in this group of patients. Like GOLDBERG et al. (1993), patients performed worse on measures of spatial memory while on clozapine.

In a crossover study of 14 patients who alternately received clozapine and risperidone for 6-week periods, clozapine improved reaction times on the Continuous Performance Test, whereas risperidone improved visual memory and the number of categories obtained on the Wisconsin Card Sort (DANIEL 1994). In a review of the cognitive literature on clozapine, GOLDBERG and WEINBERGER (1994) conclude that clozapine improves tasks which involve verbal fluency, reaction time, and attention – functions subserved by the basal ganglia or striatum. Alternatively, clozapine may impair performance on the Wisconsin Card Sort and visual memory measures because of its relatively potent D1 antagonism and strong anticholinergic profile.

Table 3. General research findings on the effects of typical neuroleptics on different areas of cognitive function and tests in schizophrenic samples

Function/tests	General findings
Attention/information processing	Either no change or improvement.
Memory	Some studies suggest improvement; neuroleptics with strong anticholinergic properties may produce impairments.
Motor	Acute administration may be related to impairments; long-term use is associated with fewer effects.
Hemispheric lateralization	Possible normalization of left hemisphere function.
Standardized batteries such as the Halstead-Reitan and Luria-Nebraska	Little differences between medicated and unmedicated groups.
Wechsler Adult Intelligence Scale	Little medication-related effects, some improvement consistent with practice effects.

J. Conclusions

Table 3 provides a general summary of the effects of standard neuroleptics on a range of cognitive and motor functions. In summary, the effects of antipsychotic drugs on the cognitive performance of schizophrenic patients are generally either negligible or positive. While acute administration may cause a temporary slowing of motor functions, long-term administration (greater than 2 weeks) generally either improves or does not alter performance on measures of attention, concentration, memory, motor speed, and higher-order intellectual and problem-solving abilities. In contrast, the effects of adjunctive anticholinergic medications and the anticholinergic activity of low-potency antipsychotics may contribute to memory dysfunction in schizophrenia. Atypical antipsychotics such as clozapine may produce unique profiles of both improvements and decrements in selected neuropsychological measures. Further work with clozapine and other candidate-atypical compounds is warranted to assess the unique properties of these new drugs. Analysis of the effects of medication must take into account that schizophrenic patients have widespread cognitive impairments as a natural part of their illness, whether because of brain structural abnormalities, neurotransmitter dysregulation, or both. Future studies which compare different neuroleptics with different neurochemical properties using double-blind or repeated measures designs will tell us much about the cognitive effects of these drugs as well as cognitive deficits of schizophrenia itself.

Acknowledgments. This work was supported in part by United States Public Health Service grant no. MH-30854 from the National Institute of Mental Health, the Department of Veterans Affairs, the California Department of Mental Health, and Napa State Hospital. The authors thank Pamela Elliott for her assistance.

References

Andreasen N, Olsen S (1982) Negative vs. positive schizophrenia: definition and reliability. Arch Gen Psychiatry 39:789–794
Angrist B, van Kammen DP (1984) CNS stimulants as tools in the study of schizophrenia. Trends Neurosci 7:388–390
Atkinson R, Shiffrin R (1968) A proposed system and its control processes. In: Spence K, Spence J (eds) Advances in the psychology of learning and motivation. Academic, New York
Bigler ED (1987) The clinical significance of cortical atrophy and ventricular enlargement in schizophrenia. Arch Clin Neuropsychology 2:385–392
Bilder RM, Lieberman JA, Kim Y, Alvir JM, Reiter G (1992a) Methylphenidate and neuroleptic effects on oral word production in schizophrenia. Neuropsychiatry Neuropsychol Behav Neurol 5:262–271
Bilder RM, Turkel E, Lipschutz-Broch L, Lieberman JA (1992b) Antipsychotic medication effects on neuropsychological functions. Psychopharmacol Bull 28:353–366
Bogerts B, Meertz E, Schönfeldt-Bausch R (1985) Basal ganglia and limbic system pathology in schizophrenia. Arch Gen Psychiatry 42:784–791
Braff DL, Saccuzzo DP (1982) Effect of antipsychotic medication on speed of information processing in schizophrenic patients. Am J Psychiatry 139:1127–1130
Braff DL, Heaton R, Kuck J, Cullum M, Moranville J, Grant I, Zisook S (1991) The generalized pattern of neuropsychological deficits in outpatients with chronic schizophrenia with heterogeneous Wisconsin Card Sorting test results. Arch Gen Psychiatry 48:891–898
Broadbent D (1959) Perception and communication. Pergamon, London
Calev A (1984) Recall and recognition in mildly disturbed schizophrenics: the use of matched tasks. Psychol Med 14:425–429
Calev A, Berlin H, Lerer B (1987) Remote and recent memory in long-hospitalized chronic schizophrenics. Biol Psychiatry 22:79–85
Caligiuri MP, Lohr JB, Jeste DV (1993) Parkinsonism in neuroleptic-naive schizophrenic patients. Am J Psychiatry 150:1343–1348
Cassens G, Inglis AK, Appelbaum PS, Gutheil TG (1990) Neuroleptics: effects on neuropsychological function in chronic schizophrenic patients. Schizophr Bull 16:477–499
Chapman LJ, Chapman JP (1973) Disordered thought in schizophrenia. Prentice-Hall, Englewood Cliffs
Conrad AJ, Abebe T, Austin R, Forsythe S, Scheibel AB (1991) Hippocampal pyramidal cell disarray in schizophrenia as a bilateral phenomenon. Arch Gen Psychiatry 45:413–417
Cooper JA, Sagar HJ, Doherty SM, Jordan N, Tidswell P, Sullivan EV (1992) Different effects of dopaminergic and anticholinergic therapies on cognitive and motor function in Parkinson's disease. A follow-up study of untreated patients. Brain 115:1701–1725
Craft S, Yurgelin-Todd D, Kaplan E, Aizley H, Levin S (1988) Concurrent deficits of frontal and spatial tasks in schizophrenics without affective symptoms. J Clin Exp Neuropsychol 10:71
Crow TJ (1980) Molecular pathology in schizophrenia: more than one disease process? Br Med J 280:66–68
Crow TJ, Done DJ (1992) Prenatal exposure to influenza does not cause schizophrenia. Br J Psychiatry 161:390–393
Csernansky JG, Murphy GM, Faustman WO (1991) Limbic/Mesolimbic connections and the pathogenesis of schizophrenia. Biol Psychiatry 30:383–400
Daniel DG (1994) Comparison of risperidone and clozapine on clinical and cognitive functions in psychotic disorders. Biol Psychiatry 35:667 [abstract]
Danion J-M, Peretti S, Grangé D, Bilik M, Imbs J-L, Singer L (1992) Effects of chlorpromazine and lorazepam on explicit memory, repetition priming and cognitive skill learning in healthy volunteers. Psychopharmacology 108:345–351

Daston PG (1959) Effects of two phenothiazine drugs on concentrative attention span of chronic schizophrenics. J Clin Psychol 15:106–109

DeLisi LE, Stritzke P, Riordan H, Holan V, Boccio A, Kushner M, McClelland J, Van Eyl O, Anand A (1992) The timing of brain morphological changes in schizophrenia and their relationship to clinical outcome. Biol Psychiatry 31:241–25

DeLisi LE, Tew W, Xie S-h, Hoff AL, Sakuma M, Kushner M, Lee G, Shedlack K, Smith AM, Grimson R (1995) A prospective follow-up study of brain morphology and cognition in first-episode schizophrenic patients: Preliminary finding. Biol Psychiatry 38:349–360

DiMascio A, Havens L, Klerman GL (1963) The psychopharmacology of phenothiazine compounds: A comparative study of the effects of chlopromazine, promethazine, trifluoperazine, and perphenazine in normal males. II. Results and discussion. J Nerv Ment Dis 136:15–28

Earle-Boyer EA, Serper MR, Davidson M, Harvey PD (1991) Continuous performance tests in schizophrenic patients: stimulus and medication effects on performance. Psychiatr Res 37:47–56

Eaton EM, Busk J, Maloney MP, Sloane RB, Whipple K, White K (1979) Hemispheric dysfunction in schizophrenia: assessment by visual perception tasks. Psychiatr Res 1:325–332

Eitan N, Levin Y, Ben-Artzi E, Levy A, Neumann M (1992) Effects of antipsychotic drugs on memory functions of schizophrenic patients. Acta Psychiatr Scand 85:74–76

Fahn S (1992) Parkinson's disease and other basal ganglion disorders. In: Asbury AK, McKhann GM, McDonald WI (eds) Diseases of the nervous system: clinical neurobiology, 2nd edn. Saunders, Philadelphia, p 1144

Faustman WO, Fowler SC (1982) An examination of methodological refinements, clozapine and fluphenazine in the anhedonia paradigm. Pharmacol Biochem Behav 17:987–993

Faustman WO, Moses JA Jr, Csernansky JG (1988) Luria-Nebraska performance and symptomatology in unmedicated schizophrenic patients. Psychiatry Res 26:29–34

Fayen M, Goldman MB, Moulthrop MA, Luchins DJ (1988) Differential memory impairment with dopaminergic versus anticholinergic treatment of drug-induced extrapyramidal symptoms. Am J Psychiatry 145:483–486

Flicker C, Serby M, Ferris SH (1990) Scopolamine effects on memory, language visuospatial praxis and psychomotor speed. Psychopharmacology 100:243–250

Fredericks RS, Finkel P (1978) Schizophrenic performance on the Halstead-Reitan Battery. J Clin Psychol 34:26–30

Gold JM, Hurt SW (1990) The effects of haloperidol on thought disorder and IQ in schizophrenia. J Pers Assess 54:390–400

Gold JM, Randolph C, Carpenter CJ, Goldberg TE, Weinberger DR (1992a) the performance of patients with schizophrenia on the Wechsler Memory Scale-Revised. The Clinical Neuropsychologist 6:367–373

Gold JM, Randolph C, Carpenter CJ, Goldberg TE, Weinberger DR (1992b) Forms of memory failure in schizophrenia. J Abn Psychology 101:487–494

Goldberg TE, Weinberger DR (1988) Probing prefrontal function in schizophrenia with neuropsychological paradigms. Schizophr Bull 14:179–183

Goldberg TE, Ragland JD, Torrey EF, Gold JM, Bigelow LB, Weinberger DR (1990) Neuropsychological assessment of monozygotic twins discordent for schizophrenia. Arch Gen Psychiatry 47:1066–1072

Goldberg TE, Bigelow LB, Weinberger DR, Daniel DG, Kleinman JE (1991) Cognitive and behavioral effects of the coadministration of dextroamphetamine and haloperidol in schizophrenia. Am J Psychiatry 148:78–84

Goldberg TE, Greenberg RD, Griffin SJ, Gold JM, Kleinman JE, Pickar D, Schultz SC, Weinberger DR (1993) The effect of clozapine on cognition and psychiatric symptoms in patients with schizophrenia. Br J Psychiatry 162:43–48

Goldberg TE, Weinberger DR (1994) The effects of clozapine on neurocognition An overview J Clin Psychiatry 55 [Suppl B]:88–90
Goldstein K (1959) Functional disturbances in brain damage. In: Arieti S (ed) American handbook of psychiatry. Basic Books, New York
Goode DJ, Manning AA, Middleton JF, Williams B (1981) Fine motor performance before and after treatment in schizophrenic and schizoaffective patients. Psychiatry Res 5:247–255
Hagger C, Buckley P, Kenny JT, Friedman L, Ubogy D, Meltzer HY (1993) Improvement in cognitive functions and psychiatric symptoms in treatment-refractory schizophrenic patients receiving clozapine. Biol Psychiatry 34:702–712
Harvey PD, Pedley M (1989) Auditory and visual distractability in schizophrenia: clinical and medication status correlations. Schizophr Res 2:295–300
Heaton R, Badde L, Johnson K (1978) Neuropsychological test results associated with psychiatric disorders in adults. Psychol Bull 85:141–162
Heaton RK, Crowley TJ (1981) Effects of psychiatric disorders and their somatic treatments on neuropsychological test results. In: Filskov SB, Boll TJ (eds) Handbook of clinical neuropsychology. Wiley, New York
Heaton R, Drexler M (1987) Clinical neuropsychological findings in schizophrenia and aging. In: Miller N, Cohen G (eds) Schizophrenia and aging. Guilford Press, New York
Heilizer F (1959) The effects of chlorpromazine upon psychomotor and psychiatric behavior of chronic schizophrenic patients. J Nerv Ment Dis 128:358–364
Hitri A, Craft RB, Fallon J, Sethi R, Sinha D (1987) Serum neuroleptic and anticholinergic activity in relationship to cognitive toxicity of antiparkinsonian agents in schizophrenic patients. Psychopharmacol Bull 23:33–37
Hoff AL, Riordan H, O'Donnell DW, Morris L, DeLisi LE (1992a) Neuropsychological functioning of first-episode schizophreniform patients. Am J Psychiatry 149:898–903
Hoff AL, Riordan H, O'Donnell D, Stritzke P, Neale C, Boccio A, Anand AK, Delisi LE (1992b) Anomalous lateral sulcus asymmetry and cognitive function in first-episode schizophrenia. Schizophr Bull 18:257–272
Hoff AL, Wieneke M, DeVilliers D, Espinoza S, Gustafson M, Mone R (1993) Effects of clozapine on cognitive function. Presented at the annual meeting of the Amer Psychiatric Assoc, San Francisco
Hollister LE, Csernansky JG (1990) Clinical pharmacology of psychotherapeutic drugs, 3rd edn. Churchill Livingstone, New York
Janowsky DS, Huey L, Storms L, Judd LL (1977) Methylphenidate hydrochloride effects on psychological tests in acute schizophrenic and nonpsychotic patients. Arch Gen Psychiatry 34:189–194
Judson AJ, MacCasland BW (1960) The effects of chlorpromazine on psychological test scores. J Consult Psychology 24:192
Kahneman D (1973) Attention and effort. Prentice-Hall, Englewood Cliffs
Killian GA, Holzman PS, Davis JM, Gibbons R (1984) Effects of psychotropic medication on selected cognitive perceptual measures. J Abn Psychology 93:58–70
King DJ (1990) The effect of neuroleptics on cognitive and psychomotor function. Br J Psychiatry 157:799–811
King DJ, Henry G (1992) The effect of neuroleptics on cognitive and psychomotor function: a preliminary study in healthy volunteers. Br J Psychiatry 160:647–653
Klorman R, Bauer LO, Coons HW, Lewis JL, Peloquin LJ, Perlmutter RA, Ryan RM, Salzman LF, Strauss J (1984) Enhancing effects of methylphenidate on normal young adults cognitive processes. Psychopharmacol Bull 20:3–9
Koh S (1978) Remembering of verbal materials by schizophrenic adults. In: Schwartz S (ed) Language and cognition in schizophrenia. Erlbaum, Hillsdale, NJ, pp 59–99
Kolb B, Whishaw I (1983) Performance of schizophrenic patients on tests sensitive to left or right frontal, temporal, or parietal function in neurological patients. J Nerv Ment Dis 171:435–443

Kornetsky C, Humphries O, Evarts Ev (1957) Comparison of psychological effects of certain centrally acting drugs in man. A.M.A. Arch Neurol Psychiatr 77:318–324
Kornetsky C, Pettit M, Wynne R, Evarts EV (1959) A comparison of the psychological effects of acute and chronic administration of chlorpromazine and secobarbitol in schizophrenic patients. J Mental Science 105:190–198
Kornetsky C, Mirsky AF (1966) On certain psychopharmacological and physiological differences between schizophrenics and normals. Psychopharmacologia 8:309–318
Kraepelin E (1913) Dementia praecox and paraphrenia. Livingston, Edinburgh.
Lee MA, Thompson PA, Meltzer HY (1994) Effects of clozapine on cognitive function in schizophrenia. J Clin Psychiatry 55 [Suppl B]:82–87
Levin BE, Liabre MM, Weiner WJ (1989a) Cognitive impairments associated with early Parkinson's disease. Neurology 39:557–561
Levin S, Yurgelun-Todd D, Craft S (1989b) Contributions of clinical neuropsyhology to the study of schizophrenia. J Abnorm Psychol 98:341–356
Lieberman JA, Koreen AR (1993) Neurochemistry and neuroendocrinology of schizophrenia. Schizophr Bull 19:371–430
Magliozzi JR, Mungas D, Laubly JN, Blunden D (1989) Effect of haloperidol on a symbol digit substitution task in normal adult males. Neuropsychopharmacology 2:29–37
Marder SR, Asarnow RF, Van Putten T (1984) Information processing and neuroleptic response in acute and stabilized schizophrenic patients. Psychiatry Res 13:41–49
Mattila MJ, Mattila ME (1990) Effects of remoxipride on psychomotor performance, alone and in combination with ethanol and diazepam. Acta Psychiatr Scand 82 [Suppl. 358]: 54–55
McGurk SR, Levin ED, Butcher LL (1992) Dopaminergic drugs reverse the impairment of radialarm maze performance caused by lesions involving the cholinergic medial pathway. Neuroscience 50:129–135
Medalia A, Gold J, Merriam A (1988) The effects of neuroleptics on neuropsychological test results of schizophrenics Arch Clinical Neuropsychology 3:249–271
Mednick SA, Machon RA, Huttunen MO, Bonett D (1988) Adult schizophrenia following prenatal exposure to an influenza epidemic. Arch Gen Psychiatry 45:189–192
Meltzer HY, Matsubara S, Lee J-C (1989) The ratios of serotonin$_2$ and dopamine$_2$ affinities differentiate atypical and typical antipsychotic drugs. Psychopharmacol Bull 25:390–392
Miller LS, Faustman WO, Moses JA Jr (1993) Effect of medication status on neuropsychological performance in schizophrenic patients. Arch Clin Neuropsychol 8:250–251 [abstract]
Molchan SE, Martinez RA, Hill JL, Weingartner HJ, Thompson K, Vitiello B, Sunderland T (1992) Increased cognitive sensitivity to scopolamine with age and a perspective on the scopolamine model. Brain Res Brain Res Rev 17:215–226
Morice R, Ingram J (1983) Language complexity and age of onset of schizophrenia. Psychiatry Res 9:233–242
Moses JA Jr (1984) The effect of presence or absence of neuroleptic medication treatment on Luria-Nebraska neuropsychological battery performance in a schizophrenic population. Int J Clin Neuropsychology 6:249–251
Moses JA Jr, Maruish ME (1988) A critical review of the Luria-Nebraska neuropsychological battery literature: IV. Cognitive deficit in schizophrenia and related disorders. Int J Clin Neuropsychol 10:51–62
Nakra BR, Margolis RB, Gfeller JD, Grossberg GT, Sata LS (1992) The effect of a single low dose of trihexyphenidyl on memory functioning in the healthy elderly. Int Psychoger 4:207–214
Nickols JE (1958) A controlled exploratory investigation into the effects of thorazine upon mental test scores of chronic hospitalized schizophrenics. The Psychological Record 8:67–76

Nuechterlein K, Dawson M (1984) Information processing and attentional functioning in the developmental course of schizophrenic disorders. Schiz Bull 10:160–203

Obiols JE, Marcos T, Salamero M (1987) Ventricular enlargement and neuropsychological testing in schizophrenia. Acta Psychiatr Scand 76:199–202

Oltmanns TF, Ohayon J, Neale JM (1978) The effect of anti-psychotic medication and diagnostic criteria on distractibility in schizophrenia. J Psychiat Res 14:81–91

Pearl D (1962) Phenothiazine effects in chronic schizophrenia. J Clin Psychology 18:86–89

Perlick D, Stastny P, Katz I, Mayer M, Mattis S (1986) Memory deficits and anticholinergic levels in chronic schizophrenia. Am J Psychiatry 143:230–232

Pfefferbaum A, Zipursky R, Lim KO, Zatz LM, Stah; SM, Jernigan TL (1988) Computed tomograpic evidence for generalized sulcal and ventricular enlargement in schizophrenia. Arch Gen Psychiatry 45:633–40

Pugh LA (1968) Response time and electrodermal measures in chronic schizophrenia: the effects of chlorpromazine. J Nerv Ment Dis 146:62–70

Pycock CJ, Kerwin RW, Carter CJ (1980) Effect of lesion of cortical dopamine terminals on subcortical dopamine in rats. Nature 286:74–77

Richardson JS, Keegan DL, Bowen RC, Blackshaw SL, Cebrian-Perez S, Dayal N, Saleh S, Shrikhande S (1994) Verbal learning by major depressive disorder patients during treatment with fluoxetine or amitriptyline. Int Clin Psychopharmacology 9:35–40

Roberts GW (1990) Schizophrenia: the cellular biology of a functional psychosis. Trends Neurosci 13:207–211

Rosofsky I, Levin S, Holtzman P (1982) Psychomotility in the functional psychoses. J Abnorm Psychol 91:71–74

Sala M, Braida D, Calcaterra P, Leone MP, Comotti FA, Gianola S, Gori E (1991) Effect of centrally administered atropine and pirenzepine on radial arm maze performance in the rat. Eur J Pharmacol 194:45–49

Saykin A, Gur RC, Gur RE, Mozley D, Mozley L, Resnick SM, Kester B, Stafiniak P (1991) Neuropsychological function in schizophrenia: selective impairment in memory and learning. Arch Gen Psychiatry 48:618–624

Saykin AJ, Shtasel DL, Gur RE, Kester DB, Mozley LH, Stafiniak P, Gur RC (1994) Neuropsychological deficits in neuroleptic-native patients with first-episode schizophrenia. Arch Gen Psychiatry 51:124–131

Schultz SC, Koller MM, Kishore PR, Hamer RM, Gehl JJ, Friedel RO (1983) Ventricular enlargement in teenage patients with schizophrenia spectrum disorder. Am J Psychiatry 140:1592–1595

Seidman L (1983) Schizophrenia and brain dysfunction: an integration of recent neurodiagnostic findings. Psychol Bull 94:195–238

Seidman LJ, Pepple JR, Faraone SV, Kremen WS, Green AI, Brown WA, Tsuang MT (1993) Neuropsychological performance in chronic schizophrenia in response to neuroleptic dose reduction. Biol Psychiatry 33:575–584

Shakow D (1962) Segmental set: a theory of formal psychological deficit in schizophrenia. Arch Gen Psychiatry 6:1–17

Shatin L, Rockmore L, Funk IC (1956) Response of psychiatric patients to massive doses of thorzine: II. Psychological test performance and comparative drug evaluation. Psychiatr Q 30:402–416

Smet IC, Goldman RS, Red Cloud S, Gupta P, Kilaru S, Tandon R, Berent S (1994) Neuropsychological performance in schizophrenic inpatients as a function of medication status. Biol Psychiatry 35:666 [abstract]

Spiegel DE, Keith-Spiegel P (1967) The effects of carphenazine, trifluoperazine, and chlorpromazine on ward behavior, physiological functioning and psychological test scores in chronic schizophrenic patients. J Nerv Ment Dis 144:111–116

Spohn HE, Lacoursiere RB, Thompson K, Coyne L (1977) Phenothiazine effects on psychological and psychophysiological dysfunction in chronic schizophrenics. Arch Gen Psychiatry 34:633–644

Spohn Re, Coyne L, Lacoursiere R, Mazur D, Hayes K (1985) Relation of neuroleptic dose and tardive dyskinesia to attention, information-processing, and psychophysiology in medicated schizophrenics. Arch Gen Psychiatry 42:849–859

Spohn HE, Strauss ME (1989) Relation of neuroleptic and anticholinergic medication to cognitive functions in schizophrenia. J Abnorm Psychol 98:367–380

Stern RG, Kahn RS, Davidson M (1993a) Predictors of response to neuroleptic treatment in schizophrenia. Psychiatr Clin North Am 16:313–338

Stern Y, Richards M, Sano M, Mayeux (1993b) Comparison of cognitive changes in patients with Alzheimer's and Parkinson's disease. Arch Neurol 50:1040–1045

Strauss ME, Lew MF, Coyle JT, Tune LE (1985) Psychopharmacologic and clinical correlates of attenion in chronic schizophrenia. Am J Psychiatry 142:497–499

Strauss ME, Reynolds KS, Jayaram G, Tune LE (1990) Effects of anticholinergic medication on memory in schizophrenia. Schizophr Res 3:127–129

Strauss WH, Klieser E (1990) Cognitive disturbances in neuroleptic therapy. Acta Psychiatr Scand 82 [Suppl 358]:56–57

Stuss DT, Benson DK (1984) Neuropsychological studies of the frontal lobes. Psychol Bull 95:3–28

Sullivan EV, Sagar HJ, Gabrieli JDE, Corkin S, Growdon JH (1989) Different cognitive profiles on standard behavioral tests in Parkinson's disease and Alzheimer's disease. J Clin Exper Neuropsychol 11:799–820

Sweeney JA, Keilp JG, Haas GL, Hill J, Weiden PJ (1991) Relationships between medication treatments and neuropsychological test performance in schizophrenia. Psychiatr Res 37:297–308

Syndulko K, Gilden ER, Hansch EC, Potvin AR, Tourtellotte WW, Potvin JH (1981) Decreased verbal memory associated with anticholinergic treatment in Parkinson's disease patients. Int J Neurosci 14:61–66

Takei N, Sham P, O'callaghan E, Murray GK, Glover G, Murray RM (1994) Prenatal exposure to influenze and the development of schizophrenia: is the effect confined to females? Am J Psychiatry 151:117–119

Tamlyn D, McKenna PJ, Mortimer AM, Lund CE, Hammond S, Baddeley AD (1992) Memory impairment in schizophrenia: its extent, affiliations and neuropsychological character. Psychol Med 22:101–115

Taylor M, Abrams R (1984) Cognitive impairment in schizophrenia. Am J Psychiatry 141:196–201

Tomer R (1989) Asymmetrical effects of neuroleptics on psychotic patients' performance on a tactile discrimination task. J Nerv Ment Dis 177:699–700

Tomer R, Flor-Henry P (1989) Neuroleptics reverse attention asymmetries in schizophrenic patients. Biol Psychiatry 25:852–860

Tomer R (1990) Neuroleptic effects on interhemispheric and intrahemispheric performance of tactile discrimination tasks by schizophrenic patients. Psychiatry Res 32:289–296

Tsuang MT, Gilbertson MW, Faraone SV (1991) The genetics of schizophrenia: current knowledge and future directions. Schizophr Res 4:157–171

Tune LE, Strauss ME, Lew MF, Breitlinger E, Coyle JT (1982) Serum levels of anticholinergic drugs and impaired recent memory in chronic schizophrenic patients. Am J Psychiatry 139:1460–1462

van Herwaarden G, Berger HJ, Horstink MW (1993) Short-term memory in Parkinson's disease after withdrawal of long-term anticholinergic therapy. Clin Neuropharmacology 16:438–443

van Spaendonck KPM, Berger HJC, Horstink MWI, Buytenhuijs EL, Cools AR (1993) Impaired cognitive shifting in parkinsonian patients on anticholinergic therapy. Neuropsychologia 31:407–411

Wagman AMI, Heinrichs DW, Carpenter WT Jr. (1987) Deficit and nondeficit forms of schizophrenia: neuropsychological evaluation. Psychiatr Res 22:319–330

Wahba M, Donlon PT, Meadow A (1981) Cognitive changes in acute schizophrenia with brief neuroleptic treatment. Am J Psychiatry 138:1307–1310

Weinberger DR, DeLisi LE, Perman GP, Targum S, Wyatt RJ (1982) Computed tomography in schizophreniform disorder and other acute psychiatric disorders. Arch Gen Psychiatry 39:778–783

Weinberger DR, Berman K, Zec R (1986) Physiological dysfunction of dorsolateral prefrontal cortex in schizophrenia: I. Regional cerebral blood flow (rCBF) evidence. Arch Gen Psychiatry 43:114–125

Weinberger DR, Berman KF, Suddath R, Torrey EF (1992) Evidence of dysfunction of a prefrontal-limbic network in schizophrenia: a magnetic resonance imaging and regional cerebral blood flow study of discordent monozygotic twins. Am J Psychiatry 149:890–897

Wynne RD, Kornetsky C (1960) The effects of chlorpromazine and secobarbital on the reaction times of chronic schizophrenics. Psychopharmacologia 1:294–302

Young AMJ, Joseph MH, Gray JA (1993) Latent inhibition of conditioned dopamine release in rat nucleus accumbens. Neuroscience 54:5–9

Zec RF, Weinberger DR (1986) Relationship between CT scan findings and neuropsychological performance in chronic schizophrenia. Psychiat Clin North Am 9:49–61

Zemishlany Z, Thorne AE (1991) Anticholinergic challenge and cognitive functions: a comparison between young and elderly normal subjects. Isr J Psychiatry Relat Sci 28:32–41

Zipursky RB, Lim KO, Sullivan EV, Browm BW, Pfefferbaum A (1992) Widespread cerebral gray matter volume deficits in schizophrenia. Arch Gen Psychiatry 49:195–205

CHAPTER 16

Antipsychotic Drugs in Children and Adolescents

EDWARD S. BRODKIN, CHRISTOPHER J. McDOUGLE, and JAMES F. LECKMAN

A. Introduction

Antipsychotic drugs are prescribed for adults to decrease the psychotic symptoms associated with schizophrenia and affective disorders. Psychosis is rare in childhood, however, and antipsychotics have been prescribed for children primarily to treat other symptoms such as severe overactivity, aggressivity towards self or others, social withdrawal, tics, and stereotypies. These symptoms are associated with a variety of neuropsychiatric disorders of childhood onset. Ideally, an antipsychotic medication will not only decrease troublesome symptoms, but by doing so will also promote the child's intellectual, emotional, and social development. However, some of the side effects of neuroleptics can interfere with such growth and development. The physician should weigh the risks and benefits of neuroleptic treatment and, if treatment is begun, carefully titrate the dose (CAMPBELL 1985). This chapter will focus on the double-blind, placebo-controlled studies of antipsychotics in children and adolescents, as well as on some significant case reports.

B. Indications

I. Autism

Autistic disorder (AD) is a neuropsychiatric syndrome with onset in childhood characterized by a serious disturbance of reciprocal social interaction, abnormalities in verbal and nonverbal communication, and a restricted repertoire of interests. Patients with this disorder may be aggressive towards themselves or others and may have prominent repetitive behavior. Ths most effective currently available treatment for autism consists of some combination of psychosocial treatment and pharmacotherapy designed to reduce the trouble some symptoms of the disorder and enhance the patient's ability to benefit from an educational program (McDOUGLE et al. 1994). Antipsychotic drugs, in particular haloperidol, have been studied extensively in children with AD. There are several double-blind, placebo-controlled studies involving autistic children which have shown haloperidol, at doses of 0.5–3.0 mg/day, to be significantly superior to placebo in reducing social withdrawal, stereotypies, hyperactivity, irritability, negativism, and labile and angry affect of autistic children aged 2–7 years (Anderson et al. 1984; Anderson et al. 1989). Haloperidol, at optimal doses, does not seem to impair the cognition of autistic children (Anderson et al. 1989). In fact, some studies (Campbell et al. 1982a; Anderson et al. 1984) suggest that haloperidol improves cognition in some autistic patients. For many autistic children, there seems to be a dose of haloperidol (usually between 0.5 and 4.0 mg/day) at which behavior is improved and cognition may be unchanged or improved without significant untoward effects (Anderson et al. 1984; Teicher and Glod 1990). Haloperidol has been found to improve language acquisition (Campbell 1985) and discrimination learning (Anderson et al. 1984) in children with AD. These therapeutic effects usually occur within 3 months of initiating therapy (Campbell 1985). Perry and coworkers found that haloperidol remained efficacious over 6 months of treatment in autistic children. They found no difference between continuous (daily) and discontinuous (5 days on, 2 days off) drug treatment either in terms of efficacy or the incidence of side effects (e.g., withdrawal dyskinesias and tardive dyskinesia) (Perry et al. 1989a). Other investigators have followed autistic children on haloperidol for as long as 2½ years and have found that haloperidol remains efficacious during this period (Campbell 1985).

Joshi et al. reported the use of haloperidol at a mean daily dose of less than 2 mg/day in childhood-onset pervasive developmental disorder (as distinguished from infantile-onset autism and schizophrenia). They found significant reductions in hyperactivity and impulsivity, improvement in social relatedness, and minimal side effects (Joshi et al. 1988). Pimozide, rather than haloperidol, may be useful in hypoactive children with autism (Ernst et al. 1992). Low-potency neuroleptics such as chlorpromazine tend to produce excessive sedation in patients with AD, whereas trifluoperazine, thiothixene, fluphenazine, and molindone may be promising alternatives to haloperidol (Campbell 1985).

Recent controlled studies indicate that serotonin uptake inhibitors (e.g., clomipramine and fluvoxamine) may be useful in reducing the repetitive behaviors and aggression and improving social relatedness in children and adults with AD, respectively (Gordon et al. 1992; Gordon et al. 1993; McDougle et al. 1994). Because of their relative safety when compared with neuroleptics, this class of drugs may be an alternative primary treatment for AD (McDougle et al. 1994).

II. Tourette's Syndrome

Tourette's syndrome (TS) is a neuropsychiatric disorder that has its onset in childhood or adolescence and is characterized by recurrent motor movements; phonic tics; variation in severity of the symptoms over months; duration of more than 1 year; and, in many patients, a premonitory tension which precedes the tics and capitulation following the tics. There is a wide range of severity of tics from virtually unnoticeable to debilitating. Many patients with TS have comorbid obsessions, compulsions, and/or attentional problems (Cohen et al. 1992).

The primary goal of treatment of TS is to help the child achieve normal developmental tasks. This does not always require medication to treat the tics. In fact, at times the side effects of certain medications may make the medication more of an impediment than a help in this regard (Cohen et al. 1992).

In those cases in which medication has been indicated, haloperidol has been the standard treatment of TS for almost three decades (Shapiro et al. 1973), and its efficacy in decreasing the frequency and severity of tics has been confirmed in double-blind, placebo-controlled studies (Shapiro et al. 1989). Patients may be started at 0.5 mg/day and slowly titrated up to 1–3 mg/day, usually divided into two daily doses. Many patients have almost complete resolution of tics at these doses with minimal side effects. If the patient does not have adequate symptomatic relief at these dosage levels, they may show further improvement at higher doses (5–10 mg/day). The risk of side effects, however, becomes more prominent at the higher doses. While up to 70% of patients benefit from haloperidol initially, only 20%–30% continue to use haloperidol over a period of years. The drug is usually discontinued due to side effects (e.g., sedation, cognitive and motivational dulling, irritability, dysphoria, and/or extrapyramidal symptoms) (Cohen et al. 1992). Dystonia can be treated with benztropine or amantadine (Borison et al. 1983). In addition to the typical side effects of haloperidol, there have been case reports of school avoidance (in children) or social phobia (in adults) associated with haloperidol treatment in patients with TS. These phobias have begun shortly after the initiation of haloperidol therapy and have resolved shortly after haloperidol has been discontinued (Mikkelsen et al. 1981). Similar phobias have been described in TS patients treated with pimozide (Linet 1985). The sedating effects of haloperidol may impair the school performance of patients with TS (Campbell and Deutsch 1985).

Pimozide has been shown to be as efficacious as haloperidol for treating the tics associated with TS, and may be less sedating (Shapiro et al. 1989; Ross and Moldofsky 1978; Shapiro and Shapiro 1984). Haloperidol may be more effective than pimozide in decreasing phonic tics, but the two drugs appear to be equally effective in decreasing motor tics (Campbell and Deutsch 1985). The initial dose of pimozide is usually 1 mg/day and the dosage is gradually increased until symptomatic relief is attained or a maximum dose of 6–10 mg/day (0.2 mg/kg) for children and 20 mg/day for adults is reached. Because pimozide has a long half-life, one daily dose may be prescribed. Pimozide has a side effect profile similar to haloperidol, and in addition pimozide may cause electrocardiogram (ECG) changes (see below). Patients should receive an ECG prior to starting pimozide and at least as frequently as every 3 months while on the medication (Cohen et al. 1992). Bruun has reported a variety of other side effects of neuroleptic treatment in some children with TS including dysphoria, akathisia (which may exacerbate tics), irritability, and aggressive behavior (Bruun 1988).

If a patient cannot tolerate haloperidol or pimozide because of side effects, a phenothiazine such as fluphenazine may be tried and found effective (Borison et al. 1983). Borison and coworkers compared haloperidol to fluphenazine and trifluoperazine in patients with TS. All three drugs had a statistically significant effect in decreasing tics when compared to placebo. Compared to each other, all three drugs were equally efficacious in suppressing tics, but fluphenazine had the least prominent side effects (Borison et al. 1983). Bruggeman and coworkers conducted an open trial of risperidone in a small number of patients with TS and found it effective in decreasing the frequency and severity of tics. Risperidone, at a mean daily dose of 3.9 mg, produced no extrapyramidal symptoms in these patients (Bruggeman et al. 1994). Dosages used are generally lower than those used to treat psychosis. Clonidine is also effective for the treatment of TS, particularly the motor tics (Leckman et al. 1991), although it is less consistently helpful than the neuroleptics. Some clinicians prefer clonidine because of its better side effect profile and its potential for enhancing attention (Cohen et al. 1992). Clonidine has a slower onset of therapeutic effects than haloperidol (Borison et al. 1983). Penfluridol, a diphenylbutylpiperidine like pimozide, has been effective in decreasing tics in some patients with TS (Shapiro et al. 1983).

Neuroleptics may be indicated in other movement disorders in children and adolescents such as juvenile Huntington's disease (Teicher and Glod 1990).

III. Obsessive-Compulsive Disorder

Obsessive-compulsive disorder (OCD) is characterized by recurrent and disturbing thoughts (obsessions) and/or repetitive, stereotyped behaviors that the person feels driven to perform (compulsions) although recognizing them as irrational or excessive. While serotonin uptake inhibitors have been shown to be effective in 40%–60% of children and adults with OCD, many patients remain clinically unchanged after treatment with these agents (McDougle et al. 1993). Recent reports indicate that the addition of low doses of high-potency neuroleptics such as pimozide (McDougle et al. 1990) and haloperidol (McDougle et al. 1993) to ongoing serotonin uptake inhibitors may further reduce obsessive-compulsive symptoms in adults with refractory OCD, particularly those with a personal history of a comorbid chronic tic disorder. Studies of neuroleptic addition to ongoing serotonin uptake inhibitor treatment in children and adolescents with refractory OCD have yet to be published.

IV. Mental Retardation

Neuroleptics have been used extensively in institutional settings for treatment of impulsivity, hyperactivity, self-injury, and aggressiveness of nonautistic mentally retarded people (Campbell and Deutsch 1985). Thioridazine has been the most frequently studied, followed by chlorpromazine, trifluoperazine, and haloperidol (Whitaker and Rao 1992). Many of these studies have been poorly designed, but the more methodologically sound studies have shown that these agents (particularly chlorpromazine and thioridazine) may impair cognitive functioning. For example, there are studies which demonstrate an increase in IQ in mentally retarded subjects following the reduction of neuroleptic dose (Campbell and Deutsch 1985; Breuning et al. 1983). Aman has carefully reviewed studies done since the 1950s on the use of chlorpromazine in mentally retarded patients. The results from studies with adequate methodologic rigor showed little or no benefit from chlorpromazine, and, in some cases, adverse effects on learning and behavior. In his review of thioridazine in mentally retarded patients, Aman found thioridazine effective in reducing stereotypies, withdrawal, and bizarre behavior in some circumstances, but possibly at the expense of learning. Aman reported that haloperidol 0.05 mg/kg per day produced improvement in ratings of stereotypy in mentally retarded patients. He found that subjects with higher degrees of stereotypy at baseline showed a more substantial decrease in the degree of stereotypy with drug treatment. In general, stereotypy seems to be the clinical variable that is most responsive to neuroleptic treatment among mentally retarded patients (Aman 1989). Stereotypy alone is usually not a sufficient reason to administer a neuroleptic. It is possible (although certainly not demonstrated) that the presence of stereotypic behavior is associated with a good response of other target psychiatric symptoms to neuroleptic drugs (Aman 1989). Finally, it is not clear whether the benefits of neuroleptics seen in mentally retarded patients are due to specific behavioral effects of neuroleptics or a nonspecific sedative action. In general, a trial of a neuroleptic drug in a patient with mental retardation with aggressive or self-injurious behavior is justified only when attempts at psychosocial intervention have failed and the behavior places the patient and/or others at risk of harm. Furthermore, if the medication is given, it should be as an adjunct to psychosocial intervention. A number of other drugs, including lithium, carbamazepine, and valproic acid, show promise in the treatment of explosive rage and may be alternatives to neuroleptics in the treatment of extreme aggressiveness in mentally retarded individuals (Campbell et al. 1992).

V. Early-Onset Schizophrenia

Schizophrenia is rare in prepubertal children. The diagnostic criteria for early-onset schizophrenia have changed significantly during the last several decades. In the 1950s, childhood schizophrenia was felt to be a disorder which is distinct from autism and the pervasive developmental disorders. In the 1960s, when DSM-II was developed, autism and all forms of childhood psychosis were lumped to together under the classification "childhood schizophrenia." Therefore, much of what was written during this era about childhood schizophrenia in fact referred to patients who would now be considered autistic. Further research demonstrated more clearly the distinction between autism and childhood-onset schizophrenia, and in DSM-III, the diagnosis of schizophrenia is reserved for children who meet essentially the same symptom criteria as adults (McClellan and Werry 1992). Recent research has validated the distinction between autistic disorder and early-onset schizophrenia. There are children who meet DSM-III criteria for schizophrenia but not for pervasive developmental disorder. Moreover, autistic children do not generally become schizophrenic in later life (Volkmar 1991; Green et al. 1984). Schizophrenia usually has an onset later in childhood than autism; schizophrenic children usually have hallucinations and/or delusions, whereas autistic children do not; and a higher percentage of autistic children are mentally retarded (Green et al. 1984). Partly because of the history of diagnostic confusion, research on schizophrenia with onset in childhood or early adolescence has been limited in both amount and in methodological rigor (McClellan and Werry 1992).

Schizophrenia is characterized by at least a 1-week period of delusions, hallucinations, thought disorder, catatonic behavior, or inappropriate affect, as well as a deterioration of social, occupational, or self-care functioning. The symptoms must last for at least 6 months (the overt psychotic symptoms may only last 1 week but either prodromal or residual symptoms last longer). Early-onset schizophrenia is usually marked by hallucinations, thought disorder, and flattened affect; systematic delusions, catatonia, and poverty of thought are less commonly found. It is important to be aware that hallucinations are not pathognomonic for schizophrenia in childhood. Mood disorders and medical or neurological disorders such as drug-induced or toxic delirium, seizure disorder, metabolic or infections disorders, or neoplastic or degenerative central nervous system disease must also be ruled out (McClellan and Werry 1992).

There has been no systematic study of the efficacy of antipsychotics in childhood-onset schizophrenia. From the work that has been done to date, it appears that antipsychotics may be less effective in reducing the symptoms of prepubertal schizophrenics than in reducing the symptoms of adults with the disorder (Campbell and Spencer 1988; Teicher and Glod 1990). For the treatment of children, high-potency neuroleptics are preferable to the more sedating, low-potency drugs (Realmuto et al. 1984). A trial of at least 4–6 weeks is necessary to determine whether a drug will be beneficial. After 4–6 months of treatment, consideration should be given to tapering and discontinuing the drug to determine whether further treatment is necessary (Campbell and Spencer 1988).

Pool et al. reported a double-blind, placebo-controlled study of the use of neuroleptics in schizophrenic adolescents. The authors compared the efficacy of loxapine succinate, haloperidol, and placebo in 75 adolescent patients with schizophrenia, ages 13–18 years. Loxapine and haloperidol were superior to placebo in reducing symptoms as measured by the Brief Psychiatric Rating Scale (BPRS), Nurses' Observation Scale for Inpatient Evaluation (NOSIE), and Clinical Global Impression (CGI) scale. The main side effects of the active agents were sedation and extrapyramidal symptoms (POOL et al. 1976). REALMUTO and coworkers assigned 21 schizophrenic adolescents to treatment with either thiothixene or thioridazine. Both drugs reduced BPRS scores and enhanced CGI scores; however, even with treatment, the adolescents remained significantly impaired (REALMUTO et al. 1984).

VI. Conduct Disorder and Explosive Rage

Conduct disorders are characterized by repetitive antisocial behaviors such as aggressiveness, stealing, lying, truancy, setting fires, and running away. In boys, the median age of onset is 7 years, and in girls, 13 years. Treatment of this disorder should always include psychosocial interventions, and if pharmacotherapy is used, it should be viewed as adjunctive. Pharmacotherapy cannot address the social factors, including chaotic households, which may contribute to the disorder. Neuroleptics may be helpful in the treatment of explosive aggression, but are less likely to be of benefit in the treatment of a conduct-disordered child without such behavior. While neuroleptics such as haloperidol, pimozide, molindone, thioridazine, and chlorpromazine have been used extensively in the treatment of aggressive children, it is not clear that neuroleptics have a specific antiaggressive effect. Neuroleptics may only cause nonspecific sedation (CAMPBELL et al. 1982b).

A double-blind, placebo-controlled study involving children aged 5 to 13 years with conduct disorder, aggressive type, compared haloperidol (1–6 mg per day) to lithium (500–2000 mg per day) and to placebo. This study found both haloperidol and lithium to be superior to placebo in decreasing symptoms (CAMPBELL et al. 1984). Haloperidol had more troublesome side effects in these patients than lithium; for example, haloperidol had mildly negative effects on tests of cognition, whereas lithium had essentially no negative effects (PLATT et al. 1984). GREENHILL and coworkers found molindone as effective as thioridazine in decreasing the symptoms of conduct disorder in a double-blind, placebo-controlled study (GREENHILL et al. 1985). Prior to prescribing medication, the child should be carefully observed for at least 2 weeks to ascertain the patient's baseline level of aggressiveness. The decision to use neuroleptics in the treatment of aggression must include a careful consideration of the risks of tardive dykinesia and other distressing side effects. Alternative drugs which are being investigated for the treatment of explosive aggression include lithium, carbamazepine, valproic acid, and propranolol (CAMPBELL et al. 1992). Lithium appears to be a more effective pharmacologic treatment for conduct disorder than haloperidol (CAMPBELL and SPENCER 1988). In any case, if a particular medication is administered to a child for the treatment of aggressive behavior, it should be tapered or discontinued after 4–6 months of treatment in order to determine whether pharmacotherapy is still necessary (CAMPBELL et al. 1992).

VII. Attention-Deficit Hyperactivity Disorder

Neuroleptics have been observed to have efficacy in reducing some symptoms of attention-deficit hyperactivity disorder (ADHD). Interestingly, neuroleptics are therapeutically efficacious in the same group of patients who are helped by stimulants. This phenomenon seems paradoxical since neuroleptics and stimulants have such different mechanisms of action (GUALTIERI and PATTERSON 1986). Because stimulants are at least as effective and generally safer to use, the use of a neuroleptic would only be justified if the patient were severely symptomatic and not responsive to stimulants (CAMPBELL 1985). Double-blind, placebo-controlled studies have found chlorpromazine (75–200 mg/day) and thioridazine (50–300 mg/day) to be effective in decreasing symptoms of ADHD (GREENBERG et al. 1972; CAMPBELL 1985). GITTELMAN-KLEIN et al. compared methylphenidate, thioridazine, a methylphenidate and thioridazine combination, and placebo in the treatment of hyperactive children. All three active treatments were superior to placebo. Methylphenidate alone and the methylphenidate/thioridazine combination were superior to thioridazine alone (GITTELMAN-KLEIN et al. 1976).

VIII. Bipolar Disorder, Manic Phase

It is uncommon for bipolar disorder to manifest itself prior to puberty (CAMPBELL 1985). However, neuroleptics can be useful adjuncts to mood stabilizers in treating the manic phase of bipolar disorder in adolescents and adults.

C. Pharmacology

I. Pharmacokinetics and Pharmacodynamics

Individuals vary in their rate of metabolizing antipsychotic drugs. For example, haloperidol plasma levels vary by ten- to 15-fold among children in the same age group receiving the same milligram (mg) per kilogram (kg) dose. In any particular individual, however, plasma levels tend to remain stable on a fixed dose of haloperidol once a steady state has been achieved, and this tends to be true for other neuroleptics as well (BALDESSARINI 1985; TEICHER and GLOD 1990). An exception to this generalization is the metabolism of chlorpromazine. Children given a constant dose of chlorpromazine will generally have declining chlorpromazine blood levels, perhaps due to autoinduction of hepatic enzymes (RIVERA-CALIMLIM et al. 1979; RIDDLE 1991).

Neuroleptics frequently raise blood levels of tricyclic antidepressants, serotonin uptake inhibitors, monoamine oxidase inhibitors (MAOIs), phenytoin, and beta blockers; and conversely, neuroleptic blood levels may be increased by tricyclic antidepressants, serotonin uptake inhibitors, MAOIs, and beta blockers. Neuroleptics may increase the cardiac depressant effects or increase the risk of arrhythmias with quinidine. When combined with other anticholinergic drugs, neuroleptics with prominent anticholinergic activity may cause delirium (TEICHER and GLOD 1990).

Infants have a decreased capacity to metabolize and eliminate antipsychotic drugs, although children eliminate them approximately twice as rapidly as adults (BALDESSARINI 1985). For example, the elimination half-life for haloperidol in children is 4–16h vs 16–24h in adults. Children's rapid hepatic biotransformation of antipsychotics is probably due to their relatively larger hepatic mass (TEICHER and GLOD 1990). The same mg per kg dose in a child will lead to lower neuroleptic blood levels than in an adult (RIVERA-CALIMLIM et al. 1979; TEICHER and GLOD 1990). Nevertheless, children are often treated with lower mg per kg dosages of neuroleptics because they demonstrate clinical improvement and develop unwanted side effects at lower blood levels of the drugs (WHITAKER and RAO 1992; RIVERA-CALIMLIM et al. 1979). Indirect evidence from preclinical and clinical studies suggests that children are more sensitive to neuroleptics because of their relatively greater density of brain D_2 and D_1 dopamine receptors (SEEMAN et al. 1987). In the rat forebrain, the levels of D_2 receptor sites increase substantially in early postnatal life followed by a decrease in receptor number during early adult life (CAMPBELL et al. 1988a; O'BOYLE and WADDINGTON 1986; MURRIN and ZENG 1986; MURRIN et al. 1985). Developing rats are more sensitive (require lower blood levels of neuroleptic to produce the same effect) to the activity-reducing, sedative, and cataleptic effects of haloperidol than adult rats. These differences do not appear to be due to pharmacokinetic effects because they also have been noted with direct administration of haloperidol or perphenazine into the brain (TEICHER and BALDESSARINI 1987). Studies of human brain also show that children have a greater density of D_1 and D_2 receptors than adults (SEEMAN et al. 1987).

The Food and Drug Administration (FDA) guidelines for the lowest age limit at which particular neuroleptics may be prescribed vary substantially. Examples include 6 months (chlorpromazine), 2 years (thioridazine, pimozide), 3 years (haloperidol), 6 years (chlorprothixene, trifluoperazine), 12 years (mesoridazine, molindone, perphenazine), and 16 years (loxitane) (TEICHER and GLOD 1990).

II. Blood Levels

There are limited data available regarding the association between neuroleptic blood levels and therapeutic efficacy in children and adolescents. Children with psychosis may show a clinical improvement at haloperidol plasma levels of 6–10ng/ml, whereas adult patients generally require levels of 14–30ng/ml. Children with tic disorders usually have a good response at lower plasma levels of haloperidol such as 0.5–4ng/ml. Plasma levels are well correlated with the development of untoward effects in children. Seventy-five percent of children with haloperidol blood levels of 6–9 ng/ml will have significant side effects, whereas only 20% with levels of 3–6 ng/ml will have significant side effects (TEICHER and GLOD 1990). Despite these associations, it is generally unnecessary to obtain neuroleptic blood levels in treating children or adolescents because blood levels correlate poorly with therapeutic effects. The clinician should titrate the dose based on a clinical assessment of therapeutic response and side effects, bearing in mind that certain individuals will require much lower than the standard doses used for most children or adolescents (TEICHER and GLOD 1990).

D. Mechanism of Action

Antipsychotic drugs have a variety of behavioral effects, some therapeutic (e.g., decreasing psychotic symptoms) and some unwanted (e.g., extrapyramidal symptoms). It is not clear whether there is a unitary mechanism or multiple mechanisms through which antipsychotics produce these diverse effects. In adults, it has been observed that the antipsychotic potency of various drugs roughly parallels their binding affinity to the dopamine D_2 receptor in vitro. Recently, two novel D_2-like receptors, the D_3 and D_4 receptors, were discovered. These receptors have similar binding affinities for typical antipsychotic drugs. Atypical antipsychotics have different binding profiles. For example, clozapine has a much higher affinity for the D_4 than for the D_2 receptor. Clozapine also binds to the D_1, D_3, serotonin 5-HT_{1c} and 5-HT_2, muscarinic cholinergic, and alpha1-adrenergic receptors. Risperidone has high affinity for the 5-HT_2 receptor, and lower affinity for the D_2 receptor. The D_2 receptor is found in the mesolimbic, mesocortical, and nigrostriatal dopamine projections in the brain, whereas the D_4 receptor is localized to limbic areas and is found in very low levels in the striatum (HYMAN and NESTLER 1993).

Generally, it takes days to weeks to achieve maximum clinical efficacy with antipsychotic drugs. Therefore, it is most likely not dopamine receptor antagonism alone which results in the therapeutic efficacy of these medications (neuroleptics have been shown to occupy dopamine receptors within hours of administration), but rather chronic adaptations to dopamine receptor blockade which occur over weeks within the brain (HYMAN and NESTLER 1993).

The firing rate of midbrain dopamine neurons increases acutely with the administration of antipsychotic drugs. After chronic administration, the neurons develop "depolarization blockade" in which they no longer produce action potentials in response to excitatory stimuli. Chronic treatment also produces changes in gene expression in particular neuronal populations. For example, there is upregulation of the D_2 receptor in various brain regions. In the striatum, the expression of genes for proenkephalin and neurotensin is increased with chronic haloperidol administration. Haloperiol also increases the expression of particular transcription factors, e.g., c-fos and zif268, in the basal ganglia and nucleus accumbens. These transcription factors, in turn, regulate the expression of other genes in these regions. None of these changes is known to be responsible for antipsychotic efficacy or for the other behavioral effects of antipsychotics (e.g., decreasing overactivity, aggressivity, and tics); however, the time course of the changes parallels that of onset of clinical response (HYMAN and NESTLER 1993).

E. Therapeutic Use

Certain generalizations can be made about the use of antipsychotic drugs in children and adolescents, regardless of the disorder being treated. Low-potency agents are usually avoided because of their sedative effects and impact on learning. The clinician should start at the lowest possible dose and slowly titrate upwards. Divided doses are initially advisable to avoid a dystonic reaction. A therapeutic response may take several weeks to emerge at a particular dose. Timing of doses in relationship to meals is usually not important because food does not significantly impair absorption, and gastrointestinal distress is uncommon (TEICHER and GLOD 1990). Monitoring during neuroleptic treatment should include a baseline medical history (including eating and sleeping patterns) and physical exam (including height, weight, blood pressure, pulse), complete blood cell count (CBC), LFTs, TFTs, electrolytes, blood urea nitrogen (BUN), Cr, glucose, ECG (for pimozide), and the administration of the Abnormal Involuntary Movements Scale (AIMS), preferably videotaped. The patient should usually be followed up weekly for the first 6 weeks of treatment to monitor the emergence of side effects. The AIMS exam should be repeated every 3 months. LFTs should be obtained at least annually. CBC with differential should be obtained should any signs of infection emerge in order to detect agranulocytosis. An ECG should be done at least annually, and every 3 months for pimozide. Informed consent must be obtained from the child's parents and must be clearly documented in the patient's chart. The duration of treatment is determined by the particular patient, the disorder in question, and the response to treatment. Patients with chronic disorders who show only a marginal clinical improvement during neuroleptic treatment might be better off without the medication, given the risk of side effects. Discontinuation should occur gradually over months (WHITAKER and RAO 1992).

F. Use in Pregnancy and Nursing

Antipsychotics cross the blood-placenta barrier, enter the fetal circulation, and cross the fetal blood-brain barrier. There is little evidence for sustained neurologic effects of in utero exposure to neuroleptics on newborn laboratory animals (BALDESSARINI 1985). There is no direct evidence of teratogenic effects of antipsychotics in humans. However, the use of chlorpromazine in late pregnancy has been associated with neonatal jaundice, and there have been case reports of infants born with persistent extrapyramidal symptoms when the mother was treated with antipsychotics in the late pregnancy (ARANA and

HYMAN 1991). Antipsychotics are secreted in human milk and can cause mild sedation followed by motor excitement in newborns who are breast-fed by a mother receiving antipsychotic medication. It is generally recommended that antipsychotics be avoided during pregnancy (especially during the first trimester) and throughout lactation (BALDESSARINI 1985).

G. Side Effects and Toxicity

I. Sedation, Fatigue, and Deterioration of Behavior

Sedation is the most common side effect of neuroleptic treatment in children, and persistent sedation may have an adverse effect on learning and development (TEICHER and GLOD 1990). Low-potency agents are more likely to cause sedation, but high-potency agents may cause initial sedation, more commonly in children than in adults. Patients must be monitored closely for impairment of schoolwork during treatment. Sedation may exacerbate the negative symptoms of schizophrenia (blunted affect, poverty of speech, social withdrawal). A subjective sense of fatigue may occur even without the decreased arousal associated with sedation (TEICHER and GLOD 1990).

There may be a worsening of preexisting symptoms (e.g., irritability, crying, apathy, hyperactivity, or hypoactivity) or the appearance of new symptoms (e.g., a dazed demeanor, decrease in activity and speech, or dysphoria) with neuroleptic treatment (CAMPBELL and DEUTSCH 1985). This deterioration is generally reversible with discontinuation of the medication.

II. Cognitive Effects

Effects on learning seem to vary according to the disorder in question, the neuroleptic used, and the dosage administered. Some studies have suggested that low doses of haloperidol may improve cognitive performance in some patients with hyperactive, aggressive, or autistic behavior, although these results have not been consistently replicated (WHITAKER and RAO 1992). CAMPBELL and DEUTSCH have shown that low doses of haloperidol may improve cognitive performance in some autistic children (CAMPBELL and DEUTSCH 1985). Higher doses of neuroleptics tend to impair cognitive performance (WHITAKER and RAO 1992). This impairment can be minimized by reducing the dose or giving most of the daily dose at bedtime (TEICHER and GLOD 1990). Most of the research on the effects of neuroleptics on cognition has been performed with mentally retarded, institutionalized patients and cannot be generalized to other patient groups (CAMPBELL 1985).

III. Acute Dystonia

All extrapyramidal symptoms (EPS) are rare in preschool children. Acute dystonia, including oculogyric crisis, torticollis, opisthotonos, and tongue protrusion, is most prevalent in 10- to 19-year-olds and is more common in boys than in girls. It is more common with high-potency than with low-potency neuroleptics (WHITAKER and RAO 1992).

Acute dystonia usually occurs within hours to days of the first dose. It can be minimized by starting at a low dose and increasing very gradually, as well as using divided daily doses rather than a single daily dose. Dystonia is treated with anticholinergic agents such as diphenhydramine or benztropine. These may be administered either orally or intramuscularly. The need for anticholinergic treatment may subside after 2–3 weeks of treatment with the neuroleptic, and anticholinergics should be tapered at that point, while monitoring for signs of reemergence of dystonia (TEICHER and GLOD 1990). It may be advisable to initiate anticholinergic therapy at the same time as starting treatment with a high-potency neuroleptic in adolescents as prophylaxis against dystonia (WHITAKER and RAO 1992). Laryngeal dystonia may be life-threatening and can be treated with IM or IV anticholinergic agents. If this is not rapidly effective, IM or IV benzodiazepine may relieve the dystonia. All acute dystonias may be terrifying to the patient and affect their subsequent compliance with drug treatment.

IV. Tardive Dyskinesia

The prevalence of withdrawal dyskinesia (WD) and tardive dyskinesia (TD) in children receiving long-term neuroleptic therapy has been reported to be from 8% to 51% in various studies (CAMPELL et al. 1983). WD lasts a maximum of 12–16 weeks, while TD lasts longer (GUALTIERI et al. 1986). WD tends to emerge approximately 14 days after neuroleptic discontinuation or dosage reduction (TEICHER and GLOD 1990). These dyskinesias may occur after as little as $3^1/_2$ months of cumulative neuroleptic treatment (CAMPBELL et al. 1983). As in adults, the mouth, cheeks, and tongue are common sites of tardive dyskinesia, but dyskinesia may also affect other parts of the face (blinking), trunk, extremities, and muscles of respiration (CAMPBELL and SPENCER 1988). Prerequisites for the diagnosis of TD include (1) at least 3 months of treatment with neuroleptics, (2) at least moderate involuntary movements in one body part or mild movements in two or more body areas, and (3) the differentiation from conditions that might mimic TD (CAMPBELL et al. 1983). Such conditions might include stereotypies associated with psychosis, autism, or mental retardation; the tics of Tourette's syndrome; facial movements related to dental problems; tardive Tourette's syndrome; rabbit syndrome; other drug-induced dystonias; Huntington's disease; Wilson's disease; Sydenham's chorea; and other systemic illnesses associated with dyskinesias (WHITAKER and RAO 1992).

Normal children may also have stereotypic-like movements (CAMPBELL and DEUTSCH 1985). The emergence of dystonia or dyskinesia upon withdrawal of neuroleptic may be due to withdrawal TD or to the reemergence of abnormal movements associated with psychosis, autism, or mental retardation in children that had been controlled by neuroleptics (Whitaker and Rao 1992). Even experienced clinicians may have great difficulty distinguishing stereotypies related to autism from drug-induced dyskinesias (Meiselas et al. 1989). Therefore it is important to obtain baseline assessment of abnormal movements prior to starting neuroleptics in children (Whitaker and Rao 1992). In the only prospective study to date employing baseline assessment, Campbell and coworkers found that 30% of autistic children treated with haloperidol developed reversible dyskinesias. Eighty percent of the dyskinesias which developed appeared upon withdrawal of neuroleptic, and the remaining 20% appeared during neuroleptic treatment. There were no cases of irreversible dyskinesia. These data suggest that tardive dyskinesia in children usually resolves spontaneously (Campbell et al. 1988b). There have, however, been reports of persistent dyskinesias in children and adolescents (Paulson et al. 1975).

In short, there is little agreement about risk factors for TD in children and adolescents. In a prospective study of the development of TD in autistic children, none of the following variables seemed to affect the risk of developing TD: IQ, chronological age, weight, the presence of stereotypies, daily dose (mg per kg), or cumulative dose. There was a trend toward greater risk for dyskinesias in the first year of haloperidol treatment for females rather than males (Campbell et al. 1988b). The American Psychiatric Association (APA) Tardive Dyskinesia Task Force Report recommends "gradual neuroleptic dose reductions to the minimal effective dose in long-term treated patients" (Whitaker and Rao 1992). At the time of this writing, only one study (Perry et al. 1989a) had compared the risk of dyskinesias in continuous versus discontinuous neuroleptic treatment. Discontinuous drug treatment did not result in a lower incidence of dyskinesias. There is no good evidence to show that drug holidays will reduce the risk of TD (Perry et al. 1989a). The treatment of TD is withdrawal of the neuroleptic whenever possible. If the neuroleptic is continued for urgent clinical reasons, the TD will not necessarily become progressively more severe. In fact, it may stabilize or even improve during maintenance on the neuroleptic (Whitaker and Rao 1992). Clozapine or risperidone may be possible alternative agents for patients who have developed tardive dyskinesia on a standard neuroleptic and who cannot be managed without a neuroleptic (Teicher and Glod 1990).

V. Neuroleptic-Induced Tics

There have been case reports of children developing multiple motor and phonic tics after long-term treatment (several months to 20 years) with neuroleptics. This has been referred to as "tardive Tourette's syndrome" (Stahl 1980). Lal and AlAnsari reported the case of a 13-year-old boy who developed such a syndrome after 2–3 months of treatment with thioridazine (Lal and AlAnsari 1986). Perry and coworkers reported the case of a $5^1/_2$-year-old autistic boy who had been treated with haloperidol for 3 years and who developed a Tourette-like syndrome during haloperidol withdrawal (Perry et al. 1989b). Karagianis and Nagpurkar described a 10-year-old boy with conduct disorder who, after being treated with haloperidol and imipramine for 17 months and propranolol for 12 months, developed motor tics and coprolalia. His tics worsened with lowering of the haloperidol dose and improved with an increase of haloperidol dose (Karagianis and Nagpurkar 1990).

There have been other reports describing the acute onset of tics within days of starting neuroleptic treatment. Gualtieri and Patterson reported a 9-year-old boy with hyperactivity and no prior history of Tourette's syndrome who developed phonic and motor tics on separate trials of *d*-amphetamine, carbamazepine, and haloperidol. They described a second 9-year-old patient with hyperactivity but without a preexisting tic disorder who developed motor tics on separate trials of methylphenidate, imipramine, and thioridazine. In both cases the tics resolved within 7–10 days of discontinuing the neuroleptic. The authors speculated that acute neuroleptic-induced tics occur through a mechanism (presynaptic dopaminergic blockade) distinct from that of tardive Tourette's syndrome (dopamine receptor supersensitivity) (Gualtieri and Patterson 1986).

VI. Parkinsonism, Catatonia, and the Rabbit Syndrome

Parkinsonian effects include tremor, rigidity, cogwheeling, bradykinesia, mask-like facies, and drooling. While rare in preschool children, these side effects are quite common in the 10- to 19-year-old age group (Whitaker and Rao 1992), and they are more likely to be produced by high-potency agents, as is the case in adults. These side effects tend to develop within the first few weeks of treatment (Campbell 1985). The preferred method of reducing such symptoms is to lower dose of the neuroleptic (Campbell and Deutsch 1985) rather than using antiparkinsonian medications. When lowering the neuroleptic dose is not feasible, anticholinergic agents such as diphenhydramine or benztropine may reduce parkinsonism (Teicher and Glod 1990).

Catatonia, which can occur with high-potency neuroleptics, usually resolves slowly after withdrawal of neuroleptic or use of an anticholinergic drug. It can resolve more rapidly after treatment with amantadine or IV lorazepam (Whitaker and Rao 1992).

The rabbit syndrome is a late-onset (after months of treatment) extrapyramidal syndrome of tremulousness of the lips which improves with the administration of anticholinergic agents (Campbell and Deutsch 1985).

VII. Akathisia

Akathisia is a distressing restlessness and need to move. This side effect may be difficult to distinguish from hyperactivity in children and it may be mistaken for a worsening of agitation secondary to the primary disorder which is being treated. It appears to be most frequent in 20- to 30-year-old patients, but does occur in childhood. High-potency agents are more likely to produce akathisia. Some patients may be relieved of this effect by switching to thiothixene (Navane) or chlorprothixene (Taractan), both of which seem to have a lower incidence of producing akathisia than the phenothiazines or butyrophenones (TEICHER and GLOD 1990). The most effective treatment for akathisia is lowering the dose of neuroleptic. There is evidence in adults that beta blockers (e.g., propranolol), benzodiazepines (e.g., lorazepam), and/or anticholinergic agents (e.g., benztropine) may have efficacy in reducing akathisia. In an open trial of clonazepam in adolescent patients with akathisia, KUTCHER et al. reported that clonazepam was fairly effective in reducing akathisia (KUTCHER et al. 1987).

VIII. Neuroleptic Malignant Syndrome

Neuroleptic malignant syndrome (NMS) is a life-threatening syndrome characterized by fever, rigidity, alterations in pulse and blood pressure, diaphoresis, and occasionally cardiac dysrhythmias. There have been only a few reported cases of NMS in adolescents and essentially none in children. GELLER and GREYDANUS reported cases of a 15-year-old female and a 12-year-old male, both of whom developed NMS during haloperidol treatment (GELLER and GREYDANUS 1979). A 16-year-old male with mental retardation became comatose after an injection of fluphenazine (DIAMOND and HAYES 1986). A 15-year-old male developed NMS during treatment with chlorpromazine and promethazine (TENENBEIN 1985–86; TEICHER and GLOD 1990). A 14-year-old male with schizophrenia treated with high-dose (60 mg or 1 mg/kg per day) trifluoperazine developed NMS (MERRY et al. 1986). NMS is treated with immediate discontinuation of the neuroleptic, as well as supportive measures such as ensuring adequate hydration and control of fever. Various pharmacologic agents have been used to treat NMS, including dantrolene, bromocriptine, and/or benzodiazepines. After an episode of NMS, a neuroleptic may be reintroduced if there is a strong clinical indication for such therapy. A different neuroleptic with a relatively low risk of NMS (molindone, thioridazine) should be used, and the dose should be increased gradually (TEICHER and GLOD 1990).

IX. Seizure Threshold

Low-potency neuroleptics, such as chlorpromazine, have a greater tendency than high potency neuroleptics to lower the seizure threshold (WHITAKER and RAO 1992). BENNETT et al. found that haloperidol treatment of a group of children aged 5–13 years with conduct disorder resulted in increased numbers of paroxysmal or focal abnormalities on the EEG in children who had abnormal EEGs at baseline (BENNETT et al. 1983). If a child has a history of a possible seizure disorder, it would be prudent to obtain a baseline EEG prior to initiation of treatment with a neuroleptic. Children with psychiatric disorders have a much higher incidence of abnormal EEGs than do children without such disorders (TEICHER and GLOD 1990). Those with a known seizure disorder may need to be treated with an anticonvulsant in conjunction with the neuroleptic.

X. Unwanted Cardiac Effects

Certain antipsychotics, particularly thioridazine (FOWLER et al. 1976) and pimozide, slow intracardiac conduction time and increase the risk for fatal ventricular dysrhythmias, particularly torsade de pointes. It has been recommended that the dose of pimozide be lowered or the drug discontinued if the Q-T interval is greater than 520ms in adults or 470ms in children (BALDESSARINI 1985). Pimozide leads to ECG changes in as many as 25% of children or adults who are treated for Tourette's syndrome. Within 1 week of treatment, T-wave inversion, U waves, QT prolongation, and bradycardia may be seen. The FDA has approved pimozide only for patients with severe Tourette's syndrome who have not responded adequately to other treatments. Pimozide should be discontinued if there is T-wave inversion or U waves. In children, the dose should not be increased if there is prologation of the corrected QT interval of greater than 470ms or more than 25% above the patient's baseline. The ECG usually returns to normal within 1 week of discontinuing the neuroleptic. A baseline ECG should be obtained prior to starting pimozide. It should be repeated every month during periods of dosage increase and every 3 months when a stable dose has been reached (WHITAKER and RAO 1992). The risk of dysrhythmia increases if the patient is dehydrated (TEICHER and GLOD 1990).

XI. Hypotension

The low-potency agents have higher alpha-adrenergic antagonist properties and have a greater tendency to cause orthostatic hypotension. Hypotension is usually not a clinically significant problem in children and adolescents except in overdose, but it may be prudent to instruct the patient not to rise too rapidly from a sitting or lying position (WHITAKER and RAO 1992). Hypotension can be minimized by administering neuroleptics in divided doses rather than a single daily dose, administering most or all of the daily dose at bedtime, or by using a high-potency agent (TEICHER and GLOD 1990).

XII. Weight Gain

Weight gain occurs frequently during treatment with neuroleptics, possibly secondary to increased appetite or decreased motor activity (TEICHER and GLOD 1990). If reasonable limits are set on caloric intake and the child is encouraged to be physically active, weight gain may be kept to a minimum.

XIII. Ocular Complications

A pigmentary retinopathy may occur in patients receiving doses of thioridazine greater than 800 mg/day. This retinopathy may result in impaired visual acuity, brownish coloration of vision, and impaired vision at night. Fundoscopic exam may reveal deposits of brown pigments in the retina. Children (aged 2 years and above) should not recevie more than 3 mg/kg per day of thioridazine (TEICHER and GLOD 1990). In addition, the low-potency neuroleptics, with their prominent anticholinergic properties, may cause cycloplegia, leading to blurred near vision.

XIV. Unwanted Cutaneous Effects

Low-potency neuroleptics are photosensitizing. Sunscreens can be used to prevent sunburn. Occassionally, the use of a neuroleptic may lead to an allergic reaction consisting of pruritic, erythematous macules and papules on the skin. In that case, the neuroleptic should be discontinued and an agent from a different chemical class used instead (WHITAKER and RAO 1992).

XV. Hypothalamic and Pituitary Functions – Adverse Effects on Growth

Antipsychotic agents may inhibit the release of growth hormone, whereas L-DOPA enhances the release of growth hormone. There is no evidence that neuroleptics impair growth of children significantly, although the growth of any young patient taking a neuroleptic should be closely monitored (BALDESSARINI 1985; SIMEON et al. 1977).

Neuroleptics cause hyperprolactinemia, which may induce amenorrhea and galactorrhea in women and may possibly cause oligospermia in men (WHITAKER and RAO 1992). GUALTIERI and coworkers have confirmed that neuroleptic treatment leads to hyperprolactinemia in children which is reversed when the drug is discontinued. These authors found no consistent effect of neuroleptics on growth hormone or thyrotropin (GUALTIERI et al. 1980). The long-term effect of hyperprolactinemia on sexual development is unknown (WHITAKER and RAO 1992). One study of adolescent males with schizophrenia found that after more than 6 months of treatment with chlorpromazine the patients had hyperprolactinemia, low basal plasma testosterone levels, and blunted luteinizing hormone (LH) response to luteinizing hormone-releasing hormone (LHRH) stimulation. The significance of these findings for sexual development and functioning is unclear (APTER et al. 1983). Clozapine has a lesser tendency than do typical neuroleptics to cause hyperprolactinemia (TEICHER and GLOD 1990).

XVI. Hepatic Complications

Low-potency neuroleptics, particularly chlorpromazine, may cause cholestatic jaundice, but this is rare in children and adolescents (Whitaker and Rao 1992). If signs and symptoms of cholestatic jaundice (e.g., nausea, malaise, fever, pruritis, abdominal pain, yellowing of the sclera and skin) occur, they usually develop during the first 3 weeks of treatment (Teicher and Glod 1990; Arana and Hyman 1991). This jaundice usually resolves within 8 weeks of discontinuing the neuroleptic (Teicher and Glod 1990). Neuroleptics may also cause transient benign elevations of transaminases (ALT and AST) and alkaline phosphatase (Campbell 1985). It is prudent to obtain liver function tests prior to initiating antipsychotic medication, and from then on a yearly basis (Whitaker and Rao 1992). If elevations of tramsaminases, alkaline phosphatase, or bilirubin occur, the neuroleptic should be discontinued. If neuroleptic treatment is still necessary, a drug from a different chemical class should be chosen (Arana and Hyman 1991).

XVII. Hematologic Complications

Most patients on neuroleptics will have a transient, benign leukopenia. One of 3000 to 4000 adults treated with chlorpromazine develops agranulocytosis within 90 days of the initiation of treatment. The incidence of agranulocytosis appears to be less common in children (Whitaker and Rao 1992). Frequent, regular measurement of white blood cell count (WBC) is not generally recommended (except in treatment with clozapine) because most of what will be seen is benign, transient leukopenia (Teicher and Glod 1990). If the patient has symptoms of sore throat, fever, malaise, or other symptoms of infection during the first 3 months of treatment with a neuroleptic, a WBC with differential should be promptly checked. The patient and patient's parents should be instructed to call the physician immediately if the patient develops signs of infection (an indication of possible leukopenia) or easy bruising (an indication of possible thrombocytopenia) after being started on a neuroleptic. If the WBC is low, the neuroleptic should be discontinued immediately (Whitaker and Rao 1992). Patients who have developed agranulocytosis should not be rechallenged with the same neuroleptic or with a neuroleptic of the same chemical class (Teicher and Glod 1990).

Clozapine use is associated with an incidence of agranulocytosis of approximately 1% per year of treatment in adults. There is not much data available yet regarding the incidence in children (Teicher and Glod 1990).

XVIII. Sexual Dysfunction

Adolescent patients may be disturbed by the altered libido, orgasmic, or ejaculatory function which may result from treatment with a low-potency neuroleptic. Administration of thioridazine has been associated with the occurrence of retrograde ejaculation. Adolescent females may experience irregular menstruation, amenorrhea, or galactorrhea due to elevation of prolactin levels by neuroleptics (TEICHER and GLOD 1990).

XIX. Enuresis

This can be avoided if the dose of neuroleptic is given in the morning or afternoon, rather than before bedtime (CAMPBELL 1985).

XX. Nondyskinetic Withdrawal Symptoms

Rebound symptoms include insomnia, nightmares, increased salivation, abdominal cramps, diarrhea, anorexia, weight loss, and diaphoresis. There may also be a rebound of hyperactivity, aggressive behavior, and irritability (WHITAKER and RAO 1992; GUALTIERI et al. 1986). GROB reported a 14.5-year-old boy who had persistent vomiting after withdrawal of haloperidol (GROB, 1986). YEPES and WINSBERG reported two cases of vomiting on neuroleptic withdrawal, one in a 9-year-old boy who had been on chlorprothixene, and another in a 9-year-old boy who was on thioridazine (YEPES and WINSBERG 1977). These nondyskinetic withdrawal symptoms tend to resolve in a few days (TEICHER and GLOD 1990). To prevent them, the neuroleptic should be tapered by no more than 5%–10% of the dose per day (WHITAKER and RAO 1992).

Tic symptoms can also increase dramatically with neuroleptic withdrawal. Families frequently report a lull in tic symptoms with cessation of neuroleptics followed by a marked increase in symptoms (at times well above premedication levels) that peaks at 4–6 weeks. Families of TS patients should be alert to this possible outcome. Gradual withdrawal may limit these symptoms.

XXI. Overdosage

Neuroleptics generally have a high therapeutic index (ratio of toxic to therapeutic dose) and are rarely fatal in overdose. Thioridazine, mesoridazine, and pimozide, however, are cardiotoxic in overdose. Complications of overdose may include hypotension, arrhythmias, severe extrapyramidal effects, coma, hyperthermia or hypothermia, and seizures (TEICHER and GLOD 1990; ARANA and HYMAN 1991). Treatment of a recent overdose may include gastric lavage, followed by administration of activated charcoal. Hypotension is treated with volume resuscitation or alpha-agonist drugs (WHITAKER and RAO 1992).

H. Conclusions and Future Directions

Neuroleptics can provide significant symptomatic relief to young patients with certain neuropsychiatric disorders, including autism, Tourette's syndrome, and schizophrenia. The use of neuroleptics to treat nonautistic mentally retarded patients has been called into question recently. Carefully conducted studies suggest that neuroleptics may reduce stereotypies in some of these patients, but may also significantly compromise cognitive functioning. Moreover, it is important to consider the many alternatives to neuroleptics in the treatment of conduct disorder. The decision to prescribe an antipsychotic drug to any child or adolescent must be informed by a thorough consideration of potential risks, benefits, and alternatives and by a discussion of these factors, whenever possible, with the patient and parents. In particular, the physician must remember the risks of acute dystonia, tardive dyskinesia, neuroleptic malignant syndrome, unwanted cardiac effects, and agranulocytosis, among others. In addition to safely treating troublesome symptoms, the physician should strive to promote the child's intellectual, emotional, and physical growth and development.

A number of atypical neuroleptics, such as clozapine and risperidone, have been developed recently. They are atypical in having a different profile of receptor blockade from the traditional neuroleptics and having a different side effect profile. In particular, these drugs produce a much lower rate of extrapyramidal symptoms. A potential problem with the use of clozapine in children is its prominent anticholinergic and sedative effects which can impair learning. SIEFEN and REMSCHMIDT treated 21 adolescent schizophrenic patients with clozapine. The patients had had an inadequate response to other neuroleptics or had had prominent extrapyramidal effects. A marked improvement in symptoms was seen in 11 of them and 6 had a partial response. Side effects included sedation, orthostatic hypotension, and hypersalivation. No cases of agranulocytosis were reported (SIEFEN and REMSCHMIDT 1986; TEICHER and GLOD 1990). FRAZIER and coworkers reported an open trial of clozapine in 11 adolescents with childhood-onset schizophrenia. Of these patients, 55% showed a greater than 30% improvement in Brief Psychiatric Rating Scale ratings compared to their baseline ratings while on another neuroleptic (FRAZIER et al. 1994). Given the risks of treatment with clozapine, its use in children and adolescents generally should be confined to the treatment of neuroleptic-resistant psychosis (TEICHER and GLOD 1990).

Risperidone is a highly potent and selective 5-HT2 receptor antagonist which also acts as an antagonist at alpha-1, histamine-1, D2, and alpha-2 receptors (LEYSEN et al. 1988; LEYSEN et al. 1992). In adults, risperidone has been shown to have efficacy in decreasing the positive and negative symptoms of schizophrenia while having minimal extrapyramidal side effects (CHOUINARD et al. 1993). VANDEN BORRE and coworkers reported a double-blind, placebo-controlled crossover trial in which 4–12 mg/day of risperidone or placebo was added to ongoing medication in 37 patients with mental retardation. Risperidone was significantly better than placebo at improving symptoms measured by the Aberrant Behavior Checklist and the Clinical Global Impression Scale. Sedation was the most frequently reported adverse effect (VANDEN BORRE et al. 1993). Risperidone may ultimately prove to be a good alternative to standard neuroleptics for the treatment of children and adolescents with mental retardation, autism, Tourette's syndrome, refractory OCD, and schizophrenia.

There is a need for long-term safety and efficacy studies of neuroleptics in children, as well as for better instruments for behavioral and psychological assessment (AMAN 1989). There is also a need for a better understanding of the neurobiological mechanisms by which antipsychotics produce their therapeutic and their untoward effects. Through such investigations, novel drugs with greater therapeutic efficacy but without serious untoward effects may be developed.

References

Aman MG (1989) Neuroleptics. In: Karasu TB (chairperson) Treatments of psychiatric disorders. A task force report of the American Psychiatric Association. vol 1. American Psychiatric Association. Washington, pp 71–77

Anderson LT, Campbell M, Grega DM, Perry R, Small AM, Green WH (1984) Haloperidol in the treatment of infantile autism: effects on learning and behavioral symptoms. Am J Psychiatry 141:1195–1202

Anderson LT, Campbell M, Adams P, Small AM, Perry R, Shell J (1989) The effects of haloperidol on discrimination learning and behavioral symptoms in autistic children. J Autism Dev Disord 19:227–239

Apter A, Dickerman Z, Gonen N, Assa S, Prager-Lewin R, Kaufman H, Tyano S, Laron Z (1983) Effect of chlorpromazine on hypothalamic-pituitary-gonadal function in 10 adolescent schizophrenic boys. Am J Psychiatry 140:1588–1591

Arana GW, Hyman SE (1991) Handbook of psychiatric drug therapy, 2nd edn. Little, Brown, Boston

Baldessarini RJ (ed) (1985) Antipsychotic agents. In: Chemotherapy in psychiatry: principles and practice. Harvard University Press, Cambridge, pp 15–92

Bennett WG, Korein J, Kalmijn M, Grega DM, Campbell M (1983) Electroencephalogram and treatment of hospitalized aggressive children with haloperidol or lithium. Biol Psychiatry 18:1427–1440

Borison RL, Ang L, Hamilton WJ, Diamond BI, Davis JM (1983) Treatment approaches in Gilles de la Tourette syndrome. Brain Res Bull 11:205–208

Breuning SE, Ferguson DG, Davidson NA, Poling AD (1983) Effects of thioridazine on the intellectual performance of mentally retarded drug responders and nonresponders. Arch Gen Psychiatry 40:309–313

Bruggeman R, van der Linden C, van Woerkom TCAM (1994) Risperidone in the treatment of Gilles de la Tourette's syndrome – an open dose finding study. Neuropsychopharmacology 10:42S

Bruun RD (1988) Subtle and underrecognized side effects of neuroleptic treatment in children with Tourette's disorder. Am J Psychiatry 145:621–624

Campbell M (1985) On the use of neuroleptics in children and adolescents. Psychiatr Ann 15:101–107

Campbell M, Deutsch SI (1985) Neuroleptics in children. In: Burrows, Norman, Davies (eds) Antipsychotics. Elsevier, New York, pp 213–238

Campbell M, Anderson LT, Small AM, Perry R, Green WH, Caplan R (1982a) The effects of haloperidol on learning and behavior in autistic children. J Autism Dev Disord 12:167–175

Campbell M, Cohen IL, Small AM (1982b) Drugs in aggressive behavior. J Am Acad Child Psychiatry 21:107–117

Campbell M, Grega DM, Green WH, Bennett WG (1983) Neuroleptic-induced dyskinesias in children. Clin Neuropharmacol 6(3):207–222

Campbell M, Small AM, Green WH, Jennings SJ, Perry R, Bennett WG, Anderson L (1984) Behavioral efficacy of haloperidol and lithium carbonate: a comparison in hospitalized aggressive children with conduct disorder. Arch Gen Psychiatry 41:650–656
Campbell M, Spencer EK (1988) Psychopharmacology in child and adolescent psychiatry: a review of the past five years. J Am Acad Child Adolesc Psychiatry 27:269–279
Campbell M, Baldessarini RJ, Teicher MH (1988a) Decreasing sensitivity to neuroleptic agents in developing rats evidence: for a pharmacodynamic factor. Psychopharmacology 94:46–51
Campbell M, Adams P, Perry R, Spencer EK, Overall JE (1988b) Tardive and withdrawal dyskinesia in autistic children: a prospective study. Psychopharmacol Bull 24:251–255
Campbell M, Gonzalez NM, Silva RR (1992) The pharmacologic treatment of conduct disorders and rage outbursts. In: Shaffer D (ed) The psychiatric clinics of North America: pediatric psychopharmacology. Saunders, Philadelphia, pp 69–85
Chouinard G, Jones B, Remington G, Bloom D, Addington D, MacEwan GW, Labelle A, Beauclair L, Arnott W (1993) A Canadian multicenter placebo-controlled study of fixed doses of risperidone and haloperidol in the treatment of chronic schizophrenic patients. J Clin Psychopharmacol 13:25–40
Cohen DJ, Riddle MA, Leckman JF (1992) Pharmacotherapy of Tourette's syndrome and associated disorders. In: Shaffer D (ed) The psychiatric clinics of North America. Pediatric Psychopharmacology. Saunders, Philadelphia, pp 109–129
Diamond JM, Hayes DD (1986) A case of neuroleptic malignant syndrome in a mentally retarded adolescent. J Adolesc Health Care 7:419–422
Ernst M, Magee HJ, Gonzalez NM, Locascio JJ, Rosenberg CR, Campbell M (1992) Pimozide in autistic children. Psychopharmacol Bull 28(2):187–191
Fowler NO, McCall D, Chou T-C, Holmes JC, Hanenson IB (1976) Electrocardiographic changes and cardiac arrhythmias in patients receiving psychotropic drugs. Am J Cardiol 37:223–230
Frazier JA, Gordon CT, McKenna K, Lenane MC, Jih D, Rapoport JL (1994) An open trial of clozapine in 11 adolescents with childhood-onset schizophrenia. J Am Acad Child Adolesc Psychiatry 33:658–663
Geller B, Greydanus DE (1979) Haloperidol-induced comatose state with hyperthermia and rigidity in adolescence: two case reports with a literature review. J Clin Psychiatry 40:102–103
Gittelman-Klein R, Klein DF, Katz S, Saraf K, Pollack E (1976) Comparative effects of methylphenidate and thioridazine in hyperkinetic children. Arch Gen Psychiatry 33:1217–1231
Gordon CT, Rapoport JL, Hamburger SD, State RC, Mannheim GB (1992) Differential response of seven subjects with autistic disorder to clomipramine and desipramine. Am J Psychiatry 149:363–366
Gordon CT, State RC, Nelson JE, Hamburger SD, Rapoport JL (1993) A double-blind comparison of clomipramine, desipramine, and placebo in the treatment of autistic disorder. Arch Gen Psychiatry 50:441–447
Green WH, Campbell M, Hardesty AS, Grega DM, Padron-Gayol M, Shell J, Erlenmeyer-Kimling L (1984) A comparison of schizophrenic and autistic children. J Am Acad Child Psychiatry 23:399–409
Greenberg LM, Deem MA, McMahon S (1972) Effects of dextroamphetamine, chlorpromazine, and hydroxyzine on behavior and performance in hyperactive children. Am J Psychiatry 129:532–539
Greenhill LL, Solomon M, Pleak R, Ambrosini P (1985) Molindone hydrochloride treatment of hospitalized children with conduct disorder. J Clin Psychiatry 46[8, sec. 2]:20–25
Grob CS (1986) Persistent supersensitivity vomiting following neuroleptic withdrawal in an adolescent. Biol Psychiatry 21:398–401

Gualtieri CT, Patterson DR (1986) Neuroleptic-induced tics in two hyperactive children. Am J Psychiatry 143:1176–1177
Gualtieri CT, Rojahn J, Staye J (1980) The influence of neuroleptic drugs on prolactin secretion in children. Develop Med Child Neurol 22:515–524
Gualtieri CT, Schroeder SR, Hicks RE, Quade D (1986) Tardive dyskinesia in young mentally retarded individuals. Arch Gen Psychiatry 43:335–340
Hyman SE, Nestler EJ (1993) The molecular foundations of psychiatry. American Psychiatric Press, Washington, DC, pp 141–150
Joshi PT, Capozzoli JA, Coyle JT (1988) Low-dose neuroleptic therapy for children with childhood-onset pervasive developmental disorder. Am J Psychiatry 145:335–338
Karagianis JL, Nagpurkar R (1990) A case of Tourette syndrome developing during haloperidol treatment. Can J Psychiatry 35:228–232
Kutcher SP, Mackenzie S, Galarraga W, Szalai J (1987) Clonazepam treatment of adolescents with neuroleptic-induced akathisia [letter to the editor]. Am J Psychiatry 144:823–824
Lal S, AlAnsari E (1986) Tourette-like syndrome following low dose short-term neuroleptic treatment. Can J Neurological Sci 13:125–128
Leckman JF, Hardin MT, Riddle MA, Stevenson J, Ort SI, Cohen DJ (1991) Clonidine treatment of Tourette's syndrome. Arch Gen Psychiatry 48:324–328
Leysen JE, Gommeren W, Eens A, De Chaffoy De Courcelles D, Stoof JC, Janssen PAJ (1988) Biochemical profile of risperidone, a new antipsychotic. J Pharmacol Expl Ther 247:661–670
Leysen JE, Janssen PMF, Gommeren W, Wynants J, Pauwels PJ, Janssen PAJ (1992) In vitro and in vivo receptor binding and effects of monoamine turnover in rat brain regions of the novel antipsychotics risperidone and ocaperidone. Mol Pharmacol 41:494–508
Linet LS (1985) Tourette syndrome, pimozide, and school phobia: the neuroleptic separation anxiety syndrome. Am J Psychiatry 142:613–615
Mandoki M (1993) Clozapine for adolescents with psychosis: literature review and two case reports. J Child Adolesc Psychopharmacol 3:213–221
McClellan JM, Werry JS (1992) Schizophrenia. In: Shaffer D (ed) The psychiatric clinics of North America: pediatric psychopharmacology. Saunders, Philadelphia, pp 131–148
McDougle CJ, Goodman WK, Price LH, Delgado PL, Krystal JH, Charney DS, Heninger GR (1990) Neuroleptic addition in fluvoxamine-refractory obsessive compulsive disorder. Am J Psychiatry 147:652–654
McDougle CJ, Goodman WK, Leckman JF, Price LH (1993) The psychopharmacology of obsessive compulsive disorder: implications for treatment and pathogenesis. In: Dunner DL (ed) Psychiatric clinics of North America, vol 16(4). Saunders, Philadelphia, pp 749–766
McDougle CJ, Price LH, Volkmar FR (1994) Recent advances in the pharmacotherapy of autism and related conditions. In: Volkmar FR (ed) Child and adolescent psychiatric clinics of North America, vol. 3(1). Saunders, Philadelphia, pp 71–89
Meiselas KD, Spencer EK, Oberfield R, Peselow ED, Angrist B, Campbell M (1989) Differentiation of stereotypies from neuroleptic-related dyskinesias in autistic children. J Clin Psychopharmacol 9:207–209
Merry SN, Werry JS, Merry AF, Birchall N (1986) The neuroleptic malignant syndrome in an adolescent. J Am Acad Child Psychiatry 25:284–286
Mikkelsen EJ, Detlor J, Cohen DJ (1981) School avoidance and social phobia triggered by haloperidol in patients with Tourette's disorder. Am J Psychiatry 138:1572–1576
Murrin CL, Gibbens DL, Ferrer JR (1985) Ontogeny of dopamine, serotonin, and spirodecanone receptors in rat forebrain – an autoradiographic study. Developmental Brain Research 23:91–109
Murrin CL, Zeng W (1986) Postnatal ontogeny of dopamine D_2 receptors in rat striatum. Biochem Pharmacol 35:1159–1162

O'Boyle KM, Waddington JL (1986) A re-evaluation of changes in rat striatal D_2 dopamine receptors during development and aging. Neurobiol Aging 7:265–267
Paulson GW, Rizvi CA, Crane GE (1975) Tardive dyskinesia as a possible sequel of long-term therapy with phenothizines. Neurology 14:953–955
Perry R, Campbell M, Adams P, Lynch N, Spencer EK, Curren EL, Overall JE (1989a) Long-term efficacy of haloperidol in autistic children: continuous versus discontinuous drug administration. J Am Acad Child Adolesc Psychiatry 28(1):87–92
Perry R, Nobler MS, Campbell M (1989b) Tourette-like syndrome associated with neuroleptic therapy in an autistic child. J Am Acad Child Adolesc Psychiatry 28:93–96
Platt JE, Campbell M, Green WH, Grega DM (1984) Cognitive effects of lithium carbonate and haloperidol in treatment-resistant aggressive children. Arch Gen Psychiatry 41:657–662
Pool D, Bloom W, Mielke DH, Roniger JJ, Gallant DM (1976) A controlled evaluation of loxitane in seventy-five adolescent schizophrenic patients. Curr Ther Res 19(1):99–104
Realmuto GM, Erickson WD, Yellin AM, Hopwood JH, Greenberg LM (1984) Clinical comparison of thiothixene and thioridazine in schizophrenic adolescents. Am J Psychiatry 141:440–442
Riddle MA (1991) Pharmacokinetics in children and adolescents. In: Lewis M (ed) Child and adolescent psychiatry: a comprehensive textbook. Wilkins, New Haven, pp 767–770
Rivera-Calimlim L, Griesbach PH, Perlmutter R (1979) Plasma chlorpromazine concentrations in children with behavioral disorders and mental illness. Clin Pharmacol Ther 26:114–121
Ross MS, Moldofsky H (1978) A comparison of pimozide and haloperidol in the treatment of Gilles de la Tourette's syndrome. Am J Psychiatry 135:585–587
Seeman P, Bzowej NH, Guan HC, Bergeron C, Becker LE, Reynolds GP, Bird ED, Riederer P, Jellinger K, Watanabe S, Tourtellotte WW (1987) Human brain dopamine receptors in children and aging adults. Synapse 1:399–404
Shapiro AK, Shapiro E, Wayne H (1973) Treatment of tourette's syndrome with haloperidol, review of 34 cases. Arch Gen Psychiatry 28:92–96
Shapiro AK, Shapiro E, Eisenkraft GJ (1983) Treatment of Tourette disorder with penfluridol. Compr Psychiatry 24:327–331
Shapiro AK, Shapiro E (1984) Controlled study of pimozide vs. placebo in Tourette's syndrome. J Am Acad Child Psychiatry 23(2):161–173
Shapiro E, Shapiro AK, Fulop G, Hubbard M, Mandeli J, Nordlie J, Phillips RA (1989) Controlled study of haloperidol, pimozide, and placebo for the treatment of Gilles de la Tourette's Syndrome. Arch Gen Psychiatry 46:722–730
Siefen G, Remschmidt H (1986) Results of treatment with clozapine in schizophrenic adolescents. Z Kinder Jungenpsychiatr 14:245–257
Simeon J, Gross M, Mueller J (1977) Neuroleptic drug effects on children's body weight and height. Psychopharmacol Bull 13:50–53
Stahl SM (1980) Tardive Tourette syndrome in an autistic patient after long-term neuroleptic administration. Am J Psychiatry 137:1267–1269
Teicher MH, Baldessarini RJ (1987) Developmental pharmacodynamics. In: Popper C (ed) Psychiatric pharmacosciences of children and adolescents. American Psychiatric Press, Washington, DC, pp 47–80
Teicher MH, Glod CA (1990) Neuroleptic drugs: indications and guidelines for their rational use in children and adolescents. J Child Adolesc Psychopharmacol (1):33–56
Tenenbein M (1985–86) The neuroleptic malignant syndrome: Occurrence in a 15-year-old boy and recovery with bromocriptine therapy. Pediatr Neurosci 12:161–164
Vanden Borre R, Vermote R, Buttiens M, Thiry P, Dierick G, Geutjens J, Sieben G, Heylen S (1993) Risperidone as add-on therapy in behavioural disturbances in

mental retardation: a double-blind placebo-controlled cross-over study. Acta Psychiatr Scand 87:167–171
Volkmar FR (1991) Childhood schizophrenia. In: Lewis M (ed) Child and adolescent psychiatry: a comprehensive textbook. Williams and Wilkins, Baltimore, pp 621–628
Whitaker A, Rao U (1992) Neuroleptics in pediatric psychiatry. In: Shaffer D (ed) The psychiatric clinics of North America: pediatric psychopharmacology. Saunders, Philadelphia, pp 243–276
Yepes LE, Winsberg BG (1977) Vomiting during neuroleptic withdrawal in children. Am J Psychiatry 134:574

CHAPTER 17

Use of Antipsychotic Drugs in the Elderly

BRUCE G. POLLOCK and BENOIT H. MULSANT

A. Introduction

Age is a major source of variation in drug response. Social, medical, and physiologic heterogeneity intertwine to complicate geriatric pharmacotherapy (WILLIAMS 1987). Inappropriate and excessive use of medication may be the most significant treatable health problem in the elderly. The elderly are especially sensitive to antipsychotics, which are disproportionately prescribed to them (BEERS and OUSLANDER 1989). Antipsychotic side effects and adverse reactions are intensified and protean in an older population, ranging from disabling to deadly. It is essential, therefore, that the indications for their use be clear, guided by knowledge of both age-related and individual determinants of drug clearance and action, and associated with prospective and frequent assessment for adverse effects (POLLOCK and MULSANT 1995). The use of antipsychotics in an older population has lately been tempered by increased awareness of adverse consequences, neurochemistry, limited efficacy, and governmental regulation.

In the United States, the OBRA-87 regulations for nursing homes limit psychotropic drug use to specific indications and require explicit documentation. Prior to the implementation of these regulations by the U.S. Health Care Financing Administration, it was estimated that as much as 50% of antipsychotic use in nursing homes would not meet these new standards (GARRARD et al. 1991). Recent data suggest substantial reductions in antipsychotic drug use since these regulations were implemented. This reduction has resulted in an increase in staffing later in the day without a compensatory increase in the use of other psychotropics or restraints (ROVNER et al. 1992; SHORR et al. 1994).

This chapter reviews current knowledge of clinical issues as well as pharmacokinetic and pharmacodynamic factors relevant to the use of antipsychotic drugs in late life.

B. Indications and Clinical Use of Antipsychotics in Late Life

The clinical awareness of the acute and long-term adverse effects associated with the prescription of traditional antipsychotics to older patients has considerably increased during the past decade. As a result, their prescription has become restricted to indications for which other psychotropic drugs are ineffective or significantly less effective. Expanding on a recently published review (Mulsant and Gershon 1993), this section briefly outlines the current [DSM-IV (American Psychiatric Association 1994)] nosology of the psychoses occurring in late life and summarizes the theoretical and empirical evidence supporting the use of antipsychotics in their treatment (Table 1).

I. Nosology of Late-Life Psychoses

It is now well recognized that a formal thought disorder, delusions, or hallucinations occurring in late life can be the manifestations of many different medical or psychiatric illnesses, leading to a diagnosis of delirium, dementia, schizophrenia, delusional disorder, a "secondary" psychotic disorder (e.g., psychosis due to a general medical condition or substance-induced psychosis), or a mood disorder (Stoudemire and Riether 1987). However, less than one century ago, psychoses in old age were considered "the darkest area of psychiatry" (Kraepelin, cited in Pearlson and Rabins 1988); the onset of psychotic symptoms in an elderly person was invariably attributed to either arteriosclerosis or senility (i.e., central nervous system [CNS] degeneration), even in the presence of associated features such as clouding of consciousness or prominent mood symptoms (Roth 1955). Following the advent of the convulsive therapies, Roth and Morissey (1952) established that late-onset "affective psychosis" and "senile psychosis" constituted two distinct nosological groups with markedly different outcomes. In a series of subsequent studies, Roth and his collaborators (Key et al. 1955; Kay and Roth

Table 1. Biological and clinical evidence supporting the use of antipsychotics in the treatment of psychoses occurring in late life

	Biological evidence	Clinical evidence in mixed-age patients	Clinical evidence in older patients
Delirium	0	D	E
Dementia	0	Not applicable	A
Early-onset schizophrenia	++++	A	B
Late-onset schizophrenia	++	Not applicable	E
Delusional disorder	+	D	E
Psychotic major depression	++	B	E
Bipolar disorder (mania)	+	B	E

Biological evidence implicating dysregulation of dopaminergic system: 0, +, ++, +++, ++++: absent to very strong.
Clinial evidence of antipsychotic efficacy: A, replicated controlled double-blind trials; B, unreplicated controlled double-blind trial(s); C, open controlled trial(s); D, naturalistic prospective trial(s), large case series; E, small case series, case reports, clinical experience.

1961; KAY 1963; ROTH 1955) characterized the clinical presentations and courses of several additional late-life psychoses: arteriosclerotic psychosis, acute confusion, and late paraphrenia. The current American nosology (AMERICAN PSYCHIATRIC ASSOCIATION 1994) remains congruent with this conceptualization of the psychotic disorders of late life, though designations have changed: senile and arteriosclerotic psychoses are now called "dementia of the Alzheimer's type" (DAT) and "vascular dementia"; acute confusion became "delirium"; most of Roth's elderly patients with "affective psychosis" would fulfill the DSM-IV criteria for "major depression" or, more rarely, "bipolar disorder"; and finally, patients with late paraphrenia would be diagnosed as suffering from "late-onset schizophrenia," "delusional disorder," or a secondary psychotic disorder (FLINT et al. 1991; GRAHAME, 1984; HOLDEN 1987; HYMAS et al. 1989; MILLER et al. 1986; MILLER and LESSER 1988).

II. Delirium

Psychotic symptoms are not considered essential features of the syndrome of delirium (AMERICAN PSYCHIATRIC ASSOCIATION 1994). Nevertheless, frank hallucinations and delusions are present in the majority of delirious patients (LIPOWSKI 1990) and they often result in the prescription of an antipsychotic, in particular in patients presenting with an hyperactive (agitated) delirium (CONN 1990; LINDESAY et al. 1990). Due to its minimal anticholinergic and hypotensive properties, haloperidol is almost universally considered as the antipsychotic of choice in the treatment of agitated delirium in young adults as well as in the elderly (ADAMS 1988; CONN 1990; LINDESAY et al. 1990; LIPOWSKI 1990; SLABY and ERLE 1993). The mode of administration and the dosage are usually guided by the clinical context. In an emergency room or on a psychiatric or medical unit, relatively low doses given intramuscularly are usually preferred (LIPOWSKI 1990; SALZMAN 1988). In the intensive care unit, case reports and naturalistic studies support the intravenous administration of moderate to very high doses – up to 1200 mg/day in one report (SANDERS et al. 1991) – given alone or in combination with lorazepam or opioids (ADAMS 1988; FERNANDEZ et al. 1989; GOLDSTEIN and HALTZMAN 1993; TESAR et al. 1985). Of note, high intravenous doses of haloperidol are reportedly associated with far fewer extrapyramidal side effects than comparable or even far lower oral doses (MENZA et al. 1987). Nevertheless, in older delirious patients, "rapid tranquilization" protocols and combinations of drugs are usually avoided; oral or intramuscular administration of low doses (e.g., 0.5–1 mg b.i.d. or t.i.d.) of haloperidol alone are recommended (CONN 1990; LINDESAY et al. 1990; LIPOWSKI 1990).

While available clinical data and clinical experience support the efficacy of high potency antipsychotics in the management of agitated delirium, current knowledge of the pathophysiology of delirium does not readily explain their efficacy. Based on a comprehensive review, LIPOWSKI (1990) proposes that delirium represents the final common pathway for a variety of processes that reduce cerebral oxidative metabolism and synthesis of acetylcholine, or bring about cholinergic blockade, or both. The association of increased cerebrospinal fluid (CSF) levels of noradrenaline and 5-hydroxyindoleacetic (5-HIAA) and decreased plasma levels of serotonin with delirium secondary to alcohol withdrawal supports the integrative hypothesis that "various patterns of imbalance of normal noradrenergic, serotonergic, and cholinergic neurotransmission appear to underlie all cases of delirium" (LIPOWSKI 1990, p. 168). Of note, a similar imbalance between cholinergic and serotonergic neurotransmission has been demonstrated recently in patients with Lewy body dementia (PERRY et al. 1990; PERRY et al. 1993), a disorder which, like delirium, is characterized by acute onset of confusion, fluctuating behavioral disturbances, and complex visual hallucinations (see below). Thus, the attribution of the beneficial effects of haloperidol (and lorazepam) in the treatment of delirium to the possible correction of a putative dopaminergic-cholinergic imbalance (ADAMS 1988) awaits supporting data.

III. Dementia

Estimates of the prevalence of delusions and hallucinations among patients with primary degenerative dementia have ranged from 10% to 73%, with most around 30% (DEUTSCH et al. 1991; DREVETS and RUBIN 1989; ROSEN and ZUBENKO 1991; WRAGG and JESTE 1989; ZUBENKO et al. 1992). In the context of dementia, psychotic symptoms have been associated with aggressiveness (DEUTSCH et al. 1991), excess patient disability, increased caregiver burden, and premature institutionalization (STEELE et al. 1990; ZUBENKO et al. 1992). Despite its high prevalence and clinical significance, there has been a dearth of methodologically sound studies that have assessed specifically the treatment of psychosis in demented patients. Even studies that are methodologically sound have not separated the response of delusions or hallucinations from the response of nonspecific agitation (BURGIO et al. 1992; COCCARO et al. 1990; DEVANAND et al. 1992; DYSKEN et al. 1994; PETRIE et al. 1982). Thus, pharmacotherapy of disruptive psychotic symptoms has been guided by the results of treatment trials in demented patients presenting with agitation.

Several reviews (RASKIND et al. 1987; SALZMAN 1987a; SALZMAN 1987b; WRAGG and JESTE 1989) and one metaanalysis (SCHNEIDER et al. 1990) have concluded that antipsychotics yield at best a modest improvement in one-third to one-half of demented patients with agitation, while a substantial proportion of these patients fails to improve or deteriorate. A variety of other psychotropic medications and placebo have been reported to produce an improvement comparable to the improvement produced by antipsychotics (HOULIHAN et al. 1994; NYTH et al. 1990; SCHNEIDER and SOBIN 1991). Thus, agitation may be the behavioral expression in demented patients of a heterogeneous group of underlying syndromes – e.g., psychosis, depression, anxiety, chronic pain, or caregiver burden (MULSANT and THORNTON 1990; ROSEN et al. 1992; SALZMAN

1987b). If this is indeed the case, antipsychotics or any single class of psychotropics would not be expected to treat agitation successfully in most demented patients; conversely, specific pharmacologic and behavioral strategies could be selected depending on the specific underlying cause of agitation (ROSEN et al. 1992; ZUBENKO et al. 1992). For instance, antipsychotics could have a much higher effectiveness when prescribed specifically to demented patients presenting with disruptive delusions or hallucinations. This would be expected to be true only if the pathophysiology of psychosis was similar in demented and nondemented patients.

Two groups of investigators have conducted postmortem neuropathologic and neurochemical studies to investigate the biological correlates of psychosis in primary dementia (PERRY et al. 1990; PERRY et al. 1993; MULSANT and ZUBENKO 1994; ZUBENKO et al. 1991). PERRY et al. (1990; 1993) evaluated neurotransmitter activities in the temporal cortex of 12 controls and 14 patients with Lewy body dementia with and without hallucinations. Patients with hallucinations were clearly distinguished by their significantly higher ratio of 5-hydroxyindolacetic (5-HIAA) level to choline acetyltransferase (ChAT) activity. By contrast, they were not distinguishable by their ratio of homovanillic acid (HVA) level to choline acetyltransferase activity. ZUBENKO et al. (1991) studied 10 controls and 27 patients with primary dementia (due to Alzheimer's disease in most patients, though some also presented with Parkinson's disease with Lewy bodies) with and without delusions or hallucinations. Psychotic symptoms were associated with a significant increase in cortical degeneration and a relative preservation of the concentration of norepinephrine in both cortical and subcortical regions. Psychosis was also associated with a reduction of serotonin and its metabolite 5-HIAA in all brain regions studied; this reduction reached statistical significance in the prosubiculum. By contrast, there was no significant difference in the dopaminergic system or in ChAT activity in any of the eight brain regions studied.

Taken together, these studies and others (e.g., PROCTER et al. 1992) suggest that, in dementia as in delirium, the pathogenesis of psychosis is associated with an imbalance between cholinergic and noradrenergic or serotonergic neurotransmission, rather than involving a disturbance in the dopaminergic system. If future studies confirm this hypothesis, cholinergic agents (PERRY et al. 1990), adrenergic agents (ZUBENKO et al. 1991), serotonergic agents (NYTH and GOTTFRIES 1990; PERRY et al. 1990), or atypical antipsychotics (CHACKO et al. 1993; STEELE et al. 1993) may then constitute the mainstay of the pharmacotherapy of psychosis occurring in the context of dementia. CHACKO et al. (1993) have highlighted the possible special role of clozapine in the pharmacologic management of Lewy body dementia that is typically associated with both Parkinsonism and psychosis. However, this role may be limited by hypotension, another frequent clinical feature of Lewy body dementia and a frequent side effect of clozapine in the elderly (FRANKENBURG and KALUNIAN 1994; NABER et al. 1992).

IV. Schizophrenia

1. Early-Onset Schizophrenia in Old Age

The majority of older schizophrenic patients seen in clinical practice have had the onset of their disease early in life (GURIAN et al. 1992; JESTE 1993; MULSANT et al. 1993). However, very little is known about the aging of patients with early-onset schizophrenia (GOLDBERG et al. 1993; MULSANT et al. 1993; PROHOVNIK et al. 1993). Given this lack of empirical data, the principles guiding the treatment of schizophrenia in late life are mostly extrapolated from the knowledge base accumulated on young and middle-aged schizophrenic patients (JESTE et al. 1993). Due to the age-associated pharmacokinetic, pharmacodynamic, and physiologic changes reviewed later in this chapter, as well as some data suggesting that a substantial proportion of elderly schizophrenic patients may present Alzheimer-type neuropathologic changes (PROHOVNIK et al. 1993), the validity of such an extrapolation is unclear. Nevertheless, since no single traditional antipsychotic has proven to be more efficacious in the treatment of schizophrenia in younger patients, the prescription of a specific antipsychotic to an older schizophrenic still depends on the drug's side effect profile. Given the sensitivity of the elderly to both anticholinergic and extrapyramidal side effects (see below), older patients tend to tolerate intermediate potency antipsychotics such as perphenazine, loxapine, or molindone better than either low- or high-potency antipsychotics (MULSANT and GERSHON 1993; PETRIE et al. 1982). In old age, as earlier in life, depot antipsychotics may be necessary to insure compliance (GOTTLIEB et al. 1988; RASKIND et al. 1979). As a general rule, the use of anticholinergic medications should be avoided or minimized since even at low dosages they are associated with cognitive impairment in the elderly (MCEVOY et al. 1987).

While little is known on the use of traditional antipsychotics in the treatment of schizophrenia in late life, even less is known about the potential role of clozapine and other "newer" antipsychotics. Due to their relative lack of action on the nigrostriatal system, some of these drugs may have a special role in late life (JESTE et al. 1993; STEELE et al. 1993). Several case series have reported the successful use of clozapine at dosages of 12.5–400 mg/day (with most patients receiving 50–150 mg/day) in older patients with treatment-resistant psychosis (FRANKENBURG and KALUNIAN 1994) or with psychosis occurring in the context of Parkinson's disease and its treatment (FRANKENBURG and KALUNIAN 1994; FRYE et al. 1993; SCHOLZ and DICHGANS 1985; WOLK and DOUGLAS 1992; WOLTERS et al. 1990). However, elderly patients may also be at higher risk for specific side effects such as agranulocytosis (ALVIR et al. 1993), hypotension (FRANKENBURG and KALUNIAN 1994; NABER et al. 1992), or delirium (SCHUSTER et al. 1977; WOLTERS et al. 1990). At the time of this writing, no report on the use of risperidone in older patients had yet been published. Before the prescription of newer antipsychotics for the treatment of schizophrenia in late life can be recommended, clinical studies need to be conducted in this population.

2. Late-Onset Schizophrenia

During the past decade, schizophrenia of late onset (i.e., after age 45) has been the subject of renewed interest (CASTLE and HOWARD 1992; CASTLE and MURRAY 1993; GRAHAME 1984; HARRIS and JESTE 1988; GOLD 1984; JESTE et al. 1988; LESTER 1982; MARNEROS and DEISTER 1984; PEARLSON et al. 1989; PEARLSON et al. 1993; RABINS et al. 1984; VOLAVKA 1985; YASSA and SURANYI-CADOTTE 1993). Typically, late-onset schizophrenia has been characterized by a high female to male ratio, premorbid personality with schizoid or paranoid traits, presence of hearing and visual loss in some patients, paranoid symptomatology, absence of formal thought disorder or negative symptoms, and a tendency towards chronicity but symptomatic improvement with antipsychotic treatment.

PEARLSON et al. (1993) have reported that, similarly to younger schizophrenic patients, drug-naive patients with late-onset schizophrenia have elevated density of dopamine D2 receptors when compared with age and gender norms. These data support the use of traditional antipsychotics in the treatment of late-onset schizophrenia. In three retrospective studies (JESTE et al. 1988; PEARLSON et al. 1989; RABINS et al. 1984b) and one ongoing uncontrolled prospective study (JESTE et al. 1993), more than 125 patients with late-onset schizophrenia have been reported to respond moderately well to various antipsychotics at relatively low dosages, typically around 200 chlorpromazine-equivalent mg/day. These dosages are about half as high as the threshold dosage reported in the treatment of younger patients with schizophrenia (BALDESSARINI 1985). This is consistent with the 50% increase in plasma level to dose ratios observed in older patients treated with neuroleptics (see Sec. C.I.3. Metabolism). Controlled studies are now needed to confirm that patients with late-onset schizophrenia do indeed respond to antipsychotics and require lower dosages than younger schizophrenic patients. These studies should also explore whether such differences are due to age-related differences in drug disposition or to differences between the pathophysiology of early- and late-onset schizophrenia.

V. Delusional Disorder

In the current American nosology, delusional disorder constitutes a discrete disorder, akin to paranoia in KRAEPELIN's classification (AMERICAN PSYCHIATRIC ASSOCIATION 1994; MANSCHRECK 1992; MUNRO 1982). The conceptualization of delusional disorder as a separate disorder is supported by the absence of familial aggregation with schizophrenia or with the mood disorders (KENDLER 1980; KENDLER and TSUANG 1981; WINOKUR 1985). Delusional disorder is a rare condition with a prevalence estimated at 20–30/

100000, a peak age of onset between 25 and 45 (KENDLER 1982), and a low likelihood of first onset in late life (MANSCHRECK 1992). However, due to a relative preservation of functioning, some patients may not come to clinical attention until late in life (GURIAN et al. 1992; KENDLER 1982). Overall, very little is known about late-life delusional disorder, though the somatic subtype has been reported to be more prevalent in older people (MUNRO 1992). The pathophysiology of delusional disorder is not known but the role of the dopaminergic system has been implicated since an indistinguishable delusional syndrome can be precipitated by cocaine, amphetamines, methyldopa, or levodopa. Furthermore, pimozide has been reported to be effective in the treatment of the somatic subtype (see OPLER and FEINBERG 1991 for a review) with symptomatic relief in as many as 64% of patients (MUNRO 1992), and in cases of erotomania (MUNRO et al. 1985), delusional jealousy (DORIAN 1979; McCOY et al. 1992; POLLOCK 1982), or persecutory and grandiose delusions (MUNRO 1992). Other antipsychotics (such as chlorpromazine, sulforidazine, thioridazine, prochlorperazine, and methotrimeprazine) have also been reported to be effective (KENDLER 1980) but the relative effectiveness of various antipsychotics has not been compared in any controlled study. Low-dose pimozide (0.5–2 mg/day) may be prescribed to an older patient presenting with delusional disorder, bearing in mind that pimozide is one of the most potent clinically available antipsychotics and that elderly patients are at very high risk for developing disabling extrapyramidal side effects with this agent. In older patients who cannot tolerate pimozide, a medium-potency antipsychotic could be used (e.g., perphenazine, 4–12 mg/day). Alternatively, some case reports suggest that tricyclic antidepressants (BROTMAN and JENIKE 1984; PYLKO and SICIGNAN 1985) and serotonergic reuptake inhibitors (LANE 1990) may be effective in the treatment of some patients with delusional disorder.

VI. Mood Disorders

1. Psychotic (Delusional) Major Depression

A quarter to one-half of elderly patients admitted to a hospital for the treatment of major depression present with delusions or hallucinations (BALDWIN and JOLLEY 1986; BURVILL et al. 1991; MEYERS and GREENBERG 1986; MULSANT et al. 1993; MURPHY 1983). Despite the significance of psychotic major depression in late life, no controlled data on its treatment have been published. Given this lack of controlled data, the treatment of psychotic major depression in late life is currently based on the results of trials involving younger or mixed-age patients, in whom close to 20 trials have demonstrated the low effectiveness (i.e., less than 40% response) of tricyclic antidepressants alone (CHAN et al. 1987; KOCSIS et al. 1990). Two small naturalistic studies in elderly patients have confirmed this poor responsiveness to a tricyclic alone (BALDWIN 1988; BROWN et al. 1984). By contrast, in mixed-age samples, 16 naturalistic studies (see

KROESSLER 1985 for a review) and two randomized double-blind controlled studies (ANTON and BURCH 1990; SPIKER et al. 1985) have shown that the combination of an antipsychotic with a tricyclic increases significantly the rate of response (to up to 70%). In younger patients, several studies suggest that this differential response to pharmacotherapy of psychotic and non-psychotic major depression is due to the activation of the dopaminergic system seen with psychotic but not with nonpsychotic depression (see SCHATZBERG and ROTHSCHILD 1992 for a review). An ongoing study is attempting to replicate this finding in older patients (MEYERS et al. 1993).

In mixed-age patients with psychotic depression, higher response rates have been associated with antipsychotic dosage above 350mg/day chlorpromazine-equivalent (NELSON et al. 1986; SPIKER et al. 1986). In an ongoing study (MULSANT et al. 1994), we have found that most elderly patients with psychotic major depression tolerate a combination of nortriptyline plus perphenazine at about half this dosage (i.e., 12–24mg/day). However, despite its relative lack of toxicity, this combination seems to be less effective in older than in younger patients. If further studies confirm these preliminary findings, they will need to explore whether this diminished responsiveness is due to aging-associated pharmacokinetic or pharmacodynamic changes.

Amoxapine is an atypical heterocyclic drug derived from loxapine. It possesses both antidepressant and antipsychotic properties (LYDIARD and GELENBERG 1981) and can induce extrapyramidal symptoms and tardive dyskinesia (COHEN et al. 1982; GAFFNEY and TUNE 1985; THORNTON and STAHL 1984). In a randomized double-blind study involving middle-age patients (ANTON and BURCH 1990), amoxapine (400mg/day) was as effective as a combination of amitriptyline (200mg/day) plus perphenazine (32mg/day). Side effects were comparable in both groups, except for extrapyramidal side effects that were significantly less prevalent with amoxapine. Despite this potential advantage of amoxapine for older patients, the lack of supporting data precludes recommending its use in the treatment of late-life psychotic major depression.

2. Bipolar Disorder (Mania)

Psychotic features, in particular grandiose delusions, commonly occur during manic episodes of bipolar or schizoaffective disorders (YOUNG et al. 1983; YOUNG and KLERMAN 1992). A significant proportion of bipolar patients experience their first manic episode late in life (STONE 1989). In mixed-age patients, older age of onset has been associated with both a lower (ROSEN et al. 1984) and a higher prevalence (ANGST et al. 1973) of psychotic symptoms. In elderly manic patients, two studies have failed to find an association between either age at index or age at onset and the presence or absence of psychosis (BROADHEAD and JACOBY 1990; YOUNG et al. 1983). Given the scarcity of empirical data on the topic, treatment of mania in late-life is patterned on its treatment in younger patients (Stone 1989; Young and Klerman 1992). Though we are not aware of data specifically relevant to the pathophysiology of psychotic symptoms in manic patients, core manic symptoms (i.e., elevation of mood, grandiose ideas, increase in activity, distractibility, and impaired judgment) have been linked to an increase in central dopaminergic activity, possibly mediated through the D1 receptors (Silverstone 1985). Furthermore, both sedating and nonsedating antipsychotics have been shown to have antimanic effects (Silverstone 1985). However, the clinical role of antipsychotics in the treatment of mania in younger patients remains a subject of debate since lithium alone or combined with clonazepam (Chouinard et al. 1983) and other benzodiazepines (Modell et al. 1985) may be effective in treating agitation and psychotic symptoms associated with mania. Some patients, however, may do better on a combination of lithium (or an anticonvulsant) plus an antipsychotic (Garfinkel et al. 1980).

Given the absence of published data supporting or challenging the relative usefulness of antipsychotics in the treatment of mania in late life, these drugs should be used in the elderly manic patients who are psychotic and who do not respond to or do not tolerate lithium, alone or combined with anticonvulsants or benzodiazepines. Some reports in mixed-age patients (McElroy et al. 1991; Naber et al. 1989) and in older patients (Frankenburg and Kalunian 1994) suggest that clozapine in combination with lithium and/or antidepressants may be effective in treatment-resistant mood disorders with psychosis. Further studies of traditional and newer antipsychotic agents in the treatment of psychosis associated with bipolar disorder in late-life are needed.

C. Pharmacokinetics

Age-related differences in response to drugs can arise from pharmacokinetic differences (i.e., differences in the way a drug is absorbed, excreted, metabolized, or distributed) or pharmacodynamic differences (i.e., differences in the response to a given plasma, or other tissue concentration of the drug).

Most differences seemingly related to age are related to conditions that, although more frequent in the elderly, can occur in patients of all ages: diminished renal or cardiac function, concurrent illness, and drug treatment. In the community the average older American is taking three prescription medications; this rises to an average of nine drugs in nursing home residents (Zimmer et al. 1985).

Knowledge of pharmacokinetic parameters such as clearance (Cl), half-life ($t_{1/2}$), bioavailability (F), and volume of distribution (Vd) are needed both for therapeutics and for the understanding of a drug's disposition. Pharmacokinetic data are essential in selecting an appropriate initial dose, a dosing rate, and the interval and route of administration. Drug–drug interactions and dose-dependent (saturation) kinetics are of considerable importance in the elderly. Undetected they can cause nonlinearity (i.e., metabolism ceases to be proportional to dose) due to drug accumulation and delays to steady-state, resulting in significant adverse effects.

I. General Changes with Age and Illness

It is now recognized that pharmacokinetic data obtained from young healthy volunteers may be different from those in elderly patients (Table 2). Altered physiology, possibly secondary to illness and inanition, may cause decreased hepatic clearance (HAMBERG et al. 1990). To this should be added medical differences and the increased possibility of interacting medications in older patients (GREENBLATT et al. 1982; DAWLING and CROME 1989; CENTER FOR DRUG EVALUATION AND RESEARCH 1989).

1. Absorption

With age there are decreases in stomach acid and in the size of the gastric absorbing surface as well as reductions in mesenteric blood flow. Although these changes may affect nutrients absorbed by active transport (e.g., calcium, iron, sugars, amino acids), antipsychotics are absorbed by passive diffusion and the rate and extent of absorption is not significantly altered. Clinicians should, however, be alert to the extensive use of compounds with huge absorbing surfaces (e.g., ispaghulahusk, Metamucil, and Maalox magnesium hydroxide), or medications with anticholinergic properties which can interfere with the rate of gastrointestinal absorption.

Table 2. Pharmacokinetic/pharmacodynamic factors affecting antipsychotic use in the elderly

Pharmacologic factor	Age-related changes
Absorption	Physiological changes of little consequence, but other medications can significantly affect.
Distribution	Increase in proportion of body fat will increase half-life of antipsychotics.
Cardiac output	Reduction with age leads to reduced hepatic first-pass effect.
Hepatic metabolism	P450 2D6 principal isozyme, not changed by age alone, but 5%–10% of population are genetically poor metabolizers and numerous medications can inhibit.
Excretion	Reduction in renal clearance will lead to increased concentrations of active, hydroxylated metabolites.
Receptor-site sensitivity	Decreased CNS cholinergic and nigrostriatal dopaminergic functioning will increase sensitivity to anticholinergics and D2 receptor blockade.

2. Distribution

Important changes in distribution occur with age. The percentage of body fat relative to lean body mass and total body water increases with age. Between the ages of 20 and 75, total body water decreases 15%–20%, extracellular fluid decreases 35%–40%, and the fraction of body weight as fat increases 25%–45% (Greenblatt et al. 1982). For lipid-soluble drugs, such as antipsychotics that distribute in body fat, the volume of distribution is increased, resulting in an increase in the half-life. Half-life is directly proportional to volume of distribution divided by clearance; thus, even if there is no change in clearance, an increase in volume of distribution will lead to a prolonged half-life.

Albumin concentrations may diminish with age-associated illness, while alpha-1-acid glycoprotein increases. This will differentially affect the protein binding of acidic and basic drugs, respectively. The antipsychotics are bound with greater affinity to alpha-1-acid glycoprotein, an acute-phase reactant protein, which increases markedly in those suffering from arthritis and cancer. The extent of protein binding will not only affect the availability of free drug but the volume of distribution as well. It is now appreciated that any increase in free drug will be buffered by redistribution and elimination, diminishing the effect of plasma protein displacement (Rolan 1994).

3. Metabolism

Data obtained in senescent rodents supported the conventional view that there is an age-associated decline in phase 1 drug-metabolizing enzyme activity (hydroxylation, demethylation). There is now evidence ranging from hepatic samples to drug-metabolic phenotyping which suggests there is not a uniform decline in P450 metabolism with age (Schmucker et al. 1990; Pollock et al. 1992). There has also been an increased appreciation of the diversity of P450 isozymes. Cytochrome P450 2D6 (debrisoquine hydroxylase) is the isozyme responsible for hydroxylation of many psychotropics. In unmedicated elders, it does not appear to undergo an age-associated decline (Pollock et al. 1992). Demethylation, mediated by other isozymes, such as P450s 2C19, 1A2 or 3A4 may be more subject to age-associated impairment as has been shown with imipramine and clomipramine (Abernethy et al. 1985; Kunik et al. 1994).

In persons older than 65, compared with those under 25, hepatic blood flow is decreased by 40%. This is at least partially the result of decreased liver size in relation to body mass as well as an age-related trend for reductions in cardiac output (decreases approximately 1%/year after age 30). For highly metabolized drugs a reduction in hepatic first-pass effect is to be expected. Although details of potentially interacting medications are not given, higher plasma level to dose ratios in older patients have been observed for thioridazine, thiotixene, and perphenazine (Axelsson and Martensson 1976; Hansen and Larsen 1985; Yesavage et al. 1981). Also, in a rare pharmacokinetic study of elderly psychiatric patients, plasma concentrations of thioridazine, mesoridazine, and sulforidazine were 1.5–2 times higher than in younger patients (COHEN and SOMMER 1988). In a larger study of haloperidol, however, the plasma level to dose ratio was not found to be increased in older patients (AOBA et al. 1985). Nevertheless, age-related changes in body composition and reductions in hepatic volume and blood flow emphasize individual differences in drug metabolism, a major factor in the variability of drug concentrations in older patients.

The discovery of the P450 2D6 polyrmorphism in oxidative metabolism is one of the most important findings in clinical pharmacology in the past two decades. Five to ten percent of the population lack P450 2D6 (EVANS et al. 1980). In addition, many drugs competitively or noncompetitively inhibit this enzyme. The list of drugs definitively shown to be metabolized by P450 2D6 continues to grow and includes perphenazine, thioridazine, and risperidone (DAHL-PUUSTINEN et al. 1989; VON BAHR et al. 1991; HUANG et al. 1993). Clozapine was believed to be a 2D6 substrate on the basis of in vitro experiments. Recent human pharmacokinetic studies conducted in poor and extensive 2D6 metabolizers have not borne this out (DAHL et al. 1994). The opposite result has occurred with haloperidol; in vitro studies suggested that its metabolism was not dependent on 2D6, while the oxidation of reduced haloperidol back to haloperidol does depend on 2D6 (TYNDALE et al. 1991). Nonetheless, poor 2D6 metabolizers demonstrate higher plasma levels of haloperidol and its active reduced metabolite compared to extensive 2D6 metabolizers (LLERENA et al. 1992). There is emerging data associating the poor debrisoquine phenotype with an increased incidence of side effects both retrospectively (SPINA et al. 1992) and prospectively POLLOCK et al. (1995).

Antipsychotics are themselves moderately potent inhibitors of P450 2D6. This has been specifically confirmed in vivo for haloperidol, perphenazine, and thioridazine (GRAM et al. 1989; GRAM and OVERØ 1972; HIRSCHOWITZ et al. 1983). Studies utilizing in vitro microsomal systems suggest that thioridazine, perphenazine, and haloperidol (approximate K_is (μM) of 1) are somewhat more potent inhibitors of P450 2D6 than is chlorpromazine ($K_i = 7$) (VON BAHR et al. 1985; INABA 1985). The potential of antipsychotics to inhibit drug metabolism would suggest a need for particular caution in elderly patients taking other 2D6 substrates such as antidepressants, β-blockers, dextromethorphan, some antiarrhythmics, and anticancer drugs (BRØSEN 1990). Moreover, the conversion of codeine to active morphine also depends on P450 2D6. Codeine has, in fact, been shown to be an ineffective analgesic when 2D6 is inhibited by quinidine (SINDRUP et al. 1991).

4. Excretion

The decrease of creatinine clearance with age is well established (POLLOCK et al. 1992). Since there is often a marked decline in muscle mass, especially with debilitation, serum creatinine may not appear elevated despite a significant decline in renal function. This will affect the elimination of the hydroxylated metabolites of antipsychotics. This has not, as yet, been systematically studied in older individuals. Active metabolites of some antipsychotics, such as reduced haloperidol, the sulfoxide metabolites of thioridazine, and 7-OH-perphenazine, may impact therapeutics to a greater degree in the elderly. In younger schizophrenic patients, increased reduced haloperidol has been associated with a less favorable outcome (Altamura et al. 1988).

5. Need for Individualized Assessment

The elderly are a very heterogeneous population, notable for wide variability in biochemical, physiological, and psychological measurements. Dependence on mean pharmacokinetic data, particularly those derived from single-dose studies in healthy volunteers, is misleading. Although age-associated trends in drug disposition identified above are of heuristic importance, there should be individualized assessment of relevant covariants. In late life, the effects of illness and other medications far outweigh the impact of age-related trends.

D. Pharmacodynamics

"Pharmacodynamics" refers to the relationship between a drug's effect and its measurable concentration, and as such, it encompasses pharmacokinetics. In older patients there is general reduction in homeostatic mechanisms (e.g., postural control, orthostatic circulatory responses, thermoregulation, visceral muscle function, higher cognitive function). This may interfere with the ability to adapt to changes in the environment and may be manifest as an adverse drug reaction. For instance, neuroleptics increase the risk of falls and hip fracture in the elderly (Ray et al. 1987; Tinetti et al. 1988).

Specific changes have been most extensively investigated for autonomic receptor-mediated effects (Pollock et al. 1990). Reduction in $\alpha 2$ (but not $\alpha 1$) adrenoceptor responsiveness might contribute to the increased risk of orthostatic hypotension in elderly patients. Orthostatic hypotension is the major cardiovascular effect of concern in the elderly, and it mitigates the use of lower-potency drugs. Nonetheless, the membrane-stabilizing properties of haloperidol, with consequent delays in ventricular conduction, have been increasingly associated with the lethal torsade de pointes arrhythmia when used in high parenteral doses in intensive care units (Goldstein and Haltzman 1993; Metzger and Friedman 1993). In older patients undergoing long-term treatment with antipsychotics, the prevalence of obesity, poor nutritional status, and increased triglycerides may contribute additional cardiovascular risk factors (Martinez et al. 1994).

Diminished cholinergic functioning in the CNS of older patients may render them more sensitive to the anticholinergic effects of antipsychotic medication (Feinberg 1993). These effects can range from blurred vision and cognitive impairment (LARSON et al. 1987) to frank delirium (SCHOR et al. 1992), which may occur at therapeutic concentrations of medication. The increased sensitivity of older patients to anticholinergic side effects also limits the use of low-potency antipsychotics in this population.

Reductions in nigrostriatal dopamine predispose the elderly to more neuroleptic-induced extrapyramidal side effects (AYD 1961; WILSON and MACLENNAN 1989) (Fig. 1).

Until recently the increased incidence of tardive dyskinesia in the elderly was believed to be due to increased exposure (SWEET et al. 1995). Two prospective studies have now demonstrated that, regardless of diagnosis, there is a very high onset of tardive dyskinesia in older patients taking antipsychotics for relatively brief periods (cumulative incidence reported as 26%–45% over 1 year and 31% after 43 weeks) (JESTE et al. 1993; SALTZ et al. 1991).

Increased attention has been given to preexisting neurological signs in Alzheimer's disease. Extrapyramidal signs are common in this illness particularly as it progresses (FUNKENSTEIN et al. 1993; WALLIN and BLENNOW 1992; RICHARDS et al. 1993). Patients suffering from Lewy body dementia have recently been recognized as being exquisitely sensitive to antipsychotics (MCKEITH et al. 1992). Unfortunately there is no way to reliably identify this condition prior to a trial of medication.

Concern with neurological damage caused by oxidative free radicals has been quickened by the success achieved in relating abnormal superoxide dismutase to amyotrophic lateral sclerosis (ROCHE and ROMERO-ALVIRA 1993). Antipsychotics are implicated in at least two ways. Acute treatment by increasing dopamine turnover could lead to increased hydrogen peroxide production (SHIVAKUMAR and RAVINDRANATH 1993). It is also possible that antipsychotics inhibit mitochondrial complex 1 enzymes. There is specific concern about haloperidol since its pyridinium metabolite bears a resemblance to MPTP (GORROD and FANG 1993). Reported effects of antipsychotics on inhibiting nitric oxide synthase (HU et al. 1994) and altering the protection afforded by the blood-brain barrier (BEN-SHACHAR et al. 1994) may be additional complexities. Given the more limited dopaminergic reserve in the eld-

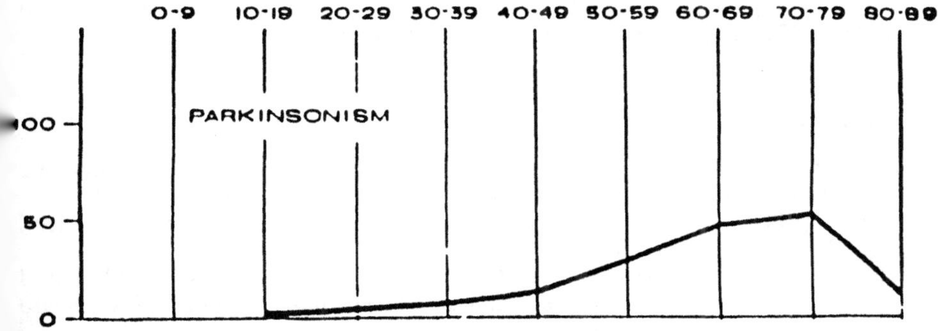

Fig. 1. Drug-induced extrapyramidal reactions incidence by age (from AYD, 1961)

erly, this putative neurotoxicity should be of particular concern when assessing the risk/benefit of antipsychotic treatment.

E. Conclusions

The elderly are more likely to be sensitive to drug effects and more likely to suffer from adverse drug interactions, especially from the increased likelihood of taking multiple medications. Before prescribing an antipsychotic to older patients, it is imperative to obtain complete records of all medications, which should be kept to a minimum. Assume impaired capacity to metabolize and/or excrete newly initiated medications, but also do not undertreat on the basis of chronological age alone. Always use higher-potency drugs (haloperidol and perphenazine) in minimal doses. Even though the current nosologic classification of late-life psychoses was established 40 years ago, controlled therapeutic trials are still direly lacking. On the basis of the scant data currently available on older patients and of extrapolations from data obtained in younger patients, a dysregulation of the dopaminergic system can be implicated in the pathophysiology of the "functional" psychoses in late life. Thus, the use of antipsychotics that exert their action mainly through dopaminergic receptor blockade can be rationally justified in the treatment of schizophrenia, delusional disorder, or mood disorders with psychotic features. By contrast, current knowledge of the pathophysiology of psychoses associated with delirium or dementia does not support the use of these antipsychotics in their treatment. One can hypothesize that the benefits attributed to antipsychotics in the treatment of behavioral disturbances associated with delirium or dementia is due either to nonspecific factors (e.g., sedation) or to the other pharmacologic properties of these drugs. However, while neuroscientists may question this practice, the prescription of antipsychotics to delirious and demented patients remains justified by the vast clinical experience with their use in this population, as opposed to the minimal empirical data supporting the use of other drugs (except for benzodiazepines) and nonpharmacologic interventions.

Acknowledgments. Preparation of this review and the authors' research described herein were supported by United States Public Health Service grant nos. K07 MH-1040 and MH-49786 from the National Institute of Mental Health.

References

Abernethy DR, Greenblatt DJ, Shader RI (1985) Imipramine and desipramine disposition in the elderly. J Pharmacol Exp Ther 232:183–188

Adams F (1988) Emergency intravenous sedation of the delirious, medically ill patient. J Clin Psychiatry 49 [Suppl 12]:22–26

Altamura C, Mauri M, Cavallaro R, Colacurcio F, Gorni A, Bareggi S (1988) Reduced haloperidol/haloperidol ratio and clinical outcome in schizophrenia. Prog Neuro psychopharmacol Biol Psychiatry 12:689–694

Alvir JMJ, Lieberman JA, Safferman AZ, Schaaf JA (1993) Clozapine-induced agranulocytosis: incidence and risk factors in the United States. N Engl J Med 329:162–167

American Psychiatric Association (1994) Diagnostic and statistical manual of mental disorders, 4th edn. American Psychiatric Association, Washington DC

Angst J, Baastrup P, Grof P, Hippius H, Pöldinger W, Weis P (1973) The course of monopolar depression and bipolar psychoses. Psychiat Neurol Neurochir 76:489–500

Anton RF, Burch EA (1990) Amoxapine versus amitriptyline combined with perphenazine in the treatment of psychotic depression. Am J Psychiatry 147:1203–1208

Aoba A, Kakita Y, Yamaguchi N, Shido M, Tsuneizumi T, Shibata M, Kitani K, Hasegawa K (1985) Absence of age effect on plasma neuroleptic levels in psychiatric patients. J Gerontol 40:303–308

Axelsson R, Martensson E (1976) Serum concentration and elimination from serum of thioridazine in psychiatric patients. Curr Ther Res 19:242–265

Ayd FJ (1961) A survey of drug-induced extrapyramidal reactions. JAMA 175:1054–1060

Baldessarini RJ (1985) Chemotherapy in Psychiatry. Principles and practice. Harvard University Press, Cambridge

Baldwin RC, Jolley DJ (1986) The prognosis of depression in old age. Br J Psychiatry 149:574–583

Baldwin RC (1988) Delusional and non-delusional depression in late life: evidence for distinct subtypes. Br J Psychiatry 152:39–44

Beers MH, Ouslander JG (1989) Risk factors in geriatric drug prescribing. Drugs 37:105–112

Ben-Shachar, Livne E, Spanier I, Leenders KL, Youdim MBH (1994) Typical and atypical neuroleptics induce alteration in blood-brain barrier and brain $^{59}FeCl_3$ uptake. J Neurochem 62:1112–1118

Broadhead J, Jacoby R (1990) Mania in old age: a first prospective study. Int J Geriatr Psychiatry 5:215–222

Brøsen K (1990) Recent developments in hepatic drug oxidation. Clin Pharmacokinet 18:220–39

Brotman AW, Jenike MA (1984) Monosymptomatic hypochondriasis treated with tricyclic antidepressants. Am J Psychiatry 141(12):1608–1609

Brown RP, Kocsis JH, Glick ID, Dhar AK (1984) Efficacy and feasibility of high dose tricyclic antidepressant treatment in elderly delusional depressives. J Clin Psychopharmacol 4:311–315

Burgio LD, Reynolds CF, Janosky JE, Perel JM, Thornton JE, Hohman MJ (1992) A behavioral microanalysis or the effects of haloperidol and oxazepam in demented psychogeriatric inpatients. Int J Geriat Psychiatry 7:253–262

Burkhardt C, Kelly JP, Lim Y-H, Filley CM, Parker WD (1993) Neuroleptic medications inhibit complex 1 of the electron transport chain. Ann Neurol 33:512–517

Burvill PW, Hall WD, Stampfer HG, Emmerson JP (1991) The prognosis of depression in old age. Br J Psychiatry 158:64–71

Caligiuri MP, Bracha HS, Lohr JB (1989) Asymmetry of neuroleptic-induced rigidity: development of quantitative methods and clinical correlates. Psychiatry Res 30:275–284

Castle DJ, Howard R (1992) What do we know about the aetiology of late-onset schizophrenia? Eur Psychiatry 7:1–9

Castle DJ, Murray RM (1993) The epidemiology of late-onset schizophrenia. Schizophr Bull 19:691–700

Center for Drug Evaluation and Research (1989) Guideline for the study of drugs likely to be used in the elderly. Food and Drug Administration, Department of Health and Human Services, Washington

Chacko R, Hurley RA, Jankovic J (1993) Clozapine use in difuse Lewy body disease. J Neuropsychiatry Clin Neurosci 5:206–208
Chan CH, Janicak PG, Davis JM, Altman E, Andriukaitis S, Hedeker D (1987) Response of psychotic and nonpsychotic depressed patients to tricyclic antidepressants. J Clin Psychiatry 48:197–200
Chouinard G, Young SN, Annable L (1983) Antimanic effect of clonazepam. Biol Psychiatry 18:451–466
Coccaro EF, Kramer E, Zemishlanhy Z, Thorne A, Rice CM, Giordani B, Duvvi K, Patel BM, Torres J, Nora R, Neufeld R, Mohs RC, Davis KL (1990) Pharmacological treatment of noncognitive behavioral disturbances in elderly demented patients. Am J Psychiatry 147:1640–1645
Cohen BM, Harris PQ, Altesman RI, Cole JO (1982) Amoxapine: neuroleptic as well as antidepressant. Am J Psychiatry 139:1165–1167
Cohen BM, Sommer BR (1988) Metabolism of thioridazine in the elderly. J Clin Psychopharmacol 8:336–339
Conn DK (1990) Delirium and other organic mental disorders. In: Sadavoy J, Lazarus LW, Jarvik LF (eds) Comprehensive review of geriatric psychiatry. American Psychiatric Press, Washington DC, p 320
Dahl M-L, LLerena A, Bondesson U, Lindström L, Bertilsson L (1994). Disposition of clozapine in man: lack of association with debrisoquine and S-mephenytoin hydroxylation polymorphisms. Br J Clin Pharmacol 37:71–4
Dahl-Puustinen M-L, Liden A, Alm C, Nordin C, Bertilsson L (1989) Disposition of perphenazine is related to polymorphic debrisoquin hydroxylation in human beings. Clin Pharmacol Ther 46:78–81
Dawling S, Crome P (1989) Clinical pharmacokinetic considerations in the elderly. Clin Pharmacokinet 17:236–263
Deutsch LH, Bylsma FW, Rovner BW, Steele C, Folstein MF (1991) Psychosis and physical aggression in probable Alzheimer's disease. Am J Psychiatry 148:1159–1163
Devanand DP, Cooper T, Sackeim HA, Taurke E, Mayeux R (1992) Low dose oral haloperidol and blood levels in Alzheimer's disease: a preliminary study. Psychopharmacol Bul 28(2):169–173
Dorian BJ (1979) Successful outcome of treatment of a case of delusional jealousy with pimozide. Can J Psychiatry 24:377
Drevets WC, Rubin EH (1989) Psychotic symptoms and the longitudinal course of senile dementia of the Alzheimer type. Biol Psychiatry 25:39–48
Dysken MW, Johnson SB, Holden L, Vatassery G, Nygren J, Jelinsky M, Kuskowski M, Schut L, McCarten JR, Knopman D, Maletta GJ, Skare S (1994) Haloperidol concentrations in patients with Alzheimer's dementia. Am J Geriatr Psychiatry 2:124–133
Evans DAP, Maghoub A, Sloan TP, Idle JR, Smith RL (1980) A family and population study of the genetic polymorphism of debrisoquin oxidation in a white British population. J Med Genet 17:102–105
Feinberg M (1993) The problems of anticholinergic adverse effects in older patients. Drugs Aging 3:335–348
Fernandez F Levy J Mansell P (1989) Management of delirium in terminally ill patients. Int J Psychiatry Med 19:165–172
Flint AJ, Rifat SL, Eastwood MR (1991) Late-onset paranoia: Distinct from paraphrenia. Int J Geriatr Psychiatry 6:103–109
Frankenburg FR, Kalunian D (1994) Clozapine in the elderly. J Geriatr Psychiatry Neurol 7:131–134
Frye MA, Wirshing WC, Ames D (1993) Clozapine as a diagnostic tool for a psychotic parkinsonian patient. J Clin Psychopharmacol 13:359–360
Funkenstein HH, Albert MS, Cook NR, West CG, Scherr PA, Chown MJ, Pilgrim D, Evans DA (1993) Extrapyramidal signs and other neurologic findings in clinically diagnosed Alzheimer's disease. Arch Neurol 50:51–56

Gaffney GR, Tune LE (1985) Serum neuroleptic levels and extrapyramidal side effects in patients treated with amoxapine. J Clin Psychiatry 46:428–429
Garfinkel PE, Stancer HC, Persad E (1980) A comparison of haloperidol, lithium carbonate, and their combination in the treatment of mania. J Affect Disord 2:279–288
Garrard J, Makris L, Dunham T, Heston LL, Cooper S, Ratner ER, Zelterman D, Kane RL (1991) Evaluation of neuroleptic drug use by nursing home elderly under proposed medicare and medicaid regulations. JAMA 265:463–467
Gold DD (1984) Late age of onset schizophrenia: present but unaccounted for. Compr Psychiatry 25:225–237
Goldberg TE, Hyde TM, Kleinman JE, Weinberger DR (1993) Course of schizophrenia: neuropsychological evidence for a static encephalopathy. Schizophr Bull 19:797–804
Goldstein MG, Haltzman SD (1993) Intensive care. In: Stoudemire A, Fogel BS (eds) Psychiatric care of the medical patient. Oxford University Press, New York
Gorrod JW, Fang J (1993) On the metabolism of haloperidol. Xenobiotica 23:495–508
Gottlieb GL, McAllister TW, Gur RC (1988) Depot neuroleptics in the treatment of behavioral disorders in patients with Alzheimer's disease. J Am Geriatr Soc 36:619–621
Grahame PS (1984) Schizophrenia in old age (late paraphrenia). Br J Psychiatry 145:493–495
Gram LF, Overø FK (1972) Drug interaction: inhibitory effect of neuroleptics on metabolism of tricyclic antidepressants in man. Br Med J 1:463–465
Gram LF, Debruyne D, Caillard V, Boulenger JP, Lacotte J, Moulin M, Zarifian E (1989) Substantial rise in sparteine metabolic ratio during haloperidol treatment. Br J Clin Pharmacol 27:272–275
Greenblatt DJ, Sellers EM, Shader RI (1982) Drug disposition in old age. N Engl J Med 306:1081–1088
Gurian BS, Wexler D, Baker EH (1992) Late-life paranoia: possible association with early trauma and infertility. Int J Geriatr Psychiatry 7:277–284
Hamberg O, Oversen L, Dorfeldt A, Loft S, Sonne J (1990) The effect of dietary energy and protein deficiency on drug metabolism. Eur J Clin Pharmacol 38:567–570
Hansen LB, Larsen N-E (1985) Therapeutic advantages of monitoring plasma concentrations of perphenazine in clinical practice. Psychopharmacology 87:16–19
Harris MJ, Jeste DV (1988) Late-onset schizophrenia: an overview. Schizophr Bull 14(1):39–55
Harris J, Ponton D, Caligiuri MP, Krull AJ, Tran-Jonhson TK, Jeste DV (1992) High incidence of tardive dyskinesia in older outpatients on low doses of neuroleptics. Psychopharmacol Bull 28:87–92
Hirschowitz J, Bennet JA, Zemlan FP, Garrer DL (1983) Thioridazine effect on desipramine plasma levels. J Clin Psychopharmacol 3:376–379
Holden NL (1987) Late paraphrenia or the paraphrenias? A descriptive study with a 10-year follow-up. Br J Psychiatry 150:635–639
Houlihan DJ, Mulsant BH, Sweet RA, Rifai AH, Pasternak R, Rosen J, Zubenko GS (1994) A naturalistic study of trazodone in the treatment of behavioral complications of dementia. Am J Geriatr Psychiatry 2:78–85
Hu J, Lee J-H, El-Fakahany EE (1994) Inhibition of neuronal nitric oxide synthase by antipsychotic drugs. Psychopharmacol 114:161–166
Huang M-L, Van Peer A, Woestenborghs R, De Coster R, Heykants J, Jansen AAI, Zylyicz Z, Visscher HW, Jonkman JHG (1993) Pharmacokinetics of the novel antipsychotic agent risperidone and the prolactin response in healthy subjects. Clin Pharmacol Ther 54:257–268
Hymas N, Naguib M, Levy R (1989) Late paraphrenia – a follow-up study. Int J Geriatr Psychiatry 4:23–29
Inaba T, Jurima M, Mahon WA, Kalow W (1985) In vitro inhibition studies of two isozymes of human liver cytochrome P-450. Drug Metab Dispos 13:443–448

Jeste DV (1993) Late-life schizophrenia. Schizophr Bull 19:687–689
Jeste DV, Harris MJ, Pearlson GD, Rabins PV, Lesser IM, Miller BL, Coles C, Yassa R (1988) Psychosis and depression in the elderly. Late-onset schizophrenia: studying clinical validity. Psychiatr Clin of North Am II(1):1–14
Jeste DV, Lacro JP, Gilbert PL, Kline J, Kline N (1993) Treatment of late-life schizophrenia with neuroleptics. Schizophr Bull 19:817–830
Kay DWK (1963) Late paraphrenia and its bearing on the aetiology of schizophrenia. Acta Psychiatr Scand 39:159–169
Kay DWK, Roth M, Hopkins B (1955) Affective disorders arising in the senium: their association with organic cerebral degeneration. Br J Psychiatry 101:302–316
Kay DWK, Roth M (1961) Environmental and heredity factors in the schizophrenias of old age ("late paraphrenias") and their bearing on the general problem of causation in schizophrenia. J Ment Sci 107:649–686
Kendler KS (1980) The nosologic validity of paranoia (simple delusional disorder). A review. Arch Gen Psychiatry 37:699–706
Kendler KS, Tsuang MT (1981) Nosology of paranoid schizophrenia and other paranoid psychoses. Schizophr Bull 7:594–610
Kendler KS (1982) Demography of paranoid psychoses (delusional disorder): a review and comparison with schizophrenia and affective illness. Arch Gen Psychiatry 39:890–902
Kocsis JH, Croughan JL, Katz MM, Butler TP, Secunda S, Bowden CL, Davis JM (1990) Response to treatment with antidepressants of patients with severe or moderate nonpsychotic depression and of patients with psychotic depression. Am J Psychiatry 147:621–624
Kroessler D (1985) Relative efficacy rates for therapies of delusional depression. Convuls Ther 1:173–182
Kunik ME, Pollock BG, Perel JM, Altieri L (1994) Clomipramine in the elderly: tolerance and plasma levels. J Geriatr Psychiatry and Neurol 7:139–143
Lane RD (1990) Successful fluoxetine treatment of pathologic jealousy. J Clin Psychiatry 51:345–346
Larson EB, Kukull WA, Buchner D, Reifler BV (1987) Adverse drug reactions associated with global cognitive impairment in elderly persons. Ann Intern Med 107:169–173
Lesser IM, Miller BL, Boone KB, Hill–Gutierrez E, Mehringer CM, Wong K, Mena I (1991) Brain injury and cognitive function in late-onset psychotic depression. J Neuropsych 3:33–40
Lester D (1982) Late-onset schizophrenia. Am J Psychiatry 139:1528
Lindesay J, Macdonald A, Starke I (1990) Delirium in the elderly. Oxford University Press, Oxford
Lipowski ZJ (1990) Delirium: acute confusional states. Oxford University Press, Oxford
LLerena A, Dahl M-L, Ekqvist B, Bertilsson L (1992) Haloperidol disposition is dependent on the debrisoquine hydroxylation phenotype: increased plasma levels of the reduced metabolite in poor metabolizers. Ther Drug Monit 14:261–264
LLerena A, Herraiz AG, Cobaleda J, Johansson I, Dahl M-L (1993) Debrisoquine and mephenytoin hydroxylation phenotypes and CYP2D6 genotype in patients treated with neuroleptic and antidepressant agents. Clin Pharmacol Ther 54:606–611
Lydiard RB, Gelenberg AJ (1981) Amoxapine – an antidepressant with some neuroleptic properties: a review of its chemistry, animal pharmacology and toxicology, human pharmacology, and clinical efficacy. Pharmacotherapy 1:163–178
Manschreck TC (1992) Delusional disorders: clinical concepts and diagnostic strategies. Psychiatric Ann 22:241–251
Marneros A, Deister A (1984) The psychopathology of "late schizophrenia". Psychopathology 17:264–274
Martinez JA, Velasco JJ, Urbistondo MD (1994) Effects of pharmacological therapy on anthropometric and biochemical status of male and female institutionalized psychiatric patients. J Am Coll Nutrition 13:192–197

McCoy LM, Schwarzkopf SB, Martin D (1992) Rapid response to pimozide in treatment resistant delusional discorder. Ann Clin Psychiatry 4:95–98
McElroy SL, Dessain EC, Pope HG, Cole JO, Keck PE, Frankenberg FR, Aizley HG, O'Brien S (1991) Clozapine in the treatment of psychotic mood disorders, schizoaffective disorder, and schizophrenia. J Clin Psychiatry 52:411–414
McEvoy JP, Mccue M, Spring B, Mohs RC, Lavori PW, Farr RM (1987) Effects of amantadine and trihexyphenidyl on memory in elderly normal volunteers. Am J Psychiatry 144:573–577
McKeith I, Fairbairn A, Perry R, Thompson P, Perry E (1992) Neuroleptic sensitivity in patients with senile dementia of Lewy body type. BMJ 305:673–678
Menza MA, Murray GB, Holmes VF, Rafuls WA (1987) Decreased extrapyramidal symptoms with intravenous haloperidol. J Clin Psychiatry 48:278–280
Metzger E, Friedman R (1993) Prolongation of the corrected QT and torsades de pointes cardiac arrhythmia associated with intravenous haloperidol in the medically ill. J Clin Psychopharmacol 13:128–132
Meyers BS, Greenberg R (1986) Late-life delusional depression. J Affect Disord 11:133–137
Meyers BS, Gabriele M, Kakuma T, Alpert S, Kalayam B, Young RC (1993) Correlates of late-life delusional depression. Proceedings of the American Psychiatric Association 146th annual meeting, p 223
Miller BL, Benson DF, Cummings JL, Neshkes R (1986) Late-life paraphrenia: an organic delusional syndrome. J Clin Psychiatry 47:204–207
Miller BL, Lesser IM (1988) Late-life psychosis and modern neuroimaging. Psychiatr Clin North Am II(1):33–45
Modell JG, Lenox RH, Weiner S (1985) Inpatient clinical trial of lorazepam for the management of manic agitation. J Clin Psychopharmacol 5:109–113
Mulsant BM, Gershon S (1993) Neuroleptics in the treatment of psychosis in late life. A rational approach. Int J Geriatr Psychiatry 8:979–992
Mulsant BH, Thornton JE (1990) Alzheimer disease and other dementias. In: Thase ME, Edelstein BA, Hersen M (eds) Handbook of outpatient treatment of adults: nonpsychotic mental disorders. Plenum, New York
Mulsant BH, Stergiou A, Keshavan MS, Sweet RA, Rifai AH, Pasternak RE, Zubenko GS (1993) Schizophrenia in late-life: a clinical study of elderly patients admitted to an acute care psychiatric hospital. Schizo Bul 19:709–721
Mulsant BH, Zubenko GS (1994) Clinical, neuropathological, and neurochemical correlates of depression and psychosis in primary dementia. In: Emery VOB, Oxam TE (eds) Dementia: presentations, differential diagnosis, and nosology. John Hopkins University Press, Baltimore
Mulsant BH, Perel JM, Sweet RA, Rosen J, Foglia JP (1994) Pharmacologic treatment of psychotic depression in late-life. Psychopharmacol Bull 30(4):634
Munro A (1982) Paranoia revisited. Br J Psychiatry 141:344–349
Munro A (1992) Psychiatric disorders characterized by delusions: treatment in relation to specific types. Psychiatric Ann 22:232–240
Munro A, O'Brien JV, Ross D (1985) Two cases of "pure" or "primary" erotomania successfully treated with pimozide. Can J Psychiatry 30:619–622
Murphy E (1983) The prognosis for depression in old age. Br J Psychiatry 142:111–119
Naber D, Leppig M, Grohmann R, Hippius H (1989) Efficacy and adverse effects of clozapine in the treatment of schizophrenia and tardive dyskinesia – a retrospective study of 387 patients. Psychopharm 99:S73–S76
Naber D, Holzbach R, Perro C, Hippius H (1992) Clinical management of clozapine patients in relation to efficacy and side-effects. Br J Psychiatry 160:54–59
Nelson JC, Price LH, Jatlow PI (1986) Neuroleptic dose and desipramine concentrations during combined treatment of unipolar delusional depression. Am J Psychiatry 143:1151–1154
Nyth AL, Gottfries CG (1990) The clinical efficacy of citalopram in treatment of emotional disturbances in dementia disorders. Brit J Psychiatry 157:894–901

Opler LA, Feinberg SS (1991) The role of pimozide in clinical psychiatry: a review. J Clin Psychiatry 52:221–233

Pearlson GD, Rabins PV (1988) Psychosis and depression in the elderly. The late-onset psychoses: possible risk factors. Psychiatr Clin North Am II(1):15–31

Pearlson GD, Kreger L, Rabins PV, Chase GA, Cohen BM, Wirth JB, Schlaepfer TB, Tune LE (1989) A chart review study of late-onset and early-onset schizophrenia. Am J Psychiatry 146:1568–1574

Pearlson GD, Tune LE, Wong DF, Aylward EH, Barta PE, Powers RE, Tien AY, Chase GA, Harris GJ, Rabins PV (1993) Quantitative D_2 dopamine receptor PET and structural MRI changes in late-onset schizophrenia. Schizophr Bull 19:783–795

Perry EK, Marshall E, Kerwin J, Smith CJ, Jabeen S, Cheng AV, Perry RH (1990) Evidence of monoaminergic-cholinergic imbalance related to visual hallucinations in Lewy body dementia. J Neurochem 55:1454–1456

Perry EK, Irving D, Kerwin JM, McKeith IG, Thompson P, Collerton D, Fairbairn AF, Ince PG, Morris CM, Cheng AV, Perry RH (1993) Cholinergic transmitter and neurotrophic activities in Lewy body dementia: similarity to Parkinson's and distinction from Alzheimer's disease. Alzheimer Dis Assoc Disord 7:69–79

Petrie WM, Ban TA, Berney S, Fujimori M, Guy W, Ragheb M, Wilson WH, Schaffer JD (1982) Loxapine in psychogeriatrics: a placebo- and standard-controlled clinical investigation. J Clin Psychopharmacol 2:122–126

Pollock BG (1982) Successful treatment of pathological jealousy with pimozide. Can J Psychiatry 27:86–87

Pollock BG, Mulsant BH (1995) Antipsychotics in older patients: a safety perspective. Drugs Aging 6(4):312–323

Pollock BG, Perel JM, Reynolds CF (1990) Pharmacodynamic issues relevant to geriatric psychopharmacology. J Geriatr Psychiatry Neurol 3:221–228

Pollock BG, Perel JM, Altieri L, Kirshner M, Yeager A, Houck P, Reynolds CF (1992) Debrisoquine hydroxylation phenotyping in geratric psychopharmacology. Psychopharmacol Bull 28:163–168

Pollock BG, Perel JM, Sweet RA, Rosen J, Mulsant BH (1995) Prospective P450 phenotyping for neuroleptic treatment in dementia. Psychopharmacol Bull 31(2):327–331

Procter AW, Francis PT, Stratmann GC, Bowen DM (1992) Serotonergic pathology is not widespread in Alzheimer patients without prominent aggressive symptoms. Neurochem Res 17:917–922

Prohovnik I, Dwork AJ, Kaufman MA, Willson N (1993) Alzheimer-type neuropathology in elderly schizophrenia patients. Schizophr Bull 19:805–816

Pylko T, Sicignan J (1985) Nortriptyline in the treatment of a monosymptomatic delusion. Am J Psychiatry 142:1223

Rabins PV, Merchant A, Nestadt G (1984a) Criteria for diagnosing reversible dementia caused by depression: validation by 2-year follow up. Br J Psychiatry 144:488–492

Rabins PV, Pauker S, Thomas J (1984b) Can schizophrenia begin after age 44. Compr Psychiatry 25:290–293

Raskind M, Alvarez C, Herlin S (1979) Fluphenazine enanthate in the outpatient treatment of late paraphrenia. J Am Geriatr Soc 27:459–463

Raskind MA, Risse SC, Lampe TH (1987) Dementia and antipsychotic drugs. J Clin Psychiatry 48:16–18

Ray WA, Griffin MR, Schaffner W, Baugh DK, Melton LJ (1987) Psychotropic drug use and the risk of hip fracture. N Engl J Med 316:363–369

Richards M, Stern Y, Mayeux R (1993) Subtle extrapyramidal signs can predict the development of dementia in elderly individuals. Neurology 43:2184–2188

Roche E, Romero-Alvira D (1993) Oxidative stress in some dementia types. Med Hypotheses 40:342–350

Rolan PE (1994) Plasma protein binding displacement interactions – Why are they still regarded as clinically important. Br J Clin Pharmacol 37:125–128

Rosen J, Zubenko GS (1991) Emergence of psychosis and depression in the longitudinal evaluation of Alzheimer's disease. Biol Psychiatry 29:224–232
Rosen LN, Rosenthal NE, VanDusen PH, Dunner DL, Fieve RR (1984) Age at onset and number of psychotic symptoms in bipolar I and schizoaffective disorder. Am J Psychiatry 140:1523–1524
Rosen J, Mulsant BH, Wright BA (1992) Agitation in severely demented patients. Ann Clin Psychiatry 4:207–215
Roth M (1955) The natural history of mental disorder in old age. Br J Psychiatry 101:281–301
Roth M, Morrisey JD (1952) Problems in the diagnosis and classification of mental disorders in old age; with a study of case material. J Ment Sci 98:66–80
Rovner BW, Edelman BA, Cox MP, Shmuely Y (1992) The impact of antipsychotic drug regulations on psychotropic prescribing practices in nursing homes. Am J Psychiatry 149:1390–1392
Saltz BL, Woerner MG, Kane JM, Lieberman JA, Alvir J, Bergmann KJ, Blank K, Koblenzer J, Kahaner K (1991) Prospective study of tardive dyskinesia incidence in the elderly. JAMA 266:2402–2406
Salzman C (1987a) Treatment of agitation in the elderly. In: Meltzer HY (ed) Psychopharmacology: the third generation of progress. Raven, New York
Salzman C (1987b) Treatment of the elderly agitated patient. J Clin Psychiatry 48:19–22
Salzman C (1988) Use of benzodiazpeines to control disruptive behavior in inpatients. J Clin Psychiatry 49 [Suppl 12]:13–15
Sanders KM, Murray GB, Cassem NH (1991) High-dose intravenous haloperidol for agitated delirium in a cardiac patient on intra-aortic balloon pump. J Clin Psychopharmacol 11:145–146
Schatzberg AF, Rothschild AJ (1992) Psychotic (delusional) major depression: should it be included as a distinct syndrome in DSM-IV. Am J Psychiatry 149:733–745
Schmucker DL, Woodhouse KW, Wang RK, Wynne H, James OF, Mcmanus M (1990) Effects of age and gender on in vitro properties of human liver microsomal monooxygenases. Clin Pharmacol Ther 48:365–374
Schneider LS, Sobin PB (1991) Non-neuroleptic medications in the management of agitation in Alzheimer's disease and other dementia: a selective review. Int J Geriatr Psychiatry 6:691–708
Schneider LS, Pollock VE, Lyness SA (1990) A metaanalysis of controlled trials of neuroleptic treatment in dementia. J Am Geriatr Soc 38:553–563
Scholz E, Dichgans J (1985) Treatment of drug-induced exogenous psychosis in Parkinsonism with clozapine and fluperlapine. Eur Arch Psychiatry Neurolo Sci 235:60–64
Schor JD, Levkoff SE, Lipsitz LA, Reilly CH, Cleary PD, Rowe JW, Evans DA (1992) Risk factors for delirium in hospitalized elderly. JAMA 267:827–831
Shorr RI, Fought RL, Ray WA (1994) Changes in antipsychotic drug use in nursing homes during implementation of the OBRA-87 regulations. JAMA 271:358–362
Schuster P, Gabriel E, Kufferle B (1977) Reversal by physostigmine of clozapine-induced delirium. Clin Toxicology 10:437–441
Shivakumar BR, Ravindranath V (1993) Oxidative stress and thiol modification induced by chronic administration of haloperidol. J Pharmacol Exp Ther 265:1137–1141
Silverstone T (1985) Dopamine in manic depressive illness: A pharmacological synthesis. J Affect Disord 8:225–231
Sindrup SH, Brøsen K, Bjerring P, Arendt-Nielsen L, Larsen U, Angelo HR, Gram LF (1991) Codeine increases pain thresholds to copper vapor laser stimuli in extensive but not poor metabolizers of sparteine. Clin Pharmacol Ther 49:686–93
Slaby AE, Erle SR (1993) Dementia and delirium. In: Stoudemire A, Fogel BS (eds) Psychiatric care of the medical patient. Oxford University Press, New York

Spiker DG, Weiss JC, Dealy RS, Griffin SJ, Hanin I, Neil JF, Perel JM, Rossi AJ, Soloff PH (1985) The pharmacological treatment of delusional depression. Am J Psychiatry 142:430–436
Spiker DG, Perel JM, Hanin I, Dealy RS, Griffin SJ, Soloff PH, Weiss JC (1986) The pharmacological treatment of delusional depression: part II. J Clin Psychopharmacol 6:339–342
Spina E, Martines C, Caputi AP, Cobaleda J, Pinas B, Carrillo JA, Benitez J (1991) Debrisoquine oxidation phenotype during neuroleptic monotherapy. Eur J Clin Pharmacol 41:467–470
Spina E, Ancione M, Di Rosa AE, Meduri M, Caputi AP (1992) Polymorphic debrisoquine oxidation and acute neuroleptic-induced adverse effects. Eur J Clin Pharmacol 42:347–8
Steele C, Rovner BW, Chase GA, Folstein MF (1990) Psychiatric symptoms and nursing home placement of patients with Alzheimer's disease. Am J Psychiatry 147:1049–1051
Steele JW, Faulds D, Sorkin E (1993) Tiapride. A review of its pharmacodynamic and pharmacokinetic properties, and therapeutic potential in geriatric agitation. Drugs and Aging 3:460–478
Stone K (1989) Mania in the elderly. Br J Psychiatry 155:220–224
Stoudemire A, Riether AM (1987) Evaluation and treatment of paranoid syndromes in the elderly: a review. Gen Hosp Psych 9:267–274
Sweet RA, Mulsant BH, Gupta B, Rifai AH, Pasternak RE, McEachran A (1995) Duration of neuroleptic treatment and prevalence of tardive dyskinesia in late life. Arch Gen Psychiatr 52(6):478–486
Tesar GE, Murray GB, Cassem NH (1985) Use of high-dose intravenous haloperidol in the treatment of agitated cardiac patients. J Clin Psychopharmacol 5:344–347
Thornton JE, Stahl SM (1984) Case report of tardive dyskinesia and parkinsonism associated with amoxapine therapy. Am J Psychiatry 141:704–705
Tinetti ME, Speechley M, Ginter SF (1988) Risk factors for falls among elderly persons living in the community. N Engl J Med 319:1701–1707
Tyndale R, Kalow W, Inaba T (1991) Oxidation of reduced haloperidol to haloperidol: Involvement of human P450IID6 (sparteine/debrisoquine monooxygenase). Br J Clin Pharmacol 31:655–60
Volavka J (1985) Late-onset schizophrenia: a review. Compr Psychiatry 26:148–156
von Bahr C, Spina E, Birgersson C, Ericsson O, Göransson M, Henthorn T, Sjöqvist F (1985) Inhibition of desmethylimipramine 2-hydroxylation by drugs in human liver microsomes. Biochem Pharmacol 34:2501–2505
von Bahr C, Movin G, Nordin C, Liden A, Hammarlund-Udenaes M, Hedberg A, Ring H, Sjöqvist F (1991) Plasma levels of thioridazine and metabolites are influenced by the debrisoquin hydroxylation phenotype. Clin Pharmacol Ther 49:234–240
Wallin A, Blennow K (1992) Neurologic motor signs in early and late onset Alzheimer's disease. Dementia 3:314–319
Williams TF (1987) Aging or disease. Clin Pharmacol Ther 42:663–665
Wilson JA, MacLennan WJ (1989) Review: Drug-induced Parkinsonism in elderly patients. Age and Ageing 18:208–210
Winokur G (1985) Familial psychopathology in delusional disorder. Compr Psychiatry 26(3):241–248
Wolk SI, Douglas CJ (1992) Clozapine treatment of psychosis in Parkinson's disease: a report of 5 consecutive cases. J Clin Psychiatry 53:373–376
Wolters EC, Hurwitz TA, Mak E, Teal P, Peppard FR, Remick R, Calne S, Calne DB (1990) Clozapine in the treatment of Parkinsonian patients with dopaminomimetic psychosis. Neurology 40:832–840
Wragg RE, Jeste DV (1989) Overview of depression and psychosis in Alzheimer's disease. Am J Psychiatry 146:577–587
Yassa R, Suranyi-Cadotte B (1993) Clinical characteristics of late-onset schizophrenia and delusional disorder. Schizophr Bull 19:701–707

Yesavage JA, Holman CA, Cohn R (1981) Correlation of thiothixene serum levels and age. Psychopharmacology 74:170–172

Young RC, Klerman GL (1992) Mania in late life: focus on age at onset. Am J Psychiatry 149:867–876

Young RC, Schreiber MT, Nysewander RW (1983) Psychotic mania. Biol Psychiatry 18:1167–1173

Zimmer AW, Calkins E, Hadley E, Ostfeld AM, Kaye JM, Kaye D (1985) Conducting clinical research in geriatric populations. Ann Intern Med 103:276–283

Zubenko GS, Moossy J, Martinez AJ, Rao G, Claassen D, Rosen J, Kopp U (1991) Neuropathologic and neurochemical correlates of psychosis in primary dementia. Arch Neurol 49:619–624

Zubenko GS, Rosen J, Sweet RA, Mulsant BH, Rifai AH (1992) Impact of psychiatric hospitalization on behavioral complications of Alzheimer's disease. Am J Psychiatry 149:1484–1491

Subject Index

Springer-Verlag and the Environment

We at Springer-Verlag firmly believe that an international science publisher has a special obligation to the environment, and our corporate policies consistently reflect this conviction.

We also expect our business partners – paper mills, printers, packaging manufacturers, etc. – to commit themselves to using environmentally friendly materials and production processes.

The paper in this book is made from low- or no-chlorine pulp and is acid free, in conformance with international standards for paper permanency.

GPSR Compliance

The European Union's (EU) General Product Safety Regulation (GPSR) is a set of rules that requires consumer products to be safe and our obligations to ensure this.

If you have any concerns about our products, you can contact us on ProductSafety@springernature.com

In case Publisher is established outside the EU, the EU authorized representative is:

Springer Nature Customer Service Center GmbH
Europaplatz 3
69115 Heidelberg, Germany

Batch number: 10844669

Printed by Printforce, the Netherlands